# MODERN CONTROL ENGINEERING

PRENTICE-HALL ELECTRICAL ENGINEERING SERIES

WILLIAM L. EVERITT, *editor*

INSTRUMENTATION AND CONTROLS SERIES

# MODERN CONTROL ENGINEERING

**KATSUHIKO OGATA**

University of Minnesota

Prentice-Hall, Inc., Englewood Cliffs, N. J.

Current printing (last digit):

27   26   25   24   23

13-590232-0
Library of Congress Catalog Card No. 72-84843
Printed in the United States of America

PRENTICE-HALL INTERNATIONAL, INC., London
PRENTICE-HALL OF AUSTRALIA, PTY. LTD., Sydney
PRENTICE-HALL OF CANADA LTD., Toronto
PRENTICE-HALL OF INDIA PRIVATE LTD., New Delhi
PRENTICE-HALL OF JAPAN, INC., Tokyo

# PREFACE

This book has been prepared as a comprehensive text for a first study of control engineering. It is written at the level of senior college students in electrical, mechanical, aeronautical, or chemical engineering.

This text covers both conventional control theory and modern control theory. The first three chapters present the fundamental concepts of feedback control systems and a basic mathematical background necessary for the understanding of the book. Chapter 4 treats the modeling of physical systems. The next six chapters (5 through 10) present conventional methods and techniques for analyzing and designing control systems. The following two chapters (11 and 12) discuss nonlinear systems. Chapter 13 provides an introduction to discrete-time systems and the $z$-transform method. The last three chapters (14 through 16) present an introduction to modern control theory based on state space concepts.

It is assumed that the reader is familiar with introductory differential equations, introductory circuit analysis, and mechanics, as is normally required for junior students in engineering curricula.

The first ten chapters can be used as the text of either a one-quarter or one-semester course of three lectures per week, depending upon how much of the material is to be covered. The remaining six chapters can be used as the text of a one-semester course. At the Department of Mechanical Engineering, University of Minnesota, the material in this book is used in two courses; the major portions of Chapters 1 through 10 are used in a one-quarter introductory course on control engineering and Chapters 13 through 16 comprise a one-quarter course on modern control theory; both courses are at the senior level.

This book provides many illustrative examples to clarify the theory presented. I believe the best approach to deepen the understanding of control theory is to digest the fundamentals through the working out of many problems, the solutions of which are available. From this viewpoint, in addition to many examples, each chapter is followed by numerous solved as well as unsolved problems. Successful solution of the unsolved problems will demonstrate that the reader understands the material presented.

I would like to express my gratitude to Dr. Richard C. Jordan, Head of the Department of Mechanical Engineering, University of Minnesota, for his continued encouragement, and convey my appreciation to many former students who solved numerous examples and problems used in the book. I would also like to express my appreciation to Mr. Richard F. Palas, Chemical Engineering Department, University of Minnesota, for making many constructive comments and for improving the presentation of the material.

KATSUHIKO OGATA

# CONTENTS

# 1

# INTRODUCTION TO CONTROL SYSTEMS

## 1-1 INTRODUCTION

Automatic control has played a vital role in the advancement of engineering and science. In addition to its extreme importance in space-vehicle, missile-guidance, and aircraft-piloting systems, etc., automatic control has become an important and integral part of modern manufacturing and industrial processes. For example, automatic control is essential in such industrial operations as controlling pressure, temperature, humidity, viscosity, and flow in the process industries; tooling, handling, and assembling mechanical parts in the manufacturing industries, among many others.

Since advances in the theory and practice of automatic control provide means for attaining optimal performance of dynamic systems, improve the quality and lower the cost of production, expand the production rate, relieve the drudgery of many routine, repetitive manual operations, etc., most engineers and scientists must now have a good understanding of this field.

**Historical review.** The first significant work in automatic control was James Watt's centrifugal governor for the speed control of a steam engine in the eighteenth century. Other significant works in the early stages of development of control theory were due to Minorsky, Hazen, and Nyquist, among many others. In 1922 Minorsky worked on automatic controllers for steering ships and showed how stability could be determined from the differential equations describing the system. In 1932 Nyquist developed a relatively simple procedure for determining the stability of closed-loop systems on the basis of open-loop response to steady-state sinusoidal inputs. In 1934 Hazen, who introduced the term "servomechanisms" for position control systems, discussed design of relay servomechanisms capable of closely following a changing input.

During the decade of the 1940's, frequency-response methods made it possible for engineers to design linear feedback control systems that satisfied performance requirements. From the end of the 1940's to early 1950's, the root-locus method in control system design was fully developed.

The frequency-response and the root-locus methods, which are the core of classical control theory, lead to systems that are stable and satisfy a set of more or less arbitrary performance requirements. Such systems are, in general, not optimal in any meaningful sense. Since the late 1950's, the emphasis in control design problems has been shifted from the design of one of many systems that work to the design of one optimal system in some meaningful sense.

As modern plants with many inputs and outputs become more and more complex, the description of a modern control system requires a large number of equations. Classical control theory, which deals only with single-input–single-output systems, becomes entirely powerless for multiple-input–multiple-output systems. Since about 1960, modern control theory has been developed to cope with the increased complexity of modern plants and the stringent requirements on accuracy, weight, and cost in military, space, and industrial applications.

Because of the readily available electronic analog, digital, and hybrid computers for use in complex computations, the use of computers in the design of control systems and the use of on-line computers in the operation of control systems are now becoming common practice.

The most recent developments in modern control theory may be said to be in the direction of the optimal control of both deterministic and stochastic systems as well as the adaptive and learning control of complex systems. Applications of modern control theory to such nonengineering fields as biology, economics, medicine, and sociology are now under way, and interesting and significant results can be expected in the near future.

## 1-2  DEFINITIONS

In this section we shall define the terminology necessary to describe control systems.

**Plants.** A plant is a piece of equipment, perhaps just a set of machine parts functioning together, the purpose of which is to perform a particular operation. In this book, we shall call any physical object to be controlled (such as a heating furnace, a chemical reactor, or a spacecraft) a *plant*.

**Processes.** The Merriam-Webster Dictionary defines a process to be a natural, progressively continuing operation or development marked by a series of gradual changes that succeed one another in a relatively fixed way and lead toward a particular result or end; or an artificial or voluntary, progressively continuing operation that consists of a series of controlled actions or movements systematically directed toward a particular result or end. In this book we shall call any operation to be controlled a *process*. Examples are chemical, economic, and biological processes.

**Systems.** A system is a combination of components that act together and perform a certain objective. A system is not limited to physical ones. The concept of the system can be applied to abstract, dynamic phenomena such as those encountered in economics. The word "system" should, therefore, be interpreted to imply physical, biological, economic, etc., systems.

**Disturbances.** A disturbance is a signal which tends to adversely affect the value of the output of a system. If a disturbance is generated within the system, it is called *internal*; while an external disturbance is generated outside the system and is an input.

**Feedback control.** Feedback control is an operation which, in the presence of disturbances, tends to reduce the difference between the output of a system and the reference input (or an arbitrarily varied, desired state) and which does so on the basis of this difference. Here, only unpredictable disturbances (i.e., those unknown beforehand) are designated for as such, since with predictable or known disturbances, it is always possible to include compensation within the system so that measurements are unnecessary.

**Feedback control systems.** A feedback control system is one which tends to maintain a prescribed relationship between the output and the reference input by comparing these and using the difference as a means of control.

Note that feedback control systems are not limited to the field of engineering but can be found in various nonengineering fields such as economics and biology. For example, the human organism, in one aspect, is analogous to an intricate chemical plant with an enormous variety of unit operations. The process control of this transport and chemical-reaction network involves a variety of control loops. In fact, the human organism is an extremely complex feedback control system.

**Servomechanisms.** A servomechanism is a feedback control system in which the output is some mechanical position, velocity, or acceleration. Therefore, the terms *servomechanism* and *position-* (or *velocity-* or *acceleration-*) *control system* are synonymous. Servomechanisms are extensively used in modern industry. For example, the completely automatic operation of machine tools, together with programmed instruction, may be accomplished by use of servomechanisms.

**Automatic regulating systems.** An automatic regulating system is a feedback control system in which the reference input or the desired output is either constant or slowly varying with time and in which the primary task is to maintain the actual output at the desired value in the presence of disturbances.

A home heating system in which a thermostat is the controller is an example of an automatic regulating system. In this system, the thermostat setting (the desired temperature) is compared with the actual room temperature. A change in outdoor temperature is a disturbance in this system. The objective is to maintain the desired room temperature despite changes in outdoor temperature. There are many other examples of automatic regulating systems, some of which are the automatic control of pressure and of electric quantities such as voltage, current, and frequency.

**Process control systems.** An automatic regulating system in which the output is a variable such as temperature, pressure, flow, liquid level, or pH is called a *process control system*. Process control is widely applied in industry. Programmed controls such as the temperature control of heating furnaces in which the furnace temperature is controlled according to a preset program are often used in such systems. For example, a preset program may be such that the furnace temperature is raised to a given temperature in a given time interval and then lowered to another given temperature in some other given time interval. In such program control the set point is varied according to the preset time schedule. The controller then functions to maintain the furnace temperature close to the varying set point. It should be noted that most process control systems include servomechanisms as an integral part.

## 1-3   CLOSED-LOOP CONTROL AND OPEN-LOOP CONTROL

We shall first define closed-loop and open-loop control systems. Then we shall make a comparison of these two types. Finally, the concepts of adaptive control and learning control will be introduced.

**Closed-loop control systems.** A closed-loop control system is one in which the output signal has a direct effect upon the control action. That is, closed-loop control systems are feedback control systems. The actuating error signal, which is the difference between the input signal and the feedback signal (which may be the output signal or a function of the output signal and its derivatives), is fed to the con-

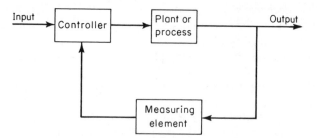

Fig. 1-1. Closed-loop control system.

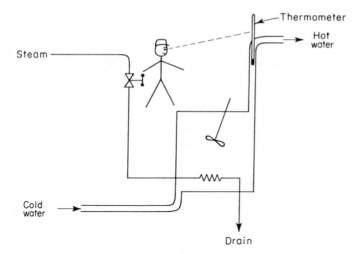

**Fig. 1-2.** Manual feedback control of a thermal system.

troller so as to reduce the error and bring the output of the system to a desired value. In other words, the term "closed loop" implies the use of feedback action in order to reduce system error. Figure 1-1 shows the input-output relationship of the closed-loop control system. Such a figure is called a *block diagram*. To illustrate the concept of closed-loop control systems, consider the thermal system shown in Fig. 1-2. Here a human being acts as the controller. He wants to maintain the temperature of the hot water at a given value. The thermometer installed in the hot water outlet pipe measures the actual temperature. This temperature is the output of the system. If the operator watches the thermometer and finds that the temperature is higher than the desired value, then he reduces the amount of steam supply in order to lower this temperature. It is quite possible that the temperature becomes too low, in which case it becomes necessary to repeat the sequence of operations in the opposite direction.

This control action is based on closed-loop operation. Since both the feedback of the output (water temperature) for comparison with the reference input and control action take place through the actions of the operator, this is a closed-loop control system. Such a system may be called a manual feedback or manual closed-loop control system.

If an automatic controller is used to replace the human operator, as shown in Fig. 1-3, the control system becomes automatic, i.e., an automatic feedback or automatic closed-loop control system. The position of the dial on the automatic controller sets the desired temperature. The output, the actual temperature of the hot water, which is measured by the temperature-measuring device, is compared with the desired temperature in order to generate an actuating error signal. In doing this, the output temperature is converted to the same units as the input (set point) by a transducer. (A transducer is a device which converts a signal from one form into another.) The error signal produced in the automatic controller is amplified, and the output of the controller is sent to the control valve in order to change the

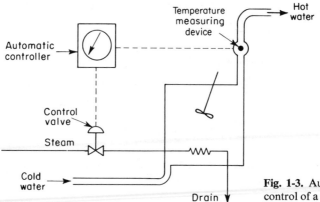

Fig. 1-3. Automatic feedback control of a thermal system.

valve opening for steam supply so as to correct the actual water temperature. If there is no error, no change in the valve opening is necessary.

In the systems considered here, variations in the ambient temperature, the inlet cold water temperature, etc., may be considered external disturbances.

The manual feedback and automatic feedback control systems cited above operate in a similar fashion. The operator's eyes are the analog of the error-measuring device; his brain, the analog of the automatic controller; and his muscles, the analog of the actuator.

The control of a complex system by a human operator is not effective because of the many interrelations among the various variables. Note that even in a simple system an automatic controller will eliminate any human errors in operation. If high precision control is necessary, control must be automatic.

Numerous closed-loop control systems may be found in industry and in homes. Some examples are all the servomechanisms, most process control systems, household refrigerators, automatic hot water heaters, and automatic home heating systems with thermostatic control.

**Open-loop control systems.** Open-loop control systems are control systems in which the output has no effect upon the control action. That is, in an open-loop control system, the output is neither measured nor fed back for comparison with the input. Figure 1-4 shows the input-output relationship of such a system. A practical example is a washing machine. Soaking, washing, and rinsing in the washing machine are operated on a time basis. The machine does not measure the output signal, namely, the cleanliness of the clothes.

In any open-loop control system the output is not compared with the reference input. Hence, for each reference input, there corresponds a fixed operating condition. Thus, the accuracy of the system depends on the calibration. (Open-loop control systems must be carefully calibrated and must maintain that calibration in order to be useful.) In the presence of disturbances an open-loop control system will not perform the desired task. Open-loop control can be used in practice only if the relationship between the input and output is known and if there are neither internal

**Fig. 1-4.** Open-loop control system.

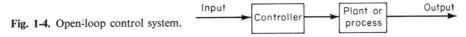

nor external disturbances. Clearly such systems are not feedback control systems. Note that any control system which operates on a time basis is open-loop. For example, traffic control by means of signals operated on a time basis is another instance of open-loop control.

**Closed-loop versus open-loop control systems.** An advantage of the closed-loop control system is that the use of feedback makes the system response relatively insensitive to external disturbances and internal variations in system parameters. It is thus possible to use relatively inaccurate and inexpensive components to obtain the accurate control of a given plant; whereas this is impossible in the open-loop case.

From the point of view of stability, the open-loop control system is easier to build since stability is not a major problem. On the other hand, stability is always a major problem in the closed-loop control system since it may tend to overcorrect errors which may cause oscillations of constant or changing amplitude.

It should be emphasized that for systems in which the inputs are known ahead of time and in which there are no disturbances it is advisable to use open-loop control. Closed-loop control systems have advantages only when unpredictable disturbances and/or unpredicatable variations in system components are present. Note that the output power rating partially determines the cost, weight, and size of a servomechanism (or capital investment, manpower, etc., in business systems). In order to decrease the required power of a system, open-loop control may be used where applicable. A proper combination of open-loop and closed-loop controls is usually less expensive and will give satisfactory overall system performance.

**Direct versus indirect controls.** Note that in order to obtain the best result, it is desirable to measure and control directly the variable which indicates the state of the system or quality of the product. In the case of process control systems we may want to measure and control directly the quality of the product. This may present a difficult problem, however, since this quality may be difficult to measure. If this is the case, it becomes necessary to control a secondary variable. For example, variables (such as temperature and pressure) which are directly related to the quality may be controlled. Because other variables may affect the relationship between the quality and the measured variable, the indirect control of a system is usually not as effective as direct control. Although it may be difficult, we should always try to control the primary variable as directly as possible.

**Adaptive control systems.** The dynamic characteristics of most control systems are not constant because of several reasons, such as the deterioration of components as time elapses or the changes in parameters and environment (for example, changes in mass and atomspheric conditions in a spacecraft control system). Although the effects of small changes on the dynamic characteristics are attenuated in a feedback control system, if changes in the system parameters and environment are significant, a satisfactory system must have the ability of adaptation. Adaptation implies the ability to self-adjust or self-modify in accordance with unpredictable changes in conditions of environment or structure. Control systems having a candid ability of adaptation are called *adaptive* control systems.

In an adaptive control system, the dynamic characteristics must be identified at all times so that the controller parameters can be adjusted in order to maintain optimal performance. This concept has a great deal of appeal to the system designer since an adaptive control system, besides accommodating environmental changes, will also accommodate moderate engineering design errors or uncertainties and will compensate for the failure of minor system components, thereby increasing overall system reliability.

**Learning control systems.** Many apparently open-loop control systems can be converted into closed-loop control systems if a human operator is considered a controller, comparing the input and output and making the corrective action based on the resulting difference or error.

If we attempt to analyze such human-operated closed-loop control systems, we encounter the difficult problem of writing equations which describe the behavior of a human being. One of the many complicating factors in this case is the learning ability of the human operator. As the operator gains more experience, he will become a better controller, and this must be taken into consideration in analyzing such a system. Control systems having an ability to learn are called *learning* control systems. This concept is quite new and has not been thoroughly explored.

Recent progress in adaptive and learning control applications has been reported in the literature, but large segments of engineering activity remain available for future study.


**1-4   ILLUSTRATIVE EXAMPLES OF CONTROL SYSTEMS**

In this section we shall present several illustrative examples of closed-loop control systems.

**Pressure control systems.** Figure 1-5 shows a pressure control system. The pressure in the furnace is controlled by the position of the damper. This pressure is measured by a pressure-measuring element. The signal thus obtained is fed to the controller for comparison with the desired value. If there is any difference or error, the controller output is fed to the actuator which positions the damper in order to reduce the error.

**Speed control systems.** The basic principle of Watt's governor for steam engines is illustrated in the schematic diagram of Fig. 1-6. The amount of steam admitted to the engine cylinder is adjusted according to the difference between the desired and the actual engine speeds.

The sequence of actions may be stated as follows: The reference input (set point) is set according to the speed desired. If the actual speed drops below the

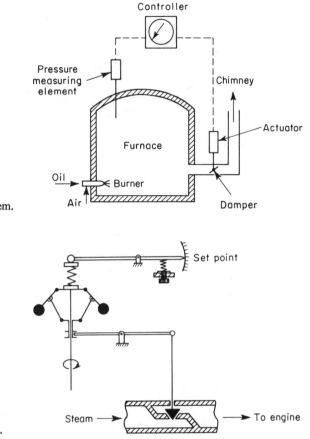

**Fig. 1-5.** Pressure control system.

**Fig. 1-6.** Speed control system.

desired value, then the decrease in the centrifugal force of the speed governor causes the control valve to move upward, supplying more steam, and the speed of the engine increases until the desired value is reached. On the other hand, if the speed of the engine increases above the desired value, then the increase in the centrifugal force of the governor causes the control valve to move downward. This decreases the supply of steam and the speed of the engine decreases until the desired value is reached.

**Numerical control systems.** Numerical control is a method of controlling the motions of machine components by use of numbers. In numerical control the motion of a workhead may be controlled by the binary information contained on a tape.

In such control, symbolic numerical values are converted into physical values (dimensions or quantities) by electrical (or other) signals that are translated into a linear or circular movement. These signals are either digital (pulses) or analog (time-varying voltages).

The system shown in Fig. 1-7 works as follows: A tape is prepared in binary

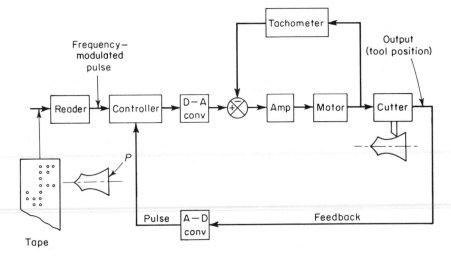

**Fig. 1-7.** Numerical control of a machine.

form representing the desired part **P**. To start the system, the tape is fed through the reader. The frequency-modulated input pulse signal is compared with the feedback pulse signal. The digital-to-analog converter converts the pulse into an analog signal which represents a certain magnitude of voltage which, in turn, causes the servomotor to rotate. The position of the cutterhead is controlled according to the input to the servomotor. The transducer attached to the cutterhead converts the motion into an electrical signal which is converted to the pulse signal by the analog-to-digital converter. Then this signal is compared with the input pulse signal. The controller carries out mathematical operations on the difference in the pulse signals. If there is any difference between these two, a signal is sent to the servomotor to reduce it.

An advantage of numerical control is that complex parts can be produced with uniform tolerances at the maximum milling speed.

**Computer control systems.** Figure 1-8 shows a schematic diagram of the computer control of a blast furnace. The blast furnace is a huge structure about 100 ft high. Modern furnaces are built to produce over 4000 tons of pig iron per day and must be kept in continuous operation because of the nature of the smelting process.

The iron ore, coke, and limestone are charged into the furnace from the top in proper proportions. (Approximately 2 tons of ore, 1 ton of coke, $\frac{1}{2}$ ton of flux, and $4\frac{1}{2}$ tons of air are needed to produce 1 ton of pig iron.) Air, which is quite important in this process, is heated in stoves and blown into the furnace. The heat in the furnace is produced by burning coke, from which carbon monoxide gas is produced by the process of partial combustion. This gas, together with the coke, reduces the iron ore in the furnace to metal, and the limestone acts as a flux, slagging the impurities. The molten iron sinks to the bottom of the furnace, and the liquid

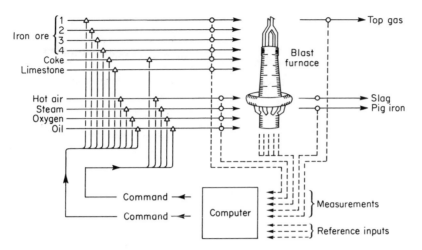

**Fig. 1-8.** Computer control of a blast furnace.

slag rises to the surface. The molten iron and liquid slag are periodically drained off through tap holes provided for this purpose.

Since the presence of carbon, manganese, silicon, sulfur, phosphorus, etc., depends largely upon the composition of the ore, coke, and limestone used, it is quite difficult for human operators to control the chemical composition of pig iron issuing from the furnace.

In the computer control of such furnaces, information about the compositions of pig iron, slag, and top gas, the temperature, and the pressure in the furnace, as well as the compositions of ore, coke, and limestone, are fed to the computer at certain time intervals. The complex computations for determining the optimal quantities of various raw materials to be charged are handled by the computer. It is then possible to keep the composition of pig iron as desired. It is also possible to maintain the steady-state operation of the blast furnace at a satisfactory condition.

Note that in the computer control of such a process it is necesssary to have mathematical models available. But deriving suitable models may be a difficult problem because not all the factors that affect the system dynamics are known. Note also that the measurement of all the variables necessary for computer control may be difficult or impossible, in which case the unmeasurable variables must be estimated by statistical methods.

**Traffic control systems.** As stated in Section 1-3, traffic control by means of traffic signals operated on a time basis constitutes an open-loop control system. If, however, the number of cars waiting at each traffic signal in a congested area of a city is continuously measured and the information fed to the central control computer which controls the traffic signals, then such a system becomes closed-loop.

Traffic movement in networks is quite complex because the variation in traffic volume depends heavily upon the hour and day of the week, as well as on many other factors. In some cases the Poisson distribution may be assumed for the arrivals at intersections, but this is not necessarily valid for all traffic problems. In fact, minimizing the average waiting time is a very complex control problem.

**Biological systems.** Consider the competition of two species of bacteria whose populations are $x_1$ and $x_2$. The two are competing in the sense that they consume the same food supply. Under certain conditions the populations $x_1$ and $x_2$ change with time according to

$$\dot{x}_1 = a_{11}x_1 - a_{12}x_1x_2$$
$$\dot{x}_2 = a_{21}x_2 - a_{22}x_1x_2$$

where $a_{11}, a_{12}, a_{21}$, and $a_{22}$ are positive constants and $x_1$ and $x_2$ are nonnegative. These equations are called Volterra's competition equations.

If a certain chemical is given to the species, the populations change according to the following equations:

$$\dot{x}_1 = a_{11}x_1 - a_{12}x_1x_2 - b_1u$$
$$\dot{x}_2 = a_{21}x_2 - a_{22}x_1x_2 - b_2u$$

where $b_1$ and $b_2$ are positive constants and $u$ is the controlling input (the amount of chemical in this example). An interesting problem is to minimize within a given time period the population $x_1$ while keeping the population $x_2$ as large as possible. This is an example of a biological system to which control theory can be applied.

**Inventory control systems.** The industrial programming of production rate and inventory level is another example of a closed-loop control system. The actual inventory level, which is the output of the system, is compared with the desired inventory level, which may change from time to time according to the market. If there is any difference between the actual inventory level and the desired inventory level, then the production rate is adjusted so that the output is always at or near the desired level which is chosen to maximize profit.

**Business systems.** A business system may consist of many groups. Each task assigned to a group will represent a dynamic element of the system. Feedback methods of reporting the accomplishments of each group must be established in such a system for proper operation. The crosscoupling between functional groups must be made a minimum in order to reduce undesirable delay times in the system. The smaller this crosscoupling, the smoother the flow of work signals and materials will be.

A business system is a closed-loop system. A good design will reduce the managerial control required. Note that disturbances in this system are the lack of manpower or materials, interruption of communication, human errors, etc.

The establishment of a well-founded estimating system based on statistics is mandatory to proper management. (Note that it is a well-known fact that the performance of such a system can be improved by use of "lead time" or "anticipation.")

To apply control theory to improve the performance of such a system, we must represent the dynamic characteristics of the component groups of the system by a relatively simple set of equations.

Although it is certainly a difficult problem to derive mathematical representations of the component groups, the application of optimization techniques to business systems represents an interesting area in which significant improvements in performance may be expected.

## 1-5  DESIGN PRINCIPLES OF CONTROL SYSTEMS

**General requirements of a control system.** Any control system must be stable. This is a primary requirement. In addition to absolute stability, a control system must have a reasonable relative stability; that is, the speed of response must be reasonably fast and this response must show reasonable damping. A control system must also be capable of reducing errors to zero or to some small tolerable value. Any useful control system must satisfy these requirements.

The requirement of reasonable relative stability and that of steady-state accuracy tend to be incompatible. In designing control systems, we therefore find it necessary to make the most effective compromise between these two requirements.

**Basic problems in control system design.** Figure 1-9 is a block diagram of a control system. The controller produces control signals based on the reference input variables and the output variables. In practical situations there are always some disturbances acting on the plant. These may be of external or internal origin, and they may be random or predictable. The controller must take into consideration any disturbances which will affect the output variables.

To determine the optimal control signal, it is necessary to define a performance index. Such an index is a quantitative measure of the performance, measuring deviation from ideal performance. The specification of the control signal over the

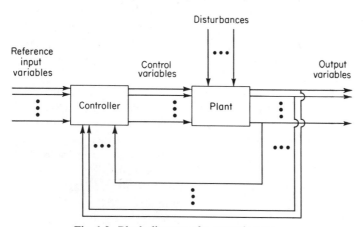

**Fig. 1-9.** Block diagram of a control system.

operating time interval is called the *control law*. Mathematically, the basic control problem is to determine the optimal control law, subject to various engineering and economic constraints, that minimizes a given performance index.

For relatively simple systems, the control law may be obtained analytically. For complex systems, it may be necessary to generate the optimal control law with an on-line digital computer.

**Analysis.** By the analysis of a control system, we mean the investigation, under specified conditions, of the performance of the system whose mathematical model is known.

Since any system is made up of components, analysis must start with a mathematical description of each component. Once a mathematical model of the complete system has been derived, the manner in which analysis is carried out is independent of whether the physical system is pneumatic, electrical, mechanical, etc.

**Design.** To design a system means to find one which accomplishes a given task. In general, the design procedure is not straightforward and will require trial-and-error methods.

**Synthesis.** By synthesis, we mean finding by a direct procedure a system that will perform in a specified way. Usually such a procedure is entirely mathematical from the start to the end of the design process. Synthesis procedures are now available for linear networks and for optimal linear systems.

**Basic approach to control system design.** The basic approach to the design of any practical control system will necessarily involve trial-and-error procedures. The synthesis of linear control systems is theoretically possible, and the control engineer can systematically determine the components necessary to perform the given objective. In practice, however, the system may be subjected to many constraints or may be nonlinear, and in such cases no synthesis methods are available at present. In addition, the characteristics of components may not be precisely known. Thus, trial-and-error procedures are always necessary.

Situations are often encountered in practice where a plant is given and the control engineer has to design the rest of the system so that the whole will meet the given specifications in accomplishing the given task. Note that the specifications must be interpreted in mathematical terms.

It is important to remember that some of the specifications may not be realistic. In such a case, the specifications must be revised in early stages of the design.

In many cases, the design of a control system proceeds as follows: The engineer begins the design procedure knowing the specifications or performance index, the dynamics of the given plant, and the dynamics of the components; the last involves design parameters. The control engineer will apply synthesis techniques if available, along with other techniques, to build a mathematical model of the system.

Once he formulates the design problem in terms of this model, he carries out a mathematical design which yields the solution to the mathematical version of the design problem. (At this stage, simulation of the mathematical model on a computer is important. Note that optimal control theory is very useful in this stage of the design because it gives the upper boundary of system performance for a given performance index.)

After the mathematical design has been completed, the control engineer simulates the model on a computer to test the behavior of the resulting system in response to various signals and disturbances. Usually, the initial system configuration is not satisfactory. Then the system must be redesigned and corresponding analysis completed. This process of design and analysis is repeated until a satisfactory system is obtained. Then a prototype physical system can be constructed.

Note that this process of constructing a prototype is the reverse of that of modeling. The prototype is a physical system that represents the mathematical model with reasonable accuracy. Once the prototype has been built, the engineer tests it to see whether or not it is satisfactory. If it is, the design is complete. If not, the prototype must be modified and tested. This process continues until the prototype is completely satisfactory.

## 1-6 OUTLINE OF THE TEXT

In designing a control system, the control engineer must be able to determine and analyze the response of systems to various signals and disturbances. Therefore, we shall start with such analysis problems. Chapters 2–9 and 11–15 are primarily related to these problems, while Chapters 10 and 16 deal mostly with design problems. To help the reader, we shall briefly discuss the contents and arrangement of the book.

Chapters 2 and 3 review the basic mathematical background necessary for understanding the control theory presented in this book.

Chapter 4 discusses the mathematical models of physical systems. We first define the transfer function and block diagram, review various basic physical laws, and derive differential equations and transfer functions for a variety of physical systems. This chapter includes a basic discussion of signal flow graphs.

Chapter 5 presents various control modes and the introduction of fluidics. Chapter 6 is involved with the time-domain analysis of control systems. The transient response of control systems is investigated in detail. Routh's stability criterion is included in this chapter. Also included is a brief discussion of analog-computer simulation techniques.

Chapter 7 presents error analyses and an introduction to system optimization. Chapter 8 introduces the root-locus method, while Chapter 9 presents classical but widely used techniques known as frequency-response methods. The Nyquist stability criterion is derived and applied to the stability analysis of control systems.

Chapter 10 presents design and compensation techniques. Examples of system compensation by use of root-locus and frequency-response techniques are presented.

Chapter 11 discusses the describing-function analysis of nonlinear control systems. Chapter 12 is concerned with the well-known phase-plane method. Chapter 13 introduces discrete-time systems and presents the z-transform method.

Chapters 14–16 present an introductory account of modern control theory. Specifically, Chapter 14 introduces the concept of state, state variables, and state space. Here we present the state-space representation of control systems and the solution of state equations. Chapter 15 is concerned with Liapunov methods of stability analysis. Finally, Chapter 16 presents the concept of controllability and observability, optimal control systems, and an introduction to adaptive control systems.

The frequency-response methods, the core of classical control theory, are widely used in industry. They are useful in treating time-invariant single-input–single-output control systems.

Modern control theory has many advantages over classical control theory. The former is applicable to the design of time-varying and multivariable systems. It enables the control engineer to take into account arbitrary initial conditions in the synthesis of optimal control systems. In such synthesis we need to deal only with the analytical aspects of the problem. A digital computer may be programmed to handle all the necessary numerical computations. This is one of the basic advantages of modern control theory.

It is important to note that modern control theory does not completely replace classical control theory. The two approaches supplement each other. Modern control engineering is based on the useful aspects of both classical and modern control theory. The aim of this book is to give the reader a good background in and the use of the tools of modern control engineering.

## Example Problems and Solutions

**PROBLEM A-1-1.** List the major advantages and disadvantages of open-loop control systems.

**Solution.** The advantages of open-loop control systems are as follows:

1. Simple construction and ease of maintenance.
2. Less expensive than a corresponding closed-loop system.
3. There is no stability problem.
4. Convenient when output is hard to measure or economically not feasible. (For example, it would be quite expensive to provide a device to measure the quality of the output of a toaster.)

The disadvantages of open-loop control systems are as follows:

1. Disturbances and changes in calibration cause errors, and the output may be different from what is desired.
2. To maintain the required quality in the output, recalibration is necessary from time to time.

**PROBLEM A-1-2.** Figure 1-10(a) is a schematic diagram of a liquid-level control system. Here the automatic controller maintains the liquid level by comparing the actual

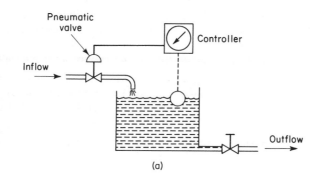

(a)

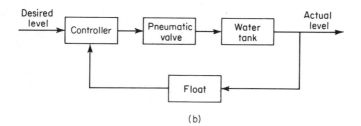

(b)

**Fig. 1-10.** (a) Liquid-level control system; (b) block diagram.

level with a desired level and correcting any error by adjusting the opening of the pneumatic valve. Figure 1-10(b) is a block diagram of the control system. Draw the corresponding block diagram for a human-operated liquid-level control system.

**Solution.** In the human-operated system, the eyes, brain, and muscles correspond to the measuring device, controller, and pneumatic valve, respectively. A block diagram is shown in Fig. 1-11.

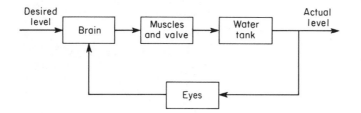

**Fig. 1-11.** Block diagram of human-operated liquid-level control system.

**PROBLEM A-1-3.** An engineering organizational system is composed of major groups, such as management, research and development, preliminary design, experiments, product design and drafting, fabrication and assembling, and testing. These groups are interconnected to make up the whole operation.

The system may be analyzed by reducing it to the most elementary set of components necessary which can provide the analytical detail required and by representing the dynamic characteristics of each component by a set of simple equations. (The dynamic performance of such a system may be determined from the relation between progressive accomplishment and time.)

Draw a functional block diagram showing an engineering organizational system.

**Solution.** A functional block diagram can be drawn by using blocks to represent the functional activities and interconnecting signal lines to represent the information or product output of the system operation. A possible block diagram is shown in Fig. 1-12.

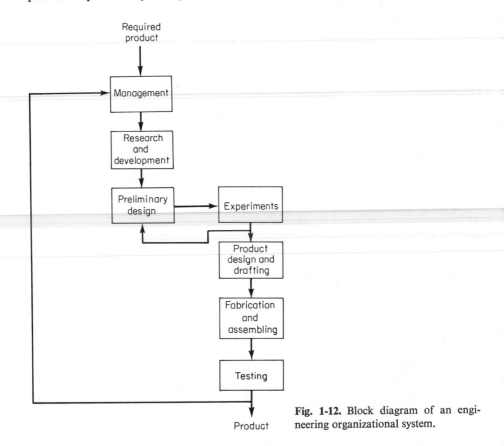

**Fig. 1-12.** Block diagram of an engineering organizational system.

PROBLEMS

**PROBLEM B-1-1.** Many closed-loop and open-loop control systems may be found in homes. List several examples and describe them.

**PROBLEM B-1-2.** Draw a block diagram of a home heating system. Note that a thermostat is the controller of this system. State what disturbances may exist in such a system.

**PROBLEM B-1-3.** Figure 1-13 shows a tension control system. Explain the sequence of control actions when the feed speed is suddenly changed for a short time.

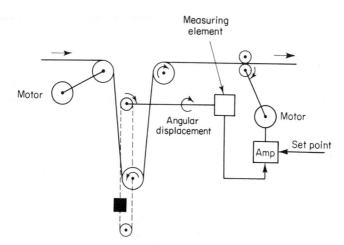

Fig. 1-13. Tension control system.

**PROBLEM B-1-4.** Figure 1-14 shows a self-operated control system. Explain its operation.

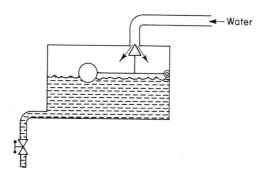

Fig. 1-14. Self-operated control system.

# 2

# MATHEMATICAL
# BACKGROUND—LAPLACE
# TRANSFORMATION

## 2-1 INTRODUCTION

The Laplace transform method is an operational method which can be used advantageously for solving linear differential equations. By use of Laplace transforms, one can convert many common functions such as sinusoidal functions, damped sinusoidal functions, and exponential functions into algebraic functions of a complex variable. Operations such as differentiation and integration can be replaced by algebraic operations in the complex plane. Thus, a linear differential equation can be transformed into an algebraic equation in a complex variable. The solution of the differential equation may then be found by use of a Laplace transform table or by use of the partial fraction expansion technique, which is presented in Section 2-4.

An advantage of the Laplace transform method is that it allows the use of graphical techniques for predicting the system performance without actually solving system differential equations. Another advantage of the Laplace transform method is that when one solves the differential equation, both the transient component and steady-state component of the solution can be obtained simultaneoulsy.

**Review of complex variables and complex functions.** Before defining the Laplace transform of the function of time $f(t)$, we present a very brief review of complex variables and complex functions. A complex variable $s$ has a real component

$\sigma$ and an imaginary component $j\omega$, or $s = \sigma + j\omega$. A complex variable $s$ may be represented by a point in the $s$ plane. Figure 2-1 illustrates the $s$ plane and a representative point $s_1 = \sigma_1 + j\omega_1$. A complex function $G(s)$, a function of $s$, has a real part and an imaginary part, or

$$G(s) = G_x + jG_y$$

where $G_x$ and $G_y$ are real quantities. The magnitude of a complex quantity $(G_x + jG_y)$ is given by $\sqrt{G_x^2 + G_y^2}$, and the angle $\theta$ of $(G_x + jG_y)$ is given by $\theta = \tan^{-1}(G_y/G_x)$. Figure 2-2 shows the complex plane and two representative complex quantities. The angle $\theta$ is measured from the positive real axis. A counterclockwise rotation is defined as the positive direction for the measurement of angles.

The complex conjugate of $G(s) = G_x + jG_y$ is defined as $\bar{G}(s) = G_x - jG_y$. A complex quantity and its conjugate have the same real part, but the imaginary part of one is the negative of the imaginary part of the other.

Complex functions $G(s)$ commonly encountered in linear control systems are single-valued functions of $s$ and are uniquely determined for a given value of $s$. A complex function $G(s)$ is said to be

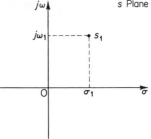

Fig. 2-1. $s$ plane and a representative point.

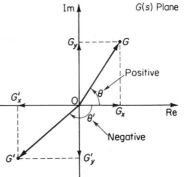

Fig. 2-2. Complex plane and two representative complex quantities.

*analytic* in a region if $G(s)$ and all its derivatives exist in that region. The derivative of an analytic function $G(s)$ is given by

$$\frac{d}{ds}G(s) = \lim_{\Delta s \to 0}\frac{G(s + \Delta s) - G(s)}{\Delta s}$$

$$= \lim_{\Delta s \to 0}\frac{\Delta G}{\Delta s}$$

The value of the derivative is independent of the choice of the path $\Delta s$. Since $\Delta s = \Delta\sigma + j\Delta\omega$, $\Delta s$ can approach zero along an infinite number of different paths. It can be shown, but is stated without a proof here, that if the derivatives taken along two particular paths, namely $\Delta s = \Delta\sigma$ and $\Delta s = j\Delta\omega$, are equal, then the derivative is unique for any other path $\Delta s = \Delta\sigma + j\Delta\omega$.

For a particular path $\Delta s = \Delta\sigma$ (which means that the path is parallel to the real axis),

$$\frac{d}{ds}G(s) = \lim_{\Delta\sigma \to 0}\left(\frac{\Delta G_x}{\Delta\sigma} + j\frac{\Delta G_y}{\Delta\sigma}\right)$$

$$= \frac{\partial G_x}{\partial\sigma} + j\frac{\partial G_y}{\partial\sigma}$$

For another particular path $\Delta s = j\,\Delta\omega$ (which means that the path is parallel to the imaginary axis),

$$\frac{d}{ds}G(s) = \lim_{j\Delta\omega \to 0}\left(\frac{\Delta G_x}{j\Delta\omega} + j\frac{\Delta G_y}{j\Delta\omega}\right)$$

$$= -j\frac{\partial G_x}{\partial\omega} + \frac{\partial G_y}{\partial\omega}$$

If these two values of the derivative are equal

$$\frac{\partial G_x}{\partial\sigma} + j\frac{\partial G_y}{\partial\sigma} = \frac{\partial G_y}{\partial\omega} - j\frac{\partial G_x}{\partial\omega}$$

or if the following two conditions

$$\frac{\partial G_x}{\partial\sigma} = \frac{\partial G_y}{\partial\omega} \quad\text{and}\quad \frac{\partial G_y}{\partial\sigma} = -\frac{\partial G_x}{\partial\omega}$$

are satisfied, then the derivative $dG(s)/ds$ is uniquely determined. These two conditions are known as the Cauchy-Riemann conditions.

As an example, consider the following $G(s)$:

$$G(s) = \frac{1}{s+1}$$

Then

$$G(\sigma + j\omega) = \frac{1}{\sigma + j\omega + 1}$$

$$= G_x + jG_y$$

where

$$G_x = \frac{\sigma + 1}{(\sigma + 1)^2 + \omega^2} \quad\text{and}\quad G_y = \frac{-\omega}{(\sigma + 1)^2 + \omega^2}$$

It can be seen that, except at $s = -1$ (namely, $\sigma = -1$, $\omega = 0$), $G(s)$ satisfies the Cauchy-Riemann conditions:

$$\frac{\partial G_x}{\partial\sigma} = \frac{\partial G_y}{\partial\omega} = \frac{\omega^2 - (\sigma + 1)^2}{[(\sigma + 1)^2 + \omega^2]^2}$$

$$\frac{\partial G_y}{\partial\sigma} = -\frac{\partial G_x}{\partial\omega} = \frac{2\omega(\sigma + 1)}{[(\sigma + 1)^2 + \omega^2]^2}$$

Hence $G(s) = 1/(s + 1)$ is analytic in the entire $s$ plane except at $s = -1$. The derivative $dG(s)/ds$ except at $s = -1$ is found to be

$$\frac{d}{ds}G(s) = \frac{\partial G_x}{\partial\sigma} + j\frac{\partial G_y}{\partial\sigma} = \frac{\partial G_y}{\partial\omega} - j\frac{\partial G_x}{\partial\omega}$$

$$= -\frac{1}{(\sigma + j\omega + 1)^2}$$

$$= -\frac{1}{(s + 1)^2}$$

Note that the derivative of an analytic function can be obtained simply by differentiating $G(s)$ with respect to $s$. In this example,

$$\frac{d}{ds}\left(\frac{1}{s+1}\right) = -\frac{1}{(s+1)^2}$$

Points in the $s$ plane at which the function $G(s)$ is analytic are called *ordinary* points, while points in the $s$ plane at which the function $G(s)$ is not analytic are called *singular* points. Singular points at which the function $G(s)$ or its derivatives approach infinity are called *poles*. For example,

$$G(s) = \frac{K(s+z)}{(s+p_1)(s+p_2)^2} \tag{2-1}$$

has poles at $s = -p_1$ and $s = -p_2$.

If $G(s)$ approaches infinity as $s$ approaches $-p$ and if the function

$$G(s)(s+p)^n \qquad (n = 1, 2, 3, \ldots)$$

has a finite, nonzero value at $s = -p$, then $s = -p$ is called a pole of order $n$. If $n = 1$, the pole is called a simple pole. If $n = 2, 3, \ldots$, the pole is called a second-order pole, a third-order pole, etc.

Points at which the function $G(s)$ equals zero are called zeros. The function given by Eq. (2-1) has a zero at $s = -z$. If points at infinity are included, $G(s)$ has the same number of poles as zeros. The function $G(s)$ given by Eq. (2-1) has two zeros at infinity in addition to a finite zero at $s = -z$ since

$$\lim_{s \to \infty} G(s) \doteq \lim_{s \to \infty} \frac{K}{s^2} \longrightarrow 0$$

Thus, $G(s)$ has three poles and three zeros (one finite zero and two infinite zeros).

**Conformal mapping.** A mapping which preserves both the size and the sense of angles is called *conformal*. Consider an analytic function $z = F(s)$. The functional relationship $z = F(s)$ can be interpreted as a mapping of points in the $s$ plane into points in the $z$ plane or the $F(s)$ plane. To any given point $P$ in the $s$ plane at which $F(s)$ is regular (meaning nonsingular), there corresponds a point $P'$ in the $F(s)$ plane. We call $P'$ the image of $P$ as given by the function $z = F(s)$.

A mapping given by an analytic function is conformal. That is, two smooth curves in the $s$ plane which intersect and form an angle $\theta$ are mapped into two smooth curves in the $F(s)$ plane which intersect and form the same angle $\theta$. (Note that distortion of the shape of two curves may occur in mapping, although the size and the sense of angles are preserved.)

If the function $G(s)$ is regular and single-valued in a domain $\Gamma$, then the image of a continuous curve in $\Gamma$ under the mapping $z = F(s)$ is also a continuous curve.

To prove that the mapping associated with an analytic function $z = F(s)$ is conformal, consider a smooth curve $s = s(\xi)$, which passes through an ordinary point $s_0$. If we write $z_0 = F(s_0)$, then

$$z - z_0 = \frac{F(s) - F(s_0)}{s - s_0}(s - s_0)$$

Therefore,

$$\underline{/z - z_0} = \underline{\left/\frac{F(s) - F(s_0)}{s - s_0}\right.} + \underline{/s - s_0}$$

where $\underline{/s - s_0}$ is the angle between the positive real axis and the vector pointing from $s_0$ to $s$. If $s$ approaches $s_0$ along the smooth curve $s(\xi)$, then $\underline{/s - s_0}$ is the angle $\theta_1$ between the positive real axis and the tangent to the curve at $s_0$. Similarly, as $z$ approaches $z_0$, $\underline{/z - z_0}$ approaches the angle $\phi_1$ which is the angle between the positive real axis and the tangent to $F(s)$ at $z_0$. Hence, we obtain

$$\phi_1 = \underline{/F'(s_0)} + \theta_1$$

provided $F'(s_0)$, the derivative of $F(s)$ evaluated at $s = s_0$, is not zero. [Note that unless $F'(s_0) \neq 0$, $\underline{/F'(s_0)}$ cannot be defined.] Hence

$$\phi_1 - \theta_1 = \underline{/F'(s_0)}$$

Note also that $\phi_1 - \theta_1$ depends not on the particular smooth curve $s = s(\xi)$ but on the point $s_0$.

Using another smooth curve $s = s_2(\xi)$ passing through the point $s_0$, we can make a similar analysis and obtain

$$\phi_2 - \theta_2 = \underline{/F'(s_0)}$$

Hence

$$\phi_1 - \theta_1 = \phi_2 - \theta_2$$

or

$$\phi_2 - \phi_1 = \theta_2 - \theta_1$$

Thus, the size and sense of the angles are preserved in this mapping. We have now seen that the mapping given by an analytic function $z = F(s)$ is conformal at all points at which $F(s)$ is regular and $F'(s) \neq 0$.

We shall use conformal mapping in discussing root-locus plots (Chapter 8) and the Nyquist stability criterion (Chapter 9).

Finally, note that it can be proved that a mapping cannot be conformal at all points of a domain $\Gamma$ unless it is the mapping given by an analytic function $z = F(s)$ which is regular in the domain. The conformality of the mapping is a characteristic property of analytic functions.

## 2-2 LAPLACE TRANSFORMATION

This section presents a definition of the Laplace transformation, a brief discussion of the condition for the existence of the Laplace transform, and examples to illustrate the derivation of Laplace transforms of several common functions.

Let us define

$f(t) =$ a function of time $t$ such that $f(t) = 0$ for $t < 0$

$s =$ a complex variable

$\mathscr{L} =$ an operational symbol indicating that the quantity which it prefixes is to be transformed by the Laplace integral $\int_0^\infty e^{-st}\, dt$

$F(s) =$ Laplace transform of $f(t)$

Then the Laplace transform of $f(t)$ is defined by

$$\mathscr{L}[f(t)] = F(s) = \int_0^\infty e^{-st}\, dt[f(t)] = \int_0^\infty f(t)\, e^{-st}\, dt$$

---

*Example 2-1. Exponential Function.* Consider the following function:

$$f(t) = 0 \qquad \text{for } t < 0$$
$$= Ae^{-\alpha t} \qquad \text{for } t \geq 0$$

where $A$ and $\alpha$ are constants. The Laplace transform of $f(t)$ is obtained as follows:

$$\mathscr{L}[f(t)] = \int_0^\infty Ae^{-\alpha t}e^{-st}\, dt$$

$$= A \int_0^\infty e^{-(\alpha+s)t}\, dt$$

$$= \frac{A}{s+\alpha}$$

It is seen that the exponential function produces a pole in the complex plane. In performing this integration, we assumed that the real part of $s$ was greater than $-\alpha$. (Note that such an assumption is necessary to make the integral absolutely convergent.)

---

The Laplace transform of a function $f(t)$ exists if $f(t)$ is sectionally continuous in every finite interval in the range $t > 0$ and if the function is of exponential order as $t$ approaches infinity. In other words, the Laplace integral must converge. A function $f(t)$ is of exponential order if a real, positive constant $\sigma$ exists such that the function

$$e^{-\sigma t}\,|f(t)|$$

approaches zero as $t$ approaches infinity. If the limit of the function $e^{-\sigma t}\,|f(t)|$ approaches zero for $\sigma$ greater than $\sigma_c$ and the limit approaches infinity for $\sigma$ less than $\sigma_c$, the value $\sigma_c$ is called the *abscissa of convergence.*

For the function $f(t) = Ae^{-\alpha t}$

$$\lim_{t\to\infty} e^{-\sigma t}\,|Ae^{-\alpha t}|$$

approaches zero if $\sigma > -\alpha$. The abscissa of convergence in this case is $\sigma_c = -\alpha$. The integral $\int_0^\infty f(t)e^{-st}\, dt$ converges only if $\sigma$, the real part of $s$, is greater than the abscissa of convergence $\sigma_c$. Thus the operator $s$ must be chosen as a constant such that this integral converges.

In terms of the poles of the function $F(s)$, the abscissa of convergence $\sigma_c$ corresponds to the real part of the pole located farthest to the right in the $s$ plane. For example, for the following function $F(s)$,

$$F(s) = \frac{K(s+3)}{(s+1)(s+2)}$$

the abscissa of convergence $\sigma_c$ is equal to $-1$. It can be seen that for such functions as $t$, $\sin \omega t$, and $t \sin \omega t$ the abscissa of convergence is equal to zero. For functions like $e^{-ct}$, $te^{-ct}$, $e^{-ct} \sin \omega t$, etc., the abscissa of convergence is equal to $-c$.

For functions which increase faster than the exponential function, however, it is impossible to find suitable values of the abscissa of convergence. Therefore, such functions as $e^{t^2}$ and $te^{t^2}$ do not possess Laplace transforms.

The reader should be cautioned that although $e^{t^2}$ (for $0 \le t \le \infty$) does not possess a Laplace transform, the time function defined by

$$f(t) = e^{t^2} \qquad \text{for } 0 \le t \le T < \infty$$
$$= 0 \qquad \text{for } t < 0, \ T < t$$

does possess a Laplace transform since $f(t) = e^{t^2}$ for only a limited time interval $0 \le t \le T$ and not for $0 \le t \le \infty$. Such a signal can physically be generated. Note that the signals that we can physically generate always have corresponding Laplace transforms.

In deriving the Laplace transform of $f(t) = Ae^{-\alpha t}$ in Example 2-1, we required that the real part of $s$ is greater than $-\alpha$ (the abscissa of convergence). A question may immediately arise as to whether or not the Laplace transform thus obtained is valid in the range where $\sigma < -\alpha$ in the $s$ plane. To answer this question, we must resort to the theory of complex variables. In the theory of complex variables, there is a theorem known as the analytic extension theorem. It states that if two analytic functions are equal for a finite length along any arc in a region in which both are analytic, then they are equal everywhere in the region. The arc of equality is usually the real axis, or a portion of it. By using this theorem the form of $F(s)$ determined by an integration in which $s$ is allowed to have any real positive value greater than the abscissa of convergence holds for any complex values of $s$ at which $F(s)$ is analytic. Thus, although we require the real part of $s$ to be greater than the abscissa of convergence to make the integral $\int_0^\infty f(t)e^{-st}\,dt$ absolutely convergent, once the Laplace transform $F(s)$ is obtained, $F(s)$ can be considered valid throughout the entire $s$ plane except at the poles of $F(s)$.

If a function $f(t)$ has a Laplace transform, then the Laplace transform of $Af(t)$, where $A$ is a constant, is given by

$$\mathscr{L}[Af(t)] = A\mathscr{L}[f(t)]$$

This is obvious from the definition of the Laplace transform. Similarly, if functions $f_1(t)$ and $f_2(t)$ have Laplace transforms, then the Laplace transform of the function $f_1(t) + f_2(t)$ is given by

$$\mathscr{L}[f_1(t) + f_2(t)] = \mathscr{L}[f_1(t)] + \mathscr{L}[f_2(t)]$$

Again the proof of this relationship is evident from the definition of the Laplace transform.

In what follows, we derive Laplace transforms of a few commonly encountered functions. Note that the Laplace transform of any Laplace-transformable function $f(t)$ is obtained by multiplying $f(t)$ by $e^{-st}$ and then integrating the product from $t = 0$ to $t = \infty$. [Note also that once we know the method of obtaining the Laplace transform, it is not necessary to derive the Laplace transform of $f(t)$ each time. Laplace transform tables can conveniently be used to find the transform of a given function $f(t)$.]

*Example 2-2. Step Function.* Consider the following step function:

$$f(t) = 0 \qquad\qquad \text{for } t < 0$$
$$= A = \text{constant} \qquad \text{for } t > 0$$

The step function here is undefined at $t = 0$. But this is immaterial since

$$\int_{0-}^{0+} Ae^{-st}\, dt = 0$$

The Laplace transform of $f(t)$ is given by

$$\mathscr{L}[f(t)] = \int_0^\infty Ae^{-st}\, dt = \frac{A}{s}$$

In performing this integration, we assume that the real part of $s$ is greater than zero (the abscissa of convergence) and therefore that $\lim_{t \to \infty} e^{-st}$ is zero. As stated earlier, the Laplace transform thus obtained is valid in the entire $s$ plane except at the pole $s = 0$.

The step function whose height is unity is called a *unit-step* function. The unit-step function which occurs at $t = t_0$ is frequently written as $u(t - t_0)$ or $1(t - t_0)$. In this book we shall use the latter notation unless otherwise stated. The step function of height $A$ can then be written as $f(t) = A1(t)$. The Laplace transform of the unit-step function which is defined by

$$1(t) = 0 \qquad \text{for } t < 0$$
$$= 1 \qquad \text{for } t > 0$$

is $1/s$, or

$$\mathscr{L}[1(t)] = \frac{1}{s}$$

Physically, a step function occurring at $t = 0$ corresponds to a constant signal suddenly applied to the system at time $t$ equals zero.

*Example 2-3. Ramp Function.* Consider the following ramp function:

$$f(t) = 0 \qquad \text{for } t < 0$$
$$= At \qquad \text{for } t \geq 0$$

where $A$ is a constant. The Laplace transform of the ramp function is given by

$$\mathscr{L}[f(t)] = A \int_0^\infty te^{-st}\, dt$$
$$= At \frac{e^{-st}}{-s}\Big|_0^\infty - \int_0^\infty \frac{Ae^{-st}}{-s}\, dt$$
$$= \frac{A}{s} \int_0^\infty e^{-st}\, dt$$
$$= \frac{A}{s^2}$$

*Example 2-4. Sinusoidal Function.* The Laplace transform of the following sinusoidal function

$$f(t) = 0 \qquad \text{for } t < 0$$
$$\phantom{f(t)} = A \sin \omega t \qquad \text{for } t \geq 0$$

where $A$ and $\omega$ are constants, is obtained as follows:

$$\mathscr{L}[f(t)] = A \int_0^\infty (\sin \omega t)e^{-st}\, dt$$

Since

$$e^{j\omega t} = \cos \omega t + j \sin \omega t$$
$$e^{-j\omega t} = \cos \omega t - j \sin \omega t$$

we obtain

$$\sin \omega t = \frac{1}{2j}(e^{j\omega t} - e^{-j\omega t})$$

Hence

$$\mathscr{L}[f(t)] = \frac{A}{2j} \int_0^\infty (e^{j\omega t} - e^{-j\omega t})e^{-st}\, dt$$

$$= \frac{A}{2j}\frac{1}{s - j\omega} - \frac{A}{2j}\frac{1}{s + j\omega}$$

$$= \frac{A\omega}{s^2 + \omega^2}$$

The Laplace transform of $\cos \omega t$ can be obtained similarly.

## 2-3 LAPLACE TRANSFORM THEOREMS

In this section we present several theorems on the Laplace transformation that are useful in the study of linear control systems.

**Translated function.** We shall obtain the Laplace transform of the translated function $f(t - \alpha)$. Here $f(t)$ is assumed to be zero for $t < 0$ or $f(t - \alpha) = 0$ for $t < \alpha$. The functions $f(t)$ and $f(t - \alpha)$ are plotted in Fig. 2-3.

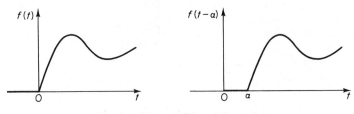

**Fig. 2-3.** Plots of $f(t)$ and $f(t - \alpha)$.

Since $f(t - \alpha) = 0$ for $0 < t < \alpha$,

$$F(s) = \int_0^\infty f(\tau)e^{-s\tau}\, d\tau$$

$$= \int_\alpha^\infty f(t - \alpha)e^{-s(t-\alpha)}\, dt$$

$$= e^{\alpha s} \int_0^\infty f(t - \alpha)e^{-st}\, dt$$

where $t - \alpha = \tau$. Thus,

$$\mathscr{L}[f(t - \alpha)] = \int_0^\infty f(t - \alpha)e^{-st}\,dt = e^{-\alpha s}\,F(s)$$

This last equation states that the translation of the time function $f(t)$ by $\alpha$ corresponds to the multiplication of the transform $F(s)$ by $e^{-\alpha s}$.

To show clearly the implication that $f(t - \alpha)$ is zero for $t < \alpha$, it is necessary to write the translated function as $f(t - \alpha)1(t - \alpha)$. With this notation,

$$\mathscr{L}[f(t - \alpha)\,1(t - \alpha)] = e^{-\alpha s}F(s)$$

---

*Example 2-5. Pulse Function.* Consider the following pulse function:

$$f(t) = A = \text{constant} \qquad \text{for } 0 < t < t_0$$
$$= 0 \qquad \text{for } t < 0,\ t_0 < t$$

The Laplace transform of $f(t)$ is obtained as follows: The pulse function $f(t)$ may be considered a step function of height $A$ which begins at $t = 0$ and which is superimposed by a negative step function of height $A$ beginning at $t = t_0$; namely,

$$f(t) = A1(t) - A1(t - t_0)$$

The Laplace transform of $f(t)$ is then obtained as

$$\mathscr{L}[f(t)] = \mathscr{L}[A1(t)] - \mathscr{L}[A1(t - t_0)]$$
$$= \frac{A}{s} - \frac{A}{s}e^{-st_0}$$
$$= \frac{A}{s}(1 - e^{-st_0})$$

---

*Example 2-6. Impulse Function.* The impulse function is a special limiting case of a pulse function. Consider the following impulse function:

$$f(t) = \lim_{t_0 \to 0} \frac{A}{t_0} \qquad \text{for } 0 < t < t_0$$
$$= 0 \qquad \text{for } t < 0,\ t_0 < t$$

Since the height of the impulse function is $A/t_0$ and the duration is $t_0$, the area under the impulse is equal to $A$. As the duration $t_0$ approaches zero, the height $A/t_0$ approaches infinity, but the area under the impulse remains equal to $A$. Note that the size of an impulse is measured by its area.

The Laplace transform of the impulse function $f(t)$ may be found as follows:

$$\mathscr{L}[f(t)] = \lim_{t_0 \to 0} \frac{A}{t_0 s}(1 - e^{-st_0})$$
$$= \lim_{t_0 \to 0} \frac{\dfrac{d}{dt_0}[A(1 - e^{-st_0})]}{\dfrac{d}{dt_0}(t_0 s)}$$
$$= \frac{As}{s}$$
$$= A$$

Thus, the Laplace transform of the impulse function is equal to the area under the impulse.

---

The impulse function whose area is equal to unity is called the *unit-impulse* function or the Dirac delta function. The unit-impulse function occurring at $t = t_0$ is usually denoted by $\delta(t - t_0)$. $\delta(t - t_0)$ satisfies the following conditions:

$$\delta(t - t_0) = 0 \qquad \text{for } t \neq t_0$$
$$\delta(t - t_0) = \infty \qquad \text{for } t = t_0$$
$$\int_{-\infty}^{\infty} \delta(t - t_0)\, dt = 1$$

An impulse which has an infinite magnitude and zero duration is a mathematical fiction and does not occur in physical systems. If, however, the magnitude of a pulse input to a system is very large and its duration is very short compared with the system time constants, then we may approximate the pulse input by an impulse function. For example, if a force or torque input $f(t)$ is applied to a system for a very short duration $0 < t < t_0$, the magnitude of $f(t)$ being sufficiently large so that the integral $\int_0^{t_0} f(t)\, dt$ is not negligible, then this input may be considered an impulse input. The impulse input supplies energy to the system in an infinitesimal time.

It should be noted that when we describe the impulse input, the size or strength of the impulse is most important but the shape of the impulse is usually immaterial. In fact, any pulse of very short duration can be considered a unit impulse if it satisfies the condition that the area under the curve remains unity as the width of the pulse approaches zero.

The concept of the impulse function is quite useful in differentiating discontinuous functions. The unit-impulse function $\delta(t)$ can be considered the time derivative of the unit-step function $1(t)$ at the point of discontinuity, or

$$\delta(t) = \frac{d}{dt} 1(t)$$

Conversely, if the unit-impulse function $\delta(t)$ is integrated, the result is the unit-step function $1(t)$. With the concept of the impulse function, we can differentiate a function containing discontinuities, giving impulses, the magnitudes of which are equal to the magnitude of each corresponding discontinuity.

**Multiplication of $f(t)$ by $e^{-\alpha t}$.** If $f(t)$ is Laplace transformable, its Laplace transform being $F(s)$, then the Laplace transform of $e^{-\alpha t} f(t)$ is obtained as follows:

$$\mathcal{L}[e^{-\alpha t} f(t)] = \int_0^{\infty} e^{-\alpha t} f(t) e^{-st}\, dt$$
$$= F(s + \alpha) \qquad (2\text{-}2)$$

We see that the multiplication of $f(t)$ by $e^{-\alpha t}$ has the effect of replacing $s$ by $(s + \alpha)$ in the Laplace transform. Conversely, changing $s$ to $(s + \alpha)$ is equivalent to multiplying $f(t)$ by $e^{-\alpha t}$. (Note that $\alpha$ may be real or complex.)

The relationship given by Eq. (2-2) is very useful for finding the Laplace transforms of such functions as $e^{-\alpha t} \sin \omega t$ and $e^{-\alpha t} \cos \omega t$. For example, since

$$\mathcal{L}[\sin \omega t] = \frac{\omega}{s^2 + \omega^2} = F(s)$$

it follows from Eq. (2-2) that the Laplace transform of $e^{-\alpha t} \sin \omega t$ is given by

$$\mathscr{L}[e^{-\alpha t}\sin\omega t] = F(s + \alpha)$$

$$= \frac{\omega}{(s + \alpha)^2 + \omega^2}$$

**Change of time scale.** In analyzing physical systems, it is sometimes desirable to change the time scale or normalize a given time function. The result obtained in terms of normalized time is useful because it can be applied directly to different systems having similar mathematical equations.

If $t$ is changed into $t/\alpha$, where $\alpha$ is a positive constant, then the function $f(t)$ is changed into $f(t/\alpha)$. If we denote the Laplace transform of $f(t)$ by $F(s)$, then the Laplace transform of $f(t/\alpha)$ may be obtained as follows:

$$\mathscr{L}\left[f\left(\frac{t}{\alpha}\right)\right] = \int_0^\infty f\left(\frac{t}{\alpha}\right)e^{-st}\,dt$$

Letting $t/\alpha = t_1$ and $\alpha s = s_1$, we obtain

$$\mathscr{L}\left[f\left(\frac{t}{\alpha}\right)\right] = \int_0^\infty f(t_1)e^{-s_1 t_1}\,d(\alpha t_1)$$

$$= \alpha \int_0^\infty f(t_1)e^{-s_1 t_1}\,dt_1$$

$$= \alpha F(s_1)$$

or

$$\mathscr{L}\left[f\left(\frac{t}{\alpha}\right)\right] = \alpha F(\alpha s)$$

As an example, consider $f(t) = e^{-t}$ and $f(t/5) = e^{-0.2t}$. We obtain

$$\mathscr{L}[f(t)] = \mathscr{L}[e^{-t}] = F(s) = \frac{1}{s+1}$$

Hence

$$\mathscr{L}\left[f\left(\frac{t}{5}\right)\right] = \mathscr{L}[e^{-0.2t}] = 5F(5s) = \frac{5}{5s+1}$$

This result can be verified easily by taking the Laplace transform of $e^{-0.2t}$ directly as follows:

$$\mathscr{L}[e^{-0.2t}] = \frac{1}{s+0.2} = \frac{5}{5s+1}$$

**Note on the lower limit of the Laplace integral.** In some cases, $f(t)$ possesses an impulse function at $t = 0$. Then the lower limit of the Laplace integral must be clearly specified as to whether it is 0- or 0+ since the Laplace transforms of $f(t)$ differ for these two lower limits. If such distinction of the lower limit of the Laplace integral is necessary, then we use the following notations:

$$\mathscr{L}_+[f(t)] = \int_{0+}^\infty f(t)e^{-st}\,dt$$

$$\mathscr{L}_-[f(t)] = \int_{0-}^\infty f(t)e^{-st}\,dt = \mathscr{L}_+[f(t)] + \int_{0-}^{0+} f(t)e^{-st}\,dt$$

If $f(t)$ involves an impulse function at $t = 0$, then

$$\mathscr{L}_+[f(t)] \neq \mathscr{L}_-[f(t)]$$

since

$$\int_{0-}^{0+} f(t)e^{-st}\, dt \neq 0$$

for such a case. Obviously, if $f(t)$ does not possess an impulse function at $t = 0$,

$$\mathscr{L}_+[f(t)] = \mathscr{L}_-[f(t)]$$

**Differentiation theorem.** The Laplace transform of the derivative of a function $f(t)$ is given by

$$\mathscr{L}\left[\frac{d}{dt}f(t)\right] = sF(s) - f(0) \qquad (2\text{-}3)$$

where $f(0)$ is the initial value of $f(t)$, evaluated at $t = 0$.

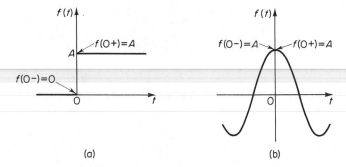

**Fig. 2-4.** (a) Step function; (b) cosine function.

For a given function $f(t)$, the values of $f(0+)$ and $f(0-)$ may be the same or different, as illustrated in Fig. 2-4. The distinction between $f(0+)$ and $f(0-)$ is important when $f(t)$ has a discontinuity at $t = 0$, because in such a case $df(t)/dt$ will involve an impulse function at $t = 0$. If $f(0+) \neq f(0-)$, Eq. (2-3) must be modified to

$$\mathscr{L}_+\left[\frac{d}{dt}f(t)\right] = sF(s) - f(0+)$$

$$\mathscr{L}_-\left[\frac{d}{dt}f(t)\right] = sF(s) - f(0-)$$

To prove the differentiation theorem, we proceed as follows: Integrating the Laplace integral by parts gives

$$\int_0^\infty f(t)e^{-st}\, dt = f(t)\frac{e^{-st}}{-s}\Big|_0^\infty - \int_0^\infty \left[\frac{d}{dt}f(t)\right]\frac{e^{-st}}{-s}\, dt$$

Hence

$$F(s) = \frac{f(0)}{s} + \frac{1}{s}\mathscr{L}\left[\frac{d}{dt}f(t)\right]$$

It follows that

$$\mathscr{L}\left[\frac{d}{dt}f(t)\right] = sF(s) - f(0)$$

Similarly, we obtain the following relationship for the second derivative of $f(t)$.

$$\mathcal{L}\left[\frac{d^2}{dt^2}f(t)\right] = s^2 F(s) - sf(0) - \dot{f}(0)$$

where $\dot{f}(0)$ is the value of $df(t)/dt$ evaluated at $t = 0$. To derive this equation, define

$$\frac{d}{dt}f(t) = g(t)$$

Then

$$\mathcal{L}\left[\frac{d^2}{dt^2}f(t)\right] = \mathcal{L}\left[\frac{d}{dt}g(t)\right]$$

$$= s\mathcal{L}[g(t)] - g(0)$$

$$= s\mathcal{L}\left[\frac{d}{dt}f(t)\right] - \dot{f}(0)$$

$$= s^2 F(s) - sf(0) - \dot{f}(0)$$

Similarly, for the $n$th derivative of $f(t)$, we obtain

$$\mathcal{L}\left[\frac{d^n}{dt^n}f(t)\right] = s^n F(s) - s^{n-1}f(0) - s^{n-2}\dot{f}(0) - \cdots - s\overset{(n-2)}{f}(0) - \overset{(n-1)}{f}(0)$$

where $f(0), \dot{f}(0), \ldots, \overset{(n-1)}{f}(0)$ represent the values of $f(t), df(t)/dt, \ldots, d^{n-1}f(t)/dt^{n-1}$, respectively, evaluated at $t = 0$. If the distinction between $\mathcal{L}_+$ and $\mathcal{L}_-$ is necessary, we substitute $t = 0+$ or $t = 0-$ into $f(t), \dot{f}(t), \ldots, \overset{(n-1)}{f}(t)$, depending on whether we take $\mathcal{L}_+$ or $\mathcal{L}_-$.

Note that, for the Laplace transform of the $n$th derivative of $f(t)$ to exist, $d^n f(t)/dt^n$ must be Laplace transformable.

Note also that if all the initial values of $f(t)$ and its derivatives are equal to zero, then the Laplace transform of the $n$th derivative of $f(t)$ is given by $s^n F(s)$.

---

*Example 2-7.* Consider the following cosine function:

$$g(t) = 0 \qquad \text{for } t < 0$$
$$= A \cos \omega t \qquad \text{for } t \geq 0$$

The Laplace transform of the cosine function can be obtained directly as in the case of the sinusoidal function. The use of the differentiation theorem will be demonstrated here, however, by deriving the Laplace transform of the cosine function from the Laplace transform of the sine function. If we define

$$f(t) = 0 \qquad \text{for } t < 0$$
$$= \sin \omega t \qquad \text{for } t \geq 0$$

then

$$F(s) = \mathcal{L}[\sin \omega t] = \frac{\omega}{s^2 + \omega^2}$$

The Laplace transform of the cosine function is obtained as

$$\mathscr{L}[A \cos \omega t] = \mathscr{L}\left[\frac{d}{dt}\frac{A}{\omega}\sin \omega t\right]$$

$$= \frac{A}{\omega}[sF(s) - f(0)]$$

$$= \frac{A}{\omega}\left(\frac{s\omega}{s^2 + \omega^2} - 0\right)$$

$$= \frac{As}{s^2 + \omega^2}$$

**Final value theorem.** If $f(t)$ and $df(t)/dt$ are Laplace transformable, if $\lim_{t\to\infty} f(t)$ exists, and if $sF(s)$ is analytic in the right-half $s$ plane including the $j\omega$ axis, except for a single pole at the origin [which means that $f(t)$ approaches a definite value as $t \longrightarrow \infty$], then

$$\lim_{t\to\infty} f(t) = \lim_{s\to0} sF(s)$$

To prove the theorem, we let $s$ approach zero in the equation for the Laplace transform of the derivative of $f(t)$, or

$$\lim_{s\to0} \int_0^\infty \left[\frac{d}{dt}f(t)\right]e^{-st}\,dt = \lim_{s\to0}[sF(s) - f(0)]$$

Since $\lim_{s\to0} e^{-st} = 1$, we obtain

$$\int_0^\infty \left[\frac{d}{dt}f(t)\right]dt = f(t)\Big|_0^\infty = f(\infty) - f(0)$$

$$= \lim_{s\to0} sF(s) - f(0)$$

Hence,

$$f(\infty) = \lim_{t\to\infty} f(t) = \lim_{s\to0} sF(s)$$

The final value theorem states that the steady-state behavior of $f(t)$ is the same as the behavior of $sF(s)$ in the neighborhood of $s = 0$. Thus, it is possible to obtain the value of $f(t)$ at $t = \infty$ directly from $F(s)$.

It is noted that when $f(t)$ is the sinusoidal function $\sin \omega t$, $sF(s)$ has poles at $s = \pm j\omega$ and $\lim_{t\to\infty} f(t)$ does not exist. Therefore, this theorem is not valid for such a function. If $f(t)$ approaches infinity as $t$ approaches infinity, then $\lim_{t\to\infty} f(t)$ does not exist, and the final value theorem does not apply to this case. We must make sure that all the conditions for the final value theorem are satisfied before applying it to a given problem.

The initial value theorem, which follows, is the counterpart of the final value theorem. The initial value theorem enables us to find the value of $f(t)$ at $t = 0+$ directly from the Laplace transform of $f(t)$. The initial value theorem does not give the value of $f(t)$ at exactly $t = 0$ but rather the value of the function $f(t)$ at a time slightly greater than zero.

**Initial value theorem.** If $f(t)$ and $df(t)/dt$ are both Laplace transformable and if $\lim_{s\to\infty} sF(s)$ exists, then

$$f(0+) = \lim_{s \to \infty} sF(s)$$

To prove this theorem, we use the equation for the $\mathscr{L}_+$ transform of $df(t)/dt$; namely,

$$\mathscr{L}_+\left[\frac{d}{dt}f(t)\right] = sF(s) - f(0+) \tag{2-4}$$

For the time interval $0+ \le t \le \infty$, as $s$ approaches infinity, $e^{-st}$ approaches zero. (Note that we must use $\mathscr{L}_+$ rather than $\mathscr{L}_-$ for this condition.) Hence

$$\lim_{s \to \infty} \int_{0+}^{\infty} \left[\frac{d}{dt}f(t)\right]e^{-st}\,dt = 0 \tag{2-5}$$

From Eqs. (2-4) and (2-5) we obtain

$$\lim_{s \to \infty} sF(s) - f(0+) = 0$$

Hence the theorem is proved.

In applying the initial value theorem, one is not limited as to the locations of the poles of $sF(s)$. (Thus, the initial value theorem is valid for the sinusoidal function.) Often the initial value theorem and the final value theorem provide a convenient check on the solution. (Note that these theorems enable one to predict the system behavior in the time domain without actually transforming functions in $s$ back to time functions.)

**Integration theorem.** The Laplace transform of the integral of $f(t)$ is given by

$$\mathscr{L}\left[\int f(t)\,dt\right] = \frac{F(s)}{s} + \frac{f^{-1}(0)}{s} \tag{2-6}$$

where $f^{-1}(0) = \int f(t)\,dt$ evaluated at $t = 0$.

Note that if $f(t)$ involves an impulse function at $t = 0$, then $f^{-1}(0+) \ne f^{-1}(0-)$. Hence, if $f(t)$ involves an impulse function at $t = 0$, then we must modify Eq. (2-6) as follows:

$$\mathscr{L}_+\left[\int f(t)\,dt\right] = \frac{F(s)}{s} + \frac{f^{-1}(0+)}{s}$$

$$\mathscr{L}_-\left[\int f(t)\,dt\right] = \frac{F(s)}{s} + \frac{f^{-1}(0-)}{s}$$

This theorem may be proved as follows: Integrating by parts yields

$$\mathscr{L}\left[\int f(t)\,dt\right] = \int_0^{\infty}\left[\int f(t)\,dt\right]e^{-st}\,dt$$

$$= \left[\int f(t)\,dt\right]\frac{e^{-st}}{-s}\Big|_0^{\infty} - \int_0^{\infty} f(t)\frac{e^{-st}}{-s}\,dt$$

$$= \frac{1}{s}\int f(t)\,dt\Big|_{t=0} + \frac{1}{s}\int_0^{\infty} f(t)e^{-st}\,dt$$

$$= \frac{f^{-1}(0)}{s} + \frac{F(s)}{s}$$

Hence the theorem is proved.

We see that integration in the time domain is converted into division in the $s$ domain. If the initial value of the integral is zero, then the Laplace transform of the integral of $f(t)$ is given by $F(s)/s$.

**Convolution integral.** Consider the Laplace transform of

$$\int_0^t f_1(t - \tau)f_2(\tau)\, d\tau$$

This integral is often written as

$$f_1(t)*f_2(t)$$

The mathematical operation $f_1(t)*f_2(t)$ is called *convolution*. Note that if we put $t - \tau = \xi$, then

$$\int_0^t f_1(t - \tau)f_2(\tau)\, d\tau = -\int_t^0 f_1(\xi)f_2(t - \xi)\, d\xi$$

$$= \int_0^t f_1(\tau)f_2(t - \tau)\, d\tau$$

Hence

$$f_1(t)*f_2(t) = \int_0^t f_1(t - \tau)f_2(\tau)\, d\tau$$

$$= \int_0^t f_1(\tau)f_2(t - \tau)\, d\tau$$

$$= f_2(t)*f_1(t)$$

Figure 2-5(a) shows curves $f_1(t)$, $f_1(t - \tau)$, and $f_2(\tau)$. Figure 2-5(b) shows the product of $f_1(t - \tau)$ and $f_2(\tau)$. The shape of the curve $f_1(t - \tau)f_2(\tau)$ depends on $t$.

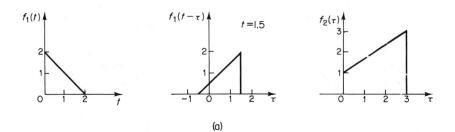

(a)

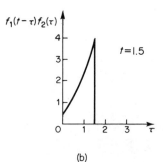

(b)

**Fig. 2-5.** (a) Plots of $f_1(t)$, $f_1(t - \tau)$, and $f_2(\tau)$; (b) plot of $f_1(t - \tau)f_2(\tau)$.

**Table 2-1.** LAPLACE TRANSFORM PAIRS

| | $f(t)$ | $F(s)$ |
|---|---|---|
| 1 | unit impulse $\delta(t)$ | $1$ |
| 2 | unit step $1(t)$ | $\dfrac{1}{s}$ |
| 3 | $t$ | $\dfrac{1}{s^2}$ |
| 4 | $e^{-at}$ | $\dfrac{1}{s+a}$ |
| 5 | $te^{-at}$ | $\dfrac{1}{(s+a)^2}$ |
| 6 | $\sin \omega t$ | $\dfrac{\omega}{s^2+\omega^2}$ |
| 7 | $\cos \omega t$ | $\dfrac{s}{s^2+\omega^2}$ |
| 8 | $t^n \quad (n=1,2,3,\ldots)$ | $\dfrac{n!}{s^{n+1}}$ |
| 9 | $t^n e^{-at} \quad (n=1,2,3,\ldots)$ | $\dfrac{n!}{(s+a)^{n+1}}$ |
| 10 | $\dfrac{1}{b-a}(e^{-at}-e^{-bt})$ | $\dfrac{1}{(s+a)(s+b)}$ |
| 11 | $\dfrac{1}{b-a}(be^{-bt}-ae^{-at})$ | $\dfrac{s}{(s+a)(s+b)}$ |
| 12 | $\dfrac{1}{ab}\left[1+\dfrac{1}{a-b}(be^{-at}-ae^{-bt})\right]$ | $\dfrac{1}{s(s+a)(s+b)}$ |
| 13 | $e^{-at}\sin \omega t$ | $\dfrac{\omega}{(s+a)^2+\omega^2}$ |
| 14 | $e^{-at}\cos \omega t$ | $\dfrac{s+a}{(s+a)^2+\omega^2}$ |
| 15 | $\dfrac{1}{a^2}(at-1+e^{-at})$ | $\dfrac{1}{s^2(s+a)}$ |
| 16 | $\dfrac{\omega_n}{\sqrt{1-\zeta^2}}e^{-\zeta\omega_n t}\sin\omega_n\sqrt{1-\zeta^2}\,t$ | $\dfrac{\omega_n^2}{s^2+2\zeta\omega_n s+\omega_n^2}$ |
| 17 | $\dfrac{-1}{\sqrt{1-\zeta^2}}e^{-\zeta\omega_n t}\sin(\omega_n\sqrt{1-\zeta^2}\,t-\phi)$ <br> $\phi=\tan^{-1}\dfrac{\sqrt{1-\zeta^2}}{\zeta}$ | $\dfrac{s}{s^2+2\zeta\omega_n s+\omega_n^2}$ |
| 18 | $1-\dfrac{1}{\sqrt{1-\zeta^2}}e^{-\zeta\omega_n t}\sin(\omega_n\sqrt{1-\zeta^2}\,t+\phi)$ <br> $\phi=\tan^{-1}\dfrac{\sqrt{1-\zeta^2}}{\zeta}$ | $\dfrac{\omega_n^2}{s(s^2+2\zeta\omega_n s+\omega_n^2)}$ |

**Table 2-2.** PROPERTIES OF LAPLACE TRANSFORMS

| | |
|---|---|
| 1 | $\mathscr{L}[Af(t)] = AF(s)$ |
| 2 | $\mathscr{L}[f_1(t) \pm f_2(t)] = F_1(s) \pm F_2(s)$ |
| 3 | $\mathscr{L}_{\pm}\left[\dfrac{d}{dt}f(t)\right] = sF(s) - f(0\pm)$ |
| 4 | $\mathscr{L}_{\pm}\left[\dfrac{d^2}{dt^2}f(t)\right] = s^2F(s) - sf(0\pm) - \dot{f}(0\pm)$ |
| 5 | $\mathscr{L}_{\pm}\left[\dfrac{d^n}{dt^n}f(t)\right] = s^nF(s) - \displaystyle\sum_{k=1}^{n} s^{n-k}\overset{(k-1)}{f}(0\pm)$ <br><br> where $\overset{(k-1)}{f}(t) = \dfrac{d^{k-1}}{dt^{k-1}}f(t)$ |
| 6 | $\mathscr{L}_{\pm}\left[\displaystyle\int f(t)\,dt\right] = \dfrac{F(s)}{s} + \dfrac{\left[\int f(t)\,dt\right]_{t=0\pm}}{s}$ |
| 7 | $\mathscr{L}_{\pm}\left[\displaystyle\iint f(t)\,dt\,dt\right] = \dfrac{F(s)}{s^2} + \dfrac{\left[\int f(t)\,dt\right]_{t=0\pm}}{s^2} + \dfrac{\left[\iint f(t)\,dt\,dt\right]_{t=0\pm}}{s}$ |
| 8 | $\mathscr{L}_{\pm}\left[\displaystyle\int\cdots\int f(t)\,(dt)^n\right] = \dfrac{F(s)}{s^n} + \displaystyle\sum_{k=1}^{n} \dfrac{1}{s^{n-k+1}}\left[\int\cdots\int f(t)\,(dt)^k\right]_{t=0\pm}$ |
| 9 | $\mathscr{L}[e^{-at}f(t)] = F(s+a)$ |
| 10 | $\mathscr{L}[f(t-a)\,1(t-a)] = e^{-as}F(s)$ |
| 11 | $\mathscr{L}[t\,f(t)] = -\dfrac{dF(s)}{ds}$ |
| 12 | $\mathscr{L}\left[\dfrac{1}{t}f(t)\right] = \displaystyle\int_s^\infty F(s)\,ds$ |
| 13 | $\mathscr{L}\left[f\left(\dfrac{t}{a}\right)\right] = a\,F(as)$ |

If $f_1(t)$ and $f_2(t)$ are piecewise continuous and of exponential order, then the Laplace transform of

$$\int_0^t f_1(t-\tau)f_2(\tau)\,d\tau$$

can be obtained as follows:

$$\mathscr{L}\left[\int_0^t f_1(t-\tau)f_2(\tau)\,d\tau\right] = F_1(s)F_2(s) \qquad (2\text{-}7)$$

where

$$F_1(s) = \int_0^\infty f_1(t)e^{-st}\,dt = \mathscr{L}[f_1(t)]$$

$$F_2(s) = \int_0^\infty f_2(t)e^{-st}\,dt = \mathscr{L}[f_2(t)]$$

To prove Eq. (2-7), note that $f_1(t-\tau)1(t-\tau) = 0$ for $\tau > t$. Hence

$$\int_0^t f_1(t - \tau)f_2(\tau)\,d\tau = \int_0^\infty f_1(t - \tau)1(t - \tau)f_2(\tau)\,d\tau$$

Then

$$\mathscr{L}\left[\int_0^t f_1(t - \tau)f_2(\tau)\,d\tau\right] = \mathscr{L}\left[\int_0^\infty f_1(t - \tau)1(t - \tau)f_2(\tau)\,d\tau\right]$$

$$= \int_0^\infty e^{-st}\left[\int_0^\infty f_1(t - \tau)1(t - \tau)f_2(\tau)\,d\tau\right]dt$$

Substituting $t - \tau = \lambda$ in this last equation and changing the order of integration, which is valid in this case because $f_1(t)$ and $f_2(t)$ are Laplace transformable, we obtain

$$\mathscr{L}\left[\int_0^t f_1(t - \tau)f_2(\tau)\,d\tau\right] = \int_0^\infty f_1(t - \tau)1(t - \tau)e^{-st}\,dt \int_0^\infty f_2(\tau)d\tau$$

$$= \int_0^\infty f_1(\lambda)e^{-s(\lambda+\tau)}\,d\lambda \int_0^\infty f_2(\tau)\,d\tau$$

$$= \int_0^\infty f_1(\lambda)e^{-s\lambda}\,d\lambda \int_0^\infty f_2(\tau)e^{-s\tau}\,d\tau$$

$$= F_1(s)F_2(s)$$

**Laplace transform tables.** Table 2-1 gives a list of Laplace transform pairs. Such a table may be used to obtain the Laplace transform of a given time function or to obtain a time function corresponding to a given Laplace transform. For solving ordinary problems in the field of control systems, Table 2-1 may be adequate. Table 2-2 summarizes a list of useful theorems and relationships in Laplace transform theory.

## 2-4 INVERSE LAPLACE TRANSFORMATION

The mathematical process of passing from the complex variable expression to the time expression is called an *inverse transformation*. The notation for the inverse Laplace transformation is $\mathscr{L}^{-1}$, so that

$$\mathscr{L}^{-1}[F(s)] = f(t)$$

In solving problems by use of the Laplace transform method, we are confronted with the question of how to find $f(t)$ from $F(s)$. Mathematically, $f(t)$ is found from $F(s)$ by the following expression:

$$f(t) = \frac{1}{2\pi j}\int_{c-j\infty}^{c+j\infty} F(s)e^{st}\,ds \qquad (t > 0) \tag{2-8}$$

where $c$, the abscissa of convergence, is a real constant and is chosen larger than the real parts of all singular points of $F(s)$. Thus, the path of integration is parallel to the $j\omega$ axis and is displaced by the amount $c$ from it. This path of integration is to the right of all singular points.

Performing the integration given by Eq. (2-8) appears complicated. Fortunately, there are simpler methods by which $f(t)$ can be found from $F(s)$ than by performing this integration directly. A convenient method for obtaining inverse Laplace transforms is to use a table of Laplace transforms. In this case, the Laplace transform must be in a form immediately recognizable in such a table. Quite often the function in question may not appear in tables of Laplace transforms available to the engineer. If a particular transform $F(s)$ cannot be found in a table, then we may expand it into partial fractions and write $F(s)$ in terms of simple functions of $s$ for which the inverse Laplace transforms are already known.

Note that these simpler methods for finding inverse Laplace transforms are based on the fact that the unique correspondence of a time function and its inverse Laplace transform holds for any continuous time function.

**Partial fraction expansion method for finding inverse Laplace transforms.** If $F(s)$, the Laplace transform of $f(t)$, is broken up into components

$$F(s) = F_1(s) + F_2(s) + \cdots + F_n(s)$$

and if the inverse Laplace transforms of $F_1(s), F_2(s), \ldots, F_n(s)$ are readily available, then

$$\mathcal{L}^{-1}[F(s)] = \mathcal{L}^{-1}[F_1(s)] + \mathcal{L}^{-1}[F_2(s)] + \cdots + \mathcal{L}^{-1}[F_n(s)]$$
$$= f_1(t) + f_2(t) + \cdots + f_n(t) \tag{2-9}$$

where $f_1(t), f_2(t), \ldots, f_n(t)$ are the inverse Laplace transforms of $F_1(s), F_2(s), \ldots, F_n(s)$, respectively.

For problems in control theory, $F(s)$ is frequently in the following form:

$$F(s) = \frac{B(s)}{A(s)}$$

where $A(s)$ and $B(s)$ are polynomials in $s$, and the degree of $B(s)$ is not higher than that of $A(s)$.

In applying the partial fraction expansion technique for finding the inverse Laplace transform of $F(s) = B(s)/A(s)$, one must know the roots of the denominator polynomial $A(s)$ in advance. (In other words, this method does not apply until the denominator polynomial has been factored.)

The advantage of the partial fraction expansion approach is that the individual terms of $F(s)$, resulting from the expansion into partial fraction form, are very simple functions of $s$; hence we need not refer to a Laplace transform table if we memorize several simple Laplace transform pairs.

Consider $F(s)$ written in the following factored form:

$$F(s) = \frac{B(s)}{A(s)} = \frac{K(s + z_1)(s + z_2) \cdots (s + z_m)}{(s + p_1)(s + p_2) \cdots (s + p_n)}$$

where $p_1, p_1, \ldots, p_n$ and $z_1, z_2, \ldots, z_m$ are either real or complex quantities; but for each complex $p$ or $z$ there will occur the complex conjugate of $p$ or $z$, respectively. Here the highest power of $s$ in $A(s)$ is assumed to be greater than that in $B(s)$.

In the expansion of $B(s)/A(s)$ into partial fraction form, it is important that the highest power of $s$ in $A(s)$ be greater than the highest power of $s$ in $B(s)$. If this is not the case, the numerator $B(s)$ must be divided by the denominator $A(s)$ to produce a polynomial in $s$ plus a remainder (a ratio of polynomials in $s$ whose numerator is of lower degree than the denominator). (For details, see Example 2-9.)

**Partial fraction expansion when $F(s)$ involves only distinct poles.** In this case $F(s)$ can always be expanded into a sum of simple partial fractions as follows:

$$F(s) = \frac{B(s)}{A(s)} = \frac{a_1}{s + p_1} + \frac{a_2}{s + p_2} + \cdots + \frac{a_n}{s + p_n} \tag{2-10}$$

where the $a_k$ are constants. Here $a_k$ is called the *residue* at the pole at $s = -p_k$. The value of $a_k$ can be found by multiplying both sides of Eq. (2-10) by $(s + p_k)$ and letting $s = -p_k$, giving

$$\left[\frac{B(s)}{A(s)}(s + p_k)\right]_{s=-p_k} = \left[\frac{a_1}{s + p_1}(s + p_k) + \frac{a_2}{s + p_2}(s + p_k)\right.$$
$$\left. + \cdots + \frac{a_k}{s + p_k}(s + p_k) + \cdots + \frac{a_n}{s + p_n}(s + p_k)\right]_{s=-p_k}$$
$$= a_k$$

We see that all the expanded terms drop out, with the exception of $a_k$. Thus, the residue $a_k$ is found from

$$a_k = \left[\frac{B(s)}{A(s)}(s + p_k)\right]_{s=-p_k} \tag{2-11}$$

Note that since $f(t)$ is a real function of time, if $p_1$ and $p_2$ are complex conjugates, then the residues $a_1$ and $a_2$ are also complex conjugates. Only one of the conjugates $a_1$ or $a_2$ need be evaluated since the other is known automatically.

Referring to Eq. (2-9) and noting that

$$\mathscr{L}^{-1}\left[\frac{a_k}{s + p_k}\right] = a_k e^{-p_k t}$$

we obtain $f(t) = \mathscr{L}^{-1}[F(s)]$ as follows:

$$f(t) = a_1 e^{-p_1 t} + a_2 e^{-p_2 t} + \cdots + a_n e^{-p_n t} \qquad (t \geq 0)$$

---

*Example 2-8.* Find the inverse Laplace transform of

$$F(s) = \frac{s + 3}{(s + 1)(s + 2)}$$

The partial fraction expansion of $F(s)$ is

$$F(s) = \frac{s + 3}{(s + 1)(s + 2)} = \frac{a_1}{s + 1} + \frac{a_2}{s + 2}$$

where $a_1$ and $a_2$ are found by use of Eq. (2-11) as follows:

$$a_1 = \left[\frac{s + 3}{(s + 1)(s + 2)}(s + 1)\right]_{s=-1} = 2$$

$$a_2 = \left[\frac{s + 3}{(s + 1)(s + 2)}(s + 2)\right]_{s=-2} = -1$$

Thus,

$$f(t) = \mathscr{L}^{-1}[F(s)]$$

$$= \mathscr{L}^{-1}\left[\frac{2}{s+1}\right] + \mathscr{L}^{-1}\left[\frac{-1}{s+2}\right]$$

$$= 2e^{-t} - e^{-2t} \quad (t \geq 0)$$

---

*Example 2-9.* Find the inverse Laplace transform of

$$G(s) = \frac{s^3 + 5s^2 + 9s + 7}{(s+1)(s+2)}$$

Dividing the numerator by the denominator, we obtain

$$G(s) = s + 2 + \frac{s+3}{(s+1)(s+2)}$$

The third term on the right-hand side of this last equation is $F(s)$ in Example 2-8. Note that the Laplace transform of the unit-impulse function $\delta(t)$ is 1 and the Laplace transform of $d\,\delta(t)/dt$ is $s$. We thus obtain the inverse Laplace transform of $G(s)$ as follows:

$$g(t) = \frac{d}{dt}\,\delta(t) + 2\,\delta(t) + 2e^{-t} - e^{-2t} \quad (t \geq 0-)$$

---

**Partial fraction expansion when $F(s)$ involves complex-conjugate poles.** If $p_1$ and $p_2$ are complex-conjugate poles, then the following expansion may be used:

$$F(s) = \frac{B(s)}{A(s)} = \frac{\alpha_1 s + \alpha_2}{(s+p_1)(s+p_2)} + \frac{a_3}{s+p_3} + \cdots + \frac{a_n}{s+p_n} \quad (2\text{-}12)$$

The values of $\alpha_1$ and $\alpha_2$ are found by multiplying both sides of Eq. (2-12) by $(s+p_1)(s+p_2)$ and letting $s = -p_1$, giving

$$\left[\frac{B(s)}{A(s)}(s+p_1)(s+p_2)\right]_{s=-p_1}$$

$$= \left[(\alpha_1 s + \alpha_2) + \frac{a_3}{s+p_3}(s+p_1)(s+p_2) + \cdots + \frac{a_n}{s+p_n}(s+p_1)(s+p_2)\right]_{s=-p_1}$$

We see that all the expanded terms drop out with the exception of the term $(\alpha_i s + \alpha_2)$. Thus,

$$(\alpha_1 s + \alpha_2)_{s=-p_1} = \left[\frac{B(s)}{A(s)}(s+p_1)(s+p_2)\right]_{s=-p_1} \quad (2\text{-}13)$$

Since $p_1$ is a complex quantity, both sides of Eq. (2-13) are complex quantities. Equating the real parts of both sides of Eq. (2-13), we obtain one equation. Similarly, equating the imaginary parts of both sides of Eq. (2-13), we obtain another equation. From these two equations it is possible to determine $\alpha_1$ and $\alpha_2$.

---

*Example 2-10.* Find the inverse Laplace transform of

$$F(s) = \frac{s+1}{s(s^2 + s + 1)}$$

$F(s)$ can be expanded as follows:

$$\frac{s+1}{s(s^2 + s + 1)} = \frac{\alpha_1 s + \alpha_2}{s^2 + s + 1} + \frac{a}{s} \quad (2\text{-}14)$$

To determine $\alpha_1$ and $\alpha_2$, note that

$$s^2 + s + 1 = (s + 0.5 + j0.866)(s + 0.5 - j0.866)$$

By multiplying both sides of Eq. (2-14) by $(s^2 + s + 1)$ and letting $s = -0.5 - j0.866$, we obtain

$$\left(\frac{s+1}{s}\right)_{s=-0.5-j0.866} = (\alpha_1 s + \alpha_2)_{s=-0.5-j0.866}$$

or

$$\frac{0.5 - j0.866}{-0.5 - j0.866} = \alpha_1(-0.5 - j0.866) + \alpha_2$$

which can be simplified as follows:

$$0.5 - j0.866 = \alpha_1(0.25 + j0.866 - 0.75) + \alpha_2(-0.5 - j0.866)$$

Equating the real and imaginary parts of both sides of this equation, respectively, gives

$$-0.5\alpha_1 - 0.5\alpha_2 = 0.5$$

$$0.866\alpha_1 - 0.866\alpha_2 = -0.866$$

or

$$\alpha_1 + \alpha_2 = -1$$

$$\alpha_1 - \alpha_2 = -1$$

It follows that

$$\alpha_1 = -1, \qquad \alpha_2 = 0$$

To determine $a$, we multiply both sides of Eq. (2-14) by $s$ and let $s = 0$, giving

$$a = \left[\frac{s(s+1)}{s(s^2 + s + 1)}\right]_{s=0} = 1$$

Hence

$$F(s) = \frac{-s}{s^2 + s + 1} + \frac{1}{s}$$

$$= \frac{1}{s} - \frac{s + 0.5}{(s + 0.5)^2 + 0.866^2} + \frac{0.5}{(s + 0.5)^2 + 0.866^2}$$

The inverse Laplace transform of $F(s)$ is then given by

$$f(t) = \mathcal{L}^{-1}[F(s)]$$

$$= 1 - e^{-0.5t} \cos 0.866t + 0.578e^{-0.5t} \sin 0.866t \qquad (t \geq 0)$$

**Partial fraction expansion when $F(s)$ involves multiple poles.** Consider $F(s) = B(s)/A(s)$, where $A(s) = 0$ has multiple roots $p_1$ of multiplicity $r$. (Other roots are assumed to be distinct.) $A(s)$ may then be written as

$$A(s) = (s + p_1)^r (s + p_{r+1})(s + p_{r+2}) \cdots (s + p_n)$$

The partial fraction expansion of $F(s)$ is

$$F(s) = \frac{B(s)}{A(s)} = \frac{b_r}{(s + p_1)^r} + \frac{b_{r-1}}{(s + p_1)^{r-1}} + \cdots + \frac{b_1}{s + p_1}$$

$$+ \frac{a_{r+1}}{s + p_{r+1}} + \frac{a_{r+2}}{s + p_{r+2}} + \cdots + \frac{a_n}{s + p_n} \qquad (2\text{-}15)$$

where $b_r, b_{r-1}, \ldots, b_1$ are given by

$$b_r = \left[\frac{B(s)}{A(s)}(s + p_1)^r\right]_{s=-p_1}$$

$$b_{r-1} = \left\{\frac{d}{ds}\left[\frac{B(s)}{A(s)}(s + p_1)^r\right]\right\}_{s=-p_1}$$

$$\vdots$$

$$b_{r-j} = \frac{1}{j!}\left\{\frac{d^j}{ds^j}\left[\frac{B(s)}{A(s)}(s + p_1)^r\right]\right\}_{s=-p_1}$$

$$\vdots$$

$$b_1 = \frac{1}{(r-1)!}\left\{\frac{d^{r-1}}{ds^{r-1}}\left[\frac{B(s)}{A(s)}(s + p_1)^r\right]\right\}_{s=-p_1}$$

The foregoing relationships for the $b$'s may be obtained as follows: By multiplying both sides of Eq. (2-15) by $(s + p_1)^r$ and letting $s$ approach $-p_1$, we obtain

$$b_r = \left[\frac{B(s)}{A(s)}(s + p_1)^r\right]_{s=-p_1}$$

If we multiply both sides of Eq. (2-15) by $(s + p_1)^r$ and then differentiate with respect to $s$,

$$\frac{d}{ds}\left[\frac{B(s)}{A(s)}(s + p_1)^r\right] = b_r\frac{d}{ds}\left[\frac{(s + p_1)^r}{(s + p_1)^r}\right] + b_{r-1}\frac{d}{ds}\left[\frac{(s + p_1)^r}{(s + p_1)^{r-1}}\right]$$
$$+ \cdots + b_1\frac{d}{ds}\left[\frac{(s + p_1)^r}{s + p_1}\right] + a_{r+1}\frac{d}{ds}\left[\frac{(s + p_1)^r}{s + p_{r+1}}\right]$$
$$+ \cdots + a_n\frac{d}{ds}\left[\frac{(s + p_1)^r}{s + p_n}\right]$$

The first term on the right-hand side of this last equation vanishes. The second term becomes $b_{r-1}$. Each of the other terms contains some power of $(s + p_1)$ as a factor, with the result that when $s$ is allowed to approach $-p_1$, these terms drop out. Hence

$$b_{r-1} = \lim_{s \to -p_1} \frac{d}{ds}\left[\frac{B(s)}{A(s)}(s + p_1)^r\right]$$
$$= \left\{\frac{d}{ds}\left[\frac{B(s)}{A(s)}(s + p_1)^r\right]\right\}_{s=-p_1}$$

Similarly, by performing successive differentiations with respect to $s$ and by letting $s$ approach $-p_1$, we obtain equations for the $b_{r-j}$.

Note that the inverse Laplace transform of $1/(s + p_1)^n$ is given by

$$\mathscr{L}^{-1}\left[\frac{1}{(s + p_1)^n}\right] = \frac{t^{n-1}}{(n-1)!}e^{-p_1 t}$$

The constants $a_{r+1}, a_{r+2}, \ldots, a_n$ in Eq. (2-15) are determined from

$$a_k = \left[\frac{B(s)}{A(s)}(s + p_k)\right]_{s=-p_k} \qquad (k = r + 1, r + 2, \ldots, n)$$

The inverse Laplace transform of $F(s)$ is then obtained as follows:

$$f(t) = \mathscr{L}^{-1}[F(s)] = \left[\frac{b_r}{(r-1)!}t^{r-1} + \frac{b_{r-1}}{(r-2)!}t^{r-2} + \cdots + b_2 t + b_1\right]e^{-p_1 t}$$
$$+ a_{r+1}e^{-p_{r+1}t} + a_{r+2}e^{-p_{r+2}t} + \cdots + a_n e^{-p_n t} \qquad (t \geq 0)$$

*Example 2-11.* Find the inverse Laplace transform of the following function $F(s)$:

$$F(s) = \frac{s^2 + 2s + 3}{(s+1)^3}$$

Expanding $F(s)$ into partial fractions, we obtain

$$F(s) = \frac{B(s)}{A(s)} = \frac{b_3}{(s+1)^3} + \frac{b_2}{(s+1)^2} + \frac{b_1}{s+1}$$

where $b_3$, $b_2$, and $b_1$ are determined as follows:

$$b_3 = \left[\frac{B(s)}{A(s)}(s+1)^3\right]_{s=-1}$$
$$= (s^2 + 2s + 3)_{s=-1} = 2$$
$$b_2 = \left\{\frac{d}{ds}\left[\frac{B(s)}{A(s)}(s+1)^3\right]\right\}_{s=-1}$$
$$= \left[\frac{d}{ds}(s^2 + 2s + 3)\right]_{s=-1}$$
$$= (2s + 2)_{s=-1} = 0$$
$$b_1 = \frac{1}{(3-1)!}\left\{\frac{d^2}{ds^2}\left[\frac{B(s)}{A(s)}(s+1)^3\right]\right\}_{s=-1}$$
$$= \frac{1}{2!}\left[\frac{d^2}{ds^2}(s^2 + 2s + 3)\right]_{s=-1}$$
$$= \frac{1}{2}(2) = 1$$

Thus, we obtain

$$f(t) = \mathscr{L}^{-1}[F(s)]$$
$$= \mathscr{L}^{-1}\left[\frac{2}{(s+1)^3}\right] + \mathscr{L}^{-1}\left[\frac{1}{s+1}\right]$$
$$= (t^2 + 1)e^{-t} \qquad (t \geq 0)$$

## 2-5 SOLUTION OF LINEAR DIFFERENTIAL EQUATIONS BY THE LAPLACE TRANSFORM METHOD

In Sections 2-1 through 2-4 we have presented some concepts and techniques of the Laplace transform method. This section presents the use of the Laplace transform method for solving linear differential equations.

The Laplace transform method yields the complete solution (the particular solution plus the complementary solution) of linear differential equations. Classical methods for finding the complete solution of a differential equation require the evaluation of the integration constants by use of the initial conditions. In the case of the Laplace transform method, however, the evaluation of the integration constants from the initial conditions is not needed since the initial conditions are automatically included in the Laplace transform of the differential equation.

If all initial conditions are zero, then the Laplace transform of the differential equation is obtained simply by replacing $d/dt$ with $s$, $d^2/dt^2$ with $s^2$, etc.

In solving linear differential equations by the Laplace transform method, we carry out the following two steps:

1. By taking the Laplace transform of each term in the given linear differential equation, convert the differential equation into an algebraic equation in $s$ and obtain the expression for the Laplace transform of the dependent variable by rearranging the algebraic equation.

2. The time solution of the differential equation is obtained by finding the inverse Laplace transform of the dependent variable.

Consider as an example the following differential equation:

$$m\ddot{x} + kx = f(t) \tag{2-16}$$

where $f(t)$ represents the forcing function. Taking the Laplace transform of each term in Eq. (2-16), we obtain

$$\mathscr{L}[m\ddot{x}] = m[s^2 X(s) - sx(0) - \dot{x}(0)]$$

$$\mathscr{L}[kx] = kX(s)$$

$$\mathscr{L}[f(t)] = F(s)$$

The Laplace transform of Eq. (2-16) can then be written as

$$(ms^2 + k)X(s) - msx(0) - m\dot{x}(0) = F(s) \tag{2-17}$$

Solving Eq. (2-17) for $X(s)$, we obtain

$$X(s) = \frac{F(s)}{ms^2 + k} + \frac{msx(0) + m\dot{x}(0)}{ms^2 + k} \tag{2-18}$$

The first term on the right-hand side of Eq. (2-18) represents the solution of the differential equation when the initial conditions are all zero (particular solution). The second term on the right-hand side of this equation represents the effect of the initial conditions (complementaly solution). The time solution of the differential equation is obtained by taking the inverse Laplace transform of $X(s)$ as follows:

$$x(t) = \mathscr{L}^{-1}\left[\frac{F(s)}{ms^2 + k}\right] + \mathscr{L}^{-1}\left[\frac{msx(0) + m\dot{x}(0)}{ms^2 + k}\right] \tag{2-19}$$

As an example, if $f(t)$ is a unit-step function, then $F(s) = 1/s$ and Eq. (2-19) becomes

$$x(t) = \mathscr{L}^{-1}\left[\frac{1}{s(ms^2 + k)}\right] + \mathscr{L}^{-1}\left[\frac{msx(0) + m\dot{x}(0)}{ms^2 + k}\right]$$

$$= \left(\frac{1}{k} - \frac{1}{k}\cos\sqrt{\frac{k}{m}}t\right) + \left[x(0)\cos\sqrt{\frac{k}{m}}t + \dot{x}(0)\sqrt{\frac{m}{k}}\sin\sqrt{\frac{k}{m}}t\right]$$

It can be seen that the initial conditions $x(0)$ and $\dot{x}(0)$ appear in the solution. Thus, $x(t)$ has no undetermined constants.

EXAMPLE PROBLEMS AND SOLUTIONS

**PROBLEM A-2-1.** Find the poles of the following $F(s)$:

$$F(s) = \frac{1}{1 - e^{-s}}$$

**Solution.** The poles are found from

$$e^{-s} = 1$$

or

$$e^{-(\sigma + j\omega)} = e^{-\sigma}(\cos \omega - j \sin \omega) = 1$$

From this it follows that $\sigma = 0$, $\omega = \pm 2n\pi$ $(n = 0, 1, 2, \ldots)$. Thus, the poles are located at

$$s = \pm j2n\pi \qquad (n = 0, 1, 2, \ldots)$$

**PROBLEM A-2-2.** The derivative of the unit-impulse function $\delta(t)$ is called a *unit-doublet function*. (Thus, the integral of the unit-doublet function is the unit-impulse function.) Mathematically, an example of the unit-doublet function $u_2(t)$ may be given by

$$u_2(t) = \lim_{t_0 \to 0} \frac{1(t) - 2[1(t - t_0)] + 1(t - 2t_0)}{t_0^2}$$

Obtain the Laplace transform of $u_2(t)$.

**Solution.** The Laplace transform of $u_2(t)$ is given by

$$\mathcal{L}[u_2(t)] = \lim_{t_0 \to 0} \frac{1}{t_0^2} \left( \frac{1}{s} - \frac{2}{s}e^{-t_0 s} + \frac{1}{s}e^{-2t_0 s} \right)$$

$$= \lim_{t_0 \to 0} \frac{1}{t_0^2 s} \left[ 1 - 2\left( 1 - t_0 s + \frac{t_0^2 s^2}{2} + \cdots \right) + \left( 1 - 2t_0 s + \frac{4t_0^2 s^2}{2} + \cdots \right) \right]$$

$$= \lim_{t_0 \to 0} \frac{1}{t_0^2 s} \left[ t_0^2 s^2 + \text{(higher order terms in } t_0 s) \right] = s$$

**PROBLEM A-2-3.** The graphical method of evaluating residues is particularly time-saving when the system involves many complex-conjugate poles and zeros.

Consider the following $F(s)$:

$$F(s) = \frac{s + z}{s(s + p_1)(s + p_2)}$$

where $p_1$ and $p_2$ are complex-conjugate quantities and $z$ is a real quantity. The function $F(s)$ has three poles, located at $s = 0$, $s = -p_1$, and $s = -p_2$, and one finite zero at $s = -z$. If these poles and zero are plotted in the complex plane, then the residues in the partial fraction expansion can be computed graphically. The partial fraction expansion of $F(s)$ is written as

$$F(s) = \frac{a_1}{s} + \frac{a_2}{s + p_1} + \frac{a_3}{s + p_2}$$

where $a_1 = z/(p_1 p_2)$ and $a_2$ is given by

$$a_2 = \left[ \frac{s + z}{s(s + p_2)} \right]_{s = -p_1} = \frac{z - p_1}{(-p_1)(p_2 - p_1)}$$

Here $(z - p_1)$ may be considered a vector drawn from $-z$ to $-p_1$. Referring to Fig. 2-6, we see that

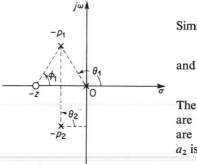

$$z - p_1 = |z - p_1| \underline{/\phi_1}$$

Similarly,

$$-p_1 = |0 - p_1| \underline{/\theta_1}$$

and

$$p_2 - p_1 = |p_2 - p_1| \underline{/\theta_2}$$

The magnitudes $|z - p_1|$, $|0 - p_1|$, and $|p_2 - p_1|$ are measured directly. The angles $\phi_1$, $\theta_1$, and $\theta_2$ are determined by a protractor. Then, the residue $a_2$ is found to be

**Fig. 2-6.** Pole-zero plot.

$$a_2 = \frac{|z - p_1|}{|0 - p_1| \cdot |p_2 - p_1|} \underline{/\phi_1 - \theta_1 - \theta_2}$$

Since $a_3$ is the complex conjugate of $a_2$, it follows that

$$a_3 = \frac{|z - p_1|}{|0 - p_1| \cdot |p_2 - p_1|} \underline{/-\phi_1 + \theta_1 + \theta_2}$$

(Note that the residue at the pole at origin can be found easily by algebraic means.)

Determine the residues at the poles when $z = 1$, $p_1 = 0.5 + j0.866$, and $p_2 = 0.5 - j0.866$.

**Solution.** The residue $a_1$ is found to be

$$a_1 = \left[ \frac{s + z}{(s + p_1)(s + p_2)} \right]_{s=0} = \frac{z}{p_1 p_2} = 1$$

Graphically $(z - p_1)$, $(0 - p_1)$, and $(p_2 - p_1)$ are found to be

$$z - p_1 = 1 \underline{/60°}$$
$$0 - p_1 = 1 \underline{/120°}$$
$$p_2 - p_1 = 1.73 \underline{/90°}$$

Therefore,

$$a_2 = \frac{1}{1 \times 1.73} \underline{/60° - 120° - 90°}$$
$$= 0.578 \underline{/-150°}$$
$$= 0.578(\cos 150° - j \sin 150°)$$
$$= -0.5 - j0.289$$

Hence, $a_3$ which is the complex conjugate of $a_2$, is obtained as

$$a_3 = -0.5 + j0.289$$

Such a graphical method of evaluating residues is quite convenient when the engineer desires a rough estimate of the values of residues without laborious computations. Note, however, that such a graphical computation does not apply to multiple poles.

**PROBLEM A-2-4.** Find the Laplace transform of the following differential equation:

$$\ddot{x} + 3\dot{x} + 6x = 0, \qquad x(0) = 0, \qquad \dot{x}(0) = 3$$

Taking the inverse Laplace transform of $X(s)$ obtain the time solution $x(t)$.

**Solution.** The Laplace transform of the differential equation is

$$s^2 X(s) - sx(0) - \dot{x}(0) + 3sX(s) - 3x(0) + 6X(s) = 0$$

Substituting the initial conditions and solving for $X(s)$,

$$X(s) = \frac{3}{s^2 + 3s + 6} = \frac{2\sqrt{3}}{\sqrt{5}} \frac{\frac{\sqrt{15}}{2}}{(s + 1.5)^2 + \left(\frac{\sqrt{15}}{2}\right)^2}$$

The inverse Laplace transform of $X(s)$ is

$$x(t) = \frac{2\sqrt{3}}{\sqrt{5}} e^{-1.5t} \sin\left(\frac{\sqrt{15}}{2}t\right)$$

**PROBLEM A-2-5.** Consider the mechanical system shown in Fig. 2-7. Suppose that the system is set into motion by a unit-impulse force. Find the resulting oscillation. Assume that the system is at rest initially.

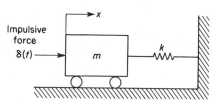

**Fig. 2-7.** Mechanical system.

**Solution.** The system is excited by an impulse input. Hence

$$m\ddot{x} + kx = \delta(t)$$

Taking the Laplace transform of both sides of this equation gives

$$m[s^2 X(s) - sx(0) - \dot{x}(0)] + kX(s) = 1$$

By substituting the initial conditions $x(0) = 0$ and $\dot{x}(0) = 0$ into this last equation and solving for $X(s)$, we obtain

$$X(s) = \frac{1}{ms^2 + k}$$

The inverse Laplace transform of $X(s)$ becomes

$$x(t) = \frac{1}{\sqrt{mk}} \sin\sqrt{\frac{k}{m}}t$$

The oscillation is a simple harmonic motion. The amplitude of the oscillation is $1/\sqrt{mk}$.

**PROBLEM A-2-6.** Consider a periodic function $f_p(t)$ with period $T$. If $f(t)$ is a time function such that $f(t) = f_p(t)$ in the interval $0 < t < T$ and $f(t) = 0$ outside the interval $0 < t < T$, then $f_p(t)$ may be expressed as

$$f_p(t) = f(t) + f(t - T) + f(t - 2T) + f(t - 3T) + \cdots$$

Obtain the Laplace transform of $f_p(t)$.

**Solution.** Define the Laplace transform of $f(t)$ to be $F(s)$, or

$$\mathscr{L}[f(t)] = F(s)$$

Then the Laplace transform of $f_p(t)$ is obtained as

$$\mathscr{L}[f_p(t)] = F_p(s) = F(s) + e^{-Ts}F(s) + e^{-2Ts}F(s) + e^{-3Ts}F(s) + \cdots$$
$$= (1 + e^{-Ts} + e^{-2Ts} + e^{-3Ts} + \cdots)F(s)$$
$$= \frac{F(s)}{1 - e^{-Ts}}$$

Thus, the Laplace transform of $f_p(t)$ can be obtained by multiplying $F(s)$ by $1/(1 - e^{-Ts})$.

PROBLEMS

**PROBLEM B-2-1.** Find the Laplace transform of the function $f(t)$ defined by

$$f(t) = \frac{1}{a^2} \qquad \text{for } 0 < t < a$$

$$= -\frac{1}{a^2} \qquad \text{for } a < t < 2a$$

$$= 0 \qquad \text{for } t < 0, 2a < t$$

Find also the limiting value of $F(s)$ as $a$ approaches zero.

**PROBLEM B-2-2.** Find the Laplace transforms of the following functions. Assume that $f(t) = 0$ for $t < 0$.

1                           $f(t) = 0.03(1 - \cos 2t)$

2.                        $f(t) = e^{-0.4t} \cos 12t$

3.                        $f(t) = \sin\left(5t + \frac{\pi}{3}\right)$

4.                        $f(t) = t^n$

5.                        $f(t) = t^n e^{at}$

**PROBLEM B-2-3.** Find the inverse Laplace transforms of the following functions:

1.                        $F(s) = \dfrac{1}{s(s + 1)}$

2.                        $F(s) = \dfrac{s + 1}{(s + 2)(s + 3)}$

3.                        $F(s) = \dfrac{e^{-s}}{s - 1}$

4.                        $F(s) = \dfrac{(s + 3)(s + 4)(s + 5)}{(s + 1)(s + 2)}$

5.                        $F(s) = \dfrac{s}{(s + 1)^2(s + 2)}$

**PROBLEM B-2-4.** By applying the final value theorem, find the final value of $f(t)$ whose Laplace transform is given by

$$F(s) = \frac{10}{s(s + 1)}$$

Verify this result by taking the inverse Laplace transform of $F(s)$ and letting $t \rightarrow \infty$.

**PROBLEM B-2-5.** Given

$$F(s) = \frac{1}{(s + 2)^2}$$

determine the values of $f(0+)$ and $\dot{f}(0+)$. (Use the initial value theorem.)

**PROBLEM B-2-6.** Solve the following differential equation:

$$a\dot{x} + bx = k, \qquad x(0) = x_0$$

where $a$, $b$, and $k$ are constants.

**PROBLEM B-2-7.** Solve the following differential equation:

$$2\ddot{x} + 7\dot{x} + 3x = 0, \qquad x(0) = x_0, \qquad \dot{x}(0) = 0$$

**PROBLEM B-2-8.** Solve the following differential equation:

$$\ddot{x} + 2\zeta\omega_n\dot{x} + \omega_n^2 x = 0, \qquad x(0) = a, \qquad \dot{x}(0) = b$$

where $a$ and $b$ are constants.

**PROBLEM B-2-9.** Consider the system shown in Fig. 2-7. The system is initially at rest. Suppose that the cart is set into motion by an impulsive force whose strength is unity. Can it be stopped by another such impulsive force?

# 3

# MATHEMATICAL
# BACKGROUND—MATRICES

## 3-1  INTRODUCTION

In deriving mathematical models of modern control systems, one finds that the differential equations involved may become quite complicated due to the multiplicity of the inputs and outputs. In fact, the number of inputs and outputs of a complex system may run into the hundreds. To simplify the mathematical expressions of the system equations, it is advantageous to use vector-matrix notation. For theoretical work, the notational simplicity gained by using vector-matrix operations is most convenient and is, in fact, essential for the analysis and synthesis of modern control systems.

By using vector-matrix notation, one can handle large and complex problems with ease by following the systematic format of representing the system equations and computing them by use of computers.

The principal objective of this chapter is to present definitions of matrices and the basic matrix algebra necessary for the subsequent analysis of control systems.

## 3-2  DEFINITIONS OF MATRICES

This section presents definitions of matrices which will be used throughout the rest of the book.

**Matrix.** A matrix is defined as a rectangular array of elements which may be real numbers, complex numbers, functions, or operators. The number of columns, in general, is not necessarily the same as the number of rows. Consider the following matrix:

$$\mathbf{A} = \begin{bmatrix} a_{11} & a_{12} & \cdots & a_{1m} \\ a_{21} & a_{22} & \cdots & a_{2m} \\ \cdot & \cdot & & \cdot \\ \cdot & \cdot & & \cdot \\ \cdot & \cdot & & \cdot \\ a_{n1} & a_{n2} & \cdots & a_{nm} \end{bmatrix}$$

where $a_{ij}$ denotes the $(i, j)$th element of the matrix $\mathbf{A}$. This matrix has $n$ rows and $m$ columns and is called an $n \times m$ matrix. The first index represents the row number and the second index the column number. The matrix $\mathbf{A}$ is sometimes written $(a_{ij})$.

**Equality of two matrices.** Two matrices are said to be equal if and only if their corresponding elements are equal. Note that equal matrices must have the same number of rows and the same number of columns.

**Vector.** A matrix having only one column such as

$$\begin{bmatrix} x_1 \\ x_2 \\ \cdot \\ \cdot \\ \cdot \\ x_n \end{bmatrix}$$

is called a *column* vector. A column vector having $n$ elements is called an $n$ vector or $n$-dimensional vector.

A matrix having only one row such as

$$[x_1 \quad x_2 \quad \cdots \quad x_n]$$

is called a *row* vector.

**Square matrix.** A square matrix is a matrix in which the number of rows is equal to that of columns. Such a square matrix is sometimes called a *matrix of order n*, where $n$ is the number of rows (or columns).

**Diagonal matrix.** If the elements other than the main diagonal elements of a square matrix $\mathbf{A}$ are zero, $\mathbf{A}$ is called a *diagonal* matrix and is written as

$$\mathbf{A} = \begin{bmatrix} a_{11} & & & 0 \\ & a_{22} & & \\ & & \cdot & \\ & & & \cdot \\ 0 & & & a_{nn} \end{bmatrix} = (a_{ij}\,\delta_{ij})$$

where the $\delta_{ij}$ are the Kronecker deltas defined by

$$\delta_{ij} = 1 \qquad \text{if } i = j$$
$$= 0 \qquad \text{if } i \neq j$$

Note that all the elements not explicitly written in the foregoing matrix are zero. The diagonal matrix is sometimes written

$$\text{diag} (a_{11}, a_{22}, \ldots, a_{nn})$$

**Identity matrix or unity matrix.** The identity matrix or unity matrix $\mathbf{I}$ is a matrix whose elements on the main diagonal are equal to unity and whose other elements are equal to zero; that is,

$$\mathbf{I} = \begin{bmatrix} 1 & 0 & \cdots & 0 \\ 0 & 1 & \cdots & 0 \\ \cdot & \cdot & & \cdot \\ \cdot & \cdot & & \cdot \\ \cdot & \cdot & & \cdot \\ 0 & 0 & \cdots & 1 \end{bmatrix} = \text{diag} (1, 1, \ldots, 1)$$

**Zero matrix.** A zero matrix is a matrix whose elements are all zero.

**Determinant of a matrix.** For each square matrix, there exists a determinant. The determinant has the following properties:

1. If any two consecutive rows or columns are interchanged, the determinant changes its sign.

2. If any row or any column consists only of zero elements, then the value of the determinant is zero.

3. If the elements of any row (or any column) are exactly $k$ times those of another row (or another column), then the value of the determinant is zero.

4. If to any row (or any column) any constant times another row (or column) is added, the value of the determinant remains unchanged.

5. If a determinant is multiplied by a constant, then only one row (or one column) is multiplied by that constant. Note, however, that the determinant of $k$ times an $n \times n$ matrix $\mathbf{A}$ is $k^n$ times the determinant of $\mathbf{A}$ or

$$|k\mathbf{A}| = k^n |\mathbf{A}|$$

6. The determinant of the product of two square matrices $\mathbf{A}$ and $\mathbf{B}$ is the product of determinants or

$$|\mathbf{AB}| = |\mathbf{A}||\mathbf{B}|$$

**Singular matrix.** A square matrix is called *singular* if the associated determinant is zero. In a singular matrix not all the rows (or not all the columns) are independent of each other.

**Nonsingular matrix.** A square matrix is called *nonsingular* if the associated determinant is nonzero.

**Transpose.** If the rows and columns of an $n \times m$ matrix $\mathbf{A}$ are interchanged, the resulting $m \times n$ matrix is called the *transpose* of the matrix $\mathbf{A}$. The transpose of the matrix $\mathbf{A}$ is denoted by $\mathbf{A}'$. Namely, if $\mathbf{A}$ is given by

$$\mathbf{A} = \begin{bmatrix} a_{11} & a_{12} & \cdots & a_{1m} \\ a_{21} & a_{22} & \cdots & a_{2m} \\ \cdot & \cdot & & \cdot \\ \cdot & \cdot & & \cdot \\ \cdot & \cdot & & \cdot \\ a_{n1} & a_{n2} & \cdots & a_{nm} \end{bmatrix}$$

then $\mathbf{A}'$ is given by

$$\mathbf{A}' = \begin{bmatrix} a_{11} & a_{21} & \cdots & a_{n1} \\ a_{12} & a_{22} & \cdots & a_{n2} \\ \cdot & \cdot & & \cdot \\ \cdot & \cdot & & \cdot \\ \cdot & \cdot & & \cdot \\ a_{1m} & a_{2m} & \cdots & a_{nm} \end{bmatrix}$$

Note that $(\mathbf{A}')' = \mathbf{A}$.

**Symmetric matrix.** If a square matrix $\mathbf{A}$ is equal to its transpose or

$$\mathbf{A} = \mathbf{A}'$$

then the matrix $\mathbf{A}$ is called a *symmetric* matrix.

**Skew-symmetric matrix.** If a square matrix $\mathbf{A}$ is equal to the negative of its transpose or

$$\mathbf{A} = -\mathbf{A}'$$

then the matrix $\mathbf{A}$ is called a *skew-symmetric* matrix.

**Conjugate matrix.** If the complex elements of a matrix $\mathbf{A}$ are replaced by their respective conjugates, then the resulting matrix is called the *conjugate* of $\mathbf{A}$. The conjugate of $\mathbf{A}$ is denoted by $\bar{\mathbf{A}} = (\bar{a}_{ij})$, where $\bar{a}_{ij}$ is the complex conjugate of $a_{ij}$. For example, if $\mathbf{A}$ is given by

$$\mathbf{A} = \begin{bmatrix} 0 & 1 & 0 \\ -1+j & -3-j3 & -1+j4 \\ -1+j & -1 & -2+j3 \end{bmatrix}$$

then

$$\bar{\mathbf{A}} = \begin{bmatrix} 0 & 1 & 0 \\ -1-j & -3+j3 & -1-j4 \\ -1-j & -1 & -2-j3 \end{bmatrix}$$

**Conjugate transpose.** The conjugate transpose is the conjugate of the transpose of a matrix. Given a matrix $\mathbf{A}$, the conjugate transpose is denoted by $\bar{\mathbf{A}}'$ or $\mathbf{A}^*$; namely,

$$\bar{\mathbf{A}}' = \mathbf{A}^* = (\bar{a}_{ji})$$

For example, if **A** is given by

$$\mathbf{A} = \begin{bmatrix} 1 & j2 & 1+j5 \\ 2+j & j & 3-j \\ 3 & 1 & 1+j3 \end{bmatrix}$$

then

$$\bar{\mathbf{A}}' = \mathbf{A}^* = \begin{bmatrix} 1 & 2-j & 3 \\ -j2 & -j & 1 \\ 1-j5 & 3+j & 1-j3 \end{bmatrix}$$

Note that

$$(\mathbf{A}^*)^* = \mathbf{A}$$

If **A** is a real matrix (a matrix whose elements are real), the conjugate transpose **A*** is the same as the transpose **A′**.

**Hermitian matrix.** A matrix whose elements are complex quantities is called a *complex* matrix. If a complex matrix **A** satisfies the relationship

$$\mathbf{A} = \mathbf{A}^* \qquad \text{or} \qquad a_{ij} = \bar{a}_{ji}$$

where $\bar{a}_{ji}$ is the complex conjugate of $a_{ji}$, then **A** is called a *Hermitian* matrix. An example is

$$\mathbf{A} = \begin{bmatrix} 1 & 4+j3 \\ 4-j3 & 2 \end{bmatrix}$$

If a Hermitian matrix **A** is written as $\mathbf{A} = \mathbf{B} + j\mathbf{C}$, where **B** and **C** are real matrices, then

$$\mathbf{B} = \mathbf{B}' \qquad \text{and} \qquad \mathbf{C} = -\mathbf{C}'$$

In the example above,

$$\mathbf{A} = \mathbf{B} + j\mathbf{C} = \begin{bmatrix} 1 & 4 \\ 4 & 2 \end{bmatrix} + j \begin{bmatrix} 0 & 3 \\ -3 & 0 \end{bmatrix}$$

**Skew-Hermitian matrix.** If a matrix **A** satisfies the relationship

$$\mathbf{A} = -\mathbf{A}^*$$

then **A** is called a *skew-Hermitian* matrix. An example is

$$\mathbf{A} = \begin{bmatrix} j5 & -2+j3 \\ 2+j3 & j \end{bmatrix}$$

If a skew-Hermitian matrix **A** is written as $\mathbf{A} = \mathbf{B} + j\mathbf{C}$, where **B** and **C** are real matrices, then

$$\mathbf{B} = -\mathbf{B}' \qquad \text{and} \qquad \mathbf{C} = \mathbf{C}'$$

In the present example,

$$\mathbf{A} = \mathbf{B} + j\mathbf{C} = \begin{bmatrix} 0 & -2 \\ 2 & 0 \end{bmatrix} + j \begin{bmatrix} 5 & 3 \\ 3 & 1 \end{bmatrix}$$

### 3-3 MATRIX ALGEBRA

This section presents matrix algebra, as well as additional definitions. It is important to remember that some of the matrix operations obey the same rules as those in ordinary algebra, but others do not.

**Addition and subtraction of matrices.** Two matrices **A** and **B** can be added if they have the same number of rows and the same number of columns. If $\mathbf{A} = (a_{ij})$ and $\mathbf{B} = (b_{ij})$, then $\mathbf{A} + \mathbf{B}$ is defined to be

$$\mathbf{A} + \mathbf{B} = (a_{ij} + b_{ij})$$

Thus, each element of **A** is added to the corresponding element of **B**. Similarly, subtraction of matrices is defined to be

$$\mathbf{A} - \mathbf{B} = (a_{ij} - b_{ij})$$

As an example, consider

$$\mathbf{A} = \begin{bmatrix} 1 & 2 & 3 \\ 4 & 5 & 6 \end{bmatrix} \quad \text{and} \quad \mathbf{B} = \begin{bmatrix} 5 & 2 & 3 \\ 1 & 4 & 1 \end{bmatrix}$$

Then $\mathbf{A} + \mathbf{B}$ and $\mathbf{A} - \mathbf{B}$ are given by

$$\mathbf{A} + \mathbf{B} = \begin{bmatrix} 6 & 4 & 6 \\ 5 & 9 & 7 \end{bmatrix} \quad \text{and} \quad \mathbf{A} - \mathbf{B} = \begin{bmatrix} -4 & 0 & 0 \\ 3 & 1 & 5 \end{bmatrix}$$

**Multiplication of a matrix by a scalar.** The product of a matrix and a scalar is a matrix in which each element is multiplied by the scalar; that is, for a matrix **A** and a scalar $k$,

$$k\mathbf{A} = \begin{bmatrix} ka_{11} & ka_{12} & \cdots & ka_{1m} \\ ka_{21} & ka_{22} & \cdots & ka_{2m} \\ \cdot & \cdot & & \cdot \\ \cdot & \cdot & & \cdot \\ \cdot & \cdot & & \cdot \\ ka_{n1} & ka_{n2} & \cdots & ka_{nm} \end{bmatrix}$$

**Multiplication of a matrix by a matrix.** Multiplication of a matrix by a matrix is possible between conformable matrices (which means that the number of columns of the first matrix must equal the number of rows of the second). Otherwise, multiplication of two matrices is not defined.

Let **A** be an $n \times m$ matrix and **B** be an $m \times p$ matrix. Then the product **AB**, which we read "**A** postmultiplied by **B**" or "**B** premultiplied by **A**," is defined as follows:

$$\mathbf{AB} = \mathbf{C} = (c_{ij}) = \left( \sum_{k=1}^{m} a_{ik} b_{kj} \right) \quad (i = 1, 2, \ldots, n; \; j = 1, 2, \ldots, p)$$

The product matrix **C** has the same number of rows as **A** and the same number of columns as **B**. Thus, the matrix **C** is an $n \times p$ matrix.

It should be noted that even if **A** and **B** are conformable for **AB**, they may not be conformable for **BA**, in which case **BA** is not defined.

The associative and distributive laws hold for matrix multiplication; namely,

$$(AB)C = A(BC)$$

$$(A + B)C = AC + BC$$

$$C(A + B) = CA + CB$$

If **AB** = **BA**, then **A** and **B** are said to commute. Note that, in general, **AB** ≠ **BA**. To show this, consider

$$A = \begin{bmatrix} 1 & 0 \\ 2 & 1 \\ 3 & 4 \end{bmatrix} \quad \text{and} \quad B = \begin{bmatrix} 2 & 1 & 5 \\ 1 & 0 & 0 \end{bmatrix}$$

**AB** and **BA** are given, respectively, by

$$AB = \begin{bmatrix} 2 & 1 & 5 \\ 5 & 2 & 10 \\ 10 & 3 & 15 \end{bmatrix} \quad \text{and} \quad BA = \begin{bmatrix} 19 & 21 \\ 1 & 0 \end{bmatrix}$$

Clearly, **AB** ≠ **BA**. As another example, consider

$$A = \begin{bmatrix} 1 & 2 \\ 2 & 4 \end{bmatrix} \quad \text{and} \quad B = \begin{bmatrix} 1 & 2 \\ 2 & 2 \end{bmatrix}$$

Then

$$AB = \begin{bmatrix} 5 & 6 \\ 10 & 12 \end{bmatrix} \quad \text{and} \quad BA = \begin{bmatrix} 5 & 10 \\ 6 & 12 \end{bmatrix}$$

Clearly, **AB** ≠ **BA**.

Because matrix multiplication is, in general, not commutative, we must preserve the order of the matrices when multiplying a matrix by another matrix. (This is the reason why we often use the terms "premultiplication" or "postmultiplication" to indicate whether the matrix is multiplied from the right or the left.)

An example of the case where **AB** = **BA** is given below:

$$A = \begin{bmatrix} 1 & 0 \\ 0 & 3 \end{bmatrix}, \quad B = \begin{bmatrix} 2 & 0 \\ 0 & 5 \end{bmatrix}$$

**AB** and **BA** are given by

$$AB = BA = \begin{bmatrix} 2 & 0 \\ 0 & 15 \end{bmatrix}$$

Clearly **A** and **B** commute in this case.

**Power of a matrix.** The $k$th power of a square matrix **A** is defined to be

$$A^k = \underbrace{AA \cdots A}_{k}$$

Note that for a diagonal matrix $A = \operatorname{diag}(a_{11}, a_{22}, \ldots, a_{nn})$

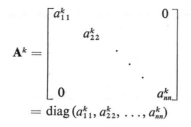

$$= \text{diag}\,(a_{11}^k, a_{22}^k, \ldots, a_{nn}^k)$$

**Further properties of matrices.** The transposes of $\mathbf{A} + \mathbf{B}$ and $\mathbf{AB}$ are given by

$$(\mathbf{A} + \mathbf{B})' = \mathbf{A}' + \mathbf{B}'$$

$$(\mathbf{AB})' = \mathbf{B}'\mathbf{A}'$$

To prove the last relationship, note that the $(i, j)$th element of $\mathbf{AB}$ is

$$\sum_{k=1}^{m} a_{ik}b_{kj} = c_{ij}$$

The $(i, j)$th element of $\mathbf{B}'\mathbf{A}'$ is

$$\sum_{k=1}^{m} b_{ki}a_{jk} = \sum_{k=1}^{m} a_{jk}b_{ki} = c_{ji}$$

which is equal to the $(j, i)$th element of $\mathbf{AB}$, or the $(i, j)$th element of $(\mathbf{AB})'$. Hence, $(\mathbf{AB})' = \mathbf{B}'\mathbf{A}'$. As an example, consider

$$\mathbf{A} = \begin{bmatrix} 1 & 4 \\ 2 & 3 \end{bmatrix} \quad \text{and} \quad \mathbf{B} = \begin{bmatrix} 4 & 2 \\ 5 & 6 \end{bmatrix}$$

then

$$\mathbf{AB} = \begin{bmatrix} 24 & 26 \\ 23 & 22 \end{bmatrix}$$

$$\mathbf{B}'\mathbf{A}' = \begin{bmatrix} 4 & 5 \\ 2 & 6 \end{bmatrix}\begin{bmatrix} 1 & 2 \\ 4 & 3 \end{bmatrix} = \begin{bmatrix} 24 & 23 \\ 26 & 22 \end{bmatrix}$$

Clearly, $(\mathbf{AB})' = \mathbf{B}'\mathbf{A}'$.

In a similar way, we obtain for the conjugate transposes of $\mathbf{A} + \mathbf{B}$ and $\mathbf{AB}$ the results

$$(\mathbf{A} + \mathbf{B})^* = \mathbf{A}^* + \mathbf{B}^*$$

$$(\mathbf{AB})^* = \mathbf{B}^*\mathbf{A}^*$$

**Rank of matrix.** A matrix is said to have rank $m$ if there exists an $m \times m$ submatrix $\mathbf{M}$ of $\mathbf{A}$ such that the determinant of $\mathbf{M}$ is nonzero and the determinant of every $r \times r$ submatrix (where $r \geq m + 1$) of $\mathbf{A}$ is zero.

As an example, consider the following matrix:

$$\mathbf{A} = \begin{bmatrix} 1 & 2 & 3 & 4 \\ 0 & 1 & -1 & 0 \\ 1 & 0 & 1 & 2 \\ 1 & 1 & 0 & 2 \end{bmatrix}$$

Note that $|\mathbf{A}| = 0$. One of the largest submatrices whose determinant is not equal to zero is

$$\begin{bmatrix} 1 & 2 & 3 \\ 0 & 1 & -1 \\ 1 & 0 & 1 \end{bmatrix}$$

Hence the rank of the matrix $\mathbf{A}$ is 3.

## 3-4  MATRIX INVERSION

This section discusses the inversion of matrices and associated subjects.

**Minor $M_{ij}$.** If the $i$th row and $j$th column are deleted from an $n \times n$ matrix $\mathbf{A}$, the resulting matrix is an $(n - 1) \times (n - 1)$ matrix. The determinant of this $(n - 1) \times (n - 1)$ matrix is called the minor $M_{ij}$ of the matrix $\mathbf{A}$.

**Cofactor $A_{ij}$.** The cofactor $A_{ij}$ of the element $a_{ij}$ of the matrix $\mathbf{A}$ is defined by the equation

$$A_{ij} = (-1)^{i+j} M_{ij}$$

That is, the cofactor $A_{ij}$ of the element $a_{ij}$ is $(-1)^{i+j}$ times the determinant of the matrix formed by deleting the $i$th row and the $j$th column from $\mathbf{A}$. Note that the cofactor $A_{ij}$ of the element $a_{ij}$ is the coefficient of the term $a_{ij}$ in the expansion of the determinant $|\mathbf{A}|$ since it can be shown that

$$a_{i1}A_{i1} + a_{i2}A_{i2} + \cdots + a_{in}A_{in} = |\mathbf{A}|$$

If $a_{i1}, a_{i2}, \ldots, a_{in}$ are replaced by $a_{j1}, a_{j2}, \ldots, a_{jn}$, then

$$a_{j1}A_{i1} + a_{j2}A_{i2} + \cdots + a_{jn}A_{in} = 0 \qquad (i \neq j)$$

since the determinant of $\mathbf{A}$ in this case possesses two identical rows. Hence we obtain

$$\sum_{k=1}^{n} a_{jk}A_{ik} = \delta_{ji}|\mathbf{A}|$$

Similarly,

$$\sum_{k=1}^{n} a_{ki}A_{kj} = \delta_{ij}|\mathbf{A}|$$

**Adjoint matrix.** The matrix $\mathbf{B}$, where the element of the $i$th row and $j$th column equals $A_{ji}$, is called the adjoint of $\mathbf{A}$ and is denoted by adj $\mathbf{A}$ or

$$\mathbf{B} = (b_{ij}) = (A_{ji}) = \text{adj } \mathbf{A}$$

That is, the adjoint of $\mathbf{A}$ is the transpose of the matrix whose elements are the cofactors of $\mathbf{A}$ or

$$\text{adj } \mathbf{A} = \begin{bmatrix} A_{11} & A_{21} & \cdots & A_{n1} \\ A_{12} & A_{22} & \cdots & A_{n2} \\ \cdot & \cdot & & \cdot \\ \cdot & \cdot & & \cdot \\ \cdot & \cdot & & \cdot \\ A_{1n} & A_{2n} & \cdots & A_{nn} \end{bmatrix}$$

Note that the element of the $j$th row and $i$th column of the product A(adj A) is

$$\sum_{k=1}^{n} a_{jk}b_{ki} = \sum_{k=1}^{n} a_{jk}A_{ik} = \delta_{ji}|\mathbf{A}|$$

Hence A(adj A) is a diagonal matrix with diagonal elements equal to $|\mathbf{A}|$. Thus,

$$\mathbf{A}(\text{adj } \mathbf{A}) = |\mathbf{A}|\,\mathbf{I}$$

Similarly, the element at the $j$th row and $i$th column of the product (adj A)A is

$$\sum_{k=1}^{n} b_{jk}a_{ki} = \sum_{k=1}^{n} A_{kj}a_{ki} = \delta_{ij}|\mathbf{A}|$$

Hence we have the relationship

$$\mathbf{A}(\text{adj } \mathbf{A}) = (\text{adj } \mathbf{A})\mathbf{A} = |\mathbf{A}|\,\mathbf{I} \qquad (3\text{-}1)$$

For example, given the matrix

$$\mathbf{A} = \begin{bmatrix} 1 & 2 & 0 \\ 3 & -1 & -2 \\ 1 & 0 & -3 \end{bmatrix}$$

we can find that the determinant of $\mathbf{A}$ is 17 and that

$$\text{adj } \mathbf{A} = \begin{bmatrix} \begin{vmatrix} -1 & -2 \\ 0 & -3 \end{vmatrix} & -\begin{vmatrix} 2 & 0 \\ 0 & -3 \end{vmatrix} & \begin{vmatrix} 2 & 0 \\ -1 & -2 \end{vmatrix} \\ -\begin{vmatrix} 3 & -2 \\ 1 & -3 \end{vmatrix} & \begin{vmatrix} 1 & 0 \\ 1 & -3 \end{vmatrix} & -\begin{vmatrix} 1 & 0 \\ 3 & -2 \end{vmatrix} \\ \begin{vmatrix} 3 & -1 \\ 1 & 0 \end{vmatrix} & -\begin{vmatrix} 1 & 2 \\ 1 & 0 \end{vmatrix} & \begin{vmatrix} 1 & 2 \\ 3 & -1 \end{vmatrix} \end{bmatrix}$$

$$= \begin{bmatrix} 3 & 6 & -4 \\ 7 & -3 & 2 \\ 1 & 2 & -7 \end{bmatrix}$$

Hence

$$\mathbf{A}(\text{adj } \mathbf{A}) = \begin{bmatrix} 1 & 2 & 0 \\ 3 & -1 & -2 \\ 1 & 0 & -3 \end{bmatrix}\begin{bmatrix} 3 & 6 & -4 \\ 7 & -3 & 2 \\ 1 & 2 & -7 \end{bmatrix}$$

$$= 17\begin{bmatrix} 1 & 0 & 0 \\ 0 & 1 & 0 \\ 0 & 0 & 1 \end{bmatrix}$$

$$= |\mathbf{A}|\,\mathbf{I}$$

**Inverse of a matrix.** If, for a square matrix $\mathbf{A}$, a matrix $\mathbf{B}$ exists such that $\mathbf{BA} = \mathbf{AB} = \mathbf{I}$, then $\mathbf{B}$ is denoted by $\mathbf{A}^{-1}$ and is called the *inverse* of $\mathbf{A}$. The inverse of a matrix $\mathbf{A}$ exists if the determinant of $\mathbf{A}$ is nonzero or $\mathbf{A}$ is nonsingular.

By definition, the inverse matrix $\mathbf{A}^{-1}$ has the property that

$$\mathbf{AA}^{-1} = \mathbf{A}^{-1}\mathbf{A} = \mathbf{I}$$

where $\mathbf{I}$ is the identity matrix. If $\mathbf{A}$ is nonsingular and $\mathbf{AB} = \mathbf{C}$, then $\mathbf{B} = \mathbf{A}^{-1}\mathbf{C}$. This can be seen from the equation

$$\mathbf{A}^{-1}\mathbf{AB} = \mathbf{IB} = \mathbf{B} = \mathbf{A}^{-1}\mathbf{C}$$

If **A** and **B** are nonsingular matrices, then the product **AB** is a nonsingular matrix. Moreover,

$$(\mathbf{AB})^{-1} = \mathbf{B}^{-1}\mathbf{A}^{-1}$$

This may be proved as follows:

$$(\mathbf{B}^{-1}\mathbf{A}^{-1})\mathbf{AB} = \mathbf{B}^{-1}(\mathbf{A}^{-1}\mathbf{A})\mathbf{B} = \mathbf{B}^{-1}\mathbf{IB} = \mathbf{B}^{-1}\mathbf{B} = \mathbf{I}$$

Similarly,

$$(\mathbf{AB})(\mathbf{B}^{-1}\mathbf{A}^{-1}) = \mathbf{I}$$

Note that

$$(\mathbf{A}^{-1})^{-1} = \mathbf{A}$$
$$(\mathbf{A}^{-1})' = (\mathbf{A}')^{-1}$$
$$(\mathbf{A}^{-1})^* = (\mathbf{A}^*)^{-1}$$

From Eq. (3-1) and the definition of the inverse matrix, we have

$$\mathbf{A}^{-1} = \frac{\text{adj }\mathbf{A}}{|\mathbf{A}|}$$

Hence the inverse of a matrix is the transpose of the matrix of its cofactors, divided by the determinant of the matrix. If **A** is given by

$$\mathbf{A} = \begin{bmatrix} a_{11} & a_{12} & \cdots & a_{1n} \\ a_{21} & a_{22} & \cdots & a_{2n} \\ \cdot & \cdot & & \cdot \\ \cdot & \cdot & & \cdot \\ \cdot & \cdot & & \cdot \\ a_{n1} & a_{n2} & \cdots & a_{nn} \end{bmatrix}$$

then

$$\mathbf{A}^{-1} = \frac{\text{adj }\mathbf{A}}{|\mathbf{A}|} = \begin{bmatrix} \dfrac{A_{11}}{|\mathbf{A}|} & \dfrac{A_{21}}{|\mathbf{A}|} & \cdots & \dfrac{A_{n1}}{|\mathbf{A}|} \\ \dfrac{A_{12}}{|\mathbf{A}|} & \dfrac{A_{22}}{|\mathbf{A}|} & \cdots & \dfrac{A_{n2}}{|\mathbf{A}|} \\ \cdot & \cdot & & \cdot \\ \cdot & \cdot & & \cdot \\ \cdot & \cdot & & \cdot \\ \dfrac{A_{1n}}{|\mathbf{A}|} & \dfrac{A_{2n}}{|\mathbf{A}|} & \cdots & \dfrac{A_{nn}}{|\mathbf{A}|} \end{bmatrix}$$

where $A_{ij}$ is the cofactor of $a_{ij}$ of the matrix **A**. Hence the terms in the $i$th column of $\mathbf{A}^{-1}$ are $1/|\mathbf{A}|$ times the cofactors of the $i$th row of the original matrix **A**. For example, if **A** is given by

$$\mathbf{A} = \begin{bmatrix} 1 & 2 & 0 \\ 3 & -1 & -2 \\ 1 & 0 & -3 \end{bmatrix}$$

then the adjoint of $\mathbf{A}$ and the determinant $|\mathbf{A}|$ are found to be

$$\text{adj } \mathbf{A} = \begin{bmatrix} 3 & 6 & -4 \\ 7 & -3 & 2 \\ 1 & 2 & -7 \end{bmatrix}, \qquad |\mathbf{A}| = 17$$

Hence the inverse of $\mathbf{A}$ is given by

$$\mathbf{A}^{-1} = \frac{\text{adj } \mathbf{A}}{|\mathbf{A}|} = \begin{bmatrix} \frac{3}{17} & \frac{6}{17} & -\frac{4}{17} \\ \frac{7}{17} & -\frac{3}{17} & \frac{2}{17} \\ \frac{1}{17} & \frac{2}{17} & -\frac{7}{17} \end{bmatrix}$$

In what follows we give formulas for finding inverse matrices for the $2 \times 2$ matrix and the $3 \times 3$ matrix. For the $2 \times 2$ matrix $\mathbf{A}$, where

$$\mathbf{A} = \begin{bmatrix} a & b \\ c & d \end{bmatrix}, \qquad ad - bc \neq 0$$

the inverse matrix is given by

$$\mathbf{A}^{-1} = \frac{1}{ad - bc} \begin{bmatrix} d & -b \\ -c & a \end{bmatrix}$$

For the $3 \times 3$ matrix $\mathbf{A}$, where

$$\mathbf{A} = \begin{bmatrix} a & b & c \\ d & e & f \\ g & h & i \end{bmatrix}, \qquad |\mathbf{A}| \neq 0$$

the inverse matrix is given by

$$\mathbf{A}^{-1} = \frac{1}{|\mathbf{A}|} \begin{bmatrix} \begin{vmatrix} e & f \\ h & i \end{vmatrix} & -\begin{vmatrix} b & c \\ h & i \end{vmatrix} & \begin{vmatrix} b & c \\ e & f \end{vmatrix} \\ -\begin{vmatrix} d & f \\ g & i \end{vmatrix} & \begin{vmatrix} a & c \\ g & i \end{vmatrix} & -\begin{vmatrix} a & c \\ d & f \end{vmatrix} \\ \begin{vmatrix} d & e \\ g & h \end{vmatrix} & -\begin{vmatrix} a & b \\ g & h \end{vmatrix} & \begin{vmatrix} a & b \\ d & e \end{vmatrix} \end{bmatrix}$$

**Remarks on cancellation of matrices.** Cancellation of matrices is not valid in matrix algebra. Consider the product of two singular matrices $\mathbf{A}$ and $\mathbf{B}$. Take, for example,

$$\mathbf{A} = \begin{bmatrix} 2 & 1 \\ 6 & 3 \end{bmatrix} \neq \mathbf{0} \qquad \text{and} \qquad \mathbf{B} = \begin{bmatrix} 1 & -2 \\ -2 & 4 \end{bmatrix} \neq \mathbf{0}$$

Then

$$\mathbf{AB} = \begin{bmatrix} 2 & 1 \\ 6 & 3 \end{bmatrix} \begin{bmatrix} 1 & -2 \\ -2 & 4 \end{bmatrix} = \begin{bmatrix} 0 & 0 \\ 0 & 0 \end{bmatrix} = \mathbf{0}$$

Clearly, $\mathbf{AB} = \mathbf{0}$ implies neither $\mathbf{A} = \mathbf{0}$ nor $\mathbf{B} = \mathbf{0}$. In fact, $\mathbf{AB} = \mathbf{0}$ implies one of the following three:

1. $\mathbf{A} = \mathbf{0}$
2. $\mathbf{B} = \mathbf{0}$
3. Both $\mathbf{A}$ and $\mathbf{B}$ are singular.

We can easily prove that if both $\mathbf{A}$ and $\mathbf{B}$ are nonzero matrices and $\mathbf{AB} = \mathbf{0}$, then both $\mathbf{A}$ and $\mathbf{B}$ are singular. For a proof, assume that if $\mathbf{A}$ and $\mathbf{B}$ are not singular, then a matrix $\mathbf{A}^{-1}$ exists with the property that

$$\mathbf{A}^{-1}\mathbf{AB} = \mathbf{B} = \mathbf{0}$$

which contradicts the assumption that $\mathbf{B}$ is a nonzero matrix. Thus, we conclude that both $\mathbf{A}$ and $\mathbf{B}$ must be singular if $\mathbf{A} \neq \mathbf{0}$ and $\mathbf{B} \neq \mathbf{0}$.

Similarly, notice that if $\mathbf{A}$ is singular, then neither $\mathbf{AB} = \mathbf{AC}$ nor $\mathbf{BA} = \mathbf{CA}$ implies $\mathbf{B} = \mathbf{C}$. If, however, $\mathbf{A}$ is a nonsingular matrix, then $\mathbf{AB} = \mathbf{AC}$ implies $\mathbf{B} = \mathbf{C}$ and $\mathbf{BA} = \mathbf{CA}$ also implies $\mathbf{B} = \mathbf{C}$.

## 3-5  DIFFERENTIATION AND INTEGRATION OF MATRICES

The derivative of an $n \times m$ matrix $\mathbf{A}(t)$ is defined to be the $n \times m$ matrix, each element of which is the derivative of the corresponding element of the original matrix, provided all the elements $a_{ij}(t)$ have derivatives with respect to $t$. That is,

$$\frac{d}{dt}\mathbf{A}(t) = \left(\frac{d}{dt}a_{ij}(t)\right) = \begin{bmatrix} \frac{d}{dt}a_{11}(t) & \frac{d}{dt}a_{12}(t) & \cdots & \frac{d}{dt}a_{1m}(t) \\ \frac{d}{dt}a_{21}(t) & \frac{d}{dt}a_{22}(t) & \cdots & \frac{d}{dt}a_{2m}(t) \\ \vdots & \vdots & & \vdots \\ \frac{d}{dt}a_{n1}(t) & \frac{d}{dt}a_{n2}(t) & \cdots & \frac{d}{dt}a_{nm}(t) \end{bmatrix}$$

Similarly, the integral of an $n \times m$ matrix $\mathbf{A}(t)$ is defined to be

$$\int \mathbf{A}(t)\,dt = \left(\int a_{ij}(t)\,dt\right) = \begin{bmatrix} \int a_{11}(t)\,dt & \int a_{12}(t)\,dt & \cdots & \int a_{1m}(t)\,dt \\ \int a_{21}(t)\,dt & \int a_{22}(t)\,dt & \cdots & \int a_{2m}(t)\,dt \\ \vdots & \vdots & & \vdots \\ \int a_{n1}(t)\,dt & \int a_{n2}(t)\,dt & \cdots & \int a_{nm}(t)\,dt \end{bmatrix}$$

**Differentiation of the product of two matrices.** If the matrices $\mathbf{A}(t)$ and $\mathbf{B}(t)$ can be differentiated with respect to $t$, then

$$\frac{d}{dt}[\mathbf{A}(t)\mathbf{B}(t)] = \frac{d\mathbf{A}(t)}{dt}\mathbf{B}(t) + \mathbf{A}(t)\frac{d\mathbf{B}(t)}{dt}$$

Here again the multiplication of $\mathbf{A}(t)$ and $d\mathbf{B}(t)/dt$ [or $d\mathbf{A}(t)/dt$ and $\mathbf{B}(t)$] is, in general, not commutative.

**Differentiation of $\mathbf{A}^{-1}(t)$.** If a matrix $\mathbf{A}(t)$ and its inverse $\mathbf{A}^{-1}(t)$ are differentiable with respect to $t$, then the derivative of $\mathbf{A}^{-1}(t)$ is given by

$$\frac{d\mathbf{A}^{-1}(t)}{dt} = -\mathbf{A}^{-1}(t)\frac{d\mathbf{A}(t)}{dt}\mathbf{A}^{-1}(t)$$

This may be obtained by differentiating $\mathbf{A}(t)\mathbf{A}^{-1}(t)$ with respect to $t$. Since

$$\frac{d}{dt}[\mathbf{A}(t)\mathbf{A}^{-1}(t)] = \frac{d\mathbf{A}(t)}{dt}\mathbf{A}^{-1}(t) + \mathbf{A}(t)\frac{d\mathbf{A}^{-1}(t)}{dt}$$

and

$$\frac{d}{dt}\mathbf{A}(t)\mathbf{A}^{-1}(t) = \frac{d}{dt}\mathbf{I} = \mathbf{0}$$

we obtain

$$\mathbf{A}(t)\frac{d\mathbf{A}^{-1}(t)}{dt} = -\frac{d\mathbf{A}(t)}{dt}\mathbf{A}^{-1}(t)$$

or

$$\frac{d\mathbf{A}^{-1}(t)}{dt} = -\mathbf{A}^{-1}(t)\frac{d\mathbf{A}(t)}{dt}\mathbf{A}^{-1}(t)$$

EXAMPLE PROBLEMS AND SOLUTIONS

**PROBLEM A-3-1.** Show that if $\mathbf{A}$ is any square matrix, then $\mathbf{A} + \mathbf{A}'$ is a symmetric matrix and $\mathbf{A} - \mathbf{A}'$ is a skew-symmetric matrix.

**Solution.** Consider the following $3 \times 3$ matrix:

$$\mathbf{A} = \begin{bmatrix} a & b & c \\ d & e & f \\ g & h & i \end{bmatrix}$$

Then

$$\mathbf{A} + \mathbf{A}' = \begin{bmatrix} 2a & b+d & c+g \\ b+d & 2e & f+h \\ c+g & f+h & 2i \end{bmatrix} = \text{symmetric matrix}$$

and

$$\mathbf{A} - \mathbf{A}' = \begin{bmatrix} 0 & b-d & c-g \\ d-b & 0 & f-h \\ g-c & h-f & 0 \end{bmatrix} = \text{skew symmetric matrix}$$

**PROBLEM A-3-2.** Obtain the product $\mathbf{AB}$, where

$$\mathbf{A} = \begin{bmatrix} 1 & 2 & 0 \\ 3 & -1 & 1 \\ 1 & 5 & 2 \end{bmatrix} \qquad \mathbf{B} = \begin{bmatrix} 1 & 0 \\ 2 & 3 \\ 1 & -1 \end{bmatrix}$$

**Solution**

$$\mathbf{AB} = \begin{bmatrix} 5 & 6 \\ 2 & -4 \\ 13 & 13 \end{bmatrix}$$

**PROBLEM A-3-3.** The Vandermonde matrix is given by

$$V = \begin{bmatrix} 1 & 1 & \cdots & 1 \\ x_1 & x_2 & \cdots & x_n \\ x_1^2 & x_2^2 & \cdots & x_n^2 \\ \cdot & \cdot & & \cdot \\ \cdot & \cdot & & \cdot \\ \cdot & \cdot & & \cdot \\ x_1^{n-1} & x_2^{n-1} & \cdots & x_n^{n-1} \end{bmatrix}$$

Find the determinant of **V** for the cases where $n = 3$ and $n = 4$.

**Solution.** For $n = 3$,

$$|V| = (x_3 - x_2)(x_3 - x_1)(x_2 - x_1)$$

For $n = 4$,

$$|V| = (x_4 - x_3)(x_4 - x_2)(x_4 - x_1)(x_3 - x_2)(x_3 - x_1)(x_2 - x_1)$$

**PROBLEM A-3-4.** Show that for the $n \times n$ matrices **A** and **B**

$$\begin{vmatrix} A & B \\ B & A \end{vmatrix} = |A + B| \cdot |A - B|$$

**Solution**

$$\begin{vmatrix} A & B \\ B & A \end{vmatrix} = \begin{vmatrix} A + B & B + A \\ B & A \end{vmatrix} = \begin{vmatrix} A + B & 0 \\ B & A - B \end{vmatrix} = |A + B| \cdot |A - B|$$

Note that for the $n \times n$ matrices **A**, **B**, **C**, and **D** the determinant

$$\begin{vmatrix} A & B \\ C & D \end{vmatrix}$$

is, in general, equal to neither $|AD - BC|$ nor $|A||D| - |B||C|$.

**PROBLEM A-3-5.** Find the inverse of

$$A = \begin{bmatrix} s + 1 & -1 \\ s + 2 & s \end{bmatrix}$$

where $s$ is a complex variable.

**Solution**

$$A^{-1} = \begin{bmatrix} \dfrac{s}{s^2 + 2s + 2} & \dfrac{1}{s^2 + 2s + 2} \\ -\dfrac{s + 2}{s^2 + 2s + 2} & \dfrac{s + 1}{s^2 + 2s + 2} \end{bmatrix}$$

**PROBLEM A-3-6.** Solve the following set of three simultaneous equations:

$$\begin{aligned} x_1 + 3x_2 + x_3 &= 1 \\ 2x_1 + x_2 \phantom{+ 2x_3} &= 2 \\ x_1 + x_2 + 2x_3 &= 3 \end{aligned}$$

**Solution.** By defining the following matrix and vectors

$$A = \begin{bmatrix} 1 & 3 & 1 \\ 2 & 1 & 0 \\ 1 & 1 & 2 \end{bmatrix} \qquad x = \begin{bmatrix} x_1 \\ x_2 \\ x_3 \end{bmatrix} \qquad b = \begin{bmatrix} 1 \\ 2 \\ 3 \end{bmatrix}$$

we obtain

$$Ax = b$$

Hence

$$x = A^{-1}b$$

where

$$A^{-1} = \frac{1}{-9} \begin{bmatrix} \begin{vmatrix} 1 & 0 \\ 1 & 2 \end{vmatrix} & -\begin{vmatrix} 3 & 1 \\ 1 & 2 \end{vmatrix} & \begin{vmatrix} 3 & 1 \\ 1 & 0 \end{vmatrix} \\ -\begin{vmatrix} 2 & 0 \\ 1 & 2 \end{vmatrix} & \begin{vmatrix} 1 & 1 \\ 1 & 2 \end{vmatrix} & -\begin{vmatrix} 1 & 1 \\ 2 & 0 \end{vmatrix} \\ \begin{vmatrix} 2 & 1 \\ 1 & 1 \end{vmatrix} & -\begin{vmatrix} 1 & 3 \\ 1 & 1 \end{vmatrix} & \begin{vmatrix} 1 & 3 \\ 2 & 1 \end{vmatrix} \end{bmatrix} = \begin{bmatrix} -\frac{2}{9} & \frac{5}{9} & \frac{1}{9} \\ \frac{4}{9} & -\frac{1}{9} & -\frac{2}{9} \\ -\frac{1}{9} & -\frac{2}{9} & \frac{5}{9} \end{bmatrix}$$

Thus

$$\begin{bmatrix} x_1 \\ x_2 \\ x_3 \end{bmatrix} = \begin{bmatrix} -\frac{2}{9} & \frac{5}{9} & \frac{1}{9} \\ \frac{4}{9} & -\frac{1}{9} & -\frac{2}{9} \\ -\frac{1}{9} & -\frac{2}{9} & \frac{5}{9} \end{bmatrix} \begin{bmatrix} 1 \\ 2 \\ 3 \end{bmatrix} = \begin{bmatrix} \frac{11}{9} \\ -\frac{4}{9} \\ \frac{10}{9} \end{bmatrix}$$

PROBLEMS

**PROBLEM B-3-1.** Show that any square matrix can be written as a sum of a Hermitian and a skew-Hermitian matrix.

**PROBLEM B-3-2.** Show that for the square matrices $A$ and $B$

1. $|A| = |A'|$
2. $|AB| = |A||B|$
3. $|(AB)'| = |A||B|$

**PROBLEM B-3-3.** In general, $(A + B)^2$ is not equal to $A^2 + 2AB + B^2$. Explain the reason. Under what condition is $(A + B)^2$ equal to $A^2 + 2AB + B^2$?

**PROBLEM B-3-4.** Find $A^{-1}$ where

$$A = \begin{bmatrix} 1 & 2 & 3 \\ 4 & 0 & 1 \\ 2 & 1 & 2 \end{bmatrix}$$

**PROBLEM B-3-5.** A matrix whose elements below (or above) the main diagonal are all zero is called a triangular matrix. An example is

$$\begin{bmatrix} a_{11} & a_{12} & a_{13} \\ 0 & a_{22} & a_{23} \\ 0 & 0 & a_{33} \end{bmatrix}$$

Using this $3 \times 3$ matrix, show that the inverse of the triangular matrix is also a triangular matrix.

**PROBLEM B-3-6.** Determine the rank of the following matrix:

$$\mathbf{A} = \begin{bmatrix} 1 & 1 & 2 & 0 \\ 2 & 0 & 1 & 3 \\ 4 & -1 & 5 & 1 \end{bmatrix}$$

**PROBLEM B-3-7.** Solve the following set of three simultaneous equations:

$$x_1 + 3x_2 + x_3 = 2$$
$$x_1 - 2x_2 + x_3 = 3$$
$$2x_1 + x_2 + 3x_3 = 1$$

**PROBLEM B-3-8.** A set of $n$ simultaneous algebraic equations

$$a_{11}x_1 + a_{12}x_2 + \cdots + a_{1n}x_n = b_1$$
$$a_{21}x_1 + a_{22}x_2 + \cdots + a_{2n}x_n = b_2$$
$$\cdots \qquad \cdots$$
$$a_{n1}x_1 + a_{n2}x_2 + \cdots + a_{nn}x_n = b_n$$

may be expressed as

$$\mathbf{A}\mathbf{x} = \mathbf{b}$$

where

$$\mathbf{A} = (a_{ij})$$

A system of equations is said to be *consistent* if it has at least one solution. If the $n$ equations are consistent but $\mathbf{A}$ is singular, how can we determine $\mathbf{x}$? Using a simple case where $n = 3$ explain how $\mathbf{x}$ may be obtained. Consider the two cases where rank $\mathbf{A} = 2$ and rank $\mathbf{A} = 1$.

# 4

# MATHEMATICAL MODELS
# OF PHYSICAL SYSTEMS

## 4-1 INTRODUCTION

Many dynamic systems, whether they are mechanical, electrical, thermal, hydraulic, economic, biological, etc., may be characterized by differential equations. The response of a dynamic system to an input (or forcing function) may be obtained if these differential equations are solved. The equations can be obtained by utilizing physical laws governing a particular system, for example, Newton's laws for mechanical systems, Kirchhoff's laws for electrical systems, etc.

**Mathematical models.** The mathematical description of the dynamic characteristics of a system is called a *mathematical model*. The first step in the analysis of a dynamic system is to derive its model. We must always keep in mind that deriving a reasonable mathematical model is the most important part of the entire analysis.

Models may assume many different forms. Depending on the particular system and the circumstances, one mathematical representation may be better suited than other representations. For example, in optimal control problems, it is often advantageous to use a set of first-order differential equations. (Refer to Chapter 14). On the other hand, for the transient-response analysis or frequency-response analysis of single-input–single-output systems, the transfer-function representation to be discussed in this chapter may be more convenient than any other.

Once a mathematical model of a system is obtained, various analytical and computer tools can be used for analysis and synthesis purposes.

**Simplicity versus accuracy.** In obtaining a model, we must make a compromise between the simplicity of the model and the accuracy of the results of the analysis. Note that the results obtained from the analysis are valid only to the extent that the model approximates a given physical system.

The rapidity with which a digital computer can perform arithmetic operations allows us to employ a new approach in formulating mathematical models. Instead of limiting models to simple ones, we may, if necessary, include hundreds of equations to describe a complete system. If extreme accuracy is not needed, however, it is preferable to obtain only a reasonably simplified model.

In deriving such a simplified model, we frequently find it necessary to ignore certain inherent physical properties of the system. In particular, if a linear lumped-parameter mathematical model (i.e., one employing ordinary differential equations) is desired, it is always necessary to ignore certain nonlinearities and distributed parameters (i.e., ones giving rise to partial differential equations) which may be present in the physical system. If the effects that these ignored properties have on the response are small, good agreement will be obtained between the results of the analysis of a mathematical model and the results of the experimental study of the physical system.

In general, in solving a new problem, we find it desirable first to build a simplified model so that we can get a general feeling for the solution. A more complete mathematical model may then be built and used for a more complete analysis.

We must be well aware of the fact that a linear lumped-parameter model, which may be valid in low-frequency operations, may not be valid at sufficiently high frequencies since the neglected property of distributed parameters may become an important factor in the dynamic behavior of the system. For example, the mass of a spring may be neglected in low-frequency operations but it becomes an important property of the system at high frequencies.

**Linear systems.** Linear systems are ones in which the equations of the model are linear. A differential equation is linear if the coefficients are constants or functions only of the independent variable. The most important property of linear systems is that the principle of superposition is applicable. The principle of superposition states that the response produced by the simultaneous application of two different forcing functions is the sum of the two individual responses. Hence, for linear systems, the response to several inputs can be calculated by treating one input at a time and adding the results. It is this principle that allows one to build up complicated solutions to the linear differential equation from simple solutions.

In an experimental investigation of a dynamic system, if cause and effect are proportional, thus implying that the principle of superposition holds, then the system can be considered linear.

**Linear time-invariant systems and linear time-varying systems.** Dynamic systems which are linear and are composed of time-invariant lumped-parameter components may be described by linear time-invariant differential equations. Such systems are called *linear time-invariant* (or *linear constant-coefficient*) systems.

Systems which are represented by differential equations whose coefficients are functions of time are called linear time-varying systems. An example of a time-varying control system is a spacecraft control system. (The mass of a spacecraft changes due to fuel consumption, and the gravity force changes as the spacecraft moves away from the earth.)

**Nonlinear systems.** Nonlinear systems are ones which are represented by nonlinear equations. Examples of nonlinear equations are

$$y = \sin x$$

$$y = x^2$$

$$z = x^2 + y^3$$

(In the last equation, $z$ is a nonlinear function of $x$ and $y$.)

A differential equation is called *nonlinear* if it is not linear. Examples of nonlinear differential equations are

$$\frac{d^2x}{dt^2} + \left(\frac{dx}{dt}\right)^2 + x = A \sin \omega t$$

$$\frac{d^2x}{dt^2} + (x^2 - 1)\frac{dx}{dt} + x = 0$$

$$\frac{d^2x}{dt^2} + \frac{dx}{dt} + x + x^3 = 0$$

Although many physical relationships are often represented by linear equations, in most cases actual relationships are not quite linear. In fact, a careful study of physical systems reveals that even so-called "linear systems" are really linear only in limited operating ranges. In practice, many electromechanical systems, hydraulic systems, pneumatic systems, etc., involve nonlinear relationships among the variables. For example, the output of a component may saturate for large input signals. There may be a dead space which affects small signals. (The dead space of a component is a small range of input variations to which the component is insensitive.) Square-law nonlinearity may occur in some components. For instance, dampers used in physical systems may be linear for low-velocity operations but may become

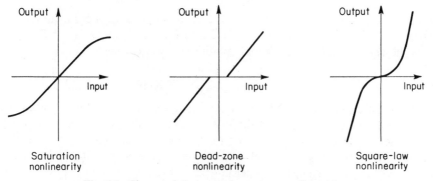

**Fig. 4-1.** Characteristic curves for various nonlinearities.

nonlinear at high velocities, and the damping force may become proportional to the square of the operating velocity. Examples of characteristic curves for these nonlinearities are shown in Fig. 4-1.

Note that some important control systems are nonlinear for signals of any size. For example, in on-off control systems, the control action is either on or off, and there is no linear relationship between the input and output of the controller.

The most important characteristic of nonlinear systems is that the principle of superposition is not applicable. Procedures for finding the solutions of problems involving such nonlinear systems, in general, are extremely complicated. Because of this mathematical difficulty attached to nonlinear systems, one often finds it necessary to introduce "equivalent" linear systems in place of nonlinear ones. Such equivalent linear systems are valid for only a limited range of operation. Once a nonlinear system is approximated by a linear mathematical model, a number of linear tools may be applied for analysis and design purposes. We shall introduce various linearization techniques in this book.

## 4-2   TRANSFER FUNCTIONS

In control theory, functions called "transfer functions" are very often used to characterize the input-output relationships of linear time-invariant systems. The concept of transfer functions applies only to linear time-invariant systems, although it can be extended to certain nonlinear control systems. (Refer to Chapter 11.)

**Transfer functions.** The transfer function of a linear time-invariant system is defined to be the ratio of the Laplace transform of the output (response function) to the Laplace transform of the input (driving function), under the assumption that all initial conditions are zero.

Consider the linear time-invariant system defined by the following differential equation:

$$
\overset{(n)}{a_0 y} + \overset{(n-1)}{a_1 y} + \cdots + a_{n-1}\dot{y} + a_n y
$$
$$
= \overset{(m)}{b_0 x} + \overset{(m-1)}{b_1 x} + \cdots + b_{m-1}\dot{x} + b_m x \qquad (n \geq m) \qquad (4\text{-}1)
$$

where $y$ is the output of the system and $x$ is the input. The transfer function of this system is obtained by taking the Laplace transforms of both sides of Eq. (4-1), under the assumption that all initial conditions are zero, or

$$
\text{Transfer function} = G(s) = \frac{Y(s)}{X(s)} = \frac{b_0 s^m + b_1 s^{m-1} + \cdots + b_{m-1}s + b_m}{a_0 s^n + a_1 s^{n-1} + \cdots + a_{n-1}s + a_n}
$$

The transfer function is an expression relating the output and input of a linear time-invariant system in terms of the system parameters and is a property of the system itself, independent of the input or driving function. The transfer function includes the units necessary to relate the input to the output; however, it does not provide any information concerning the physical structure of the system. (The transfer functions of many physically different systems can be identical.)

By using this concept, one can represent the system dynamics by algebraic equations in $s$. The highest power of $s$ in the denominator of the transfer function is equal to the order of the highest derivative term of the output. If the highest power of $s$ is equal to $n$, the system is called an $n$th-order system.

**Mechanical translational system.** Consider the spring-mass-dashpot system shown in Fig. 4-2. A dashpot is a device that provides viscous friction, or damping. It consists of a piston and oil-filled cylinder. Any relative motion between the piston rod and the cylinder is resisted by the oil because the oil must flow around the piston (or through orifices provided in the piston) from one side of the piston to the other. The dashpot essentially absorbs energy. This absorbed energy is dissipated as heat, and the dashpot does not store any kinetic or potential energy.

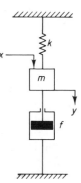

Let us obtain the transfer function of this system by assuming that the force $x(t)$ is the input and the displacement $y(t)$ of the mass is the output. We shall proceed according to the following steps:

**Fig. 4-2.** Spring-mass-dashpot system.

1. Write the differential equation of the system.

2. Take the Laplace transform of the differential equation, assuming all initial conditions to be zero.

3. Take the ratio of the output $Y(s)$ to the input $X(s)$. This ratio is the transfer function.

In order to derive a linear time-invariant differential equation, let us assume that the friction force of the dashpot is proportional to $\dot{y}$ and that the spring is a linear spring, namely, that the spring force is proportional to $y$. In this system, $m$ denotes the mass, $f$ denotes the viscous-friction coefficient, and $k$ denotes the spring constant.

The fundamental law governing mechanical systems is Newton's law. For translational systems, the law states that

$$ma = \sum F$$

where

    $m$ = mass, slug
    $a$ = acceleration, ft/sec²
    $F$ = force, lb

A slug is a unit of mass (slug = lb-sec²/ft). When acted upon by 1 lb of force, a 1-slug mass accelerates at 1 ft/sec². To obtain mass in pounds, multiply the number of slugs by 32; and to obtain mass in slugs, multiply the number of pounds by $3.1 \times 10^{-2}$.

Applying Newton's law to the present system, we obtain

$$m\frac{d^2y}{dt^2} = -f\frac{dy}{dt} - ky + x$$

or

$$m\frac{d^2y}{dt^2} + f\frac{dy}{dt} + ky = x \qquad (4\text{-}2)$$

Taking the Laplace transform of each term of Eq. (4-2) gives

$$\mathscr{L}\left[m\frac{d^2y}{dt^2}\right] = m[s^2Y(s) - sy(0) - \dot{y}(0)]$$

$$\mathscr{L}\left[f\frac{dy}{dt}\right] = f[sY(s) - y(0)]$$

$$\mathscr{L}[ky] = kY(s)$$

$$\mathscr{L}[x] = X(s)$$

If we set the initial conditions equal to zero, so that $y(0) = 0$, $\dot{y}(0) = 0$, the Laplace transform of Eq. (4-2) can be written

$$(ms^2 + fs + k)Y(s) = X(s)$$

Taking the ratio of $Y(s)$ to $X(s)$, we find that the transfer function of the system is

$$\text{Transfer function} = G(s) = \frac{Y(s)}{X(s)} = \frac{1}{ms^2 + fs + k}$$

**Mechanical rotational system.** Consider the system shown in Fig. 4-3. The system consists of a load inertia and a viscous-friction damper. Define

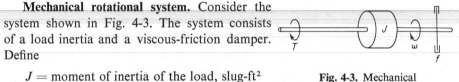

$J =$ moment of inertia of the load, slug-ft²
$f =$ viscous-friction coefficient, lb-ft/rad/sec
$\omega =$ angular velocity, rad/sec
$T =$ torque applied to the system, lb-ft

**Fig. 4-3.** Mechanical rotational system.

For mechanical rotational systems, Newton's law states that

$$J\alpha = \sum T$$

where

$J =$ moment of inertia, slug-ft²
$\alpha =$ angular acceleration, rad/sec²
$T =$ torque, lb-ft

To obtain the moment of inertia in lb-ft², multiply the number of slug-ft² by 32. Conversely, to obtain the moment of inertia in slug-ft², multiply the number of lb-ft² by $3.1 \times 10^{-2}$. The relationship between slug-ft² and gm-cm² is given by

$$1 \text{ slug-ft}^2 = 13.56 \times 10^6 \text{ gm-cm}^2$$

Sets of consistent units for mass, moment of inertia, and torque are given in the following table:

| Mass | Moment of Inertia | Torque |
|------|-------------------|--------|
| slug | slug-ft² | lb-ft |
| gram | gm-cm² | dyne-cm |
| kilogram | kg-m² | newton-m |

Applying Newton's law to the present system, we obtain

$$J\dot{\omega} + f\omega = T$$

If we assume that the applied torque $T$ is the input and that the angular velocity $\omega$ is the output, then the transfer function of this system is found to be

$$\frac{\Omega(s)}{T(s)} = \frac{1}{Js + f}$$

where

$$\Omega(s) = \mathscr{L}[\omega(t)]$$

$$T(s) = \mathscr{L}[T(t)]$$

**L-R-C circuit.** Consider the electrical circuit shown in Fig. 4-4. The circuit consists of an inductance $L$ (henry), a resistance $R$ (ohm), and a capacitance $C$ (farad). Applying Kirchhoff's law to the system, we obtain the following equations:

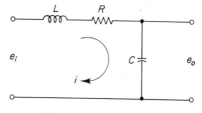

**Fig. 4-4.** Electrical circuit.

$$L\frac{di}{dt} + Ri + \frac{1}{C}\int i\,dt = e_i \qquad\qquad (4\text{-}3)$$

$$\frac{1}{C}\int i\,dt = e_0 \qquad\qquad (4\text{-}4)$$

Taking the Laplace transforms of Eqs. (4-3) and (4-4), assuming zero initial conditions, we obtain

$$LsI(s) + RI(s) + \frac{1}{C}\frac{1}{s}I(s) = E_i(s)$$

$$\frac{1}{C}\frac{1}{s}I(s) = E_0(s)$$

If $e_i$ is assumed to be the input and $e_0$ the output, then the transfer function of this system is found to be

$$\frac{E_0(s)}{E_i(s)} = \frac{1}{LCs^2 + RCs + 1} \qquad\qquad (4\text{-}5)$$

**Complex impedances.** In deriving transfer functions for electrical networks, we frequently find it convenient to write the Laplace-transformed equations directly, without writing the differential equations. Consider the system shown in Fig. 4-5(a). In this system, $Z_1$ and $Z_2$ represent complex impedances. The complex impedance $Z(s)$ of a two terminal circuit is the ratio of $E(s)$, the Laplace transform of the voltage across the terminals, to $I(s)$, the Laplace transform of the current through the element, under the assumption that the initial conditions are zero, so that $Z(s) = E(s)/I(s)$. If the two-terminal element is a resistance $R$, capacitance $C$, or inductance $L$, then the respective complex impedance is given by $R$, $1/Cs$, or $Ls$, respectively. If complex impedances are connected in series, the total impedance is the sum of the individual complex impedances.

Consider the circuit shown in Fig. 4-5 (b). Assume that the voltages $e_i$ and $e_0$ are the input and output of the circuit respectively. Then the transfer function of this circuit is

$$\frac{E_0(s)}{E_i(s)} = \frac{Z_2(s)}{Z_1(s) + Z_2(s)}$$

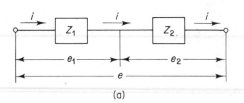

(a)

For the system shown in Fig. 4-4,

$$Z_1 = Ls + R, \qquad Z_2 = \frac{1}{Cs}$$

Hence the transfer function $E_0(s)/E_i(s)$ can be found as follows:

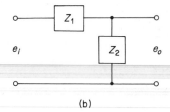

(b)

$$\frac{E_0(s)}{E_i(s)} = \frac{\dfrac{1}{Cs}}{Ls + R + \dfrac{1}{Cs}}$$

$$= \frac{1}{LCs^2 + RCs + 1}$$

**Fig. 4-5.** Electrical circuits.

which is, of course, identical to Eq. (4-5).

**Passive elements and active elements.** Some of the elements in a system (for example, capacitances and inductances in an electrical system) store energy. This energy can later be introduced into the system. The amount of energy which can be introduced cannot exceed the amount the element has stored, and unless such an element stored energy beforehand, it cannot deliver any energy to the system. Because of this, such an element is called a *passive* element. A system containing only passive elements is called a *passive* system. Examples of passive elements are capacitances, resistances, and inductances; masses, inertias, dampers, and springs. For passive systems, every term in the homogeneous system differential equation has the same sign.

A physical element which can deliver external energy into a system is called an *active* element. For example, an amplifier is an active element since it has a power source and supplies power to the system. External force, torque, or velocity sources; current or voltage sources; etc.; are also active elements.

**Force-voltage analogy.** Consider the mechanical system shown in Fig. 4-6 (a) and the electrical system shown in Fig. 4-6 (b). The differential equation for this mechanical system is

$$m\frac{d^2x}{dt^2} + f\frac{dx}{dt} + kx = p \tag{4-6}$$

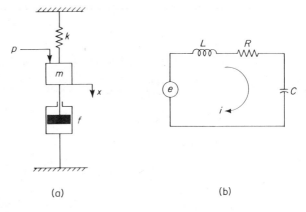

**Fig. 4-6.** (a) Mechanical system; (b) an analogous electrical system.          (a)                              (b)

while the differential equation for the electrical system is

$$L \frac{di}{dt} + Ri + \frac{1}{C} \int i \, dt = e$$

In terms of electric charge $q$, this last equation becomes

$$L \frac{d^2 q}{dt^2} + R \frac{dq}{dt} + \frac{1}{C} q = e \tag{4-7}$$

Comparing Eqs. (4-6) and (4-7), we see that the differential equations for the two systems are of identical form. Such systems are called analogous systems, and the terms which occupy corresponding positions in the differential equations are called analogous quantities. A list of analogous quantities is shown in Table 4-1. The analogy here is called the force-voltage analogy.

**Table 4-1.** ANALOGOUS QUANTITIES IN FORCE-VOLTAGE ANALOGY

| Mechanical System | Electrical System |
|---|---|
| force $p$ (torque $T$)<br>mass $m$ (moment of inertia $J$)<br>viscous-friction coefficient $f$<br>spring constant $k$<br>displacement $x$ (angular displacement $\theta$)<br>velocity $\dot{x}$ (angular velocity $\dot{\theta}$) | voltage $e$<br>inductance $L$<br>resistance $R$<br>reciprocal of capacitance $1/C$<br>charge $q$<br>current $i$ |

Figure 4-7 shows a few examples of analogous systems. Each of the electrical systems and its corresponding mechanical system have the analogous transfer function. (In Fig. 4-7, $x_i$ and $x_0$ denote displacements.) Note that in deriving the transfer functions, we made the assumption that the systems are lumped-parameter systems and that there is no loading effect at the output.

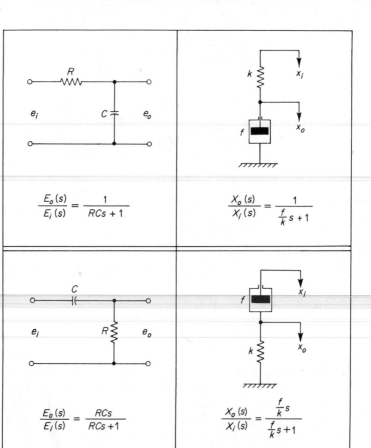

**Fig. 4-7.** Analogous systems.

**Force-current analogy.** Another useful analogy between electrical systems and mechanical systems is based on the force-current analogy. Consider the mechanical system shown in Fig. 4-8(a). The differential equation describing the system is

$$m\frac{d^2x}{dt^2} + f\frac{dx}{dt} + kx = p \qquad (4\text{-}8)$$

Consider next the electrical system shown in Fig. 4-8 (b). Applying Kirchhoff's current law, we obtain

$$i_L + i_R + i_C = i_s \qquad (4\text{-}9)$$

where

$$i_L = \frac{1}{L}\int e\,dt$$

$$i_R = \frac{e}{R}$$

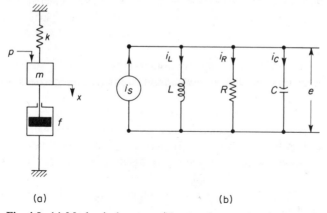

**Fig. 4-8.** (a) Mechanical system; (b) an analogous electrical system.

$$i_c = C \frac{de}{dt}$$

Equation (4-9) can be written

$$\frac{1}{L} \int e \, dt + \frac{e}{R} + C \frac{de}{dt} = i_s \qquad (4\text{-}10)$$

Note that the magnetic flux linkage $\psi$ is related to $e$ by the following equation:

$$\frac{d\psi}{dt} = e$$

In terms of $\psi$, Eq. (4-10) can be written as follows:

$$C \frac{d^2\psi}{dt^2} + \frac{1}{R} \frac{d\psi}{dt} + \frac{1}{L} \psi = i_s \qquad (4\text{-}11)$$

Comparing Eqs. (4-8) and (4-11), we find that the two systems are analogous systems. The analogous quantities are listed in Table 4-2. The analogy here is called the force-current analogy.

**Table 4-2.** ANALOGOUS QUANTITIES IN FORCE-CURRENT ANALOGY

| Mechanical System | Electrical System |
|---|---|
| force $p$ (torque $T$) <br> mass $m$ (inertia $J$) <br> viscous-friction coefficient $f$ <br> spring constant $k$ <br> displacement $x$ (angular displacement $\theta$) <br> velocity $\dot{x}$ (angular velocity $\dot{\theta}$) | current $i$ <br> capacitance $C$ <br> reciprocal of resistance $1/R$ <br> reciprocal of inductance $1/L$ <br> magnetic flux linkage $\psi$ <br> voltage $e$ |

**Analogous systems.** The concept of analogous systems is very useful in practice since one type of system may be easier to handle experimentally than another type. For example, instead of building and studying a mechanical system, we may build and study its electrical analog because, in general, electrical or electronic systems are much easier to deal with experimentally. In particular, electronic analog computers are quite useful for simulating mechanical as well as other physical systems.

It should be remembered that analogies between two systems break down when the regions of operation are extended too far. In other words, since the differential equations upon which the analogies are based are only approximations to the dynamic characteristics of physical systems, the analogy may break down if the operating region of one system is very wide. If the operating region of a given mechanical system is wide, however, it may be divided into two or more subregions, and analogous electrical systems may be built for each subregion. As a matter of fact, analogies are not limited to electrical systems and mechanical systems; they are applicable to any systems as long as their differential equations, or transfer functions, are of identical form.

**Transfer functions of cascaded elements.** Many feedback systems have components which load each other. Consider the system shown in Fig. 4-9. Assume that $e_i$ is the input and $e_0$ is the output. In this system the second stage of the circuit ($R_2C_2$ portion) produces a loading effect on the first stage ($R_1C_1$ portion). The equations for this system are

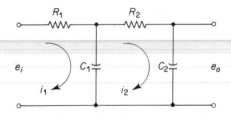

**Fig. 4-9.** Electrical system.

$$\frac{1}{C_1} \int (i_1 - i_2)\, dt + R_1 i_1 = e_i \tag{4-12}$$

and

$$\frac{1}{C_1} \int (i_2 - i_1)\, dt + R_2 i_2 = -\frac{1}{C_2} \int i_2\, dt = -e_0 \tag{4-13}$$

Taking the Laplace transforms of Eqs. (4-12) and (4-13), assuming zero initial conditions, we obtain

$$\frac{1}{C_1 s}[I_1(s) - I_2(s)] + R_1 I_1(s) = E_i(s) \tag{4-14}$$

$$\frac{1}{C_1 s}[I_2(s) - I_1(s)] + R_2 I_2(s) = -\frac{1}{C_2 s} I_2(s) = -E_0(s) \tag{4-15}$$

Eliminating $I_1(s)$ and $I_2(s)$ from Eqs. (4-14) and (4-15), we find that the transfer function between $E_0(s)$ and $E_i(s)$ is

$$\frac{E_0(s)}{E_i(s)} = \frac{1}{(R_1 C_1 s + 1)(R_2 C_2 s + 1) + R_1 C_2 s}$$

$$= \frac{1}{R_1 C_1 R_2 C_2 s^2 + (R_1 C_1 + R_2 C_2 + R_1 C_2)s + 1} \tag{4-16}$$

The term $R_1 C_2 s$ in the denominator of the transfer function represents the interaction of two simple $RC$ circuits. Since $(R_1C_1 + R_2C_2 + R_1C_2)^2 > 4R_1C_1R_2C_2$, the two roots of the denominator of Eq. (4-16) are real.

The present analysis shows that if two $RC$ circuits are connected in cascade, so that the output from the first circuit is the input to the second, the overall transfer function is not the product of $1/(R_1C_1s + 1)$ and $1/(R_2C_2s + 1)$. The reason for this is that when we derive the transfer function for an isolated circuit, we implicitly assume that the output is unloaded. In other words, the load impedance is assumed to be infinite, which means that no power is being withdrawn at the output. When the second circuit is connected to the ouput of the first, however, a certain amount of power is withdrawn, and thus the assumption of no loading is violated. Therefore, if the transfer function of this system is obtained under the assumption of no loading, then it is not valid. The degree of the loading effect determines the amount of modification of the transfer function.

**Transfer functions of nonloading cascaded elements.** The transfer function of a system consisting of two nonloading cascaded elements can be obtained by eliminating the intermediate input and output. For example, consider the system shown in Fig. 4-10 (a). The transfer function of each element is

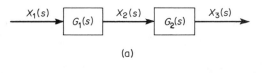

(a)

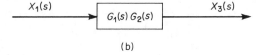

**Fig. 4-10.** (a) System consisting of two nonloading cascaded elements; (b) an equivalent system.

(b)

$$G_1(s) = \frac{X_2(s)}{X_1(s)}, \qquad G_2(s) = \frac{X_3(s)}{X_2(s)}$$

If the input impedance of the second element is infinite, the output of the first element is not affected by connecting it to the second element. Then the transfer function of the whole system is

$$G(s) = \frac{X_3(s)}{X_1(s)} = \frac{X_2(s)}{X_1(s)}\frac{X_3(s)}{X_2(s)} = G_1(s)G_2(s)$$

The transfer function of the whole system is thus the product of the transfer functions of the individual elements. This is shown in Fig. 4-10 (b).

As an example, consider the system shown in Fig. 4-11. The insertion of an isolating amplifier between the circuits to obtain nonloading characteristics is frequently used in combining electric circuits. Since both solid-state amplifiers and vacuum-tube amplifiers have very high input impedances, an isolation amplifier inserted between the two circuits justifies the nonloading assumption.

The two simple $RC$ circuits, isolated by an amplifier as shown in Fig. 4-11,

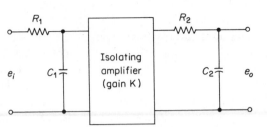

**Fig. 4-11.** Electrical system.

have negligible loading effects and the transfer function of the entire circuit equals the product of the individual transfer functions. Thus, in this case,

$$\frac{E_0(s)}{E_i(s)} = \left(\frac{1}{R_1 C_1 s + 1}\right)(K)\left(\frac{1}{R_2 C_2 s + 1}\right)$$

$$= \frac{K}{(R_1 C_1 s + 1)(R_2 C_2 s + 1)}$$

### 4-3 LINEARIZATION OF A NONLINEAR MATHEMATICAL MODEL

This section presents a linearization technique applicable to many nonlinear systems. We shall apply this technique to a hydraulic servomotor and obtain a transfer function for a linearized hydraulic servomotor.

**Linear approximation of nonlinear systems.** In order to obtain a linear mathematical model for a nonlinear system, we assume that the variables deviate only slightly from some operating condition. Consider a system whose input is $x(t)$ and output is $y(t)$. The relationship between $y(t)$ and $x(t)$ is given by

$$y = f(x) \tag{4-17}$$

If the normal operating condition corresponds to $\bar{x}$, $\bar{y}$, then Eq. (4-17) may be expanded into a Taylor series about this point as follows:

$$y = f(x)$$

$$= f(\bar{x}) + \frac{df}{dx}(x - \bar{x}) + \frac{1}{2!}\frac{d^2f}{dx^2}(x - \bar{x})^2 + \cdots \tag{4-18}$$

where the derivatives $df/dx$, $d^2f/dx^2$, ... are evaluated at $x = \bar{x}$. If the variation $x - \bar{x}$ is small, we may neglect the higher-order terms in $x - \bar{x}$. Then Eq. (4-18) may be written

$$y = \bar{y} + K(x - \bar{x}) \tag{4-19}$$

where

$$\bar{y} = f(\bar{x})$$

$$K = \frac{df}{dx}\bigg|_{x = \bar{x}}$$

Equation (4-19) may be rewritten as

$$y - \bar{y} = K(x - \bar{x}) \tag{4-20}$$

which indicates that $y - \bar{y}$ is proportional to $x - \bar{x}$. Equation (4-20) gives a linear mathematical model for the nonlinear system given by Eq. (4-17).

Next, consider a nonlinear system whose output $y$ is a function of two inputs $x_1$ and $x_2$, so that

$$y = f(x_1, x_2) \tag{4-21}$$

In order to obtain a linear approximation to this nonlinear system, we may expand Eq. (4-21) into a Taylor series about the normal operating point $\bar{x}_1, \bar{x}_2$. Then Eq. (4-21) becomes

$$y = f(\bar{x}_1, \bar{x}_2) + \left[ \frac{\partial f}{\partial x_1} (x_1 - \bar{x}_1) + \frac{\partial f}{\partial x_2} (x_2 - \bar{x}_2) \right]$$
$$+ \frac{1}{2!} \left[ \frac{\partial^2 f}{\partial x_1^2} (x_1 - \bar{x}_1)^2 + 2 \frac{\partial^2 f}{\partial x_1 \partial x_2} (x_1 - \bar{x}_1)(x_2 - \bar{x}_2) \right.$$
$$+ \left. \frac{\partial^2 f}{\partial x_2^2} (x_2 - \bar{x}_2)^2 \right] + \cdots$$

where the partial derivatives are evaluated at $x_1 = \bar{x}_1, x_2 = \bar{x}_2$. Near the normal operating point, the higher-order terms may be neglected. The linear mathematical model of this nonlinear system in the neighborhood of the normal operating condition is then given by

$$y - \bar{y} = K_1(x_1 - \bar{x}_1) + K_2(x_2 - \bar{x}_2)$$

where

$$\bar{y} = f(\bar{x}_1, \bar{x}_2)$$

$$K_1 = \left. \frac{\partial f}{\partial x_1} \right|_{x_1 = \bar{x}_1, \, x_2 = \bar{x}_2}$$

$$K_2 = \left. \frac{\partial f}{\partial x_2} \right|_{x_1 = \bar{x}_1, \, x_2 = \bar{x}_2}$$

**Hydraulic servomotor.** Figure 4-12 shows a hydraulic servomotor. It is essentially a pilot-valve-controlled hydraulic power amplifier and actuator. The pilot valve is a balanced valve, in the sense that the pressure forces acting on it are all balanced. A very large power output can be controlled by a pilot valve, which can be positioned with very little power.

The operation of this hydraulic servomotor is as follows: If the pilot valve is moved to the right, then port I is connected to the supply port, and the pressured oil enters the left-hand side of the power piston. Since port II is connected to the drain port, the oil in the right-hand side of the power piston is returned to the drain. The oil flowing into the power cylinder is at high pressure, and the oil flowing out from the power cylinder into the drain is at low pressure. The resulting difference in pressure on both sides of the power piston will cause it to move to the right. The returned oil is pressurized by a pump and is recirculated in the system. When the pilot piston is moved to the left, the power piston will move to the left.

In practice, ports $a$, $b$, and $c$ shown in Fig. 4-12 are often made wider than the corresponding valves $A$, $B$, and $C$. In this case, there is always leakage through the valve. This improves both the sensitivity and the linearity of the hydraulic

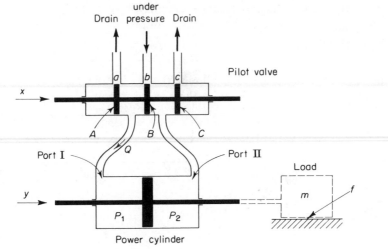

**Fig. 4-12.** Schematic diagram of a hydraulic servomotor.

servomotor. We shall make this assumption in the following analysis. [Note that sometimes a dither signal, a high-frequency signal of very small amplitude (with respect to the maximum displacement of the valve), is superimposed on the motion of the pilot valve. This also improves the sensitivity and the linearity. In this case also there is leakage through the valve.]

Let us define

$Q$ = rate of flow of oil to the power cylinder, lb/sec
$\Delta P = P_2 - P_1$ = pressure difference across the power piston, lb/in.$^2$
$x$ = displacement of pilot valve, in.

In Fig. 4-12, one can see that $Q$ is a function of $x$ and $\Delta P$. In general, the relationship among the variables $Q$, $x$, and $\Delta P$ is given by a nonlinear equation:

$$Q = f(x, \Delta P)$$

Linearizing this nonlinear equation near the normal operating point $\bar{Q}, \bar{x}, \Delta\bar{P}$, we obtain

$$Q - \bar{Q} = K_1(x - \bar{x}) - K_2(\Delta P - \Delta\bar{P}) \qquad (4\text{-}22)$$

where

$$\bar{Q} = f(\bar{x}, \Delta\bar{P})$$

$$K_1 = \frac{\partial Q}{\partial x}\bigg|_{x=\bar{x},\, \Delta P=\Delta\bar{P}}$$

$$K_2 = -\frac{\partial Q}{\partial \Delta P}\bigg|_{x=\bar{x},\, \Delta P=\Delta\bar{P}}$$

Note that for this system, the normal operating condition corresponds to $\bar{Q} = 0$, $\bar{x} = 0$, and $\Delta\bar{P} = 0$. Therefore, we obtain from Eq. (4-22)

$$Q = K_1 x - K_2 \, \Delta P \tag{4-23}$$

Figure 4-13 shows this linearized relationship among $Q$, $x$, and $\Delta P$. The straight lines shown are the characteristic curves of the linearized hydraulic servomotor. This family of curves consists of equidistant parallel straight lines, parameterized by $x$.

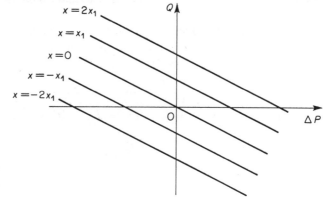

**Fig. 4-13.** Characteristic curves of the linearized hydraulic servomotor.

Referring to Fig. 4-12, we see that the rate of flow of oil $Q$ (lb/sec) times $dt$ (sec) is equal to the power piston displacement $dy$ (in.) times the piston area $A$ (in.$^2$) times the density of oil $\rho$ (lb/in.$^3$). Thus, we obtain

$$A\rho \, dy = Q \, dt$$

Notice that for a given flow rate $Q$, the larger the piston area $A$ is, the lower will be the velocity $dy/dt$. Hence, if the piston area $A$ is made smaller, the other variables remaining constant, the velocity $dy/dt$ will become higher. Also an increased flow rate $Q$ will cause an increased velocity of the power piston and will make the response time shorter.

Equation (4-23) can now be written as

$$\Delta P = \frac{1}{K_2}\left(K_1 x - A\rho \frac{dy}{dt}\right)$$

The force developed by the power piston is equal to the pressure difference $\Delta P$ times the piston area $A$ or

$$\text{Force developed by the power piston} = A \, \Delta P$$

$$= \frac{A}{K_2}\left(K_1 x - A\rho \frac{dy}{dt}\right)$$

For a given maximum force, if the pressure difference is sufficiently high, the piston area, or the volume of oil in the cylinder, can be made small. Consequently, to minimize the weight of the controller, we must make the supply pressure sufficiently high.

Assume that the power piston moves a load consisting of a mass and viscous friction. Then the force developed by the power piston is applied to the load mass and friction, and we obtain

$$m\ddot{y} + f\dot{y} = \frac{A}{K_2}(K_1 x - Ap\dot{y})$$

or

$$m\ddot{y} + \left(f + \frac{A^2 p}{K_2}\right)\dot{y} = \frac{AK_1}{K_2}x \qquad (4\text{-}24)$$

where $m$ is the mass (lb-sec$^2$/in.) of the load and $f$ is the viscous-friction coefficient (lb-sec/in.).

Assuming that the pilot valve displacement $x$ is the input and the power piston displacement $y$ is the output, we find that the transfer function for the hydraulic servomotor is, from Eq. (4-24),

$$\frac{Y(s)}{X(s)} = \frac{1}{s\left[\left(\dfrac{mK_2}{AK_1}\right)s + \dfrac{fK_2}{AK_1} + \dfrac{Ap}{K_1}\right]}$$

$$= \frac{K}{s(Ts + 1)} \qquad (4\text{-}25)$$

where

$$K = \frac{1}{\dfrac{fK_2}{AK_1} + \dfrac{Ap}{K_1}} \qquad \text{and} \qquad T = \frac{mK_2}{fK_2 + A^2 p}$$

From Eq. (4-25), one can see that this transfer function is of the second order. If the ratio $mK_2/(fK_2 + A^2 p)$ is negligibly small or the time constant $T$ is negligible, the transfer function can be simplified to give

$$\frac{Y(s)}{X(s)} = \frac{K}{s}$$

A more detailed analysis shows that if oil leakage, compressibility (including the effects of dissolved air), expansion of pipelines, etc., are taken into consideration, the transfer function becomes

$$\frac{Y(s)}{X(s)} = \frac{K}{s(T_1 s + 1)(T_2 s + 1)}$$

where $T_1$ and $T_2$ are time constants. As a matter of fact, these time constants depend on the volume of oil in the operating circuit. The smaller the volume, the smaller the time constants.

## 4-4 BLOCK DIAGRAMS

A control system may consist of a number of components. In order to show the functions performed by each component, in control engineering, we commonly use a diagram called the "block diagram."

**Block diagrams.** A block diagram of a system is a pictorial representation of the functions performed by each component and of the flow of signals. Such a diagram depicts the interrelationships which exist between the various components. Different from a purely abstract, mathematical representation, a block diagram has the advantage of indicating more realistically the signal flows of the actual system.

In a block diagram, all system variables are linked to each other through functional blocks. The "functional block," or simply "block," is a symbol for the mathematical operation on the input signal to the block which produces the output. The transfer functions of the components are usually entered in the corresponding blocks, which are connected by arrows to indicate the direction of the flow of signals. Note that the signal can pass only in the direction of the arrows. Thus, a block diagram of a control system explicitly shows a unilateral property.

Figure 4-14 shows an element of the block diagram. The arrowhead pointing toward the block indicates the input and the arrowhead leading away from the block represents the output. Such arrows are referred to as signals.

**Fig. 4-14.** Element of a block diagram.

Note that the dimensions of the output signal from the block are the dimensions of the input signal multiplied by the dimensions of the transfer function in the block.

The advantages of the block diagram representation of a system lie in the fact that it is easy to form the overall block diagram for the entire system by merely connecting the blocks of the components according to the signal flow and that it is possible to evaluate the contribution of each component to the overall performance of the system.

In general, the functional operation of the system can be visualized more readily by examination of the block diagram than by examination of the physical system itself. A block diagram contains information concerning dynamic behavior, but it does not contain any information concerning the physical construction of the system. Therefore, many dissimilar and unrelated systems can be represented by the same block diagram.

It should be noted that in a block diagram the main source of energy is not explicitly shown and, also, that a block diagram of a given system is not unique. A number of different block diagrams may be drawn for a system, depending upon the viewpoint of the analysis.

**Error detector.** The error detector produces a signal which is the difference between the reference input and the feedback signal of the control system. In design, the choice of the error detector is quite important and must be carefully decided. This is because any imperfections in the error detector will inevitably impair the performance of the entire system. The block diagram representation of the error detector is shown in Fig. 4-15.

Note that a circle with a cross is the symbol which

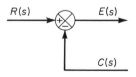

**Fig. 4-15.** Block diagram of an error detector.

indicates a summing operation. The plus or minus sign at each arrowhead indicates whether that signal is to be added or subtracted. It is important that the
quantities being added or subtracted have the same dimensions and the same units.

**Block diagram of a closed-loop system.** Figure 4-16 shows an example of a
block diagram of a closed-loop system. The output $C(s)$ is fed back to the summing
point, where it is compared with the reference input $R(s)$. The closed-loop nature
of the system is clearly indicated by the figure. The output of the block, $C(s)$ in this
case, is obtained by multiplying the transfer function $G(s)$ by the input to the block,
$E(s)$.

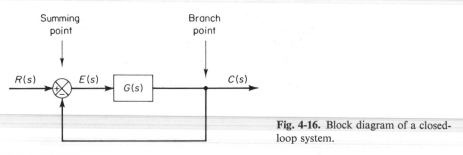

**Fig. 4-16.** Block diagram of a closed-
loop system.

Any linear control system may be represented by a block diagram consisting
of blocks, summing points, and branch points. A branch point is the point from
which the output signal from a block goes concurrently to other blocks or summing
points.

When the output is fed back to the summing point for comparison with the
input, it is necessary to convert the form of the output signal to that of the input
signal. For example, in a temperature control system, the output signal is usually
the controlled temperature. The output signal, which has the dimension of temperature, must be converted to a force or position before it can be compared with the
input signal. This conversion is accomplished by the feedback element whose transfer function is $H(s)$, as shown in Fig. 4-17. Another important role of the feedback
element is to modify the output before it is compared with the input. In the present
example, the feedback signal which is fed back to the summing point for comparison with the input is $B(s) = H(s)C(s)$.

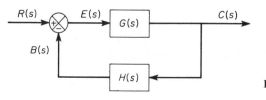

**Fig. 4-17.** Closed-loop system.

The ratio of the feedback signal $B(s)$ to the actuating error signal $E(s)$ is called
the open-loop transfer function. That is,

$$\text{Open-loop transfer function} = \frac{B(s)}{E(s)} = G(s)H(s)$$

The ratio of the output $C(s)$ to the actuating error signal $E(s)$ is called the feed-forward transfer function, so that

$$\text{Feedforward transfer function} = \frac{C(s)}{E(s)} = G(s)$$

If the feedback transfer function is unity, then the open-loop transfer function and the feedforward transfer function are the same. For the system shown in Fig. 4-17, the output $C(s)$ and input $R(s)$ are related as follows:

$$C(s) = G(s)E(s)$$
$$E(s) = R(s) - B(s)$$
$$= R(s) - H(s)C(s)$$

Eliminating $E(s)$ from these equations gives

$$C(s) = G(s)[R(s) - H(s)C(s)]$$

or

$$\frac{C(s)}{R(s)} = \frac{G(s)}{1 + G(s)H(s)} \tag{4-26}$$

The transfer function relating $C(s)$ to $R(s)$ is called the closed-loop transfer function. This transfer function relates the closed-loop system dynamics to the dynamics of the feedforward elements and feedback elements.

From Eq. (4-26) $C(s)$ is given by

$$C(s) = \frac{G(s)}{1 + G(s)H(s)} R(s)$$

Thus the output of the closed-loop system clearly depends on both the closed-loop transfer function and the nature of the input.

**Closed-loop system subjected to a disturbance.** Figure 4-18 shows a closed-loop system subjected to a disturbance. When two inputs (the reference input and disturbance) are present in a linear system, each input can be treated independently of the other; and the outputs corresponding to each input alone can be added to give the complete output. The way each input is introduced into the system is shown at the summing point by either a plus or minus sign.

Consider the system shown in Fig. 4-18. In examining the effect of the distur-

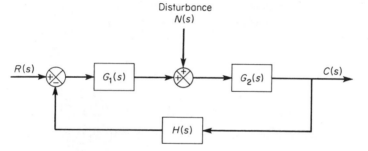

**Figure 4-18.** Closed-loop system subjected to a disturbance.

bance $N(s)$, we may assume that the system is at rest initially with zero error; we may then calculate the response $C_N(s)$ to the disturbance only. This response can be found from

$$\frac{C_N(s)}{N(s)} = \frac{G_2(s)}{1 + G_1(s)G_2(s)H(s)}$$

On the other hand, in considering the response to the reference input $R(s)$, we may assume that the disturbance is zero. Then the response $C_R(s)$ to the reference input $R(s)$ can be obtained from

$$\frac{C_R(s)}{R(s)} = \frac{G_1(s)G_2(s)}{1 + G_1(s)G_2(s)H(s)}$$

The response to the simultaneous application of the reference input and disturbance can be obtained by adding the two individual responses. In other words, the response $C(s)$ due to the simultaneous application of the reference input $R(s)$ and disturbance $N(s)$ is given by

$$C(s) = C_R(s) + C_N(s)$$
$$= \frac{G_2(s)}{1 + G_1(s)G_2(s)H(s)}[G_1(s)R(s) + N(s)]$$

Consider now the case where $|G_1(s)H(s)| \gg 1$ and $|G_1(s)G_2(s)H(s)| \gg 1$. In this case, the closed-loop transfer function $C_N(s)/N(s)$ becomes almost zero, and the effect of the disturbance is suppressed. This is an advantage of the closed-loop system.

On the other hand, the closed-loop transfer function $C_R(s)/R(s)$ approaches $1/H(s)$ as the gain of $G_1(s)G_2(s)H(s)$ increases. This means that if $|G_1(s)G_2(s)H(s)| \gg 1$, then the closed-loop transfer function $C_R(s)/R(s)$ becomes independent of $G_1(s)$ and $G_2(s)$ and becomes inversely proportional to $H(s)$ so that the variations of $G_1(s)$ and $G_2(s)$ do not affect the closed-loop transfer function $C_R(s)/R(s)$. This is another advantage of the closed-loop system. It can easily be seen that any closed-loop system with unity feedback, $H(s) = 1$, tends to equalize the input and output.

**Procedures for drawing block diagrams.** To draw a block diagram for a system, first write the equations which describe the dynamic behavior of each component, then take the Laplace transforms of these equations, assuming zero initial conditions, and represent each Laplace-transformed equation individually in block form. Finally, assemble the elements into a complete block diagram.

As an example, consider the $RC$ circuit shown in Fig. 4-19 (a). The equations for this circuit are

$$i = \frac{e_i - e_0}{R} \tag{4-27}$$

$$e_0 = \frac{\int i \, dt}{C} \tag{4-28}$$

The Laplace transforms of Eqs. (4-27) and (4-28), with an initial condition of zero, become

$$I(s) = \frac{E_i(s) - E_0(s)}{R} \qquad (4\text{-}29)$$

$$E_0(s) = \frac{I(s)}{Cs} \qquad (4\text{-}30)$$

Equation (4-29) represents a summing operation, and the corresponding diagram is shown in Fig. 4-19 (b). Equation (4-30) represents the block as shown in Fig. 4-19 (c). Assembling these two elements, we obtain the overall block diagram for the system as shown in Fig. 4-19 (d).

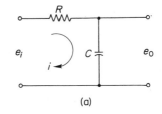

(a)

**Block diagram reduction.** It is important to note that blocks can be connected in series only if the output of one block is not affected by the next following block. If there are any loading effects between the components, it is necessary to combine these components into a single block.

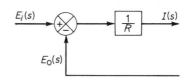

(b)

Any number of cascaded blocks representing nonloading components can be replaced by a single block, the transfer function of which is simply the product of the individual transfer functions.

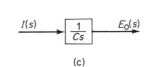

(c)

A complicated block diagram involving many feedback loops can be simplified by a step-by-step rearrangement, using rules of block diagram algebra. Some of these important rules are given in Table 4-3. They are obtained by writing the same equation in a different way. Simplification of the block diagram by rearrangements and substitutions reduces considerably the labor needed for subsequent mathematical

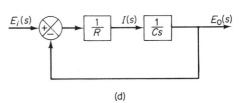

(d)

**Fig. 4-19.** (a) *RC* circuit; (b) block diagram representing Eq. (4-29); (c) block diagram representing Eq. (4-30); (d) block diagram of the *RC* circuit.

analysis. It should be noted, however, that as the block diagram is simplified, new blocks become more complex because new poles and zeros are generated.

In simplifying a block diagram, remember the following:

1. The product of the transfer functions in the feedforward direction must remain the same.

2. The product of the transfer functions around the loop must remain the same.

A general rule for simplifying a block diagram is to move branch points and summing points, interchange summing points, and then reduce internal feedback loops.

As an example of the use of the rules in Table 4-3, consider the system shown in Fig. 4-20 (a). By moving the summing point of the negative feedback loop containing $H_2$ outside the positive feedback loop containing $H_1$, we obtain Fig. 4-20 (b). Eliminating the positive feedback loop, we obtain Fig. 4-20 (c). Then, eliminating the loop containing $H_2/G_1$, we obtain Fig. 4-20 (d). Finally, by eliminating the feedback loop, we obtain Fig. 4-20 (e).

Notice that the numerator of the closed-loop transfer function $C(s)/R(s)$ is the product of the transfer functions of the feedforward path. The denominator of $C(s)/R(s)$ is equal to

**Table 4-3.** RULES OF BLOCK DIAGRAM ALGEBRA

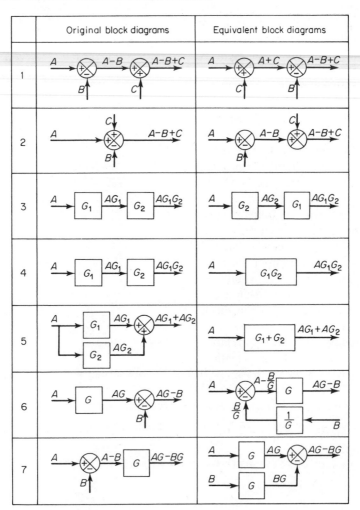

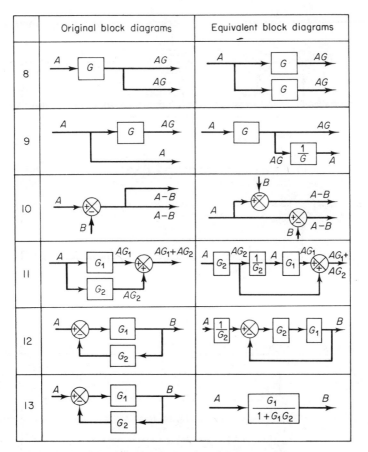

$$1 - \sum \text{(product of the transfer functions around each loop)}$$
$$= 1 - (G_1 G_2 H_1 - G_2 G_3 H_2 - G_1 G_2 G_3)$$
$$= 1 - G_1 G_2 H_1 + G_2 G_3 H_2 + G_1 G_2 G_3$$

(The positive feedback loop yields a negative term in the denominator.)

## 4-5 DERIVING TRANSFER FUNCTIONS OF PHYSICAL SYSTEMS

Control systems may consist of components of different types, such as electrical, mechanical, hydraulic, pneumatic, or thermal. A control engineer must be familiar with the fundamental laws underlying these components.

In Sections 4-2 and 4-3, we derived transfer functions for several systems. In this section we shall present additional examples showing the derivation of transfer functions for various types of physical systems.

In deriving transfer functions, note the following:

1. In approximating physical systems by linear lumped-parameter models,

(a)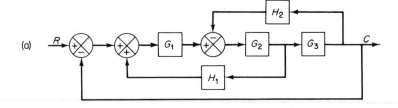

(b)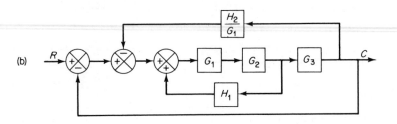

(c)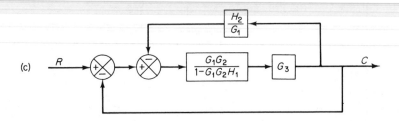

(d)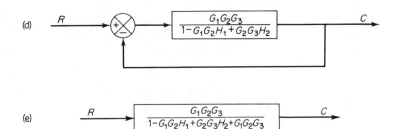

(e)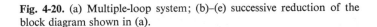

**Fig. 4-20.** (a) Multiple-loop system; (b)–(e) successive reduction of the block diagram shown in (a).

certain assumptions are necessary. In general, these assumptions depend on the operating ranges of the system. Assumptions which may be good for certain operating conditions may not be suitable for different ones. In practice, making proper assumptions is a very important part of the analysis of a system since the accuracy of the results depends on the validity of such assumptions. In this connection the engineer must realize that although the analytically obtained transfer functions show clearly the general effect of variations in system parameters on the solution of the equation, numerical predictions obtainable from the analytical study of physical systems may not be accurate. In other words, although qualitative characteristics of the system dynamics may be seen clearly from the transfer function, quantitative information may not necessarily be accurate. If quantitative accuracy is needed, it will be necessary to carry out experimental studies in addition to analytical ones.

2. It is desirable to check the validity of the transfer function by assuming that certain parameters of the system are either zero or infinity. Since this simplifies the transfer function, its validity can be checked easily.

The transfer functions obtained in the following are based on the assumption that there is no loading effect upon the output. (Remember that if there is any loading effect, it must be accounted for when the transfer function is derived.)

MECHANICAL AND ELECTROMECHANICAL SYSTEMS

**Seismograph.** Figure 4-21 shows a schematic diagram of a seismograph. A seismograph indicates the displacement of its case with respect to inertial space. It is used to measure the ground displacement during earthquakes.

Let us define

$x_i$ = displacement of the case relative to inertial space

$x_0$ = displacement of the mass $m$ relative to inertial space

$y = x_0 - x_i$ = displacement of the mass $m$ relative to the case

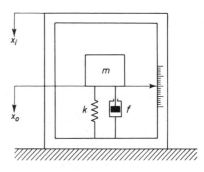

(Note that since gravity produces a steady spring deflection, we measure the displacement $x_0$ of mass $m$ from the static equilibrium position.) The equation for this system is found to be

**Fig. 4-21.** Schematic diagram of a seismograph.

$$m\ddot{x}_0 + f(\dot{x}_0 - \dot{x}_i) + k(x_0 - x_i) = 0$$

By substituting $x_0 = y + x_i$ in this last equation, we obtain a differential equation in $y$. (Note that $y$ is a signal we can actually measure.)

$$m\ddot{y} + f\dot{y} + ky = -m\ddot{x}_i \qquad (4\text{-}31)$$

Taking the Laplace transform of Eq. (4-31), assuming zero initial conditions, we obtain

$$[ms^2 + fs + k]Y(s) = -ms^2 X_i(s)$$

Considering $x_i$ as input and $y$ as output, the transfer function is

$$\frac{Y(s)}{X_i(s)} = \frac{-ms^2}{ms^2 + fs + k}$$

$$= -\frac{s^2}{s^2 + \dfrac{f}{m}s + \dfrac{k}{m}}$$

For very low frequency inputs, the mass $m$ follows the case up and down, and the spring deflection is quite small. If the input $x_i$ consists of signals whose frequencies are very much lower than $\sqrt{k/m}$, then the transfer function may be approximated by

$$\frac{Y(j\omega)}{X_i(j\omega)} = \frac{\omega^2}{\dfrac{k}{m}} \qquad (4\text{-}32)$$

[Note that in Eq. (4-32), $s$ is replaced by $j\omega$. Such a transfer function is called a sinusoidal transfer function. For details, refer to Chapter 9.] As an example, if $y(t)$ is found to be of sinusoidal form $A \sin \omega t$, where $\omega \ll \sqrt{k/m}$, then from the recording of $y(t)$ the input signal $x_i(t)$ can be determined as $(A/\omega^2)(k/m) \sin \omega t$.

Since in seismographs the undamped natural frequency $\sqrt{k/m}$ is made small, if the input frequency is much higher than $\sqrt{k/m}$, the mass $m$ will remain almost fixed in inertial space and the motion of the case will indicate the relative motion between the case and mass. If the input signal $x_i$ has a very high frequency, then the transfer function may be approximated by

$$\frac{Y(j\omega)}{X_i(j\omega)} = -1$$

This implies that if $y(t)$ is found to be $A \sin \omega t$, where $\omega \gg \sqrt{k/m}$, then the input $x_i(t)$ can be determined as $-A \sin \omega t$.

**Servomotors.** The servomotors we shall consider here are two-phase servomotors, armature-controlled dc motors, and field-controlled dc motors. We shall first consider the effect of load on the servomotor dynamics.

**Effect of load on servomotor dynamics.** Most important among the characteristics of the servomotor is the maximum acceleration obtainable. For a given available torque, the rotor moment of inertia must be a minimum. Since the servomotor operates under continuously varying conditions, acceleration and deceleration of the rotor occur from time to time. The servomotor must be able to absorb mechanical energy as well as to generate it. The performance of the servomotor when used as a brake should be satisfactory.

Let $J_m$ and $f_m$ be, respectively, the moment of inertia and friction of the rotor, and let $J_L$ and $f_L$ be, respectively, the moment of inertia and friction of the load on the output shaft. Assume that the moment of inertia and friction of the gear train are either negligible or included in $J_L$ and $f_L$, respectively. Then, the equivalent moment of inertia $J_{eq}$ referred to the motor shaft and equivalent friction $f_{eq}$ referred to the motor shaft can be written as (for details, see Problem A-4-2)

$$J_{eq} = J_m + n^2 J_L \qquad (n < 1)$$
$$f_{eq} = f_m + n^2 f_L \qquad (n < 1)$$

where $n$ is the gear ratio between the motor and load. If the gear ratio $n$ is small and $J_m \gg n^2 J_L$, then the moment of inertia of the load referred to the motor shaft is negligible with respect to the rotor moment of inertia. A similar argument applies to the load friction. In general, when the gear ratio $n$ is small, the transfer function of the electric servomotor may be obtained without taking into account the load moment of inertia and friction. If neither $J_m$ nor $n^2 J_L$ is negligibly small compared with the other, however, then the equivalent moment of inertia $J_{eq}$ must be used for evaluating the transfer function of the motor-load combination.

**Two-phase servomotors.** A two-phase servomotor, commonly used for instrument servomechanisms, is similar to a conventional two-phase induction motor except for its special design considerations. It uses a squirrel-cage rotor. This rotor has a small diameter-to-length ratio to minimize the moment of inertia and to obtain a good accelerating characteristic. The two-phase servomotor is very rugged and reliable.

In many practical applications, the power range for which two-phase servomotors are used is between a fraction of a watt and a hundred watts.

A schematic diagram of a two-phase servomotor is shown in Fig. 4-22 (a). Here one phase (fixed field) of the motor is continuously excited from the reference voltage, the frequency of which is usually 60, 400, or 1000 cycles; and the other phase (control field) is driven with the control voltage (a suppressed carrier signal) which is 90° phase-shifted in time with respect to the reference voltage. (The control voltage is of variable magnitude and polarity.)

Note that the voltage of the control phase is made 90° out of phase with respect to the voltage of the fixed phase. The stator windings for the fixed and control phases are placed 90° apart in space. These considerations are based on the fact that torque is produced most efficiently on a shaft when the phase-winding axes are in space quadrature and voltages in the two phases are in time quadrature.

The two stator windings are normally excited by a two-phase power supply. If a two-phase power supply is not available, however, then the fixed phase winding may be connected to a single-phase power supply through a capacitor, which will provide the 90° phase shift. The amplifier to which the control phase winding is connected is supplied from the same single-phase power supply.

In the two-phase servomotor, the polarity of the control voltage determines the direction of rotation. The instantaneous control voltage $e_c(t)$ is of the form

$$e_c(t) = E_c(t) \sin \omega t \qquad \text{for} \quad E_c(t) > 0$$
$$= |E_c(t)| \sin(\omega t + \pi) \qquad \text{for} \quad E_c(t) < 0$$

This means that a change in the sign of $E_c(t)$ shifts the phase by $\pi$ radians. Thus, a change in the sign of the control voltage $E_c(t)$ reverses the direction of rotation of the motor. Since the reference voltage is constant, the torque $T$ and angular speed $\dot{\theta}$ are also functions of the control voltage $E_c(t)$. If variations in $E_c(t)$ are slow compared with the ac supply frequency, the torque developed by the motor is proportional to $E_c(t)$. Figure 4-22 (b) shows the curves $e_c(t)$ versus $t$, $E_c(t)$ versus $t$, and torque $T(t)$ versus $t$. The angular speed at steady state is proportional to the control voltage $E_c(t)$.

A family of torque-speed curves, when the rated voltage is applied to the fixed phase winding and various voltages are applied to the control phase winding,

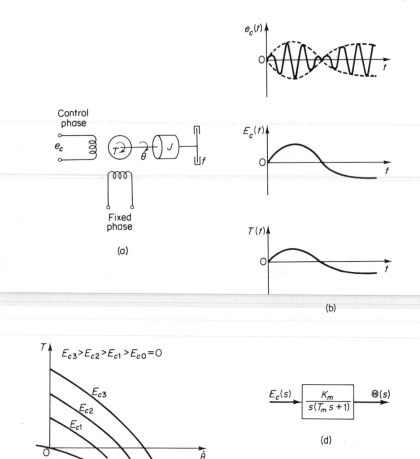

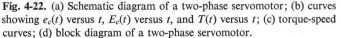

Fig. 4-22. (a) Schematic diagram of a two-phase servomotor; (b) curves showing $e_c(t)$ versus $t$, $E_c(t)$ versus $t$, and $T(t)$ versus $t$; (c) torque-speed curves; (d) block diagram of a two-phase servomotor.

gives steady-state characteristics of the two-phase servomotor. The transfer function of a two-phase servomotor may be obtained from such torque-speed curves if they are parallel and equidistant straight lines. Generally, the torque-speed curves are parallel for a relatively wide speed range but they may not be equidistant; i.e., for a given speed, the torque may not vary linearly with respect to the control voltage. In a low-speed region, however, the torque-speed curves are usually straight lines and equidistant in a region of low control voltages. Since the servomotor seldom operates at high speeds, the linear portions of the torque-speed curves may be extended to the high-speed region. If the assumption is made that they are equidistant for all control voltages, then the servomotor may be considered linear.

Figure 4-22 (c) shows a set of torque-speed curves for various values of control voltages. The torque-speed curve corresponding to zero control voltage passes through the origin. Since the slope of this curve is normally negative, if the control phase voltage becomes equal to zero, the motor develops that torque necessary to stop the rotation.

The servomotor provides a large torque at zero speed. This torque is necessary for rapid acceleration. From Fig. 4-22 (c), we see that the torque $T$ generated is a function of the motor-shaft angular speed $\dot{\theta}$ and the control voltage $E_c$. The equation for any torque-speed line is

$$T = -K_n\dot{\theta} + K_cE_c \qquad (4\text{-}33)$$

where $K_n$ and $K_c$ are positive constants. The torque-balance equation for the two-phase servomotor is

$$T = J\ddot{\theta} + f\dot{\theta} \qquad (4\text{-}34)$$

where $J$ is the moment of inertia of the motor and load referred to the motor shaft and $f$ is the viscous-friction coefficient of the motor and load referred to the motor shaft. From Eqs. (4-33) and (4-34), we obtain the following equation:

$$J\ddot{\theta} + (f + K_n)\dot{\theta} = K_cE_c$$

Noting that the control voltage $E_c$ is the input and the displacement of the motor shaft is the output, we see that the transfer function of the system is given by

$$\frac{\Theta(s)}{E_c(s)} = \frac{K_c}{Js^2 + (f + K_n)s} = \frac{K_m}{s(T_ms + 1)} \qquad (4\text{-}35)$$

where

$K_m = K_c/(f + K_n) = $ motor gain constant
$T_m = J/(f + K_n) = $ motor time constant

Figure 4-22 (d) shows a block diagram for this system. From the transfer function of this system, we can see that $(f + K_n)s$ is a viscous-friction term produced by the motor and load. Thus, $K_n$, the negative of the slope of the torque-speed curve, together with $f$, defines the equivalent viscous-friction of the motor and load combination. For steeper torque-speed curves, the damping of the motor is higher. If the rotor inertia is sufficiently low, then for most of the frequency range we have $|T_ms| \ll 1$ and the servomotor acts as an integrator.

The transfer function given by Eq. (4-35) is based on the assumption that the servomotor is linear. In practice, however, it is not quite so. For torque-speed curves not quite parallel and equidistant, the value of $K_n$ is not constant and, therefore, the values of $K_m$ and $T_m$ are also not constant; they vary with the control voltage.

**Armature-controlled dc motors.** A dc motor is often employed in a control system where an appreciable amount of shaft power is required. The dc motors are much more efficient than two-phase ac servomotors.

The dc motors have separately excited fields. They are either armature-controlled with fixed field or field-controlled with fixed armature current. For example, dc

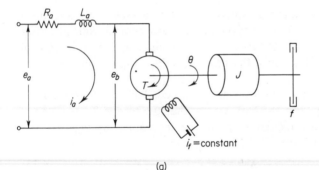

(a)

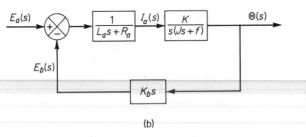

(b)

Fig. 4-23. (a) Schematic diagram of an armature-controlled dc motor; (b) block diagram.

motors used in instruments employ a fixed permanent-magnet field, and the control signal is applied to the armature terminals.

The performance characteristics of the armature-controlled dc motor resemble the idealized characteristics of the two-phase ac servomotor.

Consider the armature-controlled dc motor shown in Fig. 4-23 (a). In this system,

$R_a$ = armature-winding resistance, ohms
$L_a$ = armature-winding inductance, henrys
$i_a$ = armature-winding current, amperes
$i_f$ = field current, amperes
$e_a$ = applied armature voltage, volts
$e_b$ = back emf, volts
$\theta$ = angular displacement of the motor shaft, radians
$T$ = torque delivered by the motor, lb-ft
$J$ = equivalent moment of inertia of the motor and load referred to the motor shaft, slug-ft²
$f$ = equivalent viscous-friction coefficient of the motor and load referred to the motor shaft, lb-ft/rad/sec

The torque $T$ delivered by the motor is proportional to the product of the armature current $i_a$ and the air gap flux $\psi$, which in turn is proportional to the field current or

$$\psi = K_f i_f$$

where $K_f$ is a constant. The torque $T$ can, therefore, be written

$$T = K_f i_f K_1 i_a$$

where $K_1$ is a constant.

In the armature-controlled dc motor, the field current is held constant. For a constant field current, the flux becomes constant, and the torque becomes directly proportional to the armature current so that

$$T = K i_a$$

where $K$ is a motor-torque constant. When the armature is rotating, a voltage proportional to the product of the flux and angular veclocity is induced in the armature. For a constant flux, the induced voltage $e_b$ is directly proportional to the angular velocity $d\theta/dt$. Thus,

$$e_b = K_b \frac{d\theta}{dt} \tag{4-36}$$

where $K_b$ is a back emf constant.

The speed of an armature-controlled dc motor is controlled by the armature voltage $e_a$. The armature voltage $e_a$ is supplied by an amplifier (or by a generator, which is supplied by an amplifier). The differential equation for the armature circuit is

$$L_a \frac{di_a}{dt} + R_a i_a + e_b = e_a \tag{4-37}$$

The armature current produces the torque which is applied to the inertia and friction; hence

$$J \frac{d^2\theta}{dt^2} + f \frac{d\theta}{dt} = T = K i_a \tag{4-38}$$

Assuming that all initial conditions are zero, and taking the Laplace transforms of Eqs. (4-36), (4-37), and (4-38), we obtain the following equations:

$$K_b s \Theta(s) = E_b(s) \tag{4-39}$$

$$(L_a s + R_a) I_a(s) + E_b(s) = E_a(s) \tag{4-40}$$

$$(Js^2 + fs)\Theta(s) = T(s) = K I_a(s) \tag{4-41}$$

Considering $E_a(s)$ as the input and $\Theta(s)$ as the output, we can construct the block diagram from Eqs. (4-39), (4-40), and (4-41), as shown in Fig. 4-23 (b). The effect of the back emf is seen to be the feedback signal proportional to the speed of the motor. This back emf thus increases the effective damping of the system. The transfer function of this system is obtained as

$$\frac{\Theta(s)}{E_a(s)} = \frac{K}{s[L_a J s^2 + (L_a f + R_a J)s + R_a f + K K_b]} \tag{4-42}$$

The inductance $L_a$ in the armature circuit is usually small and may be neglected. If $L_a$ is neglected, then the transfer function given by Eq. (4-42) reduces to

$$\frac{\Theta(s)}{E_a(s)} = \frac{K_m}{s(T_m s + 1)} \tag{4-43}$$

where

$$K_m = K/(R_a f + KK_b) = \text{motor gain constant}$$
$$T_m = R_a J/(R_a f + KK_b) = \text{motor time constant}$$

From Eqs. (4-42) and (4-43), it can be seen that the transfer functions involve the term $1/s$. Thus, this system possesses an integrating property. In Eq. (4-43), notice that the time constant of the motor is smaller for a smaller $R_a$ and smaller $J$. With small $J$, as the resistance $R_a$ is reduced, the motor-time constant approaches zero, and the motor acts as an ideal integrator.

---

*Example 4-1. Positional Servomechanism.* Obtain the closed-loop transfer function for the positional servomechanism shown in Fig. 4-24. Assume that the input and output of the system are the input shaft position and the output shaft position, respectively. Assume the following numerical values for system constants:

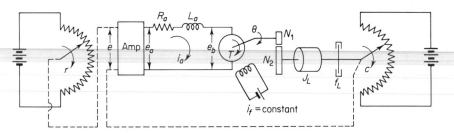

**Fig. 4-24.** Positional servomechanism.

$r =$ angular displacement of the reference input shaft, radians
$c =$ angular displacement of the output shaft, radians
$\theta =$ angular displacement of the motor shaft, radians
$K_1 =$ gain of the potentiometric error detector $= 24/\pi$ volts/rad
$K_p =$ amplifier gain $= 10$ volts/volt
$e_a =$ applied armature voltage, volts
$e_b =$ back emf, volts
$R_a =$ armature-winding resistance $= 0.2$ ohms
$L_a =$ armature-winding inductance $=$ negligible
$i_a =$ armature-winding current, amperes
$K_b =$ back emf constant $= 5.5 \times 10^{-2}$ volts-sec/rad
$K =$ motor torque constant $= 6 \times 10^{-5}$ lb-ft/amp
$J_m =$ moment of inertia of the motor $= 1 \times 10^{-5}$ lb-ft-sec$^2$
$f_m =$ viscous-friction coefficient of the motor $=$ negligible
$J_L =$ moment of inertia of the load $= 4.4 \times 10^{-3}$ lb-ft-sec$^2$
$f_L =$ viscous-friction coefficient of the load $= 4 \times 10^{-2}$ lb-ft/rad/sec
$n =$ gear ratio $N_1/N_2 = 1/10$

The equations describing the system dynamics are as follows:
*For the potentiometric error detector:*

$$E(s) = K_1[R(s) - C(s)] = 7.64[R(s) - C(s)] \qquad (4\text{-}44)$$

*For the amplifier:*

$$E_a(S) = K_p E(s) = 10E(s) \qquad (4\text{-}45)$$

*For the armature-controlled dc motor:* The equivalent moment of inertia $J$ and equivalent viscous friction $f$ referred to the motor shaft are, respectively,

$$J = J_m + n^2 J_L$$
$$= 1 \times 10^{-5} + 4.4 \times 10^{-5} = 5.4 \times 10^{-5}$$
$$f = f_m + n^2 f_L$$
$$= 4 \times 10^{-4}$$

Referring to Eq. (4-43), we obtain

$$\frac{\Theta(s)}{E_a(s)} = \frac{K_m}{s(T_m s + 1)}$$

where

$$K_m = \frac{K}{R_a f + K K_b} = \frac{6 \times 10^{-5}}{(0.2)(4 \times 10^{-4}) + (6 \times 10^{-5})(5.5 \times 10^{-2})} = 0.72$$

$$T_m = \frac{R_a J}{R_a f + K K_b} = \frac{(0.2)(5.4 \times 10^{-5})}{(0.2)(4 \times 10^{-4}) + (6 \times 10^{-5})(5.5 \times 10^{-2})} = 0.13$$

Thus

$$\frac{\Theta(s)}{E_a(s)} = \frac{10 C(s)}{E_a(s)} = \frac{0.72}{s(0.13s + 1)} \tag{4-46}$$

Using Eqs. (4-44), (4-45), and (4-46), we can draw the block diagram of the system as shown in Fig. 4-25 (a). Simplifying this block diagram, we obtain Fig. 4-25 (b). The closed-loop transfer function of this system is

$$\frac{C(s)}{R(s)} = \frac{42.3}{s^2 + 7.7s + 42.3}$$

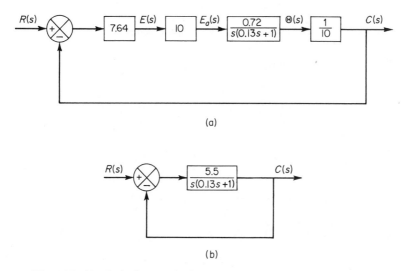

(a)

(b)

**Fig. 4-25.** (a) Block diagram of the system shown in Fig. 4-24; (b) simplified block diagram.

**Field-controlled dc motor.** Figure 4-26 (a) is a schematic diagram of a field-controlled dc motor. Let us define

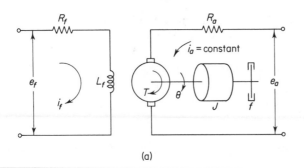

(a)

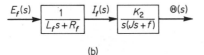

(b)

Fig. 4-26. (a) Schematic diagram of a field-controlled dc motor; (b) block diagram.

$R_f$ = field-winding resistance, ohms
$L_f$ = field-winding inductance, henrys
$i_f$ = field-winding current, amperes
$e_f$ = applied field voltage, volts
$R_a$ = sum of the armature resistance and the inserted resistance, ohms
$i_a$ = armature current, amperes
$\theta$ = angular displacement of the motor shaft, radians
$T$ = torque developed by the motor, lb-ft
$J$ = equivalent moment of inertia of the motor and load referred to the motor shaft, slug-ft²
$f$ = equivalent viscous-friction coefficient of the motor and load referred to the motor shaft, lb-ft/rad/sec

In this system, the field voltage $e_f$ is the control input. It is the output of an amplifier. The armature current $i_a$ is maintained constant; this may be accomplished by applying a constant voltage source to the armature and inserting a very large resistance in series with the armature. If the voltage drop in this resistance is large compared with the maximum back emf induced by the rotation of the armature windings in the magnetic field, the effect of the back emf is made small. Then the armature current $i_a$ can be kept approximately constant. The efficiency of the motor in such an operation is necessarily low, but such a field-controlled dc motor may be used for a speed control system. Note that maintaining a constant armature current $i_a$ is more difficult than maintaining a constant field current $i_f$, because of the back emf in the armature circuit.

The torque $T$ developed by the motor is proportional to the product of the air gap flux $\psi$ and armature current $i_a$ so that

$$T = K_1 \psi i_a \qquad (4\text{-}47)$$

where $K_1$ is a constant. Since the air gap flux $\psi$ and the field current $i_f$ are proportional for the usual operating range of the motor and $i_a$ is assumed to be constant, Eq. (4-47) can be written as

$$T = K_2 i_f$$

where $K_2$ is a constant. The equations for this system are

$$L_f \frac{di_f}{dt} + R_f i_f = e_f \tag{4-48}$$

$$J \frac{d^2\theta}{dt^2} + f \frac{d\theta}{dt} = T = K_2 i_f \tag{4-49}$$

Taking the Laplace transforms of Eqs. (4-48) and (4-49), assuming zero initial conditions, we obtain the following equations:

$$(L_f s + R_f) I_f(s) = E_f(s) \tag{4-50}$$

$$(J s^2 + f s) \Theta(s) = K_2 I_f(s) \tag{4-51}$$

Considering $E_f(s)$ as the input and $\Theta(s)$ as the output, we may construct the block diagram from Eqs. (4-50) and (4-51), as shown in Fig. 4-26(b). From this block diagram, the transfer function of this system is obtained as

$$\frac{\Theta(s)}{E_f(s)} = \frac{K_2}{s(L_f s + R_f)(J s + f)} = \frac{K_m}{s(T_f s + 1)(T_m s + 1)} \tag{4-52}$$

where

$K_m = K_2/(R_f f) = $ motor gain constant
$T_f = L_f/R_f = $ time constant of field circuit
$T_m = J/f = $ time constant of inertia-friction element

Since the field inductance $L_f$ is not negligible, the transfer function of a field-controlled dc motor is of the third order.

**Comparison of the performances of the armature-controlled dc motor with the field-controlled dc motor.** An advantage of the field control of a dc motor is that the amplifier required can be simplified because of the low power requirement in the control field. The requirement of a constant current source is, however, a serious disadvantage of field-controlled operations. Providing a constant current source is much more difficult than providing a constant voltage source. The field-controlled operation has a few more disadvantages over the armature-controlled operation of the dc motor. In the armature-controlled dc motor, the back emf acts as a damping; in the field-controlled dc motor, however, this is not the case, and the necessary damping must be provided by the motor and load. Because of the low efficiency of field-controlled operations, the heat generated in the armature may be a problem.

The time constants of the field-controlled dc motor are generally large compared with the time constants of a comparable armature-controlled motor. In making a comparison of time constants between field-controlled operations and armature-controlled operations, however, we must take the time constant of the power amplifier into consideration in the study of armature-controlled operations.

LIQUID-LEVEL SYSTEMS

**Fluid-flow laws.** In analyzing systems involving fluid flow, we find it necessary to divide flow regimes into laminar flow and turbulent flow, according to the magnitude of the Reynolds number. If the Reynolds number is greater than about 3000 ~ 4000, then the flow is turbulent. The flow is laminar if the Reynolds number is less than about 2000. In the laminar case, fluid flow occurs in streamlines with no turbulence. Systems involving turbulent flow often have to be represented by nonlinear differential equations, while systems involving laminar flow may be represented by linear differential equations. (Industrial processes often involve flow of liquids through connecting pipes and tanks. The flow in such processes is often turbulent and not laminar.)

**Resistance and capacitance of liquid-level systems.** It is convenient to introduce the concept of resistance and capacitance to describe dynamic characteristics of liquid-level systems.

Consider the flow through a short pipe connecting two tanks. The resistance for liquid flow in such a restriction is defined as the change in the level difference (the difference of the liquid levels of the two tanks) necessary to cause a unit change in flow rate; namely,

$$R = \frac{\text{change in level difference, ft}}{\text{change in flow rate, ft}^3/\text{sec}}$$

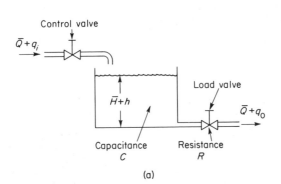

(a)

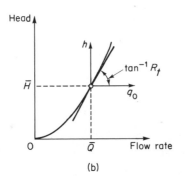

(b)

Fig. 4-27. (a) Liquid-level system; (b) head versus flow rate curve.

Since the relationship between the flow rate and level difference differs for the laminar flow and turbulent flow, we shall consider both cases in the following.

Consider the liquid-level system shown in Fig. 4-27 (a). In this system the liquid spouts through the load valve in the side of the tank. If the flow through this restriction is laminar, the relationship between the steady-state flow rate and steady-state head at the level of the restriction is given by

$$Q = KH$$

where

$Q$ = steady-state liquid flow rate, ft³/sec
$K$ = coefficient, ft²/sec
$H$ = steady-state head, ft

Notice that the law governing laminar flow is analogous to Coulomb's law, which states that the current is directly proportional to the potential difference.

For laminar flow, the resistance $R_\ell$ is obtained as

$$R_\ell = \frac{dH}{dQ} = \frac{H}{Q}$$

The laminar-flow resistance is constant and is analogous to the electrical resistance.

If the flow through the restriction is turbulent, the steady-state flow rate is given by

$$Q = K\sqrt{H} \tag{4-53}$$

where

$Q$ = steady-state liquid flow rate, ft³/sec
$K$ = coefficient, ft^{2.5}/sec
$H$ = steady-state head, ft

The resistance $R_t$ for turbulent flow is obtained from

$$R_t = \frac{dH}{dQ} = \frac{2H}{Q}$$

The value of the turbulent-flow resistance depends upon the flow rate and the head.

By use of the turbulent-flow resistance, we may linearize the nonlinear relationship between $Q$ and $H$, as given by Eq. (4-53). Such linearization is valid, provided that changes in the head and flow rate from their respective steady-state values are small. The linearized relationship is given by

$$Q = \frac{2H}{R_t}$$

The value of $R_t$ may be considered constant if the changes in head and flow rate are small.

In many practical cases, the value of the coefficient $K$ in Eq. (4-53), which depends upon the flow coefficient and the area of restriction, is not known. Then the resistance may be determined by plotting the head versus flow rate curve based on experimental data and measuring the slope of the curve at the operating condition. An example of such a plot is shown in Fig. 4-27 (b), and a steady-state operating point and resistance $R_t$ are indicated in the figure. (The resistance $R_t$

is the slope of the curve at the operating point.) The linear approximation is based on the fact that the actual curve does not differ much from its tangent line if the operating condition does not vary too much.

The capacitance $C$ of a tank is defined to be the change in quantity of stored liquid necessary to cause a unit change in the potential (head). (The potential is the quantity which indicates the energy level of the system.)

$$C = \frac{\text{change in liquid stored, ft}^3}{\text{change in head, ft}}$$

It should be noted that the capacity (ft³) and the capacitance (ft²) are different. The capacitance of the tank is equal to its cross-sectional area. If this is constant, the capacitance is constant for any head.

**Liquid-level systems.** Consider the system shown in Fig. 4-27 (a). The variables are defined as follows:

$\bar{Q}$ = steady-state flow rate (before any change has occurred), ft³/min
$q_i$ = small deviation of inflow rate from its steady-state value, ft³/min
$q_0$ = small deviation of outflow rate from its steady-state value, ft³/min
$\bar{H}$ = steady-state head (before any change has occurred), ft
$h$ = small deviation of head from its steady-state value, ft

As stated previously, a system can be considered linear if the flow is laminar. Even if the flow is turbulent, the system can be linearized if changes in the variables are kept small. Based on the assumption that the system is either linear or linearized, the differential equation of this system can be obtained as follows: Since the inflow minus outflow during the small time interval $dt$ is equal to the additional amount stored in the tank, we see that

$$C\,dh = (q_i - q_0)\,dt$$

From the definition of resistance, the relationship between $q_0$ and $h$ is given by

$$q_0 = \frac{h}{R}$$

The differential equation for this system for a constant value of $R$ becomes

$$RC\frac{dh}{dt} + h = Rq_i \tag{4-54}$$

Note that $RC$ is the time constant of the system. Taking the Laplace transforms of both sides of Eq. (4-54), assuming a zero initial condition, we obtain

$$(RCs + 1)H(s) = RQ_i(s)$$

where

$$H(s) = \mathscr{L}[h] \quad \text{and} \quad Q_i(s) = \mathscr{L}[q_i]$$

If $q_i$ is considered the input and $h$ the output, the transfer function of the system is

$$\frac{H(s)}{Q_i(s)} = \frac{R}{RCs + 1}$$

If, however, $q_0$ is taken as the output, the input being the same, then the transfer function is

$$\frac{Q_0(s)}{Q_i(s)} = \frac{1}{RCs + 1}$$

where we have used the relationship

$$Q_0(s) = \frac{1}{R}H(s)$$

**Liquid-level systems with interaction.** Consider the system shown in Fig. 4-28. In this system, the two tanks interact. Thus the transfer function of the system is not the product of two first-order transfer functions.

In the following, we shall assume only small variations of the variables from the steady-state values. Using the symbols as defined in Fig. 4-28, we can obtain the following equations for this system:

$$\frac{h_1 - h_2}{R_1} = q_1 \qquad (4\text{-}55)$$

$$C_1 \frac{dh_1}{dt} = q - q_1 \qquad (4\text{-}56)$$

$$\frac{h_2}{R_2} = q_2 \qquad (4\text{-}57)$$

$$C_2 \frac{dh_2}{dt} = q_1 - q_2 \qquad (4\text{-}58)$$

If $q$ is considered the input and $q_2$ the output, the transfer function of the system is

$$\frac{Q_2(s)}{Q(s)} = \frac{1}{R_1 C_1 R_2 C_2 s^2 + (R_1 C_1 + R_2 C_2 + R_2 C_1)s + 1} \qquad (4\text{-}59)$$

It is instructive to obtain Eq. (4-59), the transfer function of the interacted system, by block diagram reduction. From Eqs. (4-55) through (4-58), we obtain

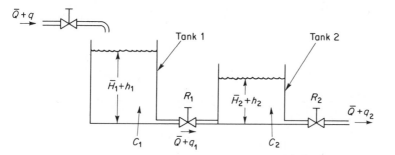

$\bar{Q}$: Steady-state flow rate
$\bar{H}_1$: Steady-state liquid level of tank 1
$\bar{H}_2$: Steady-state liquid level of tank 2

**Fig. 4-28.** Liquid-level system with interaction.

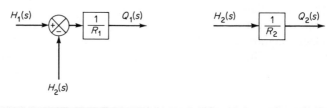

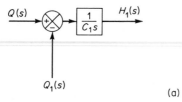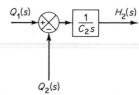

(a)

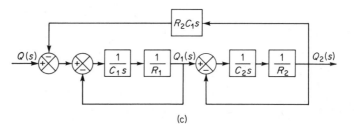

(b)

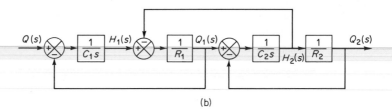

(c)

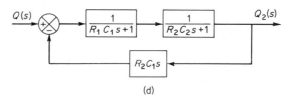

(d)

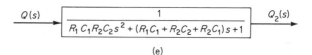

(e)

**Fig. 4-29.** (a) Elements of the block diagram of the system shown in Fig. 4-28; (b) block diagram of the system; (c)-(e) successive reduction of the block diagram.

elements of the block diagram, as shown in Fig. 4-29 (a). By co̲r̲
properly, we can construct a block diagram, as shown in Fig. 4-2̲
the rules of block diagram algebra given in Table 4-3, this block
simplified, as shown in Fig. 4-29 (c). Further simplification results
and (e). Figure 4-29 (e) is equivalent to Eq. (4-59).

PRESSURE SYSTEMS

**Resistance and capacitance of pressure systems.** Many industrial processes and pneumatic controllers involve the flow of a gas or air through connected pipelines and pressure vessels.

Consider the pressure system shown in Fig. 4-30 (a). The gas flow through the restriction is a function of the gas pressure difference $p_i - p_o$. Such a pressure system may be characterized in terms of a resistance and a capacitance.

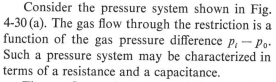

The gas flow resistance $R$ may be defined as follows:

$$R = \frac{\text{change in gas pressure difference, lb/ft}^2}{\text{change in gas flow rate, lb/sec}}$$

or

$$R = \frac{d(\Delta P)}{dq} \qquad (4\text{-}60)$$

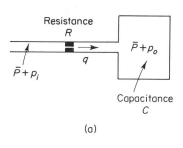

where $d(\Delta P)$ is a small change in the gas pressure difference and $dq$ is a small change in the gas flow. Computation of the value of the gas flow resistance $R$ may be quite time-consuming. Experimentally, however, it can be easily determined from a plot of the pressure difference versus flow curve by calculating the slope of the curve at a given operating condition, as shown in Fig. 4-30 (b).

**Fig. 4-30.** (a) Schematic diagram of a pressure system; (b) pressure difference versus flow curve.

The capacitance of the pressure vessel may be defined by

$$C = \frac{\text{change in gas stored, lb}}{\text{change in gas pressure, lb/ft}^2}$$

or

$$C = \frac{dm}{dp} = V\frac{dp}{dp} \qquad (4\text{-}61)$$

where

$C =$ capacitance
$m =$ mass of gas in vessel, lb
$p =$ gas pressure, lb/ft²

$V$ = volume of vessel, ft$^3$

$\rho$ = density, lb/ft$^3$

The capacitance of the pressure system depends on the type of expansion process involved. The capacitance can be calculated by use of the ideal gas law. (See Problems A-4-15 through A-4-17.) If the gas expansion process is polytropic and the change of state of the gas is between isothermal and adiabatic, then

$$p\left(\frac{V}{m}\right)^n = \frac{p}{\rho^n} = \text{constant} \tag{4-62}$$

where

$n$ = polytropic exponent

For ideal gases,

$$p\bar{v} = \bar{R}T \quad \text{or} \quad pv = \frac{\bar{R}}{M}T$$

where

$p$ = absolute pressure, lb/ft$^2$

$\bar{v}$ = the volume occupied by 1 mole of a gas, ft$^3$/lb-mole

$\bar{R}$ = universal gas constant, ft-lb/lb-mole °R

$T$ = absolute temperature, °R

$v$ = specific volume of gas, ft$^3$/lb

$M$ = molecular weight of gas per mole, lb/lb-mole

Thus

$$pv = \frac{p}{\rho} = \frac{\bar{R}}{M}T = R_{gas}T \tag{4-63}$$

where

$R_{gas}$ = gas constant, ft-lb/lb °R

The polytropic exponent $n$ is unity for isothermal expansion. For adiabatic expansion, $n$ is equal to the ratio of specific heats $c_p/c_v$, where $c_p$ is the specific heat at constant pressure and $c_v$ is the specific heat at constant volume. In many practical cases, the value of $n$ is approximately constant, and thus the capacitance may be considered constant. The value of $d\rho/dp$ is obtained from Eqs. (4-62) and (4-63) as

$$\frac{d\rho}{dp} = \frac{1}{nR_{gas}T}$$

The capacitance is then obtained as

$$C = \frac{V}{nR_{gas}T} \tag{4-64}$$

The capacitance of a given vessel is constant if the temperature stays constant. (In many practical cases, the polytropic exponent $n$ is approximately 1.0 ~ 1.2 for gases in uninsulated metal vessels.)

**Pressure systems.** Consider the system shown in Fig. 4-30 (a). If we assume only small deviations in the variables from their respective steady-state values, then this system may be considered linear.

Let us define

$\bar{P}$ = gas pressure in the vessel at steady-state (before changes in pressure have occurred), lb/ft$^2$

$p_i$ = small change in inflow gas pressure, lb/ft$^2$

$p_0$ = small change in gas pressure in the vessel, lb/ft$^2$

$V$ = volume of the vessel, ft$^3$

$m$ = mass of gas in vessel, lb

$q$ = gas flow rate, lb/sec

$\rho$ = density of gas, lb/ft$^3$

For small values of $p_i$ and $p_0$, the resistance $R$ given by Eq. (4-60) becomes constant and may be written as

$$R = \frac{p_i - p_0}{q}$$

The capacitance $C$ is given by Eq. (4-61), rewritten

$$C = \frac{dm}{dp} = V\frac{d\rho}{dp}$$

Since the pressure change $dp_0$ times the capacitance $C$ is equal to the gas added to the vessel during $dt$ seconds, we obtain

$$C\, dp_0 = q\, dt$$

or

$$C\frac{dp_0}{dt} = \frac{p_i - p_0}{R}$$

which can be written as

$$RC\frac{dp_0}{dt} + p_0 = p_i$$

If $p_i$ and $p_0$ are considered the input and output, respectively, then the transfer function of the system is

$$\frac{P_0(s)}{P_i(s)} = \frac{1}{RCs + 1}$$

where $RC$ has the dimension of time and is the time constant of the system.

THERMAL SYSTEMS

**Resistance and capacitance of thermal systems.** Thermal systems are those which involve the transfer of heat from one substance to another. Thermal systems may be analyzed in terms of resistance and capacitance, although the thermal capacitance and thermal resistance may not be represented accurately as lumped parameters since they are usually distributed throughout the substance. For precise analysis, distributed-parameter models must be used. Here, however, in order to simplify the analysis we shall assume that a thermal system can be represented by a lumped-parameter model, that substances that are characterized by resistance to heat flow have negligible heat capacitance, and that substances that are characterized by heat capacitance have negligible resistance to heat flow.

There are three different ways heat can flow from one substance to another: conduction, convection, and radiation.

For conduction or convection heat transfer,

$$q = K \, \Delta\theta$$

where

$q$ = heat flow rate, Btu/sec
$\Delta\theta$ = temperature difference, °F
$K$ = coefficient, Btu/sec °F

The coefficient $K$ is given by

$$K = \frac{kA}{\Delta X} \qquad \text{for conduction}$$

$$= HA \qquad \text{for convection}$$

where

$k$ = thermal conductivity, Btu/ft sec °F
$A$ = area normal to heat flow, ft²
$\Delta X$ = thickness of conductor, ft
$H$ = convection coefficient, Btu/ft² sec °F

For radiation heat transfer, the heat flow is given by

$$q = K_r(\theta_1^4 - \theta_2^4)$$

where

$q$ = heat flow rate, Btu/sec
$K_r$ = coefficient which depends on the emissivity, size, and configuration of the emanating surface and those of the receiving surface
$\theta_1$ = absolute temperature of emitter, °R
$\theta_2$ = absolute temperature of receiver, °R

Since the constant $K_r$ is a very small number, radiation heat transfer is appreciable only if the temperature of the emitter is very high.

The thermal resistance $R$ for heat transfer between two substances may be defined as follows:

$$R = \frac{\text{change in temperature difference, °F}}{\text{change in heat flow rate, Btu/sec}}$$

The thermal resistance for conduction or convection heat transfer is given by

$$R = \frac{d(\Delta\theta)}{dq} = \frac{1}{K}$$

Since the thermal conductivity and convection coefficient are almost constant, the thermal resistance for either conduction or convection is constant. The thermal resistance for radiation heat transfer is given by

$$R = \frac{d(\Delta\theta)}{dq} = \frac{1}{4K_r\bar{\theta}^3}$$

where

$\bar{\theta}$ = effective temperature difference of the emitter and receiver

The radiation resistance is not constant; it varies. It may be considered constant only for a small range of the operating conditions.

The thermal capacitance $C$ is defined by

$$C = \frac{\text{change in heat stored, Btu}}{\text{change in temperature, °F}}$$

or

$$C = Wc_p$$

where

$W$ = weight of substance considered, lb
$c_p$ = specific heat of substance, Btu/°F-lb

**Thermal systems.** Consider the system shown in Fig. 4-31. It is assumed that the tank is insulated to eliminate heat loss to the surrounding air. It is also assumed that there is no heat storage in the insulation and that the liquid in the tank is perfectly mixed so that it is at a uniform temperature. Thus, a single temperature is used to describe the temperature of the liquid in the tank and of the outflowing liquid.

Let us define

$\bar{\Theta}_i$ = steady-state temperature of inflowing liquid, °F
$\bar{\Theta}_0$ = steady-state temperature of outflowing liquid, °F
$G$ = steady-state liquid flow rate, lb/sec
$M$ = mass of liquid in tank, lb
$c$ = specific heat of liquid, Btu/lb °F
$R$ = thermal resistance, °F sec/Btu
$C$ = thermal capacitance, Btu/°F
$\bar{H}$ = steady-state heat input rate, Btu/sec

Assume that the temperature of the inflowing liquid is kept constant and that the heat input rate is suddenly changed from $\bar{H}$ to $\bar{H} + h_i$, where $h_i$ represents a small change in the heat input rate. The heat outflow rate will then change gradually from $\bar{H}$ to $\bar{H} + h_0$. The temperature of the outflowing liquid will also be changed from $\bar{\Theta}_0$ to $\bar{\Theta}_0 + \theta$. For this case, $h_0$, $C$, and $R$ are obtained, respectively, as

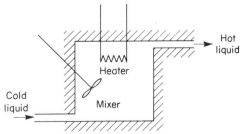

**Fig. 4-31.** Thermal system.

$$h_0 = Gc\theta$$

$$C = Mc$$

$$R = \frac{\theta}{h_0} = \frac{1}{Gc}$$

The differential equation for this system is

$$C\frac{d\theta}{dt} = h_i - h_0$$

which may be rewritten as

$$RC\frac{d\theta}{dt} + \theta = Rh_i$$

Note that the time constant of the system is equal to $RC$ or $M/G$ seconds. The transfer function relating $\theta$ and $h_i$ is given by

$$\frac{\Theta(s)}{H_i(s)} = \frac{R}{RCs + 1}$$

where

$$\Theta(s) = \mathscr{L}[\theta(t)] \quad \text{and} \quad H_i(s) = \mathscr{L}[h_i(t)]$$

In practice, the temperature of the inflowing liquid may fluctuate and may act as a load disturbance. (If a constant outflow temperature is desired, an automatic controller may be installed to adjust the heat inflow rate in order to compensate for the fluctuations in the temperature of the inflowing liquid.) If the temperature of the inflowing liquid is suddenly changed from $\bar{\Theta}_i$ to $\bar{\Theta}_i + \theta_i$ while the heat input rate $H$ and the liquid flow rate $G$ are kept constant, then the heat outflow rate will be changed from $\bar{H}$ to $\bar{H} + h_0$, and the temperature of the outflowing liquid will be changed from $\bar{\Theta}_0$ to $\bar{\Theta}_0 + \theta$. The differential equation for this case is

$$C\frac{d\theta}{dt} = Gc\theta_i - h_0$$

which may be rewritten

$$RC\frac{d\theta}{dt} + \theta = \theta_i$$

The transfer function relating $\theta$ and $\theta_i$ is given by

$$\frac{\Theta(s)}{\Theta_i(s)} = \frac{1}{RCs + 1}$$

where

$$\Theta(s) = \mathscr{L}[\theta(t)] \quad \text{and} \quad \Theta_i(s) = \mathscr{L}[\theta_i(t)]$$

If the present thermal system is subjected to changes in both the temperature of the inflowing liquid and the heat input rate, while the liquid flow rate is kept constant, the change $\theta$ in the temperature of the outflowing liquid can be given by the following equation:

$$RC\frac{d\theta}{dt} + \theta = \theta_i + Rh_i$$

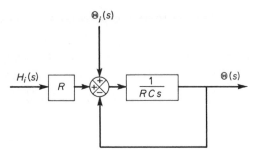

**Fig. 4-32.** Block diagram of the system shown in Fig. 4-31.

A block diagram corresponding to this case is shown in Fig. 4-32. (Notice that the system involves two inputs. We shall discuss multiple-input–multiple-output systems in Section 4-6.)

## 4-6   MULTIVARIABLE SYSTEMS AND TRANSFER MATRICES

In Section 4-2, the transfer function was defined for the single-input–single-output system. In this section we shall extend the transfer-function representation to systems with multiple inputs and multiple outputs.

**Transfer matrices.** Consider a system with $m$ inputs and $n$ outputs. We may consider the $m$ inputs to be the components of a vector. We shall call such a vector an *input vector*. Similarly we may regard the $n$ outputs as the components of an output vector. The matrix which relates the Laplace transform of the output vector to the Laplace transform of the input vector is called the *transfer matrix* between the output vector and the input vector.

Consider the system shown in Fig. 4-33. This system has two inputs and two outputs. From Fig. 4-33, the relationship between the outputs and inputs is given by

$$X_1(s) = G_{11}(s)U_1(s) + G_{12}(s)U_2(s)$$
$$X_2(s) = G_{21}(s)U_1(s) + G_{22}(s)U_2(s)$$

where $G_{ij}(s)$ is the transfer function relating the $i$th output to the $j$th input. Using vector-matrix notation, we can write this transfer relation as

$$\begin{bmatrix} X_1(s) \\ X_2(s) \end{bmatrix} = \begin{bmatrix} G_{11}(s) & G_{12}(s) \\ G_{21}(s) & G_{22}(s) \end{bmatrix} \begin{bmatrix} U_1(s) \\ U_2(s) \end{bmatrix}$$

A system having multiple inputs and multiple outputs is called a *multivariable* system. If such a system has $m$ inputs and $n$ outputs, and if the transfer function between the $i$th output and $j$th input is given by $G_{ij}(s)$, then the Laplace transform of the $i$th output is related to the Laplace transforms of the $m$ inputs by

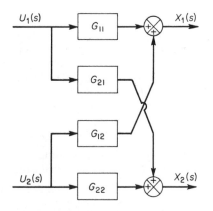

**Fig. 4-33.** Multiple-input–multiple-output system.

$$X_i(s) = G_{i1}(s)U_1(s) + G_{i2}(s)U_2(s) + \cdots + G_{im}(s)U_m(s) \qquad (i = 1, 2, \ldots, n)$$

Note that in defining $G_{ij}(s)$, only the $j$th input is considered and the other inputs are assumed to be zero. In matrix form, the Laplace transform of the output vector is related to the Laplace transform of the input vector by the following equation:

$$
\begin{bmatrix} X_1(s) \\ X_2(s) \\ \cdot \\ \cdot \\ \cdot \\ X_n(s) \end{bmatrix} =
\begin{bmatrix} G_{11}(s) & G_{12}(s) & \cdots & G_{1m}(s) \\ G_{21}(s) & G_{22}(s) & \cdots & G_{2m}(s) \\ \cdot & \cdot & & \cdot \\ \cdot & \cdot & & \cdot \\ \cdot & \cdot & & \cdot \\ G_{n1}(s) & G_{n2}(s) & \cdots & G_{nm}(s) \end{bmatrix}
\begin{bmatrix} U_1(s) \\ U_2(s) \\ \cdot \\ \cdot \\ \cdot \\ U_m(s) \end{bmatrix}
\qquad (4\text{-}65)
$$

Equation (4-65) shows the interactions between the $m$ inputs and $n$ outputs. Equation (4-65) can be rewritten as

$$\mathbf{X}(s) = \mathbf{G}(s)\mathbf{U}(s)$$

where

$$
\mathbf{X}(s) = \begin{bmatrix} X_1(s) \\ X_2(s) \\ \cdot \\ \cdot \\ \cdot \\ X_n(s) \end{bmatrix}, \qquad
\mathbf{U}(s) = \begin{bmatrix} U_1(s) \\ U_2(s) \\ \cdot \\ \cdot \\ \cdot \\ U_m(s) \end{bmatrix}
$$

$$
\mathbf{G}(s) = \begin{bmatrix} G_{11}(s) & G_{12}(s) & \cdots & G_{1m}(s) \\ G_{21}(s) & G_{22}(s) & \cdots & G_{2m}(s) \\ \cdot & \cdot & & \cdot \\ \cdot & \cdot & & \cdot \\ \cdot & \cdot & & \cdot \\ G_{n1}(s) & G_{n2}(s) & \cdots & G_{nm}(s) \end{bmatrix}
$$

$\mathbf{X}(s)$ is the Laplace transformed output vector, $\mathbf{U}(s)$ is the Laplace transformed input vector, and $\mathbf{G}(s)$ is the transfer matrix between $\mathbf{X}(s)$ and $\mathbf{U}(s)$.

---

*Example 4-2.* Consider the mechanical system shown in Fig. 4-34. We assume that the system is initially at rest. This system has two inputs $u_1(t)$ and $u_2(t)$ and two outputs $x_1(t)$ and $x_2(t)$. The equations describing the system dynamics are

$$m_1\ddot{x}_1 + f_1(\dot{x}_1 - \dot{x}_2) + k_1 x_1 = u_1$$
$$m_2\ddot{x}_2 + f_1(\dot{x}_2 - \dot{x}_1) + k_2 x_2 = u_2$$

Taking the Laplace transforms of these two equations and substituting the zero initial conditions, we obtain

$$(m_1 s^2 + f_1 s + k_1)X_1(s) - f_1 s X_2(s) = U_1(s)$$
$$(m_2 s^2 + f_1 s + k_2)X_2(s) - f_1 s X_1(s) = U_2(s)$$

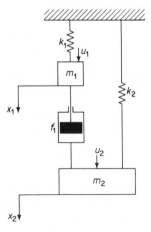

**Fig. 4-34.** Mechanical system.

In vector-matrix form, we obtain

$$\begin{bmatrix} m_1 s^2 + f_1 s + k_1 & -f_1 s \\ -f_1 s & m_2 s^2 + f_1 s + k_2 \end{bmatrix} \begin{bmatrix} X_1(s) \\ X_2(s) \end{bmatrix} = \begin{bmatrix} U_1(s) \\ U_2(s) \end{bmatrix} \tag{4-66}$$

Let us define

$$\Delta = (m_1 s^2 + f_1 s + k_1)(m_2 s^2 + f_1 s + k_2) - f_1^2 s^2 \neq 0$$

Then, by premultiplying by the inverse of the $2 \times 2$ matrix in Eq. (4-66), we obtain

$$\begin{bmatrix} X_1(s) \\ X_2(s) \end{bmatrix} = \begin{bmatrix} \dfrac{m_2 s^2 + f_1 s + k_2}{\Delta} & \dfrac{f_1 s}{\Delta} \\ \dfrac{f_1 s}{\Delta} & \dfrac{m_1 s^2 + f_1 s + k_1}{\Delta} \end{bmatrix} \begin{bmatrix} U_1(s) \\ U_2(s) \end{bmatrix}$$

The $2 \times 2$ matrix in this last equation is the transfer matrix between the outputs and the inputs. Clearly, the time responses $x_1(t)$ and $x_2(t)$ are given by

$$x_1(t) = \mathscr{L}^{-1} \left[ \frac{m_2 s^2 + f_1 s + k_2}{\Delta} U_1(s) + \frac{f_1 s}{\Delta} U_2(s) \right] \tag{4-67}$$

$$x_2(t) = \mathscr{L}^{-1} \left[ \frac{f_1 s}{\Delta} U_1(s) + \frac{m_1 s^2 + f_1 s + k_1}{\Delta} U_2(s) \right] \tag{4-68}$$

To find the responses $x_1(t)$ and $x_2(t)$ for $u_1 \neq 0$, $u_2 = 0$ (or $u_1 = 0$, $u_2 \neq 0$), we simply substitute $U_2(s) = 0$ [or $U_1(s) = 0$] into Eqs. (4-67) and (4-68).

---

*Example 4-3.* The system shown in Fig. 4-35 has two inputs, the reference input and the disturbance input, and one output. Obtain the transfer matrix between the output and the inputs.

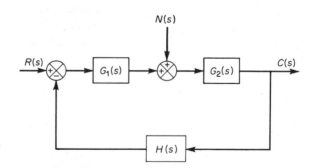

**Fig. 4-35.** System having two inputs and one output.

The Laplace transform of the output, $C(s)$, can be obtained as

$$C(s) = \frac{G_1(s)G_2(s)}{1 + G_1(s)G_2(s)H(s)} R(s) + \frac{G_2(s)}{1 + G_1(s)G_2(s)H(s)} N(s)$$

In vector matrix form,

$$C(s) = \begin{bmatrix} \dfrac{G_1(s)G_2(s)}{1 + G_1(s)G_2(s)H(s)} & \dfrac{G_2(s)}{1 + G_1(s)G_2(s)H(s)} \end{bmatrix} \begin{bmatrix} R(s) \\ N(s) \end{bmatrix} \tag{4-69}$$

The transfer matrix between the output $C(s)$ and the inputs $R(s)$ and $N(s)$ is the $1 \times 2$ matrix given in Eq. (4-69).

**Comments.** In this section, we have presented a definition of the transfer matrix and we have derived the transfer matrices for two systems. The transfer-matrix representation for multivariable systems is an extension of the transfer-function representation of single-input–single-output systems. The analysis and optimal control of multivariable systems can be carried out most conveniently by use of state variables. We shall therefore postpone the further analysis of multivariable systems to Chapters 14–16, where we shall study the state-space approach to the analysis and opimization of control systems.

## 4-7  SIGNAL FLOW GRAPHS

The block diagram is useful for graphically representing control systems. For a very complicated system, however, the block diagram reduction process becomes quite time-consuming. An alternate approach for finding the relationships among the system variables of a complicated control system is the signal flow graph approach, due to S. J. Mason.

**Signal flow graphs.** A signal flow graph is a diagram which represents a set of simultaneous linear algebraic equations. When applying the signal flow graph method to analyses of control systems, we must first transform linear differential equations into algebraic equations in $s$.

A signal flow graph consists of a network in which nodes are connected by directed branches. Each node represents a system variable, and each branch connected between two nodes acts as a signal multiplier. Note that the signal flows in only one direction. The direction of signal flow is indicated by an arrow placed on the branch, and the multiplication factor is indicated along the branch. The signal flow graph depicts the flow of signals from one point of a system to another and gives the relationships among the signals.

As might be expected, a signal flow graph contains essentially the same information as a block diagram. The advantage of using a signal flow graph to represent a control system is that a gain formula, called Mason's gain formula, is available which gives the relationships among system variables without requiring a reduction of the graph.

**Definitions.** Before we discuss signal flow graphs, we must define certain terms.

*Node.* A node is a point representing a variable or signal.

*Transmittance.* The transmittance is a gain between two nodes.

*Branch.* A branch is a directed line segment joining two nodes. The gain of a branch is a transmittance.

*Input node or source.* An input node or source is a node which has only outgoing branches. This corresponds to an independent variable.

*Output node or sink.* An output node or sink is a node which has only incoming branches. This corresponds to a dependent variable.

*Mixed node.* A mixed node is a node which has both incoming and outgoing branches.

*Path.* A path is a traversal of connected branches in the direction of the branch arrows. If no node is crossed more than once, the path is open. If the path ends at the same node from which it began and does not cross any other node more than once, it is closed. If a path crosses some node more than once but ends at a different node from which it began, it is neither open nor closed.

*Loop.* A loop is a closed path.

*Loop gain.* The loop gain is the product of the branch transmittances of a loop.

*Nontouching loops.* Loops are nontouching if they do not possess any common nodes.

*Forward path.* A forward path is a path from an input node (source) to an output node (sink) which does not cross any nodes more than once.

*Forward path gain.* A forward path gain is the product of the branch transmittances of a forward path.

Figure 4-36 shows nodes and branches, together with transmittances.

**Properties of signal flow graphs.** A few important properties of signal flow graphs are listed below.

1. A branch indicates the functional dependence of one signal upon another. A signal passes through only in the direction specified by the arrow of the branch.

2. A node adds the signals of all incoming branches and transmits this sum to all outgoing branches.

3. A mixed node, which has both incoming and outgoing branches, may be treated as an output node (sink) by adding an outgoing branch of unity transmittance. (See Fig. 4-36. Notice that a branch with unity transmittance is directed from $x_3$ to another node, also denoted by $x_3$.) Note, however, that we cannot change a mixed node to a source by this method.

4. For a given system, a signal flow graph is not unique. Many different signal flow graphs can be drawn for a given system by writing the system equations differently.

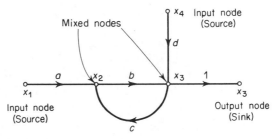

**Fig. 4-36.** Signal flow graph.

**Signal flow graph algebra.** A signal flow graph of a linear system can be drawn using the foregoing definitions. In doing so, we usually bring the input nodes (sources) to the left and the output nodes (sinks) to the right. The independent and dependent variables of the equations become the input nodes (sources) and output nodes (sinks), respectively. The branch transmittances can be obtained from the coefficients of the equations.

To determine the input-output relationship, we may use Mason's formula, which will be given later, or we may reduce the signal flow graph to a graph containing only input and output nodes. To accomplish this, we use the following rules:

1. The value of a node with one incoming branch, as shown in Fig. 4-37 (a), is $x_2 = ax_1$.

2. The total transmittance of cascaded branches is equal to the product of all the branch transmittances. Cascaded branches can thus be combined into a single branch by multiplying the transmittances, as shown in Fig. 4-37 (b).

3. Parallel branches may be combined by adding the transmittances, as shown in Fig. 4-37 (c).

4. A mixed node may be eliminated, as shown in Fig. 4-37 (d).

5. A loop may be eliminated, as shown in Fig. 4-37 (e). Note that

(a)

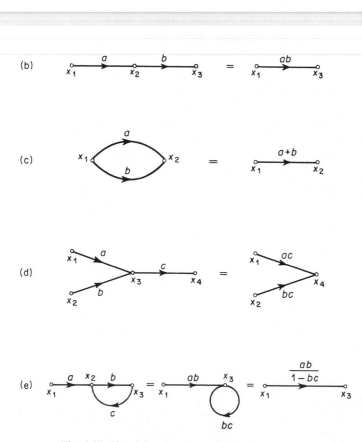

Fig. 4-37. Signal flow graphs and simplifications.

$$x_3 = bx_2, \qquad x_2 = ax_1 + cx_3$$

Hence

$$x_3 = abx_1 + bcx_3 \qquad (4\text{-}70)$$

or

$$x_3 = \frac{ab}{1 - bc} x_1 \qquad (4\text{-}71)$$

Equation (4-70) corresponds to a diagram having a self-loop of transmittance $bc$. Elimination of the self-loop yields Eq. (4-71), which clearly shows that the overall transmittance is $ab/(1 - bc)$.

**Signal flow graph representation of linear systems.** Signal flow graphs are widely applied to linear-system analysis. Here the graph can be drawn from the system

(a)

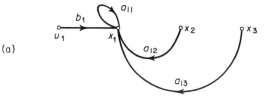

(b)

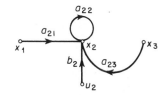

(c)

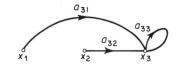

**Fig. 4-38.** (a) Signal flow graph representing Eq. (4-72); (b) signal flow graph representing Eq. (4-73); (c) signal flow graph representing Eq. (4-74); (d) complete signal flow graph for the system described by Eqs. (4-72), (4-73), and (4-74).

(d)

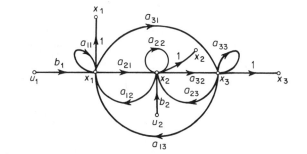

equations or, with practice, can be drawn by inspection of the physical system. Routine reduction by use of the foregoing rules gives the relation between an input and output variable.

Consider a system defined by the following set of equations:

$$x_1 = a_{11}x_1 + a_{12}x_2 + a_{13}x_3 + b_1u_1 \tag{4-72}$$

$$x_2 = a_{21}x_1 + a_{22}x_2 + a_{23}x_3 + b_2u_3 \tag{4-73}$$

$$x_3 = a_{31}x_1 + a_{32}x_2 + a_{33}x_3 \tag{4-74}$$

where $u_1$ and $u_2$ are input variables; $x_1$, $x_2$, and $x_3$ are output variables. A signal flow graph for this system, a graphical representation of these three simultaneous equations, indicating the interdependence of the variables, can be obtained as follows: First locate the nodes $x_1$, $x_2$, and $x_3$, as shown in Fig. 4-38 (a). Note that $a_{ij}$ is the transmittance between $x_j$ and $x_i$. Equation (4-72) states that $x_1$ is equal to the sum of the four signals $a_{11}x_1$, $a_{12}x_2$, $a_{13}x_3$, and $b_1u_1$. The signal flow graph representing Eq. (4-72) is shown in Fig. 4-38 (a). Equation (4-73) states that $x_2$ is equal to the sum of $a_{21}x_1$, $a_{22}x_2$, $a_{23}x_3$, and $b_2u_2$. The corresponding signal flow graph is shown in Fig. 4-38 (b). The signal flow graph representing Eq. (4-74) is shown in Fig. 4-38 (c).

The signal flow graph representing Eqs. (4-72), (4-73), and (4-74) is then obtained by combining Figs. 4-38 (a), (b), and (c). Finally the complete signal flow graph for the given simultaneous equations is shown in Fig. 4-38 (d).

In dealing with a signal flow graph, the input nodes (sources) may be considered one at a time. The output signal is then equal to the sum of the individual contributions of each input.

The overall gain from an input to an output may be obtained directly from the signal flow graph by inspection, by use of Mason's formula, or by a reduction of the graph to a simpler form.

**Signal flow graphs of control systems.** Some signal flow graphs of simple control systems are shown in Fig. 4-39. For such simple graphs, the closed-loop transfer function $C(s)/R(s)$ [or $C(s)/N(s)$] can be obtained easily by inspection. For more complicated signal flow graphs, Mason's gain formula is quite useful.

**Mason's gain formula for signal flow graphs.** In many practical cases, we wish to determine the relationship between an input variable and an output variable of the signal flow graph. The transmittance between an input node and an output node is the overall gain, or overall transmittance, between these two nodes.

Mason's gain formula, which is applicable to the overall gain, is given by

$$P = \frac{1}{\Delta} \sum_k P_k \Delta_k$$

where

    $P_k$ = path gain or transmittance of $k$th forward path
    $\Delta$ = determinant of graph
        = 1 − (sum of all different loop gains) + (sum of gain products of all

possible combinations of two nontouching loops) — (sum of gain products of all possible combinations of three nontouching loops) + ···

$$= 1 - \sum_a L_a + \sum_{b,c} L_b L_c - \sum_{d,e,f} L_d L_e L_f + \cdots$$

$\sum_a L_a$ = sum of all different loop gains

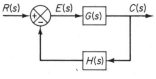

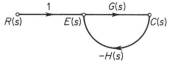

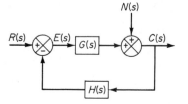

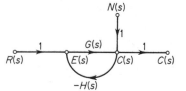

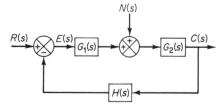

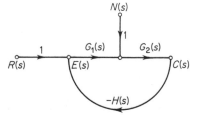

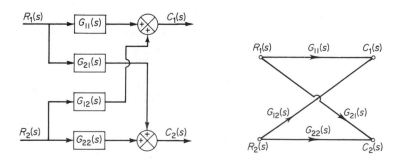

Fig. 4-39. Block diagrams and corresponding signal flow graphs.

$\displaystyle\sum_{b,c} L_b L_c =$ sum of gain products of all possible combinations of two non-touching loops

$\displaystyle\sum_{d,e,f} L_d L_e L_f =$ sum of gain products of all possible combinations of three non-touching loops

$\Delta_k =$ cofactor of the $k$th forward path determinant of the graph with the loops touching the $k$th forward path removed

(Note that the summations are taken over all possible paths from input to output.)

In the following, we shall illustrate the use of Mason's gain forumla by means of two examples.

---

*Example 4-4.* Consider the system shown in Fig. 4-40. A signal flow graph for this system is shown in Fig. 4-41. Let us obtain the closed-loop transfer function $C(s)/R(s)$ by use of Mason's gain formula.

In this system there is only one forward path between the input $R(s)$ and the output $C(s)$. The forward path gain is

$$P_1 = G_1 G_2 G_3$$

From Fig. 4-41, we see that there are three individual loops. The gains of these loops are

$$L_1 = G_1 G_2 H_1$$
$$L_2 = -G_2 G_3 H_2$$
$$L_3 = -G_1 G_2 G_3$$

Note that since all three loops have a common branch, there are no nontouching loops. Hence, the determinant $\Delta$ is given by

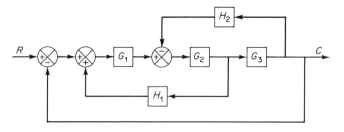

**Fig. 4-40.** Multiple-loop system.

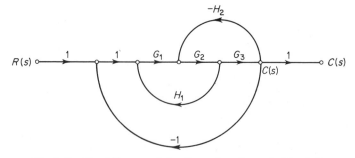

**Fig. 4-41.** Signal flow graph for the system shown in Fig. 4-40.

$$\Delta = 1 - (L_1 + L_2 + L_3)$$
$$= 1 - G_1G_2H_1 + G_2G_3H_2 + G_1G_2G_3$$

The cofactor $\Delta_1$ of the determinant along the forward path connecting the input node and output node is obtained by removing the loops that touch this path. Since path $P_1$ touches all three loops, we obtain

$$\Delta_1 = 1$$

Therefore, the overall gain between the input $R(s)$ and the output $C(s)$, or the closed-loop transfer function, is given by

$$\frac{C(s)}{R(s)} = P = \frac{P_1 \Delta_1}{\Delta}$$

$$= \frac{G_1G_2G_3}{1 - G_1G_2H_1 + G_2G_3H_2 + G_1G_2G_3}$$

which is the same as the closed-loop transfer function obtained by block diagram reduction. Mason's gain formula thus gives the overall gain $C(s)/R(s)$ without a reduction of the graph.

---

*Example 4-5.* Consider the system shown in Fig. 4-42. Obtain the closed-loop transfer function $C(s)/R(s)$ by use of Mason's gain formula.

In this system, there are three forward paths between the input $R(s)$ and the output $C(s)$. The forward path gains are

$$P_1 = G_1G_2G_3G_4G_5$$
$$P_2 = G_1G_6G_4G_5$$
$$P_3 = G_1G_2G_7$$

There are four individual loops. The gains of these loops are

$$L_1 = -G_4H_1$$
$$L_2 = -G_2G_7H_2$$
$$L_3 = -G_6G_4G_5H_2$$
$$L_4 = -G_2G_3G_4G_5H_2$$

Loop $L_1$ does not touch loop $L_2$. Hence, the determinant $\Delta$ is given by

$$\Delta = 1 - (L_1 + L_2 + L_3 + L_4) + L_1L_2 \tag{4-75}$$

The cofactor $\Delta_1$ is obtained from $\Delta$ by removing the loops that touch path $P_1$. Therefore, by removing $L_1$, $L_2$, $L_3$, $L_4$, and $L_1L_2$ from Eq. (4-75), we obtain

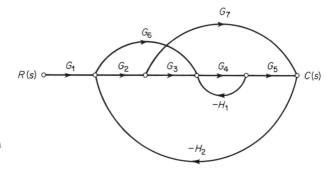

**Fig. 4-42.** Signal flow graph for a system.

$$\Delta_1 = 1$$

Similarly, the cofactor $\Delta_2$ is

$$\Delta_2 = 1$$

The cofactor $\Delta_3$ is obtained by removing $L_2$, $L_3$, $L_4$, and $L_1L_2$ from Eq. (4-75), giving

$$\Delta_3 = 1 - L_1$$

The closed-loop transfer function $C(s)/R(s)$ is then

$$\frac{C(s)}{R(s)} = P = \frac{1}{\Delta}(P_1\,\Delta_1 + P_2\,\Delta_2 + P_3\,\Delta_3)$$

$$= \frac{G_1G_2G_3G_4G_5 + G_1G_6G_4G_5 + G_1G_2G_7(1 + G_4H_1)}{1 + G_4H_1 + G_2G_7H_2 + G_6G_4G_5H_2 + G_4H_1G_2G_7H_2 + G_2G_3G_4G_5H_2}$$

**Concluding comments.** The usual application of signal flow graphs is in system diagramming. The set of equations describing a linear system is represented by a signal flow graph by establishing nodes which represent the system variables and by interconnecting the nodes with weighted, directed, transmittances, which represent the relationships among the variables. Mason's gain formula may be used to establish the relationship between an input and an output. (Alternatively, the variables in the system may be eliminated one by one with reduction techniques.) Mason's gain formula is especially useful in reducing large and complex system diagrams in one step, without requiring step-by-step reductions.

EXAMPLE PROBLEMS AND SOLUTIONS

**PROBLEM A-4-1.** Figure 4-43 shows the schematic diagram of an accelerometer. Assume that the case of the accelerometer is attached to an aircraft frame. (The accelerometer indicates the acceleration of its case with respect to inertial space.) The tilt angle $\theta$ mea-

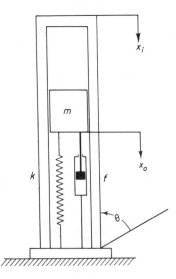

**Fig. 4-43.** Schematic diagram of an accelerometer.

sured from the horizontal line is assumed to be constant during the measurement period.

Show that for low-frequency inputs, the acceleration of the case relative to inertial space can be determined by the displacement of the mass $m$ with respect to its case.

**Solution.** Let us define

$x_i$ = displacement of the case relative to inertial space
$x_0$ = displacement of the mass $m$ relative to inertial space
$y = x_0 - x_i$ = displacement of the mass $m$ relative to the case

The equation for this system is

$$m\ddot{x}_0 + f(\dot{x}_0 - \dot{x}_i) + k(x_0 - x_i) - mg \sin \theta = 0$$

In terms of $y$, we obtain

$$m\ddot{y} + f\dot{y} + ky = -m\ddot{x}_i + mg \sin \theta$$

Since $\theta$ is assumed to be constant during the measurement period, $mg \sin \theta$ is constant, and it is possible to calibrate the displacement and define a new variable $z$ such that

$$z = y - \frac{mg}{k} \sin \theta$$

Then we obtain

$$m\ddot{z} + f\dot{z} + kz = -m\ddot{x}_i$$

If the input acceleration (the acceleration of the case relative to inertial space) $\ddot{x}_i$ is taken to be the input to the system and the displacement $z$ is taken to be the output, then the transfer function of the system becomes

$$\frac{Z(s)}{s^2 X_i(s)} = -\frac{1}{s^2 + \frac{f}{m}s + \frac{k}{m}}$$

If the input frequency is very low compared with $\sqrt{k/m}$, then

$$\frac{Z(s)}{s^2 X_i(s)} \doteq -\frac{m}{k}$$

which means that $z = y - (mg/k) \sin \theta$ is nearly proportional to the slowly varying input acceleration. Thus, for low-frequency inputs, the acceleration of the case relative to inertial space can be given by

$$\ddot{x}_i = -\frac{k}{m}\left(y - \frac{mg}{k} \sin \theta\right)$$

Note that such an accelerometer must have a sufficiently high undamped natural frequency $\sqrt{k/m}$ compared with the highest input frequency to be measured.

**PROBLEM A-4-2.** Gear trains are often used in servomechanisms to reduce speed, to magnify torque, or to obtain the most efficient power transfer by matching the driving member to the given load.

Consider the gear train system shown in Fig. 4-44. In this system, a load is driven by a motor through the gear train. Assuming that the stiffness of the shafts of the gear train is infinite (there is neither backlash nor elastic deformation) and that the number of teeth on each gear is proportional to the radius of the gear, obtain the equivalent moment of inertia and equivalent friction referred to the motor shaft and referred to the load shaft.

In Fig. 4-44, the numbers of teeth on gears 1, 2, 3, and 4 are $N_1$, $N_2$, $N_3$, and $N_4$, respectively. The angular displacements of shafts 1, 2, and 3 are $\theta_1$, $\theta_2$, and $\theta_3$, respect-

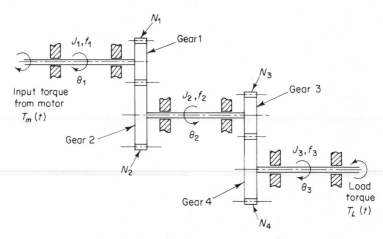

**Fig. 4-44.** Gear-train system.

ively. Thus, $\theta_2/\theta_1 = N_1/N_2$ and $\theta_3/\theta_2 = N_3/N_4$. The moment of inertia and viscous friction of each gear train component are denoted by $J_1, f_1$; $J_2, f_2$; and $J_3, f_3$; respectively. ($J_3$ and $f_3$ include the moment of inertia and friction of the load.)

**Solution.** For this gear train system, we can obtain the following three equations: For the first shaft,

$$J_1\ddot{\theta}_1 + f_1\dot{\theta}_1 + T_1 = T_m \qquad (4\text{-}76)$$

where $T_m$ is the torque developed by the motor and $T_1$ is the load torque on gear 1 due to the rest of the gear train. For the second shaft,

$$J_2\ddot{\theta}_2 + f_2\dot{\theta}_2 + T_3 = T_2 \qquad (4\text{-}77)$$

where $T_2$ is the torque transmitted to gear 2 and $T_3$ is the load torque on gear 3 due to the rest of the gear train. Since the work done by gear 1 is equal to that of gear 2,

$$T_1\theta_1 = T_2\theta_2 \qquad \text{or} \qquad T_2 = T_1\frac{N_2}{N_1}$$

If $N_1/N_2 < 1$, the gear ratio reduces the speed as well as magnifies the torque. For the third shaft,

$$J_3\ddot{\theta}_3 + f_3\dot{\theta}_3 + T_L = T_4 \qquad (4\text{-}78)$$

where $T_L$ is the load torque and $T_4$ is the torque transmitted to gear 4. $T_3$ and $T_4$ are related by

$$T_4 = T_3\frac{N_4}{N_3}$$

and $\theta_3$ and $\theta_1$ are related by

$$\theta_3 = \theta_2\frac{N_3}{N_4} = \theta_1\frac{N_1}{N_2}\frac{N_3}{N_4}$$

Elimination of $T_1$, $T_2$, $T_3$, and $T_4$ from Eqs. (4-76), (4-77), and (4-78) yields

$$J_1\ddot{\theta}_1 + f_1\dot{\theta}_1 + \frac{N_1}{N_2}(J_2\ddot{\theta}_2 + f_2\dot{\theta}_2) + \frac{N_1N_3}{N_2N_4}(J_3\ddot{\theta}_3 + f_3\dot{\theta}_3 + T_L) = T_m \qquad (4\text{-}79)$$

Eliminating $\theta_2$ and $\theta_3$ from Eq. (4-79) and writing the equation in terms of $\theta_1$ and its time derivatives, we obtain

$$\left[J_1 + \left(\frac{N_1}{N_2}\right)^2 J_2 + \left(\frac{N_1}{N_2}\right)^2\left(\frac{N_3}{N_4}\right)^2 J_3\right]\ddot{\theta}_1$$

$$+ \left[f_1 + \left(\frac{N_1}{N_2}\right)^2 f_2 + \left(\frac{N_1}{N_2}\right)^2\left(\frac{N_3}{N_4}\right)^2 f_3\right]\dot{\theta}_1 + \left(\frac{N_1}{N_2}\right)\left(\frac{N_3}{N_4}\right)T_L = T_m \qquad (4\text{-}80)$$

Thus, the equivalent moment of inertia and friction of the gear train referred to shaft 1 are given by

$$J_{1eq} = J_1 + \left(\frac{N_1}{N_2}\right)^2 J_2 + \left(\frac{N_1}{N_2}\right)^2\left(\frac{N_3}{N_4}\right)^2 J_3$$

$$f_{1eq} = f_1 + \left(\frac{N_1}{N_2}\right)^2 f_2 + \left(\frac{N_1}{N_2}\right)^2\left(\frac{N_3}{N_4}\right)^2 f_3$$

Similarly, the equivalent moment of inertia and friction of the gear train referred to the load shaft are

$$J_{3eq} = J_3 + \left(\frac{N_4}{N_3}\right)^2 J_2 + \left(\frac{N_2}{N_1}\right)^2\left(\frac{N_4}{N_3}\right)^2 J_1$$

$$f_{3eq} = f_3 + \left(\frac{N_4}{N_3}\right)^2 f_2 + \left(\frac{N_2}{N_1}\right)^2\left(\frac{N_4}{N_3}\right)^2 f_1$$

The relationship between $J_{1eq}$ and $J_{3eq}$ is thus

$$J_{1eq} = \left(\frac{N_1}{N_2}\right)^2\left(\frac{N_3}{N_4}\right)^2 J_{3eq}$$

and that between $f_{1eq}$ and $f_{3eq}$ is

$$f_{1eq} = \left(\frac{N_1}{N_2}\right)^2\left(\frac{N_3}{N_4}\right)^2 f_{3eq}$$

The effect of $J_2$ and $J_3$ on an equivalent moment of inertia is determined by the gear ratios $N_1/N_2$ and $N_3/N_4$. For speed-reducing gear trains, the ratios $N_1/N_2$ and $N_3/N_4$ are usually less than unity. If $N_1/N_2 \ll 1$ and $N_3/N_4 \ll 1$, then the effect of $J_2$ and $J_3$ on the equivalent moment of inertia $J_{1eq}$ is negligible. Similar comments apply to the equivalent friction $f_{1eq}$ of the gear train. In terms of the equivalent moment of inertia $J_{1eq}$ and equivalent friction $f_{1eq}$, Eq. (4-80) can be simplified to give

$$J_{1eq}\ddot{\theta}_1 + f_{1eq}\dot{\theta}_1 + nT_L = T_m$$

where

$$n = \frac{N_1}{N_2}\frac{N_3}{N_4}$$

Considering $T_m - nT_L$ as the input to the system and $\theta_1$ as the output, the transfer function of this system is

$$\frac{\Theta_1(s)}{T_m(s) - nT_L(s)} = \frac{1}{s(J_{1eq}s + f_{1eq})}$$

The block diagram is shown in Fig. 4-45(a). If $(T_m - nT_L)$ and $\theta_3$ are considered to be the input and output, respectively, then the transfer function becomes

$$\frac{\Theta_3(s)}{T_m(s) - nT_L(s)} = \frac{\frac{1}{n}}{s(J_{3eq}s + f_{3eq})}$$

Since

$$\frac{\Theta_3(s)}{\Theta_1(s)} = \frac{N_1 N_3}{N_2 N_4} = n$$

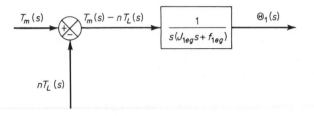

(a)

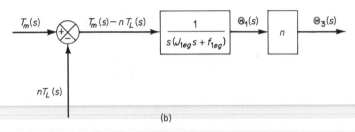

(b)

**Fig. 4-45.** Block diagrams of the gear-train system shown in Fig. 4-44.

in terms of $J_{1eq}$ and $f_{1eq}$, this transfer function can be written

$$\frac{\Theta_3(s)}{T_m(s) - nT_L(s)} = \frac{n}{s(J_{1eq}s + f_{1eq})}$$

The block diagram for this case is shown in Fig. 4-45 (b).

From the analysis above, it can be seen that in a system where a servomotor is driving load inertia and load viscous friction, the load inertia and friction are, at the motor shaft, the original values multiplied by the square of the gear ratio. If the gear ratio is a small number, the moment of inertia and friction of the servomotor give dominant effects on the dynamic behavior of the system.

**PROBLEM A-4-3.** Show that the torque-to-inertia ratios referred to the motor shaft and to the load shaft differ from each other by a factor of $n$. Show also that the torque squared-to-inertia ratios referred to the motor shaft and to the load shaft are the same.

**Solution.** Suppose that $T_{max}$ is the maximum torque which can be produced on the motor shaft. Then the torque-to-inertia ratio referred to the motor shaft is

$$\frac{T_{max}}{J_m + n^2 J_L}$$

where

$J_m$ = moment of inertia of the rotor
$J_L$ = moment of inertia of the load
$n$ = gear ratio

The torque-to-inertia ratio referred to the load shaft is

$$\frac{\dfrac{T_{max}}{n}}{J_L + \dfrac{J_m}{n^2}}$$

Clearly, they differ from each other by a factor of $n$. Hence, in comparing torque-to-inertia ratios of motors, we find it necessary to specify which shaft is the reference.

Note that the ratio of torque squared to inertia referred to the motor shaft is

$$\frac{T_{\max}^2}{J_m + n^2 J_L}$$

and that referred to the load shaft is

$$\frac{\dfrac{T_{\max}^2}{n^2}}{J_L + \dfrac{J_m}{n^2}}$$

These two ratios are clearly the same.

**PROBLEM A-4-4.** The moment of inertia $J$ of a motor may be available in the manufacturer's catalog or may be computed if the dimensions and material of the rotor are known.

The effective moment of inertia of a motor and gear train system can also be determined by experimental means. Discuss methods for determining the moment of inertia $J$ of a two-phase ac motor.

**Solution.** To determine the effective moment of inertia $J$ of a motor and gear train system, apply a step input $e_i$ and record the transient response. The transfer function of a two-phase ac motor is given by Eq. (4-35) as follows:

$$\frac{\Theta(s)}{E_i(s)} = \frac{K_c}{Js^2 + (f + K_n)s}$$

If the input voltage $e_i$ is a step input of magnitude $E_1$, then the resulting shaft velocity is

$$s\Theta(s) = \frac{K_c E_1}{s(Js + f + K_n)}$$

Hence the angular velocity $\omega(t)$ is

$$\omega(t) = \dot{\theta}(t) = \frac{K_c E_1}{f + K_n}\left[1 - \exp\left(-\frac{f + K_n}{J}t\right)\right]$$

From the record of $\omega(t)$ versus time, the time constant $J/(f + K_n)$ can be determined.

Since the steady-state angular velocity is

$$\omega(\infty) = \frac{K_c E_1}{f + K_n}$$

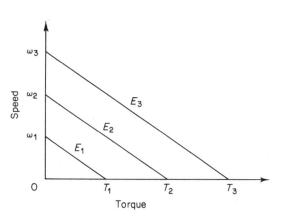

**Fig. 4-46.** Torque-speed curves.

we obtain

$$f + K_n = \frac{K_c E_1}{\omega(\infty)}$$

To determine $f + K_n$, we need to know $K_c$. [$E_1$ and $\omega(\infty)$ are known quantities.] The value of the torque constant $K_c$ can be determined from the torque-speed curves, as shown in Fig. 4-46, provided these curves are parallel. From the curves, we obtain

$$K_c = \frac{T_1}{E_1} = \frac{T_2}{E_2} = \frac{T_3}{E_3}$$

Once we know the time constant $J/(f + K_n)$, the steady-state angular velocity $\omega(\infty)$, and the torque constant $K_c$, the moment of inertia $J$ can be determined.

The moment of inertia $J$ of the rotor can also be determined experimentally by means of a torsional pendulum experiment. If the damping coefficient is very small, the period of oscillation of the pendulum is related to $J$ by the following equation:

$$J = \frac{kT^2}{4\pi^2}$$

where

$J$ = moment of inertia of the rotor
$k$ = torsional spring constant of the elastic shaft to which the motor rotor is attached
$T$ = period of the oscillation

**PROBLEM A-4-5.** Obtain the transfer function of the two-phase servomotor whose torque-speed curve is shown in Fig. 4-47. The maximum-rated fixed-phase and control-phase voltages are 115 volts. The moment of inertia $J$ of the rotor (including the effect of load) is $7.77 \times 10^{-4}$ oz-in.-sec$^2$ and the viscous-friction coefficient of the motor (includ-ing the effect of load) is 0.005 oz-in./rad/sec.

**Solution.** The equation for the torque-speed curve is

$$T + K_n \dot{\theta} = K_c E_c$$

where

$$K_n = \frac{5}{4000} \frac{60}{2\pi} = 0.0119 \text{ oz-in./rad/sec}$$

$$K_c = \frac{5}{115} = 0.0435 \text{ oz-in./volt}$$

Thus

$$T + 0.0119\dot{\theta} = 0.0435 E_c$$

The motor gain constant $K_m$ is then

$$K_m = \frac{K_c}{f + K_n}$$

$$= \frac{0.0435}{0.005 + 0.0119}$$

$$= 2.57$$

The motor time constant $T_m$ is

$$T_m = \frac{J}{f + K_n}$$

$$= \frac{7.77 \times 10^{-4}}{5 \times 10^{-3} + 11.9 \times 10^{-3}}$$

$$= 0.046$$

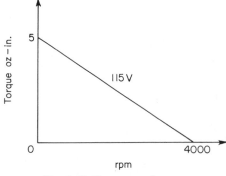

**Fig. 4-47.** Torque-speed curve.

Using the numerical values thus obtained, the transfer function given by Eq. (4-35) is as follows:

$$\frac{\Theta(s)}{E_c(s)} = \frac{K_m}{s(T_m s + 1)}$$

$$= \frac{2.57}{s(0.046s + 1)}$$

**PROBLEM A-4-6.** Assuming that a two-phase servomotor has a linear torque-speed curve so that the no-load speed is $\omega_0$ and the stall torque is $T_s$, find the maximum motor-shaft output power $P_{max}$.

**Solution.** The torque-speed curve is

$$T = T_s - \left(\frac{\omega}{\omega_0}\right)T_s$$

The shaft output power $P$ is given by

$$P = T\omega$$

$$= \left(T_s - \frac{\omega}{\omega_0}T_s\right)\omega$$

To find $P_{max}$, let us differentiate $P$ with respect to $\omega$:

$$\frac{dP}{d\omega} = T_s - 2\frac{\omega}{\omega_0}T_s$$

By setting $dP/d\omega = 0$, we obtain

$$\omega = \frac{\omega_0}{2}$$

Clearly, $d^2P/d\omega^2 = -2T_s/\omega_0 < 0$. Hence $P$ is maximum at $\omega = \omega_0/2$. $P_{max}$ is

$$P_{max} = T\omega \big|_{\omega=\omega_0/2}$$

$$= \left(T_s - \frac{T_s}{2}\right)\frac{\omega_0}{2}$$

$$= \frac{T_s\omega_0}{4}$$

For the two-phase servomotor considered in Problem A-4-5, $T_s = 5$ oz-in. and $\omega_0 = 4000$ rpm. The maximum power output $P_{max}$ is

$$P_{max} = \frac{5 \times 4000}{4} \times \frac{1}{1352}$$

$$= 3.7 \text{ watts}$$

The maximum power occurs at $\omega = 2000$ rpm.

**PROBLEM A-4-7.** The maximum acceleration which the servomotor can achieve may be indicated by the torque-to-inertia ratio, which is the ratio of the maximum torque at standstill to the rotor inertia. The higher this ratio, the better the acceleration characteristic is. Referring to the two-phase servomotor considered in Problem A-4-5, find the torque-to-inertia ratio.

**Solution.** The rotor moment of inertia $J$ of the two-phase servomotor is $7.77 \times 10^{-4}$ oz-in.-sec$^2$. The maximum torque at standstill is 5 oz-in. Hence the torque-to-inertia ratio is given by

$$\frac{T_{max}}{J} = \frac{5}{7.77 \times 10^{-4}}$$
$$= 6435 \text{ rad/sec}^2$$

In selecting a servomotor, it is important to satisfy a given requirement on accelerations.

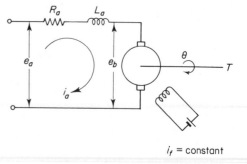

**PROBLEM A-4-8.** Consider the armature-controlled dc motor shown in Fig. 4-48. In Section 4-5 it was shown that

$$e_b = K_b \dot{\theta}$$
$$T = K_f i_f K_1 i_a = K i_a$$

**Fig. 4-48.** Armature-controlled dc motor.

where $K_b$ is the back emf constant, $i_f$ is the constant field current, and $K_f$ and $K_1$ are constants. Show that

$$\frac{K_b}{K} = 1.356 \text{ watts-sec/ft-lb}$$

**Solution.** The mechanical power $T\dot{\theta}$ must be equal to the electrical power $e_b i_a$ developed by the armature current flowing through the armature back emf. Hence from

$$e_b i_a = K_b \dot{\theta} i_a \text{ (watts)} = \frac{K_b \dot{\theta} i_a}{746} \text{ (hp)}$$

$$T\dot{\theta} = K i_a \dot{\theta} \text{ (ft-lb/sec)} = \frac{K i_a \dot{\theta}}{550} \text{ (hp)}$$

we obtain

$$\frac{K_b}{K} = 1.356 \text{ watts-sec/ft-lb}$$

**PROBLEM A-4-9.** Referring to the dc motor shown in Fig. 4-48, for fixed values of the armature voltage, the torque versus speed curves may be represented by

$$T = \frac{K}{R_a}(e_a - K_b \omega)$$

Determine $K$ and $K_b$ from these torque versus speed curves.

**Solution.** For the no-load condition, $T = 0$, Hence

$$K_b = \frac{e_a}{\omega_0}$$

where

$\omega_0 = $ no-load speed

If the motor is stalled, then $\omega = 0$, and we obtain

$$K = \frac{T_s R_a}{e_a}$$

where

$T_s = $ stall torque

**PROBLEM A-4-10.** Determine the transfer function $\Theta(s)/E_a(s)$ of the armature-controlled dc motor shown in Fig. 4-48. In the diagram

$e_a = 26$ volts
$L_a =$ negligible
$J = 2.59 \times 10^{-4}$ oz-in.-sec$^2$
$f = 3 \times 10^{-3}$ oz-in./rad/sec

Assume that the stall torque is 10 oz-in. and the no-load speed is 520 rad/sec.

**Solution.** From Eq. (4-43), the transfer function $\Theta(s)/E_a(s)$ is

$$\frac{\Theta(s)}{E_a(s)} = \frac{K_m}{s(T_m s + 1)}$$

$$= \frac{K}{s(R_a J s + R_a f + K K_b)}$$

where

$$K = \frac{T_s R_a}{e_a} \qquad (T_s = \text{stall torque})$$

$$K_b = \frac{e_a}{\omega_0} \qquad (\omega_0 = \text{no-load speed})$$

In Problem A-4-8, we found that

$$\frac{K_b}{K} = 1.356 \text{ watts-sec/ft-lb}$$

$$= 7.06 \times 10^{-3} \text{ watts-sec/in.-oz}$$

Hence

$$K_b = \frac{26}{520} = 5 \times 10^{-2} \text{ volt/rad/sec}$$

and

$$K = \frac{K_b}{7.06 \times 10^{-3}}$$

$$= \frac{5 \times 10^{-2}}{7.06 \times 10^{-3}}$$

$$= 7.08 \text{ in.-oz/amp}$$

The value of $R_a$ is then obtained as

$$R_a = \frac{K e_a}{T_s}$$

$$= 7.08 \times \frac{26}{10}$$

$$= 18.4 \text{ ohms}$$

The transfer function $\Theta(s)/E_a(s)$ is then

$$\frac{\Theta(s)}{E_a(s)} = \frac{7.08}{s[(18.4)(2.59 \times 10^{-4})s + (18.4)(3 \times 10^{-3}) + (7.08)(5 \times 10^{-2})]}$$

$$= \frac{17.3}{s(0.0116s + 1)}$$

**PROBLEM A-4-11.** Find the transfer function $\Theta(s)/E_f(s)$ of the field-controlled dc motor shown in Fig. 4-26 (a). Assume that in the diagram

$e_f = 110$ volts
$i_a = 15$ amperes
$\omega_{ss} = \dot{\theta}_{ss} = 1200$ rpm

$L_f = 20$ henrys
$R_f = 120$ ohms
$J = 1$ lb-ft-sec²
$f = 0.5$ lb-ft/rad/sec

**Solution.** From Eq. (4-52), the transfer function $\Theta(s)/E_f(s)$ is

$$\frac{\Theta(s)}{E_f(s)} = \frac{K_2}{s(L_f s + R_f)(Js + f)}$$

From this equation, we obtain the steady-state angular velocity $\omega_{ss}$ as follows: Noting that for a step input $e_f = 110$ volts, we obtain

$$\Theta(s) = \frac{K_2}{s(L_f s + R_f)(Js + f)} \frac{110}{s}$$

Hence

$$\omega_{ss} = \lim_{s \to 0} s[s\Theta(s)]$$

$$= \lim_{s \to 0} \frac{110K_2}{(L_f s + R_f)(Js + f)}$$

$$= \frac{110K_2}{R_f f}$$

Since $\omega_{ss} = 1200$ rpm $= 20 \times 2\pi$ rad/sec, $R_f = 120$ ohms, $f = 0.5$ lb-ft/rad/sec, we obtain

$$40\pi = \frac{110K_2}{120 \times 0.5}$$

or

$$K_2 = \frac{40\pi \times 60}{110} \text{ ohms lb-ft/volt}$$

$$= 68.5 \text{ lb-ft/amp}$$

The transfer function $\Theta(s)/E_f(s)$ is then obtained as follows:

$$\frac{\Theta(s)}{E_f(s)} = \frac{68.5}{s(20s + 120)(s + 0.5)}$$

$$= \frac{1.14}{s(0.167s + 1)(2s + 1)}$$

**PROBLEM A-4-12.** Consider the positional servomechanism shown in Fig. 4-49. Assume that the input to the system is the reference shaft position and the system output is the output shaft position. Assume the following numerical values for the system constants:

     $r =$ angular displacement of the reference input shaft, radians
     $c =$ angular displacement of the output shaft, radians
     $\theta =$ angular displacement of the motor shaft, radians
     $K_1 =$ gain of the potentiometer error detector $= 24/\pi$ volts/rad
     $K_p =$ amplifier gain $= 10$ volts/volt
     $R_f =$ field-winding resistence $= 2$ ohms
     $L_f =$ field-winding inductance $= 0.1$ henry
     $i_f =$ field-winding current, amperes
     $e_f =$ applied field voltage, volts
     $K_2 =$ motor torque constant $= 0.05$ newton-m/amperes
     $n =$ gear ratio $= 1/10$
     $J =$ equivalent moment of inertia of the motor and load referred to the motor
         shaft $= 0.02$ kg-m²

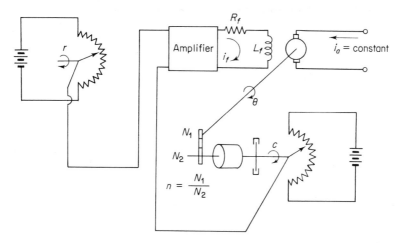

**Fig. 4-49.** Positional servomechanism.

$f$ = equivalent viscous-friction coefficient of the motor and load referred to the motor shaft = 0.02 newton-m/rad/sec

Draw a block diagram of the system. Find the transfer function of each block. Finally, simplify the block diagram.

**Solution.** The equations describing the system dynamics are as follows:
*For the potentiometric error detector:*

$$E(s) = K_1[R(s) - C(s)] = \frac{24}{\pi}[R(s) - C(s)] \tag{4-81}$$

*For the amplifier:*

$$E_f(s) = K_p E(s) = 10E(s) \tag{4-82}$$

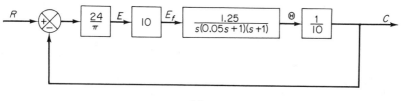

(a)

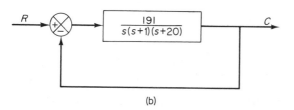

(b)

**Fig. 4-50.** (a) Block diagram of the positional servomechanism shown in Fig. 4-49; (b) simplified block diagram.

*For the field-controlled motor:* Referring to Eq. (4-52), we see that

$$\frac{\Theta(s)}{E_f(s)} = \frac{K_m}{s(T_f s + 1)(T_m s + 1)}$$

where

$$K_m = \frac{K_2}{R_f f} = \frac{0.05}{2 \times 0.02} = 1.25 \text{ rad/volt-sec}$$

$$T_f = \frac{L_f}{R_f} = \frac{0.1}{2} = 0.05 \text{ sec}$$

$$T_m = \frac{J}{f} = \frac{0.02}{0.02} = 1 \text{ sec}$$

Hence

$$\frac{\Theta(s)}{E_f(s)} = \frac{1.25}{s(0.05s + 1)(s + 1)} \tag{4-83}$$

From Eqs. (4-81), (4-82), and (4-83), we obtain the block diagram shown in Fig. 4-50(a). Simplification of the block diagram of Fig. 4-50(a) yields Fig. 4-50(b).

**PROBLEM A-4-13.** In feedback control systems, dc generators are used for producing large power amplification. (Vacuum-tube amplifiers, transistor amplifiers, etc., are not suitable for use in directly controlling a dc motor of more than 100 watts if small time constants and good linearity are desired.) In the diagram shown in Fig. 4-51 (a), the dc generator drives an armature-controlled dc motor. In the schematic diagram of Fig. 4-51 (b),

    $R_f$ = field-winding resistance of the dc generator, ohms
    $L_f$ = field-winding inductance of the dc generator, henrys
    $i_g$ = field-winding current of the dc generator, amperes
    $e_f$ = applied field-winding voltage of the dc generator, volts
    $R_a$ = armature-winding resistance, ohms
    $L_a$ = armature-winding inductance, henrys
    $i_a$ = armature-winding current, amperes
    $e_a$ = armature-generated voltage, volts
    $\dot{\theta}_g$ = constant speed of the dc generator, rad/sec
    $i_f$ = field current of the dc motor (constant), amperes
    $e_b$ = back emf, volts
    $\theta$ = angular displacement of the motor shaft, radians
    $T$ = torque delivered by the motor, lb-ft
    $J$ = moment of inertia of the motor and load referred to the motor shaft, slug-ft²
    $f$ = viscous-friction coefficient of the motor and load referred to the motor shaft, lb-ft/rad/sec

The generator is rotated at a constant speed $\dot{\theta}_g$. The armature generated voltage $e_a$ is determined by the voltage $e_f$ applied to the generator field. Derive the transfer function between $\theta$ and $e_f$.

**Solution.** Equations for this system are

$$L_f \frac{di_g}{dt} + R_f i_g = e_f \tag{4-84}$$

and

$$e_a = K\dot{\theta}_g \psi_g$$

where $K$ is a constant and $\psi_g$ is the air gap flux. Since the field-winding current and the air gap flux are proportional and $\dot{\theta}_g$ is a constant, $e_a$ can be written

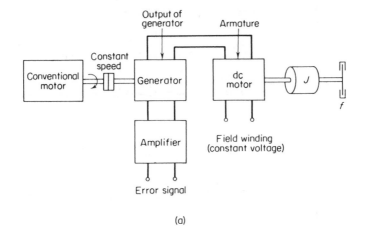

(a)

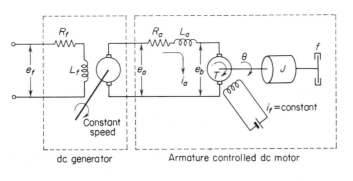

(b)

**Fig. 4-51.** (a) A dc generator-motor system; (b) schematic diagram.

$$e_a = K_a i_g \tag{4-85}$$

where $K_a$ is a constant. Eliminating $i_g$ from Eqs. (4-84) and (4-85), we obtain

$$L_f \frac{de_a}{dt} + R_f e_a = K_a e_f$$

The transfer function between the armature-generated voltage $e_a$ and field-winding voltage $e_f$ is then

$$\frac{E_a(s)}{E_f(s)} = \frac{K_a}{L_f s + R_f} \tag{4-86}$$

In the armature-controlled dc motor, the armature is energized by the output of the generator and the field current is held constant. The transfer function between $\theta$ and $e_a$ was obtained in Section 4-5 and given by Eq. (4-42). Rewritten, it is

$$\frac{\Theta(s)}{E_a(s)} = \frac{K}{s[L_a J s^2 + (L_a f + R_a J)s + R_a f + KK_b]} \tag{4-87}$$

Hence the transfer function between $\theta$ and $e_f$ is obtained from Eqs. (4-86) and (4-87) as

$$\frac{\Theta(s)}{E_f(s)} = \frac{KK_a}{(L_f s + R_f)s[L_a J s^2 + (L_a f + R_a J)s + R_a f + KK_b]} \quad (4\text{-}88)$$

The inductance $L_a$ in the armature circuit is usually small and may be neglected. If $L_a$ is neglected, then the transfer function given by Eq. (4-88) reduces to

$$\frac{\Theta(s)}{E_f(s)} = \frac{K_g K_m}{s(T_f s + 1)(T_m s + 1)}$$

where

$K_g = K_a/R_f$ = generator gain constant
$T_f = L_f/R_f$ = generator time constant
$K_m = K/(R_a f + KK_b)$ = motor gain constant
$T_m = R_a J/(R_a f + KK_b)$ = motor time constant

**PROBLEM A-4-14.** Considering small deviations from steady-state operation, draw the block diagram of the air heating system shown in Fig. 4-52. Assume that the heat loss to the surroundings and the heat capacitance of the metal parts of the heater are negligible.

**Solution.** Let us define

$\bar{\Theta}_i$ = steady-state temperature of inlet air, °F
$\bar{\Theta}_0$ = steady-state temperature of outlet air, °F
$G$ = flow rate of air through the heating chamber, lb/sec
$M$ = air contained in the heating chamber, lb
$c$ = specific heat of air, Btu/lb °F
$R$ = thermal resistance, °F sec/Btu
$C$ = thermal capacitance of air contained in the heating chamber = $Mc$, Btu/°F
$\bar{H}$ = steady-state heat input, Btu/sec

Let us assume that the heat input is suddenly changed from $\bar{H}$ to $\bar{H} + h$ and the inlet air temperature is suddenly changed from $\bar{\Theta}_i$ to $\bar{\Theta}_i + \theta_i$. Then the outlet air temperature will be changed from $\bar{\Theta}_0$ to $\bar{\Theta}_0 + \theta_0$.

The equation describing the system behavior is

$$C\, d\theta_0 = [h + Gc(\theta_i - \theta_0)]\, dt$$

or

$$C\frac{d\theta_0}{dt} = h + Gc(\theta_i - \theta_0)$$

Noting that

$$Gc = \frac{1}{R}$$

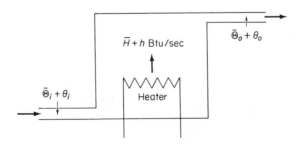

**Fig. 4-52.** Air heating system.

we obtain

$$C\frac{d\theta_0}{dt} = h + \frac{1}{R}(\theta_i - \theta_0)$$

or

$$RC\frac{d\theta_0}{dt} + \theta_0 = Rh + \theta_i$$

Taking the Laplace transforms of both sides of this last equation and substituting the initial condition that $\theta_0(0) = 0$, we obtain

$$\Theta_0(s) = \frac{R}{RCs + 1}H(s) + \frac{1}{RCs + 1}\Theta_i(s)$$

The block diagram of the system corresponding to this equation is shown in Fig. 4-53.

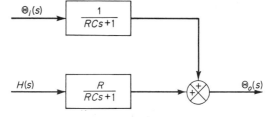

**Fig. 4-53.** Block diagram of the air heating system shown in Fig. 4-52.

**PROBLEM A-4-15.** Consider an ideal gas changing from a state represented by $(p_1, v_1, T_1)$ to a state represented by $(p_2, v_2, T_2)$. If we keep the temperature constant at $T$ but change the pressure from $p_1$ to $p_2$, then the volume of gas will change from $v_1$ to $v'$ such that

$$p_1v_1 = p_2v' \tag{4-89}$$

Now keep the pressure constant but change the temperature to $T_2$. Then the volume of gas reaches $v_2$. Thus

$$\frac{v'}{T_1} = \frac{v_2}{T_2} \tag{4-90}$$

By eliminating $v'$ between Eqs. (4-89) and (4-90), we obtain

$$\frac{p_1v_1}{T_1} = \frac{p_2v_2}{T_2}$$

This means that, for a fixed quantity of gas, no matter what physical changes occur, $pv/T$ will be constant. We may therefore write

$$pv = kT$$

where the value of constant $k$ depends on the quantity and nature of the gas considered.

In dealing with gas systems, we find it convenient to work in molar quantities since 1 mole of any gas contains the same number of molecules. Thus 1 mole occupies the same volume if measured under the same conditions of temperature and pressure.

If we consider 1 mole of gas, then

$$p\bar{v} = \bar{R}T \tag{4-91}$$

The value of $\bar{R}$ is the same for all gases under all conditions. The constant $\bar{R}$ is called the universal gas constant. At standard temperature and pressure (that is, at 492°R and 14.7 psia), 1 lb-mole of any gas is found to occupy 359 ft³. [For example, at 492°R(= 32°F) and 14.7 psia, the volume occupied by 2 lb of hydrogen, 32 lb of oxygen, or 28 lb of nitrogen is the same, 359 ft³.] This volume is called the molal volume and is denoted by $\bar{v}$.

Obtain the value of the universal gas constant.

**Solution.** By substituting $p = 14.7$ lb/in.², $\bar{v} = 359$ ft³/lb-mole, and $T = 492°R$ into Eq. (4-91), we obtain

$$\bar{R} = \frac{14.7 \times 144 \times 359}{492}$$

$$= 1545 \text{ ft-lb/lb-mole } °R$$

$$= 1.987 \text{ Btu/lb-mole } °R$$

**PROBLEM A-4-16.** The molecular weight of a pure substance is the weight of one molecule of the substance compared to the weight of one oxygen atom, which is taken to be 16. That is, the molecular weight of carbon dioxide ($CO_2$) is $12 + (16 \times 2) = 44$. The molecular weights of oxygen (molecular) and water vapor are 32 and 18, respectively.

Obtain the specific volume of a mixture which consists of 100 ft³ of oxygen, 5 ft³ of carbon dioxide, and 20 ft³ of water vapor. Assume the temperature and pressure to be 70°F and 14.7 psia, respecively.

**Solution.** The mean molecular weight of the mixture is

$$M = \left(32 \times \frac{100}{125}\right) + \left(44 \times \frac{5}{125}\right) + \left(18 \times \frac{20}{125}\right) = 25.6 + 1.76 + 2.88 = 30.24$$

Thus

$$v = \frac{\bar{R}}{M} \frac{T}{p}$$

$$= \frac{1545 \times 530}{30.24 \times 14.7 \times 144} = 12.8 \text{ ft}^3/\text{lb}$$

**PROBLEM A-4-17.** The value of the gas constant for any gas may be determined from accurate experimental observations of simultaneous values of $p$, $v$, and $T$.

Obtain the gas constant $R_{\text{air}}$ for air. Note that at 32°F and 14.7 psia the specific volume of air is 12.39 ft³/lb. Then obtain the capacitance of a 20-ft³ pressure vessel which contains air at 160°F. Assume that the expansion process is isothermal.

**Solution**

$$R_{\text{air}} = \frac{pv}{T} = \frac{14.7 \times 144 \times 12.39}{460 + 32} = 53.3 \text{ ft-lb/lb } °R$$

Referring to Eq. (4-64) the capacitance of a 20-ft³ pressure vessel is

$$C = \frac{V}{nR_{\text{air}}T} = \frac{20}{1 \times 53.3 \times 620} = 6.05 \times 10^{-4} \frac{\text{lb}}{\text{lb/ft}^2}$$

**PROBLEM A-4-18.** Consider the system shown in Fig. 4-54. Obtain the closed-loop transfer function $H(s)/Q(s)$.

**Solution.** In the given system, there is only one forward path that connects the input $Q(s)$ and the output $H(s)$. Thus,

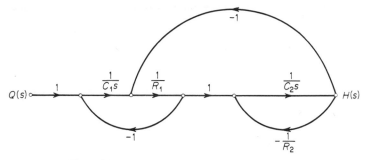

**Fig. 4-54.** Signal flow graph of a control system.

$$P_1 = \frac{1}{C_1 s}\frac{1}{R_1}\frac{1}{C_2 s}$$

There are three individual loops. Thus,

$$L_1 = -\frac{1}{C_1 s}\frac{1}{R_1}$$

$$L_2 = -\frac{1}{C_2 s}\frac{1}{R_2}$$

$$L_3 = -\frac{1}{R_1}\frac{1}{C_2 s}$$

Loop $L_1$ does not touch loop $L_2$. (Loop $L_1$ touches loop $L_3$, and loop $L_2$ touches loop $L_3$). Hence the determinant $\Delta$ is given by

$$\Delta = 1 - (L_1 + L_2 + L_3) + (L_1 L_2)$$
$$= 1 + \frac{1}{R_1 C_1 s} + \frac{1}{R_2 C_2 s} + \frac{1}{R_1 C_2 s} + \frac{1}{R_1 C_1 R_2 C_2 s^2}$$

Since all three loops touch the forward path $P_1$, we remove $L_1$, $L_2$, and $L_3$ from $\Delta$ and evaluate the cofactor $\Delta_1$ as follows:

$$\Delta_1 = 1$$

Thus we obtain the closed-loop transfer function as shown:

$$\frac{H(s)}{Q(s)} = \frac{P_1 \Delta_1}{\Delta}$$

$$= \frac{\dfrac{1}{R_1 C_1 C_2 s^2}}{1 + \dfrac{1}{R_1 C_1 s} + \dfrac{1}{R_2 C_2 s} + \dfrac{1}{R_1 C_2 s} + \dfrac{1}{R_1 C_1 R_2 C_2 s^2}}$$

$$= \frac{R_2}{R_1 C_1 R_2 C_2 s^2 + (R_1 C_1 + R_2 C_2 + R_2 C_1)s + 1}$$

## PROBLEMS

**PROBLEM B-4-1.** Obtain the transfer functions of the mechanical systems shown in Figs. 4-55 (a) and (b). Also obtain electrical analogs of the mechanical systems.

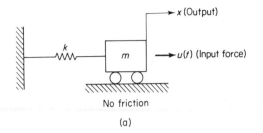

(a)

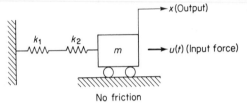

(b)

**Fig. 4-55.** Mechanical systems.

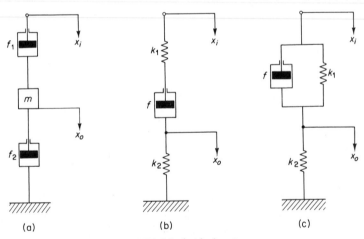

(a)                    (b)                    (c)

**Fig. 4-56.** Mechanical systems.

**PROBLEM B-4-2.** Obtain the transfer function of each of the three mechanical systems shown in Fig. 4-56. In the diagrams, $x_i$ denotes the input displacement and $x_0$ denotes the output displacement. Assume that the systems are lumped-parameter ones (no distributed mass) and that the loading effect at the output is negligible.

**PROBLEM B-4-3.** Obtain the transfer functions of the systems shown in Figs. 4-57 (a), (b), and (c).

**PROBLEM B-4-4.** Show that the systems in Figs. 4-58 (a) and (b) are analogous systems. (Show that the transfer functions of the two systems are of identical form.)

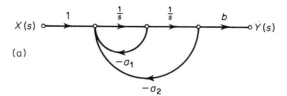

(a)

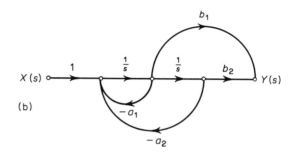

(b)

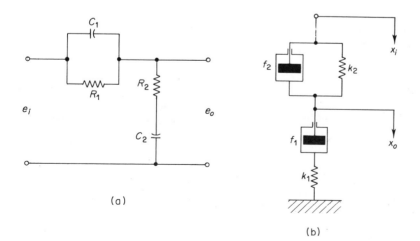

(c)

**Fig. 4-57.** Signal flow graphs of systems.

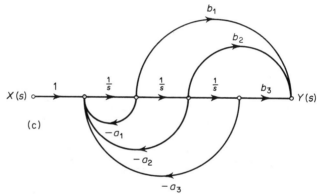

(a)

(b)

**Fig. 4-58.** (a) Electrical system; (b) mechanical analog of the electrical system.

**PROBLEM B-4-5.** Derive the transfer function of the electrical network shown in Fig. 4-59. Draw a schematic diagram of an equivalent mechanical network.

**PROBLEM B-4-6.** Derive the transfer function of the mechanical network shown in Fig. 4-60. Draw a schematic diagram of an equivalent electrical network.

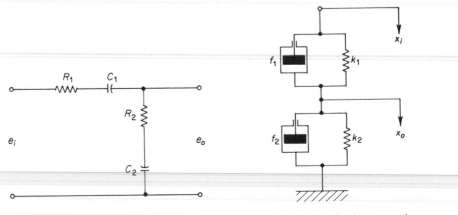

Fig. 4-59. Electrical network.       Fig. 4-60. Mechanical network.

**PROBLEM B-4-7.** Obtain the transfer function $\Theta(s)/E_f(s)$ of the field-controlled dc motor shown in Fig. 4-61. In the system, assume that $J = 0.5$ lb-ft-sec$^2$, $f = 0.2$ lb-ft/rad/sec, and $K_2 =$ motor torque constant $= 27.4$ lb-ft/amp.

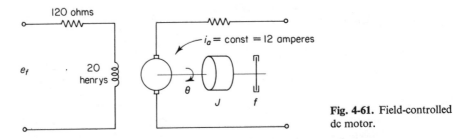

Fig. 4-61. Field-controlled dc motor.

**PROBLEM B-4-8.** Draw a block diagram of the control system shown in Fig. 4-62. Assume the transfer functions of the synchro error detector and the amplifier to be constants ($K_1$ and $K_2$, repsectively) and the transfer function of the motor plus load to be

$$\frac{C(s)}{E_a(s)} = \frac{K_3}{s(Ts + 1)}$$

The velocity feedback coefficient is $K_4$. The gear ratio is $n = N_1/N_2 < 1$.

**PROBLEM B-4-9.** Consider the liquid-level system shown in Fig. 4-27 (a). Assuming that $\bar{H} = 10$ ft, $\bar{Q} = 1$ ft$^3$/sec and the cross-sectional area of the tank is equal to 5 ft$^2$,

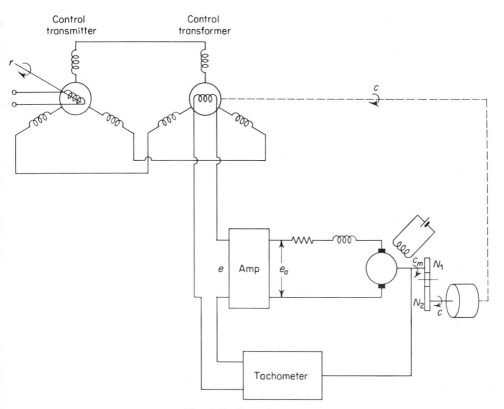

Control
transmitter

Control
transformer

**Fig. 4-62.** Control system.

obtain the time constant of the system at the operating point $(\bar{H}, \bar{Q})$.

**PROBLEM B-4-10.** Obtain the capacitance of the pressure vessel which contains 2500 in.³ of air at 100°F and 34.7 psia. Assume that the expansion process is isothermal.

**PROBLEM B-4-11.** Obtain the transfer function of the air pressure system shown in Fig. 4-63 (a). Assume that $p_1$ is the input and $p_2$ is the output. The volume of the vessel is 5 in.³ and the $\Delta p$ versus flow curve is given in Fig. 4-63(b). The temperature of air is 75°F and the expansion process is assumed to be isothermal. Assume that the operating pressure $\Delta p = p_1 - p_2$ is 0~10 lb/in.². (Use an average resistance.)

**PROBLEM B-4-12.** Obtain the transfer function of the thermometer shown in Fig. 4-64. Obtain an electrical analog of the thermometer.

**PROBLEM B-4-13.** A thermocouple has a time constant of 2 sec. A thermal well has a time constant of 30 sec. When the thermocouple is inserted into the well, this temperature-measuring device can be considered a two-capacity system.

Determine the time constants of the combined thermocouple-thermal well system. Assume that the weight of the thermocouple is 0.5 oz and the weight of the thermal

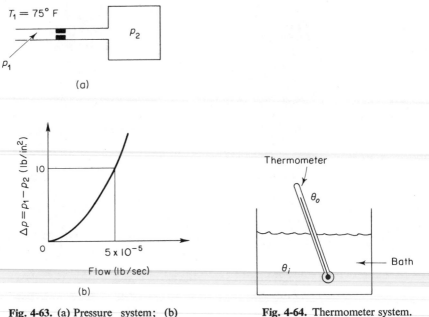

Fig. 4-63. (a) Pressure system; (b)
$\Delta p$ versus flow curve.

Fig. 4-64. Thermometer system.

well is 3.2 oz. Assume also that the specific heats of the thermocouple and thermal well are the same.

**PROBLEM B-4-14.** Figure 4-65 is the block diagram of an engine-speed control system. The speed is measured by a set of flyweights. Draw a signal flow graph for this system.

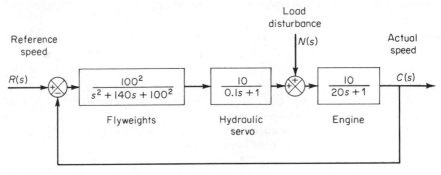

Fig. 4-65. Block diagram of an engine-speed control system.

# 5

# BASIC CONTROL ACTIONS
# AND INDUSTRIAL
# AUTOMATIC CONTROLS

## 5-1 INTRODUCTION

An automatic controller compares the actual value of the plant output with the desired value, determines the deviation, and produces a control signal which will reduce the deviation to zero or to a small value. The manner in which the automatic controller produces the control signal is called the *control action*.

In this chapter, we shall present the basic control actions commonly used in industrial automatic controllers. First we shall introduce the principle of operation of automatic controllers and the methods for generating various control signals, such as the use of the derivative and integral of the error signal. Next we shall discuss the effects of particular control modes on the system performance. Then we shall give a brief discussion of methods for reducing the effects of external disturbances on the system performance. Finally, we shall introduce fluid amplifiers, present basic principles of fluidics, and discuss applications of fluidic devices.

**Classifications of industrial automatic controllers.** Industrial automatic controllers may be classified according to their control action as

1. two-position or on-off controllers
2. proportional controllers
3. integral controllers

4. proportional-plus-integral controllers

5. proportional-plus-derivative controllers

6. proportional-plus-derivative-plus-integral controllers

Most industrial automatic controllers use electricity or pressurized fluid such as oil or air as power sources. Automatic controllers may also be classified according to the kind of power employed in the operation, such as pneumatic controllers, hydraulic controllers, or electronic controllers. What kind of controller to use must be decided by the nature of the plant and the operating conditions, including such considerations as safety, cost, availability, reliability, accuracy, weight, and size.

**Elements of industrial automatic controllers.** An automatic controller must detect the actuating error signal, which is usually at a very low power level, and amplify it to a sufficiently high level. Thus, an amplifier is necessary. The output of an automatic controller is fed to a power device, such as a pneumatic motor or valve, a hydraulic motor, or an electric motor.

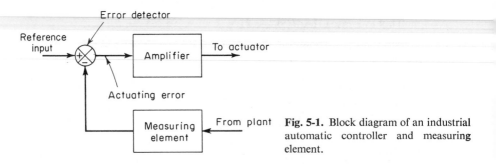

**Fig. 5-1.** Block diagram of an industrial automatic controller and measuring element.

Figure 5-1 shows a block diagram of an industrial automatic controller together with a measuring element. The controller consists of an error detector and amplifier. The measuring element is a device which converts the output variable into another suitable variable, such as a displacement, pressure, or electric signal, which can be used for comparing the output to the reference input signal. This element is in the feedback path of the closed-loop system. The set point of the controller must be converted to a reference input of the same units as the feedback signal from the measuring element. The amplifier amplifies the power of the actuating error signal, which in turn operates the actuator. (Quite often an amplifier, together with a suitable feedback circuit, is used to alter the actuating error signal by amplifying and sometimes by differentiating and/or integrating it to produce a better control signal.) The actuator is an element which alters the input to the plant according to the control signal so that the feedback signal may be brought into correspondence with the reference input signal.

**Self-operated controllers.** In most industrial automatic controllers, separate units are used for the measuring element and for the actuator. In a very simple one, however, such as a self-operated controller, these elements are assembled in one unit. Self-operated controllers utilize power developed by the measuring element and are very simple and inexpensive. An example of such a self-operated controller is shown in Fig. 5-2. The set point is determined by the adjustment of the spring force. The controlled pressure is measured by the diaphragm. The actuating error signal is the net force acting on the diaphragm. Its position determines the valve opening.

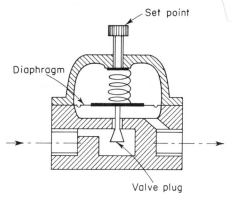

Fig. 5-2. Self-operated controller.

The operation of the self-operated controller is as follows: Suppose that the output pressure is lower than the reference pressure, as determined by the set point. Then the downward spring force is greater than the upward pressure force, resulting in a downward movement of the diaphragm. This increases the flow rate and raises the output pressure. When the upward pressure force equals the downward spring force, the valve plug stays stationary and the flow rate is constant. Conversely, if the output pressure is higher than the reference pressure, the valve opening becomes small and reduces the flow rate through the valve opening. Such a self-operated controller is widely used for water and gas pressure control. In such a controller, the flow rate through the valve opening is approximately proportional to the actuating error signal.

**Control actions.** The following six basic control actions are very common among industrial auomatic controllers: two-position or on-off, proportional, integral, proportional-plus-integral, proportional-plus-derivative, and proportional-plus-derivative-plus-integral control action. These six will be discussed in this chapter. Note that an understanding of the basic characteristics of the various actions is necessary in order for the control engineer to select the one best suited to his particular application.

**Two-position or on-off control action.** In a two-position control system, the actuating element has only two fixed positions which are, in many cases, simply on and off. Two-position or on-off control is relatively simple and inexpensive and, for this reason, is very widely used in both industrial and domestic control systems.

Let the output signal from the controller be $m(t)$ and the actuating error signal be $e(t)$. In two-position control, the signal $m(t)$ remains at either a maximum or minimum value, depending on whether the actuating error signal is positive or negative, so that

$$m(t) = M_1 \qquad \text{for } e(t) > 0$$
$$= M_2 \qquad \text{for } e(t) < 0$$

where $M_1$ and $M_2$ are constants. The minimum value $M_2$ is usually either zero or

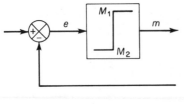

(a)

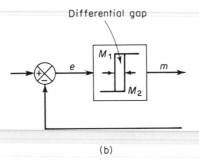

(b)

**Fig. 5-3.** (a) Block diagram of an on-off controller; (b) block diagram of an on-off controller with differential gap.

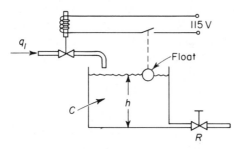

**Fig. 5-4.** Liquid-level control system.

$-M_1$. Two-position controllers are generally electrical devices, and an electric solenoid-operated valve is widely used in such controllers. Pneumatic proportional controllers with very high gains act as two-position controllers and are sometimes called pneumatic two-position controllers.

Figures 5-3(a) and (b) show the block diagrams for two-position controllers. The range through which the actuating error signal must move before the switching occurs is called the differential gap. A differential gap is indicated in Fig. 5-3(b). Such a differential gap causes the controller output $m(t)$ to maintain its present value until the actuating error signal has moved slightly beyond the zero value. In some cases, the differential gap is a result of unintentional friction and lost motion; however, quite often it is intentionally provided in order to prevent too frequent operation of the on-off mechanism.

Consider the liquid-level control system shown in Fig. 5-4. With two position control, the valve is either open or closed. Thus the water inflow rate is either a positive constant or zero. As shown in Fig. 5-5, the output signal continuously moves between the two limits required to cause the actuating element to move from one fixed position to the other. Notice that the output curve follows one of two exponential curves, one corresponding to the filling curve and the other to the emptying curve. Such output oscillation between two limits is a typical response characteristic of a system under two-position control.

From Fig. 5-5, we notice that the amplitude of the output oscillation can be reduced by decreasing the differential gap. This, however, increases the number of on-off switchings per minute and reduces the useful life of the component. The magnitude of the differential gap must be determined from such considerations as the accuracy required and the life of the component. Further analysis of control systems with two-position controllers is deferred to Chapter 11.

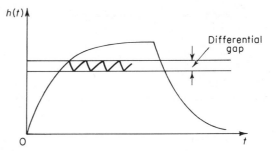

**Fig. 5-5.** Level $h(t)$ versus $t$ curve for the system shown in Fig. 5-4.

**Proportional control action.** For a controller with proportional control action, the relationship between the output of the controller $m(t)$ and the actuating error signal $e(t)$ is

$$m(t) = K_p e(t)$$

or, in Laplace-transformed quantities,

$$\frac{M(s)}{E(s)} = K_p$$

where $K_p$ is termed the proportional sensitivity or the gain.

Whatever the actual mechanism may be and whatever the form of the operating power, the proportional controller is essentially an amplifier with an adjustable gain. A block diagram of such a controller is shown in Fig. 5-6.

**Integral control action.** In a controller with integral control action, the value of the controller output $m(t)$ is changed at a rate proportional to the actuating error signal $e(t)$. Namely,

$$\frac{dm(t)}{dt} = K_i e(t)$$

or

$$m(t) = K_i \int_0^t e(t)\, dt$$

where $K_i$ is an adjustable constant. The transfer function of the integral controller is

$$\frac{M(s)}{E(s)} = \frac{K_i}{s}$$

If the value of $e(t)$ is doubled, then the value of $m(t)$ varies twice as fast. For zero actuating error, the value of $m(t)$ remains stationary. The integral control action is sometimes called reset control. Figure 5-7 shows a block diagram of such a controller.

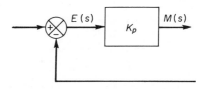

**Fig. 5-6.** Block diagram of a proportional controller.

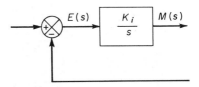

**Fig. 5-7.** Block diagram of an integral controller.

**Proportional-plus-integral control action.** The control action of a proportional-plus-integral controller is defined by the following equation:

$$m(t) = K_p e(t) + \frac{K_p}{T_i} \int_0^t e(t)\, dt$$

or the transfer function of the controller is

$$\frac{M(s)}{E(s)} = K_p\left(1 + \frac{1}{T_i s}\right)$$

where $K_p$ represents the proportional sensitivity or gain, and $T_i$ represents the integral time. Both $K_p$ and $T_i$ are adjustable. The integral time adjusts the integral control action, while a change in the value of $K_p$ affects both the proportional and integral parts of the control action. The inverse of the integral time $T_i$ is called the reset rate. The reset rate is the number of times per minute that the proportional part of the control action is duplicated. Reset rate is measured in terms of repeats per minute. Figure 5-8(a) shows a block diagram of a proportional-plus-integral

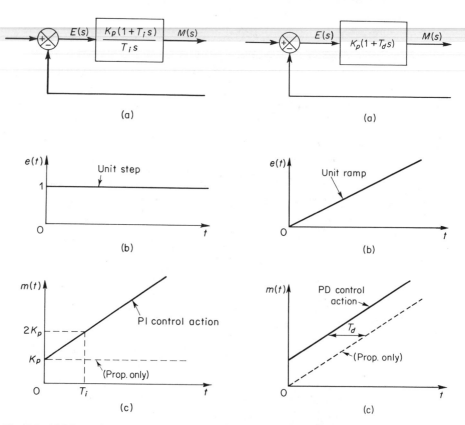

(a)                                                      (a)

(b)                                                      (b)

(c)                                                      (c)

**Fig. 5-8.** (a) Block diagram of a proportional-plus-integral controller; (b) and (c) diagrams depicting a unit-step input and the controller output.

**Fig. 5-9.** (a) Block diagram of a porportional-plus-derivative controller; (b) and (c) diagrams depicting a unit-ramp input and the controller output.

controller. If the actuating error signal $e(t)$ is a unit-step function as shown in Fig. 5-8(b), then the controller output $m(t)$ becomes as shown in Fig. 5-8(c).

**Proportional-plus-derivative control action.** The control action of a proportional-plus-derivative controller is defined by the following equation:

$$m(t) = K_p e(t) + K_p T_d \frac{de(t)}{dt}$$

and the transfer function is

$$\frac{M(s)}{E(s)} = K_p(1 + T_d s)$$

where $K_p$ represents the proportional sensitivity and $T_d$ represents the derivative time. Both $K_p$ and $T_d$ are adjustable. The derivative control action, sometimes called rate control, is where the magnitude of the controller output is proportional to the rate of change of the actuating error signal. The derivative time $T_d$ is the time interval by which the rate action advances the effect of the proportional control action. Figure 5-9(a) shows a block diagram of a proportional-plus-derivative controller. If the actuating error signal $e(t)$ is a unit-ramp function as shown in Fig. 5-9(b), then the controller output $m(t)$ becomes as shown in Fig. 5-9(c). As may be seen from Fig. 5-9(c), the derivative control action has an anticipatory

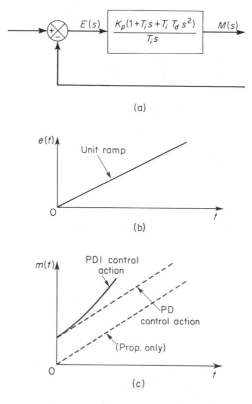

**Fig. 5-10.** (a) Block diagram of a proportional-plus-derivative-plus-integral controller; (b) and (c) diagrams depicting a unit-ramp input and the controller output.

character. As a matter of course, however, derivative control action can never anticipate any action that has not yet taken place.

While derivative control action has an advantage of being anticipatory, it has the disadvantages that it amplifies noise signals and may cause a saturation effect in the actuator.

Note that derivative control action can never be used alone because this control action is effective only during transient periods.

**Proportional-plus-derivative-plus-integral control action.** The combination of proportional control action, derivative control action, and integral control action is termed proportional-plus-derivative-plus-integral control action. This combined action has the advantages of each of the three individual control actions. The equation of a controller with this combined action is given by

$$m(t) = K_p e(t) + K_p T_d \frac{de(t)}{dt} + \frac{K_p}{T_i} \int_0^t e(t)\, dt$$

or the transfer function is

$$\frac{M(s)}{E(s)} = K_p \left( 1 + T_d s + \frac{1}{T_i s} \right)$$

where $K_p$ represents the proportional sensitivity, $T_d$ represents the derivative time, and $T_i$ represents the integral time. The block diagram of a proportional-plus-derivative-plus-integral controller is shown in Fig. 5-10(a). If $e(t)$ is a unit-ramp function as shown in Fig. 5-10(b), then the controller output $m(t)$ becomes as shown in Fig. 5-10(c).

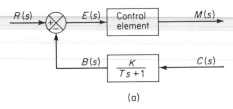

(a)

**Effects of the measuring element on system performance.** Since the dynamic and static characteristics of the measuring element affect the indication of the actual value of the output variable, the measuring element plays an important role in determining the overall performance of the control system. The measuring element usually determines the transfer function in the feedback path. If the time constants of a measuring element are negligibly small compared with other time constants of the control system, the transfer function of the measuring element simply becomes a constant. Figures 5-11(a), (b), and (c) show block

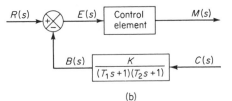

(b)

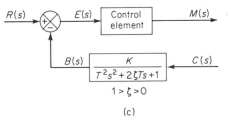

(c)

**Fig. 5-11.** Block diagrams of automatic controllers with (a) first-order measuring element; (b) overdamped second-order measuring element; (c) underdamped second-order measuring element.

diagrams of automatic controllers having a first-order, an overdamped second-order, and an underdamped second-order measuring element, respectively. The response of a thermal measuring element is often of the overdamped second-order type.

**Block diagrams of automatic control systems.** A block diagram of a simple automatic control system may be obtained by connecting the plant to the automatic controller, as shown in Fig. 5-12. Feedback of the output signal is accom-

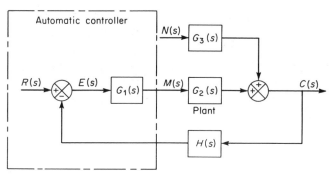

**Fig. 5-12.** Block diagram of a control system.

plished by the measuring element. The equation relating the output variable $C(s)$ to the reference input $R(s)$ and disturbance variable $N(s)$ may be obtained as follows:

$$C(s) = \frac{G_1(s)G_2(s)}{1 + G_1(s)G_2(s)H(s)} R(s) + \frac{G_3(s)}{1 + G_1(s)G_2(s)H(s)} N(s)$$

In process control systems, we are usually interested in the response to the load disturbance $N(s)$. In servomechanisms, however, the response to a varying input $R(s)$ is of most interest. We shall postpone the analysis of the system response to changes in load disturbances to Section 5-4. The system response to changes in the reference input will be studied in detail in Chapter 6.

### 5-2　PROPORTIONAL CONTROLLERS

In this section, we shall illustrate the fact that proportional controllers utilize the principle of negative feedback in themselves. We shall give a detailed discussion of the principle by which proportional controllers operate by considering pneumatic ones. We shall then show that the same principle applies to hydraulic and electronic controllers. Throughout this discussion, we shall place emphasis on the fundamental principles rather than on the details of the operation of the actual mechanisms.

**Pneumatic systems.** Low-pressure pneumatic controllers have been well developed for industrial control systems and have been used extensively in industrial processes. Reasons for the widespread use of pneumatic controllers are their explosion-proof characteristics, simplicity, and ease of maintenance.

**Pneumatic nozzle-flapper amplifiers.** A schematic diagram of a pneumatic nozzle-flapper amplifier is shown in Fig. 5-13(a). The power source for this amplifer is a supply of air at constant pressure. The nozzle-flapper amplifier converts small changes in the position of the flapper into large changes in the back pressure in the nozzle. Thus a large power output can be controlled by the very little power that is needed to position the flapper.

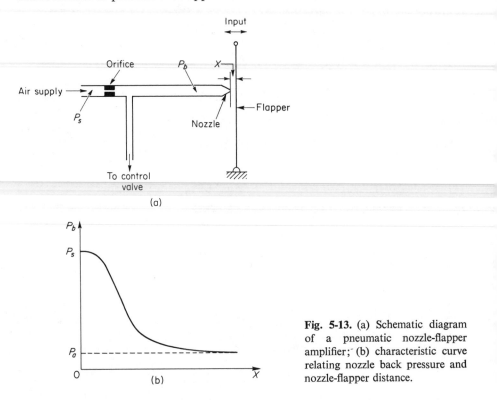

Fig. 5-13. (a) Schematic diagram of a pneumatic nozzle-flapper amplifier; (b) characteristic curve relating nozzle back pressure and nozzle-flapper distance.

In Fig. 5-13(a), pressurized air is fed through the orifice, and the air is ejected from the nozzle toward the flapper. Usually the supply pressure $P_s$ for such a controller is 20 psig. The diameter of the orifice is of the order of 0.010 in. and that of the nozzle is of the order of 0.015 in. The nozzle diameter must be larger than the orifice diameter for the proper functioning of the amplifier. The flapper is positioned against the nozzle opening, and the nozzle back pressure $P_b$ is controlled by the nozzle-flapper distance $X$. As the flapper approaches the nozzle, the opposition to the flow of air through the nozzle increases, with the result that the nozzle back pressure $P_b$ increases. If the nozzle is completely closed by the flapper, the nozzle back pressure $P_b$ becomes equal to the supply pressure $P_s$. If the flapper is moved away from the nozzle, so that the nozzle-flapper distance is wide (of the order of 0.01 in.), then there is practically no restriction to flow and the nozzle back pressure $P_b$ takes on a minimum value which depends on the nozzle-flapper device. (The lowest possible pressure will be the ambient pressure $P_a$.)

Note that because the air jet puts a force against the flapper, it is necessary to make the nozzle diameter as small as possible.

A typical curve relating the nozzle back pressure $P_b$ to the nozzle-flapper distance $X$ is shown in Fig. 5-13(b). The steep and almost linear part of the curve is utilized in the actual operation of the nozzle-flapper amplifier. Because the range of flapper displacements is restricted to a small value, the change in output pressure is also small, unless the curve happens to be very steep.

The nozzle-flapper amplifier converts displacement into a pressure signal. Since industrial process control systems require large output power to operate large pneumatic actuating valves, the power amplification of the nozzle-flapper amplifier is usually not sufficient. Therefore, a pneumatic relay is often employed as a power amplifier in connection with the nozzle-flapper amplifier.

**Pneumatic relays.**  In a practical pneumatic controller, a nozzle-flapper amplifier acts as the first-stage amplifier and a pneumatic relay as the second-stage amplifier. The pneumatic relay is capable of handling a large quantity of air flow.

A schematic diagram of a pneumatic relay is shown in Fig. 5-14(a). As the nozzle back pressure $P_b$ increases, the ball valve is forced toward the lower seat, thus decreasing the control pressure $P_c$. Such a relay is called a reverse acting relay.

When the ball valve is at the top of the seat, the atmospheric opening is closed and the control pressure $P_c$ becomes equal to the supply pressure $P_s$. When the ball

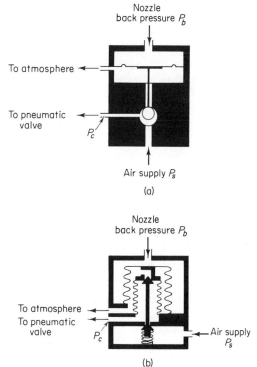

**Fig. 5-14.** (a) Schematic diagram of a bleed-type relay; (b) schematic diagram of a nonbleed-type relay.

valve is at the bottom of its seat, it shuts off the air supply and the control pressure $P_c$ drops to the ambient pressure. The control pressure $P_c$ can thus be made to vary from 0 psig to full supply pressure, usually 20 psig.

The total movement of the ball valve between the upper and lower seat is very small (of the order of 0.01 in.). In all positions of the ball valve, except at the top seat, air continues to bleed into the atmosphere, even after the equilibrium condition is attained between the nozzle back pressure and the control pressure. Thus the relay shown in Fig. 5-14(a) is called a bleed-type relay.

There is another type of relay, the nonbleed type. In this one the air bleed stops when the equilibrium condition is obtained and, therefore, there is no loss of pressurized air at steady-state operation. Note, however, that the nonbleed type relay must have an atmospheric relief to release the control pressure $P_c$ from the pneumatic actuating valve. A schematic diagram of a nonbleed-type relay is shown in Fig. 5-14(b).

In either type of relay, the air supply is controlled by a valve, which is in turn controlled by the nozzle back pressure. Thus, the nozzle back pressure is converted into the control pressure with power amplification.

Since the control pressure $P_c$ changes almost instantaneously with changes in the nozzle back pressure $P_b$, the time constant of the pneumatic relay is negligible compared with the other larger time constants of the pneumatic controller and the plant.

**Pneumatic proportional controllers (force-distance type).** Two types of pneumatic controllers, one called the force-distance type and the other the force-balance type, are used extensively in industry. Regardless of how differently industrial pneumatic controllers may appear, careful study will show the close similarity in the functions of the pneumatic circuit. Here we shall consider only force-distance pneumatic controllers.

Figure 5-15(a) shows a schematic diagram of such a proportional controller. The nozzle-flapper amplifier constitutes the first-stage amplifier, and the nozzle back pressure is controlled by the nozzle-flapper distance. The relay-type amplifier constitutes the second-stage amplifier. The nozzle back pressure determines the position of the ball valve for the second-stage amplifier, which is capable of handling a large quantity of air flow.

In most pneumatic controllers, some type of pneumatic feedback is employed. Feedback of the pneumatic output reduces the amount of actual movement of the flapper. Instead of mounting the flapper on a fixed point, as shown in Fig. 5-15(b), it is often pivoted on the feedback bellows, as shown in Fig. 5-15(c). The amount of feedback can be regulated by introducing a variable linkage between the feedback bellows and the flapper connecting point. The flapper then becomes a floating link. It can be moved by both the error signal and the feedback signal.

The operation of the controller shown in Fig. 5-15(a) is as follows: The input signal to the two-stage pneumatic amplifier is the actuating error signal. Increasing the actuating error signal moves the flapper to the right. This will, in turn, decrease the nozzle back pressure, and the bellows $B$ will contract, which results in an upward

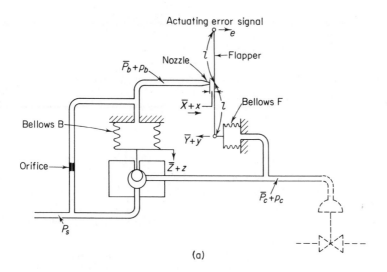

(a)

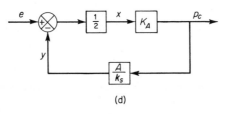

(b)            (c)

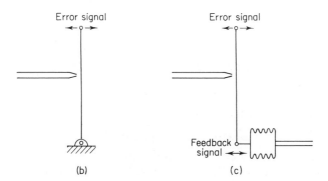

(d)

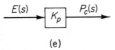

(e)

**Fig. 5-15.** (a) Schematic diagram of a force-distance type pneumatic proportional controller; (b) flapper mounted on a fixed point; (c) flapper mounted on a feedback bellows; (d) block diagram for the controller; (e) simplified block diagram for the controller.

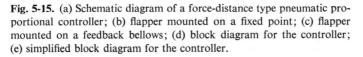

movement of the ball valve. This will cause more flow to the pneumatic valve, and the control pressure will increase. This increase will cause the bellows $F$ to expand and move the flapper to the left, closing the nozzle.

The nozzle-flapper displacement is very small because of this feedback, but the change in the control pressure can be large. In the case where the actuating error decreases, the nozzle back pressure increases and the ball valve moves down, resulting in a decrease in supply flow to the valve and an increase in bleeding to the atmosphere. This will cause the control pressure to decrease.

It is important to note that the feedback bellows should move the flapper less than that movement caused by the error signal alone. If these two movements were equal, no control action would result.

Equations for this controller can be derived as follws: When the actuating error is zero, or $e = 0$, and equilibrium state exists with the nozzle-flapper distance equal to $\bar{X}$, the displacement of the bellows $F$ equal to $\bar{Y}$, the displacement of the bellows $B$ equal to $\bar{Z}$, the nozzle back pressure equal to $\bar{P}_b$, and the control pressure equal to $\bar{P}_c$. When there is any actuating error, the nozzle-flapper distance, the displacements of the bellows $F$ and $B$, the nozzle back pressure, and the control pressure deviate from their respective equilibrium values. Let these deviations be $x, y, z, p_b$, and $p_c$, respectively. (The positive direction for each displacement variable is indicated by an arrowhead.)

Assuming that the relationship between the variation in the nozzle back pressure and the variation in the nozzle-flapper distance is linear, we have

$$p_b = -K_1 x \qquad (5\text{-}1)$$

where $K_1$ is a constant. For the bellows $B$

$$p_b = K_2 z \qquad (5\text{-}2)$$

where $K_2$ is a constant. The position of the ball valve which depends upon the displacement of the bellows $B$ determines the control pressure. If the ball valve is such that the relationship between $p_c$ and $z$ is linear, then

$$p_c = -K_3 z \qquad (5\text{-}3)$$

where $K_3$ is a constant. From Eqs. (5-1), (5-2), and (5-3), we obtain

$$p_c = -\frac{K_3}{K_2} p_b = K_A x \qquad (5\text{-}4)$$

where $K_A = K_1 K_3 / K_2$ is a constant. For the flapper movement, we have

$$x = \frac{e - y}{2} \qquad (5\text{-}5)$$

The bellows $F$ acts like a spring, and the follwing equation holds:

$$Ap_c = k_s y \qquad (5\text{-}6)$$

where $A$ is the effective area of the bellows $F$, and $k_s$ is the equivalent spring constant or the stiffness due to the action of the corrugated side of the bellows.

Assuming that all variations in the variables are within the linear range, we may obtain a block diagram for this system from Eqs. (5-4), (5-5), and (5-6), as shown in Fig. 5-15(d). From Fig. 5-15(d), it can clearly be seen that the pneumatic controller shown in Fig. 5-15(a) itself is a feedback system. The transfer function between $p_c$ and $e$ is given by

$$\frac{P_c(s)}{E(s)} = \frac{\dfrac{1}{2}K_A}{1 + \dfrac{K_A A}{2k_s}} = K_p \tag{5-7}$$

A simplified block diagram is shown in Fig. 5-15(e). Since $p_c$ and $e$ are proportional, the pneumatic controller shown in Fig. 5-15(a) is called a pneumatic proportional controller.

Note that since the value of $K_A A/k_s$ is generally very much greater than unity in actual controllers, the transfer function given by Eq. (5-7) can be simplified to give

$$\frac{P_c(s)}{E(s)} = \frac{\dfrac{1}{2}K_A}{\dfrac{K_A A}{2k_s}} = \frac{k_s}{A} \tag{5-8}$$

As seen from Eq. (5-7) or (5-8), the gain of the pneumatic proportional controller can be widely varied by adjusting the effective value of $k_s$. This can be accomplished easily by adjusting the flapper connecting linkage. [The flapper connecting linkage is not shown in Fig. 5-15(a).] In most commercial proportional controllers an adjusting knob or other mechanism is provided for varying the gain by adjusting this linkage.

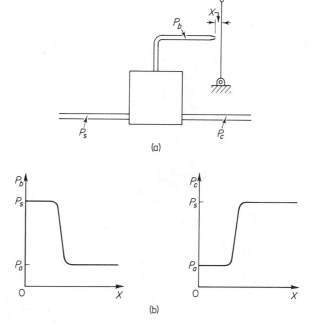

**Fig. 5-16.** (a) Pneumatic controller without a feedback mechanism; (b) curves $P_b$ versus $X$ and $P_c$ versus $X$.

As we have seen previously, the actuating error signal moved the flapper in one direction, and the feedback bellows moved the flapper in the opposite direction, but to a smaller degree. The effect of the feedback bellows is thus to reduce the sensitivity of the controller. The principle of feedback is commonly used to obtain wide proportional-band controllers.

Pneumatic controllers which do not have feedback mechanisms [which means that one end of the flapper is fixed, as shown in Fig. 5-16(a)] have high sensitivity and are called narrow-band proportional controllers or two-position controllers. In such a controller, only a small motion between the nozzle and the flapper is required to give a complete change from the maximum to the minimum control pressure. The curves relating $P_b$ to $X$ and $P_c$ to $X$ are shown in Fig. 5-16(b). Notice that a small change in $X$ can cause a large change in $P_b$, which causes the ball valve to be completely open or completely closed.

**Pneumatic actuating valves.** One characteristic of pneumatic controls is that they almost exclusively employ pneumatic actuating valves. A pneumatic actuating valve can provide a large power output. (Since a pneumatic actuator requires a large power input to produce a large power output, it is necessary that a sufficient quantity of pressurized air be available.) In practical pneumatic actuating valves, the valve characteristics may not be linear; i.e., the flow may not be directly proportional to the valve stem position, and also there may be other nonlinear effects, such as hysteresis.

Consider the schematic diagram of a pneumatic actuating valve shown in Fig. 5-17. Assume that the area of the diaphragm is $A$. Assume also that when the actuating error is zero, the control pressure is equal to $\bar{P}_c$ and the valve displacement is equal to $\bar{X}$.

Control pressure

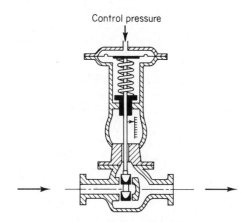

Fig. 5-17. Schematic diagram of a pneumatic actuating valve.

In the following analysis ,we shall consider small variations in the variables and linearize the pneumatic actuating valve. Let us define the small variation in the control pressure and the corresponding valve displacement to be $p_c$ and $x$, respectively. Since a small change in the pneumatic pressure force applied to the diaphragm repositions the load, consisting of the spring, viscous friction, and mass, the force balance equation becomes

$$Ap_c = m\ddot{x} + f\dot{x} + kx \qquad (5\text{-}9)$$

where

$m$ = mass of the valve and valve stem
$f$ = viscous-friction coefficient
$k$ = spring constant

If the force due to the mass and viscous friction are negligibly small, then Eq. (5-9) can be simplified to

$$Ap_c = kx$$

The transfer function between $x$ and $p_c$ thus becomes

$$\frac{X(s)}{P_c(s)} = \frac{A}{k} = K_c$$

where $X(s) = \mathscr{L}[x]$ and $P_c(s) = \mathscr{L}[p_c]$. If $q_i$, the change in flow through the pneumatic actuating valve, is proportional to $x$, the change in the valve-stem displacement, then

$$\frac{Q_i(s)}{X(s)} = K_q$$

where $Q_i(s) = \mathscr{L}[q_i]$ and $K_q$ is a constant. The transfer function between $q_i$ and $p_c$ becomes

$$\frac{Q_i(s)}{P_c(s)} = K_c K_q = K_v$$

where $K_v$ is a constant.

The standard control pressure for this kind of a pneumatic actuating valve is between 3 and 15 psig. The valve-stem displacement is limited by the allowable stroke of the diaphragm and is only a few inches. If a longer stroke is needed, a piston-spring combination may be employed.

In pneumatic actuating valves, the static-friction force must be limited to a low value so that excessive hysteresis does not result. Because of the compressibility of air, the control action may not be positive; i.e., an error may exist in the valve-stem position. The use of a valve positioner results in improvements in the performance of a pneumatic actuating valve.

A schematic diagram of a valve postioner is shown in Fig. 5-18. The principle of operation of this device is that if the valve position does not correspond to the

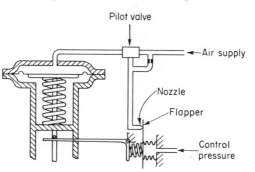

Fig. 5-18. Schematic diagram of a valve positioner.

control pressure, then the pilot valve will operate until the valve position corresponds exactly to the control pressure.

**Liquid-level control systems.** Consider the liquid-level control system shown in Fig. 5-19. It is desired to maintain the liquid level at a constant value, regardless of the change in the opening of the load valve.

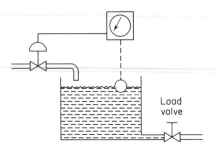

Let us assume that the controller is of the proportional type, as shown in Fig. 5-15(a), and that the control valve is that shown in Fig. 5-17. If the liquid level rises as a result of a change in the load valve, the float moves upward, causing the flapper to move into closer contact with the nozzle, increasing the nozzle back pressure. Since the relay is reverse acting, this will result in a reduction in the control pressure and will cause a decrease in the opening of the pneumatic actuating valve. This is in the proper direction to correct for the rising level.

**Fig. 5-19.** Liquid-level control system.

**Proportional control of a first-order system.** Consider the liquid-level control system shown in Fig. 5-20(a). [The controller is assumed to be the proportional controller shown in Fig. 5-15(a).] We assume that all the variables $r$, $q_i$, $h_1$, and $q_0$ are measured from their respective steady-state values $\bar{R}$, $\bar{Q}$, $\bar{H}$, and $\bar{Q}$. We also assume that the magnitudes of the variables $r$, $q_i$, $h_1$, and $q_0$ are sufficiently small so that the system can be approximated by a linear mathematical model, i.e., a transfer function.

Referring to Section 4-5, we can obtain the transfer function of the liquid-level system as

$$\frac{H_1(s)}{Q_i(s)} = \frac{R}{RCs + 1}$$

Since the controller is a proportional controller, the change in inflow $q_i$ is proportional to the actuating error $e$ so that $q_i = K_p K_v e$, where $K_p$ is the gain of the controller and $K_v$ is the gain of the control valve. In terms of Laplace-transformed quantities,

$$Q_i(s) = K_p K_v E(s)$$

A block diagram of this system is shown in Fig. 5-20(b). A simplified block diagram is given in Fig. 5-20(c), where $X(s) = (1/K_b)R(s)$, $K = K_p K_v RK_b$, and $T = RC$.

In what follows we shall investigate the response $h_1(t)$ to a change in the reference input. We shall assume a unit-step change in $x(t)$, where $x(t) = (1/K_b)r(t)$. The closed-loop transfer function between $H_1(s)$ and $X(s)$ is given by

$$\frac{H_1(s)}{X(s)} = \frac{K}{Ts + 1 + K} \tag{5-10}$$

Since the Laplace transform of the unit-step function is $1/s$, substituting $X(s) = 1/s$ into Eq. (5-10) gives

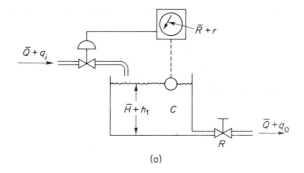

(a)

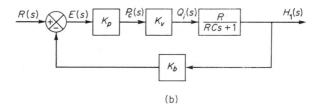

(b)

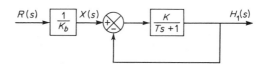

(c)

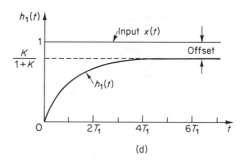

(d)

**Fig. 5-20.** (a) Liquid-level control system; (b) block diagram; (c) simplified block diagram; (d) curve $h_1(t)$ versus $t$.

$$H_1(s) = \frac{K}{Ts + 1 + K} \frac{1}{s}$$

Expanding $H_1(s)$ into partial fractions gives

$$H_1(s) = \frac{K}{1+K} \frac{1}{s} - \frac{TK}{1+K} \frac{1}{Ts + 1 + K} \tag{5-11}$$

Taking the inverse Laplace transforms of both sides of Eq. (5-11), we obtain the following time solution $h_1(t)$:

$$h_1(t) = \frac{K}{1+K}(1 - e^{-t/T_1}) \qquad (t \geq 0) \tag{5-12}$$

where

$$T_1 = \frac{T}{1+K}$$

The response curve $h_1(t)$ is plotted in Fig. 5-20(d). From Eq. (5-12), notice that the time constant $T_1$ of the closed-loop system is different from the time constant $T$ of the feedforward block.

From Eq. (5-12), we see that as $t$ approaches infinity, the value of $h_1(t)$ approaches $K/(1+K)$, or

$$h_1(\infty) = \frac{K}{1+K}$$

Since $x(\infty) = 1$, there is a steady-state error of $1/(1+K)$. Such an error is called *offset*. The value of the offset becomes smaller as the gain $K$ becomes larger.

Offset is a characteristic of the proportional control of a plant whose transfer function does not possess an integrating element. (Clearly, we need a nonzero error in order to provide a nonzero output.) To eliminate such offset, we must add integral control action. (Refer to Section 5-3.)

**Pneumatic proportional controllers (force-balance type).** Figure 5-21 shows a schematic diagram of a force-balance pneumatic proportional controller. Force-balance controllers are in extensive use in industry. Such controllers are sometimes called stack controllers. The basic principle of operation does not differ from that of the force-distance controller. The main advantage of the force-balance controller is that it eliminates many mechanical linkages and pivot joints, thereby reducing the effects of friction.

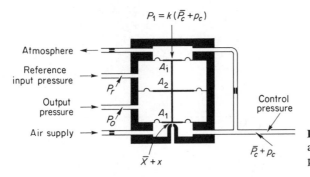

Fig. 5-21. Schematic diagram of a force-balance type pneumatic proportional controller.

In what follows, we shall consider the principle of the force-balance controller. In the controller shown in Fig. 5-21, the reference input pressure $P_r$ and the output pressure $P_0$ are fed to large diaphragm chambers. Note that a force-balance pneumatic controller operates only on pressure signals. Therefore, it is necessary to convert the reference input and system output to corresponding pressure signals.

As in the case of the force-distance controller, this controller employs a flapper, nozzle, and orifices. In Fig. 5-21, the drilled opening in the bottom chamber is the nozzle. The diaphragm just above the nozzle acts as a flapper.

The operation of the force-balance controller shown in Fig. 5-21 may be summarized as follows: 20 psig air from an air supply flows through an orifice, causing a reduced pressure in the bottom chamber. Air in this chamber escapes to the atmosphere through the nozzle. The flow through the nozzle depends upon the gap and the pressure drop across it. An increase in the reference input pressure $P_r$, while the output pressure $P_0$ remains the same, causes the valve stem to move down, decreasing the gap between the nozzle and the flapper diaphragm. This causes the control pressure $P_c$ to increase. Let

$$p_e = P_r - P_0$$

If $p_e = 0$, there is an equilibrium state with the nozzle-flapper distance equal to $\bar{X}$ and the control pressure equal to $\bar{P}_c$. At this equilibrium state, $P_1 = \bar{P}_c k$ (where $k < 1$) and

$$\bar{X} = \alpha(\bar{P}_c A_1 - \bar{P}_c k A_1) \tag{5-13}$$

where $\alpha$ is a constant.

Let us assume that $p_e \neq 0$ and define small variations in the nozzle-flapper distance and control pressure as $x$ and $p_c$, respectively. Then we obtain the following equation:

$$\bar{X} + x = \alpha[(\bar{P}_c + p_c)A_1 - (\bar{P}_c + p_c)kA_1 - p_e(A_2 - A_1)] \tag{5-14}$$

From Eqs. (5-13) and (5-14), we obtain

$$x = \alpha[p_c(1 - k)A_1 - p_e(A_2 - A_1)] \tag{5-15}$$

At this point, we must examine the quantity $x$. In the design of pneumatic controllers, the nozzle-flapper distance is quite small. In view of the fact that $x/\alpha$ is a higher-order term than $p_c(1 - k)A_1$ or $p_e(A_2 - A_1)$: that is, for $p_e \neq 0$,

$$\frac{x}{\alpha} \ll p_c(1 - k)A_1$$

$$\frac{x}{\alpha} \ll p_e(A_2 - A_1)$$

we may neglect the term $x$ in our analysis. Equation (5-15) can then be rewritten to reflect this assumption as follows:

$$p_c(1 - k)A_1 = p_e(A_2 - A_1)$$

and the transfer function between $p_c$ and $p_e$ becomes

$$\frac{P_c(s)}{P_e(s)} = \frac{A_2 - A_1}{A_1} \frac{1}{1 - k} = K_p$$

Thus, the controller shown in Fig. 5-21 is a proportional controller. The value of gain $K_p$ increases as $k$ approaches unity. Note that the value of $k$ depends upon the diameters of the orifices in the inlet and outlet pipes of the feedback chamber. (The value of $k$ approaches unity as the resistance to flow in the orifice of the inlet pipe is made smaller.)

**Hydraulic proportional controllers.** Except for low-pressure pneumatic controllers, compressed air has seldom been applied to the continuous control of the motion of devices having significant mass under external load forces. For such a case,

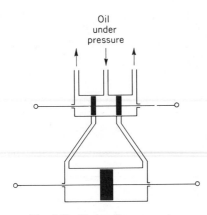

**Fig. 5-22.** Hydraulic servomotor.

hydraulic controllers are generally preferred. Hydraulic controllers are also used extensively in industry. With high-pressure hydraulic systems, very large force can be obtained. Rapid-acting accurate positioning of heavy loads is possible with hydraulic systems. A combination of electronic and hydraulic systems is widely used because it combines the advantages of both electronic control and hydraulic power.

A brief description of the operation of a hydraulic servomotor was given in Section 4-3. It was shown that for negligibly small load mass, the servomotor shown in Fig. 4-12 acts as an integrator or an integral controller. [Refer to Eq. (4-25).] Such a servomotor constitutes the basis of the hydraulic control circuit.

The servomotor shown in Fig. 5-22 acts as an integral controller. We can modify this servomotor to a proportional controller by including a feedback mechanism. Figure 5-23(a) is identical with Fig. 5-22, with the exception of the link attached to the left side of the power piston joining feedback link *ABC* at *C*. Link *AC* is a floating link rather than one moving about a fixed pivot. We shall see that the servomotor shown in Fig. 5-23(a) acts as a proportional controller.

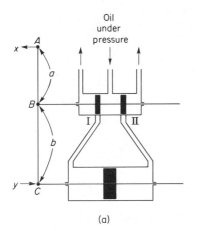

(a)

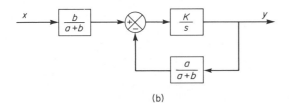

(b)

**Fig. 5-23.** (a) Servomotor which acts as a proportional controller; (b) block diagram of the servomotor.

The hydraulic system shown in Fig. 5-23(a) operates as follows: If the input $x$ moves the pilot piston to the left, this will uncover port I so that high-pressure oil flows through port I into the left side of the power piston and forces this piston to the right. The power piston, in moving to the right, will carry the feedback link $AC$ with it, thus moving the pilot piston to the right. This action continues until the pilot piston again covers ports I and II. A block diagram of the system can be drawn as in Fig. 5-23(b). The transfer function between $y$ and $x$ is given by

$$\frac{Y(s)}{X(s)} = \frac{\dfrac{b}{a+b}\dfrac{K}{s}}{1 + \dfrac{K}{s}\dfrac{a}{a+b}}$$

$$= \frac{bK}{s(a+b) + Ka}$$

Noting that under normal operation $|Ka/s(a+b)| \gg 1$, we obtain

$$\frac{Y(s)}{X(s)} = \frac{b}{a} = K_p$$

The transfer function between $y$ and $x$ becomes a constant, and thus the hydraulic system shown in Fig. 5-23(a) acts as a proportional controller, the gain of which is $K_p$. The gain $K_p$ can be adjusted by effectively changing the lever ratio, $b/a$.

We have thus seen that the addition of a feedback lever will cause the hydraulic servomotor to act as a proportional controller.

**Electronic proportional controllers.** An electronic proportional controller is an amplifier which receives a small voltage signal and produces a voltage output at a higher power level. A schematic diagram of such a controller is shown in Fig. 5-24. For this controller

$$e_0 = K\left(e_i - e_o\frac{R_2}{R_1}\right), \qquad K\frac{R_2}{R_1} \gg 1$$

Thus the transfer function $G(s)$ of this controller is

$$G(s) = \frac{E_0(s)}{E_i(s)} = \frac{R_1}{R_2} = K_p$$

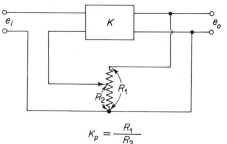

$$K_p = \frac{R_1}{R_2}$$

**Fig. 5-24.** Schematic diagram of an electronic proportional controller.

$K_p$ is the gain of the proportional controller. The gain $K_p$ can be adjusted by changing the ratio of resistances $(R_1/R_2)$ in the feedback circuit.

## 5-3 OBTAINING DERIVATIVE AND INTEGRAL CONTROL ACTION

In this section, we shall present methods for obtaining derivative and integral control action. We shall again place the emphasis on the principle and not on the details of the actual meachanisms.

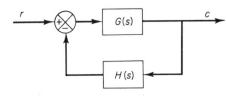

**Fig. 5-25.** Control system.

The basic principle for generating a desired control action is to insert the inverse of the desired transfer function in the feedback path. For the system shown in Fig. 5-25, the closed-loop transfer function is

$$\frac{C(s)}{R(s)} = \frac{G(s)}{1 + G(s)H(s)}$$

If $|G(s)H(s)| \gg 1$, then $C(s)/R(s)$ can be modified to

$$\frac{C(s)}{R(s)} = \frac{1}{H(s)}$$

Thus, if proportional-plus-derivative control action is desired, we insert an element having the transfer function $1/(Ts + 1)$ in the feedback path.

**Obtaining pneumatic proportional-plus-derivative control action.** Consider the pneumatic controller shown in Fig. 5-26 (a). This is a narrow-band proportional controller or a pneumatic two-position controller. Suppose that the zero actuating error $e = 0$ corresponds to the nozzle-flapper distance $\bar{X}$ and control pressure $\bar{P}_c$. If a small change in the actuating error from the zero position produces a small change in $x$ in the nozzle-flapper distance, then a small change $p_c$ is produced in the control pressure. A block diagram of the system (under the assumption of small variations) can be drawn, as shown in Fig. 5-26(b). The transfer function between $p_c$ and $e$ becomes

$$\frac{P_c(s)}{E(s)} = \frac{bK}{a + b} = K_p$$

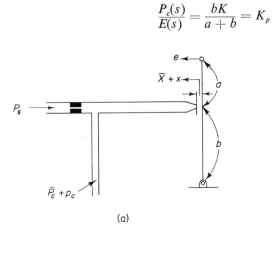

(a)

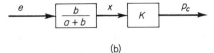

(b)

**Fig. 5-26.** (a) A pneumatic proportional controller; (b) block diagram of the controller.

We shall next show that the addition of delayed negative feedback to the controller shown in Fig. 5-26 (a) will modify the narrow-band proportional controller to a proportional-plus-derivative controller.

Consider the pneumatic controller shown in Fig. 5-27 (a). Assuming again small changes in the actuating error, nozzle-flapper distance, and control pressure, we can summarize the operation of this controller as follows: Let us first assume a small step change in $e$. Then the change in the control pressure $p_c$ will be instan-

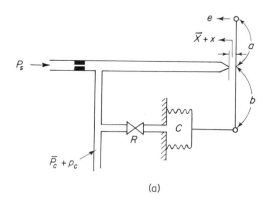

(a)

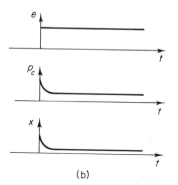

(b)

**Fig. 5-27.** (a) A pneumatic proportional-plus-derivative controller; (b) step change in $e$ and the corresponding changes in $p_c$ and $x$ plotted versus $t$; (c) block diagram of the controller.

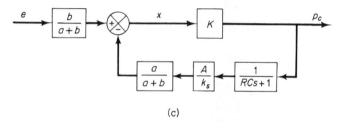

(c)

taneous. The restriction $R$ will momentarily prevent the feedback bellows from sensing the pressure change $p_c$. Thus the feedback bellows will not respond momentarily, and the pneumatic actuating valve will feel the full effect of the movement of the flapper. As time goes on the feedback bellows will expand or contract. The change in the nozzle-flapper distance $x$ and the change in the control pressure $p_c$ can be plotted against time $t$, as shown in Fig. 5-27 (b). At steady state, the feedback bellows acts like an ordinary feedback mechanism. The curve $p_c$ versus $t$ clearly shows that this controller is of the proportional-plus-derivative type.

A block diagram corresponding to this pneumatic controller is shown in Fig. 5-27 (c). In the block diagram, $K$ is a constant, $A$ is the area of the bellows, and $k_s$ is the equivalent spring constant of the bellows. The transfer function between $p_c$ and $e$ can be obtained from the block diagram as follows:

$$\frac{P_c(s)}{E(s)} = \frac{\dfrac{b}{a+b}K}{1 + \dfrac{Ka}{a+b}\dfrac{A}{k_s}\dfrac{1}{RCs+1}}$$

In such a controller the loop gain $|KaA/[(a+b)k_s(RCs+1)]|$ is normally very much greater than unity. Thus the transfer function $P_c(s)/E(s)$ can be simplified to give

$$\frac{P_c(s)}{E(s)} = K_p(1 + T_d s)$$

where

$$K_p = \frac{bk_s}{aA}, \qquad T_d = RC$$

Thus, delayed negative feedback, or the transfer function $1/(RCs+1)$ in the feedback path, modifies the proportional controller to a proportional-plus-derivative controller.

Note that if the feedback valve is fully opened, the control action becomes proportional. If the feedback valve is fully closed, the control action becomes narrow-band proportional (on-off).

**Obtaining pneumatic proportional-plus-integral control action.** Consider the proportional controller shown in Fig. 5-28 (a). Considering small changes in the variables, we can draw a block diagram of this controller as in Fig. 5-28 (b). We shall show that the addition of delayed positive feedback will modify this proportional controller to a proportional-plus-integral controller.

Consider the pneumatic controller shown in Fig. 5-29 (a). The operation of this controller is as follows: The bellows denoted by I is connected to the control pressure source without any restriction. The bellows denoted by II is connected to the control pressure source through a restriction. Let us assume a small step change in the actuating error. This will cause the back pressure in the nozzle to change instantaneously. Thus a change in the control pressure $p_c$ also occurs instantaneously. Due to the restriction of the valve in the path to bellows II, there will be a pressure drop across the valve. As time goes on, air will flow across the valve

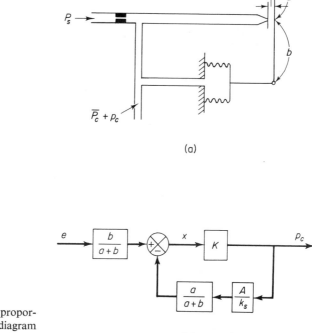

**Fig. 5-28.** (a) A pneumatic proportional controller; (b) block diagram of the controller.

(b)

in such a way that the change in pressure in bellows II attains the value $p_c$. Thus bellows II will expand or contract as time elapses in such a way as to move the flapper an additional amount in the direction of the original displacement $e$. This will cause the back pressure $p_e$ in the nozzle to change continuously, as shown in Fig. 5-29 (b).

Note that the integral control action in the controller takes the form of slowly canceling the feedback that the proportional control originally provided.

A block diagram of this controller under the assumption of small variations in the variables is shown in Fig. 5-29 (c). A simplification of this block diagram yields Fig. 5-29 (d). The transfer function of this controller is

$$\frac{P_c(s)}{E(s)} = \frac{\dfrac{b}{a+b}K}{1 + \dfrac{Ka}{a+b}\dfrac{A}{k_s}\left(1 - \dfrac{1}{RCs+1}\right)}$$

where $K$ is a constant, $A$ is the area of the bellows, and $k_s$ is the equivalent spring constant of the combined bellows. If $|\,KaARCs/[(a+b)k_s(RCs+1)]\,| \gg 1$, which is usually the case, the transfer function can be simplified to

$$\frac{P_c(s)}{E(s)} = K_p\left(1 + \frac{1}{T_i s}\right)$$

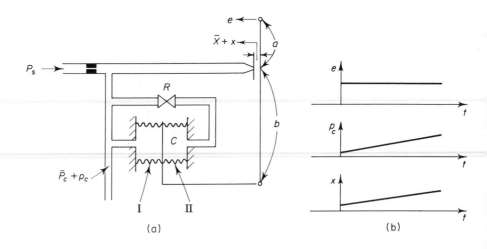

(a)                                                    (b)

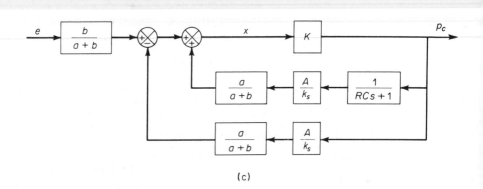

(c)

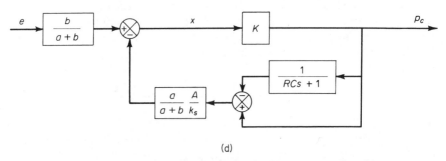

(d)

**Fig. 5-29.** (a) A pneumatic proportional-plus-integral controller; (b) step change in $e$ and the corresponding changes in $p_c$ and $x$ plotted versus $t$; (c) block diagram of the controller; (d) simplified block diagram.

where

$$K_p = \frac{bk_s}{aA}, \qquad T_i = RC$$

**Obtaining pneumatic proportional-plus-derivative-plus-integral control action.** A combination of the pneumatic control actions shown in the systems of Figs. 5-27 (a) and 5-29 (a) yields a proportional-plus-derivative-plus-integral control action. Figure 5-30 (a) shows a schematic diagram of such a controller. Figure 5-30 (b) shows a block diagram of this controller under the assumption of small variations in the variables.

The transfer function of this controller is

$$\frac{P_c(s)}{E(s)} = \frac{\dfrac{bK}{a+b}}{1 + \dfrac{Ka}{a+b}\dfrac{A}{k_s}\dfrac{(R_iC - R_dC)s}{(R_dCs + 1)(R_iCs + 1)}}$$

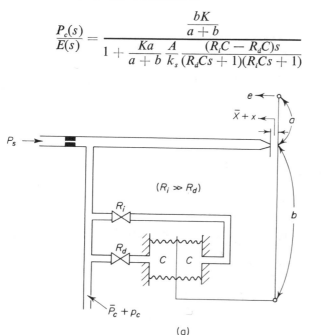

(a)

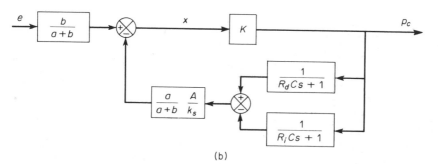

(b)

**Fig. 5-30.** (a) A pneumatic proportional-plus-derivative-plus-integral controller; (b) block diagram of the controller.

By defining

$$T_i = R_iC, \qquad T_d = R_dC$$

and noting that under normal operation $|KaA(T_i-T_d)s/[(a+b)k_s(T_ds+1)(T_is+1)]| \gg 1$ and $T_i \gg T_d$, we obtain

$$\frac{P_c(s)}{E(s)} \doteq \frac{bk_s}{aA}\frac{(T_ds + 1)(T_is + 1)}{(T_i - T_d)s}$$

$$\doteq \frac{bk_s}{aA}\frac{T_dT_is^2 + T_is + 1}{T_is}$$

$$= K_p\left(1 + T_ds + \frac{1}{T_is}\right) \tag{5-16}$$

where

$$K_p = \frac{bk_s}{aA}$$

Equation (5-16) indicates that the controller shown in Fig. 5-30 (a) is a proportional-plus-derivative-plus-integral controller.

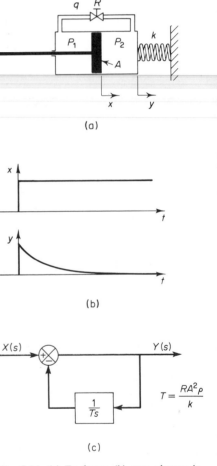

(a)

(b)

(c)

**Fig. 5-31.** (a) Dashpot; (b) step change in $x$ and the corresponding change in $y$ plotted versus $t$; (c) block diagram of the dashpot.

**Dashpots.** The dashpot shown in Fig. 5-31 (a) acts as a differentiating element. Suppose we introduce a step displacement to the piston position $x$. Then the displacement $y$ becomes equal to $x$ momentarily. Because of the spring force, however, the oil will flow through the resistance $R$ and the cylinder will come back to the original position. The curves $x$ versus $t$ and $y$ versus $t$ are shown in Fig. 5-31 (b).

Let us derive the transfer function between the displacement $y$ and displacement $x$. Define the pressures existing on the left and right sides of the piston as $P_1$ (lb/in.$^2$) and $P_2$ (lb/in.$^2$), respectively. Suppose that the inertia force involved is negligible. Then the force acting on the piston must balance the spring force. Thus

$$A(P_2 - P_1) = ky$$

where

$A$ = piston area, in.$^2$
$k$ = spring constant, 1b/in.

The flow rate $q$ is given by

$$q = \frac{P_2 - P_1}{R}$$

where

$q$ = flow rate through the restriction, lb/sec
$R$ = resistance to flow at the restriction, lb-sec/in.²-lb

Since the flow through the restriction during $dt$ seconds must equal the change in the mass of oil to the left of the piston during the same $dt$ seconds, we obtain

$$q\, dt = A\rho(dx - dy)$$

where

$\rho$ = density, lb/in.³

(We assume that the fluid is incompressible or $\rho$ = constant.) This last equation can be rewritten as

$$\frac{dx}{dt} - \frac{dy}{dt} = \frac{q}{A\rho} = \frac{P_2 - P_1}{RA\rho} = \frac{ky}{RA^2\rho}$$

or

$$\frac{dx}{dt} = \frac{dy}{dt} + \frac{ky}{RA^2\rho}$$

Taking the Laplace transforms of both sides of this last equation, assuming zero initial conditions, we obtain

$$sX(s) = sY(s) + \frac{k}{RA^2\rho}Y(s)$$

The transfer function of this system thus becomes

$$\frac{Y(s)}{X(s)} = \frac{s}{s + \dfrac{k}{RA^2\rho}}$$

Let us define $RA^2\rho/k = T$. Then

$$\frac{Y(s)}{X(s)} = \frac{Ts}{Ts + 1} = \frac{1}{1 + \dfrac{1}{Ts}}$$

Figure 5-31 (c) shows a block diagram representation for this system.

**Obtaining hydraulic proportional-plus-integral control action.** Figure 5-32 (a) shows a schematic diagram of a hydraulic proportional-plus-integral controller. A block diagram of this controller is shown in Fig. 5-32 (b). The transfer function $Y(s)/E(s)$ is given by

$$\frac{Y(s)}{E(s)} = \frac{\dfrac{b}{a+b}\dfrac{K}{s}}{1 + \dfrac{Ka}{a+b}\dfrac{T}{Ts+1}}$$

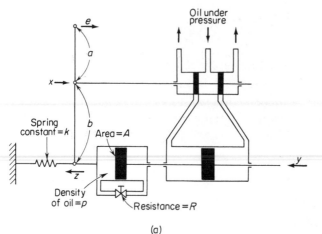

(a)

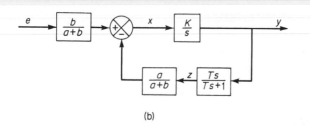

(b)

**Fig. 5-32.** (a) Schematic diagram of a hydraulic proportional-plus-integral controller; (b) block diagram of the controller.

In such a controller, under normal operation $|KaT/[(a + b)(Ts + 1)]| \gg 1$, with the result that

$$\frac{Y(s)}{E(s)} = K_p\left(1 + \frac{1}{T_i s}\right)$$

where

$$K_p = \frac{b}{a}, \qquad T_i = T = \frac{RA^2 p}{k}$$

Thus the controller shown in Fig. 5-32 (a) is a proportional-plus-integral controller.

**Obtaining derivative and integral control action in electronic controllers.** Figure 5-33 shows the principle of obtaining derivative and integral control action in electronic controllers. Essentially we insert an appropriate circuit in the feedback path to generate the desired control action. The transfer functions of the controllers may be obtained as follows: For the controller shown in Fig. 5-33 (a),

$$\frac{E_f(s)}{E_0(s)} = \frac{1}{R_d C_d s + 1}$$

$$[E_i(s) - E_f(s)]K = E_0(s)$$

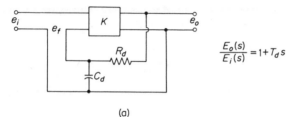

(a)

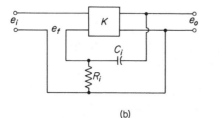

(b)

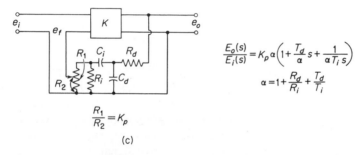

(c)

**Fig. 5-33.** Electronic controllers. (a) Proportional-plus-derivative controller; (b) proportional-plus-integral controller; (c) proportional-plus-derivative-plus-integral controller.

Hence, for $|K/(R_dC_ds + 1)| \gg 1$,

$$\frac{E_0(s)}{E_i(s)} = \frac{K(R_dC_ds + 1)}{R_dC_ds + 1 + K} \doteq R_dC_ds + 1 = T_ds + 1$$

where $T_d = R_dC_d$.

Similarly, for the controller shown in Fig. 5-33 (b),

$$\frac{E_f(s)}{E_0(s)} = \frac{R_iC_is}{R_iC_is + 1}$$

$$[E_i(s) - E_f(s)]K = E_0(s)$$

Hence, for $|KR_iC_is/(R_iC_is + 1)| \gg 1$,

$$\frac{E_0(s)}{E_i(s)} = \frac{K(R_iC_is + 1)}{KR_iC_is + R_iC_is + 1} \doteq \frac{R_iC_is + 1}{R_iC_is} = 1 + \frac{1}{T_is}$$

where $T_i = R_iC_i$.

For the controller shown in Fig. 5-33 (c), if the loop gain is very much greater than unity the transfer function may be derived as follows:

$$\frac{E_0(s)}{E_i(s)} = K_p\alpha\left(1 + \frac{T_d}{\alpha}s + \frac{1}{\alpha T_i s}\right)$$

where

$$\alpha = 1 + \frac{R_d}{R_i} + \frac{T_d}{T_i}$$

The derivation of this transfer function is given in Problem A-5-5.

## 5-4 EFFECTS OF INTEGRAL AND DERIVATIVE CONTROL ACTION ON SYSTEM PERFORMANCE

In this section, we shall investigate the effects of integral and derivative control action on the system performance, but we shall consider only simple systems. (In later chapters we shall study more about integral and derivative control.)

**Integral control action.** In the proportional control of a plant whose transfer function does not possess an integrator $1/s$, there is a steady-state error, or offset, in the response to a step input. Such an offset can be eliminated if the integral control action is included in the controller.

In the integral control of a plant, the control signal, the output signal from the controller, at any instant is the area under the actuating error signal curve up to that instant. The control signal $m(t)$ can have a nonzero value when the actuating error signal $e(t)$ is zero, as shown in Fig. 5-34 (a). This is impossible in the case of the proportional controller since a nonzero control signal requires a nonzero actuating error signal. (A nonzero actuating error signal at steady state means that there is an offset.) Figure 5-34(b) shows the curve $e(t)$ versus $t$ and the corresponding curve $m(t)$ versus $t$ when the controller is of the proportional type.

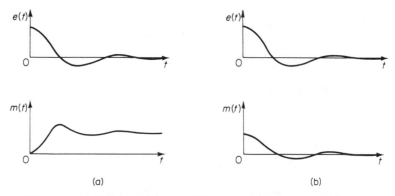

**Fig. 5-34.** (a) Plots of $e(t)$ and $m(t)$ curves showing nonzero control signal when the actuating error signal is zero; (b) plots of $e(t)$ and $m(t)$ curves showing zero control signal when the actuating error signal is zero.

Note that integral control action, while removing offset or steady-state error, may lead to oscillatory response of slowly decreasing amplitude or even increasing amplitude, both of which are usually undesirable. (For details, see Chapter 6.)

**Integral control of liquid-level control systems.** In Section 5-2, we found that the proportional control of a liquid-level system will result in a steady-state error with a step input. We shall now show that such an error can be eliminated if integral control action is included in the controller.

Figure 5-35 (a) shows a liquid-level control system. We assume that the controller is an integral controller. We also assume that the variables $x$, $q_i$, $h$, and $q_0$, which are measured from their respective steady-state values $\bar{X}$, $\bar{Q}$, $\bar{H}$, and $\bar{Q}$, are small quantities so that the system can be considered linear. Under these assumptions, the block diagram of the system can be obtained as shown in Fig. 5-35 (b). From Fig. 5-35 (b), the closed-loop transfer function between $H(s)$ and $X(s)$ is

$$\frac{H(s)}{X(s)} = \frac{KR}{RCs^2 + s + KR}$$

Hence

$$\frac{E(s)}{X(s)} = \frac{X(s) - H(s)}{X(s)}$$

$$= \frac{RCs^2 + s}{RCs^2 + s + KR}$$

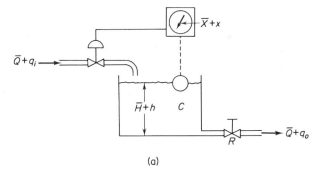

(a)

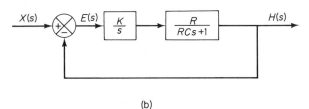

**Fig. 5-35.** (a) Liquid-level control system; (b) block diagram of the system.

(b)

Since the system is a stable one, the steady-state error for the unit-step response is obtained by applying the final value theorem as follows:

$$e_{ss} = \lim_{s \to 0} s E(s)$$

$$= \lim_{s \to 0} \frac{s(RCs^2 + s)}{RCs^2 + s + KR} \frac{1}{s}$$

$$= 0$$

Integral control of the liquid-level system thus eliminates the steady-state error in the response to the step input. This is an important improvement over the proportional control alone which gives offset.

**Response to torque disturbances (proportional control).** Let us investigate the effect of a torque disturbance occurring at the load element. Consider the system shown in Fig. 5-36. The proportional controller delivers torque $T$ to position the load element, which consists of moment of inertia and viscous friction. Torque disturbance is denoted by $N$.

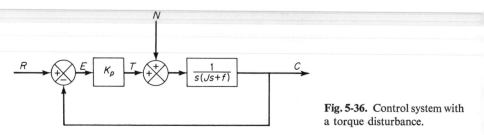

**Fig. 5-36.** Control system with a torque disturbance.

Assuming that the reference input is zero or $R(s) = 0$, the transfer function between $C(s)$ and $N(s)$ is given by

$$\frac{C(s)}{N(s)} = \frac{1}{Js^2 + fs + K_p}$$

Hence

$$\frac{E(s)}{N(s)} = -\frac{C(s)}{N(s)} = -\frac{1}{Js^2 + fs + K_p}$$

The steady-state error due to a step disturbance torque of magnitude $T_n$ is given by

$$e_{ss} = \lim_{s \to 0} s E(s)$$

$$= \lim_{s \to 0} \frac{-s}{Js^2 + fs + K_p} \frac{T_n}{s}$$

$$= -\frac{T_n}{K_p}$$

At steady state, the proportional controller provides the torque $-T_n$, which is equal in magnitude but opposite in sign to the disturbance torque $T_n$. The steady-state output due to the step disturbance torque is

$$c_{ss} = -e_{ss} = \frac{T_n}{K_p}$$

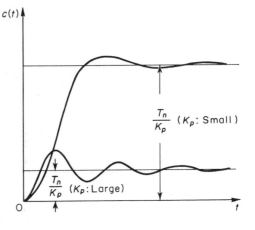

**Fig. 5-37.** Typical response curves to a step torque disturbance.

The steady-state error can be reduced by increasing the value of the gain $K_p$. Increasing this value, however, will cause the system response to be more oscillatory. Typical response curves for a small value of $K_p$ and a large value of $K_p$ are shown in Fig. 5-37.

Since the value of the gain $K_p$ cannot be increased too much, it is desirable to modify the proportional controller to a proportional-plus-integral controller.

**Response to torque disturbances (proportional-plus-integral control).** In order to eliminate offset due to torque disturbance, the proportional controller may be replaced by a proportional-plus-integral controller.

If integral control action is added to the controller, then as long as there is an error signal, there is a torque developed by the controller to reduce this error, provided the control system is a stable one.

Figure 5-38 shows the proportional-plus-integral control of the load element, consisting of moment of inertia and viscous friction.

The closed-loop transfer function between $C(s)$ and $N(s)$ is

$$\frac{C(s)}{N(s)} = \frac{s}{Js^3 + fs^2 + K_ps + \dfrac{K_p}{T_i}}$$

In the absence of the reference input, or $r(t) = 0$, the error signal is obtained from

$$E(s) = -\frac{s}{Js^3 + fs^2 + K_ps + \dfrac{K_p}{T_i}} N(s)$$

**Fig. 5-38.** Proportional-plus-integral control of a load element consisting of moment of inertia and viscous friction.

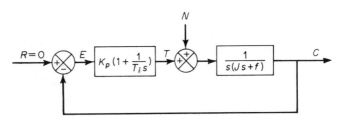

If this control system is a stable one, i.e., if the roots of the characteristic equation

$$Js^3 + fs^2 + K_ps + \frac{K_p}{T_i} = 0$$

have negative real parts, then the steady-state error in the response to a step disturbance torque of magnitude $T_n$ is obtained by applying the final value theorem as follows:

$$e_{ss} = \lim_{s \to 0} sE(s)$$

$$= \lim_{s \to 0} \frac{-s^2}{Js^3 + fs^2 + K_ps + \frac{K_p}{T_i}} \frac{T_n}{s}$$

$$= 0$$

Thus steady-state error to the torque disturbance can be eliminated if the controller is of the proportional-plus-integral type.

Note that the integral control action added to the proportional controller has converted the originally second-order system to a third-order one. Hence the control system may become unstable for a large value of $K_p$ since the roots of the characteristic equation may have positive real parts. (The second-order system is always stable if the coefficients in the system differential equation are all positive.)

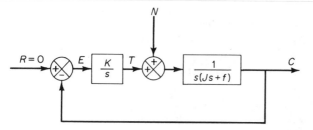

Fig. 5-39. Integral control of a load element consisting of moment of inertia and viscous friction.

It is important to point out that if the controller were an integral controller, as in Fig. 5-39, then the system always becomes unstable because the characteristic equation

$$Js^3 + fs^2 + K = 0$$

will have roots with positive real parts. Such an unstable system cannot be used in practice.

Note that in the system of Fig. 5-38, the proportional control action tends to stabilize the system, while the integral control action tends to eliminate or reduce steady-state error in response to various inputs.

**Derivative control action.** Derivative control action, when added to a proportional controller, provides a means of obtaining a controller with high sensitivity. An advantage of using derivative control action is that it responds to the rate of change of the actuating error and can produce a significant correction before the magnitude of the actuating error becomes too large. Derivative control thus anticipates the actuating error, initiates an early corrective action, and tends to increase the stability of the system.

Although derivative control does not affect the steady-state error directly, it adds damping to the system and thus permits the use of a larger value of the gain $K$, which will result in an improvement in the steady-state accuracy.

Because derivative control operates on the rate of change of the actuating error and not the actuating error itself, this mode is never used alone. It is always used in combination with proportional or proportional-plus-integral action.

**Proportional control of systems with inertia load.** Before we discuss the effect of derivative action on system performance, we shall consider the proportional control of an inertia load.

Consider the system shown in Fig. 5-40 (a). The closed-loop transfer function is obtained as

$$\frac{C(s)}{R(s)} = \frac{K_p}{Js^2 + K_p}$$

Since the roots of the characteristic equation

$$Js^2 + K_p = 0$$

are imaginary, the response to a unit-step input continues to oscillate indefinitely, as shown in Fig. 5-40 (b).

Control systems exhibiting such response characteristics are not desirable. We shall see that the addition of derivative control will stabilize the system.

**Proportional-plus-derivative control of a system with inertia load.** Let us modify the proportional controller to a proportional-plus-derivative controller whose transfer function is $K_p(1 + T_d s)$. The torque developed by the controller is proportional to $K_p(e + T_d \dot{e})$. Derivative control is essentially anticipatory, measures the instantaneous error velocity, and predicts the large overshoot ahead of time and produces an appropriate counteraction before too large an overshoot occurs.

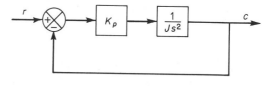

(a)

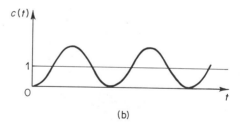

**Fig. 5-40.** (a) Proportional control of a system with inertia load; (b) response to a unit step-input.

(b)

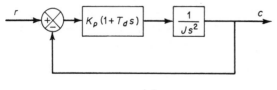

(a)

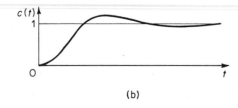

(b)

**Fig. 5-41.** (a) Proportional-plus-derivative control of a system with inertia load; (b) response to a unit step-input.

Consider the system shown in Fig. 5-41 (a). The closed-loop transfer function is given by

$$\frac{C(s)}{R(s)} = \frac{K_p(1 + T_d s)}{Js^2 + K_p T_d s + K_p}$$

The characteristic equation

$$Js^2 + K_p T_d s + K_p = 0$$

now has two roots with negative real parts for positive values of $J$, $K_p$, and $T_d$. Thus derivative control introduces a damping effect. A typical response curve $c(t)$ to a unit-step input is shown in Fig. 5-41 (b). Clearly, the response curve shows a marked improvement over the original response curve shown in Fig. 5-40 (b).

## 5-5 REDUCTION OF PARAMETER VARIATIONS BY USE OF FEEDBACK

The primary purpose of using feedback in control systems is to reduce the sensitivity of the system to parameter variations and unwanted disturbances.

If we are to construct a suitable open-loop control system, we must select all the components of the open-loop transfer function $G(s)$ very carefully so that they respond accurately. In the case of constructing a closed-loop control system, however, the components can be less accurate since the sensitivity to parameter variations in $G(s)$ is reduced by a factor of $1 + G(s)$.

To illustrate this, consider the open-loop and the closed-loop system shown in Figs. 5-42 (a) and (b), respectively. Suppose that, due to parameter variations, $G(s)$ is changed to $G(s) + \Delta G(s)$, where $|G(s)| \gg |\Delta G(s)|$. Then, in the open-loop system shown in Fig. 5-42 (a), the output is given by

$$C(s) + \Delta C(s) = [G(s) + \Delta G(s)]R(s)$$

Hence the change in the output is given by

$$\Delta C(s) = \Delta G(s)R(s)$$

In the closed-loop system shown in Fig. 5-42 (b),

$$C(s) + \Delta C(s) = \frac{G(s) + \Delta G(s)}{1 + G(s) + \Delta G(s)}R(s)$$

or

$$\Delta C(s) \doteq \frac{\Delta G(s)}{1 + G(s)}R(s)$$

Thus, the change in the output of the closed-loop system, due to the parameter variations in $G(s)$, is reduced by a factor of $1 + G(s)$. In many practical cases, the magnitude of $1 + G(s)$ is generally much greater than one.

Note that in reducing the effects of the parameter variations of the components, we very often bridge the offending component with a feedback loop.

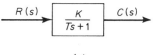

(a)

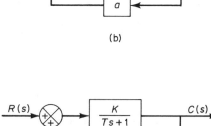

(b)

(a)

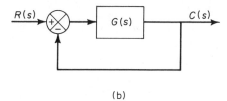

(b)

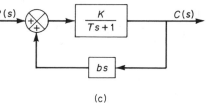

(c)

**Fig. 5-42.** (a) Open-loop system; (b) closed-loop system.

**Fig. 5-43.** (a) Open-loop system; (b) closed-loop system with time constant $T/(1 + Ka)$; (c) closed-loop system with time constant $T - bK$.

**Changing time constants by use of feedback.** Consider the system shown in Fig. 5-43 (a). The time constant of the system is $T$. The addition of a negative feedback loop around this element reduces the time constant. Figure 5-43 (b) shows the system with the same feedforward transfer function as that shown in Fig. 5-43 (a), with the exception that a negative feedback loop has been added. The time constant of this system has been reduced to $T/(1 + Ka)$. Note also that the gain constant for this system has also been reduced from $K$ to $K/(1 + Ka)$.

If, instead of a negative feedback loop, a positive feedback loop is added around the transfer function $K/(Ts + 1)$ and if the feedback transfer function is properly chosen, then the time constant can be made zero or a very small value. Consider the system shown in Fig. 5-43 (c). Since the closed-loop transfer function is

$$\frac{C(s)}{R(s)} = \frac{K}{(T - bK)s + 1}$$

the time constant can be reduced by properly choosing the value of $b$. If $b$ is set equal to $T/K$, then the time constant becomes zero. Note, however, that if disturbances cause $T - bK$ to be negative instead of zero, the system becomes unstable. Hence if positive feedback is employed to reduce the time constant to a small value, we must be very careful so that $T - bK$ never becomes negative.

**Increasing loop gains by use of positive feedback.** The system shown in Fig. 5-44 (a) has the transfer function $C(s)/R(s) = G(s)$. Consider now the system shown in Fig. 5-44(b). The closed-loop transfer function for this system is

(a)

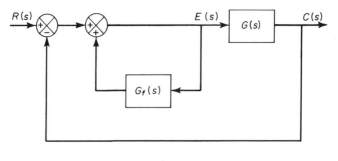

(b)

**Fig. 5-44.** (a) Open-loop system; (b) closed-loop system whose transfer function is nearly unity.

$$\frac{C(s)}{R(s)} = \frac{G(s)}{1 - G_f(s) + G(s)}$$

If $G_f(s)$ is chosen nearly unity, or $G_f(s) \doteq 1$, then

$$\frac{C(s)}{R(s)} \doteq 1$$

Essentially, this means that the inner loop, using positive feedback, has increased the feedforward gain to a very large value. As we stated earlier, when the loop gain is very large, the closed-loop transfer function $C(s)/R(s)$ becomes equal to the inverse of the transfer function of the feedback element. Since the system shown in Fig. 5-44 (b) has unity feedback, $C(s)/R(s)$ becomes almost equal to unity. [Thus $C(s)/R(s)$ is not sensitive to the parameter variations in $G(s)$.]

**Elimination of integration.** Addition of a minor loop around an integrator modifies it to a first-order delay element. Consider the system shown in Fig. 5-45 (a). Negative feedback of the output, as shown in Fig. 5-45 (b), modifies the integrator $K/s$ to a first-order delay element $K/(s + K)$.

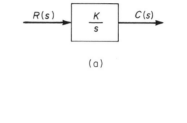

(a)

**Comments on the use of feedback loops.** As we have seen in the previous discussion, feedback control, or closed-loop control, reduces the sensitivity of a system to parameter variations and therefore decreases the effects of gain variations in the feedforward path in response to variations of supply pressure, supply voltage, temperature, etc. In the study of controllers

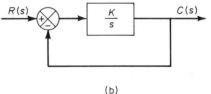

(b)

Fig. 5-45. (a) Integrating element; (b) first-order delay element.

made in Sections 5-2 and 5-3, we have also seen that the elements which perform the various control actions are in the feedback part of the controller mechanisms and that the feedback elements in a controller essentially increase the linearity of the amplifier and increase the range of the proportional sensitivity.

The use of feedback loops in control systems, however, will increase the number of components of the systems, will thereby increase the complexity, and also will introduce the possibility of instability.

## 5-6 FLUIDICS

**Introduction.** Fluid devices through which air, gases, or liquids flow in intricate and precise channels are called *fluidic devices*. Such devices use solid components, or circuits, to perform sensing, logic, amplification, and control functions. Fluidics, the general study of fluidic devices and systems, is one of the newest and most interesting fields to be developed in recent years. The advantages of fluidic devices are that they are quite rugged (because of the absence of mechanical moving parts) and can perform control and computational functions in adverse environments. Fluidic devices can withstand wide temperature changes, extremely high temperature, shock, vibration, G-force (acceleration), and radiation. (Thus fluidic devices are ideal for operation in hazardous locations where other devices fail.) In addition, the simplicity of basic fluidic devices assures high reliability, long life, and very low maintenance.

In certain applications of fluidic devices, the controlling medium is also the flowing or power medium. In such a case, power consumption of fluidic devices is not a problem. In applications where available power is limited, however, such as in a satellite system, relatively high power consumption of fluidic devices becomes a problem. By reducing jet sizes and lowering operating pressures, power consumption can be reduced to a certain extent; however, the minimum values of jet sizes and operating pressures are determined by signal noises and manufacturing reproducibilities.

As far as the speed of response of fluidic devices is concerned, they are comparable to conventional pneumatic and hydraulic devices or electromechanical relays. [This means that the speed of response of fluidic devices is very much slower than that of electronic devices. Electronic responses are generally expressed in terms of microseconds ($10^{-6}$ sec) or nanoseconds ($10^{-9}$ sec), while fluidic responses are expressed in terms of milliseconds ($10^{-3}$ sec).]

In this section we shall first discuss those fluidic devices called wall attachment amplifiers, and then we shall present digital fluidics, one of the most promising applications of fluidics. Finally, we shall discuss an application of digital fluidics.

**Wall attachment phenomena.** The operating principle of fluidic devices is known as jet interaction or wall attachement phenomenon. A fluid jet can be diverted from its normal path of flow by introducing another jet perpendicular to the first, as shown in Fig. 5-46. If the fluid jet enters a relatively narrow chamber and touches one wall, it attaches itself to the wall, as shown in Fig. 5-47. This phenomenon is called the "wall effect." It is possible to break such wall effects if we apply a stream or jet to the low-pressure region below the point where the jet touches the wall. This fact makes it possible to design a bistable device, or flip-flop, by providing control jets on either side of the main jet, as shown in Fig. 5-48. (A bistable device is one which has two output possibilities and which will alternate from one output to another upon receipt of correctly phased input signals.) Such a bistable device is suitable for logic operations using binary signals.

Fluidic amplifiers based on wall attachment phenomena are called wall attachment amplifiers. These amplifiers are usually made of glass, plastic, aluminum, brass, or stainless steel.

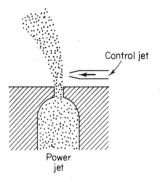

**Fig. 5-46.** Jet interaction in a fluidic device.

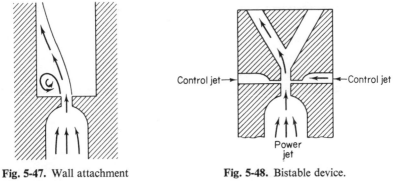

**Fig. 5-47.** Wall attachment phenomenon.

**Fig. 5-48.** Bistable device.

**Bistable fluidic amplifiers (flip-flops).** Wall attachment amplifiers are basically bistable devices. The fluid will continue to flow in one channel until acted on by a signal (a jet of fluid) which directs the main stream of fluid toward another channel wall.

Figure 5-49 shows schematic diagrams of a bistable amplifier or flip-flop. In this element, the output may be switched from one output channel to the other by applying air pressure at the opposite control port. Output will continue through this channel, even after the control signal has been removed. The element continues to operate from a single output channel until a change signal is issued. Thus this device has a memory. Flip-flop elements serve as the memory bank in a fluidic system.

**Fig. 5-49.** Schematic diagrams of a bistable amplifier. (a) Control signal applied from left; (b) control signal removed; (c) control signal applied from right.

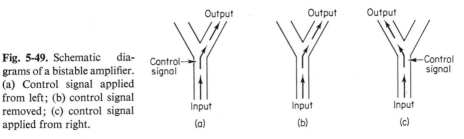

In fluidic devices, a percentile representation of output capture as related to supply, such as output pressure versus input pressure, is commonly used to indicate the degree of recovery. For the flip-flop element considered here, the maximum output pressure is about 35% of the supply pressure, while the maximum flow out is about 50% of the supply flow. The remaining 50% of the supply flow will exit through the exhaust vents. The minimum control pressure to cause switching is about 10% of the supply pressure.

**Digital amplifiers.** A digital amplifier is similar to a flip-flop element except that the former has no memory. Flow from an output channel will continue only as long as there is a control signal present. A signal applied at either control port will produce an output at the opposite output channel.

Operating independently of other fluidic elements, the digital amplifier may be used for simple functions such as operating cylinders, air relays, or other logic units.

In a digital amplifier, control pressure that equals 10% of the supply pressure will change and maintain the output flow. The maximum output pressure is about 50% of the supply pressure, and the maximum flow out is about 70% of the supply flow. (The remaining 30% of supply flow will exit through the exhaust vents.)

**Proportional amplifiers.** Although wall attachment amplifiers are basically bistable devices, they can be changed to proportional devices by widening the passages following the nozzle where wall attachment occurs, as shown in Fig. 5-50. In this amplifier, the main jet flow is distributed between the two outlet passages according to the balance of control streams.

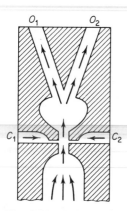

The proportional amplifier is a differential input-output device. Referring to Fig. 5-50, we see that the input signal is the differential pressure existing across ports $C_1$ and $C_2$. The output is the differential pressure existing across channels $O_1$ and $O_2$.

In this device the jet is deflected by the input signal pressure difference which modulates the fraction of the jet flow that each output channel receives. Figure 5-51 shows the characteristic curves of a wall attachment proportional amplifier.

**Fig. 5-50.** Wall attachment proportional amplifier.

To give some idea of the gains of such amplifiers, we note that pressure gains of 10 and power gains of 100 per

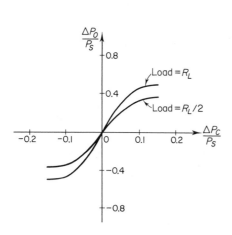

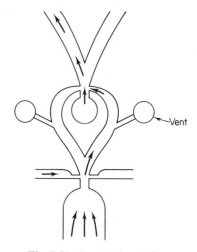

**Fig. 5-51.** Characteristic curves of a wall attachment proportional amplifier. ($P_s =$ supply pressure, $\Delta P_c =$ differential pressure existing across control ports, $\Delta P_0 =$ differential pressure existing across output channels, and $R_L =$ load resistance.)

**Fig. 5-52.** Cascaded amplifier.

stage may be obtained. If two such devices are cascaded, as shown in Fig. 5-52, then the gains are multiplied.

As in the case of electrical and mechanical systems, the total opposition to circuit flow, represented by resistance, capacitance, and inductance combined in a resultant, is called *impedance*. The number of fluidic elements which can be operated in parallel from a single similar element is represented by a fan-out capability. ("Similar" refers to the impedance and does not refer to devices performing the same function.) The fan-out capability of wall attachment amplifiers is about four.

**Digital fluidics.** Digital fluidic devices are those fluidic components which perform logic functions, such as FLIP-FLOP, OR/NOT, and AND.

Fluidic logic gates can be built into familiar digital circuits. (A gate is a device

**Table 5-1.** FLUIDIC LOGIC GATES AND THEIR RESPECTIVE TRUTH TABLES

| Truth table | | Fluidic device | Symbol |
|---|---|---|---|
| Present state    Next state | | FLIP-FLOP | |
| $C_1$ $C_2$ $O_1$ $O_2$    $O_1$ $O_2$<br>0   0   0   1     0   1<br>0   0   1   0     1   0<br>0   1   0   1     1   0<br>0   1   1   0     1   0<br>1   0   0   1     0   1<br>1   0   1   0     0   1 | | | |
| | | OR/NOT | |
| $C_1$   $C_2$   $C_3$   $O_1$   $O_2$<br>0    0    0    1    0<br>All other     0    1<br>states | | | |
| | | AND | |
| $C_1$   $C_2$   $O_1$   $O_2$<br>0    0    0    1<br>0    1    0    1<br>1    0    0    1<br>1    1    1    0 | | | |

or circuit which allows passage of a signal only if certain control requirements have been satisfied.) Table 5-1 shows a few fluidic logic gates and their respective truth tables. (A truth table is a tabular correlation of input and output relationships for logic elements.) As seen from Table 5-1, digital fluidic devices can gate, or inhibit, signal transmission by the application, removal, or other combinations of input signals.

Digital fluidics perform the same logic functions as their electronic counterpart. There are, however, some areas where each has distinct advantages. (Usually, the choice is clear.) For example, in applications where reliability in extreme environments (e.g., high temperature or radiation) is more important than speed of operations, the digital fluidics should be chosen. If high-speed logic operation in normal environments is required, electronic devices are preferable. Applications of digital fluidics are most often found in automatic warehousing, machine feeding, sequencing, handling, etc.

Many functions of control relays can be performed by fluidic devices, and many logic functions of relays and electronic systems can be implemented by fluidic devices.

**A typical fluidic application\*** Fluidic components can sense positions, provide operating commands, perform logic interlocking, and control the valving of a machine's control system.

Consider the mechanical system shown in Fig. 5-53. It is desired to design a control logic system to sequence a drill head so that when the head is retracted and a pushbutton is actuated, the drill will advance downward. The head drills in the downward position and then returns to the up position to complete the cycle. For emergency return, a pushbutton is required to retract the head at any time.

We shall present a solution to this problem using digital fluidics. Figure 5-54 shows a logic circuit which can perform the

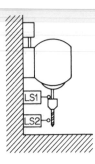

**Fig. 5-53.** Mechanical system.

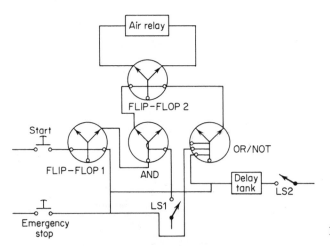

**Fig. 5-54.** Logic circuit.

\*This example is taken from "Fluidics-Plugging the Control Capability Gap" by The General Purpose Control Department, General Electric Company, Bloomington, Illinois.

desired sequence of operations. FLIP-FLOP element 1 is set by the start button. The start button is released, and the element will continue to give the set output. The output of FLIP-FLOP element 1 is fed to AND element, which combines this with the limit switch (LS 1) signal. This signal sets FLIP-FLOP 2, which shifts the air relay and starts the head moving down. The drill moves down and drills until the signal from LS 2 fills the delay tank. This shifts OR/NOT element, which in turn switches FLIP-FLOP 2, which shifts the air relay back to its original state. The drill head now moves up. The cycle will repeat when the start button is pushed and LS 1 is activated.

If the emergency stop button is pushed at any time, it will shift the OR/NOT and FLIP-FLOP 1. The OR/NOT will shift FLIP-FLOP 2 and immediately cause the drill head to retract.

## EXAMPLE PROBLEMS AND SOLUTIONS

**PROBLEM A-5-1.** The term commonly used to define the gain or sensitivity of a proportional controller is the *proportional band*. This is the percentage change in the input to the controller (error signal) required to cause 100% change in the output of the actuator. Thus small proportional band corresponds to high gain or high proportional sensitivity.

What is the proportional band if the controller and actuator have an overall gain of 4%/%? (Note that the total changes in the input to the controller and the output of the actuator are given as 100%. Thus a gain of 4%/% means that there is a change of 4% in the output if the change in input is 1%.)

**Solution**

$$\text{Proportional band} = \frac{100\%}{\text{gain in }\%/\%} = \frac{100\%}{4\%/\%} = 25\%$$

**PROBLEM A-5-2.** Consider the liquid-level control system shown in Fig. 5-55. Assume that the set point of the controller is fixed. Assuming a step disturbance of magnitude $n_0$, determine the error. Assume that $n_0$ is small and the variations in the variables from their respective steady-state values are also small. The controller is a proportional one.

If the controller is not a proportional one, but integral, what is the steady-state error?

**Solution.** Figure 5-56 is a block diagram of the system when the controller is proportional with gain $K_p$. (We assume the transfer function of the pneumatic valve to be unity.) Since

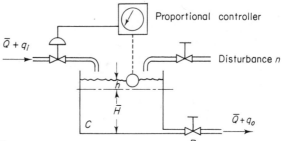

**Fig. 5-55.** Liquid-level control system.

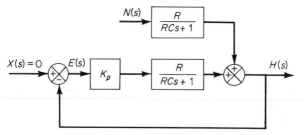

**Fig. 5-56.** Block diagram of the liquid-level control system shown in Fig. 5-55.

the set point is fixed, the variation in the set point is zero, or $X(s) = 0$. The Laplace transform of $h(t)$ is

$$H(s) = \frac{K_p R}{RCs + 1} E(s) + \frac{R}{RCs + 1} N(s)$$

Then

$$E(s) = -H(s) = -\frac{K_p R}{RCs + 1} E(s) - \frac{R}{RCs + 1} N(s)$$

Hence

$$E(s) = -\frac{R}{RCs + 1 + K_p R} N(s)$$

Since

$$N(s) = \frac{n_0}{s}$$

we obtain

$$E(s) = -\frac{R}{RCs + 1 + K_p R} \frac{n_0}{s}$$

$$= \frac{Rn_0}{1 + K_p R} \left( \frac{1}{s + \dfrac{1 + K_p R}{RC}} \right) - \frac{Rn_0}{1 + K_p R} \frac{1}{s}$$

The time solution for $t > 0$ is

$$e(t) = \frac{Rn_0}{1 + K_p R} \left[ \exp\left( -\frac{1 + K_p R}{RC} t \right) - 1 \right]$$

Thus, the time constant is $RC/(1 + K_p R)$. (In the absence of the controller, the time constant is equal to $RC$.) As the gain of the controller is increased, the time constant is decreased. The steady-state error is

$$e(\infty) = -\frac{Rn_0}{1 + K_p R}$$

As the gain $K_p$ of the controller is increased, the steady-state error, or offset, is reduced. Thus, mathematically, the larger the gain $K_p$ is, the smaller the offset and time constant are. In practical systems, however, if the gain $K_p$ of the proportional controller is increased to a very large value, oscillation may result in the output since in our analysis all the small lags and small time constants which may exist in the actual control system are neglected. (If these small lags and time constants are included in the analysis, the transfer function becomes higher order and for very large values of $K_p$ the possibility of oscillation or even instability may occur.)

If the controller is an integral one, then assuming the transfer function of the controller to be

$$G_c = \frac{K}{s}$$

we obtain

$$E(s) = -\frac{Rs}{RCs^2 + s + KR}N(s)$$

The steady-state error for a step disturbance $N(s) = n_0/s$ is

$$e(\infty) = \lim_{s \to 0} sE(s)$$

$$= \lim_{s \to 0} \frac{-Rs^2}{RCs^2 + s + KR}\frac{n_0}{s}$$

$$= 0$$

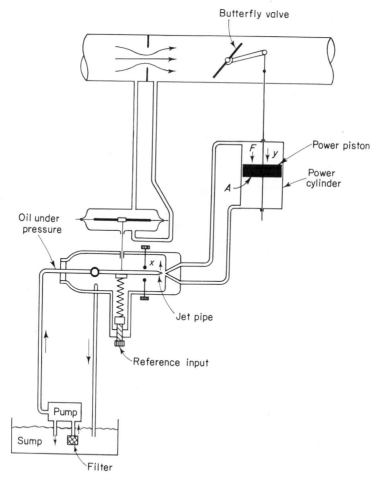

**Fig. 5-57.** Schematic diagram of a flow control system using a hydraulic jet pipe controller.

Thus, an integral controller eliminates steady-state error or offset due to the step disturbance. (The value of $K$ must be chosen so that the transient response due to the command input and/or disturbance damps out with a reasonable speed. See Chapter 6 for transient-response analysis.)

**PROBLEM A-5-3.** Figure 5-57 shows a hydraulic jet pipe controller applied to a flow control system. The jet pipe controller governs the position of the butterfly valve. Discuss the operation of this system. Plot a possible curve relating the displacement $x$ of the nozzle to the total force $F$ acting on the power piston.

**Solution.** The operation of this system is as follows: The flow rate is measured by the orifice, and the pressure difference produced by this orifice is transmitted to the diaphragm of the pressure-measuring device. The diaphragm is connected to the free swinging nozzle, or jet pipe, through a linkage. High-pressure oil ejects from the nozzle all the time. When the nozzle is at a neutral position, no oil flows through either of the pipes to move the power piston. If the nozzle is displaced by the motion of the balance arm to one side, the high-pressure oil flows through the corresponding pipe, and the oil in the power cylinder flows back to the sump through the other pipe.

Assume that the system is initially at rest. If the reference input is changed suddenly to a higher flow rate, then the nozzle is moved in such a direction as to move the power piston and open the butterfly valve. Then the flow rate will increase, the pressure difference

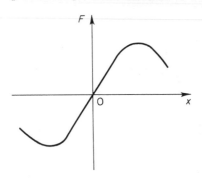 across the orifice becomes larger, and the nozzle will move back to the neutral position. The movement of the power piston stops when $x$, the displacement of the nozzle, comes back to and stays at the neutral position. (The jet pipe controller thus possesses an integrating property.)

**Fig. 5-58.** Force versus displacement curve.

The relationship between the total force $F$ acting on the power piston and the displacement $x$ of the nozzle is shown in Fig. 5-58. The total force is equal to the pressure difference $\Delta P$ across the piston times the area $A$ of the power piston. For a small displacement $x$ of the nozzle, the total force $F$ and displacement $x$ may be considered proportional.

**PROBLEM A-5-4.** Draw a block diagram of the pneumatic controller shown in Fig. 5-59. Then derive the transfer function of this controller.

If the resistance $R_d$ is removed (replaced by the line-sized tubing), what control action do we get? If the resistance $R_i$ is removed (replaced by the line-sized tubing), what control action do we get?

**Solution.** Let us assume that when $e = 0$, the nozzle-flapper distance is equal to $\bar{X}$ and the control pressure is equal to $\bar{P}_c$. In the present analysis, we shall assume small deviations from the respective reference values as follows:

$e$ = small error signal
$x$ = small change in the nozzle-flapper distance
$p_c$ = small change in the control pressure
$p_I$ = small pressure change in bellows I due to small change in the control pressure

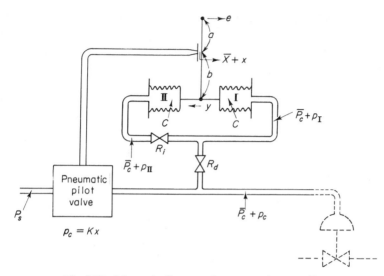

**Fig. 5-59.** Schematic diagram of a pneumatic controller.

$p_{II}$ = small pressure change in bellows II due to small change in the control pressure

$y$ = small displacement at the lower end of the flapper

In this controller, $p_c$ is transmitted to bellows I through the resistance $R_d$. Similarly, $p_c$ is transmitted to bellows II through the series of resistances $R_d$ and $R_i$. An approximate relationship between $p_I$ and $p_c$ is

$$\frac{P_I(s)}{P_c(s)} = \frac{1}{R_d Cs + 1} = \frac{1}{T_d s + 1}$$

where

$T_d = R_d C$ = derivative time

Similarly $p_{II}$ and $p_I$ are related by the transfer function

$$\frac{P_{II}(s)}{P_I(s)} = \frac{1}{R_i Cs + 1} = \frac{1}{T_i s + 1}$$

where

$T_i = R_i C$ = integral time

The force-balance equation for the two bellows is

$$(p_I - p_{II})A = k_s y$$

where $k_s$ is the stiffness of the two connected bellows and $A$ is the cross-sectional area of the bellows. The relationship among the variables $e$, $x$, and $y$ is

$$x = \frac{b}{a+b}e - \frac{a}{a+b}y$$

The relationship between $p_c$ and $x$ is

$$p_c = Kx$$

From the equations just derived, a block diagram of the controller can be drawn, as shown in Fig. 5-60(a). Simplification of this block diagram results in Fig. 5-60(b).

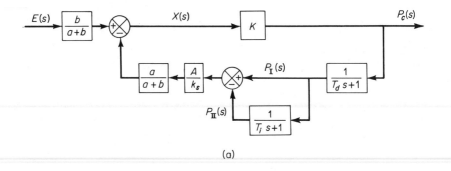

(a)

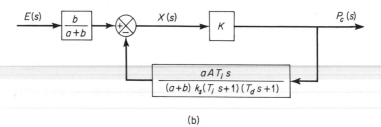

(b)

**Fig. 5-60.** (a) Block diagram of the pneumatic controller shown in Fig. 5-59; (b) simplified block diagram.

The transfer function between $P_c(s)$ and $E(s)$ is

$$\frac{P_c(s)}{E(s)} = \frac{\dfrac{b}{a+b}K}{1 + K\dfrac{a}{a+b}\dfrac{A}{k_s}\left(\dfrac{T_i s}{T_i s + 1}\right)\left(\dfrac{1}{T_d s + 1}\right)}$$

For a practical controller, under normal operation $|KaAT_i s/[(a+b)k_s(T_i s + 1)(T_d s + 1)]|$ is very much greater than unity and $T_i \gg T_d$. Therefore, the transfer function can be simplified as follows:

$$\frac{P_c(s)}{E(s)} \doteqdot \frac{bk_s(T_i s + 1)(T_d s + 1)}{aAT_i s}$$

$$= \frac{bk_s}{aA}\left(\frac{T_i + T_d}{T_i} + T_d s + \frac{1}{T_i s}\right)$$

$$\doteqdot K_p\left(1 + T_d s + \frac{1}{T_i s}\right)$$

where

$$K_p = \frac{bk_s}{aA}$$

Thus the controler shown in Fig. 5-59 is a proportional-plus-derivative-plus-integral one.

If the resistance $R_d$ is removed, or $R_d = 0$, the action becomes that of a proportional-plus-integral controller.

If the resistance $R_i$ is removed, or $R_i = 0$, the action becomes that of a narrow-band proportional, or two-position, controller. (Note that the actions of two feedback bellows cancel each other, and there is no feedback.)

**PROBLEM A-5-5.** Obtain the transfer function of the controller shown in Fig. 5-33 (c).

**Solution.** Figure 5-61 shows the feedback circuit. The equations for this feedback circuit are

$$\frac{1}{C_d s}[I_1(s) - I_2(s)] + R_d I_1(s) = E_0(s)$$

$$\frac{1}{C_d s}[I_2(s) - I_1(s)] + \frac{1}{C_i s}I_2(s) + R_i I_2(s) = 0$$

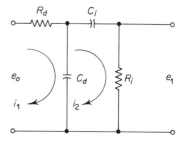

Hence

$$\frac{I_2(s)}{E_0(s)} = \frac{C_i s}{R_i C_i R_d C_d s^2 + (R_i C_i + R_d C_i + R_d C_d)s + 1}$$

or

$$\frac{E_1(s)}{E_0(s)} = \frac{R_i C_i s}{R_i C_i R_d C_d s^2 + (R_i C_i + R_d C_i + R_d C_d)s + 1}$$

Noting that in Fig. 5-33 (c)

**Fig. 5-61.** Feedback circuit used in the controller shown in Fig. 5-33 (c).

$$(e_i - e_f)K = e_0, \qquad e_f = e_1 \frac{R_2}{R_1}$$

we obtain

$$\left[ E_i(s) - \frac{R_2}{R_1}\left( \frac{R_i C_i s E_0(s)}{R_i C_i R_d C_d s^2 + (R_i C_i + R_d C_i + R_d C_d)s + 1} \right) \right] K = E_0(s)$$

The transfer function $E_0(s)/E_i(s)$ is

$$\frac{E_0(s)}{E_i(s)} = \frac{KR_1[R_i C_i R_d C_d s^2 + (R_i C_i + R_d C_i + R_d C_d)s + 1]}{KR_2 R_i C_i s + R_1[R_i C_i R_d C_d s^2 + (R_i C_i + R_d C_i + R_d C_d)s + 1]}$$

If the loop gain is very much greater than unity, then this last equation may be simplified to give

$$\frac{E_0(s)}{E_i(s)} = \frac{R_1[R_i C_i R_d C_d s^2 + (R_i C_i + R_d C_i + R_d C_d)s + 1]}{R_2 R_i C_i s}$$

$$= K_p\left[ T_d s + \left( 1 + \frac{R_d}{R_i} + \frac{T_d}{T_i} \right) + \frac{1}{T_i s} \right]$$

where

$$K_p = R_1/R_2, \qquad T_d = R_d C_d, \qquad T_i = R_i C_i$$

Define

$$\alpha = 1 + \frac{R_d}{R_i} + \frac{T_d}{T_i}$$

Then

$$\frac{E_0(s)}{E_i(s)} = K_p\alpha\left( 1 + \frac{T_d}{\alpha}s + \frac{1}{\alpha T_i s} \right)$$

Thus, the controller is a proportional-plus-derivative-plus-integral one.

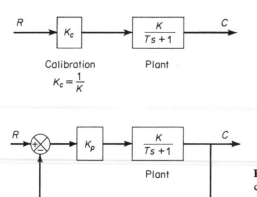

Plant

**Fig. 5-62.** Block diagrams of an open-loop control system and a closed-loop control system.

**PROBLEM A-5-6.** Consider the open-loop and closed-loop control system shown in Fig. 5-62. In the open-loop one, gain $K_c$ is calibrated so that $K_c = 1/K$. Thus, the transfer function of the open-loop control system is

$$G_0(s) = \frac{1}{K} \frac{K}{Ts+1} = \frac{1}{Ts+1}$$

In the closed-loop control system, gain $K_p$ of the controller is set so that $K_p K \gg 1$.

Assuming a unit-step input, compare the steady-state errors for these control systems.

**Solution.** For the open-loop control system, the error signal is

$$e(t) = r(t) - c(t)$$

or

$$E(s) = R(s) - C(s)$$
$$= [1 - G_0(s)]R(s)$$

The steady-state error for the unit-step response is

$$e_{ss} = \lim_{s \to 0} sE(s)$$
$$= \lim_{s \to 0} s[1 - G_0(s)]\frac{1}{s}$$
$$= 1 - G_0(0)$$

If $G_0(0)$, the dc gain of the open-loop control system, is equal to unity, then the steady-state error is zero. Due to environmental changes and aging of components, however, the dc gain $G_0(0)$ will drift from unity as time elapses, and the steady-state error will no longer be equal to zero. Such steady-state error in an open-loop control system will remain until the system is recalibrated.

For the closed-loop control system, the error signal is

$$E(s) = R(s) - C(s)$$
$$= \frac{1}{1 + G(s)}R(s)$$

where

$$G(s) = \frac{K_p K}{Ts+1}$$

The steady-state error for a unit-step input is

$$e_{ss} = \lim_{s \to 0} s \left[ \frac{1}{1 + G(s)} \right] \frac{1}{s}$$

$$= \frac{1}{1 + G(0)}$$

$$= \frac{1}{1 + K_p K}$$

In the closed-loop control system, gain $K_p$ is set at a large value compared with $1/K$. Thus the steady-state error can be made small, although not exactly zero.

Let us assume the following variation in the transfer function of the plant, assuming $K_c$ and $K_p$ constant:

$$\frac{K + \Delta K}{Ts + 1}$$

For simplicity, let us assume $K = 10$, $\Delta K = 1$, or $\Delta K/K = 0.1$. Then the steady-state error for a unit-step input in the open-loop control system becomes

$$e_{ss} = 1 - \frac{1}{K}(K + \Delta K)$$

$$= 1 - 1.1$$

$$= -0.1$$

In the closed-loop control system, if $K_p$ is set at $100/K$, then the steady-state error for a unit-step input becomes

$$e_{ss} = \frac{1}{1 + G(0)}$$

$$= \frac{1}{1 + \frac{100}{K}(K + \Delta K)}$$

$$= \frac{1}{1 + 110}$$

$$= 0.009$$

Thus, the closed-loop control system is superior to the open-loop control system in the presence of environmental changes, aging of components, etc., which definitely affect the steady-state performance.

**PROBLEM A-5-7.** The block diagram of Fig. 5-63 shows a speed control system in which the output member of the system is subject to a torque disturbance. In the diagram,

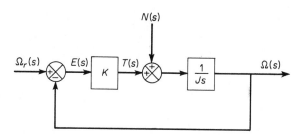

**Fig. 5-63.** Block diagram of a speed control system.

$\Omega_r(s)$, $\Omega(s)$, $T(s)$, and $N(s)$ are the Laplace transforms of the reference speed, output speed, driving torque, and disturbance torque, respectively. In the absence of a disturbance torque, the output speed is equal to the reference speed.

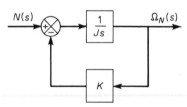

Investigate the response of this system to a unit step disturbance torque. Assume that the reference input is zero, or $\Omega_r(s) = 0$.

**Solution.** Figure 5-64 is a modified block diagram convenient for the present analysis. The closed-loop transfer function is

**Fig. 5-64.** Block diagram of the speed control system of Fig. 5-63 when $\Omega_r(s) = 0$.

$$\frac{\Omega_N(s)}{N(s)} = \frac{1}{Js + K}$$

where $\Omega_N(s)$ is the Laplace transform of the output speed due to the disturbance torque. For a unit-step disturbance torque, the steady-state output velocity is

$$\omega_N(\infty) = \lim_{s \to 0} s\Omega_N(s)$$

$$= \lim_{s \to 0} \frac{s}{Js + K} \frac{1}{s}$$

$$= \frac{1}{K}$$

From this analysis, we conclude that if a step disturbance torque is applied to the output member of the system, an error speed will result so that the ensuing motor torque will exactly cancel the disturbance torque. To develop this motor torque, it is necessary that there be an error in speed so that nonzero torque will result.

**PROBLEM A-5-8.** In the system considered in Problem A-5-7, it is desired to eliminate as much as possible the speed errors due to torque disturbances.

Is it possible to cancel the effect of a disturbance torque at steady state so that a constant disturbance torque applied to the output member will cause no speed change at steady state?

**Solution.** Suppose we choose a suitable controller whose transfer function is $G_c(s)$, as shown in Fig. 5-65. Then in the absence of the reference input, the closed-loop transfer function between the output velocity $\Omega_N(s)$ and the disturbance torque $N(s)$ is

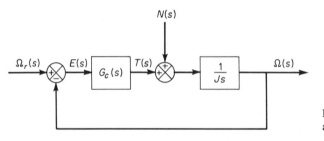

**Fig. 5-65.** Block diagram of a speed control system.

$$\frac{\Omega_N(s)}{N(s)} = \frac{\dfrac{1}{Js}}{1 + \dfrac{1}{Js}G_c(s)}$$

$$= \frac{1}{Js + G_c(s)}$$

The steady-state output speed due to a unit-step disturbance torque is

$$\omega_N(\infty) = \lim_{s \to 0} s\Omega_N(s)$$

$$= \lim_{s \to 0} \frac{s}{Js + G_c(s)} \frac{1}{s}$$

$$= \frac{1}{G_c(0)}$$

In order to satisfy the requirement that

$$\omega_N(\infty) = 0$$

we must choose $G_c(0) = \infty$. This can be realized if we choose

$$G_c(s) = \frac{K}{s}$$

Integral control action will continue to correct until the error is zero. This controller, however, presents a stability problem because the characteristic equation will have two imaginary roots.

One method of stabilizing such a system is to add a proportional mode to the controller or choose

$$G_c(s) = K_p + \frac{K}{s}$$

With this controller, the block diagram of Fig. 5-65 in the absence of the reference input can be modified to that of Fig. 5-66. The closed-loop transfer function $\Omega_N(s)/N(s)$ becomes

$$\frac{\Omega_N(s)}{N(s)} = \frac{s}{Js^2 + K_p s + K}$$

For a unit-step disturbance torque, the steady-state output speed is

$$\omega_N(\infty) = \lim_{s \to 0} s\Omega_N(s)$$

$$= \lim_{s \to 0} \frac{s^2}{Js^2 + K_p s + K} \frac{1}{s}$$

$$= 0$$

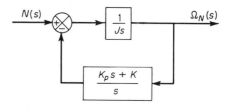

**Fig. 5-66.** Block diagram of the speed control system of Fig. 5-65 when $G_c(s) = K_p + (K/s)$ and $\Omega_r(s) = 0$.

Thus, we see that the proportional-plus-integral controller eliminates speed error at steady state.

The use of integral control action has increased the order of the system by one. (This tends to produce an oscillatory response.)

In the present system, a step disturbance torque will cause a transient error in the output speed, but the error will become zero at steady state. The integrator provides a nonzero output with zero error. (The nonzero output of the integrator produces a motor torque which exactly cancels the disturbance torque.)

Note that the integrator in the transfer function of the plant does not eliminate the steady-state error due to a disturbance torque. To eliminate this, we must have an integrator before the point where the disturbance torque enters.

**PROBLEM A-5-9.** Figure 5-67(a) is a schematic diagram of a pneumatic control device. It consists of four chambers separated by three diaphragms which are rigidly linked, as shown in the diagram. This device has two input pipes where two control pressures $P_1$ and $P_2$ are connected. If such pressures are applied, the diaphragm assembly distorts and shuts off either one of the two supply inputs. The output pressure $P_0$ is then equal to either $P_s$ or 0. The output pressure $P_0$ can be plotted versus the pressure difference $P_1 - P_2$, as shown in Fig. 5-67(b). Such a device can be used as a logic device.

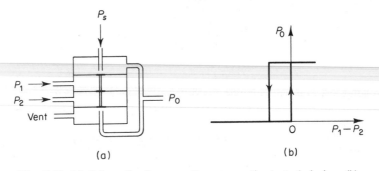

(a)                                        (b)

**Fig. 5-67.** (a) Schematic diagram of a pneumatic control device; (b) characteristic curve.

If a bias pressure is applied to one of the chambers, as shown in Fig. 5-68(a), the characteristic curve is shifted, as shown in Fig. 5-68(b).

Figure 5-69 shows a combination of two such devices. What logic operation can we get from this combination? Assuming the supply pressure is 1 and the input pressures at points $A$ and $B$ are either 0 or 1, find the output pressure at point $F$.

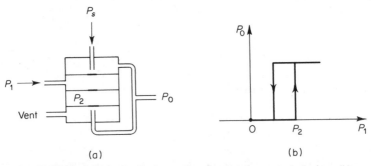

(a)                                        (b)

**Fig. 5-68.** (a) Schematic diagram of a pneumatic control device; (b) characteristic curve.

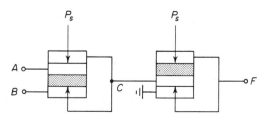

**Fig. 5-69.** Combination of two pneumatic
control devices.

**Solution.** From Fig. 5-69, we obtain the following table:

| $A$ | $B$ | $C$ | $F$ |
|---|---|---|---|
| 0 | 0 | 0 | 1 |
| 1 | 0 | 1 | 0 |
| 0 | 1 | 1 | 0 |
| 1 | 1 | 1 | 0 |

Thus, the logic operation obtained is that of **NOR**; that is, $F = \overline{A} \cdot \overline{B}$.

## PROBLEMS

**PROBLEM B-5-1.** The schematic diagram of a pneumatic controller is shown in Fig. 5-70. Draw a block diagram of the controller, and then derive the transfer function.

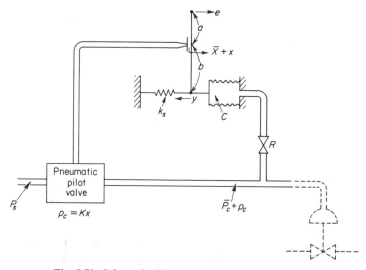

**Fig. 5-70.** Schematic diagram of a pneumatic controller.

**PROBLEM B-5-2.** Figure 5-71 shows the schematic diagram of a hydraulic servomotor in which the error signal is amplified in two stages. Draw a block diagram of the system and then find the transfer function between $y$ and $x$.

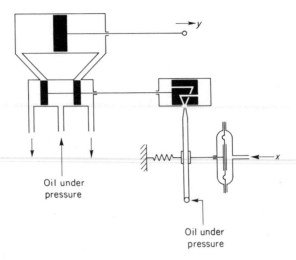

**Fig. 5-71.** Schematic diagram of a hydraulic servomotor.

**PROBLEM B-5-3.** Figure 5-72 shows the schematic diagram of a hydraulic servomotor. What type of control action does this servomotor produce?

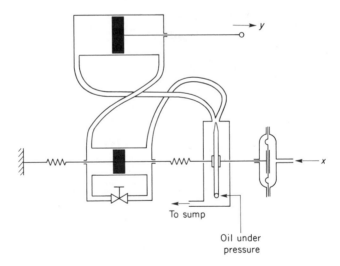

**Fig. 5-72.** Schematic diagram of a hydraulic servomotor.

**PROBLEM B-5-4.** Figure 5-73 shows the schematic diagram of a speed control system. The governor shaft, through reduction gears, is driven at speed $\omega$. The rotating flyweights result in a centrifugal force which is opposed by the spring force. The desired speed is set by the spring preload. What type of control action does the controller produce?

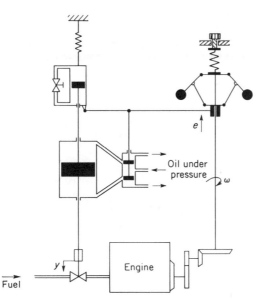

**Fig. 5-73.** Schematic diagram of a speed control system.

**PROBLEM B-5-5.** Consider the controller shown in Fig. 5-74. What type of control action does this controller produce?

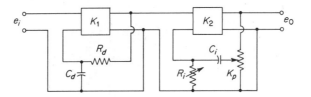

**Fig. 5-74.** Schematic diagram of an electronic controller.

**PROBLEM B-5-6.** Figure 5-75(a) is the schematic diagram of a satellite attitude control system. Small jets apply reaction forces to rotate the satellite body into the desired attitude. The two skew symmetrically placed jets denoted by $A$ or $B$ operate in pairs. Assume that each jet thrust is $F/2$ and a torque of $Fl$ is applied to the system. The moment of inertia about the center of mass is $J$.

Suppose that the attitude controller is a proportional-plus-derivative type. The block diagram representation of the system is shown in Fig. 5-75(b). Determine the value of the derivative time so that the damping ratio $\zeta$ is 0.7.

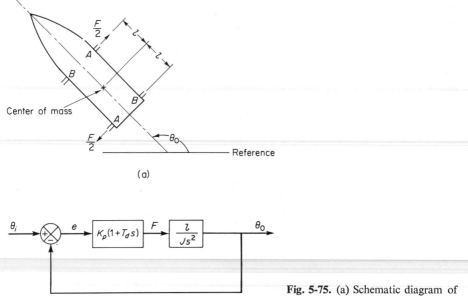

(a)

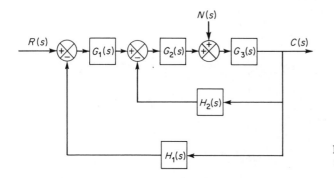

(b)

Fig. 5-75. (a) Schematic diagram of a satellite attitude control system; (b) block diagram.

**PROBLEM B-5-7.** Consider the multiple-loop system shown in Fig. 5-76. Obtain the closed-loop transfer function between $C(s)$ and $N(s)$.

Fig. 5-76. Multiple-loop system.

**PROBLEM B-5-8.** If the feedforward path of a control system contains at least one integrating element, then the output continues to change as long as an error is present. The output stops when the error is precisely zero. If an external disturbance enters the system, it is desirable to have an integrating element between the error-measuring element and the point where the disturbance enters so that the effect of the external disturbance may be made zero at steady state.

Show that if the disturbance is a ramp function, then the steady-state error due to this ramp disturbance may be eliminated only if two integrators precede the point where the disturbance enters.

**PROBLEM B-5-9.** Consider the system shown in Fig. 5-77(a) where $K$ is an adjustable gain and $G(s)$ and $H(s)$ are fixed components. The closed-loop transfer function for the disturbance is

$$\frac{C(s)}{N(s)} = \frac{1}{1 + KG(s)H(s)}$$

To minimize the effect of disturbances, the adjustable gain $K$ should be chosen as large as possible.

Is this true for the system shown in Fig. 5-77(b) too?

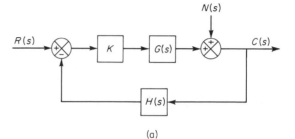

(a)

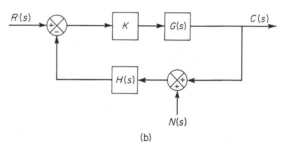

**Fig. 5-77.** (a) Control system with disturbance entering in the feed-forward path; (b) control system with disturbance entering in the feedback path.

(b)

**PROBLEM B-5-10.** Figure 5-78 shows the schematic diagram of a fluidic device. What function does this device perform?

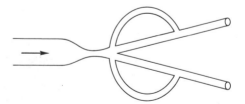

**Fig. 5-78.** Schematic diagram of a fluidic device.

**PROBLEM B-5-11.** A fluidic resistor is a passive fluidic element which, because of viscous losses, produces a pressure drop as a function of the flow through it and has a transfer function with essentially real components over the frequency range of interest.

What is the electrical analog of a passive fluidic element which, because of fluid compressibility, produces pressure which lags flow into it by essentially 90°?

What is the electrical analog of a passive fluidic element which, because of fluid inertness, has a pressure drop which leads flow by essentially 90°?

# 6

# TRANSIENT-RESPONSE ANALYSIS

## 6-1 INTRODUCTION

It was stated in Chapter 4 that the first step in analyzing a control system was to derive a mathematical model of the system. Once such a model is obtained, various methods are available for the analysis of system performance.

In practice, the input signal to a control system is not known ahead of time but is random in nature, and the instantaneous input cannot be expressed analytically. Only in some special cases is the input signal known in advance and expressible analytically or by curves, such as in the case of the automatic control of cutting tools.

In analyzing and designing control systems, we must have a basis of comparison of performance of various control systems. This basis may be set up by specifying particular test input signals and by comparing the responses of various systems to these input signals.

Many design criteria are based on such signals or on the response of systems to changes in initial conditions (without any test signals). The use of test signals can be justified because of a correlation existing between the characteristics of a system to a typical test input signal and the capability of the system to cope with actual input signals.

**Typical test signals.** The commonly used test input signals are those of step functions, ramp functions, acceleration functions, impulse functions, sinusoidal functions, etc. With these test signals, mathematical and experimental analyses of control systems can be carried out easily since the signals are very simple functions of time.

Which one or ones of these typical input signals to use for analyzing system characteristics may be determined by the form of the input that the system will be subjected to most frequently under normal operation. If the inputs to a control system are gradually changing functions of time, then a ramp function of time may be a good test signal. Similarly, if a system is subjected to sudden disturbances, a step function of time may be a good test signal; and for a system subjected to shock inputs, an impulse function may be best. Once a control system is designed on the basis of test signals, the performance of the system in response to actual inputs is generally satisfactory. The use of such test signals enables one to compare the performance of all systems on the same basis.

This chapter is concerned with system response to aperiodic signals (such as step, ramp, acceleration, and impulse functions of time). (System analysis and design based on sinusoidal test signals are given in Chapters 9 and 10.)

**Transient response and steady-state response.** The time response of a control system consists of two parts: the transient and the steady-state response. By transient response, we mean that which goes from the initial state to the final state. By steady-state response, we mean the manner in which the system output behaves as $t$ approaches infinity.

**Absolute stability, relative stability, and steady-state error.** In designing a control system, we must be able to predict the dynamic behavior of the system from a knowledge of the components. The most important characteristic of the dynamic behavior of a control system is absolute stability, that is, whether the system is stable or unstable. A control system is in equilibrium if, in the absence of any disturbance or input, the output stays in the same state. A linear time-invariant control system is stable if the output eventually comes back to its equilibrium state when the system is subjected to a disturbance. A linear time-invariant control system is unstable if either oscillation of the output continues forever or the output diverges without bound from its equilibrium state when the system is subjected to a disturbance. Actually, the output of a physical system may increase to a certain extent but may be limited by mechanical "stops," or the system may break down or become nonlinear after the output exceeds a certain magnitude so that the linear differential equations no longer apply. We shall not discuss the stability of nonlinear systems in this chapter. (Refer to such analyses of nonlinear systems in Chapters 11 and 15.)

Important system behavior (other than absolute stability) to which we must give careful consideration includes relative stability and steady-state error. Since a physical control system involves energy storage, the output of the system, when subjected to an input, cannot follow the input immediately but exhibits a transient response before a steady state can be reached. The transient response of a practical control system often exhibits damped oscillations before reaching a steady state. If the output of a system at steady state does not exactly agree with the input, the system is said to have steady-state error. This error is indicative of the accuracy of the system. In analyzing a control system, we must examine transient-response behavior, such as the time required to reach a new steady state and the value of the error while following an input signal, as well as the steady-state behavior.

## 6-2 IMPULSE-RESPONSE FUNCTIONS

For a linear time-invariant system, the transfer function $G(s)$ is

$$G(s) = \frac{Y(s)}{X(s)}$$

where $X(s)$ is the Laplace transform of the input and $Y(s)$ is the Laplace transform of the output. It follows that the output $Y(s)$ can be written as the product of $G(s)$ and $X(s)$, or

$$Y(s) = G(s)X(s) \tag{6-1}$$

Note that multiplication in the complex domain is equivalent to convolution in the time domain, so that the inverse Laplace transform of Eq. (6-1) is given by the following convolution integral:

$$y(t) = \int_0^t x(\tau)g(t - \tau)\, d\tau$$

$$= \int_0^t g(\tau)x(t - \tau)\, d\tau$$

where

$$g(t) = x(t) = 0 \qquad \text{for } t < 0$$

**Impulse-response functions.** Consider the output (response) of a system to a unit-impulse input when the initial conditions are zero. Since the Laplace transform of the unit-impulse function is unity, the Laplace transform of the output of the system is just

$$Y(s) = G(s) \tag{6-2}$$

The inverse Laplace transform of the output given by Eq. (6-2) is the impulse-response function, or

$$y(t) = g(t) = \text{impulse-response function}$$

This function is also called the weighting function of the system.

The impulse-response function $g(t)$ is thus the response of a linear system to a unit-impulse input when the initial conditions are zero. The Laplace transform of this function gives the transfer function. Therefore, the transfer function and impulse-response function of a linear time-invariant system contain the same information about system dynamics. It is, hence, possible to obtain complete information about the dynamic characteristics of the system by exciting it with an impulse input and measuring the response. In practice, a pulse input with a very short duration compared with the significant time constants of the system can be considered an impulse.

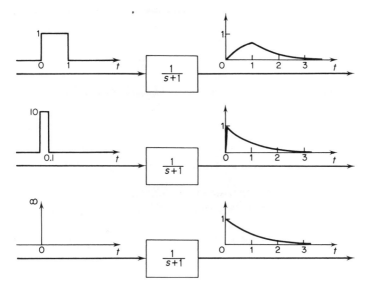

**Fig. 6-1.** Response curves of a first-order system subjected to pulse inputs and impulse input.

Consider the response of a first-order system to a pulse input of amplitude $1/t_1$ and duration $t_1$. If the time duration $0 < t < t_1$ of this input is sufficiently small compared with the system time constant $T$, then the response is approximately a unit-impulse response. One method of testing to determine whether $t_1$ is sufficiently small is to apply the pulse input of magnitude $2/t_1$ and duration $t_1/2$. If the response to the original pulse input and to the modified pulse input are essentially the same, $t_1$ can be considered sufficiently small. Figure 6-1 shows the response curves of a first-order system to pulse inputs and impulse input. Note that if $t_1 < 0.1T$, the response of the system is almost identical to the unit-impulse response.

**Convolution integrals.** In practice we can approximate an impulse function by a pulse function of large amplitude and narrow width, the area of which is equal to the strength, or area, of the impulse function. If the input function $x(t)$ is a pulse function with amplitude $1/t_1$ and width $t_1$, then the convolution integral

$$y(t) = \int_0^t x(\tau)g(t - \tau)\, d\tau$$

becomes

$$y(t) = \int_0^{t_1} \frac{1}{t_1} g(t - \tau)\, d\tau$$

$$\doteq g(t) \qquad \text{for sufficiently small } t_1$$

Thus, the response of a system to a large-amplitude–narrow-width pulse having unity area is almost equal to the impulse-response function.

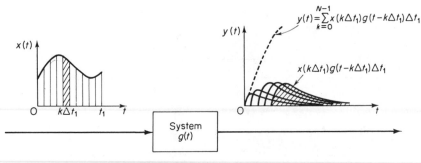

Fig. 6-2. System output as convolution summation.

**Approximation to convolution integrals.** Consider the system shown in Fig. 6-2. Assume that the impulse-response function is $g(t)$. {Clearly, the transfer function of the system is $\mathscr{L}[g(t)] = G(s)$}. The input $x(t)$ starts at $t = 0$ and lasts until $t = t_1$. Let us find the response of this system to $x(t)$ by approximating the convolution integral:

$$y(t) = \int_0^t x(t)g(t - \tau)\, d\tau$$

by the sum of the responses to $N$ pulse functions.

The input $x(t)$ can be approximated by a sequence of $N$ pulse functions whose width is $\Delta t_1$, where $\Delta t_1 = t_1/N$. If $\Delta t_1$ is sufficiently small compared with the smallest time constant of the system, then the $k$th pulse can be considered an impulse whose magnitude is the area $x(k\,\Delta t_1)\,\Delta t_1$. The response to the $k$th pulse is then

$$x(k\,\Delta t_1)\,\Delta t_1\, g(t - k\,\Delta t_1)$$

which is the product of the area of the impulse and the impulse-response function delayed by $k\,\Delta t_1$.

Since the system under consideration is linear, the principle of superposition holds. Hence the response $y(t)$ of the system to the sequence of $N$ pulse functions is given by convolution summation:

$$y(t) = \sum_{k=0}^{N-1} x(k\,\Delta t_1)g(t - k\,\Delta t_1)\,\Delta t_1 \tag{6-3}$$

where

$$g(\tau) = 0 \quad \text{for } \tau < 0$$

Equation (6-3) gives the response at time $t$. Note that since $g(\tau) = 0$ for $\tau < 0$, the response does not precede the input. Thus

$$y(0 \le t < \Delta t_1) = x(0)g(t)\,\Delta t_1$$
$$y(\Delta t_1 \le t < 2\,\Delta t_1) = [x(0)g(t) + x(\Delta t_1)g(t - \Delta t_1)]\,\Delta t_1$$
$$y(2\,\Delta t_1 \le t < 3\,\Delta t_1) = [x(0)g(t) + x(\Delta t_1)g(t - \Delta t_1)$$
$$+ x(2\,\Delta t_1)g(t - 2\,\Delta t_1)]\,\Delta t_1$$
$$\cdots$$
$$y(N\,\Delta t_1 - \Delta t_1 \le t \le N\,\Delta t_1) = \left[\sum_{k=0}^{N-1} x(k\,\Delta t_1)g(t - k\,\Delta t_1)\right]\Delta t_1$$

## 6-3  FIRST-ORDER SYSTEMS

Consider the first-order system shown in Fig. 6-3 (a). Physically, this system may represent an $R$-$C$ circuit, thermal system, etc. A simplified block diagram is shown in Fig. 6-3(b). The input-output relationship is given by

$$\frac{C(s)}{R(s)} = \frac{1}{Ts + 1} \qquad (6\text{-}4)$$

In the following, we shall analyze the system responses to such inputs as the unit-step, unit-ramp, and unit-impulse functions. The initial conditions will be zero.

Note that all systems having the same transfer function will exhibit the same output in response to the same input. For any given physical system, the mathematical response can be given a physical interpretation.

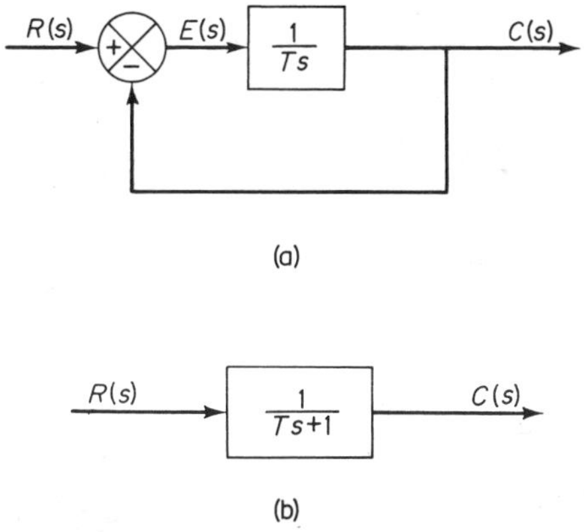

(a)

(b)

**Fig. 6-3.** (a) Block diagram of a first-order system; (b) simplified block diagram.

**Unit-step response of first-order systems.** Since the Laplace transform of the unit-step function is $1/s$, substituting $R(s) = 1/s$ into Eq. (6-4), we obtain

$$C(s) = \frac{1}{Ts + 1}\frac{1}{s}$$

Expanding $C(s)$ into partial fractions gives

$$C(s) = \frac{1}{s} - \frac{T}{Ts + 1} \qquad (6\text{-}5)$$

Taking the inverse Laplace transform of Eq. (6-5), we obtain

$$c(t) = 1 - e^{-t/T} \qquad (t \geq 0) \qquad (6\text{-}6)$$

Equation (6-6) states that initially the output $c(t)$ is zero and finally it becomes unity. One of the important characteristics of such an exponential response curve $c(t)$ is that at $t = T$ the value of $c(t)$ is 0.632, or the response $c(t)$ has reached 63.2 % of its total change. This may be easily seen by substituting $t = T$ in $c(t)$. Namely,

$$c(T) = 1 - e^{-1} = 0.632$$

It is well-known that $T$ is the time constant of the system. The smaller the time constant, the faster the system response. Another important characteristic of the exponential response curve is that the slope of the tangent line at $t = 0$ is $1/T$ since

$$\frac{dc}{dt} = \frac{1}{T}e^{-t/T}\bigg|_{t=0} = \frac{1}{T} \qquad (6\text{-}7)$$

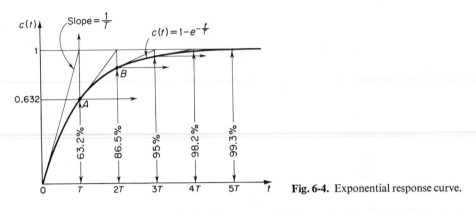

**Fig. 6-4.** Exponential response curve.

The output would reach the final value at $t = T$ if it maintained its initial speed of response. From Eq. (6-7) we see that the slope of the response curve $c(t)$ decreases monotonically from $1/T$ at $t = 0$ to zero at $t = \infty$.

The exponential response curve $c(t)$ given by Eq. (6-6) is shown in Fig. 6-4. In one time constant, the exponential response curve has gone from 0 to 63.2% of the final value. In two time constants, the response reaches 86.5% of the final value. At $t = 3T$, $4T$, and $5T$, the response reaches 95, 98.2, and 99.3%, respectively, of the final value. Thus, for $t \geq 4T$, the response remains within 2% of the final value. As seen from Eq. (6-6), the steady state is reached mathematically only after an infinite time. In practice, however, a reasonable estimate of the response time is the length of time the response curve needs to reach the 2% line of the final value, or four time constants.

**Fig. 6-5.** A general system.

Consider the system shown in Fig. 6-5. In order to determine experimentally whether or not the system is of first order, plot the curve $\log |c(t) - c(\infty)|$, where $c(t)$ is the system output, as a function of $t$. If the curve turns out to be a straight line, the system is of first order. The time constant $T$ can be read from the graph as the time $T$ which satisfies the following equation:

$$c(T) - c(\infty) = 0.368\,[c(0) - c(\infty)]$$

Note that instead of plotting $\log |c(t) - c(\infty)|$ versus $t$, it is convenient to plot $|c(t) - c(\infty)|/|c(0) - c(\infty)|$ versus $t$ on semilog paper, as shown in Fig. 6-6.

**Unit-ramp response of first-order systems.** Since the Laplace transform of the unit-ramp function is $1/s^2$, we obtain the output of the system of Fig. 6-3 (a) as

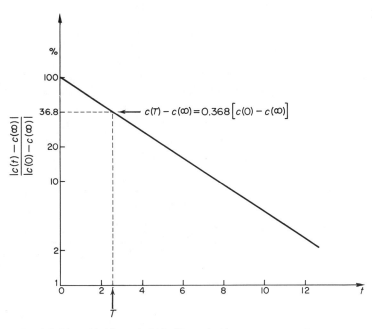

**Fig. 6-6.** Plot of $|c(t) - c(\infty)|/|c(0) - c(\infty)|$ versus $t$ on semilog paper.

$$C(s) = \frac{1}{Ts + 1} \frac{1}{s^2}$$

Expanding $C(s)$ into partial fractions gives

$$C(s) = \frac{1}{s^2} - \frac{T}{s} + \frac{T^2}{Ts + 1} \tag{6-8}$$

Taking the inverse Laplace transform of Eq. (6-8), we obtain

$$c(t) = t - T + Te^{-t/T} \qquad (t \geq 0)$$

The error signal $e(t)$ is then

$$e(t) = r(t) - c(t)$$
$$= T(1 - e^{-t/T})$$

As $t$ approaches infinity, $e^{-t/T}$ approaches zero, and thus the error signal $e(t)$ approaches $T$ or

$$e(\infty) = T$$

The unit-ramp input and the system output are shown in Fig. 6-7. The error in following the unit-ramp input is equal to $T$ for sufficiently large $t$. The smaller the time constant $T$, the smaller the steady-state error in following the ramp input.

**Unit-impulse response of first-order systems.** For the unit-impulse input, $R(s) = 1$ and the output of the system of Fig. 6-3 (a) can be obtained as

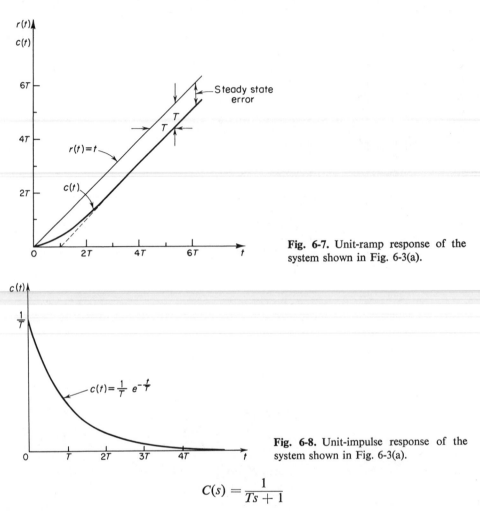

**Fig. 6-7.** Unit-ramp response of the system shown in Fig. 6-3(a).

**Fig. 6-8.** Unit-impulse response of the system shown in Fig. 6-3(a).

$$C(s) = \frac{1}{Ts + 1}$$

or

$$c(t) = \frac{1}{T}e^{-t/T} \qquad (t \geq 0) \tag{6-9}$$

The response curve given by Eq. (6-9) is shown in Fig. 6-8.

**An important property of linear time-invariant systems.** In the analysis above, it has been shown that for the unit-ramp input the output $c(t)$ is

$$c(t) = t - T + Te^{-t/T} \qquad (t \geq 0)$$

For the unit-step input, which is the derivative of unit-ramp input, the output $c(t)$ is

$$c(t) = 1 - e^{-t/T} \qquad (t \geq 0)$$

Finally, for the unit-impulse input, which is the derivative of unit-step input, the output $c(t)$ is

$$c(t) = \frac{1}{T} e^{-t/T} \qquad (t \geq 0)$$

Comparison of the system response to these three inputs clearly indicates that the response to the derivative of an input signal can be obtained by differentiating the response of the system to the original signal. It can also be seen that the response to the integral of the original signal can be obtained by integrating the response of the system to the original signal and by determining the integration constants from the zero output initial condition. As stated in Chapter 4, this is the property of linear time-invariant systems. Linear time-varying systems and nonlinear systems do not possess this property.

## 6-4 SECOND-ORDER SYSTEMS

In this section, we shall first obtain the response of a specific second-order control system to a step input and then extend the analysis to the solution of other second-order systems.

**A servomechanism.** Consider the servomechanism shown in Fig. 6-9 (a). The objective of this system is to control the position of the mechanical load in accordance with the reference position. The operation of this system is as follows: A pair of potentiometers acts as an error-measuring device. They convert the input and output positions into proportional electric signals. The command input signal determines the angular position $r$ of the wiper arm of the input potentiometer. The angular position $r$ is the reference input to the system, and the electric potential of the arm is proportional to the angular position of the arm. The output shaft position determines the angular position $c$ of the wiper arm of the output potentiometer. The potential difference $e_r - e_c = e$ is the error signal, where $e_r$ is proportional to $r$ and $e_c$ is proportional to $c$; namely, $e_r = K_0 r$ and $e_c = K_0 c$, where $K_0$ is a proportionality constant. The error signal which appears at the potentiometer terminals is amplified by the amplifier whose gain constant is $K_1$. The output voltage of this amplifier is applied to the armature circuit of the dc motor. (The amplifier must have very high input impedance because the potentiometers are essentially high impedance circuits and do not tolerate current drain. At the same time, the amplifier must have low output impedance since it feeds into the armature circuit of the motor.) A fixed voltage is applied to the field winding. If an error exists, the motor develops a torque to rotate the output load in such a way as to reduce the error to zero. For constant field current, the torque developed by the motor is

$$T = K_2 i_a$$

where $K_2$ is the motor torque constant and $i_a$ is the armature current. For the armature circuit,

$$L_a \frac{di_a}{dt} + R_a i_a + K_3 \frac{d\theta}{dt} = K_1 e \qquad (6\text{-}10)$$

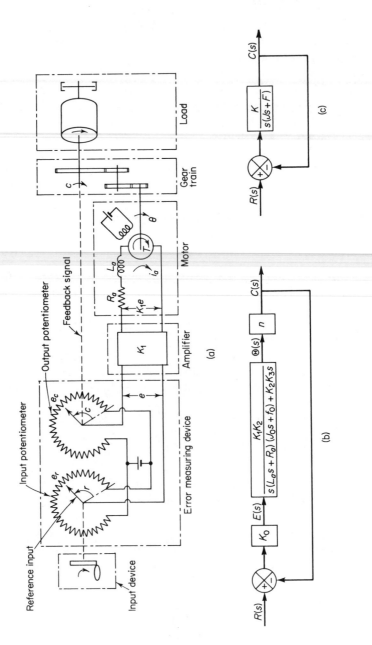

**Fig. 6-9.** (a) Schematic diagram of a servomechanism; (b) and (c) are block diagrams.

where $K_3$ is the back emf constant of the motor and $\theta$ is the angular displacement of the motor shaft. The equation for torque equilibrium is

$$J_0 \frac{d^2\theta}{dt^2} + f_0 \frac{d\theta}{dt} = T = K_2 i_a \tag{6-11}$$

where $J_0$ is the inertia of the combination of the motor, load, and gear train referred to the motor shaft and $f_0$ is the viscous friction coefficient of the combination of the motor, load, and gear train referred to the motor shaft. The transfer function between the motor shaft displacement and the error signal is obtained from Eqs. (6-10) and (6-11) as follows:

$$\frac{\Theta(s)}{E(s)} = \frac{K_1 K_2}{s(L_a s + R_a)(J_0 s + f_0) + K_2 K_3 s} \tag{6-12}$$

where $\Theta(s) = \mathscr{L}[\theta(t)]$ and $E(s) = \mathscr{L}[e(t)]$. We assume that the gear ratio of the gear train is such that the output shaft rotates $n$ times for each revolution of the motor shaft. Thus,

$$C(s) = n\Theta(s) \tag{6-13}$$

where $C(s) = \mathscr{L}[c(t)]$ and $c(t)$ is the angular displacement of the output shaft. The relationship among $E(s)$, $R(s)$, and $C(s)$ is

$$E(s) = K_0[R(s) - C(s)] \tag{6-14}$$

where $R(s) = \mathscr{L}[r(t)]$. The block diagram of this system can be constructed from Eqs. (6-12), (6-13), and (6-14), as shown Fig. 6-9 (b). The transfer function in the feedforward path of this system is

$$G(s) = \frac{K_0 K_1 K_2 n}{s[(L_a s + R_a)(J_0 s + f_0) + K_2 K_3]}$$

Since $L_a$ is usually small, it can be neglected, and the transfer function $G(s)$ in the feedforward path becomes

$$G(s) = \frac{K_0 K_1 K_2 n}{s[R_a(J_0 s + f_0) + K_2 K_3]}$$

$$= \frac{K_0 K_1 K_2 n / R_a}{J_0 s^2 + \left(f_0 + \frac{K_2 K_3}{R_a}\right)s} \tag{6-15}$$

The term $[f_0 + (K_2 K_3/R_a)]s$ indicates that the back emf of the motor effectively increases the viscous friction of the system. The inertia $J_0$ and viscous friction coefficient $f_0 + (K_2 K_3/R_a)$ are referred to the motor shaft. When $J_0$ and $f_0 + (K_2 K_3/R_a)$ are multiplied by $1/n^2$, the inertia and viscous friction coefficient are expressed in terms of the output shaft. Introducing new parameters defined by

$J = J_0/n^2 =$ moment of inertia referred to the output shaft
$F = [f_0 + (K_2 K_3/R_a)]/n^2 =$ viscous-friction coefficient referred to the output shaft
$K = K_0 K_1 K_2/nR_a$

the transfer function $G(s)$ given by Eq. (6-15) can be simplified, yielding

$$G(s) = \frac{K}{Js^2 + Fs}$$

The block diagram of the system shown in Fig. 6-9 (b) is then simplified as shown in Fig. 6-9 (c).

In the following, we shall investigate the dynamic responses of this system to unit-step, unit-ramp, and unit-impulse inputs.

**Step response of second-order systems.** The closed-loop transfer function of the system shown in Fig. 6-9 (c) is

$$\frac{C(s)}{R(s)} = \frac{K}{Js^2 + Fs + K}$$

$$= \frac{\dfrac{K}{J}}{\left[s + \dfrac{F}{2J} + \sqrt{\left(\dfrac{F}{2J}\right)^2 - \dfrac{K}{J}}\right]\left[s + \dfrac{F}{2J} - \sqrt{\left(\dfrac{F}{2J}\right)^2 - \dfrac{K}{J}}\right]} \qquad (6\text{-}16)$$

The closed loop poles are complex if $F^2 - 4JK < 0$, and they are real if $F^2 - 4JK \geq 0$. In transient-response analysis, it is convenient to write

$$\frac{K}{J} = \omega_n^2, \qquad \frac{F}{J} = 2\zeta\omega_n = 2\sigma$$

where $\sigma$ is called the attenuation; $\omega_n$, the undamped natural frequency; and $\zeta$, the damping ratio of the system. The damping ratio $\zeta$ is the ratio of the actual damping $F$ to the critical damping $F_c = 2\sqrt{JK}$ or

$$\zeta = \frac{F}{F_c} = \frac{F}{2\sqrt{JK}}$$

With this notation, the system shown in Fig. 6-9 (c) can be modified to that shown in Fig. 6-10, and the closed-loop transfer function $C(s)/R(s)$ given by Eq. (6-16) can be written

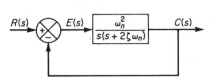

$$\frac{C(s)}{R(s)} = \frac{\omega_n^2}{s^2 + 2\zeta\omega_n s + \omega_n^2} \qquad (6\text{-}17)$$

**Fig. 6-10.** Second-order system.

The dynamic behavior of second-order systems can then be described in terms of two parameters $\zeta$ and $\omega_n$. If $0 < \zeta < 1$, the closed-loop poles are complex conjugates and lie in the left-half $s$ plane. The system is then called underdamped, and the transient response is oscillatory. If $\zeta = 1$, the system is called critically damped. Overdamped systems correspond to $\zeta > 1$. The transient response of critically damped and overdamped systems do not oscillate. If $\zeta = 0$, the transient response does not die out.

We shall now solve for the response of the system shown in Fig. 6-10 to a unit-step input. We shall consider three different cases: the underdamped ($0 < \zeta < 1$), critically damped ($\zeta = 1$), and overdamped ($\zeta > 1$) cases.

(1) Underdamped case ($0 < \zeta < 1$): In this case, $C(s)/R(s)$ can be written

$$\frac{C(s)}{R(s)} = \frac{\omega_n^2}{(s + \zeta\omega_n + j\omega_d)(s + \zeta\omega_n - j\omega_d)}$$

where $\omega_d = \omega_n\sqrt{1 - \zeta^2}$. The frequency $\omega_d$ is called the damped natural frequency. For a unit-step input, $C(s)$ can be written

$$C(s) = \frac{\omega_n^2}{(s^2 + 2\zeta\omega_n s + \omega_n^2)s} \tag{6-18}$$

The inverse Laplace transform of Eq. (6-18) can be obtained easily if $C(s)$ is written in the following form:

$$C(s) = \frac{1}{s} - \frac{s + 2\zeta\omega_n}{s^2 + 2\zeta\omega_n s + \omega_n^2}$$

$$= \frac{1}{s} - \frac{s + \zeta\omega_n}{(s + \zeta\omega_n)^2 + \omega_d^2} - \frac{\zeta\omega_n}{(s + \zeta\omega_n)^2 + \omega_d^2}$$

In Chapter 2 it was shown that

$$\mathscr{L}^{-1}\left[\frac{s + \zeta\omega_n}{(s + \zeta\omega_n)^2 + \omega_d^2}\right] = e^{-\zeta\omega_n t}\cos\omega_d t$$

$$\mathscr{L}^{-1}\left[\frac{\omega_d}{(s + \zeta\omega_n)^2 + \omega_d^2}\right] = e^{-\zeta\omega_n t}\sin\omega_d t$$

Hence the inverse Laplace transform of Eq. (6-18) is obtained as

$$\mathscr{L}^{-1}[C(s)] = c(t)$$

$$= 1 - e^{-\zeta\omega_n t}\left(\cos\omega_d t + \frac{\zeta}{\sqrt{1 - \zeta^2}}\sin\omega_d t\right)$$

$$= 1 - \frac{e^{-\zeta\omega_n t}}{\sqrt{1 - \zeta^2}}\sin\left(\omega_d t + \tan^{-1}\frac{\sqrt{1 - \zeta^2}}{\zeta}\right) \qquad (t \geq 0) \tag{6-19}$$

This result can, of course, be obtained directly by using a table of Laplace transforms. From Eq. (6-19), it can be seen that the frequency of transient oscillation is the damped natural frequency $\omega_d$ and thus varies with the damping ratio $\zeta$. The error signal for this system is the difference between the input and output and is

$$e(t) = r(t) - c(t)$$

$$= e^{-\zeta\omega_n t}\left(\cos\omega_d t + \frac{\zeta}{\sqrt{1 - \zeta^2}}\sin\omega_d t\right) \qquad (t \geq 0)$$

This error signal exhibits a damped sinusoidal oscillation. At steady state, or at $t = \infty$, no error exists between the input and output.

If the damping ratio $\zeta$ is equal to zero, the response becomes undamped and oscillations continue indefinitely. The response $c(t)$ for the zero damping case may be obtained by substituting $\zeta = 0$ in Eq. (6-19), yielding

$$c(t) = 1 - \cos\omega_n t \qquad (t \geq 0) \tag{6-20}$$

Thus, from Eq. (6-20), we see that $\omega_n$ represents the undamped natural frequency of the system. That is, $\omega_n$ is that frequency at which the system would oscillate if the damping were decreased to zero. If the linear system has any amount of damping, the undamped natural frequency can not be observed experimentally. The frequency which may be observed is the damped natural frequency $\omega_d$, which is equal to $\omega_n\sqrt{1 - \zeta^2}$. This frequency is always lower than the undamped natural frequency. An increase in $\zeta$ would reduce the damped natural frequency $\omega_d$. If $\zeta$ is increased beyond unity, the response becomes overdamped and will not oscillate.

(2) Critically damped case ($\zeta = 1$): If the two poles of $C(s)/R(s)$ are nearly equal, the system may be approximated by a critically damped one.

For a unit-step input, $R(s) = 1/s$ and $C(s)$ can be written

$$C(s) = \frac{\omega_n^2}{(s + \omega_n)^2 s} \tag{6-21}$$

The inverse Laplace transform of Eq. (6-21) may be found as

$$c(t) = 1 - e^{-\omega_n t}(1 + \omega_n t) \qquad (t \geq 0) \tag{6-22}$$

This result can, of course, be obtained by letting $\zeta$ approach unity in Eq. (6-19) and by using the following limit:

$$\lim_{\zeta \to 1} \frac{\sin \omega_d t}{\sqrt{1 - \zeta^2}} = \lim_{\zeta \to 1} \frac{\sin \omega_n \sqrt{1 - \zeta^2}\, t}{\sqrt{1 - \zeta^2}} = \omega_n t$$

(3) Overdamped case ($\zeta > 1$): In this case, the two poles of $C(s)/R(s)$ are negative real and unequal. For a unit-step input, $R(s) = 1/s$ and $C(s)$ can be written

$$C(s) = \frac{\omega_n^2}{(s + \zeta\omega_n + \omega_n\sqrt{\zeta^2 - 1})(s + \zeta\omega_n - \omega_n\sqrt{\zeta^2 - 1})s} \tag{6-23}$$

The inverse Laplace transform of Eq. (6-23) is

$$\begin{aligned}
c(t) &= 1 + \frac{1}{2\sqrt{\zeta^2 - 1}(\zeta + \sqrt{\zeta^2 - 1})} e^{-(\zeta + \sqrt{\zeta^2 - 1})\omega_n t} \\
&\quad - \frac{1}{2\sqrt{\zeta^2 - 1}(\zeta - \sqrt{\zeta^2 - 1})} e^{-(\zeta - \sqrt{\zeta^2 - 1})\omega_n t} \\
&= 1 + \frac{\omega_n}{2\sqrt{\zeta^2 - 1}}\left(\frac{e^{-s_1 t}}{s_1} - \frac{e^{-s_2 t}}{s_2}\right) \qquad (t \geq 0)
\end{aligned} \tag{6-24}$$

where $s_1 = (\zeta + \sqrt{\zeta^2 - 1})\omega_n$ and $s_2 = (\zeta - \sqrt{\zeta^2 - 1})\omega_n$. Thus, the response $c(t)$ includes two decaying exponential terms.

When $\zeta$ is appreciably greater than unity, one of the two decaying exponentials decreases much faster than the other, so that the faster decaying exponential term (which corresponds to a smaller time constant) may be neglected. That is, if $-s_2$ is located very much closer to the $j\omega$ axis than $-s_1$ (which means $|s_2| \ll |s_1|$), then for an approximate solution we may neglect $-s_1$. This is permissible because the effect of $-s_1$ on the response is much smaller than that of $-s_2$ since the term involving $s_1$ in Eq. (6-24) decays much faster than the term involving $s_2$. Once the faster decaying exponential term has disappeared, the response is similar to that of a first-order system and $C(s)/R(s)$ may be approximated by

$$\frac{C(s)}{R(s)} = \frac{\zeta\omega_n - \omega_n\sqrt{\zeta^2 - 1}}{s + \zeta\omega_n - \omega_n\sqrt{\zeta^2 - 1}} = \frac{s_2}{s + s_2}$$

This approximate form is a direct consequence of the fact that the initial values and final values of both the original $C(s)/R(s)$ and the approximate one agree with each other.

With the approximate transfer function $C(s)/R(s)$, the unit-step response can be obtained as

$$C(s) = \frac{\zeta\omega_n - \omega_n\sqrt{\zeta^2 - 1}}{(s + \zeta\omega_n - \omega_n\sqrt{\zeta^2 - 1})s}$$

The time response $c(t)$ is then

$$c(t) = 1 - e^{-(\zeta - \sqrt{\zeta^2 - 1})\omega_n t} \qquad (t \geq 0)$$

This gives an approximate unit-step response when one of the poles of $C(s)/R(s)$ can be neglected. An example of the approximate time-response function $c(t)$ with $\zeta = 2$, $\omega_n = 1$ is shown in Fig. 6-11, together with the exact solution for $c(t)$. The approximate solution is

$$c(t) = 1 - e^{-0.27t} \qquad (t \geq 0)$$

and the exact solution for this case is

$$c(t) = 1 + 0.077e^{-3.73t} - 1.077e^{-0.27t} \qquad (t \geq 0)$$

The major difference between the exact and approximate response curves is in the starting part of the response curves.

A family of curves $c(t)$ with various values of $\zeta$ is shown in Fig. 6-12, where the abscissa is the dimensionless variable $\omega_n t$. The curves are functions only of $\zeta$. These curves are obtained from Eqs. (6-19), (6-22), and (6-24). The system described by these equations was initially at rest.

Note that two second-order systems having the same $\zeta$ but different $\omega_n$ will

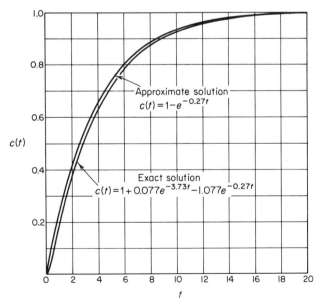

**Fig. 6-11.** Unit-step response curves of the system shown in Fig. 6-10. (Overdamped case.)

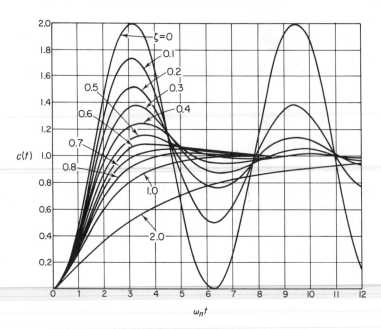

**Fig. 6-12.** Unit-step response curves of the system shown in Fig. 6-10.

exhibit the same overshoot and the same oscillatory pattern. Such systems are said to have the same relative stability.

It is important to note that for second-order systems whose closed-loop transfer functions are different from that given by Eq. (6-17), the step-response curves may look quite different from those shown in Fig. 6-12.

From Fig. 6-12, we see that an underdamped system with $\zeta$ between 0.5 and 0.8 gets close to the final value more rapidly than a critically damped or overdamped system. Among the systems responding without oscillation, a critically damped system exhibits the fastest response. An overdamped system is always sluggish in responding to any inputs.

**Definitions of transient-response specifications.** In many practical cases, the desired performance characteristics of control systems are specified in terms of time-domain quantities. Systems with energy storage cannot respond instantaneously and will exhibit transient responses whenever they are subjected to inputs or disturbances.

Frequently, the performance characteristics of a control system are specified in terms of the transient response to a unit-step input since it is easy to generate and is sufficiently drastic. (If the response to a step input is known, it is mathematically possible to compute the response to any input.)

The transient response of a system to a unit-step input depends on the initial conditions. For convenience in comparing transient responses of various systems, it is a common practice to use the standard initial condition that the system is at rest initially with output and all time derivatives thereof zero. Then the response characteristics can easily be compared.

The transient response of a practical control system often exhibits damped oscillations before reaching steady state. In specifying the transient-response characteristics of a control system to a unit-step input, it is common to specify the following:

1. delay time, $t_d$
2. rise time, $t_r$
3. peak time, $t_p$
4. maximum overshoot, $M_p$
5. settling time, $t_s$

These specifications are defined in what follows and are shown graphically in Fig. 6-13.

1. Delay time, $t_d$: The delay time is the time required for the response to reach half the final value the very first time.

2. Rise time, $t_r$: The rise time is the time required for the response to rise from 10 to 90%, 5 to 95%, or 0 to 100% of its final value. For underdamped second-order systems, the 0–100% rise time is normally used. For overdamped systems, the 10–90% rise time is commonly used.

3. Peak time, $t_p$: The peak time is the time required for the response to reach the first peak of the overshoot.

4. Maximum (per cent) overshoot, $M_p$: The maximum overshoot is the maximum peak value of the response curve measured from unity. If the final steady-state value of the response differs from unity, then it is common to use the maximum per cent overshoot. It is defined by

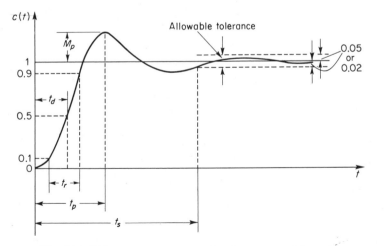

**Fig. 6-13.** Unit-step response curve showing $t_d$, $t_r$, $t_p$, $M_p$, and $t_s$.

$$\text{Maximum per cent overshoot} = \frac{c(t_p) - c(\infty)}{c(\infty)} \times 100\%$$

The amount of the maximum (per cent) overshoot directly indicates the relative stability of the system.

5. Settling time, $t_s$: The settling time is the time required for the response curve to reach and stay within a range about the final value of size specified by absolute percentage of the final value (usually 5% or 2%). The settling time is related to the largest time constant of the control system. Which percentage error criterion to use may be determined from the objectives of the system design in question.

The time-domain specifications just given are quite important since most control systems are time-domain systems; that is, they must exhibit acceptable time responses. (This means that the control system must be modified until the transient response is satisfacotry.) Note that if we specify the values of $t_d$, $t_r$, $t_p$, $t_s$, and $M_p$, then the shape of the response curve is virtually determined. This may be seen clearly from Fig. 6-14.

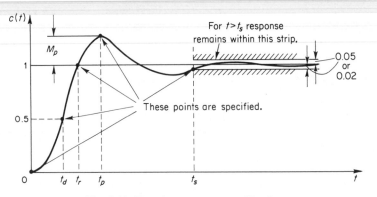

**Fig. 6-14.** Transient-response specifications.

Note that not all these specifications necessarily apply to any given case. For example, for an overdamped system, the terms peak time and maximum overshoot do not apply. (For systems which yield steady-state errors for step inputs, this error must be kept within a specified percentage level. Detailed discussions of steady-state errors are postponed until Chapter 7.)

**A few comments on transient-response specifications.** Except for certain applications where oscillations cannot be tolerated, it is desirable that the transient response be sufficiently fast and be sufficiently damped. Thus for a desirable transient response of a second-order system, the damping ratio must be between 0.4 and 0.8. Small values of $\zeta$ ($\zeta < 0.4$) yield excessive overshoot in the transient response, and a system with a large value of $\zeta$ ($\zeta > 0.8$) responds sluggishly.

We shall see later that the maximum overshoot and the rise time conflict with each other. In other words, both the maximum overshoot and the rise time cannot be made smaller simultaneously. If one of them is made smaller, the other necessarily becomes larger.

**Second-order systems and transient-response specifications.** In the following, we shall obtain the rise time, peak time, maximum overshoot, and settling time of the second-order system given by Eq. (6-17). These values will be obtained in terms of $\zeta$ and $\omega_n$. The system is assumed to be underdamped.

Rise time $t_r$: Referring to Eq. (6-19), we obtain the rise time $t_r$ by letting $c(t_r) = 1$ or

$$c(t_r) = 1 = 1 - e^{-\zeta \omega_n t_r}\left(\cos \omega_d t_r + \frac{\zeta}{\sqrt{1-\zeta^2}} \sin \omega_d t_r\right) \tag{6-25}$$

Since $e^{-\zeta \omega_n t_r} \neq 0$, we obtain from Eq. (6-25) the following equation:

$$\cos \omega_d t_r + \frac{\zeta}{\sqrt{1-\zeta^2}} \sin \omega_d t_r = 0$$

or

$$\tan \omega_d t_r = -\frac{\sqrt{1-\zeta^2}}{\zeta} = -\frac{\omega_d}{\sigma}$$

Thus, the rise time $t_r$ is

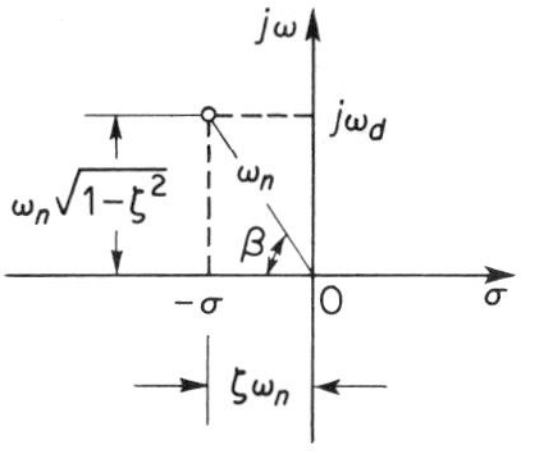

**Fig. 6-15.** Definition of the angle $\beta$.

$$t_r = \frac{1}{\omega_d} \tan^{-1}\left(\frac{\omega_d}{-\sigma}\right) = \frac{\pi - \beta}{\omega_d} \tag{6-26}$$

where $\beta$ is defined in Fig. 6-15. It can be seen that the value of $\tan^{-1}(-\sqrt{1-\zeta^2}/\zeta)$ lies between $\pi/2$ and $\pi$. If $\zeta = 0+$, then $\tan^{-1}(-\sqrt{1-\zeta^2}/\zeta) = \pi+/2$; and if $\zeta = 1-$, then $\tan^{-1}(-\sqrt{1-\zeta^2}/\zeta) = \pi-$. Clearly, for a small value of $t_r$, $\omega_n$ must be large.

Peak time $t_p$: Referring to Eq. (6-19), we may obtain the peak time by differentiating $c(t)$ with respect to time and letting this derivative equal zero, or

$$\left.\frac{dc}{dt}\right|_{t=t_p} = (\sin \omega_d t_p)\frac{\omega_n}{\sqrt{1-\zeta^2}} e^{-\zeta \omega_n t_p} = 0$$

This yields the following equation:

$$\sin \omega_d t_p = 0$$

or

$$\omega_d t_p = 0, \pi, 2\pi, 3\pi, \ldots$$

Since the peak time corresponds to the first peak overshoot, $\omega_d t_p = \pi$. Hence

$$t_p = \frac{\pi}{\omega_d} \tag{6-27}$$

The peak time $t_p$ corresponds to one-half cycle of the frequency of damped oscillation.

Maximum overshoot $M_p$: The maximum overshoot occurs at the peak time or at $t = t_p = \pi/\omega_d$. Thus, from Eq. (6-19), $M_p$ is obtained as

$$M_p = c(t_p) - 1$$

$$= -e^{-\zeta\omega_n(\pi/\omega_d)}\left(\cos\pi + \frac{\zeta}{\sqrt{1-\zeta^2}}\sin\pi\right)$$

$$= e^{-(\sigma/\omega_d)\pi} = e^{-(\zeta/\sqrt{1-\zeta^2})\pi} \qquad (6\text{-}28)$$

The maximum per cent overshoot is $e^{-(\sigma/\omega_d)\pi} \times 100\%$.

Settling time $t_s$: For an underdamped second-order system, the transient response is obtained from Eq. (6-19) as

$$c(t) = 1 - \frac{e^{-\zeta\omega_n t}}{\sqrt{1-\zeta^2}}\sin\left(\omega_d t + \tan^{-1}\frac{\sqrt{1-\zeta^2}}{\zeta}\right) \qquad (t \geq 0)$$

The curves $1 \pm (e^{-\zeta\omega_n t}/\sqrt{1-\zeta^2})$ are the envelope curves of the transient response for a unit-step input. The response curve $c(t)$ always remains within a pair of the envelope curves, as shown in Fig. 6-16. The time constant of these envelope curves is $1/\zeta\omega_n$.

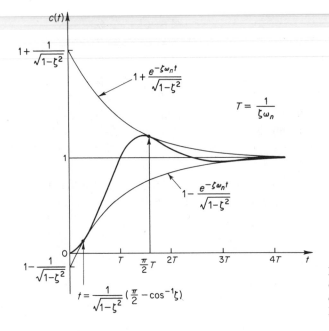

**Fig. 6-16.** A pair of envelope curves for the unit-step response curve of the system shown in Fig. 6-10.

The speed of decay of the transient response depends on the value of the time constant $1/\zeta\omega_n$. For a given $\omega_n$, the settling time $t_s$ is a function of the damping ratio $\zeta$. From Fig. 6-12, we see that for the same $\omega_n$ and for a range of $\zeta$ between 0 and 1 the settling time $t_s$ for a very lightly damped system is larger than that for a properly damped system. For an overdamped system, the settling time $t_s$ becomes large because of the sluggish start of the response.

The settling time corresponding to a $\pm 2\%$ or $\pm 5\%$ tolerance band may be measured in terms of the time constant $T = 1/\zeta\omega_n$ from the curves of Fig. 6-12

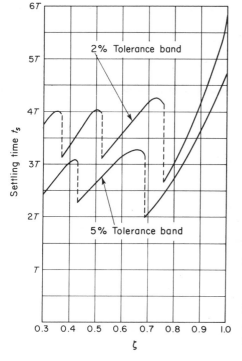

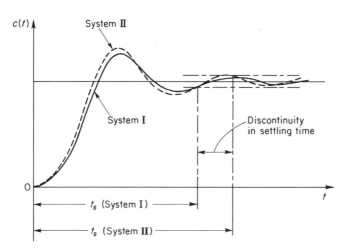

for different values of $\zeta$. The results are shown in Fig. 6-17. For $0 < \zeta < 0.9$, if the 2% criterion is used, $t_s$ is approximately four times the time constant of the system or

$$t_s = 4T = \frac{4}{\sigma} = \frac{4}{\zeta\omega_n} \qquad (6\text{-}29)$$

If the 5% criterion is used, then $t_s$ is approximately three times the time constant or

$$t_s = 3T = \frac{3}{\sigma} = \frac{3}{\zeta\omega_n} \qquad (6\text{-}30)$$

The settling time given by Eq. (6-29) [or Eq. (6-30)] reaches a minimum value around $\zeta = 0.76$ [or $\zeta = 0.68$] and then increases almost linearly for large values of $\zeta$. The discontinuities in the curves of Fig. 6-17 arise because an infinitesimal change in the value of $\zeta$ can cause a finite change in the settling time, as seen in Fig. 6-18.

Note that the settling time is inversely proportional to the product of the damping ratio and the undamped natural frequency of the system. Since the value of $\zeta$ is usually determined from the requirement of permissible maximum overshoot, the settling time is determined

**Fig. 6-17.** Settling time $t_s$ versus $\zeta$ curves.

**Fig. 6-18.** Unit-step response curves showing a discontinuity in the settling time.

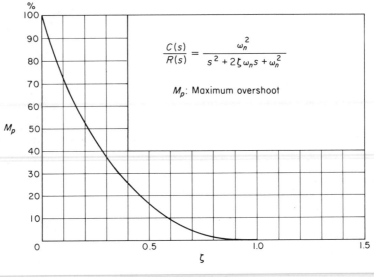

**Fig. 6-19.** $M_p$ versus $\zeta$ curve.

primarily by the undamped natural frequency $\omega_n$. This means that the duration of the transient period may be varied, without changing the maximum overshoot, by adjusting the undamped natural frequency $\omega_n$.

From the preceding analysis, it is evident that for rapid response $\omega_n$ must be large. In order to limit the maximum overshoot $M_p$ and to make the settling time small, the damping ratio $\zeta$ should not be too small. The relationship between the maximum per cent overshoot $M_p$ and the damping ratio $\zeta$ is presented in Fig. 6-19. Note that if the damping ratio is between 0.4 and 0.8, then the maximum per cent overshoot for step response is between 25% and 2.5%.

*Example 6-1.* Consider the system shown in Fig. 6-10, where $\zeta = 0.6$ and $\omega_n = 5$ rad/sec. Let us obtain the rise time $t_r$, peak time $t_p$, maximum overshoot $M_p$, and settling time $t_s$ when the system is subjected to a unit-step input.

From the given values of $\zeta$ and $\omega_n$, we obtain $\omega_d = \omega_n\sqrt{1-\zeta^2} = 4$ and $\sigma = \zeta\omega_n = 3$.

Rise time $t_r$: The rise time is

$$t_r = \frac{\pi - \beta}{\omega_d} = \frac{3.14 - \beta}{4}$$

where $\beta$ is given by

$$\beta = \tan^{-1}\frac{\omega_d}{\sigma} = \tan^{-1}\frac{4}{3} = 0.93 \text{ rad}$$

The rise time $t_r$ is thus

$$t_r = \frac{3.14 - 0.93}{4} = 0.55 \text{ sec}$$

Peak time $t_p$: The peak time is

$$t_p = \frac{\pi}{\omega_d} = \frac{3.14}{4} = 0.785 \text{ sec}$$

Maximum overshoot $M_p$: The maximum overshoot is

$$M_p = e^{-(\sigma/\omega_d)\pi} = e^{-(3/4) \times 3.14} = 0.095$$

The maximum per cent overhsoot is thus 9.5%.

Settling time $t_s$: For the 2% criterion, the settling time is

$$t_s = \frac{4}{\sigma} = \frac{4}{3} = 1.33 \text{ sec}$$

For the 5% criterion,

$$t_s = \frac{3}{\sigma} = \frac{3}{3} = 1 \text{ sec}$$

**Impulse response of second-order systems.** For a unit-impulse input $r(t)$, the corresponding Laplace transform is unity, or $R(s) = 1$. The unit-impulse response $C(s)$ of the second-order system shown in Fig. 6-10 is

$$C(s) = \frac{\omega_n^2}{s^2 + 2\zeta\omega_n s + \omega_n^2}$$

The inverse Laplace transform of this equation yields the time solution for the response $c(t)$ as follows:

For $0 \leq \zeta < 1$,

$$c(t) = \frac{\omega_n}{\sqrt{1 - \zeta^2}} e^{-\zeta\omega_n t} \sin \omega_n \sqrt{1 - \zeta^2} t \qquad (t \geq 0) \qquad (6\text{-}31)$$

For $\zeta = 1$,

$$c(t) = \omega_n^2 t e^{-\omega_n t} \qquad (t \geq 0) \qquad (6\text{-}32)$$

For $\zeta > 1$,

$$c(t) = \frac{\omega_n}{2\sqrt{\zeta^2 - 1}} e^{-(\zeta - \sqrt{\zeta^2 - 1})\omega_n t} - \frac{\omega_n}{2\sqrt{\zeta^2 - 1}} e^{-(\zeta + \sqrt{\zeta^2 - 1})\omega_n t} \qquad (t \geq 0) \qquad (6\text{-}33)$$

Note that without taking the inverse Laplace transform of $C(s)$, we can also obtain the time response $c(t)$ by differentiating the corresponding unit-step response since the unit-impulse function is the time derivative of the unit-step function. A family of unit-impulse response curves given by Eqs. (6-31) and (6-32) with various values of $\zeta$ is shown in Fig. 6-20. The curves $c(t)/\omega_n$ are plotted against the dimensionless variable $\omega_n t$, and thus they are functions only of $\zeta$. For the critically damped and overdamped cases, the unit-impulse response is always positive or zero; i.e., $c(t) \geq 0$. This can be seen from Eqs. (6-32) and (6-33). For the underdamped case, the unit-impulse response $c(t)$ oscillates about zero and takes both positive and negative values.

From the foregoing analysis, we may conclude that if the impulse response $c(t)$ does not change sign, the system is either critically damped or overdamped,

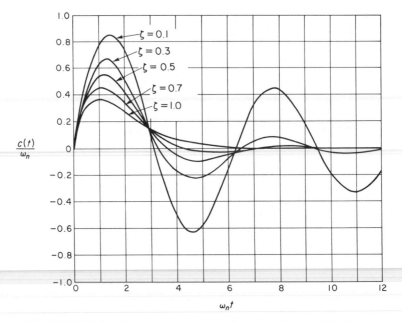

**Fig. 6-20.** Unit-impulse response curves of the system shown in Fig. 6-10.

in which case the corresponding step response does not overshoot but increases or decreases monotonically and approaches a constant value.

The maximum overshoot for the unit-impulse response of the underdamped system occurs at

$$t = \frac{\tan^{-1}\frac{\sqrt{1-\zeta^2}}{\zeta}}{\omega_n\sqrt{1-\zeta^2}} \qquad (0 < \zeta < 1)$$

and the maximum overshoot is

$$c(t)_{\max} = \omega_n \exp\left(-\frac{\zeta}{\sqrt{1-\zeta^2}}\tan^{-1}\frac{\sqrt{1-\zeta^2}}{\zeta}\right) \qquad (0 < \zeta < 1)$$

Since the unit-impulse response function is the time derivative of the unit-step response function, the maximum overshoot $M_p$ for the unit-step response can be found from the corresponding unit-impulse response. Namely, the area under the unit-impulse response curve from $t = 0$ to the time of the first zero, as shown in Fig. 6-21, is $1 + M_p$, where $M_p$ is the maximum overshoot (for the unit-step response) given by Eq. (6-28). The peak time $t_p$ (for the unit-step response) given by Eq. (6-27) corresponds to the time that the unit-impulse response first crosses the axis.

**Steady-state error for ramp responses.** The transient response of a second-order system when subjected to a ramp input can be obtained by a straightforward

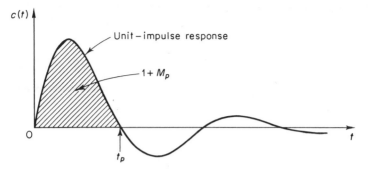

**Fig. 6-21.** Unit-impulse response curve of the system shown in Fig. 6-10.

method. (See Problem A-6-5.) In the present analysis, we shall examine the steady-state error when a second-order system is subjected to such an input.

For the system shown in Fig. 6-22, we obtain

$$E(s) = \frac{Js^2 + Fs}{Js^2 + Fs + K} R(s)$$

The steady-state error for the unit-ramp response can be obtained as follows: For a unit-ramp input $r(t) = t$, we obtain $R(s) = 1/s^2$. The steady-state error $e_{ss}$ is

$$
\begin{aligned}
e_{ss} &= \lim_{s \to 0} sE(s) \\
&= \lim_{s \to 0} s \frac{Js^2 + Fs}{Js^2 + Fs + K} \frac{1}{s^2} \\
&= \frac{F}{K} \\
&= \frac{2\zeta}{\omega_n}
\end{aligned}
$$

where

$$\zeta = \frac{F}{2\sqrt{KJ}}, \qquad \omega_n = \sqrt{\frac{K}{J}}$$

In order to assure acceptable transient response and acceptable steady-state error in following a ramp input, $\zeta$ must not be too small and $\omega_n$ must be sufficiently large. It is possible to make the steady-state error $e_{ss}$ small by making the value of the gain $K$ large. (A large value of $K$ has an additional advantage of suppressing

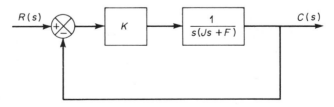

**Fig. 6-22.** Control system.

undesirable effects caused by dead zone, backlash, coulomb friction, etc.) A large value of $K$ would, however, make the value of $\zeta$ small and increase the maximum overshoot, which is undesirable.

It is therefore necessary to compromise between the magnitude of the steady-state error to a ramp input and the maximum overshoot to a unit-step input. In the system shown in Fig. 6-22, a reasonable compromise may not be reached easily. It is then desirable to consider other types of control action which may improve both the transient-response and steady-state behavior. In what follows, we shall consider two schemes to improve the behavior, one using a proportional-plus-derivative controller and the other using tachometer feedback.

**Proportional-plus-derivative control of second-order systems.** A compromise between acceptable transient-response behavior and acceptable steady-state behavior may be achieved by use of proportional-plus-derivative control action.

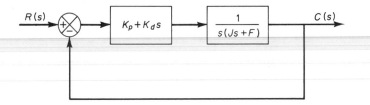

**Fig. 6-23.** Control system.

Consider the system shown in Fig. 6-23. The closed-loop transfer function is

$$\frac{C(s)}{R(s)} = \frac{K_p + K_d s}{Js^2 + (F + K_d)s + K_p}$$

The steady-state error for a unit-ramp input is

$$e_{ss} = \frac{F}{K_p}$$

The characteristic equation is

$$Js^2 + (F + K_d)s + K_p = 0 \qquad (6\text{-}34)$$

The effective damping coefficient of this system is thus $F + K_d$ rather than $F$. Since the damping ratio $\zeta$ of this system is

$$\zeta = \frac{F + K_d}{2\sqrt{K_p J}}$$

it is possible to make both the steady-state error $e_{ss}$ for a ramp input and the maximum overshoot for a step input small by making $F$ small, $K_p$ large, and $K_d$ large enough so that $\zeta$ is between 0.4 and 0.7.

In the following, we shall examine the unit-step response of the system shown in Fig. 6-23. Let us define

$$\omega_n = \sqrt{\frac{K_p}{J}}, \qquad z = \frac{K_p}{K_d}$$

The closed-loop transfer function can then be written

$$\frac{C(s)}{R(s)} = \frac{\omega_n^2}{z}\frac{s+z}{s^2 + 2\zeta\omega_n s + \omega_n^2}$$

When a second-order system has a zero near the closed-loop poles, the transient-response behavior becomes considerably different from that of a second-order system without a zero.

If the zero at $s = -z$ is located close to the $j\omega$ axis, the effect of the zero on the unit-step response is quite significant. Typical step-response curves of this system with $\zeta = 0.5$ and various values of $z/(\zeta\omega_n)$ are shown in Fig. 6-24.

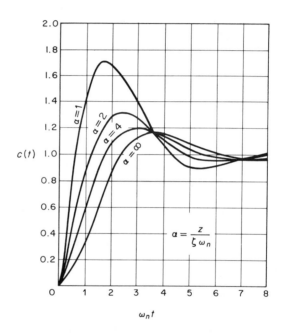

**Fig. 6-24.** Unit-step response curves of the second-order system

$$\frac{C(s)}{R(s)} = \left(\frac{\omega_n^2}{z}\right)\left(\frac{s+z}{s^2 + 2\zeta\omega_n s + \omega_n^2}\right)$$
$$\zeta = 0.5$$

**Tachometers.** Another method for improving servomechanism performance is to add tachometer feedback. This method is easily adjustable and relatively economical. A dc tachometer is a generator which produces a voltage proportional to its rotating speed. It is used as a transducer, converting the velocity of the rotating shaft into a proportional dc voltage. In a dc tachometer, a permanent magnet supplies a constant air gap flux. Thus, the tachometer-induced voltage $e$ can be written

$$e = K\psi\dot{\theta} = K_1\dot{\theta}$$

where $K$ is a constant, $\psi$ is the air gap flux, and $\dot{\theta}$ is the rotating speed. Since $\psi$ is a constant, $K_1 = K\psi$ is also a constant. The transfer function of the dc tachometer is

$$\frac{E(s)}{\Theta(s)} = K_1 s$$

A schematic diagram of a dc tachometer is shown in Fig. 6-25 (a). Although

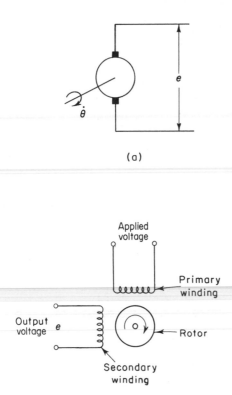

(a)

Applied voltage

Primary winding

Output voltage  e

Rotor

Secondary winding

(b)

**Fig. 6-25.** (a) Schematic diagram of a dc tachometer; (b) schematic diagram of an ac tachometer.

its output is a dc voltage, if this is converted into an ac voltage, the dc tachometer can also be used in ac servomechanisms.

The ac tachometer is a device which is similar to a two-phase induction motor. (Two-phase motors generally make good ac tachometers.) Figure 6-25(b) shows the schematic diagram of an ac tachometer. An ac voltage of rated value is applied to the primary winding of the tachometer. The secondary winding is placed 90 degrees apart in space from the primary winding. Thus, when the rotor shaft is stationary, the output voltage is zero. When the rotor shaft is rotating, the output voltage at the secondary winding is proportional to the rotor velocity. The polarity of the output voltage is determined by the direction of rotation. The transfer function of an ac tachometer is

$$\frac{E(s)}{\Theta(s)} = Ks$$

where $E(s)$ is the Laplace transform of the output voltage, $\Theta(s)$ is the Laplace transform of the rotor position, and $K$ is a constant. Although the output of an ac tachometer is an ac voltage, this tachometer can be used in a dc servomechanism if the output ac voltage is converted into a dc voltage by use of a demodulator.

Note that a tachometer used for damping purposes is usually built as an integral part of the servomotor.

**Servomechanisms with velocity feedback.** The derivative of the output signal can be used to improve system performance. In obtaining the derivative of the output position signal, it is desirable to use a tachometer instead of physically differentiating the output signal. (Note that differentiation amplifies noise effects. In fact, if discontinuous noises are present, differentiation amplifies the discontinuous noises more than the useful signal. For example, the output of a potentiometer is a discontinuous voltage signal because, as the potentiometer brush is moving on the windings, voltages are induced in the switchover turns and thus generate transients. The output of the potentiometer therefore should not be followed by a differentiating element.)

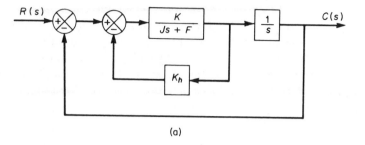

(a)

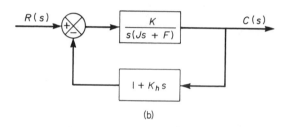

(b)

**Fig. 6-26.** (a) Block diagram of a servomechanism; (b) simplified block diagram.

Consider the servomechanism shown in Fig. 6-26 (a). In this device, the velocity signal, together with the positional signal, is fed back to the input to produce the actuating error signal. In any servomechanism, such a velocity signal can be easily generated by a tachometer. The block diagram shown in Fig. 6-26 (a) can be simplified, as shown in Fig. 6-26 (b), giving

$$\frac{C(s)}{R(s)} = \frac{K}{Js^2 + (F + KK_h)s + K}$$

The characteristic equation is

$$Js^2 + (F + KK_h)s + K = 0 \tag{6-35}$$

Comparing Eq. (6-35) with Eq. (6-34), we see that they are of the same form. If $KK_h$ and $K$ were equal to $K_d$ and $K_p$, respectively, then the two equations would be identical. Hence we can expect that velocity feedback gives a similar improvement on system performance as does proportional-plus-derivative control action.

The steady-state error for a unit-ramp input is

$$e_{ss} = \frac{F}{K}$$

The damping ratio $\zeta$ is

$$\zeta = \frac{F + KK_h}{2\sqrt{KJ}} \tag{6-36}$$

The undamped natural frequency $\omega_n = \sqrt{K/J}$ is not affected by velocity feedback. Noting that the maximum overshoot for a unit-step input can be controlled by controlling the value of the damping ratio $\zeta$, we can reduce both the steady-state error $e_{ss}$ and the maximum overshoot by making $F$ small and $K$ large and then adjusting the velocity feedback constant $K_h$ so that $\zeta$ is between 0.4 and 0.7.

Remember that velocity feedback has the effect of increasing the damping ratio without affecting the undamped natural frequency of the system.

---

*Example 6-2.* For the system shown in Fig. 6-27, determine the values of gain $K$ and velocity feedback constant $K_h$ so that the maximum overshoot in the unit step response is 0.2 and the peak time is 1 sec. With these values of $K$ and $K_h$, obtain the rise time and settling time.

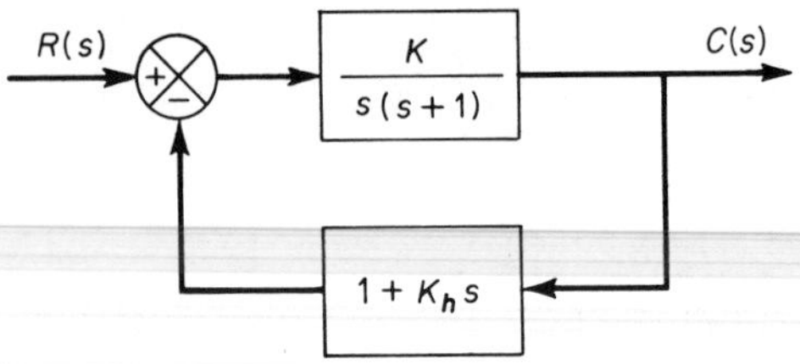

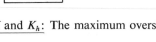

**Fig. 6-27.** Control system.

Determination of the values of $K$ and $K_h$: The maximum overshoot $M_p$ is given by Eq. (6-28) as

$$M_p = e^{-(\zeta/\sqrt{1-\zeta^2})\pi}$$

This value must be 0.2. Thus,

$$e^{-(\zeta/\sqrt{1-\zeta^2})\pi} = 0.2$$

or

$$\frac{\zeta\pi}{\sqrt{1-\zeta^2}} = 1.61$$

which yields

$$\zeta = 0.456$$

The peak time $t_p$ is specified as 1 sec; therefore, from Eq. (6-27),

$$t_p = \frac{\pi}{\omega_d} = 1$$

or

$$\omega_d = 3.14$$

Since $\zeta$ is 0.456, $\omega_n$ is

$$\omega_n = \frac{\omega_d}{\sqrt{1-\zeta^2}} = 3.53$$

Since the natural frequency $\omega_n$ is equal to $\sqrt{K}$ in this example,

$$K = \omega_n^2 = 12.5$$

Then, $K_h$ is, from Eq. (6-36),

$$K_h = \frac{2\sqrt{K}\zeta - 1}{K} = 0.178$$

Rise time $t_r$: From Eq. (6-26), the rise time $t_r$ is

$$t_r = \frac{\pi - \beta}{\omega_d}$$

where

$$\beta = \tan^{-1} \frac{\omega_d}{\sigma} = \tan^{-1} 1.95 = 1.10$$

Thus, $t_r$ is

$$t_r = 0.65 \text{ sec}$$

Settling time $t_s$: For the 2% criterion,

$$t_s = \frac{4}{\sigma} = 2.48 \text{ sec}$$

For the 5% criterion,

$$t_s = \frac{3}{\sigma} = 1.86 \text{ sec}$$

## 6-5 HIGHER-ORDER SYSTEMS

In this section, we shall first discuss the unit-step response of a particular type of third-order system. We shall then present a transient response analysis of higher-order systems in general terms. Finally, we shall present a discussion of stability analysis in the complex plane.

**Unit-step response of third-order systems.** We shall discuss the unit-step response of a commonly encountered third-order system whose closed-loop transfer function is

$$\frac{C(s)}{R(s)} = \frac{\omega_n^2 p}{(s^2 + 2\zeta\omega_n s + \omega_n^2)(s + p)}$$

The unit-step response of this system can be obtained as follows:

$$c(t) = 1 - \frac{e^{-\zeta\omega_n t}}{\beta\zeta^2(\beta - 2) + 1} \left\{ \beta\zeta^2(\beta - 2) \cos \sqrt{1 - \zeta^2}\, \omega_n t \right.$$
$$\left. + \frac{\beta\zeta[\zeta^2(\beta - 2) + 1]}{\sqrt{1 - \zeta^2}} \sin \sqrt{1 - \zeta^2}\, \omega_n t \right\} - \frac{e^{-pt}}{\beta\zeta^2(\beta - 2) + 1} \qquad (t \geq 0)$$

where

$$\beta = \frac{p}{\zeta\omega_n}$$

Note that since

$$\beta\zeta^2(\beta - 2) + 1 = \zeta^2(\beta - 1)^2 + (1 - \zeta^2) > 0$$

the coefficient of the term $e^{-pt}$ is always negative.

The effect of the real pole $s = -p$ on the unit-step response is that of reducing the maximum overshoot and increasing the settling time. Figure 6-28 shows unit-step response curves of this third-order system with $\zeta = 0.5$. The ratio $\beta = p/(\zeta\omega_n)$ is a parameter in the family of curves.

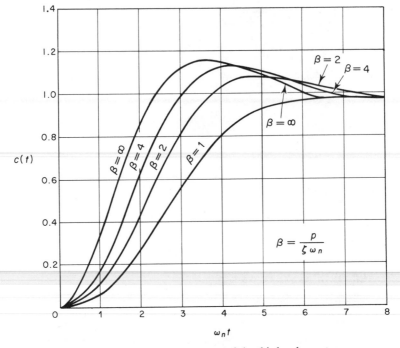

**Fig. 6-28.** Unit-step response curves of the third-order system

$$\frac{C(s)}{R(s)} = \frac{\omega_n^2 p}{(s^2 + 2\zeta\omega_n s + \omega_n^2)(s + p)}, \qquad \zeta = 0.5$$

If the real pole is located to the right of the complex-conjugate poles, then there is a tendency for sluggish response. The system will behave like an overdamped system. The complex-conjugate poles will add ripple to the response curve.

**Transient response of higher-order systems.** Consider the system shown in Fig. 6-29. The closed-loop transfer function is

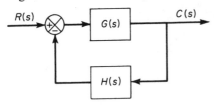

**Fig. 6-29.** Control system.

$$\frac{C(s)}{R(s)} = \frac{G(s)}{1 + G(s)H(s)} \qquad (6\text{-}37)$$

In general, $G(s)$ and $H(s)$ are given as ratios of polynomials in $s$, or

$$G(s) = \frac{p(s)}{q(s)} \qquad \text{or} \qquad H(s) = \frac{n(s)}{d(s)}$$

where $p(s)$, $q(s)$, $n(s)$, and $d(s)$ are polynomials in $s$. The closed-loop transfer function given by Eq. (6-37) may then be written

$$\frac{C(s)}{R(s)} = \frac{p(s)d(s)}{q(s)d(s) + p(s)n(s)}$$

$$= \frac{b_0 s^m + b_1 s^{m-1} + \cdots + b_{m-1}s + b_m}{a_0 s^n + a_1 s^{n-1} + \cdots + a_{n-1}s + a_n}$$

In order to determine the transient response of this system to any given input, it is necessary to factor the denominator polynomial. (Various methods are available for factoring polynomials. In this book we shall present a convenient technique called the root-locus method. See Chapter 8.) Once the denominator polynomial has been factored, $C(s)/R(s)$ can be written

$$\frac{C(s)}{R(s)} = \frac{K(s + z_1)(s + z_2) \cdots (s + z_m)}{(s + p_1)(s + p_2) \cdots (s + p_n)} \tag{6-38}$$

Let us examine the response behavior of this system to a unit-step input. We assume that the closed-loop poles are all distinct. (This is usually the case in practical systems.) For a unit-step input, Eq. (6-38) can be written

$$C(s) = \frac{a}{s} + \sum_{i=1}^{n} \frac{a_i}{s + p_i} \tag{6-39}$$

where $a_i$ is the residue of the pole at $s = -p_i$.

If all closed-loop poles lie in the left-half $s$ plane, the relative magnitudes of the residues determine the relative importance of the components in the expanded form of $C(s)$. If there is a closed-loop zero close to a closed-loop pole, then the residue at this pole is small and the coefficient of the transient-response term corresponding to this pole becomes small. A pair of closely located poles and zeros will effectively cancel each other. If a pole is located very far from the origin, the residue at this pole may be small. The transients corresponding to such a remote pole are small and last a short time. Terms in the expanded form of $C(s)$ having very small residues contribute little to the transient response, and these terms may be neglected. If this is done, the higher-order system may be approximated by a lower-order one. (Such an approximation often enables us to estimate the response characteristics of a higher-order system from those of a simplified one.)

The poles of $C(s)$ consist of real poles and pairs of complex-conjugate poles. A pair of complex-conjugate poles yields a second-order term in $s$. Since the factored form of the higher-order characteristic equation consists of first- and second-order terms, Eq. (6-39) can be rewritten

$$C(s) = \frac{K \prod_{i=1}^{m} (s + z_i)}{s \prod_{j=1}^{q} (s + p_j) \prod_{k=1}^{r} (s^2 + 2\zeta_k \omega_k s + \omega_k^2)} \tag{6-40}$$

where $q + 2r = n$. If the closed-loop poles are distinct, Eq. (6-40) can be expanded into partial fractions as follows:

$$C(s) = \frac{a}{s} + \sum_{j=1}^{q} \frac{a_j}{s + p_j} + \sum_{k=1}^{r} \frac{b_k(s + \zeta_k \omega_k) + c_k \omega_k \sqrt{1 - \zeta_k^2}}{s^2 + 2\zeta_k \omega_k s + \omega_k^2}$$

From this last equation, we see that the response of a higher-order system is composed of a number of terms involving the simple functions found in the responses of first- and second-order systems. The unit-step response $c(t)$, the inverse Laplace transform of $C(s)$, is then

$$c(t) = a + \sum_{j=1}^{q} a_j e^{-p_j t} + \sum_{k=1}^{r} b_k e^{-\zeta_k \omega_k t} \cos \omega_k \sqrt{1 - \zeta_k^2}\, t$$

$$+ \sum_{k=1}^{r} c_k e^{-\zeta_k \omega_k t} \sin \omega_k \sqrt{1 - \zeta_k^2}\, t \qquad (t \geq 0) \tag{6-41}$$

If all closed-loop poles lie in the left-half $s$ plane, then the exponential terms and the damped exponential terms in Eq. (6-41) will approach zero as time $t$ increases. The steady-state output is then $c(\infty) = a$.

Let us assume that the system considered is a stable one. Then the closed-loop poles which are located far from the $j\omega$ axis have large negative real parts. The exponential terms which correspond to these poles decay very rapidly to zero. (Note that the horizontal distance from a closed-loop pole to the $j\omega$ axis determines the settling time of transients due to that pole. The smaller the distance, the longer the settling time.)

The response curve of a stable higher-order system is the sum of a number of exponential curves and damped sinusoidal curves. Examples of step-response curves of higher-order systems are shown in Fig. 6-30. A particular characteristic of such response curves is that small oscillations are superimposed upon larger oscillations or upon exponential curves. Fast-decaying components have significance only in the initial part of the transient response.

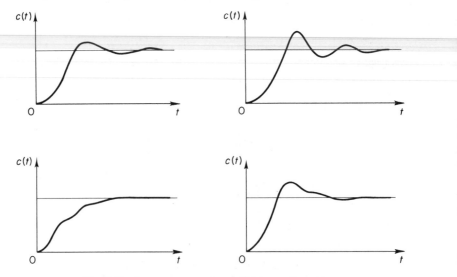

**Fig. 6-30.** Step-response curves of higher-order systems.

Remember that the type of transient response is determined by the closed-loop poles, while the shape of the transient response is primarily determined by the closed-loop zeros. As we have seen earlier, the poles of the input $R(s)$ yield the steady-state response terms in the solution, while the poles of $C(s)/R(s)$ enter into the exponential transient-response terms. (The closed-loop poles always appear in the time solution as exponential response terms.) The zeros of $C(s)/R(s)$ do not affect the exponents in the exponential response terms, but they do affect the magnitudes and signs of the residues. (The actual time-response curve depends on the coefficients of the individual exponential terms, as well as on their exponents.)

**Dominant closed-loop poles.** The relative dominance of closed-loop poles is determined by the ratio of the real parts of the closed-loop poles, as well as by the relative magnitudes of the residues evaluated at the closed-loop poles. The magnitudes of the residues depend upon both the closed-loop poles and zeros.

If the ratios of the real parts exceed five, and there are no zeros nearby, then the closed-loop poles nearest the $j\omega$ axis will dominate in the transient-response behavior because these poles correspond to transient-response terms which decay slowly. Those closed-loop poles which have dominant effects upon the transient-response behavior are called *dominant* poles. Quite often the dominant closed-loop poles occur in the form of a complex-conjugate pair. The dominant closed-loop poles are most important among all closed-loop poles.

The gain of a higher-order system is often adjusted so that there will exist a pair of dominant complex-conjugate closed-loop poles. The presence of such poles in a stable system reduces the effect of such nonlinearities as dead zone, backlash, and coulomb friction.

Remember that although the concept of dominant closed-loop poles is useful for estimating the dynamic behavior of a closed-loop system, we must be careful to see that the underlying assumptions are met before using it.

**Nonoscillatory response of higher-order systems.** If the closed-loop system does not possess complex-conjugate poles, then the transient response is nonoscillatory. Nonoscillatory transients may be plotted on semilog paper, and if the largest time constant is dominant, the plot will approach a straight line whose slope is determined by the time constant. Subtraction of the part of the transient due to this time constant will yield other smaller time constants.

**Stability analysis in the complex plane.** The stability of a linear closed-loop system can be determined from the location of the closed-loop poles in the $s$ plane. If any of these poles lie in the right-half $s$ plane, then with increasing time they give rise to the dominant mode, and the transient response increases monotonically or oscillates with increasing amplitude. This represents an unstable system. For such a system, as soon as the power is turned on, the output may increase with time. If no saturation takes place in the system and no mechanical stop is provided, then the system may eventually be subjected to damage and fail since the response of a real physical system cannot increase indefinitely. Therefore, closed-loop poles in the right-half $s$ plane are not permissible in the usual linear control system. If all closed-loop poles lie to the left of the $j\omega$ axis, any transient response eventually reaches equilibrium. This represents a stable system.

Whether a linear system is stable or unstable is a property of the system itself and does not depend on the input or driving function of the system. The poles of the input, or driving function, do not affect the property of stability of the system but they contribute only to steady-state response terms in the solution. Thus, the problem of absolute stability can be solved readily by choosing no closed-loop poles in the right-half $s$ plane, including the $j\omega$ axis. (Mathematically, closed-loop

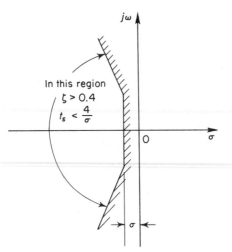

**Fig. 6-31.** Region in the complex plane satisfying the conditions $\zeta > 0.4$ and $t_s < 4/\sigma$.

poles on the $j\omega$ axis will yield oscillations, the amplitude of which is neither decaying nor growing with time. In practical cases, where noise is present, however, the amplitude of oscillations may increase at a rate determined by the noise power level. Therefore, a control system should not have closed-loop poles on the $j\omega$ axis.)

Note that the mere fact that all closed-loop poles lie in the left-half $s$ plane does not guarantee satisfactory transient-response characteristics. If dominant complex-conjugate closed-loop poles lie close to the $j\omega$ axis, the transient response may exhibit excessive oscillations or may be very slow. Therefore, to guarantee fast, yet well-damped, transient-response characteristics, it is necessary that the closed-loop poles of the system lie in a particular region in the complex plane, such as the region bounded by the shaded area in Fig. 6-31.

Since relative stability and transient performance of a closed-loop control system are directly related to the closed-loop pole-zero configuration in the $s$ plane, it is frequently necessary to adjust one or more system parameters in order to obtain suitable configurations. The effects of varying system parameters on the closed-loop poles will be discussed in detail in Chapter 8.

## 6-6 ROUTH'S STABILITY CRITERION

The most important problem in linear control systems concerns stability. Namely, under what conditions will a system become unstable? If it is unstable, how should we stabilize the system? In Section 6-5 it was stated that a control system is stable if and only if all closed-loop poles lie in the left-half $s$ plane. Since most linear closed-loop systems have closed-loop transfer functions of the form

$$\frac{C(s)}{R(s)} = \frac{b_0 s^m + b_1 s^{m-1} + \cdots + b_{m-1}s + b_m}{a_0 s^n + a_1 s^{n-1} + \cdots + a_{n-1}s + a_n} = \frac{B(s)}{A(s)}$$

where the $a$'s and $b$'s are constants and $m \leq n$, we must first factor the polynomial $A(s)$ in order to find the closed-loop poles. This process is very time-consuming for a polynomial of degree greater than second. A simple criterion, known as Routh's stability criterion, enables us to determine the number of closed-loop poles which lie in the right-half $s$ plane without having to factor the polynomial.

**Routh's stability criterion.** Routh's stability criterion tells us whether or not there are positive roots in a polynomial equation without actually solving for them. This stability criterion applies to polynomials with only a finite number of terms. When the criterion is applied to a control system, information about absolute stability can be obtained directly from the coefficients of the characteristic equation.

The procedure in Routh's stability criterion is as follows:

1. Write the polynomial in $s$ in the following form:

$$a_0 s^n + a_1 s^{n-1} + \cdots + a_{n-1} s + a_n = 0 \qquad (6\text{-}42)$$

where the coefficients are real quantities. We assume that $a_n \neq 0$; that is, any zero root has been removed.

2. If any of the coefficients are zero or negative in the presence of at least one positive coefficient, there is a root or roots which are imaginary or which have positive real parts. Therefore, in such a case, the system is not stable. If we are interested in only the absolute stability, there is no need to follow the procedure further. Note that all the coefficients must be positive. This is a necessary condition, as may be seen from the following argument: A polynomial in $s$ having real coefficients can always be factored into linear and quadratic factors, such as $(s + a)$ and $(s^2 + bs + c)$, where $a$, $b$, and $c$ are real. The linear factors yield real roots and the quadratic factors yield complex roots of the polynomial. The factor $(s^2 + bs + c)$ yields roots having negative real parts only if $b$ and $c$ are both positive. In order for all roots to have negative real parts, the constants $a$, $b$, $c$, etc., in all factors must be positive. The product of any number of linear and quadratic factors containing only positive coefficients always yields a polynomial with positive coefficients. It is important to note that the condition that all the coefficients are positive is not sufficient to assure stability. The necessary but not sufficient condition for stability is that the coefficients of Eq. (6-42) all be present and all have a positive sign. (If all $a$'s are negative, they can be made positive by multiplying both sides of the equation by $-1$)

3. If all coefficients are positive, arrange the coefficients of the polynomial in rows and columns according to the following pattern:

$$
\begin{array}{cccccccc}
s^n & a_0 & a_2 & a_4 & a_6 & \cdot & \cdot & \cdot \\
s^{n-1} & a_1 & a_3 & a_5 & a_7 & \cdot & \cdot & \cdot \\
s^{n-2} & b_1 & b_2 & b_3 & b_4 & \cdot & \cdot & \cdot \\
s^{n-3} & c_1 & c_2 & c_3 & c_4 & \cdot & \cdot & \cdot \\
s^{n-4} & d_1 & d_2 & d_3 & d_4 & \cdot & \cdot & \cdot \\
& \cdot & \cdot & \cdot & & & & \\
& \cdot & \cdot & \cdot & & & & \\
& \cdot & \cdot & \cdot & & & & \\
s^2 & e_1 & e_2 & & & & & \\
s^1 & f_1 & & & & & & \\
s^0 & g_1 & & & & & & \\
\end{array}
$$

The coefficients $b_1, b_2, b_3$, etc., are evaluated as follows:

$$b_1 = \frac{a_1 a_2 - a_0 a_3}{a_1}$$

$$b_2 = \frac{a_1 a_4 - a_0 a_5}{a_1}$$

$$b_3 = \frac{a_1 a_6 - a_0 a_7}{a_1}$$

. . .

The evaluation of the $b$'s is continued until the remaining ones are all zero. The same pattern of cross multiplying the coefficients of the two previous rows is followed in evaluating the $c$'s, $d$'s, $e$'s, etc. Namely,

$$c_1 = \frac{b_1 a_3 - a_1 b_2}{b_1}$$

$$c_2 = \frac{b_1 a_5 - a_1 b_3}{b_1}$$

$$c_3 = \frac{b_1 a_7 - a_1 b_4}{b_1}$$

. . .

and

$$d_1 = \frac{c_1 b_2 - b_1 c_2}{c_1}$$

$$d_2 = \frac{c_1 b_3 - b_1 c_3}{c_1}$$

. . .

This process is continued until the $n$th row has been completed. The complete array of coefficients is triangular. Note that in developing the array, an entire row may be divided or multiplied by a positive number in order to simplify the subsequent numerical calculation without altering the stability conclusion.

Routh's stability criterion states that the number of roots of Eq. (6-42) with positive real parts is equal to the number of changes in sign of the coefficients of the first column of the array. It should be noted that the exact values of the terms in the first column need not be known; instead only the signs are needed. The necessary and sufficient condition that all roots of Eq. (6-42) lie in the left-half $s$ plane is that all the coefficients of Eq. (6-42) be positive and all terms in the first column of the array have positive signs.

---

*Example 6-3.* Let us apply Routh's stability criterion to the following third-order polynomial:

$$a_0 s^3 + a_1 s^2 + a_2 s + a_3 = 0$$

where all the coefficients are positive numbers. The array of coefficients becomes

$$
\begin{array}{ccc}
s^3 & a_0 & a_2 \\
s^2 & a_1 & a_3 \\
s^1 & \dfrac{a_1 a_2 - a_0 a_3}{a_1} & \\
s^0 & a_3 &
\end{array}
$$

The condition that all roots have negative real parts is given by

$$a_1 a_2 > a_0 a_3$$

---

*Example 6-4.* Consider the following polynomial:

$$s^4 + 2s^3 + 3s^2 + 4s + 5 = 0$$

Let us follow the procedure just presented and construct the array of coefficients. (The first two rows can be obtained directly from the given polynomial. The remaining terms are obtained from these. If any coefficients are missing, they may be replaced by zeros in the array.)

$$
\begin{array}{cccc}
s^4 & 1 & 3 & 5 \\
s^3 & 2 & 4 & 0 \\
& & & \\
s^2 & 1 & 5 & \\
s^1 & -6 & & \\
s^0 & 5 & &
\end{array}
\qquad
\begin{array}{cccc}
s^4 & 1 & 3 & 5 \\
s^3 & \cancel{2} & \cancel{4} & \cancel{0} \\
& 1 & 2 & 0 \\
s^2 & 1 & 5 & \\
s^1 & -3 & & \\
s^0 & 5 & &
\end{array}
$$

The second row is divided by two.

In this example, the number of changes in sign of the coefficients in the first column is two. This means that there are two roots with positive real parts. Note that the result is unchanged when the coefficients of any row are multiplied or divided by a positive number in order to simplify the computation.

**Special cases.** If a first-column term in any row is zero, but the remaining terms are not zero or there is no remaining term, then the zero term is replaced by a very small positive number $\epsilon$ and the rest of the array is evaluated. For example, consider the following equation:

$$s^3 + 2s^2 + s + 2 = 0 \tag{6-43}$$

The array of coefficients is

$$
\begin{array}{ccc}
s^3 & 1 & 1 \\
s^2 & 2 & 2 \\
s^1 & 0 \approx \epsilon & \\
s^0 & 2 &
\end{array}
$$

If the sign of the coefficient above the zero ($\epsilon$) is the same as that below it, it indicates that there are a pair of imaginary roots. Actually Eq. (6-43) has two roots at $s = \pm j$.

If, however, the sign of the coefficient above the zero ($\epsilon$) is opposite that below it, it indicates that there is one sign change. For example, for the following equation,

$$s^3 - 3s + 2 = (s - 1)^2(s + 2) = 0$$

the array of coefficients is

|     | $s^3$ | 1 | $-3$ |
|-----|-------|---|------|
| One sign change: | $s^2$ | $0 \approx \epsilon$ | 2 |
| One sign change: | $s^1$ | $-3 - \dfrac{2}{\epsilon}$ | |
|     | $s^0$ | 2 | |

There are two sign changes of the coefficients in the first column. This agrees with the correct result indicated by the factored form of the polynomial equation.

If all the coefficients in any derived row are zero, it indicates that there are roots of equal magnitude lying radially opposite in the $s$ plane, i.e., two real roots with equal magnitudes and opposite signs and/or two conjugate imaginary roots. In such a case, the evaluation of the rest of the array can be continued by forming an auxiliary polynomial with the coefficients of the last row and by using the coefficients of the derivative of this polynomial in the next row. Such roots with equal magnitudes and lying radially opposite in the $s$ plane can be found by solving the auxiliary polynomial, which is always even. For a $2n$-degree auxiliary polynomial, there are $n$ pairs of equal and opposite roots. For example, consider the following equation:

$$s^5 + 2s^4 + 24s^3 + 48s^2 - 25s - 50 = 0$$

The array of coefficients is

| $s^5$ | 1 | 24 | $-25$ | |
|-------|---|----|-------|--|
| $s^4$ | 2 | 48 | $-50$ | $\leftarrow$ Auxiliary polynomial $P(s)$ |
| $s^3$ | 0 | 0 | | |

The terms in the $s^3$ row are all zero. The auxiliary polynomial is then formed from the coefficients of the $s^4$ row. The auxiliary polynomial $P(s)$ is

$$P(s) = 2s^4 + 48s^2 - 50$$

which indicates that there are two pairs of roots of equal magnitude and opposite sign. These pairs are obtained by solving the auxiliary polynomial equation $P(s) = 0$. The derivative of $P(s)$ with respect to $s$ is

$$\frac{dP(s)}{ds} = 8s^3 + 96s$$

The terms in the $s^3$ row are replaced by the coefficients of the last equation, namely 8 and 96. The array of coefficients then becomes

| $s^5$ | 1 | 24 | $-25$ | |
|-------|---|----|-------|--|
| $s^4$ | 2 | 48 | $-50$ | |
| $s^3$ | 8 | 96 | | $\leftarrow$ Coefficients of $dP(s)/ds$ |
| $s^2$ | 24 | $-50$ | | |
| $s^1$ | 112.7 | 0 | | |
| $s^0$ | $-50$ | | | |

We see that there is one change in sign in the first column of the new array. Thus, the original equation has one root with a positive real part. By solving for roots of the auxiliary polynomial equation,

$$2s^4 + 48s^2 - 50 = 0$$

we obtain

$$s^2 = 1, \qquad s^2 = -25$$

or

$$s = \pm 1, \qquad s = \pm j5$$

These two pairs of roots are a part of the roots of the original equation. As a matter of fact, the original equation can be written in factored form as follows:

$$(s + 1)(s - 1)(s + j5)(s - j5)(s + 2) = 0$$

Clearly, the original equation has one root with a positive real part.

**Relative stability analysis.** Routh's stability criterion provides the answer to the question of absolute stability. This, in many practical cases, is not sufficient. We usually require information about the relative stability of the system. A useful approach for examining relative stability is to shift the $s$-plane axis and apply Routh's stability criterion. Namely, we substitute

$$s = z - \sigma \qquad (\sigma = \text{constant})$$

into the characteristic equation of the system, write the polynomial in terms of $z$, and apply Routh's stability criterion to the new polynomial in $z$. The number of changes of sign in the first column of the array developed for the polynomial in $z$ is equal to the number of roots which are located to the right of the vertical line $s = -\sigma$. Thus, this test reveals the number of roots which lie to the right of the vertical line $s = -\sigma$.

**Application of Routh's stability criterion to control system analysis.** Routh's stability criterion is of limited usefulness in linear control system analysis mainly because it does not suggest how to improve relative stability or how to stabilize an unstable system. It is possible, however, to determine the effects of changing one or two parameters of a system by examining the values that cause instability. In the following, we shall consider the problem of determining the stability range of a parameter value.

Consider the system shown in Fig. 6-32. Let us determine the range of $K$ for stability.

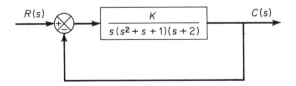

**Fig. 6-32.** Control system.

The closed-loop transfer function is

$$\frac{C(s)}{R(s)} = \frac{K}{s(s^2 + s + 1)(s + 2) + K}$$

The characteristic equation is

$$s^4 + 3s^3 + 3s^2 + 2s + K = 0$$

The array of coefficients becomes

| | | | |
|---|---|---|---|
| $s^4$ | 1 | 3 | $K$ |
| $s^3$ | 3 | 2 | 0 |
| $s^2$ | $\frac{7}{3}$ | $K$ | |
| $s^1$ | $2 - \frac{9}{7}K$ | | |
| $s^0$ | $K$ | | |

For stability, $K$ must be positive, and all coefficients in the first column must be positive. Therefore,

$$\frac{14}{9} > K > 0$$

When $K = \frac{14}{9}$, the system becomes oscillatory and, mathematically, the oscillation is sustained at constant amplitude.

## 6-7  ANALOG COMPUTERS

This section presents the principle of operation of electronic analog computers and the techniques of setting up computer diagrams for solving differential equations and for simulating physical systems.

It is well-known that the analog computer is one of the most useful engineering tools available for the analysis and design of both linear and nonlinear systems. We shall begin our discussion by describing the elements of an electronic analog computer.

**Operational amplifiers.** An operational amplifier is one used to produce various mathematical operations, such as sign inversion, summation, and integration. It is a dc amplifier and has a very high gain, approximately $10^6 \sim 10^8$. The current drawn at the input of an operational amplifier is negligibly small. The output voltage is usually limited to $\pm 100$ volts. Figure 6-33(a) shows a schematic diagram of an operational amplifier. The output voltage $e_0$ and input grid voltage $e_g$ are related by the following equation:

$$e_0 = -Ke_g$$

where $K = 10^6 \sim 10^8$. An operational amplifier thus changes algebraic sign.

Figure 6-33 (b) shows an operational amplifier in series with an input impedance $Z_i$, and with an impedance $Z_0$ inserted in the feedback path. Because the internal impedance of the amplifier is very high, there is essentially negligible input current, or

$$i_g \doteq 0$$

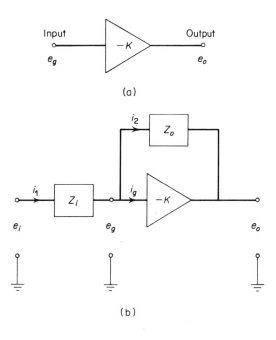

**Fig. 6-33.** (a) Schematic diagram of an operational amplifier; (b) schematic diagram of an operational amplifier with input and feedback impedances.

(b)

Then by Kirchhoff's law,

$$i_1 = i_2$$

Taking the Laplace transform of both sides of this last equation, we obtain

$$I_1(s) = I_2(s)$$

where

$$I_1(s) = \frac{E_i(s) - E_g(s)}{Z_i(s)}$$

$$I_2(s) = \frac{E_g(s) - E_0(s)}{Z_0(s)}$$

Hence we obtain

$$\frac{E_i(s)}{Z_i(s)} = \frac{E_g(s)}{Z_i(s)} + \frac{E_g(s)}{Z_0(s)} - \frac{E_0(s)}{Z_0(s)} \qquad (6\text{-}44)$$

Noting that

$$E_0(s) = -KE_g(s)$$

and that $K$ is a very large number (between $10^6$ and $10^8$), we obtain the following equation by neglecting the first two terms in the right-hand side of Eq. (6-44):

$$E_0(s) = -\frac{Z_0(s)}{Z_i(s)} E_i(s) \qquad (6\text{-}45)$$

Equation (6-45) is the basic equation relating $e_i$ and $e_0$ of the high-gain dc operational amplifier shown in Fig. 6-33 (b).

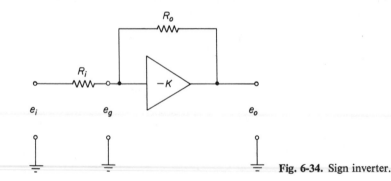

Fig. 6-34. Sign inverter.

**Sign inversion.** If we use resistors as the input and feedback impedances, as shown in Fig. 6-34, then Eq. (6-45) becomes

$$E_0(s) = -\frac{R_0}{R_i}E_i(s)$$

The output voltage is equal to the input voltage times a constant $(-R_0/R_i)$, which is negative. Thus, Fig. 6-34 shows a circuit giving a sign inversion.

**Summation.** Summation of several inputs can be accomplished by using resistors as the input and feedback impedances of the high-gain dc operational amplifier. Figure 6-35 (a) shows a schematic diagram of a summer which adds $n$ inputs. The output $e_0$ is

$$e_0 = -\left(\frac{R_0}{R_1}e_1 + \frac{R_0}{R_2}e_2 + \cdots + \frac{R_0}{R_n}e_n\right)$$

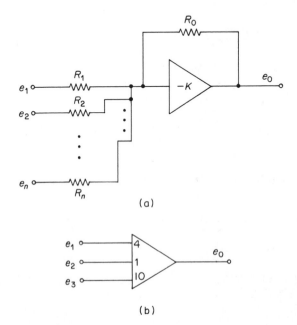

(a)

(b)

Fig. 6-35. (a) Schematic diagram of a summer; (b) symbol for the summer.

If, for example, $n = 3$ and $R_0 = 1$ megohm $= 10^6$ ohms, $R_1 = 0.25$ megohm, $R_2 = 1$ megohm, and $R_3 = 0.1$ megohm, then

$$e_0 = -(4e_1 + e_2 + 10e_3)$$

Figure 6-35(b) is the commonly used symbol for such a case.

**Integration.** If a capacitor is used as the feedback impedance and a resistor as the input impedance, as shown in Fig. 6-36(a), then Eq. (6-45) becomes

$$E_0(s) = -\frac{1}{R_i C_0 s} E_i(s)$$

or

$$e_0 = -\frac{1}{R_i C_0} \int e_i \, dt \tag{6-46}$$

Note that in integrating the input $e_i$ to obtain the output $e_0$, the initial condition must be given. Indicating the initial condition $e_0(0)$, we can write Eq. (6-46) as

$$e_0(t) = -\frac{1}{R_i C_0} \int_0^t e_i(t) \, dt + e_0(0)$$

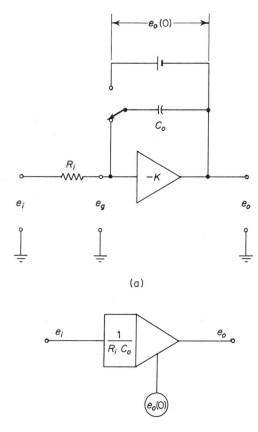

(a)

(b)

**Fig. 6-36.** (a) Schematic diagram of an integrator; (b) symbol for the integrator.

Each integrator must be initially biased by a dc voltage to give the necessary initial condition.

Figure 6-36 (b) shows the commonly used symbol for an integrator. If, for example, $R_i = 0.25$ megohm and $C_0 = 1 \ \mu f$, then $1/(R_i C_0) = 4$. The initial condition $e_0(0)$ is indicated in the circle. (Note that some operational amplifiers use standard resistors of 1, 0.25, and 0.10 megohm and a standard capacitor of 1 microfarad. In such a case, the values of $1/R_i C_0$ are equal only to 1, 4, or 10.)

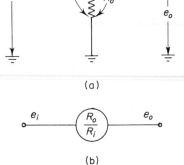

(a)

**Multiplication by a fraction.** The multiplication of $e_i$ by a constant $\alpha$, where $0 < \alpha < 1$, can be accomplished by the potentiometer shown in Fig. 6-37 (a). The output $e_0$ is

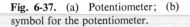

(b)

$$e_0 = \frac{R_0}{R_i} e_i$$

**Fig. 6-37.** (a) Potentiometer; (b) symbol for the potentiometer.

Figure 6-37 (b) shows the commonly used symbol for a potentiometer.

**Solving differential equations.** We shall now illustrate the procedure for solving differential equations by use of the electronic analog computer. Consider, as an example, the following differential equation:

$$\ddot{x} + 10\dot{x} + 16x = 0, \qquad x(0) = 0, \qquad \dot{x}(0) = 8 \qquad (6\text{-}47)$$

The first step in setting up a computer diagram is to assume that the highest-order derivative is available. Then solve the differential equation for this highest-order derivative. In the present differential equation,

$$\ddot{x} = -10\dot{x} - 16x$$

The variable $\dot{x}$ can be obtained by integrating $\ddot{x}$, and $x$ can be obtained by integrating $\dot{x}$. Figure 6-38 shows a computer diagram for this system. Note that the sum of the inputs to the first integrator is the highest derivative term that originally assumed available. Note also that the initial conditions are indicated in the circles.

It is important to remember that there is a sign change associated with each operational amplifier. Hence if the number of operational amplifiers (integrators and summers) in a loop is even, the output voltages will increase until they saturate. Hence, the number of operational amplifiers in any loop must be odd. (In the computer diagram shown in Fig. 6-38, the inner loop has one operational amplifier

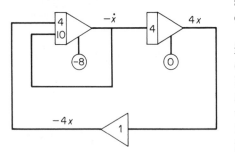

**Fig. 6-38.** Analog-computer diagram for solving Eq. (6-47).

and the outer loop three.) This serves as a convenient check to see if there is any mistake in setting up the computer diagram.

**Generating sinusoidal functions.** We shall next illustrate the generation of a sinusoidal signal, such as $10 \sin 3t$. In order to set up the analog computer diagram, let us obtain the differential equation whose solution is $10 \sin 3t$.

Let

$$x(t) = 10 \sin 3t$$

Then

$$\ddot{x}(t) = -90 \sin 3t$$

Hence the required differential equation is

$$\ddot{x} + 9x = 0, \qquad x(0) = 0, \qquad \dot{x}(0) = 30$$

Solving this differential equation for the highest-order derivative, we obtain

$$\ddot{x} = -9x$$

Assuming that $\ddot{x}$ is available, $x$ can be obtained by integrating $\ddot{x}$ twice. A computer diagram for this system is given in Fig. 6-39.

Note that the output of the second integrator is $9x$ and that the output voltage oscillates between $\pm 90$ volts. (The initial conditions determine the amplitude of the voltage oscillation.)

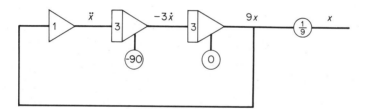

**Fig. 6-39.** Analog-computer diagram for generating the sinusoidal signal $x(t) = 10 \sin 3t$.

**Magnitude scale factors.**   The output voltage of any amplifier should not exceed $\pm 100$ volts to avoid saturation. Voltage saturation will cause errors in the solution. On the other hand, the maximum voltage at any amplifier should not be too small. To assure proper accuracy, it is desirable that the maximum swing of output voltage at any operational amplifier be about $\pm 70 \sim \pm 90$ volts.

In setting up the computer diagram, it is desirable that the maximum swing of output voltage of any amplifier be about the same. In this respect, choosing proper magnitude scale factors is most important. The magnitude scale factors relate the output voltages of amplifiers to the corresponding physical quantities.

Consider the system given by Eq. (6-47). Let us assume that $x$ has the units of feet. Then $x(0) = 0$ ft, and $\dot{x}(0) = 8$ ft/sec. Solving Eq. (6-47) for the highest-order derivative, we obtain

$$\ddot{x} = -10\dot{x} - 16x \qquad\qquad (6-48)$$

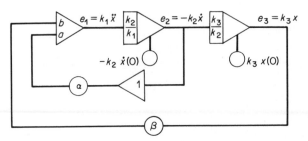

**Fig. 6-40.** Unscaled analog-computer diagram for solving Eq. (6-47).

Let us define $k_1$, $k_2$, and $k_3$ as magnitude scale factors and rewrite Eq. (6-48) as

$$k_1\ddot{x} = -\frac{10k_1}{k_2}(k_2\dot{x}) - \frac{16k_1}{k_3}(k_3x) \qquad (6\text{-}49)$$

Referring to Fig. 6-40, the output voltage of the first amplifier (summer) is $e_1 = k_1\ddot{x}$. The output voltage of the second amplifier (first integrator) is $e_2 = -k_2\dot{x}$, and finally the output voltage of the third amplifier (second integrator) is $e_3 = k_3x$. These output voltages must be limited to $\pm100$ volts.

The undamped natural frequency of this system is 4 rad/sec. An approximate solution for $0 < t \ll 1$ is

$$x(t) = 2\sin 4t$$

Hence, for $0 < t \ll 1$,

$$\dot{x}(t) = 8\cos 4t$$
$$\ddot{x}(t) = -32\sin 4t$$

Note that this solution is valid only for very small values of $t$. From this solution, we obtain

$$\text{Maximum value of } |x(t)| = 2$$
$$\text{Maximum value of } |\dot{x}(t)| = 8$$
$$\text{Maximum value of } |\ddot{x}(t)| = 32$$

[Note that in more complex systems (especially in nonlinear systems) the maximum values of $x(t)$, $\dot{x}(t)$, $\ddot{x}(t)$, ... are not available. Then we must estimate these maximum values. If the estimated values are way off, then the computer diagram must be revised until it is acceptable.] We would like to choose $k_1$, $k_2$, and $k_3$ so that $e_1 = -e_2 = e_3 = 90$ volts for the maximum values of $\ddot{x}$, $\dot{x}$, and $x$, respectively. (We have chosen 90 volts rather than 100 volts as a conservative estimate.) Hence

$$k_1 = \frac{90}{|\ddot{x}|_{\max}} = \frac{90}{32} = \frac{45}{16} \text{ volts/ft/sec}^2$$

$$k_2 = \frac{90}{|\dot{x}|_{\max}} = \frac{90}{8} = \frac{45}{4} \text{ volts/ft/sec}$$

$$k_3 = \frac{90}{|x|_{\max}} = \frac{90}{2} = 45 \text{ volts/ft}$$

Therefore

$$\frac{k_2}{k_1} = 4, \qquad \frac{k_3}{k_2} = 4$$

Note that from Fig. 6-40 we obtain

$$k_1\ddot{x} = e_1 = -(-a\alpha e_2 + b\beta e_3) \qquad (6\text{-}50)$$

From Eq. (6-49),

$$k_1\ddot{x} = -\frac{10k_1}{k_2}(-e_2) - \frac{16k_1}{k_3}e_3 \qquad (6\text{-}51)$$

Equating Eqs. (6-50) and (6-51), we obtain

$$a\alpha = \frac{10k_1}{k_2}, \qquad b\beta = \frac{16k_1}{k_3}$$

Hence

$$a\alpha = 2.5, \qquad b\beta = 1$$

Let us choose $a = 4$, $\alpha = 2.5/4$, and $b = \beta = 1$. Then all the unknown constants in Fig. 6-40 are determined. The properly scaled computer diagram is shown in Fig. 6-41. The initial conditions are

$$e_2(0) = -k_2\dot{x}(0) = -\frac{45}{4} \times 8 = -90 \text{ volts}$$

$$e_3(0) = k_3 x(0) = 45 \times 0 = 0 \text{ volts}$$

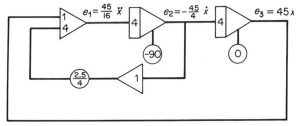

**Fig. 6-41.** Properly scaled analog-computer diagram for solving Eq. (6-47).

Note that the diagram shown in Fig. 6-41 is not the final diagram we would like to use as the computer diagram for solving Eq. (6-48). In order to minimize the effect of noise and keep the accuracy high, it is desirable to use the minimal number of amplifiers necessary. (In any analog computer, the number of integrators and summers is limited. In solving a complex problem requiring many integrators and summers, it is necessary to use the minimal number for each equation in order to save components.)

In the case of the diagram of Fig. 6-41, it is possible to eliminate one summer. The simplified computer diagram using the same magnitude scale factors is shown in Fig. 6-42. Note that since the gain of the inner loop is $4 \times 1 \times (2.5/4) \times 4 = 10$, the potentiometer has been eliminated. Comparing the computer diagram shown in Fig. 6-38 with that in Fig. 6-42, we see that the latter is better from the standpoint of accuracy since all the amplifiers are driven over a larger portion of their linearity region.

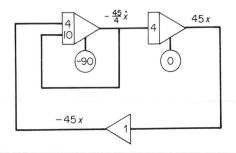

**Time scale factors.** Time scaling relates the independent variable of the physical problem to the independent variable of the analog computer. For phenomena which take place very rapidly, it is necessary to slow down the speed at which such problems are simulated on the computer.

**Fig. 6-42.** Simplified version of the computer diagram shown in Fig. 6-41.

Let the following equation relate the real time $t$ in seconds to the computer time (or machine time) $\tau$ in seconds:

$$\tau = \lambda t$$

where $\lambda$ is the time scale factor. If $\lambda$ is chosen as 0.1, then 10 sec of real time is equivalent to 1 computer second. This means that if the actual response takes 10 sec of real time to complete, then the response is completed in 1 sec on the computer. Conversely, if $\lambda$ is chosen as 10, then 1 sec of real time is equivalent to 10 sec of computer time.

As an example, consider the following differential equation:

$$\frac{d^2x}{dt^2} + 10\frac{dx}{dt} + 16x = 20 \sin \omega t, \qquad x(0) = 0, \qquad \frac{dx}{dt}\bigg|_{t=0} = 8 \qquad (6\text{-}52)$$

Let us convert the independent variable $t$ to $\tau$. Since

$$\tau = \lambda t$$

we obtain

$$\frac{dx}{dt} = \frac{dx}{d\tau}\frac{d\tau}{dt} = \lambda\frac{dx}{d\tau}$$

$$\frac{d^2x}{dt^2} = \lambda^2\frac{d^2x}{d\tau^2}$$

Equation (6-52) then becomes

$$\lambda^2\frac{d^2x}{d\tau^2} + 10\lambda\frac{dx}{d\tau} + 16x = 20 \sin \frac{\omega}{\lambda}\tau \qquad (6\text{-}53)$$

To slow down the solution by a factor of 5, we substitute $\lambda = 5$ into Eq. (6-53). The computer equation is then

$$25\frac{d^2x}{d\tau^2} + 50\frac{dx}{d\tau} + 16x = 20 \sin \frac{\omega}{5}\tau$$

The initial conditions are transformed to

$$x(0) = 0, \qquad \frac{dx}{d\tau}\bigg|_{\tau=0} = \frac{1}{\lambda}\frac{dx}{dt}\bigg|_{t=0} = \frac{1}{5}(8) = 1.6$$

As another example, consider the following equation:

$$100\frac{d^2x}{dt^2} + 2\frac{dx}{dt} + 0.5x = \sin t, \qquad x(0) = 2, \qquad \frac{dx}{dt}\bigg|_{t=0} = 5 \qquad (6\text{-}54)$$

Because of a large difference in the magnitudes of the coefficients, setting up Eq. (6-54) on the computer is not desirable. If we choose a value for $\lambda$ such that the coefficients of all the terms are of the same order of magnitude, then the computer solution will generally have less error. In the present example, if $\lambda$ is chosen to be 0.1, then

$$\tau = 0.1t$$

$$\frac{dx}{dt} = 0.1\frac{dx}{d\tau}$$

$$\frac{d^2x}{dt^2} = 0.01\frac{d^2x}{d\tau^2}$$

Equation (6-54) now becomes

$$\frac{d^2x}{d\tau^2} + 0.2\frac{dx}{d\tau} + 0.5x = \sin 10\tau$$

The initial conditions are

$$x(0) = 2, \qquad \frac{dx}{d\tau}\bigg|_{\tau=0} = 50$$

Note that in a specific problem, in addition to time scaling, it may be necessary to choose magnitude scale factors so that the magnitude of the forcing function is neither too large nor too small and that the amplifier outputs are driven over a large portion of the linearity range ($\pm 100$ volts).

**Analog computer simulation of physical systems.** The simulation of dynamic systems is a very important application of the analog computer. In particular, the analog computer is quite useful in determining the effects of parameter variations on system performance.

Consider the system shown in Fig. 6-43. We wish to simulate this system on an analog computer and investigate the effect of variations in $a$ and $b$ on the unit-step response.

The closed-loop transfer function is

$$\frac{X(s)}{U(s)} = \frac{as + b}{s^3 + 3s^2 + (2 + a)s + b}$$

The differential equation corresponding to this transfer function is

$$\dddot{x} + 3\ddot{x} + (2 + a)\dot{x} + bx = a\dot{u} + bu \qquad (6\text{-}55)$$

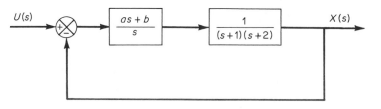

Fig. 6-43. Control system.

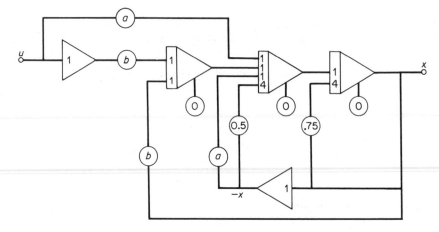

**Fig. 6-44.** Analog-computer diagram for solving Eq. (6-56).

Solving Eq. (6-55) for the highest-order derivative in $x$, we obtain

$$\dddot{x} = -3\ddot{x} - (2\dot{x} + a\dot{x} - a\dot{u}) - (bx - bu) \tag{6-56}$$

where $x(0) = \dot{x}(0) = \ddot{x}(0) = 0$. An analog-computer diagram corresponding to Eq. (6-56) is shown in Fig. 6-44. This diagram shows only one possibility. If the ranges of variation of the values of $a$ and $b$ are specified, proper magnitude scale factors may be determined.

The effects of the constants $a$ and $b$ on the unit-step response can be investigated by varying the potentiometer settings involving $a$ and $b$. (If $a, b > 1$, we may need to add amplifiers to the computer diagram.) The initial conditions are set equal to zero.

As another example of the simulation of a dynamic system, consider the mechanical system shown in Fig. 6-45 (a). The equations are

$$m_1\ddot{x}_1 + f_1\dot{x}_1 + k_1x_1 + k_2(x_1 - x_2) = 0$$
$$m_2\ddot{x}_2 + f_2\dot{x}_2 + k_3x_2 + k_2(x_2 - x_1) = 0$$

An analog computer simulation diagram for this system is shown in Fig. 6-45 (b).

**Concluding comments.** In the course of the analysis and design of complicated systems, analog computer simulation plays an important role. The effects of change in system parameters on the performance of the system can be easily determined. The advantage of analog (or digital) simulation is that any convenient time scale may be used.

In general, the precise mathematical representation of a complicated component is very difficult. It is likely that some of the important features of the component may be overlooked in the simulation. This may cause serious errors in the solution. In order to avoid such errors, the simulator may include actual system components. If such components are included, no important characteristics of the actual components are lost. The solution, however, must be obtained in real time.

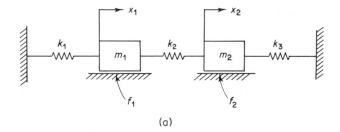

(a)

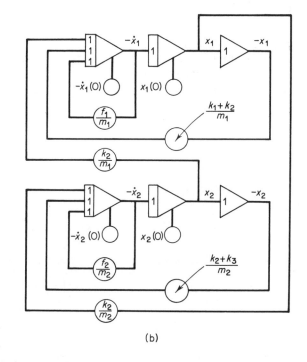

**Fig. 6-45.** (a) Mechanical system; (b) analog-computer simulation diagram for the mechanical system.

(b)

Although we have not discussed nonlinear components yet, the analog computer is quite versatile and convenient to use in simulating control systems involving such components. Such nonlinear operations as the multiplication of two variables can be carried out easily with the electronic analog computer. Standard electronic circuits are available for simulating such commonly encountered non-linearities as dead zone, saturation, backlash, and coulomb friction.

EXAMPLE PROBLEMS AND SOLUTIONS

**PROBLEM A-6-1.** Explain why the proportional control of a plant which does not possess an integrating property (which means that the plant transfer function does not include the factor $1/s$) suffers offset in response to step inputs.

**Solution.** Consider, for example, the system shown in Fig. 6-46. At steady state, if $c$ were equal to a nonzero constant $r$, then $e = 0$ and $m = Ke = 0$, resulting in $c = 0$, which contradicts the assumption that $c = r =$ nonzero constant.

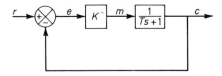

**Fig. 6-46.** Control system.

A nonzero offset must exist for proper operation of such a control system. In other words, at steady state, if $e$ were equal to $r/(1 + K)$, then $m = Kr/(1 + K)$ and $c = Kr/(1 + K)$, which results in the assumed error signal $e = r/(1 + K)$. Thus the offset is $r/(1 + K)$.

**PROBLEM A-6-2.** Consider the liquid-level control system shown in Fig. 6-47. The controller is of the proportional type. The set point of the controller is fixed.

Draw a block diagram of the system, assuming that changes in the variables are small. Investigate the response of the level of the second tank subjected to a step disturbance $u$.

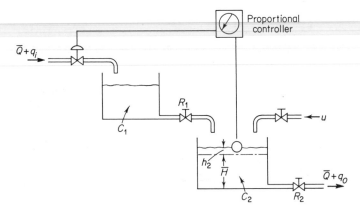

**Fig. 6-47.** Liquid-level control system.

**Solution.** Figure 6-48(a) is a block diagram of this system when changes in the variables are small. Since the set point of the controller is fixed, $r = 0$. (Note that $r$ is the change in set point.)

To investigate the response of the level of the second tank subjected to a step disturbance $u$, we find it convenient to modify the block diagram of Fig. 6-48(a) to the one shown in Fig. 6-48(b).

The transfer function between $H_2(s)$ and $U(s)$ can be obtained as

$$\frac{H_2(s)}{U(s)} = \frac{R_2(R_1 C_1 s + 1)}{(R_1 C_1 s + 1)(R_2 C_2 s + 1) + KR_2}$$

From this equation, the response $H_2(s)$ to a disturbance $U(s)$ can be found. The effect of the controller is seen by the presence of $K$ in the denominator of this last equation.

For a step disturbance of magnitude $U_0$, we obtain

$$h_2(\infty) = \frac{R_2}{1 + KR_2} U_0$$

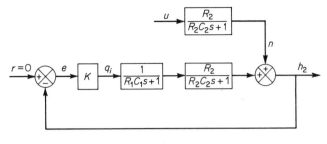

(a)

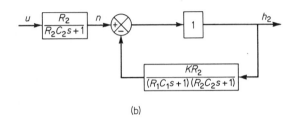

**Fig. 6-48.** (a) Block diagram of the system shown in Fig. 6-47; (b) modified block diagram.

(b)

or

$$\text{Steady-state error} = -\frac{R_2}{1 + KR_2} U_0$$

The system exhibits offset for a step disturbance. In the response to a step disturbance, both the damping ratio and the undamped natural frequency depend upon the value of the gain $K$. This gain must be adjusted so that the transient response to disturbances shows reasonable damping and a reasonable speed.

**PROBLEM A-6-3.** Consider the unit-step response of the second-order system

$$\frac{C(s)}{R(s)} = \frac{\omega_n^2}{s^2 + 2\zeta\omega_n s + \omega_n^2}$$

The amplitude of the exponentially damped sinusoid changes as a geometric series. At time $t = t_p = \pi/\omega_d$, the amplitude is equal to $e^{-(\sigma/\omega_d)\pi}$. After one oscillation, or at $t = t_p + 2\pi/\omega_d = 3\pi/\omega_d$, the amplitude is equal to $e^{-(\sigma/\omega_d)3\pi}$, after another cycle of oscillation, the amplitude is $e^{-(\sigma/\omega_d)5\pi}$. The logarithm of this amplitude ratio is called the logarithmic decrement. Determine the logarithmic decrement for this second-order system.

**Solution.** The amplitude ratio per one period of damped oscillation is

$$\frac{c(t_1)}{c\left(t_1 + \dfrac{2\pi}{\omega_d}\right)} = \frac{e^{-(\sigma/\omega_d)\pi}}{e^{-(\sigma/\omega_d)3\pi}} = e^{2(\sigma/\omega_d)\pi} = e^{2\zeta\pi/\sqrt{1-\zeta^2}}$$

where

$$t_1 = t_p + n\frac{2\pi}{\omega_d} \qquad (n = 0, 1, 2, \ldots)$$

Thus, the logarithmic decrement $\delta$ is

$$\delta = \ln \left| \frac{c(t_1)}{c\left(t_1 + \frac{2\pi}{\omega_d}\right)} \right| = \frac{2\zeta\pi}{\sqrt{1 - \zeta^2}}$$

It is a function only of the damping ratio $\zeta$.

**PROBLEM A-6-4.** Consider a unity-feedback control system whose open-loop transfer function is

$$G(s) = \frac{K}{s(Js + F)}$$

Discuss the effects that varying the values of $K$ and $F$ has on the steady-state error in unit-ramp response. Sketch typical unit-ramp response curves for a small value, medium value, and large value of $K$.

**Solution.** The closed-loop transfer function is

$$\frac{C(s)}{R(s)} = \frac{K}{Js^2 + Fs + K}$$

For a unit-ramp input, $R(s) = 1/s^2$. Thus

$$\frac{E(s)}{R(s)} = \frac{R(s) - C(s)}{R(s)} = \frac{Js^2 + Fs}{Js^2 + Fs + K}$$

or

$$E(s) = \frac{Js^2 + Fs}{Js^2 + Fs + K} \frac{1}{s^2}$$

The steady-state error is

$$e_{ss} = e(\infty) = \lim_{s \to 0} sE(s) = \frac{F}{K}$$

We see that we can reduce the steady-state error $e_{ss}$ by increasing the gain $K$ or decreasing the viscous-friction coefficient $F$. Increasing the gain or decreasing the viscous-friction coefficient, however, causes the damping ratio to decrease, with the result that the transient response of the system will become more oscillatory. Doubling $K$ decreases $e_{ss}$ to half of its original value, while $\zeta$ is decreased to 0.707 of its original value since $\zeta$ is inversely proportional to the square root of $K$. On the other hand, decreasing $F$ to half of its original value decreases both $e_{ss}$ and $\zeta$ to the halves of their original values, respectively. Therefore, it is advisable to increase the value of $K$ rather than to decrease the value of $F$. After the

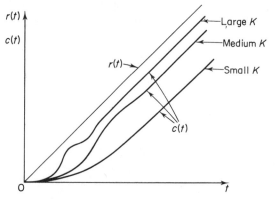

Fig. 6-49. Unit-ramp response curves of the system considered in Problem A-6-4.

transient response has died out and a steady-state reached, the output velocity becomes the same as the input velocity. However, there is a steady-state positional error between the input and the output. Examples of the unit-ramp response of the system for three different values of $K$ are illustrated in Fig. 6-49.

**PROBLEM A-6-5.** Obtain the unit-ramp response of the system given by

$$\frac{C(s)}{R(s)} = \frac{\omega_n^2}{s^2 + 2\zeta\omega_n s + \omega_n^2}$$

Obtain also the steady-state error.

**Solution.** For a unit-ramp input, $R(s) = 1/s^2$. The output $C(s)$ may be written

$$C(s) = \frac{\omega_n^2}{s^2 + 2\zeta\omega_n s + \omega_n^2} \frac{1}{s^2}$$

For the underdamped case ($0 \le \zeta < 1$), $c(t)$, the inverse Laplace transform of $C(s)$, may be obtained as

$$c(t) = t - \frac{2\zeta}{\omega_n} + e^{-\zeta\omega_n t}\left(\frac{2\zeta}{\omega_n}\cos\omega_d t + \frac{2\zeta^2 - 1}{\omega_n\sqrt{1 - \zeta^2}}\sin\omega_d t\right)$$

$$= t - \frac{2\zeta}{\omega_n} + \frac{e^{-\zeta\omega_n t}}{\omega_n\sqrt{1 - \zeta^2}}\sin\left(\omega_d t + \tan^{-1}\frac{2\zeta\sqrt{1 - \zeta^2}}{2\zeta^2 - 1}\right) \qquad (t \ge 0)$$

where

$$\omega_d = \omega_n\sqrt{1 - \zeta^2}$$

$$\tan^{-1}\frac{2\zeta\sqrt{1 - \zeta^2}}{2\zeta^2 - 1} = 2\tan^{-1}\frac{\sqrt{1 - \zeta^2}}{\zeta}$$

For the critically damped case ($\zeta = 1$), $c(t)$ is

$$c(t) = t - \frac{2\zeta}{\omega_n} + \frac{2}{\omega_n}e^{-\omega_n t}\left(1 + \frac{\omega_n t}{2}\right) \qquad (t \ge 0)$$

For the overdamped case ($\zeta > 1$), $c(t)$ is

$$c(t) = t - \frac{2\zeta}{\omega_n} - \frac{2\zeta^2 - 1 - 2\zeta\sqrt{\zeta^2 - 1}}{2\omega_n\sqrt{\zeta^2 - 1}}e^{-(\zeta+\sqrt{\zeta^2-1})\omega_n t}$$

$$+ \frac{2\zeta^2 - 1 + 2\zeta\sqrt{\zeta^2 - 1}}{2\omega_n\sqrt{\zeta^2 - 1}}e^{-(\zeta-\sqrt{\zeta^2-1})\omega_n t} \qquad (t \ge 0)$$

(The unit-ramp response of this system can also be obtained by integrating the unit-step response with respect to time and determining the integration constants, using the condition that the initial conditions on the output are zero.) The error signal $e(t)$ in this case is

$$e(t) = r(t) - c(t) = t - c(t)$$

The steady-state error, $e_{ss}$, for $\zeta > 0$, is then

$$e_{ss} = e(\infty) = \frac{2\zeta}{\omega_n}$$

**PROBLEM A-6-6.** Consider the system shown in Fig. 6-50 (a). The steady-state error to a unit-ramp input is $e_{ss} = 2\zeta/\omega_n$. Show that the steady-state error for following a ramp input may be eliminated if the input is introduced to the system through a proportional-plus-derivative element, as shown in Fig. 6-50 (b), and the value of $k$ is properly set.

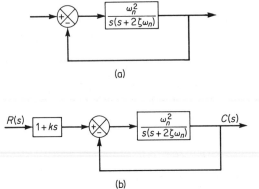

(a)

(b)                        **Fig. 6-50.** Control systems.

**Solution.** The closed-loop transfer function of the system shown in Fig. 6-50 (b) is

$$\frac{C(s)}{R(s)} = \frac{(1 + ks)\omega_n^2}{s^2 + 2\zeta\omega_n s + \omega_n^2}$$

Then

$$R(s) - C(s) = \left(\frac{s^2 + 2\zeta\omega_n s - \omega_n^2 ks}{s^2 + 2\zeta\omega_n s + \omega_n^2}\right) R(s)$$

If the input is a unit ramp, then the steady-state error is

$$e(\infty) = r(\infty) - c(\infty)$$
$$= \lim_{s \to 0} s\left(\frac{s^2 + 2\zeta\omega_n s - \omega_n^2 ks}{s^2 + 2\zeta\omega_n s + \omega_n^2}\right)\frac{1}{s^2}$$
$$= \frac{2\zeta\omega_n - \omega_n^2 k}{\omega_n^2}$$

Therefore, if $k$ is chosen as

$$k = \frac{2\zeta}{\omega_n}$$

then the steady-state error for following a ramp input can be made equal to zero. Note that if there are any variations in the values of $\zeta$ and/or $\omega_n$ due to environmental changes or aging, then a nonzero steady-state error for a ramp response may result.

**PROBLEM A-6-7.** Consider the servomechanism shown in Fig. 6-51. Determine the values of $K$ and $k$ so that the maximum overshoot in unit-step response is 25% and the peak time is 2 sec.

**Solution.** The maximum overshoot $M_p$ is

$$M_p = e^{-(\zeta/\sqrt{1-\zeta^2})\pi}$$

which is specified as 25%. Hence

$$e^{-(\zeta/\sqrt{1-\zeta^2})\pi} = 0.25$$

from which

$$\frac{\zeta}{\sqrt{1-\zeta^2}}\pi = 1.39$$

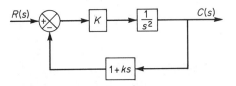

**Fig. 6-51.** Block diagram of a servomechanism.

or

$$\zeta = 0.4$$

The peak time $t_p$ is specified as 2 sec. Hence

$$t_p = \frac{\pi}{\omega_d} = 2$$

or

$$\omega_d = 1.57$$

Then the undamped natural frequency $\omega_n$ is

$$\omega_n = \frac{\omega_d}{\sqrt{1 - \zeta^2}} = \frac{1.57}{\sqrt{1 - 0.4^2}} = 1.71$$

From the block diagram, we obtain

$$\frac{C(s)}{R(s)} = \frac{K}{s^2 + Kks + K}$$

Noting that

$$\omega_n = \sqrt{K}, \qquad \zeta = \frac{Kk}{2\omega_n}$$

we obtain

$$K = \omega_n^2 = 1.71^2 = 2.93$$

$$k = \frac{2\zeta\omega_n}{K} = \frac{2 \times 0.4 \times 1.71}{2.93} = 0.47$$

**PROBLEM A-6-8.** Figure 6-52 (a) shows a mechanical vibratory system. When 2 lb of force (step input) is applied to the system, the mass oscillates, as shown in Fig. 6-52 (b). Determine $m$, $f$, and $k$ of the system from this response curve.

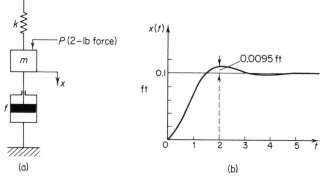

Fig. 6-52. (a) Mechanical vibratory system; (b) step-response curve.        (a)        (b)

**Solution.** The transfer function of this system is

$$\frac{X(s)}{P(s)} = \frac{1}{ms^2 + fs + k}$$

Since

$$P(s) = \frac{2}{s}$$

we obtain

$$X(s) = \frac{2}{s(ms^2 + fs + k)}$$

It follows that the steady-state value of $x$ is

$$x(\infty) = \lim_{s \to 0} sX(s) = \frac{2}{k} = 0.1 \text{ ft}$$

Hence

$$k = 20 \text{ lb/ft}$$

Note that $M_p = 9.5\%$ corresponds to $\zeta = 0.6$. The peak time $t_p$ is given by

$$t_p = \frac{\pi}{\omega_d} = \frac{\pi}{\omega_n \sqrt{1 - \zeta^2}} = \frac{\pi}{0.8\omega_n}$$

The experimental curve shows that $t_p = 2$ sec. Therefore

$$\omega_n = \frac{3.14}{2 \times 0.8} = 1.96 \text{ rad/sec}$$

Since $\omega_n^2 = k/m = 20/m$, we obtain

$$m = \frac{20}{\omega_n^2} = \frac{20}{1.96^2} = 5.2 \text{ slugs}$$
$$= 166 \text{ lb}$$

Then $f$ is determined from

$$2\zeta\omega_n = \frac{f}{m}$$

or

$$f = 2\zeta\omega_n m = 2 \times 0.6 \times 1.96 \times 5.2 = 12.2 \text{ lb/ft/sec}$$

**PROBLEM A-6-9.** Consider a system whose closed-loop poles and closed-loop zero are located in the $s$ plane on a line parallel to the $j\omega$ axis, as shown in Fig. 6-53. Show that the impulse response of such a system is a damped cosine function.

**Solution.** The closed-loop transfer function is

$$\frac{C(s)}{R(s)} = \frac{K(s + \sigma)}{(s + \sigma + j\omega_d)(s + \sigma - j\omega_d)}$$

For a unit-impulse input, $R(s) = 1$ and

$$C(s) = \frac{K(s + \sigma)}{(s + \sigma)^2 + \omega_d^2}$$

The inverse Laplace transform of $C(s)$ is

$$c(t) = Ke^{-\sigma t} \cos \omega_d t \qquad (t \geq 0)$$

which is a damped cosine function.

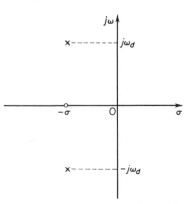

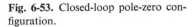

**Fig. 6-53.** Closed-loop pole-zero configuration.

**PROBLEM A-6-10.** Consider the system shown in Fig. 6-54. This system is subjected to two signals, one the reference input and the other the external disturbance. Show that the characteristic equation of this system is the same regardless of which signal is chosen as input.

**Solution.** The transfer function which relates the reference input and the corresponding output, without considering the external disturbance, is

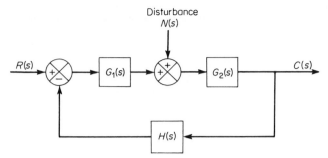

**Fig. 6-54.** Control system.

$$\frac{C(s)}{R(s)} = \frac{G_1 G_2}{1 + G_1 G_2 H} \qquad (6\text{-}57)$$

The transfer function which relates the external disturbance and the corresponding output in the absence of the reference input is

$$\frac{C(s)}{N(s)} = \frac{G_2}{1 + G_1 G_2 H} \qquad (6\text{-}58)$$

Note that the denominators of Eqs. (6-57) and (6-58) are the same. The characteristic equation is

$$1 + G_1 G_2 H = 0$$

It contains the information necessary to determine basic characteristics of the system response. (Remember that for a given system there is only one characteristic equation. This means that the characteristic equation of a given transfer function is the same regardless of which signal is chosen as input.)

**PROBLEM A-6-11.** Are the open-loop and closed-loop zeros identical in a closed-loop system?

**Solution.** No. Consider a closed-loop system whose feedforward transfer function is $G(s) = p(s)/q(s)$ and feedback transfer function is $H(s) = n(s)/d(s)$, where $p(s)$, $q(s)$, $n(s)$, and $d(s)$ are polynomials in $s$. Then

$$\frac{C(s)}{R(s)} = \frac{G(s)}{1 + G(s)H(s)}$$

$$= \frac{p(s)d(s)}{q(s)d(s) + p(s)n(s)}$$

We see that the zeros of the closed-loop transfer function are those values of $s$ which make $p(s)d(s) = 0$ and those of the open-loop transfer function are those values of $s$ which make $p(s)n(s) = 0$. (Thus, some of the zeros of the closed-loop transfer function are the same as those of the open-loop transfer function.)

**PROBLEM A-6-12\*** Consider the inverted pendulum mounted on the motor-driven cart shown in Fig. 6-55. This is a model of the attitude control of a space booster on take-off. The objective of the attitude control problem is to keep the space booster in a vertical position. In the present problem, we wish to keep the pendulum in a vertical position.

*\*The problem of controlling unstable systems such as the inverted pendulum mounted on the motor-driven cart is treated in detail in Reference H-6.*

The actual space booster (or the inverted pendulum in this problem) is unstable and may fall over any time and in any direction.

In this problem, we consider only a two-dimensional problem so that the pendulum shown in Fig. 6-55 moves only in the plane of the page. In order to keep the inverted pendulum vertical, suppose we measure $\theta$ and $\dot{\theta}$ continuously and employ a proportional-plus-derivative controller to produce the control force $u$, or

$$u = M(a\theta + b\dot{\theta})$$

Determine the conditions on $a$ and $b$ so that the system is stable. Assume that there is neither friction at the pivot nor slip at the wheels of the cart. Assume also that $\theta$ and $\dot{\theta}$ are small.

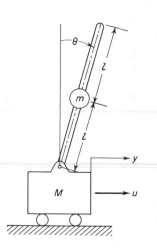

**Fig. 6-55.** Inverted pendulum mounted on a motor-driven cart.

**Solution.** Using the symbols defined in Fig. 6-55, the differential equations describing the system dynamics may be obtained as follows: Noting that $\theta$ and $\dot{\theta}$ are small, so that $\sin \theta \doteq \theta$, $\cos \theta \doteq 1$, we have

$$(J + ml^2)\ddot{\theta} + ml\ddot{y} - mgl\theta = 0 \tag{6-59}$$

$$ml\ddot{\theta} + (M + m)\ddot{y} = u \tag{6-60}$$

$$M(a\theta + b\dot{\theta}) = u \tag{6-61}$$

where $J$ is the moment of inertia of the inverted pendulum about the center of mass. Since $J$ can be given as

$$J = \frac{ml^2}{3}$$

Eq. (6-59) may be written

$$\frac{4}{3}ml^2\ddot{\theta} + ml\ddot{y} - mgl\theta = 0$$

from which

$$\ddot{y} = -\frac{4}{3}l\ddot{\theta} + g\theta \tag{6-62}$$

Substituting Eqs. (6-62) and (6-61) into Eq. (6-60), we obtain

$$ml\ddot{\theta} + (M + m)\left(-\frac{4}{3}l\ddot{\theta} + g\theta\right) = M(a\theta + b\dot{\theta}) \tag{6-63}$$

Let us define

$$\frac{m}{M} = \mu$$

Then Eq. (6-63) becomes

$$\ddot{\theta} + \frac{3b}{(\mu + 4)l}\dot{\theta} + \frac{3[a - (1 + \mu)g]}{(\mu + 4)l}\theta = 0$$

For stability, the coefficients of $\theta$ and $\dot{\theta}$ must be positive. Hence, the stability conditions are

$$b > 0, \qquad a > (1 + \mu)g$$

If the values of $a$ and $b$ are chosen to satisfy these conditions, any small tilt may be recovered without having the pendulum fall over.

**PROBLEM A-6-13.** Obtain the unit-step response of a unity-feedback system whose open-loop transfer function is

$$G(s) = \frac{5(s + 20)}{s(s + 4.59)(s^2 + 3.41s + 16.35)}$$

**Solution.** The closed-loop transfer function is

$$\frac{C(s)}{R(s)} = \frac{5(s + 20)}{s(s + 4.59)(s^2 + 3.41s + 16.35) + 5(s + 20)}$$

$$= \frac{5s + 100}{s^4 + 8s^3 + 32s^2 + 80s + 100}$$

$$= \frac{5(s + 20)}{(s^2 + 2s + 10)(s^2 + 6s + 10)}$$

The unit-step response of this system is then

$$C(s) = \frac{5(s + 20)}{s(s^2 + 2s + 10)(s^2 + 6s + 10)}$$

$$= \frac{1}{s} + \frac{\frac{3}{8}(s + 1) - \frac{17}{8}}{(s + 1)^2 + 3^2} + \frac{-\frac{11}{8}(s + 3) - \frac{13}{8}}{(s + 3)^2 + 1^2}$$

The time response $c(t)$ can be found by taking the inverse Laplace transform of $C(s)$ as follows:

$$c(t) = 1 + \tfrac{3}{8}e^{-t}\cos 3t - \tfrac{17}{24}e^{-t}\sin 3t - \tfrac{11}{8}e^{-3t}\cos t - \tfrac{13}{8}e^{-3t}\sin t \qquad (t \geq 0)$$

## PROBLEMS

**PROBLEM B-6-1.** A thermometer requires 1 min to indicate 98% of the response to a step input. Assuming the thermometer to be a first-order system, find the time constant.

If the thermometer is placed in a bath, the temperature of which is changing linearly at a rate 10°/min, how much error does the thermometer show?

**PROBLEM B-6-2.** What would be the error if the network shown in Fig. 6-56 is used as an integrator?

**PROBLEM B-6-3.** Obtain the unit-step response of a unity-feedback system whose open-loop transfer function is

$$G(s) = \frac{4}{s(s + 5)}$$

**PROBLEM B-6-4.** Consider the unit-step response of a unity-feedback control system whose open-loop transfer function is

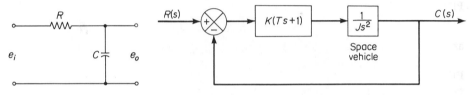

**Fig. 6-56.** Network system.          **Fig. 6-57.** Space-vehicle attitude-control system.

$$G(s) = \frac{1}{s(s + 1)}$$

Obtain the rise time, peak time, maximum overshoot, and settling time.

**PROBLEM B-6-5.** Consider the closed-loop system given by

$$\frac{C(s)}{R(s)} = \frac{\omega_n^2}{s^2 + 2\zeta\omega_n s + \omega_n^2}$$

Determine the values of $\zeta$ and $\omega_n$ so that the system responds to a step input with approximately 5% overshoot and with a settling time of 2 sec. (Use the 2% criterion.)

**PROBLEM B-6-6.** Figure 6-57 is the block diagram of a space-vehicle attitude-control system. Assuming the time constant $T$ of the controller to be 3 sec and the ratio of torque to inertia $K/J$ to be $\frac{2}{9}$ rad²/sec², find the damping ratio of the system.

**PROBLEM B-6-7.** Consider a unity-feedback control system whose open-loop transfer function is

$$G(s) = \frac{0.4s + 1}{s(s + 0.6)}$$

Obtain the response to a unit-step input. What is the rise time for this system? What is the maximum overshoot?

**PROBLEM B-6-8.** Obtain the unit-impulse response and the unit-step response of a unity-feedback system whose open-loop transfer function is

$$G(s) = \frac{2s + 1}{s^2}$$

**PROBLEM B-6-9.** Consider the system shown in Fig. 6-58. Show that the transfer function $Y(s)/X(s)$ has a zero in the right-half $s$ plane. Then obtain $y(t)$ when $x(t)$ is a unit step. Plot $y(t)$ versus $t$.

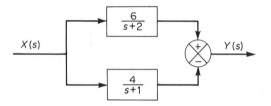

**Fig. 6-58.** System with a zero in the right-half $s$ plane. (Nonminimum phase system.)

**PROBLEM B-6-10.** Apply Routh's stability criterion to the following equation:

$$s^4 + Ks^3 + s^2 + s + 1 = 0$$

and determine the range of $K$ for stability.

**PROBLEM B-6-11.** Determine the range of $K$ for stability of a unity-feedback control system whose open-loop transfer function is

$$G(s) = \frac{K}{s(s + 1)(s + 2)}$$

**PROBLEM B-6-12.** Are the following two systems stable?

1. $$G(s) = \frac{10(s + 1)}{s(s - 1)(s + 5)}, \qquad H(s) = 1$$

2. $$G(s) = \frac{10}{s(s - 1)(2s + 3)}, \qquad H(s) = 1$$

**PROBLEM B-6-13.** Consider the system shown in Fig. 6-59 (a). The damping ratio of this system is 0.137 and the undamped natural frequency is 3.16 rad/sec. To improve the relative stability, we employ tachometer feedback. Figure 6-59 (b) shows such a tachometer-feedback system.

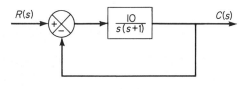

(a)

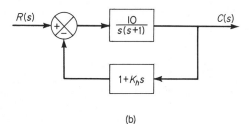

**Figure 6-59.** (a) Control system; (b) control system with tachometer feedback.

(b)

Determine the value of $K_h$ so that the damping ratio of the system is 0.5. Draw unit-step response curves of both the original and tachometer-feedback systems. Also draw the error-versus-time curves for the unit-ramp response of both systems.

**PROBLEM B-6-14.** Draw an analog computer diagram for the spring-mass-dashpot system described by

$$\ddot{x} + 2\dot{x} + 3x = u(t)$$

where

$$x(0) = 2 \text{ ft}, \qquad \dot{x}(0) = 5 \text{ ft/sec}$$

The input $u(t)$ is a step function and is given by

$$u(t) = 10 \text{ lb}$$

**PROBLEM B-6-15.** Suppose that the dynamic characteristic of a process is given by

$$\frac{C(s)}{R(s)} = \frac{\omega_n^2}{s^2 + 2\zeta\omega_n s + \omega_n^2} \qquad (0 < \zeta < 1)$$

The response of this system to a step input $a1(t)$ and that to a delayed step input $b1(t - t_1)$ are shown in Figs. 6-60 (a) and (b), respectively. From these response curves we can see that if we choose the values of $a$, $b$, and $t_1$ properly, then we can obtain the finite-time settling response without oscillation as shown in Fig. 6-60 (c). The control method of applying a step input and a delayed step input to obtain a finite-time settling response without oscillation is called *posicast control*. (This name was coined by O. J. M. Smith.)

If the dynamic characteristic of a batch process can be represented by an underdamped second-order transfer function, then this method may conveniently be applied to adjust the set point of the controller in starting the process cycle.

Suppose we desire to start the batch process and bring the output $c(t)$ to $h$ by posicast control. Determine the values of $a$, $b$, and $t_1$ shown in Fig. 6-60 (c) in terms of $h$, $\zeta$, and $\omega_n$.

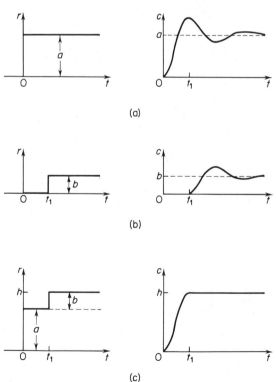

(a)

(b)

(c)

**Fig. 6-60.** (a) Step input and response curve; (b) delayed step input and response curve; (c) input and response curve showing characteristic of posicast control.

# 7

# ERROR ANALYSIS AND
# INTRODUCTION TO
# SYSTEM OPTIMIZATION

## 7-1 STATIC ERROR COEFFICIENTS

The transient-response characteristics discussed in Chapter 6 are important performance features of control systems. Another important feature is concerned with system error. Errors in a control system can be attributed to many factors. Changes in the reference input will cause unavoidable errors during transient periods and may also cause steady-state errors. Imperfections in the system components, such as static friction, backlash, and amplifier drift, as well as aging or deterioration, will cause errors at steady state.

In previous chapters, it was stated that steady-state error is a measure of the accuracy of a control system. The steady-state performance of a stable control system is generally judged by the steady-state error due to step, ramp, or acceleration inputs. In this section, we shall investigate such performance. Namely, we shall investigate a type of steady-state error which is caused by the incapability of a system to follow particular types of inputs.

Any physical control system inherently suffers steady-state error in response to certain types of inputs. A system may have no steady-state error to a step input, but the same system may exhibit nonzero steady-state error to a ramp input. (The only way we may be able to eliminate this error is to modify the system structure.) Whether or not a given system will exhibit steady-state error for a given type of input depends upon the type of open-loop transfer function of the system, to be discussed in what follows.

**Classification of control systems.** Control systems may be classified according to their ability to follow step inputs, ramp inputs, parabolic inputs, etc. This is a reasonable classification scheme because actual inputs may frequently be considered combinations of such inputs. The magnitudes of the steady-state errors due to these individual inputs are indicative of the "goodness" of the system.

Consider the following open-loop transfer function $G(s)H(s)$:

$$G(s)H(s) = \frac{K(T_a s + 1)(T_b s + 1) \cdots (T_m s + 1)}{s^N(T_1 s + 1)(T_2 s + 1) \cdots (T_p s + 1)}$$

It involves the term $s^N$ in the denominator, representing a pole of multiplicity $N$ at the origin. The present classification scheme is based on the number of integrations indicated by the open-loop transfer function. A system is called type 0, type 1, type 2, ... if $N = 0$, $N = 1$, $N = 2$, ..., respectively. Note that this classification is different from that of the order of a system. As the type number is increased, accuracy is improved; however, increasing the type number aggravates the stability problem. A compromise between steady-state accuracy and relative stability is always necessary. In practice, it is rather exceptional to have type 3 or higher systems because we find it generally difficult to design stable systems having more than two integrations in the feedforward path.

We shall see later that if $G(s)H(s)$ is written so that each term in the numerator and denominator, except the term $s^N$, approaches unity as $s$ approaches zero, then the open-loop gain $K$ is directly related to the steady-state error.

**Steady-state errors.** Consider the system shown in Fig. 7-1. The closed-loop transfer function is

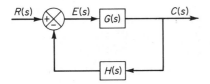

Fig. 7-1. Control system.

$$\frac{C(s)}{R(s)} = \frac{G(s)}{1 + G(s)H(s)}$$

The transfer function between the actuating error signal $e(t)$ and the input signal $r(t)$ is

$$\frac{E(s)}{R(s)} = 1 - \frac{C(s)H(s)}{R(s)} = \frac{1}{1 + G(s)H(s)}$$

where the actuating error $e(t)$ is the difference between the input signal and the feedback signal.

The final value theorem provides a convenient way to find the steady-state performance of a stable system. Since $E(s)$ is

$$E(s) = \frac{1}{1 + G(s)H(s)} R(s)$$

the steady-state actuating error is

$$e_{ss} = \lim_{t \to \infty} e(t) = \lim_{s \to 0} \frac{s R(s)}{1 + G(s)H(s)}$$

The static error coefficients defined in the following are figures of merit of control systems. The higher the coefficients, the smaller the steady-state error. In a given system, the output may be the position, velocity, pressure, temperature, etc. The physical form of the output, however, is immaterial to the present analysis. Therefore, in what follows, we shall call the output "position," the rate of change of the output "velocity," etc. This means that in a temperature control system "position" represents the output temperature, "velocity" represents the rate of change of the output temperature, etc.

**Static position error coefficient** $K_p$. The steady-state actuating error of the system for a unit-step input is

$$e_{ss} = \lim_{s \to 0} \frac{s}{1 + G(s)H(s)} \frac{1}{s}$$

$$= \frac{1}{1 + G(0)H(0)}$$

The static position error coefficient $K_p$ is defined by

$$K_p = \lim_{s \to 0} G(s)H(s) = G(0)H(0)$$

Thus, the steady-state actuating error in terms of the static position error coefficient $K_p$ is given by

$$e_{ss} = \frac{1}{1 + K_p}$$

For a type 0 system,

$$K_p = \lim_{s \to 0} \frac{K(T_a s + 1)(T_b s + 1) \cdots}{(T_1 s + 1)(T_2 s + 1) \cdots} = K$$

For a type 1 or higher system,

$$K_p = \lim_{s \to 0} \frac{K(T_a s + 1)(T_b s + 1) \cdots}{s^N(T_1 s + 1)(T_2 s + 1) \cdots} = \infty \qquad (N \geq 1)$$

Hence, for a type 0 system, the static position error coefficient $K_p$ is finite, while for a type 1 or higher system $K_p$ is infinite.

For a unit-step input, the steady-state actuating error $e_{ss}$ may be summarized as follows:

$$e_{ss} = \frac{1}{1 + K} \qquad \text{for type 0 systems}$$

$$e_{ss} = 0 \qquad \text{for type 1 or higher systems}$$

From the foregoing analysis, it is seen that the response of a unity-feedback control system to a step input involves a steady-state error if there is no integration in the feedforward path. (If small errors for step inputs can be tolerated, then a type 0 system may be permissible, provided that the gain $K$ is sufficiently large. If the gain $K$ is too large, however, it is difficult to obtain reasonable relative stability.) If zero steady-state error for a step input is desired, the type of the system must be one or higher.

**Static velocity error coefficient** $K_v$. The steady-state actuating error of the system with a unit-ramp input (unit-velocity input) is given by

$$e_{ss} = \lim_{s \to 0} \frac{s}{1 + G(s)H(s)} \frac{1}{s^2}$$

$$= \lim_{s \to 0} \frac{1}{sG(s)H(s)}$$

The static velocity error coefficient $K_v$ is defined by

$$K_v = \lim_{s \to 0} sG(s)H(s)$$

Thus, the steady-state actuating error in terms of the static velocity error coefficient $K_v$ is given by

$$e_{ss} = \frac{1}{K_v}$$

The term *velocity error* is used here to express the steady-state error for a ramp input. The dimension of the velocity error is the same as the system error. That is, velocity error is not an error in velocity, but it is an error in position due to a ramp input.

For a type 0 system,

$$K_v = \lim_{s \to 0} \frac{sK(T_a s + 1)(T_b s + 1) \cdots}{(T_1 s + 1)(T_2 s + 1) \cdots} = 0$$

For a type 1 system,

$$K_v = \lim_{s \to 0} \frac{sK(T_a s + 1)(T_b s + 1) \cdots}{s(T_1 s + 1)(T_2 s + 1) \cdots} = K$$

For a type 2 or higher system,

$$K_v = \lim_{s \to 0} \frac{sK(T_a s + 1)(T_b s + 1) \cdots}{s^N(T_1 s + 1)(T_2 s + 1) \cdots} = \infty \qquad (N \geq 2)$$

The steady-state actuating error $e_{ss}$ for the unit-ramp input can be summarized as follows:

$$e_{ss} = \frac{1}{K_v} = \infty \qquad \text{for type 0 systems}$$

$$e_{ss} = \frac{1}{K_v} = \frac{1}{K} \qquad \text{for type 1 systems}$$

$$e_{ss} = \frac{1}{K_v} = 0 \qquad \text{for type 2 or higher systems}$$

The foregoing analysis indicates that a type 0 system is incapable of following a ramp input in the steady state. The type 1 system with unity feedback can follow the ramp input with a finite error. In steady-state operation, the output velocity is exactly the same as the input velocity, but there is a positional error. This error is proportional to the velocity of the input and is inversely proportional to the gain $K$. Figure 7-2 shows an example of the response of a type 1 system with unity feedback to a ramp input. The type 2 or higher system can follow a ramp input with zero actuating error at steady state.

**Static acceleration error coefficient** $K_a$**.** The steady-state actuating error of the system with a unit-parabolic input (acceleration input) which is defined by

$$r(t) = \frac{t^2}{2} \qquad \text{for } t \geq 0$$

$$= 0 \qquad \text{for } t < 0$$

is given by

$$e_{ss} = \lim_{s \to 0} \frac{s}{1 + G(s)H(s)} \frac{1}{s^3}$$

$$= \frac{1}{\lim_{s \to 0} s^2 G(s)H(s)}$$

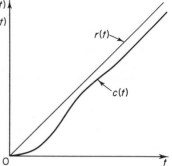

**Fig. 7-2.** Response of a type 1 unity-feedback system to a ramp input.

The static acceleration error coefficient $K_a$ is defined by the equation

$$K_a = \lim_{s \to 0} s^2 G(s)H(s)$$

The steady-state actuating error is then

$$e_{ss} = \frac{1}{K_a}$$

Note that the acceleration error, the steady-state error due to a parabolic input, is an error in position.

The values of $K_a$ are obtained as follows:

For a type 0 system,

$$K_a = \lim_{s \to 0} \frac{s^2 K(T_a s + 1)(T_b s + 1) \cdots}{(T_1 s + 1)(T_2 s + 1) \cdots} = 0$$

For a type 1 system,

$$K_a = \lim_{s \to 0} \frac{s^2 K(T_a s + 1)(T_b s + 1) \cdots}{s(T_1 s + 1)(T_2 s + 1) \cdots} = 0$$

For a type 2 system,

$$K_a = \lim_{s \to 0} \frac{s^2 K(T_a s + 1)(T_b s + 1) \cdots}{s^2(T_1 s + 1)(T_2 s + 1) \cdots} = K$$

For a type 3 or higher system,

$$K_a = \lim_{s \to 0} \frac{s^2 K(T_a s + 1)(T_b s + 1) \cdots}{s^N(T_1 s + 1)(T_2 s + 1) \cdots} = \infty \qquad (N \geq 3)$$

Thus, the steady-state actuating error for the unit parabolic input is

$$e_{ss} = \infty \qquad \text{for type 0 and type 1 systems}$$

$$e_{ss} = \frac{1}{K} \qquad \text{for type 2 systems}$$

$$e_{ss} = 0 \qquad \text{for type 3 or higher systems}$$

Note that both type 0 and type 1 systems are incapable of following a parabolic input in the steady state. The type 2 system with unity feedback can follow a parabolic input with a finite actuating error signal. Figure 7-3 shows an example of the response of a type 2 system with unity feedback to a parabolic input. The type 3 or higher system with unity feedback follows a parabolic input with zero actuating error at steady state.

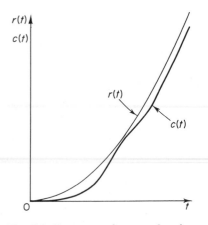

**Summary.** Table 7-1 summarizes the steady-state errors for type 0, type 1, and type 2 systems when they are subjected to various inputs. The finite values for steady-state errors appear on the diagonal line. Above the diagonal, the steady-state errors are infinity; below the diagonal, they are zero.

**Fig. 7-3.** Response of a type 2 unity-feedback system to a parabolic input.

Remember that the terms *position error, velocity error,* and *acceleration error* mean steady-state deviations in the output position. A finite velocity error implies that after transients have died out, the input and output move at the same velocity but have a finite position difference.

The error coefficients $K_p$, $K_v$, and $K_a$ describe the ability of a system to reduce or eliminate steady-state error. Therefore, they are indicative of the steady-state performance. It is generally desirable to increase the error coefficients, while maintaining the transient response within an acceptable range. If there is any conflict between the static velocity error coefficient and the static acceleration error coefficient, then the latter may be considered less important than the former. It is noted that to improve the steady-state performance, we can increase the type of the system by adding an integration or integrations to the feedforward path. This, however, introduces an additional stability problem. The design of a satisfactory system with more than two integrations in the feedforward path is generally difficult.

**Table 7-1.** STEADY-STATE ERROR IN TERMS OF GAIN $K$

|  | Step Input $r(t) = 1$ | Ramp Input $r(t) = t$ | Acceleration Input $r(t) = \frac{1}{2}t^2$ |
|---|---|---|---|
| Type 0 System | $\dfrac{1}{1 + K}$ | $\infty$ | $\infty$ |
| Type 1 System | $0$ | $\dfrac{1}{K}$ | $\infty$ |
| Type 2 System | $0$ | $0$ | $\dfrac{1}{K}$ |

**Correlation between integral of error in step response and steady-state error in ramp response.** We shall next discuss the correlation between the integral of the error in the unit-step response of a system and the steady-state error in the unit-ramp response of the same system.

In a unity-feedback control system, the total area under the error curve $\int_0^\infty e(t)\,dt$ as a result of unit-step response gives the steady-state error in the response to the unit-ramp input. To prove this, let us define

$$\mathscr{L}[e(t)] = \int_0^\infty \varepsilon^{-st}e(t)\,dt = E(s)$$

Then

$$\lim_{s\to 0}\int_0^\infty \varepsilon^{-st}e(t)\,dt = \int_0^\infty e(t)\,dt = \lim_{s\to 0} E(s)$$

Note that

$$\frac{E(s)}{R(s)} = 1 - \frac{C(s)}{R(s)} = \frac{1}{1 + G(s)}$$

Hence

$$\int_0^\infty e(t)\,dt = \lim_{s\to 0}\left[\frac{R(s)}{1 + G(s)}\right]$$

For a unit-step input,

$$\int_0^\infty e(t)\,dt = \lim_{s\to 0}\left[\frac{1}{1 + G(s)}\frac{1}{s}\right]$$

$$= \lim_{s\to 0}\frac{1}{sG(s)}$$

$$= \frac{1}{K_v}$$

$$= \text{steady-state error in unit-ramp response}$$

Hence in a unity-feedback control system, we have

$$\int_0^\infty e(t)\,dt = e_{ssr} \tag{7-1}$$

where

$e(t) = $ error in the unit-step response
$e_{ssr} = $ steady-state error in the unit-ramp response

If $e_{ssr}$ is zero, then $e(t)$ must change its sign at least once. This means that a zero-velocity-error system (a system having $K_v = \infty$) will exhibit at least one overshoot, when the system is subjected to a step input.

***

*Example 7-1.* Consider the system shown in Fig. 7-4. Let us obtain the unit-step response of the system, sketch the response curve, graphically compute the integral of the error signal, $\int_0^\infty e(t)\,dt$, and verify Eq. (7-1), (The upper limit of integration must, of course, be approximated by some finite time $T$, where $T$ is the time beyond which the output practically remains unity.)

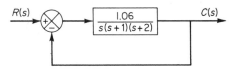

**Fig. 7-4.** Control system.

The closed-loop transfer function for the present system is given by

$$\frac{C(s)}{R(s)} = \frac{G(s)}{1 + G(s)}$$

$$= \frac{1.06}{s(s + 1)(s + 2) + 1.06}$$

$$= \frac{1.06}{(s + 2.33)(s + 0.33 + j0.58)(s + 0.33 - j0.58)}$$

For a unit-step input,

$$R(s) = \frac{1}{s}$$

The unit-step response is thus obtained as

$$c(t) = \mathscr{L}^{-1}[C(s)]$$

$$= \mathscr{L}^{-1}\left[\frac{1.06}{s(s + 2.33)[(s + 0.33)^2 + 0.58^2]}\right]$$

Referring to the formula

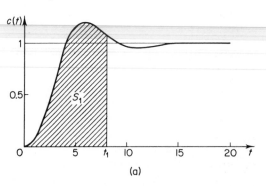

(a)

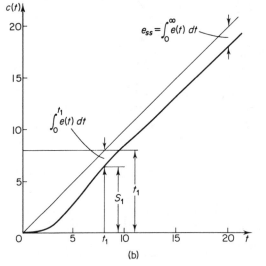

(b)

**Fig. 7-5.** (a) Unit-step response of the system shown in Fig. 7-4; (b) unit-ramp response of the system shown in Fig. 7-4.

$$\mathcal{L}^{-1}\left[\frac{1}{s(s+a)[(s+\zeta\omega_n)^2+\omega_d^2]}\right]=\frac{1}{a\omega_n^2}-\frac{e^{-at}}{a[(a-\zeta\omega_n)^2+\omega_d^2]}$$

$$+\frac{e^{-\zeta\omega_n t}}{\omega_n[(a-\zeta\omega_n)^2+\omega_d^2]}\left[\left(2\zeta-\frac{a}{\omega_n}\right)\cos\omega_d t+\frac{1}{\sqrt{1-\zeta^2}}\left(2\zeta^2-\frac{a\zeta}{\omega_n}-1\right)\sin\omega_d t\right]$$

and noting that in the present system

$$a=2.33,\qquad \zeta=0.5,\qquad \omega_n=0.667,\qquad \omega_d=0.58$$

we obtain $c(t)$ as follows:

$$c(t)=1-0.103\,e^{-2.33t}-e^{-0.33t}(0.897\cos 0.58t+0.933\sin 0.58t)\qquad (t\geq 0)\qquad (7\text{-}2)$$

Equation (7-2) gives the response to a unit-step input; see Fig. 7-5(a).

Now we shall examine the response of the system to a unit-ramp input. When all initial conditions are zero, the unit-ramp response of the system can be obtained by integrating the unit-step response curve. Figure 7-5(b) shows the unit-ramp response curve for the present system. [Note that the area $S_1$ under the unit-step response curve from $t=0$ to $t=t_1$ gives the value of the output at $t=t_1$ with a unit-ramp input. See Figs. 7-5(a) and (b).]

In the present system, the unit-step response settles at approximately $t=20$. Hence, the difference between 20 and the area under the unit-step response curve from $t=0$ to $t=20$ gives the steady-state error in the unit-ramp response. This value of the error should be equal to the value given by $1/\lim_{s\to 0} sG(s)$.

From the response curve of Fig. 7-5(a), we obtain the following data:

| $t$ | Area $S$ under Unit-Step Response Curve from 0 to $t$ | Error Area $1\times t-S$ |
|---|---|---|
| 0 | 0 | 0 |
| 2 | 0.16 | 1.84 |
| 4 | 1.50 | 2.50 |
| 6 | 3.80 | 2.20 |
| 8 | 6.10 | 1.90 |
| 10 | 8.15 | 1.85 |
| 12 | 10.10 | 1.90 |
| 14 | 12.06 | 1.94 |
| 16 | 14.07 | 1.93 |
| 18 | 16.08 | 1.92 |
| 20 | 18.08 | 1.92 |

The error area at steady state is found as 1.92.

The static velocity error coefficient for this system is obtained from the open-loop transfer function as

$$K_v=\lim_{s\to 0}sG(s)$$
$$=\frac{1.06}{2}$$
$$=0.53$$

Hence the steady-state error in the unit-ramp response is given by

$$e_{ss}=\frac{1}{K_v}=1.89$$

The graphically obtained value and the analytically obtained value of the steady-state error in the unit-ramp response are close to each other. Thus, we verified Eq. (7-1). (Note that if these two values differ considerably, then we should check to see where the discrepancy comes from, find the errors, and correct them.)

## 7-2. DYNAMIC ERROR COEFFICIENTS

A characteristic of the definition of static error coefficients is that only one of the coefficients assumes a finite value for a given system. The other coefficients are either zero or infinity. The steady-state error obtained through the static error coefficients is either zero, a finite nonzero value, or infinity. Thus, the variation of the error with time cannot be obtained through the use of such coefficients. The dynamic error coefficients presented in this section provide some information about how the error varies with time; namely, whether or not the steady-state error of the system with a given input increases in proportion to $t$, $t^2$, etc.

**Systems with different dynamic errors but identical static error coefficients.** We shall first demonstrate that two systems having different dynamic errors can have identical static error coefficients. Consider the following two systems:

$$G_1(s) = \frac{10}{s(s+1)}, \qquad G_2(s) = \frac{10}{s(5s+1)}$$

The static error coefficients are given by

$$K_{p1} = \infty, \qquad K_{p2} = \infty$$
$$K_{v1} = 10, \qquad K_{v2} = 10$$
$$K_{a1} = 0, \qquad K_{a2} = 0$$

Thus, the two systems have the same steady-state error for the same step input. Similar comments apply to the steady-state errors for ramp and parabolic inputs. This analysis indicates that it is impossible to estimate the system dynamic error from the static error coefficients.

**Dynamic error coefficients.** We shall now introduce dynamic error coefficients to describe dynamic error. We shall limit our systems to unity-feedback ones. By dividing the numerator polynomial of $E(s)/R(s)$ by its denominator polynomial, $E(s)/R(s)$ can be expanded into a series in ascending powers of $s$ as follows:

$$\frac{E(s)}{R(s)} = \frac{1}{1+G(s)} = \frac{1}{k_1} + \frac{1}{k_2}s + \frac{1}{k_3}s^2 + \cdots$$

The coefficients $k_1, k_2, k_3, \ldots$ of the power series are defined to be the *dynamic error coefficients*. Namely,

$k_1 =$ dynamic position error coefficient
$k_2 =$ dynamic velocity error coefficient
$k_3 =$ dynamic acceleration error coefficient

In a given system, the dynamic error coefficients are related to the static error coefficients. Consider the following type 0 system with unity feedback:

$$G(s) = \frac{K}{Ts + 1}$$

The static position error, static velocity error, and static acceleration error coefficients are, respectively,

$$K_p = K$$
$$K_v = 0$$
$$K_a = 0$$

Since $E(s)/R(s)$ can be expanded as

$$\frac{E(s)}{R(s)} = \frac{1 + Ts}{1 + K + Ts} = \frac{1}{1 + K} + \frac{TK}{(1 + K)^2} s + \cdots$$

the dynamic error coefficients are given in terms of the static error coefficients as follows:

The dynamic position error coefficient is

$$k_1 = 1 + K = 1 + K_p$$

The dynamic velocity error coefficient is

$$k_2 = \frac{(1 + K)^2}{TK}$$

As another example, consider the unity-feedback control system with the following feedforward transfer function:

$$G(s) = \frac{\omega_n^2}{s^2 + 2\zeta\omega_n s}$$

The static error coefficients are given by

$$K_p = \lim_{s \to 0} G(s) = \infty$$
$$K_v = \lim_{s \to 0} sG(s) = \frac{\omega_n}{2\zeta}$$
$$K_a = \lim_{s \to 0} s^2 G(s) = 0$$

Since $E(s)/R(s)$ can be expanded as

$$\begin{aligned}
\frac{E(s)}{R(s)} &= \frac{s^2 + 2\zeta\omega_n s}{s^2 + 2\zeta\omega_n s + \omega_n^2} \\
&= \frac{\dfrac{2\zeta}{\omega_n} s + \dfrac{1}{\omega_n^2} s^2}{1 + \dfrac{2\zeta}{\omega_n} s + \dfrac{s^2}{\omega_n^2}} \\
&= \frac{2\zeta}{\omega_n} s + \left(\frac{1 - 4\zeta^2}{\omega_n^2}\right) s^2 + \cdots
\end{aligned}$$

the dynamic velocity error coefficient is equal to the static velocity error coefficient; namely,

$$k_2 = \frac{\omega_n}{2\zeta} = K_v$$

The dynamic acceleration error coefficient is given by

$$k_3 = \frac{\omega_n^2}{1 - 4\zeta^2}$$

If a similar analysis is made for higher-order systems, we can show that for a type $N$ system the dynamic error coefficients are given by

$$k_{n+1} = \infty \qquad \text{for } n < N$$
$$k_{n+1} = \lim_{s \to 0} s^N G(s) \qquad \text{for } n = N$$

where $n = 0, 1, 2, \ldots$. The values of $k_{n+1}$ for $n > N$ are determined by the results of the expansion of $E(s)/R(s)$ near the origin.

**Advantage of dynamic error coefficients.** An advantage of the dynamic error coefficients becomes clear when $E(s)$ is written in the following form:

$$E(s) = \frac{1}{k_1} R(s) + \frac{1}{k_2} s R(s) + \frac{1}{k_3} s^2 R(s) + \cdots$$

The region of convergence of this series is the neighborhood of $s = 0$. This corresponds to $t = \infty$ in the time domain. The corresponding time solution or the steady-state error is given, assuming all initial conditions are zero and neglecting impulses at $t = 0$, as follows:

$$\lim_{t \to \infty} e(t) = \lim_{t \to \infty} \left[ \frac{1}{k_1} r(t) + \frac{1}{k_2} \dot{r}(t) + \frac{1}{k_3} \ddot{r}(t) + \cdots \right]$$

The steady-state error due to the input function and its derivatives can thus be given in terms of the dynamic error coefficients. This is an advantage of the dynamic error coefficients.

---

*Example 7-2.* Find the dynamic error coefficients of the unity-feedback control system whose feedforward transfer function is given by

$$G(s) = \frac{10}{s(s+1)}$$

Find also the steady-state error to the input defined by

$$r(t) = a_0 + a_1 t + a_2 t^2$$

For the present system,

$$\frac{E(s)}{R(s)} = \frac{1}{1 + G(s)}$$

$$= \frac{s + s^2}{10 + s + s^2}$$

$$= 0.1s + 0.09s^2 - 0.019s^3 + \cdots$$

or

$$E(s) = 0.1 s R(s) + 0.09 s^2 R(s) - 0.019 s^3 R(s) + \cdots$$

In the time domain, the steady-state error is given by

$$\lim_{t \to \infty} e(t) = \lim_{t \to \infty} [0.1\dot{r}(t) + 0.09\ddot{r}(t) - 0.019\dddot{r}(t) + \cdots]$$

The dynamic error coefficients are

$$k_1 = \infty$$

$$k_2 = \frac{1}{0.1} = 10$$

$$k_3 = \frac{1}{0.09} = 11.1$$

Since $r(t)$ is given by

$$r(t) = a_0 + a_1 t + a_2 t^2$$

we obtain

$$\dot{r}(t) = a_1 + 2a_2 t, \qquad \ddot{r}(t) = 2a_2, \qquad \dddot{r}(t) = 0$$

The steady-state error is then

$$\lim_{t \to \infty} e(t) = \lim_{t \to \infty} [0.1(a_1 + 2a_2 t) + 0.09(2a_2)]$$

$$= \lim_{t \to \infty} (0.1a_1 + 0.18a_2 + 0.2a_2 t)$$

Unless $a_2 = 0$, the steady-state error becomes infinity.

___

From the foregoing analysis, it can be seen that if $E(s)/R(s)$ is expanded around the origin into a power series, successive coefficients of the series indicate the dynamic error of the system when it is subjected to a slowly varying input. The dynamic error coefficients provide a simple way of estimating the error signal to arbitrary inputs and the steady-state error, without requiring us actually to solve the system differential equation.

**A remark on dynamic velocity error coefficients.** It is important to point out that the dynamic velocity error coefficient $k_2$ can be estimated from the time constant of the first-order system which approximates the given closed-loop transfer function in the neighborhood of $s = 0$.

Consider a unity-feedback control system with the following closed-loop transfer function:

$$\frac{C(s)}{R(s)} = \frac{1 + T_a s + T_b s^2 + \cdots}{1 + T_1 s + T_2 s^2 + \cdots}$$

In the neighborhood of $s = 0$,

$$\lim_{s \to 0} \frac{C(s)}{R(s)} = \frac{1}{1 + (T_1 - T_a)s + \cdots}$$

The dynamic velocity error coefficient $k_2$ is then given by

$$k_2 = \frac{1}{T_1 - T_a} \tag{7-3}$$

This can be verified by expanding the function $E(s)/R(s)$ about the origin as follows:

$$\frac{E(s)}{R(s)} = 1 - \frac{C(s)}{R(s)}$$

$$= 1 - \frac{1}{1 + (T_1 - T_a)s}$$

$$= 1 - 1 + (T_1 - T_a)s + \cdots$$

$$= (T_1 - T_a)s + \cdots$$

The dynamic velocity error coefficient $k_2$ is thus given by Eq. (7-3).

From the preceding analysis, we may conclude that if $C(s)/R(s)$ is approximated by

$$\frac{C(s)}{R(s)} = \frac{1}{1 + T_{eq}s} \qquad \text{for } s \ll 1$$

then the dynamic velocity error coefficient $k_2$ is

$$k_2 = \frac{1}{T_{eq}}$$

Note that the dynamic velocity error coefficient $k_2$ thus obtained is the same as the static velocity error coefficient. Since $C(s)/R(s)$ can be written

$$\frac{C(s)}{R(s)} = \frac{1}{1 + T_{eq}s}$$

$$= \frac{\dfrac{1}{T_{eq}s}}{1 + \dfrac{1}{T_{eq}s}}$$

$$= \frac{G(s)}{1 + G(s)}$$

the static velocity error coefficient $K_v$ is

$$K_v = \lim_{s \to 0} sG(s)$$

$$= \lim_{s \to 0} s \frac{1}{T_{eq}s}$$

$$= \frac{1}{T_{eq}}$$

## 7-3 ERROR CRITERIA

In the design of a control system, it is important that the system meet given performance specifications. Since control systems are dynamic, the performance specifications may be given in terms of the transient-response behavior to specific inputs, such as step inputs, ramp inputs, etc., or the specifications may be given in terms of a performance index.

**Performance indexes.** A performance index is a number which indicates the "goodness" of system performance. A control system is considered optimal if the values of the parameters are chosen so that the selected performance index is minimum or maximum. The optimal values of the parameters depend directly upon the performance index selected.

**Requirements of performance indexes.** A performance index must offer selectively; that is, an optimal adjustment of the parameters must clearly distinguish nonoptimal adjustments of the parameters. In addition, a performance index must yield a single positive number or zero, the latter being obtained if and only if the measure of the deviation is identically zero. To be useful, a performance index must be a function of the parameters of the system, and it must exhibit a maximum or minimum. Finally, to be practical, a performance index must be easily computed, analytically or experimentally.

**Error performance indexes.** In what follows, we shall discuss several error criteria in which the corresponding performance indexes are integrals of some function or weighted function of the deviation of the actual system output from the desired output. Since the values of the integrals can be obtained as functions of the system parameters, once a performance index is specified, the optimal system can be designed by adjusting the parameters to yield, say, the smallest value of the integral.

Various error performance indexes have been proposed in the literature. We shall discuss the following four in this section.

$$\int_0^\infty e^2(t)\, dt, \qquad \int_0^\infty te^2(t)\, dt, \qquad \int_0^\infty |e(t)|\, dt, \qquad \int_0^\infty t\,|e(t)|\, dt$$

Consider a control system whose desired output and actual output are $x(t)$ and $y(t)$, respectively. We shall define the error $e(t)$ as

$$e(t) = x(t) - y(t)$$

Note that unless

$$\lim_{t \to \infty} e(t) = 0$$

the performance indexes approach infinity. If $\lim_{t \to \infty} e(t)$ does not approach zero, we may define

$$e(t) = y(\infty) - y(t)$$

With this definition of the error, the performance indexes will yield finite numbers.

**Integral square-error criterion.** According to the integral square-error (ISE) criterion, the quality of system performance is evaluated by the following integral:

$$\int_0^\infty e^2(t)\, dt$$

where the upper limit $\infty$ may be replaced by $T$ which is chosen sufficiently large so that $e(t)$ for $T < t$ is negligible. The optimal system is the one which minimizes this integral. This performance index has been used extensively for both deterministic inputs (such as step inputs) and statistical inputs because of the ease of

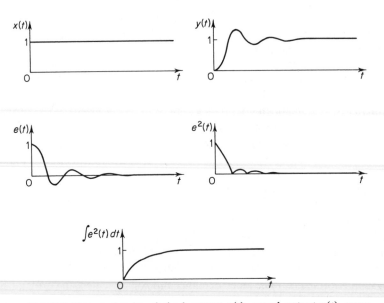

**Fig. 7-6.** Curves showing desired output $x(t)$, actual output $y(t)$, error $e(t)$, squared error $e^2(t)$, and integral square error $\int e^2(t)\,dt$ as functions of $t$.

computing the integral both analytically and experimentally. Figure 7-6 shows $x(t)$, $y(t)$, $e(t)$, $e^2(t)$, and $\int e^2(t)\,dt$ when the desired output $x(t)$ is a unit step. The integral of $e^2(t)$ from 0 to $T$ is the total area under the curve $e^2(t)$.

A characteristic of this performance index is that it weighs large errors heavily and small errors lightly. This criterion is not very selective since, for the following second-order system,

$$\frac{C(s)}{R(s)} = \frac{1}{s^2 + 2\zeta s + 1}$$

a change in $\zeta$ from 0.5 to 0.7 does not yield much change in the value of the integral.

A system designed by this criterion tends to show a rapid decrease in a large initial error. Hence the response is fast and oscillatory. Thus the system has poor relative stability.

Note, however, that the integral square-error criterion is often of practical significance because the minimization of the performance index results in the minimization of power consumption for some systems, such as spacecraft systems.

**Integral-of-time-multiplied square-error criterion.** The performance index based on the integral-of-time-multiplied square-error (ITSE) criterion is

$$\int_0^\infty t e^2(t)\,dt$$

The optimal system is the one which minimizes this integral.

This criterion has a characteristic that in the unit-step response of the system a large initial error is weighed lightly, while errors occurring late in the transient response are penalized heavily. This criterion has a better selectivity than the integral square-error criterion.

**Integral absolute-error criterion.** The performance index defined by the integral absolute-error (IAE) criterion is

$$\int_0^\infty |e(t)|\, dt$$

This is one of the most easily applied performance indexes. If this criterion is used, both highly underdamped and highly over-damped systems cannot be made optimum. An optimum system based on this criterion is a system which has reasonable damping and a satisfactory transient-response characteristic; however, the selectivity of this performance index is not too good. Although this performance index cannot easily be evaluated by analytical means, it is particularly convenient for analog computer study. Figure 7-7 shows the $e(t)$ versus $t$ curve and the $|e(t)|$ versus $t$ curve.

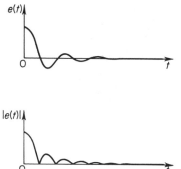

Note that minimization of the integral absolute error is directly related to the minimization of fuel consumption of spacecraft systems.

**Fig. 7-7.** Curves showing $e(t)$ versus $t$ and $|e(t)|$ versus $t$.

**Integral-of-time-multiplied absolute-error criterion.** According to this criterion, the ITAE criterion, the optimum system is the one which minimizes the following performance index:

$$\int_0^\infty t\,|e(t)|\, dt$$

As in the preceding criteria, a large initial error in a unit-step response is weighed lightly, and errors occurring late in the transient response are penalized heavily.

A system designed by use of this criterion has a characteristic that the overshoot in the transient response is small and oscillations are well damped. This criterion possesses good selectivity and is an improvement over the integral absolute-error criterion. It is, however, very difficult to evaluate analytically, although it can be easily measured experimentally.

**Comparison of various error criteria.** Figure 7-8 shows several error performance curves. The system considered is

$$\frac{C(s)}{R(s)} = \frac{1}{s^2 + 2\zeta s + 1}$$

Inspection of these curves reveals that the integral square-error criterion does not have good selectivity because the curve is rather flat near the point where the performance index is minimum. Addition of the square of error rate to the integrand

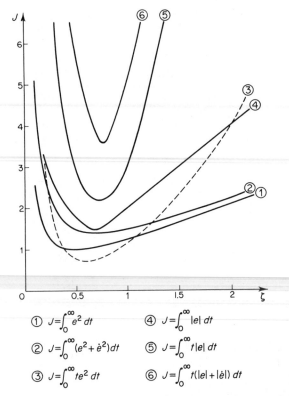

① $J = \int_0^\infty e^2\, dt$       ④ $J = \int_0^\infty |e|\, dt$

② $J = \int_0^\infty (e^2 + \dot{e}^2)\, dt$    ⑤ $J = \int_0^\infty t|e|\, dt$

③ $J = \int_0^\infty t e^2\, dt$      ⑥ $J = \int_0^\infty t(|e| + |\dot{e}|)\, dt$

**Fig. 7-8.** Error performance curves.

of the performance index, or

$$\int_0^\infty [e^2(t) + \dot{e}^2(t)]\, dt$$

shifts the optimal value of $\zeta$ to 0.707, but the selectivity is again rather poor. From Fig. 7-8, it is clearly seen that other performance indexes have good selectivity.

The curves in Fig. 7-8 indicate that in the second-order system considered $\zeta = 0.7$ corresponds to the optimal, or near optimal, value with respect to each of the performance indexes employed. When the damping ratio $\zeta$ is 0.7, the second-order system results in swift response to a step input with approximately 5% overshoot.

**Application of the ITAE criterion to $n$th-order systems.*** The ITAE criterion has been applied to the following $n$th-order system:

$$\frac{C(s)}{R(s)} = \frac{a_n}{s^n + a_1 s^{n-1} + \cdots + a_{n-1}s + a_n}$$

and the coefficients that will minimize

$$\int_0^\infty t\,|e|\, dt$$

have been determined. Clearly, this transfer function yields zero steady-state error to step inputs.

*Reference G-5.

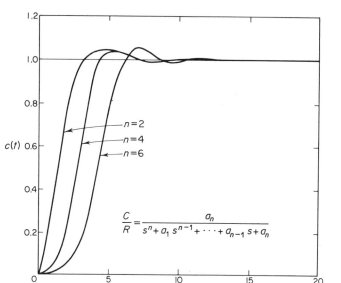

**Fig. 7-9.** Step-response curves of systems with optimal transfer functions based on the ITAE criterion.

Table 7-2 shows the optimal form of the closed-loop transfer function based on this ITAE criterion. Figure 7-9 shows the step-response curves of the optimal systems.

**Table 7-2.** OPTIMAL FORM OF THE CLOSED-LOOP TRANSFER FUNCTION BASED ON THE ITAE CRITERION (ZERO-STEP ERROR SYSTEMS)

$$\frac{C(s)}{R(s)} = \frac{a_n}{s^n + a_1 s^{n-1} + \cdots + a_{n-1}s + a_n}, \qquad a_n = \omega_n^n$$

$$s + \omega_n$$

$$s^2 + 1.4\omega_n s + \omega_n^2$$

$$s^3 + 1.75\omega_n s^2 + 2.15\omega_n^2 s + \omega_n^3$$

$$s^4 + 2.1\omega_n s^3 + 3.4\omega_n^2 s^2 + 2.7\omega_n^3 s + \omega_n^4$$

$$s^5 + 2.8\omega_n s^4 + 5.0\omega_n^2 s^3 + 5.5\omega_n^3 s^2 + 3.4\omega_n^4 s + \omega_n^5$$

$$s^6 + 3.25\omega_n s^5 + 6.60\omega_n^2 s^4 + 8.60\omega_n^3 s^3 + 7.45\omega_n^4 s^2 + 3.95\omega_n^5 s + \omega_n^6$$

## 7-4. INTRODUCTION TO SYSTEM OPTIMIZATION

In this section, we shall solve a simple optimization problem in which we minimize certain error performance indexes. We shall introduce the direct approach and the Laplace transform approach to the solution of this problem. We shall present an alternative (Liapunov) approach to the solution of similar optimization problems in Chapter 16.

**Direct computation of ISE performance indexes.** Consider the control system shown in Fig. 7-10. Suppose that a unit-step input $r(t)$ is applied to the system at $t = 0$. We assume that the system is initially at rest. The problem is to determine the optimal value of $\zeta > 0$ such that

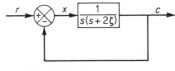

Fig. 7-10. Control system.

$$J = \int_0^\infty x^2(t)\, dt \qquad (7\text{-}4)$$

is minimum, where $x(t)$ is the error signal.

From Fig. 7-10, the system equation is obtained as follows:

$$\ddot{c} + 2\zeta\dot{c} + c = r$$

In terms of the error signal $x = r - c$,

$$\ddot{x} + 2\zeta\dot{x} + x = 0 \qquad (t > 0) \qquad (7\text{-}5)$$

The solution of Eq. (7-5) can be written

$$x(t) = k_1 e^{\lambda_1 t} + k_2 e^{\lambda_2 t} \qquad (7\text{-}6)$$

where

$$\lambda_1 = -\zeta + \sqrt{\zeta^2 - 1}, \qquad \lambda_2 = -\zeta - \sqrt{\zeta^2 - 1}$$

$k_1$ and $k_2$ are determined from the initial conditions.

Substituting Eq. (7-6) into Eq. (7-4) and noting that $\lambda_1$ and $\lambda_2$ have negative real parts, we obtain

$$J = \int_0^\infty x^2(t)\, dt$$

$$= \int_0^\infty (k_1^2 e^{2\lambda_1 t} + 2k_1 k_2 e^{(\lambda_1 + \lambda_2)t} + k_2^2 e^{2\lambda_2 t})\, dt$$

$$= -\frac{k_1^2}{2\lambda_1} - \frac{2k_1 k_2}{\lambda_1 + \lambda_2} - \frac{k_2^2}{2\lambda_2}$$

Since $\lambda_1 + \lambda_2 = -2\zeta$, $\lambda_1 \lambda_2 = 1$, we can simplify $J$ as follows:

$$J = -\frac{k_1^2 \lambda_2 + k_2^2 \lambda_1}{2} + \frac{k_1 k_2}{\zeta} \qquad (7\text{-}7)$$

The constants $k_1$ and $k_2$ are determined from the initial conditions as follows:

$$1 = k_1 + k_2, \qquad 0 = k_1 \lambda_1 + k_2 \lambda_2$$

or

$$k_1 = \frac{\lambda_2}{\lambda_2 - \lambda_1}, \qquad k_2 = -\frac{\lambda_1}{\lambda_2 - \lambda_1}$$

Substituting the values of $k_1$ and $k_2$ into Eq. (7-7), we obtain

$$J = -\frac{1}{2(\lambda_2 - \lambda_1)^2}\left(\lambda_2^3 + \lambda_1^3 + \frac{2\lambda_1 \lambda_2}{\zeta}\right)$$

$$= -\frac{1}{2(\lambda_2 - \lambda_1)^2}\left\{(\lambda_1 + \lambda_2)[(\lambda_1 + \lambda_2)^2 - 3\lambda_1 \lambda_2] + \frac{2\lambda_1 \lambda_2}{\zeta}\right\}$$

$$= \zeta + \frac{1}{4\zeta}$$

The optimal value of $\zeta$ can be obtained by differentiating $J$ with respect to $\zeta$, equating $dJ/d\zeta$ to zero, and solving for $\zeta$. Since

$$\frac{dJ}{d\zeta} = 1 - \frac{1}{4\zeta^2} = 0$$

the optimal value of $\zeta > 0$ is

$$\zeta = 0.5$$

(Clearly, $\zeta = 0.5$ corresponds to a minimum since $d^2J/d\zeta^2 > 0$.) The minimum value of $J$ is

$$\min J = 0.5 + 0.5 = 1$$

**Real integration theorem.** We shall next compute $\int_0^\infty f(t)\, dt$ by use of the Laplace transform method. In doing this, we shall use the following real integration theorem:

*Real Integration Theorem:* If $f(t)$ is of exponential order, then the Laplace transform of $\int_0^t f(t)\, dt$ exists and is

$$\mathscr{L}\left[\int_0^t f(t)\, dt\right] = \frac{F(s)}{s}$$

where

$$F(s) = \mathscr{L}[f(t)]$$

To prove this theorem, we use integration by parts.

$$\mathscr{L}\left[\int_0^t f(t)\, dt\right] = \int_0^\infty \left[\int_0^t f(t)\, dt\right] e^{-st}\, dt$$

$$= -\frac{e^{-st}}{s}\int_0^t f(t)\, dt\bigg|_0^\infty + \frac{1}{s}\int_0^\infty f(t) e^{-st}\, dt$$

$$= \lim_{t\to\infty}\left[-\frac{e^{-st}}{s}\int_0^t f(t)\, dt\right] + \frac{F(s)}{s}$$

Since $f(t)$ is of exponential order, when the real part of $s$ is sufficiently large, we obtain

$$\lim_{t\to\infty}\frac{e^{-st}}{s}\int_0^t f(t)\, dt \longrightarrow 0$$

Hence

$$\mathscr{L}\left[\int_0^t f(t)\, dt\right] = \frac{F(s)}{s}$$

This proves the theorem.

**Laplace transform approach to the computation of ISE performance indexes.** By use of the real integration and the final value theorems, we obtain

$$\int_0^\infty f(t)\, dt = \lim_{t\to\infty}\int_0^t f(t)\, dt$$

$$= \lim_{s\to 0} s\frac{F(s)}{s}$$

$$= \lim_{s\to 0} F(s)$$

This is the basic equation for computing the ISE performance index. In the present problem

$$f(t) = x^2(t)$$

Hence

$$\lim_{t \to \infty} \int_0^t x^2(t)\, dt = \lim_{s \to 0} F(s)$$

where

$$F(s) = \mathcal{L}[x^2(t)]$$

Consider again the control system shown in Fig. 7-10. For this system

$$\frac{X(s)}{R(s)} = \frac{s^2 + 2\zeta s}{s^2 + 2\zeta s + 1}$$

Hence, for the unit-step input,

$$X(s) = \frac{s^2 + 2\zeta s}{s^2 + 2\zeta s + 1} \frac{1}{s}$$

$$= \frac{s + \zeta}{s^2 + 2\zeta s + 1} + \frac{\zeta}{s^2 + 2\zeta s + 1}$$

The inverse Laplace transform is

$$x(t) = e^{-\zeta t}\left(\cos\sqrt{1 - \zeta^2}\,t + \frac{\zeta}{\sqrt{1 - \zeta^2}} \sin\sqrt{1 - \zeta^2}\,t\right) \qquad (t \ge 0)$$

Hence

$$x^2(t) = e^{-2\zeta t}\left[\frac{1}{2(1 - \zeta^2)} + \frac{1}{2}\frac{1 - 2\zeta^2}{1 - \zeta^2} \cos 2\sqrt{1 - \zeta^2}\,t + \frac{\zeta}{\sqrt{1 - \zeta^2}} \sin 2\sqrt{1 - \zeta^2}\,t\right]$$

Then

$$F(s) = \mathcal{L}[x^2(t)] = \frac{1}{2(1 - \zeta^2)}\frac{1}{s + 2\zeta} + \frac{1 - 2\zeta^2}{2(1 - \zeta^2)}\frac{s + 2\zeta}{(s + 2\zeta)^2 + 4(1 - \zeta^2)}$$

$$+ \frac{\zeta}{\sqrt{1 - \zeta^2}}\frac{2\sqrt{1 - \zeta^2}}{(s + 2\zeta)^2 + 4(1 - \zeta^2)} \tag{7-8}$$

Taking the limit as $s$ approaches zero gives

$$\lim_{s \to 0} F(s) = \frac{1}{2(1 - \zeta^2)}\frac{1}{2\zeta} + \frac{1 - 2\zeta^2}{2(1 - \zeta^2)}\frac{\zeta}{2} + \frac{\zeta}{2}$$

$$= \zeta + \frac{1}{4\zeta}$$

Hence

$$J = \int_0^\infty x^2(t)\, dt$$

$$= \zeta + \frac{1}{4\zeta} \tag{7-9}$$

This value of $J$, as a matter of course, is the same as that obtained by the direct approach, and thus the optimal value of $\zeta$ that minimizes $J$ is 0.5.

**Complex differentiation theorem.** In evaluating the performance index $\int_0^\infty tf(t)\,dt$ we find it convenient to use the following complex differentiation theorem:

*Complex differentiation theorem:* If $f(t)$ is Laplace transformable, then, except at singular points of $F(s)$,

$$\mathscr{L}[tf(t)] = -\frac{d}{ds}F(s)$$

where

$$\mathscr{L}[f(t)] = F(s)$$

To prove this, note that

$$\frac{d}{ds}F(s) = \frac{d}{ds}\int_0^\infty f(t)e^{-st}\,dt$$
$$= -\int_0^\infty tf(t)e^{-st}\,dt$$

Hence the theorem.

**Laplace transform approach to the computation of ITSE performance indexes.**
Consider next the system shown in Fig. 7-10, with the following performance index:

$$J = \int_0^\infty tx^2(t)\,dt$$

We again assume that the system is at rest before a unit-step input is applied. Let us find the optimal value of $\zeta$ that minimizes this performance index.

By use of the real integration and the complex differentiation theorems, we obtain

$$\mathscr{L}[tx^2(t)] = -\frac{d}{ds}F(s)$$

where

$$F(s) = \mathscr{L}[x^2(t)]$$

In the present problem, $F(s)$ is given by Eq. (7-8).

$$\mathscr{L}[tx^2(t)] = -\frac{d}{ds}F(s)$$

$$= \frac{1}{2(1-\zeta^2)}\frac{1}{(s+2\zeta)^2} + \frac{1-2\zeta^2}{2(1-\zeta^2)}\frac{(s+2\zeta)^2 - 4(1-\zeta^2)}{[(s+2\zeta)^2 + 4(1-\zeta^2)]^2}$$
$$+ \frac{4\zeta(s+2\zeta)}{[(s+2\zeta)^2 + 4(1-\zeta^2)]^2}$$

By use of the final value theorem,

$$\int_0^\infty tx^2(t)\,dt = \lim_{s\to 0}\left[-\frac{d}{ds}F(s)\right]$$

Taking the limit of $-dF(s)/ds$ as $s$ approaches zero, we obtain

$$\lim_{s \to 0} \left[ -\frac{d}{ds} F(s) \right] = \frac{1}{2(1 - \zeta^2)} \frac{1}{4\zeta^2} + \frac{1 - 2\zeta^2}{2(1 - \zeta^2)} \frac{2\zeta^2 - 1}{4} + \frac{\zeta^2}{2}$$

$$= \zeta^2 + \frac{1}{8\zeta^2}$$

Hence

$$J = \int_0^\infty t x^2(t) \, dt$$

$$= \zeta^2 + \frac{1}{8\zeta^2} \tag{7-10}$$

To obtain the optimal value of $\zeta$ that minimizes $J$ we differentiate $J$ with respect to $\zeta$:

$$\frac{dJ}{d\zeta} = 2\zeta - \frac{1}{4\zeta^3} = 0$$

which yields

$$\zeta = 0.595$$

Clearly, $d^2J/d\zeta^2 > 0$. The optimal value of $\zeta$ is 0.595 and the minimum value of $J$ is

$$\min J = 0.595^2 + \frac{1}{8 \times 0.595^2} = 0.71$$

## EXAMPLE PROBLEMS AND SOLUTIONS

**PROBLEM A-7-1.** Find the steady-state error in the output of a linear control system with unity feedback when the input is given by

$$r(t) = r_0 + r_1 t + \frac{r_2}{2} t^2$$

**Solution.** The steady-state error is the sum of the errors due to each input component. Therefore, the steady-state error is found as

$$e_{ss} = \frac{r_0}{1 + K_p} + \frac{r_1}{K_v} + \frac{r_2}{K_a}$$

The steady-state error becomes infinity unless the system is of type 2 or higher.

**PROBLEM A-7-2.** Consider the unity-feedback control system with feedforward transfer function $G(s)$. Suppose that the closed-loop transfer function can be written

$$\frac{C(s)}{R(s)} = \frac{G(s)}{1 + G(s)} = \frac{(T_a s + 1)(T_b s + 1) \cdots (T_m s + 1)}{(T_1 s + 1)(T_2 s + 1) \cdots (T_n s + 1)}$$

Show that

$$\int_0^\infty e(t) \, dt = (T_1 + T_2 + \cdots + T_n) - (T_a + T_b + \cdots + T_m)$$

where $e(t)$ is the error in the unit-step response. Show also that

$$\frac{1}{\lim_{s \to 0} sG(s)} = (T_1 + T_2 + \cdots + T_n) - (T_a + T_b + \cdots + T_m)$$

**Solution.** Let us define

$$(T_a s + 1)(T_b s + 1) \cdots (T_m s + 1) = P(s)$$

and

$$(T_1 s + 1)(T_2 s + 1) \cdots (T_n s + 1) = Q(s)$$

Then

$$\frac{C(s)}{R(s)} = \frac{P(s)}{Q(s)}$$

and

$$E(s) = \frac{Q(s) - P(s)}{Q(s)} R(s)$$

For a unit-step input, $R(s) = 1/s$. Hence

$$E(s) = \frac{Q(s) - P(s)}{s Q(s)}$$

Using the fact that

$$\int_0^\infty e(t)\, dt = \lim_{s \to 0} s \frac{E(s)}{s} = \lim_{s \to 0} E(s)$$

we obtain

$$\int_0^\infty e(t)\, dt = \lim_{s \to 0} \frac{Q(s) - P(s)}{s Q(s)}$$

$$= \lim_{s \to 0} \frac{Q'(s) - P'(s)}{Q(s) + s Q'(s)}$$

$$= \lim_{s \to 0} [Q'(s) - P'(s)]$$

Since

$$\lim_{s \to 0} P'(s) = T_a + T_b + \cdots + T_m$$

$$\lim_{s \to 0} Q'(s) = T_1 + T_2 + \cdots + T_n$$

we have

$$\int_0^\infty e(t)\, dt = (T_1 + T_2 + \cdots + T_n) - (T_a + T_b + \cdots + T_m)$$

Note that since

$$\int_0^\infty e(t)\, dt = \frac{1}{K_v} = \frac{1}{\lim_{s \to 0} s G(s)}$$

we immediately obtain the following relationship:

$$\frac{1}{\lim_{s \to 0} s G(s)} = (T_1 + T_2 + \cdots + T_n) - (T_a + T_b + \cdots + T_m)$$

Note that zeros will improve $K_v$. Poles close to the origin cause low velocity error coefficients unless there are zeros nearby.

**PROBLEM A-7-3.** Compare the errors of the unity-feedback control systems with the following feedforward transfer functions:

$$G_1(s) = \frac{10}{s(s+1)}, \qquad G_2(s) = \frac{10}{s(2s+1)}$$

The input is assumed to be

$$r(t) = 5 + 2t + 3t^2$$

**Solution.** Since

$$\frac{E_1(s)}{R(s)} = \frac{1}{1 + G_1(s)} = \frac{s + s^2}{10 + s + s^2}$$

$$= 0.1s + 0.09s^2 - 0.019s^3 + \cdots$$

and

$$\frac{E_2(s)}{R(s)} = \frac{1}{1 + G_2(s)} = \frac{s + 2s^2}{10 + s + 2s^2}$$

$$= 0.1s + 0.19s^2 - 0.039s^3 + \cdots$$

the error time functions are given by

$$\lim_{t \to \infty} e_1(t) = \lim_{t \to \infty} [0.1(2 + 6t) + 0.09 \times 6]$$

and

$$\lim_{t \to \infty} e_2(t) = \lim_{t \to \infty} [0.1(2 + 6t) + 0.19 \times 6]$$

It follows that

$$\lim_{t \to \infty} e_1(t) < \lim_{t \to \infty} e_2(t)$$

**PROBLEM A-7-4.** Consider the system whose closed-loop transfer function is

$$\frac{C(s)}{R(s)} = \frac{1}{s^2 + 2\zeta s + 1} \qquad (0 < \zeta < 1)$$

Compute $\int_0^\infty c^2(t)\, dt$ for a unit-impulse input.

**Solution.** For a unit-impulse input, $R(s) = 1$. Hence

$$C(s) = \frac{1}{s^2 + 2\zeta s + 1}$$

$$= \frac{1}{\sqrt{1 - \zeta^2}} \frac{\sqrt{1 - \zeta^2}}{(s + \zeta)^2 + 1 - \zeta^2}$$

The inverse Laplace transform of $C(s)$ is

$$c(t) = \frac{1}{\sqrt{1 - \zeta^2}} e^{-\zeta t} \sin \sqrt{1 - \zeta^2}\, t \qquad (t \geq 0)$$

Since $c^2(t)$ is

$$c^2(t) = \frac{1}{1 - \zeta^2} e^{-2\zeta t} \sin^2 \sqrt{1 - \zeta^2}\, t$$

$$= \frac{1}{1 - \zeta^2} e^{-2\zeta t} \frac{1}{2} (1 - \cos 2\sqrt{1 - \zeta^2}\, t)$$

we get

$$\mathcal{L}[c^2(t)] = \frac{1}{2(1 - \zeta^2)} \left[ \frac{1}{s + 2\zeta} - \frac{s + 2\zeta}{(s + 2\zeta)^2 + 4(1 - \zeta^2)} \right]$$

Then

$$\int_0^\infty c^2(t)\, dt = \lim_{s \to 0} \mathscr{L}[c^2(t)]$$

$$= \lim_{s \to 0} \frac{1}{2(1 - \zeta^2)} \left[ \frac{1}{s + 2\zeta} - \frac{s + 2\zeta}{(s + 2\zeta)^2 + 4(1 - \zeta^2)} \right]$$

$$= \frac{1}{4\zeta}$$

**PROBLEM A-7-5.** Consider the following system:

$$\ddot{x} + 2\zeta\dot{x} + x = 0 \qquad (0 < \zeta \le 0.9) \qquad\qquad (7\text{-}11)$$

Find the value of $\zeta$ which minimizes the following performance index $J$:

$$J = \int_0^\infty [x^2(t) + \dot{x}^2(t)]\, dt$$

**Solution.** The solution of Eq. (7-11) is

$$x(t) = k_1 e^{\lambda_1 t} + k_2 e^{\lambda_2 t}$$

where

$$\lambda_1 = -\zeta + \sqrt{\zeta^2 - 1}, \qquad \lambda_2 = -\zeta - \sqrt{\zeta^2 - 1}$$

$k_1$ and $k_2$ are determined from the initial conditions. Hence

$$\dot{x}(t) = k_1 \lambda_1 e^{\lambda_1 t} + k_2 \lambda_2 e^{\lambda_2 t}$$

Then $x^2(t)$ and $\dot{x}^2(t)$ are obtained as

$$x^2(t) = k_1^2 e^{2\lambda_1 t} + 2k_1 k_2 e^{(\lambda_1 + \lambda_2)t} + k_2^2 e^{2\lambda_2 t}$$

and

$$\dot{x}^2(t) = k_1^2 \lambda_1^2 e^{2\lambda_1 t} + 2k_1 k_2 \lambda_1 \lambda_2 e^{(\lambda_1 + \lambda_2)t} + k_2^2 \lambda_2^2 e^{2\lambda_2 t}$$

The performance index then becomes

$$J = \int_0^\infty [x^2(t) + \dot{x}^2(t)]\, dt$$

$$= \int_0^\infty [k_1^2(1 + \lambda_1^2)e^{2\lambda_1 t} + 2k_1 k_2(1 + \lambda_1 \lambda_2)e^{(\lambda_1 + \lambda_2)t} + k_2^2(1 + \lambda_2^2)e^{2\lambda_2 t}]\, dt$$

Performing the integration and evaluating at $t = \infty$ and $t = 0$, and noting that the real parts of $\lambda_1$ and $\lambda_2$ are negative, we obtain

$$J = -\left[ \frac{k_1^2(1 + \lambda_1^2)}{2\lambda_1} + \frac{2k_1 k_2(1 + \lambda_1 \lambda_2)}{\lambda_1 + \lambda_2} + \frac{k_2^2(1 + \lambda_2^2)}{2\lambda_2} \right]$$

Since

$$\lambda_1 + \lambda_2 = -2\zeta, \qquad \lambda_1 \lambda_2 = 1$$

we obtain

$$J = \zeta(k_1^2 + k_2^2) + \frac{2k_1 k_2}{\zeta}$$

$$= \zeta(k_1 + k_2)^2 - 2k_1 k_2\left(\zeta - \frac{1}{\zeta}\right) \qquad\qquad (7\text{-}12)$$

From the initial conditions,

$$x(0) = c_1 = k_1 + k_2, \qquad \dot{x}(0) = c_2 = k_1 \lambda_1 + k_2 \lambda_2$$

Since $0 < \zeta \leq 0.9$, $\lambda_1 \neq \lambda_2$ and $k_1$ and $k_2$ are obtained as

$$k_1 = \frac{c_1\lambda_2 - c_2}{\lambda_2 - \lambda_1}, \qquad k_2 = \frac{c_2 - \lambda_1 c_1}{\lambda_2 - \lambda_1}$$

Substitution of these values of $k_1$ and $k_2$ into Eq. (7-12) yields

$$J = \zeta c_1^2 + c_1 c_2 + \frac{c_1^2 + c_2^2}{2\zeta} \tag{7-13}$$

To find the minimum value of $J$, first differentiate $J$ with respect to $\zeta$ and then set the result to zero.

$$\frac{\partial J}{\partial \zeta} = c_1^2 - \frac{c_1^2 + c_2^2}{2\zeta^2} = 0$$

Solving this equation

$$\zeta = \sqrt{\frac{c_1^2 + c_2^2}{2c_1^2}} \tag{7-14}$$

Since $\partial^2 J / \partial \zeta^2 > 0$, Eq. (7-14) gives the optimal value of $\zeta$, provided

$$\sqrt{\frac{c_1^2 + c_2^2}{2c_1^2}} \leq 0.9$$

or

$$c_2^2 \leq 0.62 c_1^2$$

In the special case where $c_2 = 0$,

$$\zeta = \frac{1}{\sqrt{2}}$$

is the optimal value of $\zeta$. Hence if $c_1 = 1$ and $c_2 = 0$, then

$$\min J = \zeta + \frac{1}{2\zeta} = 1.414 \tag{7-15}$$

If $c_2^2 > 0.62 c_1^2$, then Eq. (7-14) becomes

$$\zeta = \sqrt{\frac{c_1^2 + c_2^2}{2c_1^2}} > \sqrt{\frac{1}{2} + \frac{0.62}{2}} = 0.9$$

But $\zeta$ is limited between 0 and 0.9. Hence Eq. (7-14) will not give the optimal value of $\zeta$. To find the optimal value of $\zeta$, define

$$c_2^2 = (0.62 + 2h)c_1^2, \qquad h > 0$$

Then we can rewrite Eq. (7-13) as follows:

$$J = \left(\zeta + \frac{0.81 + h}{\zeta}\right)c_1^2 + c_1 c_2$$

The optimal value of $\zeta$ which minimizes $J$ for given values of $c_1$ and $c_2$ can be obtained by minimizing the following quantity $I$ with respect to $\zeta$.

$$I = \zeta + \frac{0.81 + h}{\zeta}$$

The value of $I$ is minimum when $\zeta^2 = 0.81 + h$. This value of $\zeta$, however, cannot be attained because of the constraint that $0 < \zeta \leq 0.9$. Since the value of $I$ increases as $\zeta$ decreases from $\sqrt{0.81 + h}$, the optimal value of $\zeta$ when $c_2^2 > 0.62 c_1^2$ is 0.9.

In summary, the optimal value of $\zeta$ is

$$\zeta = \sqrt{\frac{c_1^2 + c_2^2}{2c_1^2}} \qquad \text{if } c_2^2 \le 0.62c_1^2$$

$$= 0.9 \qquad \text{if } c_2^2 > 0.62c_1^2$$

**PROBLEM A-7-6.** Compute the following performance indexes:

$$\int_0^\infty |e(t)|\, dt, \qquad \int_0^\infty t\, |e(t)|\, dt, \qquad \int_0^\infty t[|e(t)| + |\dot{e}(t)|]\, dt$$

for the system

$$\frac{C(s)}{R(s)} = \frac{1}{s^2 + 2\zeta s + 1} \qquad (\zeta \ge 1)$$

$$E(s) = \mathcal{L}[e(t)] = R(s) - C(s)$$

Assume that the system is initially at rest and is subjected to a unit-step input.

**Solution.** For a unit-step input, $R(s) = 1/s$ and

$$E(s) = \frac{s^2 + 2\zeta s}{s^2 + 2\zeta s + 1}\, \frac{1}{s}$$

$$= \frac{-\zeta + \sqrt{\zeta^2 - 1}}{2\sqrt{\zeta^2 - 1}}\, \frac{1}{s + \zeta + \sqrt{\zeta^2 - 1}} + \frac{\zeta + \sqrt{\zeta^2 - 1}}{2\sqrt{\zeta^2 - 1}}\, \frac{1}{s + \zeta - \sqrt{\zeta^2 - 1}}$$

For $\zeta \ge 1$, the system does not overshoot. Hence $|e(t)| = e(t)$ for all $t \ge 0$. The given performance indexes then can be computed as follows:

1.    $\displaystyle \int_0^\infty |e(t)|\, dt = \int_0^\infty e(t)\, dt$

$$= \lim_{s \to 0} E(s)$$

$$= \frac{-\zeta + \sqrt{\zeta^2 - 1}}{2\sqrt{\zeta^2 - 1}}\, \frac{1}{\zeta + \sqrt{\zeta^2 - 1}} + \frac{\zeta + \sqrt{\zeta^2 - 1}}{2\sqrt{\zeta^2 - 1}}\, \frac{1}{\zeta - \sqrt{\zeta^2 - 1}}$$

$$= 2\zeta \qquad (\zeta \ge 1)$$

2.    $\displaystyle \int_0^\infty t\, |e(t)|\, dt = \int_0^\infty t e(t)\, dt$

$$= \lim_{s \to 0}\left[ -\frac{d}{ds} E(s) \right]$$

$$= \lim_{s \to 0}\left[ \frac{-\zeta + \sqrt{\zeta^2 - 1}}{2\sqrt{\zeta^2 - 1}}\, \frac{1}{(s + \zeta + \sqrt{\zeta^2 - 1})^2} \right.$$

$$\left. + \frac{\zeta + \sqrt{\zeta^2 - 1}}{2\sqrt{\zeta^2 - 1}}\, \frac{1}{(s + \zeta - \sqrt{\zeta^2 - 1})^2} \right]$$

$$= 4\zeta^2 - 1 \qquad (\zeta \ge 1)$$

3. Noting that $\dot{e}(t) < 0$ and $|\dot{e}(t)| = -\dot{e}(t)$ for $t \ge 0$, we obtain

$$\int_0^\infty t(|e(t)| + |\dot{e}(t)|)\, dt = \int_0^\infty [t e(t) - t\dot{e}(t)]\, dt$$

$$= \lim_{s \to 0}\left[ -\frac{d}{ds} E(s) + \frac{d}{ds} s E(s) \right]$$

$$= \lim_{s \to 0}\left[ (-1 + s)\frac{d}{ds} E(s) + E(s) \right]$$

$$= 4\zeta^2 - 1 + 2\zeta \qquad (\zeta \ge 1)$$

For purposes of comparison, we shall list a few other performance indexes computed for the same system. [Refer to Eqs. (7-9), (7-10), and (7-15).]

$$\int_0^\infty e^2(t)\, dt = \zeta + \frac{1}{4\zeta} \qquad (\zeta > 0)$$

$$\int_0^\infty t e^2(t)\, dt = \zeta^2 + \frac{1}{8\zeta^2} \qquad (\zeta > 0)$$

$$\int_0^\infty [e^2(t) + \dot{e}^2(t)]\, dt = \zeta + \frac{1}{2\zeta} \qquad (\zeta > 0)$$

Plots of these six performance indexes are shown in Fig. 7-8 as functions of $\zeta$. (Note that the curves shown in Fig. 7-8 are valid for $\zeta > 0$. The three performance indexes computed in this problem are valid only for $\zeta \geqslant 1$.)

## PROBLEMS

**PROBLEM B-7-1.** Show that the steady-state error in the response to ramp inputs can be made zero if the closed-loop transfer function is given by

$$\frac{C(s)}{R(s)} = \frac{a_{n-1}s + a_n}{s^n + a_1 s^{n-1} + \cdots + a_{n-1}s + a_n}$$

**PROBLEM B-7-2.** Obtain the dynamic velocity error coefficients of the following systems:

1. $$\frac{C(s)}{R(s)} = \frac{10}{(s+1)(5s^2 + 2s + 10)}$$

2. $$\frac{C(s)}{R(s)} = \frac{3s + 10}{5s^2 + 2s + 10}$$

**PROBLEM B-7-3.** Consider the unity-feedback control system whose open-loop transfer function is

$$G(s) = \frac{100}{s(0.1s + 1)}$$

Determine the steady-state error when the input is

$$r(t) = 1 + t + at^2, \qquad a \geq 0$$

**PROBLEM B-7-4.** Consider the system shown in Fig. 7-11. Assume that the input is a ramp input, or

$$r(t) = at$$

where $a$ is an arbitrary constant.

Show that by properly adjusting the value of $K_i$, the steady-state error in the response to ramp inputs can be made zero.

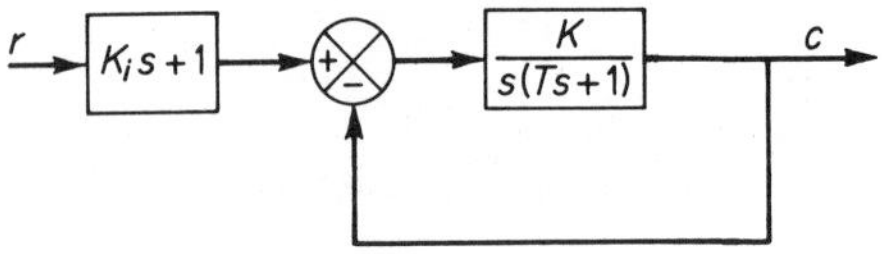

Fig. 7-11. Control system.

**PROBLEM B-7-5.** Consider a system described by

$$\ddot{x} + 2\zeta\dot{x} + x = 0$$

Determine the value of the damping ratio $\zeta$ so that

$$\int_0^\infty (x^2 + \mu \dot{x}^2)\, dt \qquad (\mu \geq 0)$$

is minimum.

**PROBLEM B-7-6.** The closed-loop transfer function for a disturbance input $n(t)$ is given by

$$\frac{C(s)}{N(s)} = \frac{s(s+a)}{s^2 + as + 10}$$

Determine the value of the constant $a$ so that the integral square-error due to a step disturbance is minimized.

**PROBLEM B-7-7.** Consider a unity-feedback control system with the open-loop transfer function

$$G(s) = \frac{(a+b)s + ab}{s^2}$$

For a unit-step input, compute the following performance indexes:

$$J_1 = \int_0^\infty e^2(t)\, dt$$

$$J_2 = \int_0^\infty t e^2(t)\, dt$$

If $a = b$, what are the values of $J_1$ and $J_2$?

# 8

# THE ROOT-LOCUS
# METHOD

## 8-1  INTRODUCTION

The basic characteristic of the transient response of a closed-loop system is determined from the closed-loop poles. Thus, in analysis problems, it is important to locate the closed-loop poles in the $s$ plane. In the design of closed-loop systems, we want to adjust the open-loop poles and zeros so as to place the closed-loop poles and zeros at desirable locations in the $s$ plane.

The closed-loop poles are the roots of the characteristic equation. Finding them requires factoring the characteristic polynomial. This is, in general, laborious if the degree of the characteristic polynomial is three or higher. The classical techniques of factoring polynomials are not convenient because as the gain of the open-loop transfer function varies, the computations must be repeated.

A simple method for finding the roots of the characteristic equation has been developed by W. R. Evans and used extensively in control engineering. This method, called the *root-locus method*, is one in which the roots of the characteristic equation are plotted for all values of a system parameter. The roots corresponding to a particular value of this parameter can then be located on the resulting graph. Note that the parameter is usually the gain but any other variable of the open-loop transfer function may be used. Unless otherwise stated, we shall assume that the gain of the open-loop transfer function is the parameter to be varied through all values, i.e., from zero to infinity.

314

**Root-locus method.** The basic idea behind the root-locus method is that the values of $s$ which make the transfer function around the loop equal $-1$ must satisfy the characteristic equation of the system.

The locus of roots of the characteristic equation of the closed-loop system as the gain is varied from zero to infinity gives the method its name. Such a plot clearly shows the contributions of each open-loop pole or zero to the locations of the closed-loop poles.

The root-locus method enables us to find the closed-loop poles from the open-loop poles and zeros with the gain as parameter. It avoids inherent difficulties existing in classical techniques by providing a graphical display of all closed-loop poles for all values of the gain of the open-loop transfer function.

In designing a linear control system, we find that the root-locus method proves quite useful since it indicates the manner in which the open-loop poles and zeros should be modified so that the response meets system performance specifications. This method is particularly suited to obtaining approximate results very quickly.

Since the method is a graphical one for finding the roots of the characteristic equation, it provides an effective graphical procedure for finding the roots of any polynomial equation arising in the study of physical systems.

**Outline of the chapter.** In Section 8-2 we shall introduce the concepts underlying the root-locus method. Section 8-3 presents the general procedure for sketching root loci. Following this presentation, we shall, in Section 8-4, analyze closed-loop systems by use of the root-locus method. (The root-locus approach to design problems will be given in Chapter 10.)

## 8-2 ROOT-LOCUS PLOTS

**Angle and magnitude conditions.** Consider the system shown in Fig. 8-1. The closed-loop transfer function is

$$\frac{C(s)}{R(s)} = \frac{G(s)}{1 + G(s)H(s)} \tag{8-1}$$

The characteristic equation for this closed-loop system is obtained by setting the denominator of the right-hand side of Eq. (8-1) equal to zero. Namely,

$$1 + G(s)H(s) = 0$$

or

$$G(s)H(s) = -1 \tag{8-2}$$

Since $G(s)H(s)$ is a complex quantity, Eq. (8-2) can be split into two equations by equating the angles and magnitudes of both sides, respectively, to obtain

Angle condition:

$$\underline{/G(s)H(s)} = \pm 180°(2k + 1) \qquad (k = 0, 1, 2, \ldots) \tag{8-3}$$

Magnitude condition:

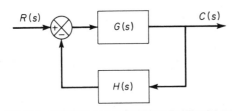

Fig. 8-1. Control system.

$$|G(s)H(s)| = 1 \qquad (8\text{-}4)$$

The values of $s$ which fulfill the angle and magnitude conditions are the roots of the characteristic equation, or the closed-loop poles. A plot of the points of the complex plane satisfying the angle condition alone is the root locus. The roots of the characteristic equation (the closed-loop poles) corresponding to a given value of the gain can be determined from the magnitude condition. The details of applying the angle and magnitude conditions to obtain the closed-loop poles are presented in Section 8-3.

**Root-locus plots of second-order systems.** Before presenting a method for constructing such plots in detail, a root-locus plot of a simple second-order system is illustrated.

Consider the system shown in Fig. 8-2. The open-loop transfer function $G(s)H(s)$ is

$$G(s)H(s) = \frac{K}{s(s + 1)}$$

The closed-loop transfer function is then

$$\frac{C(s)}{R(s)} = \frac{K}{s^2 + s + K}$$

The characteristic equation is

$$s^2 + s + K = 0 \qquad (8\text{-}5)$$

We wish to find the locus of roots of this equation as $K$ is varied from zero to infinity.

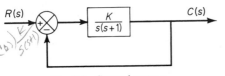

Fig. 8-2. Control system.

In order to give a clear idea of what a root-locus plot looks like for this system, we shall first obtain the roots of the characteristic equation analytically in terms of $K$ and then vary $K$ from zero to infinity. It should be noted that this is not the proper way to construct the root-locus plot. The proper way is through a graphical trial-and-error approach, and the graphical work can be simplified greatly by applying the general rules to be presented in Section 8-3. (Obviously, if an analytical solution for the characteristic roots can be found easily, there is no need for the root-locus method.) The roots of the characteristic equation, Eq. (8-5), are

$$s_1 = -\tfrac{1}{2} + \tfrac{1}{2}\sqrt{1 - 4K}, \qquad s_2 = -\tfrac{1}{2} - \tfrac{1}{2}\sqrt{1 - 4K}$$

The roots are real for $K \leq \tfrac{1}{4}$, and are complex for $K > \tfrac{1}{4}$.

The loci of the roots corresponding to all values of $K$ are plotted in Fig. 8-3 (a). The root loci are graduated with $K$ as parameter. (The motion of the roots with

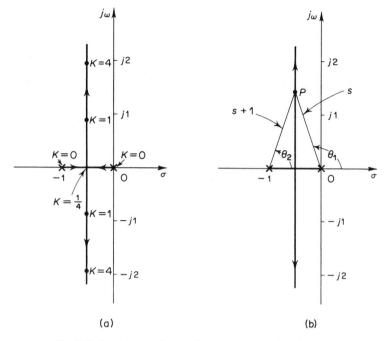

**Fig. 8-3.** Root-locus plots of the system shown in Fig. 8-2.

increasing $K$ is shown by arrows.) Once we draw such a plot, we can immediately determine the value of $K$ that will yield a root, or a closed-loop pole, at a desired point. From this analysis, it is clear that the closed-loop poles corresponding to $K = 0$ are the same as the poles of $G(s)H(s)$. As the value of $K$ is increased from zero to $\frac{1}{4}$, the closed-loop poles move toward point $(-\frac{1}{2}, 0)$. For values of $K$ between zero and $\frac{1}{4}$, all the closed-loop poles are on the real axis. This corresponds to an overdamped system, and the impulse response is nonoscillatory. At $K = \frac{1}{4}$, the two real closed-loop poles coalesce. This corresponds to the case of a critically damped system. As $K$ is increased from $\frac{1}{4}$, the closed-loop poles break away from the real axis, becoming complex, and since the real part of the closed-loop pole is constant for $K > \frac{1}{4}$, the closed-loop poles move along the line $s = -\frac{1}{2}$. Hence, for $K > \frac{1}{4}$, the system becomes underdamped. For a given value of $K$, one of the conjugate closed-loop poles moves toward $s = -\frac{1}{2} + j\infty$ while the other moves toward $s = -\frac{1}{2} - j\infty$.

We shall show that any point on the root locus satisfies the angle condition. The angle condition given by Eq. (8-3) is

$$\Big/\frac{K}{s(s+1)} = -\big/s - \big/s+1 = \pm 180°(2k+1) \qquad (k = 0, 1, 2, \ldots)$$

Consider point $P$ on the root locus shown in Fig. 8-3(b). The complex quantities $s$ and $s + 1$ have angles $\theta_1$ and $\theta_2$, respectively, and magnitudes $|s|$ and $|s+1|$, respectively. (Note that all angles are considered positive when they are measured in counterclockwise direction.) The sum of the angles $\theta_1$ and $\theta_2$ is clearly 180°.

If point $P$ is located on the real axis between 0 and $-1$, then $\theta_1 = 180°$ and $\theta_2 = 0°$. Thus, we see that any point on the root locus satisfies the angle condition. We also see that if point $P$ is not a point on the root locus, then the sum of $\theta_1$ and $\theta_2$ is not equal to $\pm 180°(2k + 1)$ where $k = 0, 1, 2, \ldots$. Thus, points that are not on the root locus do not satisfy the angle condition and, therefore, cannot be closed-loop poles for any value of $K$.

If the closed-loop poles are specified on the root locus, then the corresponding value of $K$ is determined from the magnitude condition, Eq. (8-4). If, for example, the closed-loop poles selected are $s = -\frac{1}{2} \pm j2$, then the corresponding value of $K$ is found from

$$|G(s)H(s)| = \left| \frac{K}{s(s + 1)} \right|_{s = -(1/2)+j2} = 1$$

or

$$K = |s(s + 1)|_{s=-(1/2)+j2} = \frac{17}{4}$$

Since complex poles are conjugates, if one of them, for example $s = -\frac{1}{2} + j2$,

**Table 8-1.** COLLECTION OF SIMPLE ROOT-LOCUS PLOTS

| $G(s)H(s)$ | Open–loop pole–zero locations and root loci | $G(s)H(s)$ | Open–loop pole–zero locations and root loci |
|---|---|---|---|
| $\dfrac{K}{s}$ | | $\dfrac{K}{s^2}$ | |
| $\dfrac{K}{s+p}$ | | $\dfrac{K}{s^2 + \omega_1^2}$ | |
| $\dfrac{K(s+z)}{s+p}$ $(z > p)$ | | $\dfrac{K}{(s+\sigma)^2 + \omega_1^2}$ | |
| $\dfrac{K(s+z)}{s+p}$ $(z < p)$ | | $\dfrac{K}{(s+p_1)(s+p_2)}$ | |

is specified, then the other is automatically fixed. In evaluating the value of $K$, either pole can be used.

From the root-locus plot of Fig. 8-3 (a), we clearly see the effects of changes in the value of $K$ on the transient-response behavior of the second-order system. An increase in the value of $K$ will decrease the damping ratio $\zeta$, resulting in an increase in the overshoot of the response. An increase in the value of $K$ will also result in increases in the damped and undamped natural frequencies. (If $K$ is greater than the critical value, that which corresponds to a critically damped system, increasing $K$ will have no effect on the value of the real part of the closed-loop poles.) From the root-locus plot, it is clear that the closed-loop poles are always in the left half of the $s$ plane; thus no matter how much $K$ is increased, the system always remains stable. Hence, the second-order system is always stable. (Note, however, that if the gain is set at a very high value, the effects of some of the neglected time constants may become important and the system which is supposedly of second order, but actually of higher order, may become unstable.)

Table 8-1 shows a collection of simple root-locus plots.

**Constant gain locus.** Figure 8-4 shows a plot of the constant gain loci of the system shown in Fig. 8-2. The constant gain loci for this system have been obtained from the magnitude condition:

$$|G(s)H(s)| = \left| \frac{K}{s(s+1)} \right| = 1$$

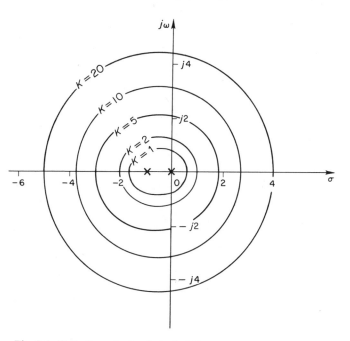

**Fig. 8-4.** Plot of constant gain loci of the system shown in Fig. 8-2.

or

$$|s(s + 1)| = K \qquad (8\text{-}6)$$

The points in the complex plane that satisfy Eq. (8-6) for a given $K$ constitute a constant gain locus.

**Orthogonality of root loci and constant gain loci.** Consider the system shown in Fig. 8-1. In the $G(s)H(s)$ plane, the loci of $|G(s)H(s)| = $ constant are circles centered at the origin, and the loci corresponding to $\underline{/G(s)H(s)} = \pm180°(2k + 1)$ ($k = 0, 1, 2, \ldots$) lie on the negative real axis of the $G(s)H(s)$ plane, as shown in Fig. 8-5. [Note that the complex plane employed here is not the $s$ plane but the $G(s)H(s)$ plane.]

The root loci and constant gain loci in the $s$ plane are conformal mappings of the loci of $\underline{/G(s)H(s)} = \pm180°(2k + 1)$ and of $|G(s)H(s)| = $ constant in the $G(s)H(s)$ plane.

Since the constant phase and constant gain loci in the $G(s)H(s)$ plane are orthogonal, the root loci and constant gain loci in the $s$ plane are orthogonal. Figure 8-6 (a) shows the root and constant gain loci for the following system:

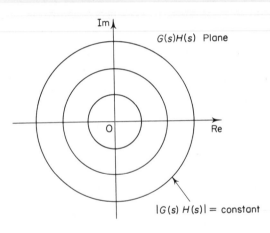

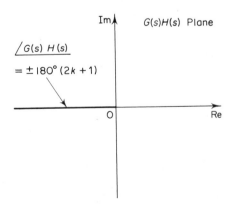

**Fig. 8-5.** Plots of constant gain and constant phase loci in the $G(s)H(s)$ plane.

$$G(s) = \frac{K(s+2)}{s^2 + 2s + 3}, \qquad H(s) = 1$$

Note that since the pole-zero configuration is symmetric about the real axis, the constant gain loci are also symmetric about the real axis.

Figure 8-6 (b) shows the root and constant gain loci for the system:

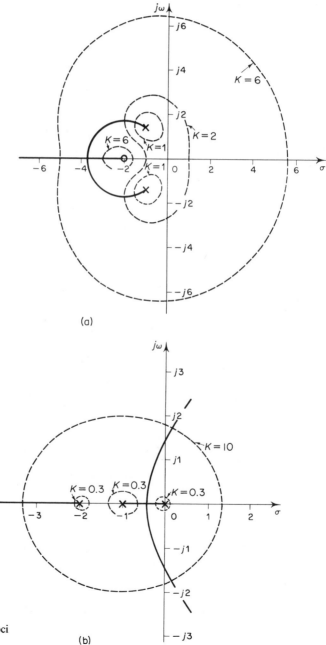

(a)

(b)

**Fig. 8-6.** Plots of root loci and constant gain loci.

$$G(s) = \frac{K}{s(s + 1)(s + 2)}, \qquad H(s) = 1$$

Note that since the configuration of the poles in the $s$ plane is symmetric about the real axis and the line parallel to the imaginary axis passing through point ($\sigma = -1, \omega = 0$), the constant gain loci are symmetric about the $\omega = 0$ line (real axis) and the $\sigma = -1$ line.

## 8-3  TWO ILLUSTRATIVE EXAMPLES

In this section, we shall present two illustrative examples for constructing root-locus plots. We shall use graphical computation, combined with inspection, to determine the root loci upon which the roots of the characteristic equation of the closed-loop system must lie. Although we employ only simple systems for illustrating purposes, the procedure for finding the root loci is no more complicated for higher-order systems.

The first step in the procedure for constructing a root-locus plot is to seek out the loci of possible roots using the angle condition. Then the loci are scaled, or graduated, in gain using the magnitude condition.

Because graphical measurements of angles and magnitudes are involved in the analysis, in sketching the root locus on graph paper, we find it necessary to use the same divisions on the abscissa as on the ordinate axis.

---

*Example 8-1.* Consider the system shown in Fig. 8-7. (We assume that the value of gain $K$ is positive.) For this system,

$$G(s) = \frac{K}{s(s + 1)(s + 2)}, \qquad H(s) = 1$$

Let us sketch the root-locus plot and then determine the value of $K$ so that the damping ratio $\zeta$ of a pair of dominant complex-conjugate closed-loop poles is 0.5.

For the given system, the angle condition becomes

$$
\begin{aligned}
\underline{/G(s)} &= \underline{\bigg/ \frac{K}{s(s + 1)(s + 2)}} \\
&= -\underline{/s} - \underline{/s + 1} - \underline{/s + 2} \\
&= \pm 180°(2k + 1) \qquad (k = 0, 1, 2, \ldots)
\end{aligned}
$$

The magnitude condition is

$$|G(s)| = \left| \frac{K}{s(s + 1)(s + 2)} \right| = 1$$

A typical procedure for sketching the root-locus plot is as follows:

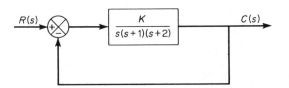

Fig. 8-7. Control system.

1. Determine the root loci on the real axis. The first step in constructing a root-locus plot is to locate the open-loop poles, $s = 0$, $s = -1$, and $s = -2$ in the complex plane. (There are no open-loop zeros in this system.) The locations of the open-loop poles are indicated by crosses. (The locations of the open-loop zeros will be indicated by small circles in this book.) Note that the starting points of the root loci (the points corresponding to $K = 0$) are open-loop poles. The number of individual root loci for this system is three, which is the same as the number of open-loop poles.

To determine the root loci on the real axis, we select a test point, $s$. If the test point is on the positive real axis, then

$$\underline{/s} = \underline{/s+1} = \underline{/s+2} = 0°$$

This shows that the angle condition cannot be satisfied. Hence, there is no root locus on the positive real axis. Next, select a test point on the negative real axis between 0 and $-1$. Then

$$\underline{/s} = 180°, \qquad \underline{/s+1} = \underline{/s+2} = 0°$$

Thus

$$-\underline{/s} - \underline{/s+1} - \underline{/s+2} = -180°$$

and the angle condition is satisfied. Therefore, the portion of the negative real axis between 0 and $-1$ forms a portion of the root locus. If a test point is selected between $-1$ and $-2$, then

$$\underline{/s} = \underline{/s+1} = 180°, \qquad \underline{/s+2} = 0°$$

It can be seen that the angle condition will not be satisfied. Therefore, the negative real axis from $-1$ to $-2$ is not a part of the root locus. Similarly, if a test point is located on the negative real axis from $-2$ to $-\infty$, the angle condition is satisfied. Thus, root loci exist on the negative real axis between 0 and $-1$ and between $-2$ and $-\infty$.

2. Determine the asymptotes of the root loci. The asymptotes of the root loci as $s$ approaches infinity can be determined as follows: If a test point $s$ is selected very far from the origin, then

$$\lim_{s \to \infty} G(s) = \lim_{s \to \infty} \frac{K}{s(s+1)(s+2)} = \lim_{s \to \infty} \frac{K}{s^3}$$

and the angle condition becomes

$$-3\underline{/s} = \pm 180°(2k+1) \qquad (k = 0, 1, 2, \ldots)$$

or

$$\text{Angle of asymptote} = \frac{\pm 180°(2k+1)}{3}$$

Since the angle repeats itself as $k$ is varied, the distinct angles for the asymptotes are determined as $60°$, $-60°$, and $180°$. Thus, there are three asymptotes. The one having the angle of $180°$ is the negative real axis.

Before we can draw these asymptotes in the complex plane, we must find where they intersect the real axis. Since the characteristic equation of this system is

$$\frac{K}{s(s+1)(s+2)} = -1$$

or

$$s^3 + 3s^2 + 2s = -K$$

if $s$ is assumed to be very large, then we can approximate this characteristic equation by the following equation:

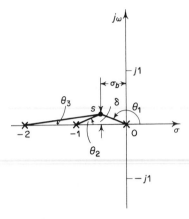

**Fig. 8-8.** Determination of breakaway point.

$$(s + 1)^3 = 0$$

The abscissa of the intersection of the asymptotes and the real axis can be obtained by letting $s = \sigma_a$ and solving for $\sigma_a$. Thus

$$\sigma_a = -1$$

and the point of origin of the asymptotes is $(-1, 0)$. The asymptotes are almost part of the root loci in regions very far from the origin.

3. Determine the breakaway point. In order to plot root loci accurately, we must find the breakaway point, where the root-locus branches originating from the poles at 0 and $-1$ break away (as $K$ is increased) from the real axis and move into the complex plane. The breakaway point corresponds to a point in the $s$ plane where multiple roots of the characteristic equation are found.

The breakaway point $(-\sigma_b, 0)$ may be found as follows: Assume a test point which is near the negative real axis between 0 and $-1$, as shown in Fig. 8-8. If we denote the vertical distance of the test point measured from the negative real axis by $\delta$, then the angles $\theta_1$, $\theta_2$, and $\theta_3$ of the complex quantities $s, s + 1,$ and $s + 2$, respectively, can be written as follows:

$$\theta_1 = 180° - \tan^{-1}\left(\frac{\delta}{\sigma_b}\right)$$

$$\theta_2 = \tan^{-1}\left[\frac{\delta}{-\sigma_b - (-1)}\right]$$

$$\theta_3 = \tan^{-1}\left[\frac{\delta}{-\sigma_b - (-2)}\right]$$

For a small value of $\delta$,

$$\theta_1 = 180° - \frac{\delta}{\sigma_b}$$

$$\theta_2 = \frac{\delta}{-\sigma_b + 1}$$

$$\theta_3 = \frac{\delta}{-\sigma_b + 2}$$

The angle condition

$$\theta_1 + \theta_2 + \theta_3 = 180°$$

becomes

$$-\frac{\delta}{\sigma_b} + \frac{\delta}{-\sigma_b + 1} + \frac{\delta}{-\sigma_b + 2} = 0°$$

or

$$\frac{1}{\sigma_b} + \frac{1}{\sigma_b - 1} + \frac{1}{\sigma_b - 2} = 0 \qquad (8-7)$$

By solving Eq. (8-7), we obtain

$$3\sigma_b^2 - 6\sigma_b + 2 = 0$$

which yields

$$\sigma_b = 0.423, \qquad \sigma_b = 1.577$$

Since $0 > -\sigma_b > -1$, the breakaway point must be $(-0.423, 0)$. If the number of terms of Eq. (8-7) were four or more, a trial-and-error approach would be used for finding the value of $\sigma_b$. Such an approach can be facilitated if we find an approximate value for $\sigma_b$ by neglecting the poles and zeros located far from the breakaway point, and then we make a small correction to this approximate value for $\sigma_b$.

An alternate method for finding the breakaway point is available. We shall present this method in the following: Let us write the characteristic equation as

$$f(s) = A(s) + KB(s) = 0 \qquad (8\text{-}8)$$

where $A(s)$ and $B(s)$ do not contain $K$. Note that $f(s) = 0$ has multiple roots at points where

$$\frac{df(s)}{ds} = 0$$

From Eq. (8-8), we obtain

$$\frac{df(s)}{ds} = A'(s) + KB'(s) = 0 \qquad (8\text{-}9)$$

where

$$A'(s) = \frac{dA(s)}{ds}, \qquad B'(s) = \frac{dB(s)}{ds}$$

The particular value of $K$ which will yield multiple roots of the characteristic equation is obtained from Eq. (8-9) as

$$K = -\frac{A'(s)}{B'(s)}$$

If we substitute this value of $K$ into Eq. (8-8), we get

$$f(s) = A(s) - \frac{A'(s)}{B'(s)}B(s) = 0$$

or

$$A(s)B'(s) - A'(s)B(s) = 0 \qquad (8\text{-}10)$$

If Eq. (8-10) is solved for $s$, the points where multiple roots occur can be obtained. On the other hand, from Eq. (8-8) we obtain

$$K = -\frac{A(s)}{B(s)}$$

and

$$\frac{dK}{ds} = -\frac{A'(s)B(s) - A(s)B'(s)}{B^2(s)}$$

If $dK/ds$ is set equal to zero, we get the same equation as Eq. (8-10). Therefore, the breakaway points can be simply determined from the roots of

$$\frac{dK}{ds} = 0$$

It should be noted that not all the solutions of Eq. (8-10) or of $dK/ds = 0$ correspond to actual breakaway points. We shall demonstrate this using the present example where

$$f(s) = s^3 + 3s^2 + 2s + K = 0 \qquad (8\text{-}11)$$

Referring to the plot of $f(\sigma)$ versus $\sigma$ shown in Fig. 8-9, notice that points $P$ and $Q$ correspond to $df(\sigma)/d\sigma = 0$. Clearly, point $Q$ corresponds to $K < 0$, which means that point $Q$ cannot be a breakaway point of the system under consideration. [If at a point at which

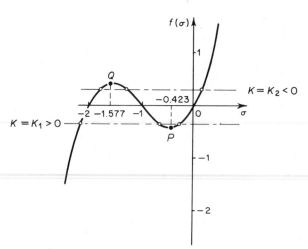

**Fig. 8-9.** Plot of $f(\sigma)$ versus $\sigma$.

$df(s)/ds = 0$ the value of $K$ takes a positive real value, then that point is an actual breakaway point.]

For the present example, we have from Eq. (8-11)

$$K = -s(s + 1)(s + 2) = -(s^3 + 3s^2 + 2s)$$

By setting $dK/ds = 0$, we obtain

$$\frac{dK}{ds} = -(3s^2 + 6s + 2) = 0$$

or

$$s = -0.423, \qquad s = -1.577$$

Since the breakaway point must lie between 0 and $-1$, it is clear that $s = -0.423$ corresponds to the actual breakaway point. In fact, the evaluation of the values of $K$ corresponding to $s = -0.423$ and $s = -1.577$ yields

$$K = 0.385 \qquad \text{for } s = -0.423$$
$$K = -0.385 \qquad \text{for } s = -1.577$$

4. Determine the points where the root loci cross the imaginary axis. These points can be found by use of Routh's stability criterion as follows: Since the characteristic equation for the present system is

$$s^3 + 3s^2 + 2s + K = 0$$

the Routh array becomes

$$
\begin{array}{c|cc}
s^3 & 1 & 2 \\
s^2 & 3 & K \\
s^1 & \dfrac{6 - K}{3} & \\
s^0 & K &
\end{array}
$$

The value of $K$ which makes the $s^1$ term in the first column equal zero is $K = 6$. The crossing points on the imaginary axis can then be found by solving the auxiliary equation obtained from the $s^2$ row; that is,

$$3s^2 + K = 3s^2 + 6 = 0$$

which yields

$$s = \pm j\sqrt{2}$$

The frequencies at the crossing points on the imaginary axis are thus $\omega = \pm\sqrt{2}$. The gain value corresponding to the crossing points is $K = 6$.

An alternate approach is to let $s = j\omega$ in the characteristic equation, equate both the real part and the imaginary part to zero, and then solve for $\omega$ and $K$. For the present system, the characteristic equation, with $s = j\omega$, is

$$(j\omega)^3 + 3(j\omega)^2 + 2(j\omega) + K = 0$$

or

$$(K - 3\omega^2) + j(2\omega - \omega^3) = 0$$

Equating both the real and imaginary parts of this last equation to zero, we obtain

$$K - 3\omega^2 = 0, \qquad 2\omega - \omega^3 = 0$$

from which

$$\omega = \pm\sqrt{2}, \qquad K = 6$$

Thus, the root loci cross the imaginary axis at $\omega = \pm\sqrt{2}$, and the value of $K$ at the crossing points is 6.

5. Choose a test point in the broad neighborhood of the $j\omega$ axis and the origin, as shown in Fig. 8-10, and apply the angle condition. If a test point is on the root loci, then the sum of the three angles, $\theta_1 + \theta_2 + \theta_3$, must be 180°. If the test point does not satisfy the angle condition, select another test point until it satisfies the condition. (The sum of the angles at the test point will indicate which direction the test point should be moved.) Continue this process and locate a sufficient number of points satisfying the angle condition.

6. Based on the information obtained in the foregoing steps, draw the complete root-locus plot, as shown in Fig. 8-11.

7. Determine a pair of dominant complex-conjugate closed-loop poles such that the damping ratio $\zeta$ is 0.5. Closed-loop poles with $\zeta = 0.5$ lie on lines passing through the origin and making the angles $\beta = \pm\cos^{-1}\zeta = \pm\cos^{-1}0.5 = \pm60°$ with the negative real axis. From Fig. 8-11, such closed-loop poles having $\zeta = 0.5$ are obtained as follows:

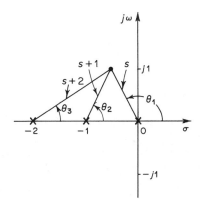

**Fig. 8-10.** Construction of root locus.

$$s_1 = -0.33 + j0.58, \qquad s_2 = -0.33 - j0.58$$

The value of $K$ which yields such poles is found from the magnitude condition as follows:

$$K = |s(s + 1)(s + 2)|_{s=-0.33+j0.58}$$
$$= 1.06$$

Using this value of $K$, the third pole is found at $s = -2.33$.

Note that, from Step 4, it can be seen that for $K = 6$ the dominant closed-loop poles lie on the imaginary axis at $s = \pm j\sqrt{2}$. With this value of $K$, the system will exhibit sustained oscillations. For $K > 6$ the dominant closed-loop poles lie in the right-half $s$ plane, resulting in an unstable system.

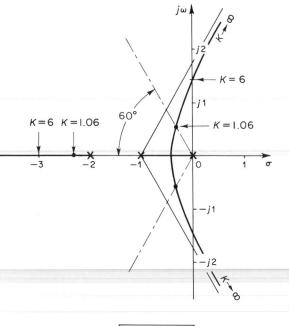

**Fig. 8-11.** Root-locus plot.

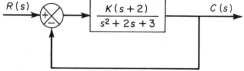

**Fig. 8-12.** Control system.

Finally, note that, if necessary, the root loci can be easily graduated in terms of $K$ by use of the magnitude condition. We simply pick out a point on a root locus and measure the magnitudes of the three complex quantities $s$, $s + 1$, and $s + 2$, multiply these three magnitudes, and the product is equal to the gain value $K$ at that point, or

$$|s| \cdot |s + 1| \cdot |s + 2| = K$$

*Example 8-2.* In this example, we shall sketch the root-locus plot of a system with complex-conjugate open-loop poles. Consider the system shown in Fig. 8-12. For this system,

$$G(s) = \frac{K(s + 2)}{s^2 + 2s + 3}, \qquad H(s) = 1$$

It is seen that $G(s)$ has a pair of complex-conjugate poles at

$$s = -1 + j\sqrt{2}, \qquad s = -1 - j\sqrt{2}$$

A typical procedure for sketching the root-locus plot is as follows:

1. Determine the root loci on the real axis. For any test point $s$ on the real axis, the sum of the angular contributions of the complex-conjugate poles is $360°$, as shown in Fig. 8-13. Thus, the net effect of the complex-conjugate poles is zero on the real axis. The location of the root locus on the real axis is determined from the open-loop zero on the negative real axis. A simple test reveals that a section of the negative real axis, that between $-2$ and $-\infty$, is a part of the root locus. It is noted that since this locus lies

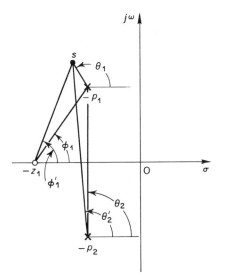

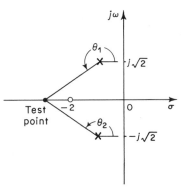

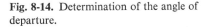

**Fig. 8-13.** Determination of the root locus on the real axis.

**Fig. 8-14.** Determination of the angle of departure.

between two zeros (at $s = -2$ and $s = -\infty$), it is actually a part of two root loci, each of which starts from one of the two complex-conjugate poles. In other words, two root loci break in the part of the negative real axis between $-2$ and $-\infty$.

Since there are two open-loop poles and one zero, there is one asymptote, which coincides with the negative real axis.

2. Determine the angle of departure from the complex-conjugate open-loop poles. The presence of a pair of complex-conjugate open-loop poles requires determination of the angle of departure from these poles. Knowledge of this angle is important since the root locus near a complex pole yields information as to whether the locus originating from the complex pole migrates toward the real axis or extends toward the asymptote.

Referring to Fig. 8-14, if we choose a test point and move it in the very vicinity of the complex open-loop pole at $s = -p_1$, we find that the sum of the angular contributions from the pole at $s = -p_2$ and zero at $s = -z_1$ to the test point can be considered remaining the same. The sum of $\phi_1'$ and $-\theta_2'$ remains the same as the test point is moved in the very vicinity of pole at $s = -p_1$. If the test point is to be on the root locus, then the sum of $\phi_1'$, $-\theta_1$, and $-\theta_2'$ must be $\pm 180°(2k + 1)$, where $k = 0, 1, 2, \ldots$. Thus, in this example,

$$\phi_1' - (\theta_1 + \theta_2') = \pm 180°(2k + 1)$$

or

$$\theta_1 = 180° - \theta_2' + \phi_1' = 180° - \theta_2 + \phi_1$$

The angle of departure is then

$$\theta_1 = 180° - \theta_2 + \phi_1 = 180° - 90° + 55° = 145°$$

Since the root locus is symmetric about the real axis, the angle of departure from the pole at $s = -p_2$ is $-145°$.

3. Determine the break-in point. The break-in point is that where a pair of root-locus branches coalesce as $K$ is increased. The break-in point located between $-2$ and $-\infty$

on the negative real axis can be determined as follows: If a test point is chosen on the real axis, then the sum of the angular contributions of a pair of complex-conjugate poles is always 360°. If the test point is slightly off the real axis, however, then the angular contribution is not equal to 360°.

To determine the angular contribution, assume that a pair of complex-conjugate poles are located as in Fig. 8-15. The total angular contribution of the poles may be written $360° - \Delta\psi$, where $\Delta\psi$ is obtained as follows: From Fig. 8-15,

$$\psi_1 = \tan^{-1}\frac{a}{b}, \qquad \psi_2 = \tan^{-1}\frac{a-\delta}{b}$$

Thus

$$\psi_1 - \psi_2 = \tan^{-1}\frac{a}{b} - \tan^{-1}\frac{a-\delta}{b}$$

or

$$\tan(\psi_1 - \psi_2) = \frac{\dfrac{a}{b} - \dfrac{a-\delta}{b}}{1 + \dfrac{a(a-\delta)}{b^2}} = \frac{b\delta}{b^2 + a^2 - a\delta}$$

If $\delta/a \ll 1$, then

$$\psi_1 - \psi_2 \doteq \tan(\psi_1 - \psi_2) = \frac{b\delta}{a^2 + b^2}$$

The sum of the angular contributions of a pair of complex-conjugate poles is equal to twice $(\psi_1 - \psi_2)$, or

$$2(\psi_1 - \psi_2) = \frac{2b\delta}{a^2 + b^2}$$

Therefore, at a test point near the real axis, the sum of the angular contributions of a pair of complex-conjugate poles is given by

$$\Delta\psi = \pm\left|\frac{2b\delta}{a^2 + b^2}\right|$$

where the plus sign is used if the poles lie to the right of the test point $s$, while the negative sign is used if they are located to the left. Referring to Fig. 8-15, we see that at the test point $s$ the sum of the angles of the three complex quantities $s + p_1$, $s + p_2$, and $s + z$ is

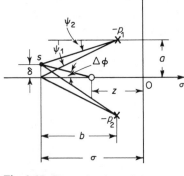

**Fig. 8-15.** Determination of the break-in point.

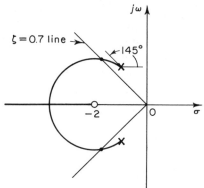

**Fig. 8-16.** Root-locus plot.

$$(180° - \Delta\phi) - \left(360° - \left|\frac{2b\delta}{a^2 + b^2}\right|\right)$$

where

$$\Delta\phi = \left|\frac{\delta}{\sigma - z}\right|$$

If the test point lies on the root locus, then the sum of the angles of the three complex quantities must be $\pm 180°$; that is,

$$\left(180° - \left|\frac{\delta}{\sigma - z}\right|\right) - \left(360° - \left|\frac{2b\delta}{a^2 + b^2}\right|\right) = \pm 180°$$

or

$$-\left|\frac{1}{\sigma - z}\right| + \left|\frac{2b}{a^2 + b^2}\right| = 0$$

Substituting the values of $a$, $b$, and $z$ into this last equation, we obtain

$$-\frac{1}{\sigma - 2} + \frac{2(\sigma - 1)}{2 + (\sigma - 1)^2} = 0$$

The value of $\sigma$ must be greater than two because the break-in point must be located between $-2$ and $-\infty$. Solving this last equation, we obtain

$$\sigma = 3.73$$

Hence, the break-in point is $(-3.73, 0)$.

4. Based on the information obtained in the foregoing steps, sketch a root-locus plot. In order to determine accurate root loci, several points must be found by trial and error between the break-in point and the complex open-loop poles. (In order to facilitate sketching the root-locus plot, we should find the direction in which the test point should be moved by mentally summing up the changes on the angles of the poles and zeros.) Figure 8-16 shows a complete root-locus plot for the system considered.

The value of the gain $K$ at any point on the root locus can be found by applying the magnitude condition. For example, the value of $K$ at which the closed-loop complex-conjugate poles have the damping ratio $\zeta = 0.7$ can be found by locating the roots, as shown in Fig. 8-16, and computing the value of $K$ as follows:

$$K = \left|\frac{(s + 1 - j\sqrt{2})(s + 1 + j\sqrt{2})}{s + 2}\right|_{s=-1.67+j1.70}$$

$$= 1.33$$

From the preceding examples, it may be clear that it is possible to sketch a reasonably accurate root-locus plot for a given system by following simple rules. At preliminary design stages, we may not need the precise locations of the closed-loop poles. Often their approximate locations are all that is needed to make an estimate of system performance.

## 8-4   SUMMARY OF GENERAL RULES FOR CONSTRUCTING ROOT LOCI

For a complicated system with many open-loop poles and zeros, constructing a root-locus plot may seem complicated but actually it is not difficult if the rules for constructing the root loci are applied. By locating particular points and asymp-

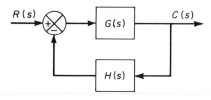

Fig. 8-17. Control system.

totes, and by computing angles of departure from complex poles and angles of arrival at complex zeros, we can construct the general form of the root loci without difficulty. As a matter of fact, the full advantage of the root-locus method can be realized in the case of higher-order systems for which other methods of finding closed-loop poles are extremely laborious.

Some of the rules for constructing root loci were given in Section 8-3. The purpose of this section is to summarize the general rules for constructing root loci of the system shown in Fig. 8-17. While the root-locus method is essentially based on a trial-and-error technique, the number of trials required can be greatly reduced if we use these rules.

Before we summarise the general rules, remember that the angles of the complex quantities originating from the open-loop poles and open-loop zeros to the test point $s$ are measured in the counterclockwise direction. For example, for the system

$$G(s)H(s) = \frac{K(s + z_1)}{(s + p_1)(s + p_2)(s + p_3)(s + p_4)}$$

where $-p_2$ and $-p_3$ are complex-conjugate poles, the angle of $G(s)H(s)$ is

$$\underline{/G(s)H(s)} = \phi_1 - \theta_1 - \theta_2 - \theta_3 - \theta_4$$

where $\phi_1, \theta_1, \theta_2, \theta_3,$ and $\theta_4$ are measured counterclockwise as shown in Figs. 8-18 (a) and (b). The magnitude of $G(s)H(s)$ for this system is

$$|G(s)H(s)| = \frac{KB_1}{A_1 A_2 A_3 A_4}$$

where $A_1, A_2, A_3, A_4,$ and $B_1$ are the magnitudes of the complex quantities $s + p_1, s + p_2, s + p_3, s + p_4,$ and $s + z_1$, respectively, as shown in Fig. 8-18 (a).

Note that because the open-loop complex-conjugate poles and complex-conjugate zeros, if any, are always located symmetrically about the real axis, the root loci are always symmetric with respect to this axis. Therefore, we only need to construct the upper half of the root loci and draw the mirror image of the upper half in the lower-half $s$ plane.

**General rules for constructing root loci.** We shall now summarize the general rules and procedure for constructing the root loci of the system shown in Fig. 8-17.

1. First obtain the characteristic equation

$$1 + G(s)H(s) = 0$$

and rearrange this equation so that the parameter of interest appears as the multiplying factor in the form

$$1 + \frac{K(s + z_1)(s + z_2) \cdots (s + z_m)}{(s + p_1)(s + p_2) \cdots (s + p_n)} = 0$$

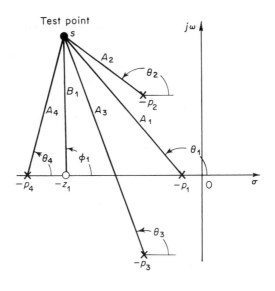

(a)

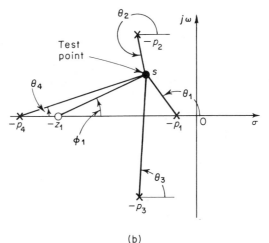

**Fig. 8-18.** Construction of the root locus.

(b)

In the present discussions, we assume that the parameter of interest is the gain $K$, where $K > 0$. (If $K < 0$ which corresponds to the positive feedback case, the angle condition must be modified. See Problem A-8-6.) Note, however, that the method is still applicable to systems with parameters of interest other than gain.

From the factored form of the open-loop transfer function, locate the open-loop poles and zeros in the $s$ plane. [Note that the open-loop zeros are the zeros of $G(s)H(s)$, while the closed-loop zeros consist of the zeros of $G(s)$ and the poles of $H(s)$.]

2. Find the starting points and terminating points of the root loci and find also the number of separate root loci. The points on the root loci corresponding to $K = 0$ are open-loop poles. This can be seen from the magnitude condition by letting $K$ approach zero, or

$$\lim_{K \to 0} \left| \frac{(s + z_1)(s + z_2) \cdots (s + z_m)}{(s + p_1)(s + p_2) \cdots (s + p_n)} \right| = \lim_{K \to 0} \frac{1}{K} = \infty$$

This last equation implies that the value of $s$ must approach one of the open-loop poles. Each root locus thus originates at a pole of the open-loop transfer function $G(s)H(s)$. As $K$ is increased to infinity, each root-locus approaches either a zero of the open-loop transfer function or infinity in the complex plane. This can be seen as follows: If we let $K$ approach infinity in the magnitude condition, then

$$\lim_{K \to \infty} \left| \frac{(s + z_1)(s + z_2) \cdots (s + z_m)}{(s + p_1)(s + p_2) \cdots (s + p_n)} \right| = \lim_{K \to \infty} \frac{1}{K} = 0$$

Hence the value of $s$ must approach one of the open-loop zeros or an open-loop zero at infinity. [If the zeros at infinity are included in the count, $G(s)H(s)$ has the same number of zeros as poles.]

A root-locus plot will have just as many branches as there are roots of the characteristic equation. Since the number of open-loop poles generally exceeds that of zeros, the number of branches equals that of poles. If the number of closed-loop poles is the same as the number of open-loop poles, then the number of individual root-locus branches terminating at finite open-loop zeros is equal to the number $m$ of the open-loop zeros. The remaining $n - m$ branches terminate at infinity ($n - m$ implicit zeros at infinity) along asymptotes.

It is important to note, however, that if we consider a purely mathematical problem, then the number of the closed-loop poles can be made the same as the number of open-loop zeros rather than the number of open-loop poles. In such a case, the number of root-locus branches is equal to the number of open-loop zeros. For example, consider the following polynomial equation:

$$s^2 + s + 1 = 0$$

This equation may be rewritten

$$1 + \frac{s^2}{s + 1} = 0$$

Then the transfer function $s^2/(s + 1)$ may be considered the open-loop transfer function. It has two zeros and one pole. Hence the number of finite zeros is greater than that of finite poles. The number of root-locus branches is equal to that of open-loop zeros.

If we include poles and zeros at infinity, the number of open-loop poles is equal to that of open-loop zeros. Hence we can always state that the root loci start at the poles of $G(s)H(s)$ and end at the zeros of $G(s)H(s)$, as $K$ increases from zero to infinity, where the poles and zeros include both those in the finite $s$ plane and those at infinity.

3. Determine the root loci on the real axis. Root loci on the real axis are determined by open-loop poles and zeros lying on it. The complex-conjugate poles and

zeros of the open-loop transfer function have no effect on the location of the root loci on the real axis because the angle contribution of a pair of complex-conjugate poles or zeros is 360° on the real axis. Each portion of the root locus on the real axis extends over a range from a pole or zero to another pole or zero. In constructing the root loci on the real axis, choose a test point on it. If the total number of real poles and real zeros to the right of this test point is odd, then this point lies on a root locus. The root locus and its complement form alternate segments along the real axis.

4. Determine the asymptotes of the root loci. If the test point $s$ is located far from the origin, then the angle of each complex quantity may be considered the same. One open-loop zero and one open-loop pole then cancel the effects of the other. Therefore, the root loci for very large values of $s$ must be asymptotic to straight lines whose angles (slopes) are given by

$$\text{Angle of asymptote} = \frac{\pm 180°(2k + 1)}{n - m} \qquad (k = 0, 1, 2, \ldots)$$

where

$n = $ number of finite poles of $G(s)H(s)$
$m = $ number of finite zeros of $G(s)H(s)$

Here, $k = 0$ corresponds to the asymptotes with the smallest angle with the real axis. Although $k$ assumes an infinite number of values, as $k$ is increased, the angle repeats itself, and the number of distinct asymptotes is $n - m$.

All the asymptotes intersect on the real axis. The point at which they do so is obtained as follows: If both the numerator and denominator of the open-loop transfer function are expanded, the result is

$$G(s)H(s) = \frac{K[s^m + (z_1 + z_2 + \cdots + z_m)s^{m-1} + \cdots + z_1 z_2 \cdots z_m]}{s^n + (p_1 + p_2 + \cdots + p_n)s^{n-1} + \cdots + p_1 p_2 \cdots p_n}$$

If a test point is located very far from the origin, then $G(s)H(s)$ may be written

$$G(s)H(s) = \frac{K}{s^{n-m} + [(p_1 + p_2 + \cdots + p_n) - (z_1 + z_2 + \cdots + z_m)]s^{n-m-1} + \cdots}$$

Since the characteristic equation is

$$G(s)H(s) = -1$$

it may be written

$$s^{n-m} + [(p_1 + p_2 + \cdots + p_n) - (z_1 + z_2 + \cdots + z_m)]s^{n-m-1} + \cdots = -K \tag{8-12}$$

For a large value of $s$, Eq. (8-12) may be approximated by

$$\left[s + \frac{(p_1 + p_2 + \cdots + p_n) - (z_1 + z_2 + \cdots + z_m)}{n - m}\right]^{n-m} = 0$$

If the abscissa of the intersection of the asymptotes and the real axis is denoted by $-\sigma_a$, then

$$-\sigma_a = -\frac{(p_1 + p_2 + \cdots + p_n) - (z_1 + z_2 + \cdots + z_m)}{n - m} \tag{8-13}$$

Because all the complex poles and zeros occur in conjugate pairs, $-\sigma_a$ is always a real quantity. Note that $-\sigma_a$ is the centroid of the poles and zeros of $G(s)H(s)$, where the centroid is computed on the basis of all poles having mass $+1$ and all zeros mass $-1$. Once the intersection of the asymptotes and the real axis is found, they can be readily drawn in the complex plane.

5. Find the breakaway and break-in points. Because of the conjugate symmetry of the root loci, the breakaway points and break-in points either lie on the real axis or occur in complex-conjugate pairs.

If a root locus lies between two adjacent open-loop poles on the real axis, then there exists at least one breakaway point between the two poles. Similarly, if the root locus lies between two adjacent zeros (one zero may be located at $-\infty$) on the real axis, then there always exists at least one break-in point between the two zeros. If the root locus lies between an open-loop pole and a zero (finite or infinite) on the real axis, then there may exist no breakaway or break-in points or there may exist both breakaway and break-in points.

If the characteristic equation is given by

$$A(s) + KB(s) = 0$$

the breakaway and break-in points can be determined from the roots of

$$\frac{dK}{ds} = -\frac{A'(s)B(s) - A(s)B'(s)}{B^2(s)} = 0$$

where the prime indicates differentiation with respect to $s$. If the value of $K$ corresponding to a root $s = s_1$ of $dK/ds = 0$ is positive, point $s = s_1$ is an actual breakaway or break-in point. (Since $K$ is assumed to be nonnegative, if the value of $K$ thus obtained is negative, then point $s = s_1$ is not a breakaway or break-in point.) Note that this approach can, of course, be used when there are complex poles and/or zeros.

6. Find the angles of departure (or angles of arrival) of root loci from the complex poles (or at complex zeros). In order to sketch the root loci with reasonable accuracy, we must find the directions of the root loci near the complex poles and zeros. If a test point is chosen and moved in the very vicinity of a complex pole (or complex zero), the sum of the angular contributions from all other poles and zeros can be considered remaining the same. Therefore, the angle of departure (or angle of arrival) of the root locus from a complex pole (or at a complex zero) can be found by subtracting from $180°$ the sum of all the angles of complex quantities from all other poles and zeros to the complex pole (or complex zero) in question, with appropriate signs included. The angle of departure is shown in Fig. 8-19.

7. Find the points where the root loci cross the imaginary axis. The points where the root loci intersect the $j\omega$ axis can be found easily by use of Routh's stability criterion, by a trial-and-error approach, or by letting $s = j\omega$ in the characteristic equation, equating both the real part and the imaginary part to zero, and solving for $\omega$ and $K$. The value of $\omega$ thus found gives the frequency at which root loci cross the imaginary axis, and the value of $K$ corresponds to a critical gain for stability.

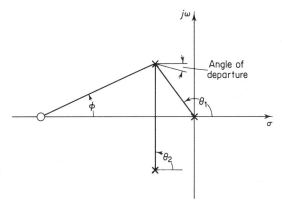

**Fig. 8-19.** Construction of the root locus. [Angle of departure $= 180° - (\theta_1 + \theta_2 - \phi_1)$]

8. The characteristic equation of the system whose open-loop transfer function is

$$G(s)H(s) = \frac{K(s^m + b_1 s^{m-1} + \cdots + b_m)}{s^n + a_1 s^{n-1} + \cdots + a_n} \qquad (n \geq m)$$

is an $n$th-degree algebraic equation in $s$. If the order of the numerator of $G(s)H(s)$ is lower than that of the denominator by two or more (which means that there are two or more zeros at infinity), then the coefficient $a_1$ is the negative sum of the roots of the equation and is independent of $K$. In such a case, if some of the roots move on the locus toward the left as $K$ is increased, then the other roots must move toward the right as $K$ is increased. This information is helpful in finding the general shape of the root loci.

9. Determine the root loci in the broad neighborhood of the $j\omega$ axis and the origin. The most important part of the root loci is on neither the real axis nor the asymptotes but the part in the broad neighborhood of the $j\omega$ axis and the origin. The shape of the root loci in this important region in the $s$ plane must be obtained with sufficient accuracy.

In constructing a root-locus plot, we need only a protractor and a ruler. A device called the Spirule, consisting of a protractor and a turning arm, is particularly convenient, however, for checking the angle condition of test points. (A spiral logarithmic scale printed on the Spirule can be used for determining the gain $K$ from the magnitude condition.)

**A few comments on root-locus plots.** Any point on the root loci is a possible closed-loop pole. A particular point will be a closed-loop pole when the value of $K$ satisfies the magnitude condition. Thus, the magnitude condition enables us to determine the value of the gain $K$ at any specific root location on the locus. (If necessary, the root loci may be graduated in terms of $K$. The root loci are continuous with $K$.)

If the gain $K$ of the open-loop transfer function is given in the problem, then by applying the magnitude condition we can find the correct locations of the closed-loop poles for a given $K$ on each branch of the root loci by a trial-and-error approach.

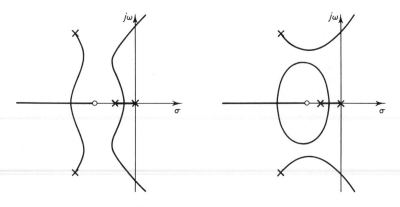

**Fig. 8-20.** Root-locus plots.

Once the dominant closed-loop poles (or the closed-loop poles closest to the $j\omega$ axis) are found by the root-locus method, then the remaining closed-loop poles may be found by dividing the characteristic equation by the factors corresponding to the dominant closed-loop poles. The remainder corresponds to the remaining closed-loop poles. Often this division may not be done exactly. This is unavoidable because of inaccuracies introduced in graphical analysis.

As a final comment, note that a slight change in the pole-zero configuration may cause significant changes in the root-locus configuration. Figure 8-20 demonstrates the fact that a slight change in the location of the zero will make the root-locus configuration look quite different.

**Cancellation of poles of $G(s)$ with zeros of $H(s)$.** It is important to note that if the denominator of $G(s)$ and the numerator of $H(s)$ involve common factors, then the corresponding open-loop poles and zeros will cancel each other, reducing the degree of the characteristic equation by one or more. For example, consider the system shown in Fig. 8-21 (a). (This system has velocity feedback.) The closed-loop transfer function $C(s)/R(s)$ is

$$\frac{C(s)}{R(s)} = \frac{K}{s(s+1)(s+2) + K(s+1)}$$

The characteristic equation is

$$[s(s+2) + K](s+1) = 0$$

Because of the cancellation of the terms $(s+1)$ appearing in $G(s)$ and $H(s)$, however, we have

$$1 + G(s)H(s) = 1 + \frac{K(s+1)}{s(s+1)(s+2)}$$

$$= \frac{s(s+2) + K}{s(s+2)}$$

The reduced characteristic equation is

$$s(s+2) + K = 0$$

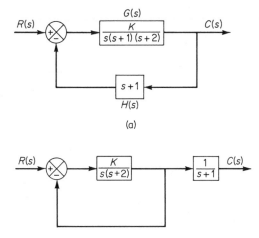

**Fig. 8-21.** Control systems.                    (b)

The root-locus plot of $G(s)H(s)$ does not show all the roots of the characteristic equation, only the roots of the reduced equation.

To obtain the complete set of closed-loop poles, we must add the canceled pole of $G(s)H(s)$ to those closed-loop poles obtained from the root-locus plot of $G(s)H(s)$. The important thing to remember is that the canceled pole of $G(s)H(s)$ is a closed-loop pole of the system, as seen from Fig. 8-21 (b).

## 8-5 ROOT-LOCUS ANALYSIS OF CONTROL SYSTEMS

In this section, we first examine the effects of a zero on the configuration of the root loci of a second-order system. Next, we compare the effects of derivative control and velocity feedback on the performance of a positional servomechanism. Analyses of a conditionally stable system and nonminimum phase systems follow. Finally, our attention is turned to an analysis of the variations of two system parameters.

**Effects of a zero on the root loci of second-order systems.** Consider the system shown in Fig. 8-22 (a). The open-loop transfer function $G(s)H(s)$ is

$$G(s)H(s) = \frac{K(s+z)}{(s+p_1)(s+p_2)} \qquad (K \geq 0, \, p_1 > p_2 > 0, \, z > 0)$$

We shall examine the root loci of this system for the following three cases:

Case 1:     $z > p_1 > p_2$

Case 2:     $p_1 > p_2 > z$

Case 3:     $p_1 > z > p_2$

For Case 1, the break points can be obtained as follows: Since the characteristic equation for this system is

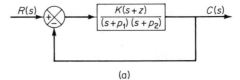

(a)

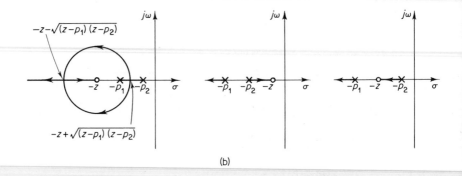

(b)

**Fig. 8-22.** (a) Control system; (b) root-locus plots.

$$(s + p_1)(s + p_2) + K(s + z) = 0$$

or

$$K = - \frac{(s + p_1)(s + p_2)}{s + z}$$

the roots of $dK/ds = 0$ may be found from

$$(s + p_1)(s + p_2) - (2s + p_1 + p_2)(s + z) = 0$$

as follows:

$$s = -z \pm \sqrt{(z - p_1)(z - p_2)} \qquad (8\text{-}14)$$

Since $z > p_1 > p_2$, the quantity inside the square root sign is positive. The value of $K$ corresponding to $s = -z + \sqrt{(z - p_1)(z - p_2)}$ is

$$K = (\sqrt{z - p_1} - \sqrt{z - p_2})^2 > 0$$

Similarly, the value of $K$ corresponding to $s = -z - \sqrt{(z - p_1)(z - p_2)}$ is

$$K = (\sqrt{z - p_1} + \sqrt{z - p_2})^2 > 0$$

Thus, both $s = -z + \sqrt{(z - p_1)(z - p_2)}$ and $s = -z - \sqrt{(z - p_1)(z - p_2)}$ are break points.

For Case 2, although the argument of the square root of Eq. (8-14) is positive and thus the values of $s$ given by Eq. (8-14) are real, the corresponding values of $K$ become negative since

$$K = -(\sqrt{p_1 - z} \pm \sqrt{p_2 - z})^2 < 0$$

This means that for Case 2 the points given by Eq. (8-14) are not break points and thus no break points exist.

For Case 3, the argument of the square root of Eq. (8-14) becomes negative, meaning that the points given by Eq. (8-14) are complex conjugates. Since break-away and break-in points, if they exist, must be on the real axis in the present example, the points given by Eq. (8-14) are not break points. Therefore, there exist no break points in Case 3, and the root loci are simply two segments of the negative real axis.

Figure 8-22 (b) shows the root locus plots corresponding to the three cases considered.

**Comparison of the effects of derivative control and velocity feedback (tachometer feedback) on the performance of positional servomechanisms.** System I shown in Fig. 8-23 is a positional servomechanism. (The output is position.) System II shown in Fig. 8-23 is a positional servomechanism utilizing proportional-plus-derivative control action. System III shown in Fig. 8-23 is a positional servomechanism utilizing velocity feedback or tachometer feedback. Let us compare the relative merits of derivative control and velocity feedback.

The root-locus plot for System I is shown in Fig. 8-24 (a). The closed-loop poles are located at $s = -0.1 \pm j0.995$.

The open-loop transfer function of System II is

$$G_{\mathrm{II}}(s)H_{\mathrm{II}}(s) = \frac{5(1 + 0.8s)}{s(5s + 1)}$$

The open-loop transfer function of System III is

$$G_{\mathrm{III}}(s)H_{\mathrm{III}}(s) = \frac{5(1 + 0.8s)}{s(5s + 1)}$$

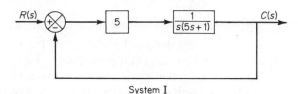

System I

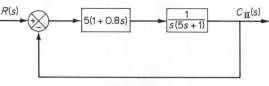

System II

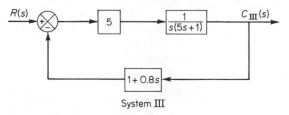

**Fig. 8-23.** Positional servomechanisms.

System III

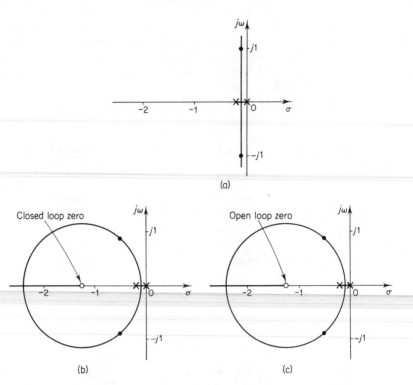

**Fig. 8-24.** Root-locus plots of the systems shown in Fig. 8-23.

Thus Systems II and III have identical open-loop transfer functions. (Both systems have the same open-loop poles and zero.) The root-locus plots of Systems II and III are therefore identical and are shown in Figs. 8-24 (b) and (c), respectively.

Note, however, that the closed-loop transfer function of System II and that of System III are clearly different. System II has two closed-loop poles and one finite closed-loop zero, while System III has two closed-loop poles and no finite closed-loop zero. (Velocity feedback, or tachometer feedback, possesses an open-loop zero but not a closed-loop zero.) The closed-loop poles of Systems II and III are $s = -0.5 \pm j0.866$.

The closed-loop transfer function of System II is

$$\frac{C_{\mathrm{II}}(s)}{R(s)} = \frac{1 + 0.8s}{(s + 0.5 + j0.866)(s + 0.5 - j0.866)}$$

For a unit-impulse input,

$$C_{\mathrm{II}}(s) = \frac{0.4 + j0.346}{s + 0.5 + j0.866} + \frac{0.4 - j0.346}{s + 0.5 - j0.866}$$

The residue at the closed-loop pole $s = -0.5 - j0.866$ is $0.4 + j0.346$ and that at the closed-loop pole $s = -0.5 + j0.866$ is $0.4 - j0.346$. The inverse Laplace transform of $C_{\mathrm{II}}(s)$ yields

$$c_{\mathrm{II}}(t)_{\mathrm{impulse}} = e^{-0.5t}(0.8 \cos 0.866t + 0.693 \sin 0.866t) \qquad (t \geq 0)$$

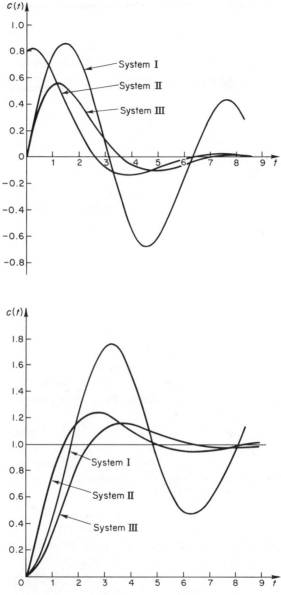

**Fig. 8-25.** Unit-impulse response curves for Systems I, II, and III shown in Fig. 8-23.

**Fig. 8-26.** Unit-step response curves for Systems I, II, and III shown in Fig. 8-23.

The closed-loop transfer function of System III is

$$\frac{C_{\mathrm{III}}(s)}{R(s)} = \frac{1}{(s + 0.5 + j0.866)(s + 0.5 - j0.866)}$$

For a unit-impulse input,

$$C_{\mathrm{III}}(s) = \frac{j0.577}{s + 0.5 + j0.866} + \frac{-j0.577}{s + 0.5 - j0.866}$$

The residue at the closed-loop pole $s = -0.5 - j0.866$ is $j0.577$ and that at the closed-loop pole $s = -0.5 + j0.866$ is $-j0.577$. The inverse Laplace transform of $C_{III}(s)$ yields

$$c_{III}(t)_{impulse} = 1.155e^{-0.5t}\sin 0.866t \qquad (t \geq 0)$$

The unit-impulse responses of Systems II and III are clearly different because the residues at the same pole are different for the two systems. (Remember that the residues depend upon both the closed-loop poles and zeros.) Figure 8-25 shows the unit-impulse response curves for the three systems.

Note that the unit-step response can be obtained either directly or by integrating the unit-impulse response. For example, for System III, the unit-step response is obtained as follows:

$$c_{III}(t)_{step} = \int_0^t c_{III}(t)_{impulse}\, dt$$

$$= \int_0^t 1.155e^{-0.5t}\sin 0.866t\, dt$$

$$= 1 - e^{-0.5t}(\cos 0.866t + 0.577\sin 0.866t)$$

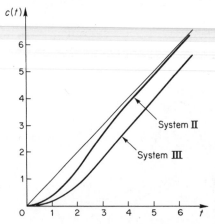

Figure 8-26 shows unit-step response curves for the three systems. The system utilizing proportional-plus-derivative control action exhibits the shortest rise time. The system with velocity feedback has the least maximum overshoot, or the best relative stability, of the three systems.

Figure 8-27 shows the unit-ramp response curves for Systems II and III. System II has the advantage of quicker response and less steady-state error to a ramp input.

The main reason why the system utilizing proportional-plus-derivative control action has superior response characteristics is that derivative control responds to the rate of change of the error signal and can produce early corrective action before the magnitude of the error becomes large.

**Fig. 8-27.** Unit-ramp response curves for Systems II and III shown in Fig. 8-23.

Notice that the output of System III is the output of System II delayed by a first-order lag term $1/(1 + 0.8s)$. Figure 8-28 shows the relationship between the output of System II and that of System III.

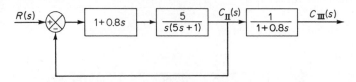

**Fig. 8-28.** Block diagram showing the relationship between the output of System II and that of System III, shown in Fig. 8-23.

**Conditionally stable systems.** Consider the system shown in Fig. 8-29 (a). The root loci for this system can be plotted by applying the general rules and pro-

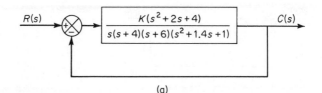

(a)

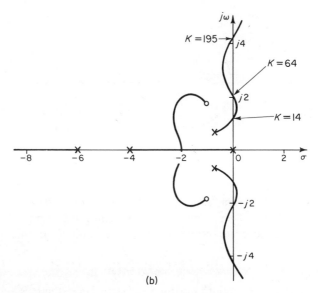

**Fig. 8-29.** (a) Control system;
(b) root-locus plot.                    (b)

cedure for constructing root loci. A root locus plot for this system is shown in Fig. 8-29 (b). It can be seen that this system is stable only for limited ranges of the value of $K$; namely, $14 > K > 0$ and $195 > K > 64$. The system becomes unstable for $64 > K > 14$ and $K > 195$. If $K$ assumes a value corresponding to unstable operation, the system may break down or may become nonlinear due to a saturation nonlinearity which may exist. Such a system is called *conditionally stable*.

In practice, conditionally stable systems are not desirable. Conditional stability is dangerous but does occur in certain systems, in particular, a system which has an unstable feedforward path. Such a feedforward path may occur if the system has a minor loop. It is advisable to avoid such conditional stability since if the gain drops beyond the critical value for some reason or other, the system becomes unstable. Note that the addition of a proper compensating network will eliminate conditional stability. [An addition of a zero will cause the root loci to bend to the left. (See Section 10-3.) Hence conditional stability may be eliminated by adding proper compensation.]

**Nonminimum phase systems.** If all the poles and zeros of a system lie in the left-half $s$ plane, then the system is called minimum phase. If a system has at least one pole or zero in the right-half $s$ plane, then the system is called nonminimum

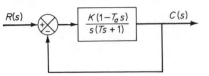

(a)

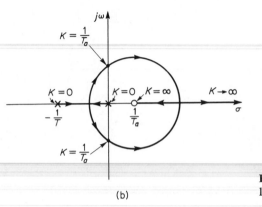

(b)

**Fig. 8-30.** (a) Feedback system; (b) root-locus plot.

phase. The term "nonminimum phase" comes from the phase shift characteristics of such a system when subjected to sinusoidal inputs. (See Section 9-2.)

Consider the system shown in Fig. 8-30 (a). In this system

$$G(s) = \frac{K(1 - T_a s)}{s(Ts + 1)} \quad (T_a > 0), \quad H(s) = 1$$

This is a nonminimum phase system since there is one zero in the right-half $s$ plane. For this system, the angle condition becomes

$$\underline{/G(s)} = \left| -\frac{K(T_a s - 1)}{s(Ts + 1)} \right.$$

$$= \left| \frac{K(T_a s - 1)}{s(Ts + 1)} + 180° \right.$$

$$= \pm 180°(2k + 1) \quad (k = 0, 1, 2, \ldots)$$

or

$$\left| \frac{K(T_a s - 1)}{s(Ts + 1)} \right. = 0° \tag{8-15}$$

The root loci can be obtained from Eq. (8-15). Figure 8-30 (b) shows a root-locus plot for this system. From the plot, we see that the system is stable if the gain $K$ is less than $1/T_a$.

**Systems with transportation lag or dead time.** Figure 8-31 shows a thermal system in which hot air is circulated to keep the temperature of a chamber constant. In this system, the measuring element is placed downstream a distance $L$ ft from the furnace, the air velocity is $v$ ft/sec, and $T = L/v$ sec would elapse before any

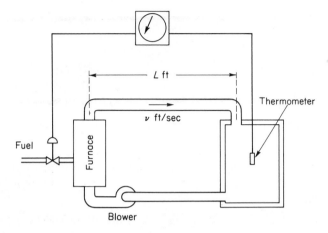

**Fig. 8-31.** Thermal system.

change in the furnace temperature was sensed by the thermometer. Such a delay in measuring, delay in controller action, or delay in actuator operation, etc., is called *dead time* or *transportation lag*. Dead time is present in most process control systems.

The input $x(t)$ and the output $y(t)$ of a dead time element are related by

$$y(t) = x(t - T)$$

where $T$ is dead time. The transfer function of dead time is given by

$$\text{Transfer function of dead time or transportation lag} = \frac{\mathscr{L}[x(t - T)1(t - T)]}{\mathscr{L}[x(t)1(t)]}$$

$$= \frac{X(s)e^{-Ts}}{X(s)} = e^{-Ts}$$

Suppose that the feedforward transfer function of this thermal system can be approximated by

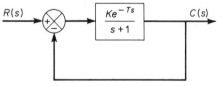

$$G(s) = \frac{Ke^{-Ts}}{s + 1}$$

as shown in Fig. 8-32. Let us construct a root-locus plot for this system. The characteristic equation for this closed-loop sys-

**Fig. 8-32.** Block diagram of the system shown in Fig. 8-31.

tem is

$$1 + \frac{Ke^{-Ts}}{s + 1} = 0 \qquad (8\text{-}16)$$

From Eq. (8-16), we obtain

$$\frac{Ke^{-Ts}}{s + 1} = -1$$

Thus, the angle condition becomes

$$\left/\frac{Ke^{-Ts}}{s + 1}\right. = \left/e^{-Ts}\right. - \left/s + 1\right. = \pm 180°(2k + 1) \qquad (k = 0, 1, 2, \ldots) \qquad (8\text{-}17)$$

To find the angle of $e^{-Ts}$, we write $s = \sigma + j\omega$. Then we obtain

$$e^{-Ts} = e^{-T\sigma - j\omega T}$$

Since $e^{-T\sigma}$ is a real quantity, the angle of $e^{-T\sigma}$ is zero. Hence

$$\underline{/e^{-Ts}} = \underline{/e^{-j\omega T}} = \underline{/\cos \omega T - j \sin \omega T}$$
$$= -\omega T \quad \text{(radians)}$$
$$= -57.3\omega T \quad \text{(degrees)}$$

The angle condition, Eq. (8-17), then becomes

$$-57.3\omega T - \underline{/s + 1} = \pm 180°(2k + 1)$$

Since $T$ is a given constant, the angle of $e^{-Ts}$ is a function of $\omega$ only.

We shall next determine the angle contribution due to $e^{-Ts}$. For $k = 0$, the angle condition may be written

$$\underline{/s + 1} = \mp 180° - 57.3°\omega T \qquad (8\text{-}18)$$

Since the angle contribution of $e^{-Ts}$ is zero for $\omega = 0$, the real axis from $-1$ to $-\infty$ forms a part of the root loci. Now assume a value $\omega_1$ for $\omega$ and compute

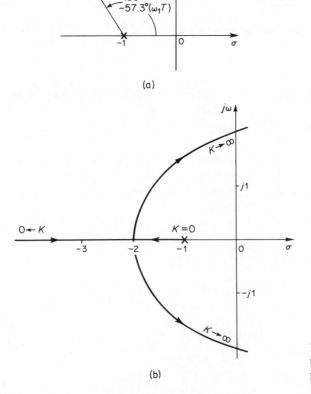

(a)

(b)

**Fig. 8-33.** (a) Construction of the root locus; (b) root-locus plot.

$57.3°\omega_1 T$. At point $-1$ on the negative real axis, draw a line which makes an angle of $180° - 57.3°\omega_1 T$ with the real axis. Find the intersection of this line and the horizontal line $\omega = \omega_1$. This intersection, point $P$ in Fig. 8-33 (a), is a point satisfying Eq. (8-18) and hence is on a root locus. Continuing the same process, we obtain the root-locus plot as shown in Fig. 8-33 (b).

Note that as $s$ approaches minus infinity, the open-loop transfer function

$$\frac{Ke^{-Ts}}{s+1}$$

approaches minus infinity since

$$\lim_{s=-\infty} \frac{Ke^{-Ts}}{s+1} = \frac{\frac{d}{ds}(Ke^{-Ts})}{\frac{d}{ds}(s+1)}\bigg|_{s=-\infty}$$

$$= -KTe^{-Ts}\big|_{s=-\infty}$$

$$= -\infty$$

Therefore, $s = -\infty$ is a pole of the open-loop transfer function. Thus, root loci start from $s = -1$ or $s = -\infty$ and terminate at $s = \infty$, as $K$ increases from zero to infinity. Since the right-hand side of the angle condition given by Eq. (8-17) has an infinite number of values, there are an infinite number of root loci, as the value of $k$ ($k = 0, 1, 2, \ldots$) goes from zero to infinity. For example, if $k = 1$, the angle condition becomes

$$\underline{/s+1} = \mp 540° - 57.3°\omega T \quad \text{(degrees)}$$
$$= \mp 3\pi - \omega T \quad \text{(radians)}$$

The construction of the root loci for $k = 1$ is the same as that for $k = 0$. A plot of root loci for $k = 0, 1,$ and $2$ when $T = 1$ is shown in Fig. 8-34.

The magnitude condition states that

$$\left|\frac{Ke^{-Ts}}{s+1}\right| = 1$$

Since the magnitude of $e^{-Ts}$ is equal to that of $e^{-T\sigma}$ or

$$|e^{-Ts}| = |e^{-T\sigma}| \cdot |e^{-j\omega T}| = e^{-T\sigma}$$

the magnitude condition becomes

$$|s+1| = Ke^{-T\sigma}$$

The root loci shown in Fig. 8-34 are graduated in terms of $K$ when $T = 1$. From this plot, it is clear that the root-locus branch corresponding to $k = 0$ is the dominant one; other branches corresponding to $k = 1, 2, 3, \ldots$ are not so important and may be neglected.

This example illustrates the fact that dead time can cause instability even in the first-order system because the root loci enter the right-half $s$ plane for large values of $K$. Therefore, although the gain $K$ of the first-order system can be set at a high value in the absence of dead time, it cannot be set too high if dead time is present. (For the system considered here, the value of the gain $K$ must be considerably less than 2 for a satisfactory operation.)

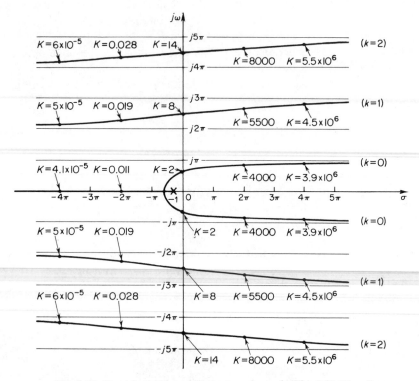

**Fig. 8-34.** Root-locus plot of the system shown in Fig. 8-32 ($T = 1$).

**Approximation of transportation lag or dead time.** If the dead time $T$ is very small, then $e^{-Ts}$ may be approximated by

$$e^{-Ts} \doteq 1 - Ts$$

or

$$e^{-Ts} \doteq \frac{1}{Ts + 1}$$

Such approximations are good if the dead time is very small and, in addition, the input time function $f(t)$ to the dead-time element is a smooth and continuous one. [This means that the second- and higher-order derivatives of $f(t)$ are small.]

**Effects of parameter variations on closed-loop poles.** In many design problems, the effects on the closed-loop poles need to be investigated when system parameters other than the gain $K$ are varied. The effects of these other parameters can be easily investigated by the root-locus method. When two (or more) parameters are varied, the corresponding root loci are sometimes called *root contours.*

We shall use an example to illustrate the construction of the root contours when two parameters are varied, respectively, from zero to infinity.

Consider the system shown in Fig. 8-35. We wish to investigate the effect of varying the parameter $a$ as well as the gain $K$. The closed-loop transfer function of this system is

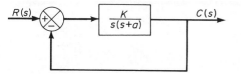

$$\frac{C(s)}{R(s)} = \frac{K}{s^2 + as + K}$$

The characteristic equation is

$$s^2 + as + K = 0 \qquad (8\text{-}19)$$

**Fig. 8-35.** Control system.

which may be rewritten

$$1 + \frac{as}{s^2 + K} = 0$$

or

$$\frac{as}{s^2 + K} = -1 \qquad (8\text{-}20)$$

In Eq. (8-20), the parameter $a$ is written as a multiplying factor. For a given value of $K$, the effect of $a$ on the closed-loop poles can be investigated from Eq. (8-20). The root contours for this system can be constructed by following the usual procedure for constructing root loci.

We shall now construct the root contours as $K$ and $a$ vary, respectively, from zero to infinity. The root contours start and terminate at the poles and zeros of $as/(s^2 + K)$.

We shall first construct the locus of roots when $a = 0$. This can be done easily as follows: Substitute $a = 0$ into Eq. (8-19). Then

$$s^2 + K = 0$$

or

$$\frac{K}{s^2} = -1 \qquad (8\text{-}21)$$

The open-loop poles are thus a double pole at the origin. The root-locus plot of Eq. (8-21) is shown in Fig. 8-36 (a).

In order to construct the root contours, let us first assume that $K$ is a constant; for example, $K = 4$. Then Eq. (8-20) becomes

$$\frac{as}{s^2 + 4} = -1 \qquad (8\text{-}22)$$

The open-loop poles are $s = \pm j2$. The finite open-loop zero is at the origin. The root-locus plot corresponding to Eq. (8-22) is shown in Fig. 8-36 (b). For different values of $K$, Eq. (8-22) yields similar root loci.

The root contour, the diagram showing the root loci corresponding to $0 \leq K \leq \infty$, $0 \leq a \leq \infty$, can be sketched as in Fig. 8-36 (c). Clearly, the root contours start at the poles of and end at the zeros of the transfer function $as/(s^2 + K)$. The arrowheads on the root contours indicate the direction of increase in the value of $a$.

The root contours show the effects of the variations of system parameters on the closed-loop poles. From the root-contour plot shown in Fig. 8-36 (c), we see that, for $0 < K < \infty$, $0 < a$, the closed-loop poles lie in the left-half $s$ plane and the system is stable.

Note that if the value of $K$ is fixed, say $K = 4$, then the root contours become simply the root loci, as shown in Fig. 8-36 (b).

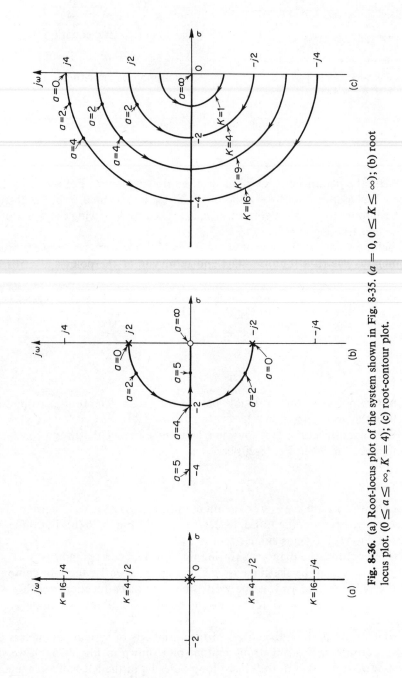

**Fig. 8-36.** (a) Root-locus plot of the system shown in Fig. 8-35. ($a = 0, 0 \leq K \leq \infty$); (b) root locus plot. ($0 \leq a \leq \infty, K = 4$); (c) root-contour plot.

We have illustrated a method of constructing root contours when the gain $K$ and parameter $a$ are varied, respectively, from zero to infinity.

**Typical pole-zero configurations and corresponding root loci.** In concluding this section, we show several open-loop pole-zero configurations and their corresponding root loci in Table 8-2. The pattern of the root loci depends only on the relative

**Table 8-2.** OPEN-LOOP POLE-ZERO CONFIGURATIONS
AND THE CORRESPONDING ROOT LOCI

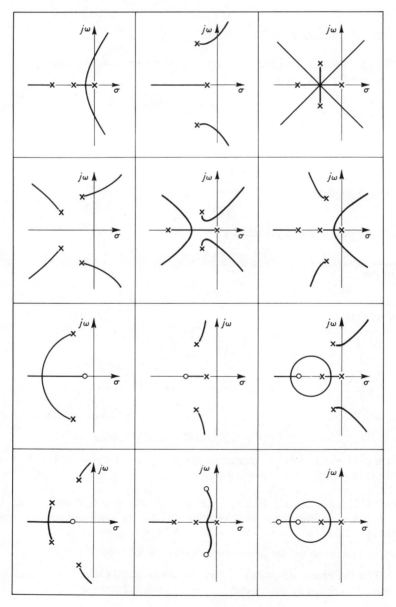

separation of the open-loop poles and zeros. If the number of open-loop poles exceeds the number of finite zeros by three or more, there is a value of the gain $K$ beyond which root loci enter the right-half $s$ plane, and thus the system can become unstable. A stable system must have all its closed-loop poles in the left-half $s$ plane.

Note that once we have some experience with the method, we can easily evaluate the changes in the root loci due to the changes in the number and location of the open-loop poles and zeros by visualizing the root-locus plots resulting from various pole-zero configurations.

EXAMPLE PROBLEMS AND SOLUTIONS

**PROBLEM A-8-1.** Sketch constant gain loci for the unity-feedback system whose feedforward transfer function is

$$G(s) = \frac{K}{s(s+1)}$$

**Solution.** The constant gain loci of this system for various values of $K$ are given by the following mathematical relationship:

$$|G(s)| = \left| \frac{K}{s(s+1)} \right| = 1$$

which can be rewritten

$$|s(s+1)| = K \tag{8-23}$$

Let us substitute $s = \sigma + j\omega$ into Eq. (8-23). Then

$$\sqrt{\sigma^2 + \omega^2} \sqrt{(\sigma + 1)^2 + \omega^2} = K$$

Let us put

$$\sigma + \frac{1}{2} = \sigma_1$$

Then we obtain

$$
\begin{aligned}
K^2 &= (\sigma^2 + \omega^2)[(\sigma + 1)^2 + \omega^2] \\
&= \left[ \left( \sigma_1 - \frac{1}{2} \right)^2 + \omega^2 \right] \left[ \left( \sigma_1 + \frac{1}{2} \right)^2 + \omega^2 \right] \\
&= \left( \sigma_1^2 - \frac{1}{4} \right)^2 + 2\omega^2 \left( \sigma_1^2 - \frac{1}{4} \right) + \omega^4 + \omega^2
\end{aligned}
$$

or

$$\left[ \left( \sigma_1^2 - \frac{1}{4} \right) + \omega^2 \right]^2 = K^2 - \omega^2 \tag{8-24}$$

The constant gain loci can be plotted by use of Eq. (8-24). For $K = 1, 2, 5, 10,$ and $20$, the constant gain loci are shown in Fig. 8-4.

**PROBLEM A-8-2.** Sketch the root-locus plot for the system shown in Fig. 8-37 (a). (The gain $K$ is assumed to be positive.) Observe that for small or large values of $K$ the system is overdamped and for medium values of $K$ it is underdamped.

**Solution.** The procedure for plotting the root loci is as follows:

1. Plot the open-loop poles and zeros on the complex plane. Root loci exist on the negative real axis between 0 and $-1$ and between $-2$ and $-3$.

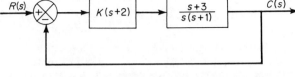

(a)

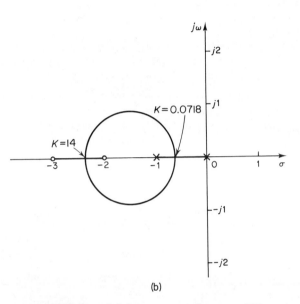

Fig. 8-37. (a) Control system;
(b) root-locus plot.                                            (b)

2. The number of open-loop poles and that of finite zeros are the same. This means that there are no asymptotes in the complex region of the $s$ plane.

3. Determine the breakaway and break-in points. The characteristic equation for the system is

$$1 + \frac{K(s + 2)(s + 3)}{s(s + 1)} = 0$$

or

$$K = -\frac{s(s + 1)}{(s + 2)(s + 3)}$$

The breakaway and break-in points are determined from

$$\frac{dK}{ds} = -\frac{(2s + 1)(s + 2)(s + 3) - s(s + 1)(2s + 5)}{[(s + 2)(s + 3)]^2}$$

$$= -\frac{4(s + 0.634)(s + 2.366)}{[(s + 2)(s + 3)]^2}$$

$$= 0$$

as follows:

$$s = -0.634, \qquad s = -2.366$$

At point $s = -0.634$, the value of $K$ is

$$K = -\frac{(-0.634)(0.366)}{(1.366)(2.366)} = 0.0718$$

Similarly, at $s = -2.366$,

$$K = -\frac{(-2.366)(-1.366)}{(-0.366)(0.634)} = 14$$

Note that the values of $K$ at $s = -0.634$ and $s = -2.366$ are positive and these points are actual breakaway and break-in points. Because point $s = -0.634$ ($s = -2.366$) lies between two poles (zeros), it is a breakaway (break-in) point.

4. Determine a sufficient number of points that satisfy the angle condition. (It can be found that the root locus is a circle with center at $-1.5$ that passes through the breakaway and break-in points.) The root-locus plot for this system is shown in Fig. 8-37 (b).

5. Using the magnitude condition, calibrate the root loci in terms of $K$. For a given value of $K$, the closed-loop poles, which satisfy both the angle and magnitude conditions, can be found from the root-locus plot.

Note that this system is stable for any positive value of $K$ since all the root loci lie in the left-half $s$ plane.

Small values of $K$ ($0 < K < 0.0718$) correspond to an overdamped system. Medium values of $K$ ($0.0718 < K < 14$) correspond to an underdamped system. Finally, large values of $K$ ($14 < K$) correspond to an overdamped system. With a large value of $K$, the steady state can be reached in much shorter time than with a small value of $K$.

The value of $K$ should be adjusted so that system performance is optimum according to a given performance index.

**PROBLEM A-8-3.** Find the roots of the following polynomial by use of the root-locus method:

$$3s^4 + 10s^3 + 21s^2 + 24s - 16 = 0 \qquad (8\text{-}25)$$

**Solution.** First rearrange the polynomial and put it into the form

$$\frac{P(s)}{Q(s)} = -1$$

where $P(s)$ and $Q(s)$ are factored polynomials. Then apply the general rules presented in Section 8-4 to locate the roots of the polynomial.

Equation (8-25) may be rearranged in a convenient way as follows:

$$3s^4 + 10s^3 + 21s^2 = -24s + 16$$

In this case the polynomial can be rewritten

$$\frac{8(s - \frac{2}{3})}{s^2(s^2 + \frac{10}{3}s + 7)} = -1 \qquad (8\text{-}26)$$

This form has two poles at the origin, two complex poles, and one zero on the positive real axis. Since Eq. (8-26) has the form $G(s)H(s) = -1$, the root-locus method can be applied to find the roots of the polynomial.

Equation (8-25) can, of course, be rearranged in different ways. For example, it can be rewritten

$$3s^4 + 10s^3 = -21s^2 - 24s + 16$$

or

$$\frac{7(s^2 + \frac{8}{7}s - \frac{16}{21})}{s^3(s + \frac{10}{3})} = -1 \qquad (8\text{-}27)$$

In this case, however, the system involves three poles at the origin, one pole and one zero on the negative real axis, and one zero on the positive real axis. The amount of graphical work needed to sketch the root-locus plot of Eq. (8-27) is almost the same as that of Eq. (8-26).

Note that if two root-locus plots, one corresponding to Eq. (8-26) and the other corresponding to Eq. (8-27), are sketched on the same diagram, the intersections of two give the roots for the polynomial. (If the magnitude condition is used, only one root-locus plot need be sketched.)

In this problem, we shall sketch only one root-locus plot based on Eq. (8-26) and utilize the magnitude condition to determine the roots of the polynomial. Equation (8-26) can be rewritten

$$\frac{8(s - \frac{2}{3})}{s^2(s + 1.67 + j2.06)(s + 1.67 - j2.06)} = -1 \tag{8-28}$$

To determine the root loci, replace the constant 8 in the numerator of Eq. (8-28) by $K$ and write

$$\frac{K(s - \frac{2}{3})}{s^2(s + 1.67 + j2.06)(s + 1.67 - j2.06)} = -1$$

To sketch the root-locus plot, we follow this procedure:

1. Plot the poles and zero in the complex plane. Root loci exist on the real axis between $\frac{2}{3}$ and 0 and between 0 and $-\infty$.

2. Determine the asymptotes of the root loci. There are three asymptotes which make angles of

$$\frac{\pm 180°(2k + 1)}{4 - 1} = 60°, -60°, 180°$$

with the positive real axis. Referring to Eq. (8-13) the abscissa of the intersection of the asymptotes is given by

$$-\sigma_a = -\frac{(0 + 0 + \frac{5}{3} + j2.06 + \frac{5}{3} - j2.06) + \frac{2}{3}}{4 - 1} = -\frac{4}{3}$$

3. Using Routh's stability criterion, determine the value of $K$ at which root loci cross the imaginary axis. The characteristic equation is

$$s^2(s^2 + \tfrac{10}{3}s + 7) = -K(s - \tfrac{2}{3})$$

or

$$s^4 + \tfrac{10}{3}s^3 + 7s^2 + Ks - \tfrac{2}{3}K = 0$$

The Routh array becomes

| | | | |
|---|---|---|---|
| $s^4$ | $1$ | $7$ | $-\frac{2}{3}K$ |
| $s^3$ | $\frac{10}{3}$ | $K$ | $0$ |
| $s^2$ | $7 - \frac{3}{10}K$ | $-\frac{2}{3}K$ | |
| $s^1$ | $\dfrac{-\frac{3}{10}K^2 + \frac{83}{9}K}{7 - \frac{3}{10}K}$ | $0$ | |
| $s^0$ | $-\frac{2}{3}K$ | | |

The values of $K$ which make the $s^1$ term in the first column equal zero are $K = 30.7$ and $K = 0$. The crossing points on the imaginary axis can be found by solving the auxiliary equation obtained from the $s^2$ row, or

$$(7 - \tfrac{3}{10}K)s^2 - \tfrac{2}{3}K = 0$$

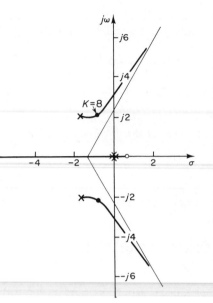

**Fig. 8-38.** Root-locus plot.

where $K = 30.7$. The result is

$$s = \pm j3.04$$

The crossing points on the imaginary axis are, therefore, $s = \pm j3.04$.

4. See if there are any breakaway or break-in points. From Step 3 we know that there are only two crossing points on the $j\omega$ axis. Hence, in this particular pole-zero configuration, there cannot exist any breakway or break-in points.

5. Find the angle of departure of the root loci from the complex poles. At the pole $s = -1.67 + j2.06$, the angle of departure $\theta$ is found from

$$110° - 106° - 106° - 90° - \theta$$
$$= \pm 180°(2k + 1)$$

as follows:

$$\theta = -12°$$

(The angle of departure from the pole $s = -1.67 - j2.06$ is $12°$.)

6. In the broad neighborhood of the $j\omega$ axis and the origin, locate a sufficient number of points that satisfy the angle condition.

Based on the information obtained thus far, the root-locus plot for this system can be sketched as shown in Fig. 8-38.

7. Use the magnitude condition

$$K = \left| \frac{s^2(s + 1.67 + j2.06)(s + 1.67 - j2.06)}{s - \frac{2}{3}} \right|$$

to determine the points on the root locus at which $K = 8$. By use of a trial-and-error procedure, we find

$$s = -0.79 \pm j2.16$$

A trial-and-error procedure can be used to locate the two remaining roots. It may be simpler, however, to factor the known roots from the given polynomial.

$$3s^4 + 10s^3 + 21s^2 + 24s - 16$$
$$= (s + 0.79 + j2.16)(s + 0.79 - j2.16)(3s^2 + 5.28s - 3.06)$$
$$= 3(s + 0.79 + j2.16)(s + 0.79 - j2.16)(s + 2.22)(s - 0.46)$$

Thus, the roots of the given polynomial are

$$s_1 = -0.79 - j2.16, \qquad s_2 = -0.79 + j2.16, \qquad s_3 = -2.22, \qquad s_4 = 0.46$$

**PROBLEM A-8-4.** A simplified form of the open-loop transfer function of an airplane with an autopilot in the longitudinal mode is

$$G(s)H(s) = \frac{K(s + a)}{s(s - b)(s^2 + 2\zeta\omega_n s + \omega_n^2)}, \qquad a > 0, \qquad b > 0$$

Such a system involving an open-loop pole in the right-half $s$ plane may be conditionally stable. Sketch the root-locus plot when $a = b = 1$, $\zeta = 0.5$, and $\omega_n = 4$. Find the range of gain $K$ for stability.

**Solution.** The open-loop transfer function for the system is

$$G(s)H(s) = \frac{K(s + 1)}{s(s - 1)(s^2 + 4s + 16)}$$

To sketch the root-locus plot, we follow this procedure:

1. Plot the open-loop poles and zero in the complex plane. Root loci exist on the real axis between 1 and 0 and between $-1$ and $-\infty$.

2. Determine the asymptotes of the root loci. There are three asymptotes whose angles can be determined as

$$\text{Angles of asymptotes} = \frac{180°(2k + 1)}{4 - 1} = 60°, -60°, 180°$$

The abscissa of the intersection of the asymptotes and the real axis is

$$-\sigma_a = -\frac{(0 - 1 + 2 + j2\sqrt{3} + 2 - j2\sqrt{3}) - 1}{4 - 1} = -\frac{2}{3}$$

3. Determine the breakaway and break-in points. Since the characteristic equation is

$$1 + \frac{K(s + 1)}{s(s - 1)(s^2 + 4s + 16)} = 0$$

we obtain

$$K = -\frac{s(s - 1)(s^2 + 4s + 16)}{s + 1}$$

By differentiating $K$ with respect to $s$, we get

$$\frac{dK}{ds} = -\frac{3s^4 + 10s^3 + 21s^2 + 24s - 16}{(s + 1)^2}$$

In Problem A-8-3, it was shown that

$$3s^4 + 10s^3 + 21s^2 + 24s - 16$$
$$= 3(s + 0.79 + j2.16)(s + 0.79 - j2.16)(s + 2.22)(s - 0.46)$$

Hence the breakaway and break-in points are $s = 0.46$ and $s = -2.22$, respectively. (Points $s = -0.79 \pm j2.16$ do not satisfy the angle condition.)

4. Using Routh's stability criterion, determine the value of $K$ at which the root loci cross the imaginary axis. Since the characteristic equation is

$$s^4 + 3s^3 + 12s^2 + (K - 16)s + K = 0$$

the Routh array becomes

| | | | |
|---|---|---|---|
| $s^4$ | 1 | 12 | $K$ |
| $s^3$ | 3 | $K - 16$ | 0 |
| $s^2$ | $\dfrac{52 - K}{3}$ | $K$ | 0 |
| $s^1$ | $\dfrac{-K^2 + 59K - 832}{3(52 - K)}$ | 0 | |
| $s^0$ | $K$ | | |

The values of $K$ which make the $s^1$ term in the first column equal zero are $K = 35.7$ and $K = 23.3$.

The crossing points on the imaginary axis can be found by solving the auxiliary equation obtained from the $s^2$ row, that is, by solving the following equation for $s$:

$$\frac{52 - K}{3}s^2 + K = 0$$

The results are

$$s = \pm j2.56 \qquad \text{for } K = 35.7$$

$$s = \pm j1.56 \qquad \text{for } K = 23.3$$

The crossing points on the imaginary axis are thus $s = \pm j2.56$ and $s = \pm j1.56$.

5. Find the angles of departure of the root loci from the complex poles. For the open-loop pole at $s = -2 + j2\sqrt{3}$, the angle of departure $\theta$ is

$$106° - 120° - 130.5° - 90° - \theta = \pm 180°(2k + 1)$$

or

$$\theta = -54.5°$$

(The angle of departure from the open-loop pole at $s = -2 - j2\sqrt{3}$ is 54.5°.)

6. Choose a test point in the broad neighborhood of the $j\omega$ axis and the origin, and apply the angle condition. If the test point does not satisfy the angle condition, select another test point until it does. Continue the same process and locate a sufficient number of points which satisfy the angle condition.

Figure 8-39 shows the root-locus plot for this system. From Step 4 the system is stable for $23.3 < K < 35.7$. Otherwise it is unstable.

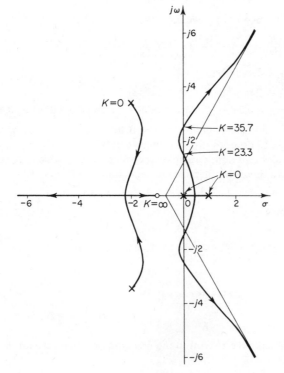

**Fig. 8-39.** Root-locus plot.

**PROBLEM A-8-5.** Consider the system with transportation lag shown in Fig. 8-40. Sketch the root-locus plot and find the two pairs of closed-loop poles nearest the $j\omega$ axis.

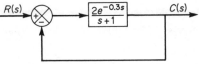

Using only the dominant closed-loop poles, obtain the unit-step response and sketch the response curve.

**Fig. 8-40.** Control system with transportation lag.

**Solution.** The characteristic equation is

$$\frac{2e^{-0.3s}}{s+1} + 1 = 0$$

which is equivalent to the following angle and magnitude conditions:

$$\left/\frac{2e^{-0.3s}}{s+1}\right. = \pm 180°(2k+1)$$

$$\left|\frac{2e^{-0.3s}}{s+1}\right| = 1$$

The angle condition reduces to

$$\underline{/s+1} = \mp\pi(2k+1) - 0.3\omega \qquad \text{(radians)}$$

For $k = 0$,

$$\underline{/s+1} = \mp\pi - 0.3\omega \qquad \text{(radians)}$$
$$= \mp 180° - 17.2°\omega \qquad \text{(degrees)}$$

For $k = 1$,

$$\underline{/s+1} = \mp 3\pi - 0.3\omega \qquad \text{(radians)}$$
$$= \mp 540° - 17.2°\omega \qquad \text{(degrees)}$$

The root-locus plot for this system is shown in Fig. 8-41.

Let us set $s = \sigma + j\omega$ in the magnitude condition and replace 2 by $K$. Then we obtain

$$\frac{\sqrt{(1+\sigma)^2 + \omega^2}}{e^{-0.3\sigma}} = K$$

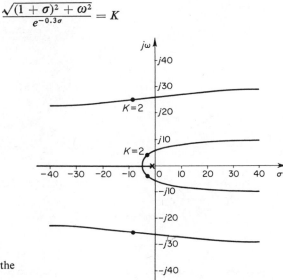

**Fig. 8-41.** Root-locus plot for the system shown in Fig. 8-40.

By evaluating $K$ at different points on the root loci, the points may be found for which $K = 2$. These points are closed-loop poles. The dominant pair of closed-loop poles is

$$s = -2.5 \pm j3.9$$

The next pair of closed-loop poles is

$$s = -8.6 \pm j25.1$$

Using only the pair of dominant closed-loop poles, the closed-loop transfer function may be approximated as follows: Noting that

$$\frac{C(s)}{R(s)} = \frac{2e^{-0.3s}}{1 + s + 2e^{-0.3s}}$$

$$= \frac{2e^{-0.3s}}{1 + s + 2\left(1 - 0.3s + \frac{0.09s^2}{2} + \cdots\right)}$$

$$= \frac{2e^{-0.3s}}{3 + 0.4s + 0.09s^2 + \cdots}$$

and

$$(s + 2.5 + j3.9)(s + 2.5 - j3.9) = s^2 + 5s + 21.46$$

we may approximate $C(s)/R(s)$ by

$$\frac{C(s)}{R(s)} = \frac{\frac{2}{3}(21.46)e^{-0.3s}}{s^2 + 5s + 21.46}$$

or

$$\frac{C(s)}{R(s)} = \frac{14.31e^{-0.3s}}{(s + 2.5)^2 + 3.9^2}$$

For a unit-step input,

$$C(s) = \frac{14.31e^{-0.3s}}{[(s + 2.5)^2 + 3.9^2]s}$$

Note that

$$\frac{14.31}{[(s + 2.5)^2 + 3.9^2]s} = \frac{\frac{2}{3}}{s} + \frac{-\frac{2}{3}s - \frac{10}{3}}{(s + 2.5)^2 + 3.9^2}$$

Hence

$$C(s) = \left(\frac{\frac{2}{3}}{s}\right)e^{-0.3s} + \left[\frac{-\frac{2}{3}s - \frac{10}{3}}{(s + 2.5)^2 + 3.9^2}\right]e^{-0.3s}$$

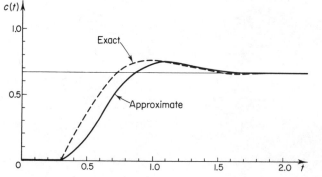

**Fig. 8-42.** Unit-step response curves for the system shown in Fig. 8-40.

The inverse Laplace transform of $C(s)$ gives

$$c(t) = \tfrac{2}{3}[1 - e^{-2.5(t-0.3)} \cos 3.9(t-0.3) - 0.64e^{-2.5(t-0.3)} \sin 3.9(t-0.3)]1(t-0.3)$$

where $1(t - 0.3)$ is the unit-step function occurring at $t = 0.3$.

Figure 8-42 shows the approximate response curve thus obtained, together with the exact unit-step response curve obtained by numerical computation. Note that in this system a fairly good approximation can be obtained by use of only the dominant closed-loop poles.

**PROBLEM A-8-6.**\* In a complex control system, there may be a positive feedback inner loop as shown in Fig. 8-43. Such a loop is usually stabilized by the outer loop.

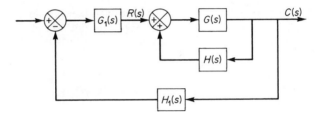

**Fig. 8-43.** Control system.

In this problem we shall be concerned only with the positive feedback inner loop. The closed-loop transfer function of the inner loop is

$$\frac{C(s)}{R(s)} = \frac{G(s)}{1 - G(s)H(s)}$$

The characteristic equation is

$$1 - G(s)H(s) = 0 \qquad\qquad (8\text{-}29)$$

This equation can be solved in a manner similar to the development of the root-locus method in Sections 8-3 and 8-4. The angle condition, however, must be altered.

Equation (8-29) can be rewritten

$$G(s)H(s) = 1$$

which is equivalent to the following two equations:

$$\underline{/G(s)H(s)} = 0° \pm k360° \qquad (k = 0, 1, 2, \ldots)$$
$$|G(s)H(s)| = 1$$

The total sum of all angles from the open-loop poles and zeros must be equal to $0° \pm k360°$. Thus, the root locus follows a $0°$ locus in contrast to the $180°$ locus considered previously. The magnitude condition remains unaltered.

Sketch the root-locus plot for the positive feedback system with the following transfer functions:

$$G(s) = \frac{K(s + 2)}{(s + 3)(s^2 + 2s + 2)}, \qquad H(s) = 1$$

The gain $K$ is assumed to be positive.

**Solution.** The general rules for constructing root loci given in Section 8-4 must be modified in the following way:

Rule 3 is modified as follows: If the total number of real poles and real zeros to the right of a test point on the real axis is even, then this test point lies on the root locus.

\*Reference W-4.

Rule 4 is modified as follows:

$$\text{Angles of asymptotes} = \frac{\pm k 360°}{n - m}$$

where

$n$ = number of finite poles of $G(s)H(s)$
$m$ = number of finite zeros of $G(s)H(s)$

Rule 6 is modified as follows: When calculating the angle of departure (or angle of arrival) from a complex open-loop pole (or at a complex zero), subtract from $0°$ the sum of all the angles of the complex quantities from all the other poles and zeros to the complex pole (or complex zero) in question, with appropriate signs included.

Other rules for constructing the root-locus plot remain the same. We shall now apply the modified rules to construct the root-locus plot. The closed-loop transfer function for the positive feedback system is given by

$$\frac{C(s)}{R(s)} = \frac{G(s)}{1 - G(s)H(s)}$$
$$= \frac{K(s + 2)}{(s + 3)(s^2 + 2s + 2) - K(s + 2)}$$

1. Plot the open-loop poles ($s = -1 + j$, $s = -1 - j$, $s = -3$) and zero ($s = -2$) in the complex plane. As $K$ is increased from 0 to $\infty$, the closed-loop poles start at the open-loop poles and terminate at the open-loop zeros (finite or infinite), just as in the case of negative feedback systems.

2. Determine the root loci on the real axis. Root loci exist on the real axis between $-2$ and $+\infty$ and between $-3$ and $-\infty$.

3. Determine the asymptotes of the root loci. For the present system,

$$\text{Angle of asymptote} = \frac{\pm k 360°}{3 - 1} = \pm 180°$$

This simply means that root-locus branches are on the real axis.

4. Determine the breakaway and break-in points. Since the characteristic equation is

$$(s + 3)(s^2 + 2s + 2) - K(s + 2) = 0$$

we obtain

$$K = \frac{(s + 3)(s^2 + 2s + 2)}{s + 2}$$

By differentiating $K$ with respect to $s$, we obtain

$$\frac{dK}{ds} = \frac{2s^3 + 11s^2 + 20s + 10}{(s + 2)^2}$$

Note that

$$2s^3 + 11s^2 + 20s + 10 = 2(s + 0.8)(s^2 + 4.7s + 6.24)$$
$$= 2(s + 0.8)(s + 2.35 + j0.77)(s + 2.35 - j0.77)$$

Points $s = -2.35 \pm j0.77$ do not satisfy the angle condition. At point $s = -0.8$, the value of $K$ becomes positive. The break-in point is thus $s = -0.8$.

5. Find the angle of departure of the root locus from a complex pole. For the complex pole at $s = -1 + j$, the angle of departure $\phi$ is

$$45° - 27° - 90° - \phi = 0°$$

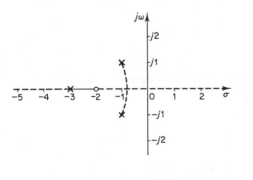

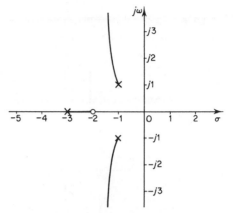

**Fig. 8-44.** Root-locus plot for the positive feedback system with
  $G(s) = K(s + 2)/[(s + 3)(s^2 + 2s + 2)]$,
  $H(s) = 1$.

**Fig. 8-45.** Root-locus plot for the negative feedback system with
  $G(s) = K(s + 2)/[(s + 3)(s^2 + 2s + 2)]$,
  $H(s) = 1$.

or
$$\phi = -72°$$
(The angle of departure from the complex pole at $s = -1 - j$ is 72°.)

6. Choose a test point in the broad neighborhood of the $j\omega$ axis and the origin and apply the angle condition. Locate a sufficient number of points that satisfy the angle condition.

Figure 8-44 shows the root-locus plot for the given positive feedback system. The root loci are shown with dotted lines and curve.

Note that if
$$K > \frac{(s + 3)(s^2 + 2s + 2)}{s + 2}\bigg|_{s=0} = 3$$

one real root enters the right-half $s$ plane. Hence for values of $K$ greater than 3, the system becomes unstable. (For $K > 3$, the system must be stabilized with an outer loop.)

To compare this root-locus plot with that of the corresponding negative feedback system, we show in Fig. 8-45 the root-locus plot of the negative feedback system whose closed-loop transfer function is

$$\frac{C(s)}{R(s)} = \frac{K(s + 2)}{(s + 3)(s^2 + 2s + 2) + K(s + 2)}$$

Table 8-3 shows various root-locus plots of negative-feedback and positive-feedback systems. The closed-loop transfer functions are given by

$$\frac{C}{R} = \frac{G}{1 + GH} \qquad \text{for negative feedback systems}$$

$$\frac{C}{R} = \frac{G}{1 - GH} \qquad \text{for positive feedback systems}$$

where $GH$ is the open-loop transfer function. In Table 8-3, the root loci for negative feedback systems are drawn with heavy lines and curves and those for positive feedback systems are drawn with dotted lines and curves.

**Table 8-3.** ROOT-LOCUS PLOTS OF NEGATIVE FEEDBACK
AND POSITIVE FEEDBACK SYSTEMS

Heavy lines and curves correspond to negative feedback systems;
dotted lines and curves correspond to positive feedback systems.

**PROBLEM A-8-7.** Consider the system shown in Fig. 8-46, which has an unstable feedforward transfer function. Sketch the root-locus plot and locate the closed-loop poles.
Show that although the closed-loop poles lie on the negative real axis and the system is
**not oscillatory,** the unit-step response curve will exhibit overshoot.

**Solution.** The rool-locus plot for this system is shown in Fig. 8-47. The closed-loop poles are located at $s = -2$ and $s = -5$.

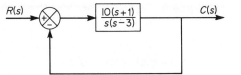

The closed-loop transfer function becomes

$$\frac{C(s)}{R(s)} = \frac{10(s + 1)}{s^2 + 7s + 10}$$

**Fig. 8-46.** Control system.

The unit-step response of this system is

$$C(s) = \frac{10(s + 1)}{s(s + 2)(s + 5)}$$

The inverse Laplace transform of $C(s)$ gives

$$c(t) = 1 + 1.666e^{-2t} - 2.666e^{-5t} \qquad (t \geq 0)$$

The unit-step response curve is shown in Fig. 8-48. Although the system is not oscillatory, the unit-step response curve exhibits overshoot. (This is due to the presence of a zero at $s = -1$.)

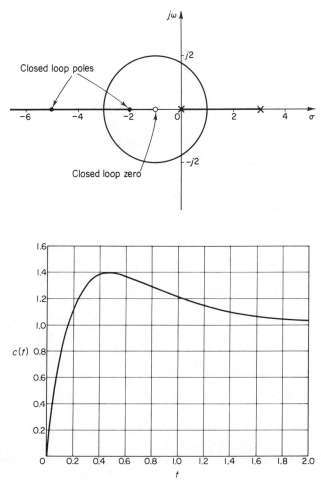

**Fig. 8-47.** Root-locus plot of the system shown in Fig. 8-46.

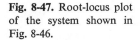

**Fig. 8-48.** Unit-step response curve for the system shown in Fig. 8-46.

**PROBLEM A-8-8.** Consider a system with

$$G(s) = \frac{s+3}{s(s+2)}, \qquad H(s) = 1$$

Show that a part of the root locus is circular.

**Solution.** On the root locus, the angle condition becomes

$$\underline{/G(s)} = \underline{/s+3} - \underline{/s} - \underline{/s+2} = 180°$$

If $s = \sigma + j\omega$ is substituted into this last equation, we obtain

$$\underline{/\sigma + j\omega + 3} - \underline{/\sigma + j\omega} - \underline{/\sigma + j\omega + 2} = 180°$$

which can be rewritten

$$\tan^{-1}\frac{\omega}{\sigma+3} - \tan^{-1}\frac{\omega}{\sigma} = 180° + \tan^{-1}\frac{\omega}{\sigma+2}$$

Taking the tangents of both sides of this last equation using the relationship

$$\tan(x \pm y) = \frac{\tan x \pm \tan y}{1 \mp \tan x \tan y}$$

we obtain

$$\tan\left(\tan^{-1}\frac{\omega}{\sigma+3} - \tan^{-1}\frac{\omega}{\sigma}\right) = \frac{\dfrac{\omega}{\sigma+3} - \dfrac{\omega}{\sigma}}{1 + \dfrac{\omega}{\sigma+3}\dfrac{\omega}{\sigma}} = \frac{-3\omega}{\sigma(\sigma+3) + \omega^2}$$

$$\tan\left(180° + \tan^{-1}\frac{\omega}{\sigma+2}\right) = \frac{0 + \dfrac{\omega}{\sigma+2}}{1 - 0 \times \dfrac{\omega}{\sigma+2}} = \frac{\omega}{\sigma+2}$$

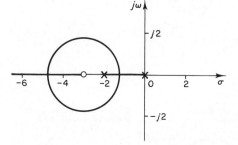

Hence

$$\frac{-3\omega}{\sigma(\sigma+3) + \omega^2} = \frac{\omega}{\sigma+2}$$

or

$$(\sigma+3)^2 + \omega^2 = (\sqrt{3})^2$$

This represents an equation of a circle with center at $\sigma = -3$, $\omega = 0$ and with radius $\sqrt{3}$. Note that the center is at the zero of the open-loop transfer function. Figure 8-49 shows the root-locus plot for the system.

**Fig. 8-49.** Root-locus plot of the system with $G(s) = (s+3)/[s(s+2)]$, $H(s) = 1$.

PROBLEMS

**PROBLEM B-8-1.** Sketch the root-locus plots for the open-loop pole-zero configurations shown in Fig. 8-50.

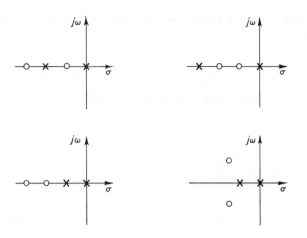

**Fig. 8-50.** Pole-zero systems.

**PROBLEM B-8-2.** Sketch the root-locus plots for the open-loop pole-zero configurations shown in Fig. 8-51.

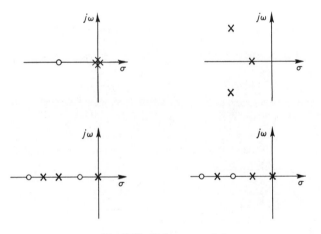

**Fig. 8-51.** Pole-zero systems.

**PROBLEM B-8-3.** Consider the system shown in Fig. 8-52. Investigate the effect of increasing the value of $K_h$ on the root loci. Sketch typical unit-step response curves for the

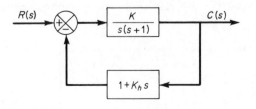

**Fig. 8-52.** Control system.

cases $K_h = 0$, $0 < K_h < 1$, and $K_h > 1$. Then sketch the root-locus plot when $K_h = 0.5$. Locate the closed-loop poles on the root loci when $K = 10$.

**PROBLEM B-8-4.** Show that the root loci for a control system with

$$G(s) = \frac{K(s^2 + 6s + 10)}{s^2 + 2s + 10}, \qquad H(s) = 1$$

are arcs of the circle centered at the origin with radius equal to $\sqrt{10}$.

**PROBLEM B-8-5.** A control system with

$$G(s) = \frac{K}{s^2(s + 1)}, \qquad H(s) = 1$$

is unstable for all positive values of the gain $K$.

Sketch the root-locus plot of the system. By using this plot, show that this system can be stabilized by adding a zero on the negative real axis or by modifying $G(s)$ to $G_1(s)$, where

$$G_1(s) = \frac{K(s + a)}{s^2(s + 1)} \qquad (0 \leq a < 1)$$

**PROBLEM B-8-6.** Sketch the root-locus plot of a system with

$$G(s) = \frac{K}{(s^2 + 2s + 2)(s^2 + 2s + 5)}, \qquad H(s) = 1$$

Determine the exact points where the root loci cross the $j\omega$ axis.

**PROBLEM B-8-7.** Sketch the root-locus plot for the system with

$$G(s) = \frac{K}{s(s + 0.5)(s^2 + 0.6s + 10)}, \qquad H(s) = 1$$

**PROBLEM B-8-8.** Sketch the root-locus plot for the system shown in Fig. 8-53 and show that it may become unstable for large values of $K$.

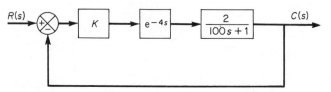

Fig. 8-53. Control system.

**PROBLEM B-8-9.** Sketch the root contours for the system shown in Fig. 8-54 when the gain $K$ and parameter $a$ vary, respectively, from zero to infinity.

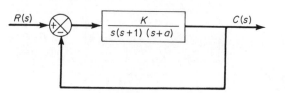

Fig. 8-54. Control system.

**PROBLEM B-8-10.** Consider the system shown in Fig. 8-55. Determine the values of the gain $K$ and the velocity feedback coefficient $K_h$ so that the closed-loop poles are $s = -1 \pm j\sqrt{3}$. Then, using the value of $K_h$ so determined, sketch the root-locus plot.

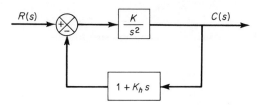

**Fig. 8-55.** Control system.

# 9

# FREQUENCY-RESPONSE METHODS

## 9-1 INTRODUCTION

**Frequency response.** By the term "frequency response," we mean the steady-state response of a system to a sinusoidal input. In the frequency-response methods, the most conventional methods available to control engineers for analysis and design of control systems, we vary the frequency of the input signal over a certain range and study the resulting frequency response.

The Nyquist stability criterion, to be given in Section 9-5, enables us to investigate both the absolute and relative stabilities of linear closed-loop systems from a knowledge of their open-loop frequency-response characteristics. In using this stability criterion, we do not have to determine the roots of the characteristic equation. This is one advantage of the frequency-response approach. Another advantage of this approach is that frequency-response tests are, in general, simple and can be made accurately by use of readily available sinusoidal signal generators and precise measurement equipments. Often the transfer functions of complicated components can be determined experimentally by frequency-response tests. In addition, the frequency-response approach has the advantages that a system may be designed so that the effects of undesirable noise are negligible and that such analysis and design can be extended to certain nonlinear control systems.

Although the frequency response of a control system presents a qualitative picture of the transient response, the correlation between frequency and transient responses is indirect, except for the case of second-order systems. In designing a closed-loop system, we may adjust the frequency-response characteristic by using several design criteria (to be given in Section 9-7) in order to obtain acceptable transient-response characteristics.

Once we understand the indirect correlation between several measures of the transient response and the frequency response, the frequency-response approach may be used to advantage. The design of a control system through this approach is based on the interpretation of the desired dynamic characteristics in terms of the frequency-response characteristics. Such analysis of a control system indicates graphically what changes have to be made in the open-loop transfer function in order to obtain the desired transient-response characteristics.

**Obtaining steady-state solutions to sinusoidal inputs.** We shall first prove the basic fact that the frequency-response characteristics of a system can be obtained directly from the sinusoidal transfer function, i.e., the transfer function in which $s$ is replaced by $j\omega$, where $\omega$ is frequency.

Consider the linear time-invariant system shown in Fig. 9-1. The input and output of the system, whose transfer function is $G(s)$, are denoted by $x(t)$ and $y(t)$, respectively. The input $x(t)$ is sinusoidal and is given by

**Fig. 9-1.** Linear time-invariant system.

$$x(t) = X \sin \omega t$$

Suppose that the transfer function $G(s)$ can be written as a ratio of two polynomials in $s$; namely,

$$G(s) = \frac{p(s)}{q(s)} = \frac{p(s)}{(s + s_1)(s + s_2) \cdots (s + s_n)}$$

The Laplace-transformed output $Y(s)$ is then

$$Y(s) = G(s)X(s) = \frac{p(s)}{q(s)} X(s) \tag{9-1}$$

where $X(s)$ is the Laplace transform of the input $x(t)$.

Let us limit our discussion only to stable systems. For such systems, the real parts of the $-s_i$ are negative. The steady-state response of a stable linear time-invariant system to a sinusoidal input does not depend upon the initial conditions. (Thus, we can assume the zero initial condition.)

If $Y(s)$ has only distinct poles, then the partial fraction expansion of Eq. (9-1) yields

$$Y(s) = \frac{p(s)}{q(s)} \frac{\omega X}{s^2 + \omega^2}$$

$$= \frac{a}{s + j\omega} + \frac{\bar{a}}{s - j\omega} + \frac{b_1}{s + s_1} + \frac{b_2}{s + s_2} + \cdots + \frac{b_n}{s + s_n} \tag{9-2}$$

where $a$ and the $b_i$ (where $i = 1, 2, \ldots, n$) are constants and $\bar{a}$ is the complex conjugate of $a$. The inverse Laplace transform of Eq. (9-2) gives

$$y(t) = ae^{-j\omega t} + \bar{a}e^{j\omega t} + b_1 e^{-s_1 t} + b_2 e^{-s_2 t} + \cdots + b_n e^{-s_n t} \qquad (t \geq 0) \tag{9-3}$$

For a stable system, $-s_1, -s_2, \ldots, -s_n$ have negative real parts. Therefore, as

$t$ approaches infinity, the terms $e^{-s_1t}, e^{-s_2t}, \ldots,$ and $e^{-s_nt}$ approach zero. Thus, all the terms on the right-hand side of Eq. (9-3), except the first two, drop out at steady state.

If $Y(s)$ involves multiple poles $s_j$ of multiplicity $m_j$, then $y(t)$ will involve terms such as $t^{h_j}e^{-s_jt}$ ($h_j = 0, 1, 2, \ldots, m_j - 1$). Since the real parts of the $-s_j$ are negative for a stable system, the terms $t^{h_j}e^{-s_jt}$ approach zero as $t$ approaches infinity.

Thus, regardless of whether or not the system is of the distinct-pole type, the steady-state response becomes

$$y(t) = ae^{-j\omega t} + \bar{a}e^{j\omega t} \tag{9-4}$$

where the constant $a$ can be evaluated from Eq. (9-2) as follows:

$$a = G(s)\frac{\omega X}{s^2 + \omega^2}(s + j\omega)\bigg|_{s=-j\omega} = -\frac{XG(-j\omega)}{2j}$$

Note that

$$\bar{a} = G(s)\frac{\omega X}{s^2 + \omega^2}(s - j\omega)\bigg|_{s=j\omega} = \frac{XG(j\omega)}{2j}$$

Since $G(j\omega)$ is a complex quantity, it can be written in the following form:

$$G(j\omega) = |G(j\omega)|e^{j\phi}$$

where $|(Gj\omega)|$ represents the magnitude and $\phi$ represents the angle of $G(j\omega)$; namely

$$\phi = \underline{/G(j\omega)} = \tan^{-1}\left[\frac{\text{imaginary part of } G(j\omega)}{\text{real part of } G(j\omega)}\right]$$

The angle $\phi$ may be negative, positive, or zero. Similarly, we obtain the following expression for $G(-j\omega)$:

$$G(-j\omega) = |G(-j\omega)|e^{-j\phi} = |G(j\omega)|e^{-j\phi}$$

Then, Eq. (9-4) can be written

$$\begin{aligned}
y(t) &= X|G(j\omega)|\frac{e^{j(\omega t+\phi)} - e^{-j(\omega t+\phi)}}{2j} \\
&= X|G(j\omega)|\sin(\omega t + \phi) \\
&= Y\sin(\omega t + \phi)
\end{aligned} \tag{9-5}$$

where $Y = X|G(j\omega)|$. We see that a stable linear time-invariant system subjected to a sinusoidal input will, at steady state, have a sinusoidal output of the same frequency as the input. But the amplitude and phase of the output will, in general, be different from those of the input. In fact, the amplitude of the output is given by the product of that of the input and $|G(j\omega)|$, while the phase angle differs from that of the input by the amount $\phi = \underline{/G(j\omega)}$. An example of input and output sinusoidal signals is shown in Fig. 9-2.

On the basis of this, we obtain this important result: For sinusoidal inputs,

$$|G(j\omega)| = \left|\frac{Y(j\omega)}{X(j\omega)}\right| = \text{amplitude ratio of the output sinusoid to the input sinusoid}$$

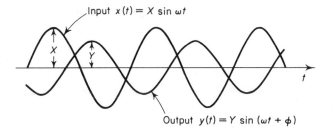

Fig. 9-2. Input and output
sinusoidal signals.

$$\underline{/G(j\omega)} = \underline{\Big/\frac{Y(j\omega)}{X(j\omega)}} = \text{phase shift of the output sinusoid with respect to the input sinusoid}$$

Hence, the response characteristics of a system to a sinusoidal input can be obtained directly from

$$\frac{Y(j\omega)}{X(j\omega)} = G(j\omega)$$

The sinusoidal transfer function $G(j\omega)$, the ratio of $Y(j\omega)$ to $X(j\omega)$, is a complex quantity and can be represented by the magnitude and phase angle with frequency as a parameter. (A negative phase angle is called *phase lag*, and a positive phase angle is called *phase lead*.) The sinusoidal transfer function of any linear system is obtained by substituting $j\omega$ for $s$ in the transfer function of the system. To completely characterize a linear system in the frequency domain, we must specify both the amplitude ratio and the phase angle as functions of the frequency $\omega$.

---

*Example 9-1.* Consider the system shown in Fig. 9-3. The transfer function $G(s)$ is

$$G(s) = \frac{K}{Ts + 1}$$

For the sinusoidal input $x(t) = X \sin \omega t$, the output $y(t)$ can be found as follows: Substituting $j\omega$ for $s$ in $G(s)$ yields

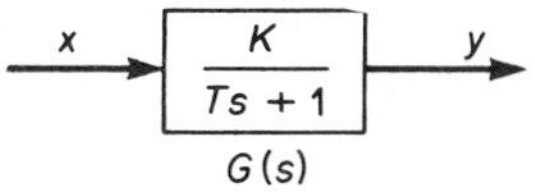

Fig. 9-3. First-order system.

$$G(j\omega) = \frac{K}{jT\omega + 1}$$

The amplitude ratio of the output to input is

$$|G(j\omega)| = \frac{K}{\sqrt{1 + T^2\omega^2}}$$

while the phase angle $\phi$ is

$$\phi = \underline{/G(j\omega)} = -\tan^{-1} T\omega$$

Thus, for the input $x(t) = X \sin \omega t$, the output $y(t)$ can be obtained from Eq. (9-5) as follows:

$$y(t) = \frac{XK}{\sqrt{1 + T^2\omega^2}} \sin (\omega t - \tan^{-1} T\omega) \qquad (9\text{-}6)$$

From Eq. (9-6), it can be seen that for small $\omega$ the amplitude of the output $y(t)$ is almost equal to $K$ times the amplitude of the input. The phase shift of the output is small for small $\omega$. For large $\omega$, the amplitude of the output is small and almost inversely proportional to $\omega$. The phase shift approaches $-90°$ as $\omega$ approaches infinity.

**Frequency response from pole-zero plots.** The frequency response can be determined graphically from the pole-zero plot of the transfer function. Consider the following transfer function:

$$G(s) = \frac{K(s + z)}{s(s + p)}$$

where $p$ and $z$ are real. The frequency response of this transfer function can be obtained from

$$G(j\omega) = \frac{K(j\omega + z)}{j\omega(j\omega + p)}$$

The factors $j\omega + z$, $j\omega$, and $j\omega + p$ are complex quantities, as shown in Fig. 9-4. The magnitude of $G(j\omega)$ is

$$|G(j\omega)| = \frac{K|j\omega + z|}{|j\omega||j\omega + p|}$$

$$= \frac{K|\overline{AP}|}{|\overline{OP}| \cdot |\overline{BP}|}$$

and the phase angle of $G(j\omega)$ is

$$\underline{/G(j\omega)} = \underline{/j\omega + z} - \underline{/j\omega} - \underline{/j\omega + p}$$

$$= \tan^{-1}\frac{\omega}{z} - 90° - \tan^{-1}\frac{\omega}{p}$$

$$= \phi - \theta_1 - \theta_2$$

where the angles $\phi$, $\theta_1$, and $\theta_2$ are defined in Fig. 9-4. Note that a counterclockwise rotation is defined as the positive direction for the measurement of angles.

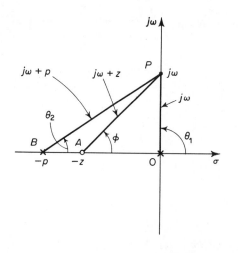

**Fig. 9-4.** Determination of the frequency response in the complex plane.

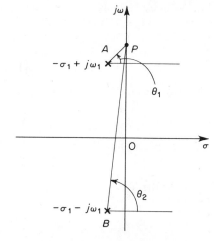

**Fig. 9-5.** Determination of the frequency response in the complex plane.

From the transient-response analysis of closed-loop systems, we know that a complex-conjugate pair of poles near the $j\omega$ axis will produce a highly oscillatory mode of transient response. In the case of the frequency response, such a pair of poles will produce a highly peaked response.

Consider, for example, the following transfer function:

$$G(s) = \frac{K}{(s + p_1)(s + p_2)}$$

where $p_1$ and $p_2$ are complex conjugates, as shown in Fig. 9-5. The frequency response of this transfer function can be found from

$$|G(j\omega)| = \frac{K}{|j\omega + p_1||j\omega + p_2|}$$

$$= \frac{K}{|\overline{AP}||\overline{BP}|}$$

$$\underline{/G(j\omega)} = -\theta_1 - \theta_2$$

where the angles $\theta_1$ and $\theta_2$ are defined in Fig. 9-5. Since $|\overline{AP}||\overline{BP}|$ is very small near $\omega = \omega_1$, $|G(j\omega_1)|$ is very large. Thus a pair of complex-conjugate poles near the $j\omega$ axis will cause a highly peaked frequency response.

Conversely, if the frequency response is not highly peaked, then the transfer function will not have complex-conjugate poles near the $j\omega$ axis. Such a transfer function wil not exhibit highly oscillatory transient response either. Since the frequency response indirectly describes the location of the poles and zeros of the transfer function, we can estimate the transient-response characteristics of a system from frequency-response characteristics. We shall present a detailed discussion of this subject in Section 9-7.

## 9-2 LOGARITHMIC PLOTS

This section and the following two are concerned primarily with presenting frequency-response characteristics of linear control systems.

The sinusoidal transfer function, a complex function of the frequency $\omega$, is characterized by its magnitude and phase angle, with frequency as the parameter. There are three commonly used representations of sinusoidal transfer functions. They are

1. logarithmic plot or Bode diagram
2. polar plot
3. log-magnitude versus phase plot

This section presents logarithmic plots of sinusoidal transfer functions. Polar plots and log-magnitude versus phase plots are presented, respectively, in Sections 9-3 and 9-4.

**Logarithmic plots or Bode diagrams.** A sinusoidal transfer function may be represented by two separate plots, one giving the magnitude versus frequency and the other the phase angle versus frequency. A logarithmic plot or Bode diagram consists of two graphs: One is a plot of the logarithm of the magnitude of a sinusoidal transfer function; the other is a plot of the phase angle; both are plotted against the frequency in logarithmic scale.

The standard representation of the logarithmic magnitude of $G(j\omega)$ is 20 log $|G(j\omega)|$, where the base of the logarithm is 10. The unit used in this representation of the magnitude is the decibel, usually abbreviated db. In the logarithmic representation, the curves are drawn on semilog paper, using the log scale for frequency and the linear scale for either magnitude (but in db) or phase angle (in degrees). (The frequency range of interest determines the number of logarithmic cycles required on the abscissa.)

The main advantage of using the logarithmic plot is that multiplication of magnitudes can be converted into addition. Furthermore, a simple method for sketching an approximate log-magnitude curve is available. It is based on asymptotic approximations. Such approximation by straight-line asymptotes is sufficient if only rough information on the frequency-response characteristics is needed. Should exact curves be desired, corrections can be made easily to these basic asymptotic ones. The phase-angle curves can be drawn easily if a template for the phase-angle curve of $1 + j\omega$ is available.

Note that the experimental determination of a transfer function can be made simple if frequency-response data are presented in the form of a logarithmic plot.

The logarithmic representation is useful in that it shows both the low-frequency and high-frequency characteristics of the transfer function in one diagram. Expanding the low-frequency range by use of a logarithmic scale for the frequency is very advantageous since characteristics at low frequencies are most important in practical systems. (Note that because of the logarithmic frequency scale, it is impossible to plot the curves right down to zero frequency; however, this does not create any serious problems.)

**Basic factors of $G(j\omega)H(j\omega)$.** As stated earlier, the main advantage in using the logarithmic plot is in the relative ease of plotting frequency-response curves. The basic factors which very frequently occur in an arbitrary transfer function $G(j\omega)H(j\omega)$ are

1. gain $K$
2. integral and derivative factors $(j\omega)^{\mp 1}$
3. first-order factors $(1 + j\omega T)^{\mp 1}$
4. quadratic factors $[1 + 2\zeta(j\omega/\omega_n) + (j\omega/\omega_n)^2]^{\mp 1}$

Once we become familiar with the logarithmic plots of these basic factors, it is possible to utilize them in constructing a composite logarithmic plot for any general form of $G(j\omega)H(j\omega)$ by sketching the curves for each factor and adding individual curves graphically because adding the logarithms of the gains corresponds to multiplying them together.

The process of obtaining the logarithmic plot can be further simplified by using asymptotic approximations to the curves for each factor. (If necessary, corrections can be made easily to an approximate plot to obtain an accurate one.)

**The gain $K$.** A number greater than unity has a positive value in decibels, while a number smaller than unity has a negative value. The log-magnitude curve for a constant gain $K$ is a horizontal straight line at the magnitude of $20 \log K$ db. The phase angle of the gain $K$ is zero. The effect of varying the gain $K$ in the transfer function is that it raises or lowers the log-magnitude curve of the transfer function by the corresponding constant amount, but it has no effect on the phase angle.

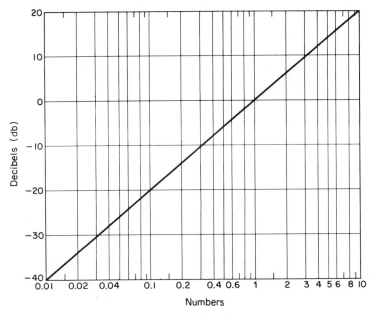

**Fig. 9-6.** Number-decibel conversion line.

A number-decibel conversion line is given in Fig. 9-6. The decibel value of any number can be obtained from this line. As a number increases by a factor of 10, the corresponding decibel value increases by a factor of 20. This may be seen from the following:

$$20 \log (K \times 10^n) = 20 \log K + 20n$$

Note that, when expressed in db, the reciprocal of a number differs from its value only in sign; i.e., for the number $K$,

$$20 \log K = -20 \log \frac{1}{K}$$

**Integral and derivative factors** $(j\omega)^{\mp 1}$**.** The logarithmic magnitude of $1/j\omega$ in db is

$$20 \log \left| \frac{1}{j\omega} \right| = -20 \log \omega \text{ db}$$

The phase angle of $1/j\omega$ is a constant and equal to $-90°$.

In logarithmic plots, frequency ratios are expressed in terms of octaves or decades. An octave is a frequency band from $\omega_1$ to $2\omega_1$, where $\omega_1$ is any frequency value. A decade is a frequency band from $\omega_1$ to $10\omega_1$ where again $\omega_1$ is any frequency. (On the logarithmic scale of semilog paper, any given frequency ratio can be represented by the same horizontal distance. For example, the horizontal distance from $\omega = 1$ to $\omega = 10$ is equal to that from $\omega = 3$ to $\omega = 30$.)

If the log magnitude $-20 \log \omega$ db is plotted against $\omega$ on a logarithmic scale, it is a straight line. Since

$$(-20 \log 10\omega) \, \text{db} = (-20 \log \omega - 20) \, \text{db}$$

the slope of the line is $-20$ db/decade (or $-6$ db/octave).

Similarly, the log magnitude of $j\omega$ in db is

$$20 \log |j\omega| = 20 \log \omega \, \text{db}$$

The phase angle of $j\omega$ is constant and equal to $90°$. The log-magnitude curve is a straight line with a slope of 20 db/decade. Figures 9-7(a) and (b) show frequency-response curves for $1/j\omega$ and $j\omega$, respectively. We can clearly see that the differences in the frequency responses of the factors $1/j\omega$ and $j\omega$ lie in the signs of the slopes of the log-magnitude curves and in the signs of the phase angles. Both log magnitudes become equal to 0 db at $\omega = 1$.

If the transfer function contains the factor $(1/j\omega)^n$ or $(j\omega)^n$, the log magnitude becomes, respectively,

$$20 \log \left| \frac{1}{(j\omega)^n} \right| = -n \times 20 \log |j\omega| = -20n \log \omega \, \text{db}$$

or

$$20 \log |(j\omega)^n| = n \times 20 \log |j\omega| = 20n \log \omega \, \text{db}$$

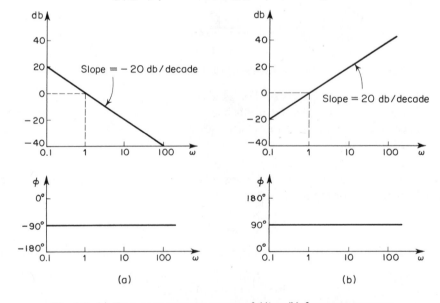

(a)                                            (b)

**Fig. 9-7.** (a) Frequency-response curves of $1/j\omega$; (b) frequency-response curves of $j\omega$.

The slopes of the log-magnitude curves for the factors $(1/j\omega)^n$ and $(j\omega)^n$ are then $-20n$ db/decade and $20n$ db/decade, respectively. The phase angle of $(1/j\omega)^n$ is equal to $-90° \times n$ over the entire frequency range, while that of $(j\omega)^n$ is equal to $90° \times n$ over the entire frequency range.

**First-order factors** $(1 + j\omega T)^{\mp 1}$. The log magnitude of the first-order factor $1/(1 + j\omega T)$ is

$$20 \log \left| \frac{1}{1 + j\omega T} \right| = -20 \log \sqrt{1 + \omega^2 T^2} \text{ db}$$

For low frequencies, such that $\omega \ll 1/T$, the log magnitude may be approximated by

$$-20 \log \sqrt{1 + \omega^2 T^2} \doteq -20 \log 1 = 0 \text{ db}$$

Thus, the log-magnitude curve at low frequencies is the constant 0-db line. For high frequencies, such that $\omega \gg 1/T$,

$$-20 \log \sqrt{1 + \omega^2 T^2} \doteq -20 \log \omega T \text{ db}$$

This is an approximate expression for the high-frequency range. At $\omega = 1/T$, the log magnitude equals 0 db; at $\omega = 10/T$, the log magnitude is $-20$ db. Thus, the value of $-20 \log \omega T$ db decreases by 20 db for every decade of $\omega$. For $\omega \gg$

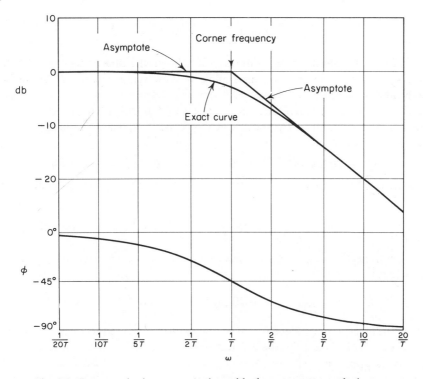

**Fig. 9-8.** Log-magnitude curve together with the asymptotes and phase angle curve of $1/(1 + j\omega T)$.

$1/T$, the log-magnitude curve is thus a straight line with a slope of $-20$ db/decade (or $-6$ db/octave.)

The analysis above shows that the logarithmic representation of the frequency-response curve of the factor $1/(1 + j\omega T)$ can be approximated by two straight-line asymptotes, one a straight line at 0 db for the frequency range $0 < \omega < 1/T$, and the other a straight line with slope $-20$ db/decade (or $-6$ db/octave) for the frequency range $1/T < \omega < \infty$. The exact log-magnitude curve, the asymptotes, and the exact phase-angle curve are shown in Fig. 9-8.

The frequency at which the two asymptotes meet is called the *corner* frequency or *break* frequency. For the factor $1/(1 + j\omega T)$, the frequency $\omega = 1/T$ is the corner frequency since at $\omega = 1/T$ the two asymptotes have the same value. (The low-frequency asymptotic expression at $\omega = 1/T$ is 20 log 1 db = 0 db and the high-frequency asymptotic expression at $\omega = 1/T$ is also 20 log 1 db = 0 db.) The corner frequency divides the frequency-response curve into two regions, a curve for the low-frequency region and a curve for the high-frequency region. The corner frequency is very important in sketching logarithmic frequency-response curves.

The exact phase angle $\phi$ of the factor $1/(1 + j\omega T)$ is

$$\phi = -\tan^{-1} \omega T$$

At zero frequency, the phase angle is 0°. At the corner frequency, the phase angle is

$$\phi = -\tan^{-1} \frac{T}{T} = -\tan^{-1} 1 = -45°$$

At infinity, the phase angle becomes $-90°$. Since the phase angle is given by an inverse-tangent function, the phase angle is skew symmetric about the inflection point at $\phi = -45°$.

The error in the magnitude curve caused by the use of asymptotes can be calculated. The maximum error occurs at the corner frequency and is approximately equal to $-3$ db since

$$-20 \log \sqrt{1 + 1} + 20 \log 1 = -10 \log 2 = -3.03 \text{ db}$$

The error at the frequency one octave below the corner frequency, namely at $\omega = 1/2T$, is

$$-20 \log \sqrt{\frac{1}{4} + 1} + 20 \log 1 = -20 \log \frac{\sqrt{5}}{2} = -0.97 \text{ db}$$

The error at the frequency one octave above the corner frequency, namely at $\omega = 2/T$, is

$$-20 \log \sqrt{2^2 + 1} + 20 \log 2 = -20 \log \frac{\sqrt{5}}{2} = -0.97 \text{ db}$$

Thus, the error at one octave below or above the corner frequency is approximately equal to $-1$ db. Similarly, the error at one decade below or above the corner frequency is approximately $-0.04$ db. The error in decibels involved in using the asymptotic expression for the frequency-response curve of $1/(1 + j\omega T)$ is shown in Fig. 9-9. The error is symmetric with respect to the corner frequency.

Since the asymptotes are quite easy to draw and are sufficiently close to the exact curve, the use of such approximations in drawing Bode diagrams is convenient in establishing the general nature of the frequency-response characteristics quickly with a minimum amount of calculation and may be used for most preliminary design work. If accurate frequency-response curves are desired, corrections may easily be made by referring to the curve given in Fig. 9-9. In practice, an accurate frequency-response curve can be drawn by locating the $-3$-db point at the corner frequency and the $-1$-db points one octave below and above the corner frequency and then connecting these points by a smooth curve.

Note that varying the time constant $T$ shifts the corner frequency to the left or to the right, but the shapes of the log-magnitude and the phase-angle curves remain the same.

The transfer function $1/(1 + j\omega T)$ has the characteristics of a low-pass filter. For frequencies above $\omega = 1/T$, the log magnitude falls off rapidly toward $-\infty$. This is essentially due to the presence of the time constant. In the low-pass filter, the output can follow a sinusoidal input faithfully at low frequencies. But as the input frequency is increased, the output cannot follow the input because a certain amount of time is required for the system to build up in magnitude. Thus, at high frequencies, the amplitude of the output approaches zero and the phase angle of the output approaches $-90°$. Therefore, if the input function contains many harmonics, then the low-frequency components are reproduced faithfully at the output, while the high-frequency components are attenuated in amplitude and shifted in phase. Thus, a first-order element yields exact, or almost exact, duplication only for constant or slowly varying phenomena.

An advantage of the logarithmic representation is that for reciprocal factors, for example, the factor $1 + j\omega T$, the log-magnitude and the phase-angle curves need only be changed in sign. Since

$$20 \log |1 + j\omega T| = -20 \log \left| \frac{1}{1 + j\omega T} \right|$$

$$\underline{/1 + j\omega T} = \tan^{-1} \omega T = -\underline{\left/ \frac{1}{1 + j\omega T} \right.}$$

the corner frequency is the same for both cases. The slope of the high-frequency

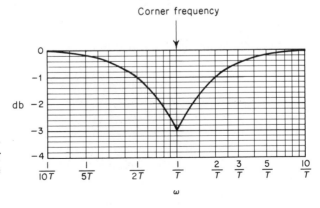

**Fig. 9-9.** Log-magnitude error in the asymptotic expression of the frequency-response curve of $1/(1 + j\omega T)$.

asymptote of $1 + j\omega T$ is 20 db/decade, and the phase angle varies from 0 to 90° as the frequency $\omega$ is increased from zero to infinity. The log-magnitude curve together with the asymptotes and the phase-angle curve for the factor $1 + j\omega T$ are shown in Fig. 9-10.

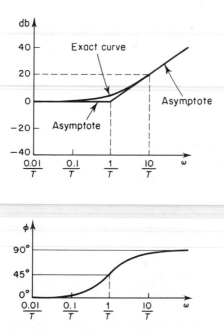

The shapes of phase-angle curves are the same for any factor of the form $(1 + j\omega T)^{\mp 1}$. Hence, it is convenient to have a template for the phase-angle curve on cardboard. Then such a template may be used repeatedly for constructing phase-angle curves for any function of the form $(1 + j\omega T)^{\mp 1}$. If such a template is not available, we have to locate several points on the curve. The phase angles of $(1 + j\omega T)^{\mp 1}$ are

$$\mp 45° \quad \text{at} \quad \omega = \frac{1}{T}$$

$$\mp 26.6° \quad \text{at} \quad \omega = \frac{1}{2T}$$

$$\mp 5.7° \quad \text{at} \quad \omega = \frac{1}{10T}$$

$$\mp 63.4° \quad \text{at} \quad \omega = \frac{2}{T}$$

$$\mp 84.3° \quad \text{at} \quad \omega = \frac{10}{T}$$

**Fig. 9-10.** Log-magnitude curve together with the asymptotes and phase-angle curve of $1 + j\omega T$.

For the case where a given transfer function involves terms like $(1 + j\omega T)^{\mp n}$, a similar asymptotic construction may be made. The corner frequency is still at $\omega = 1/T$, and the asymptotes are straight lines. The low-frequency asymptote is a horizontal straight line at 0 db, while the high-frequency asymptote has the slope of $-20n$ db/decade or $20n$ db/decade. The error involved in the asymptotic expressions is $n$ times that for $(1 + j\omega T)^{\mp 1}$. The phase angle is $n$ times that of $(1 + j\omega T)^{\mp 1}$ at each frequency point.

**Quadratic factors** $[1 + 2\zeta(j\omega/\omega_n) + (j\omega/\omega_n)^2]^{\mp 1}$. Control systems often possess quadratic factors of the form

$$\frac{1}{1 + 2\zeta\left(j\dfrac{\omega}{\omega_n}\right) + \left(j\dfrac{\omega}{\omega_n}\right)^2} \tag{9-7}$$

If $\zeta > 1$, this quadratic factor can be expressed as a product of two first-order ones with real poles. If $0 < \zeta < 1$, this quadratic factor is the product of two complex-conjugate factors. Asymptotic approximations to the frequency-response curves are not accurate for a factor with low values of $\zeta$. This is because the magnitude and phase of the quadratic factor depend upon both the corner frequency and the damping ratio $\zeta$.

The asymptotic frequency-response curve may be obtained as follows: Since

$$20 \log \left| \frac{1}{1 + 2\zeta\left(j\dfrac{\omega}{\omega_n}\right) + \left(j\dfrac{\omega}{\omega_n}\right)^2} \right|$$

$$= -20 \log \sqrt{\left(1 - \frac{\omega^2}{\omega_n^2}\right)^2 + \left(2\zeta\frac{\omega}{\omega_n}\right)^2}$$

for low frequencies such that $\omega \ll \omega_n$, the log magnitude becomes

$$-20 \log 1 = 0 \text{ db}$$

The low-frequency asymptote is thus a horizontal line at 0 db. For high frequencies such that $\omega \gg \omega_n$, the log magnitude becomes

$$-20 \log \frac{\omega^2}{\omega_n^2} = -40 \log \frac{\omega}{\omega_n} \text{ db}$$

The equation for the high-frequency asymptote is a straight line having the slope $-40$ db/decade since

$$-40 \log \frac{10\omega}{\omega_n} = -40 - 40 \log \frac{\omega}{\omega_n}$$

The high-frequency asymptote intersects the low-frequency one at $\omega = \omega_n$ since at this frequency

$$-40 \log \frac{\omega}{\omega_n} = -40 \log 1 = 0 \text{ db}$$

This frequency is the corner frequency for the quadratic factor considered.

The two asymptotes just derived are independent of the value of $\zeta$. Near the frequency $\omega = \omega_n$, a resonant peak occurs, as may be expected from (9-7). The damping ratio $\zeta$ determines the magnitude of this resonant peak. Errors obviously exist in the approximation by straight-line asymptotes. The magnitude of the error depends on the value of $\zeta$. It is large for small values of $\zeta$. Figure 9-11 shows the exact log-magnitude curves together with the straight-line asymptotes and the exact phase-angle curves for the quadratic factor given by (9-7) with several values of $\zeta$. If corrections are desired in the asymptotic curves, the necessary amounts of correction at a sufficient number of frequency points may be obtained from Fig. 9-11.

The phase angle of the quadratic factor $[1 + 2\zeta(j\omega/\omega_n) + (j\omega/\omega_n)^2]^{-1}$ is

$$\phi = \underline{\left| \frac{1}{1 + 2\zeta\left(j\dfrac{\omega}{\omega_n}\right) + \left(j\dfrac{\omega}{\omega_n}\right)^2} \right.} = -\tan^{-1}\left[\frac{2\zeta\dfrac{\omega}{\omega_n}}{1 - \left(\dfrac{\omega}{\omega_n}\right)^2}\right] \qquad (9\text{-}8)$$

The phase angle is a function of both $\omega$ and $\zeta$. At $\omega = 0$, the phase angle equals $0°$. At the corner frequency $\omega = \omega_n$, the phase angle is $-90°$ regardless of $\zeta$ since

$$\phi = -\tan^{-1}\left(\frac{2\zeta}{0}\right) = -\tan^{-1}\infty = -90°$$

At $\omega = \infty$, the phase angle becomes $-180°$. The phase-angle curve is skew symmetric about the inflection point, the point where $\phi = -90°$.

The frequency-response curves for the factor

$$1 + 2\zeta \left(j\frac{\omega}{\omega_n}\right) + \left(j\frac{\omega}{\omega_n}\right)^2$$

can be obtained by merely reversing the sign of the log magnitude and that of the phase angle of the factor

$$\frac{1}{1 + 2\zeta \left(j\frac{\omega}{\omega_n}\right) + \left(j\frac{\omega}{\omega_n}\right)^2}$$

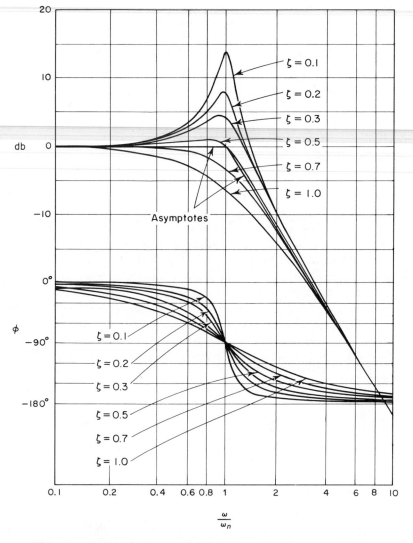

**Fig. 9-11.** Log-magnitude curves together with the asymptotes and phase-angle curves of the quadratic transfer function given by (9-7).

To obtain the frequency-response curves of a given quadratic transfer function, we must first determine the value of the corner frequency $\omega_n$ and that of the damping ratio $\zeta$. Then, by using the family of curves given in Fig. 9-11, the frequency-response curves can be plotted.

**The resonant frequency $\omega_r$ and the resonant peak value $M_r$.** The magnitude of

$$G(j\omega) = \frac{1}{1 + 2\zeta\left(j\dfrac{\omega}{\omega_n}\right) + \left(j\dfrac{\omega}{\omega_n}\right)^2}$$

is

$$|G(j\omega)| = \frac{1}{\sqrt{\left(1 - \dfrac{\omega^2}{\omega_n^2}\right)^2 + \left(2\zeta\dfrac{\omega}{\omega_n}\right)^2}} \tag{9-9}$$

If $|G(j\omega)|$ has a peak value at some frequency, this frequency is called the *resonant frequency*. Since the numerator of $|G(j\omega)|$ is constant, a peak value of $|G(j\omega)|$ will occur when

$$g(\omega) = \left(1 - \frac{\omega^2}{\omega_n^2}\right)^2 + \left(2\zeta\frac{\omega}{\omega_n}\right)^2 \tag{9-10}$$

is a minimum. Since Eq. (9-10) can be written

$$g(\omega) = \left[\frac{\omega^2 - \omega_n^2(1 - 2\zeta^2)}{\omega_n^2}\right]^2 + 4\zeta^2(1 - \zeta^2) \tag{9-11}$$

the minimum value of $g(\omega)$ occurs at $\omega = \omega_n\sqrt{1 - 2\zeta^2}$. Thus the resonant frequency $\omega_r$ is

$$\omega_r = \omega_n\sqrt{1 - 2\zeta^2} \qquad (0 \leq \zeta \leq 0.707) \tag{9-12}$$

As the damping ratio $\zeta$ approaches zero, the resonant frequency approaches $\omega_n$. For $0 < \zeta \leq 0.707$, the resonant frequency $\omega_r$ is less than the damped natural frequency $\omega_d = \omega_n\sqrt{1 - \zeta^2}$, which is exhibited in the transient response. From Eq. (9-11) it can be seen that for $\zeta > 0.707$ there is no resonant peak. For $\zeta > 0.707$, the magnitude $|G(j\omega)|$ decreases monotonically with increasing frequency $\omega$. This means that there is no peak in the magnitude curve for $\zeta > 0.707$. (The magnitude is less than 0 db for all values of $\omega > 0$. Recall that, for $0.7 < \zeta < 1$, the step response is oscillatory but the oscillations are well damped and are hardly perceptible.)

The magnitude of the resonant peak $M_r$ can be found by substituting Eq. (9-12) into Eq. (9-9). For $0 \leq \zeta \leq 0.707$,

$$M_r = |G(j\omega)|_{\max} = |G(j\omega_r)| = \frac{1}{2\zeta\sqrt{1 - \zeta^2}} \tag{9-13}$$

For $\zeta > 0.707$,

$$M_r = 1 \tag{9-14}$$

As $\zeta$ approaches zero, $M_r$ approaches infinity. This means that if the undamped system is excited at its natural frequency, the magnitude of $G(j\omega)$ becomes infinity. The relationship between $M_r$ and $\zeta$ is shown in Fig. 9-12.

The phase angle of $G(j\omega)$ at the frequency where the resonant peak occurs can be obtained by substituting Eq. (9-12) into Eq. (9-8). Thus, at the resonant frequency $\omega_r$,

$$\underline{/G(j\omega_r)} = -\tan^{-1}\frac{\sqrt{1-2\zeta^2}}{\zeta}$$

$$= -90° + \sin^{-1}\frac{\zeta}{\sqrt{1-\zeta^2}}$$

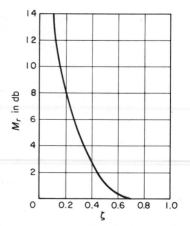

**Fig. 9-12.** $M_r$ versus $\zeta$ curve for the second-order system $1/[1 + 2\zeta\,(j\omega/\omega_n) + (j\omega/\omega_n)^2]$.

**General procedure for plotting logarithmic frequency-response curves.** First rewrite the sinusoidal transfer function $G(j\omega)H(j\omega)$ as a product of basic factors discussed above. Then identify the corner frequencies associated with these basic factors. Finally, draw the asymptotic log-magnitude curves with proper slopes between the corner frequencies. The exact curve, which lies close to the asymptotic curve, can be obtained by adding proper corrections.

The phase-angle curve of $G(j\omega)H(j\omega)$ can be drawn by adding the phase-angle curves of individual factors.

The use of logarithmic plots employing asymptotic approximations requires much less time than other methods which may be used for computing the frequency response of a transfer function. The ease of plotting the frequency-response curves for a given transfer function and the ease of modification of the frequency-response curve as compensation is added are the main reasons why logarithmic plots are very frequently used in practice.

---

*Example 9-2.* Draw the Bode diagram for the following transfer function.

$$G(j\omega) = \frac{10(j\omega + 3)}{(j\omega)(j\omega + 2)[(j\omega)^2 + j\omega + 2]}$$

Make corrections so that the log-magnitude curve is accurate.

In order to avoid any possible mistakes in drawing the log-magnitude curve, it is desirable to put $G(j\omega)$ in the following normalized form, where the low-frequency asymptotes for the first-order factors and the second-order factor are the 0-db line.

$$G(j\omega) = \frac{7.5\left(\dfrac{j\omega}{3} + 1\right)}{(j\omega)\left(\dfrac{j\omega}{2} + 1\right)\left[\dfrac{(j\omega)^2}{2} + \dfrac{j\omega}{2} + 1\right]}$$

This function is composed of the following factors:

$$7.5, \quad (j\omega)^{-1}, \quad 1 + j\frac{\omega}{3}, \quad \left(1 + j\frac{\omega}{2}\right)^{-1} \quad \left[1 + j\frac{\omega}{2} + \frac{(j\omega)^2}{2}\right]^{-1}$$

The corner frequencies of the third, fourth, and fifth terms are $\omega = 3$, $\omega = 2$, and $\omega = \sqrt{2}$, respectively.

To plot the Bode diagram, the separate asymptotic curves for each of the factors are shown in Fig. 9-13. The composite curve is then obtained by adding algebraically the

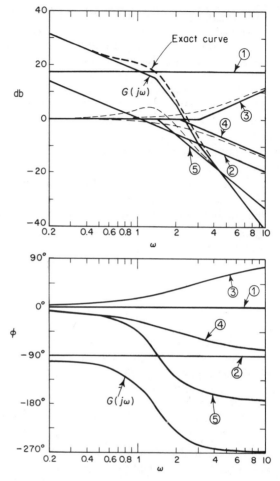

**Fig. 9-13.** Bode diagram of the system considered in Example 9-2.

individual curves, also shown in Fig. 9-13. Note that when the individual asymptotic curves are added at each frequency, the slope of the composite curve is cumulative. Below $\omega = \sqrt{2}$, the plot has the slope of $-20$ db/decade. At the first corner frequency $\omega = \sqrt{2}$, the slope changes to $-60$ db/decade and continues to the next corner frequency $\omega = 2$, where the slope becomes $-80$ db/decade. At the last corner frequency $\omega = 3$, the slope changes to $-60$ db/decade.

Once such an approximate log-magnitude curve has been drawn, the actual curve can be obtained by adding corrections at each corner frequency and at frequencies one octave below and above the corner frequencies. For first-order factors $(1 + j\omega T)^{\pm 1}$, the corrections are $\pm 3$ db at the corner frequency and $\pm 1$ db at the frequencies one octave below and above the corner frequency. Corrections necessary for the quadratic factor are obtained from Fig. 9-11. The exact log-magnitude curve for $G(j\omega)$ is shown by a dotted curve in Fig. 9-13.

For plotting the complete phase-angle curve, the phase-angle curves for all factors have to be sketched. The algebraic sum of all phase-angle curves provides the complete phase-angle curve, as shown in Fig. 9-13.

The simplicity in plotting the Bode diagram should now be apparent. Once we become used to drawing Bode diagrams, it is not necessary to draw the asymptotes for each of the factors involved in the transfer function. We may draw directly the composite, asymptotic, log-magnitude curve by mentally summing up the asymptotes for each of the factors.

**Minimum phase systems and nonminimum phase systems.** Transfer functions having no poles or zeros in the right-half $s$ plane are minimum-phase transfer functions; whereas those having poles and/or zeros in the right-half $s$ plane are nonminimum phase transfer functions. Systems with minimum phase transfer functions are called minimum phase systems; whereas those with nonminimum phase transfer functions are called nonminimum phase systems.

For systems with the same magnitude characteristic, the range in phase angle of the minimum phase transfer function is minimum for all such systems, while the range in phase angle of any nonminimum phase transfer function is greater than this minimum.

Consider as an example the two systems whose sinusoidal transfer functions are, respectively,

$$G_1(j\omega) = \frac{1 + j\omega T}{1 + j\omega T_1}, \qquad G_2(j\omega) = \frac{1 - j\omega T}{1 + j\omega T_1} \qquad (0 < T < T_1)$$

The pole-zero configurations of these systems are shown in Fig. 9-14. The two sinusoidal transfer functions have the same magnitude characteristics, but they have different phase-angle characteristics, as shown in Fig. 9-15. These two systems differ from each other by the factor

$$G(j\omega) = \frac{1 - j\omega T}{1 + j\omega T}$$

The magnitude of the factor $(1 - j\omega T)/(1 + j\omega T)$ is always unity. But the phase angle equals $-2 \tan^{-1} \omega T$ and varies from 0 to $-180°$ as $\omega$ is increased from zero to infinity.

For a minimum phase system, the magnitude and phase-angle characteristics

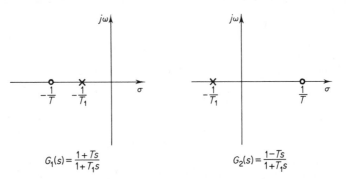

**Fig. 9-14.** Pole-zero configurations of a minimum phase system $G_1(s)$ and nonminimum phase system $G_2(s)$.

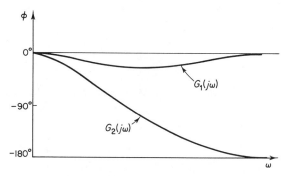

**Fig. 9-15.** Phase angle characteristics of the systems $G_1(s)$ and $G_2(s)$ shown in Fig. 9-14.

are directly related. This means that if the magnitude curve of a system is specified over the entire frequency range from zero to infinity, then the phase-angle curve is uniquely determined and vice versa. This, however, does not hold for a nonminimum phase system.

Nonminimum phase situations may arise in two different ways. One is simply when a system includes a nonminimum phase element or elements. The other situation may arise in the case where a minor loop is unstable.

For a minimum phase system, the phase angle at $\omega = \infty$ becomes $-90°$ ($q - p$), where $p$ and $q$ are the degrees of the numerator and denominator polynomials of the transfer function, respectively. For a nonminimum phase system, the phase angle at $\omega = \infty$ differs from $-90°(q - p)$. In either system, the slope of the log-magnitude curve at $\omega = \infty$ is equal to $-20(q - p)$ db/decade. It is therefore possible to detect whether or not the system is a minimum phase one by examining both the slope of the high-frequency asymptote of the log-magnitude curve and the phase angle at $\omega = \infty$. If the slope of the log-magnitude curve as $\omega$ approaches infinity is $-20(q - p)$ db/decade and the phase angle at $\omega = \infty$ is equal to $-90°(q - p)$, then the system is a minimum phase one.

Nonminimum phase systems are slow in response because of their faulty behavior at the start of response. In most practical control systems, excessive phase lag should be carefully avoided. In designing a system, if fast speed of response is of primary importance, we should not use nonminimum phase components. (A common example of nonminimum phase elements which may be present in control systems is transportation lag.)

It is noted that the techniques of frequency-response anlysis and design to be presented in this and the next chapter are valid for both minimum phase and nonminimum phase systems.

**Transportation lag.** Transportation lag is of nonminimum phase behavior and has an excessive phase lag with no attenuation at high frequencies. Consider the transportation lag given by

$$G(j\omega) = e^{-j\omega T}$$

The magnitude is always equal to unity since

$$|G(j\omega)| = |\cos \omega T - j \sin \omega T| = 1$$

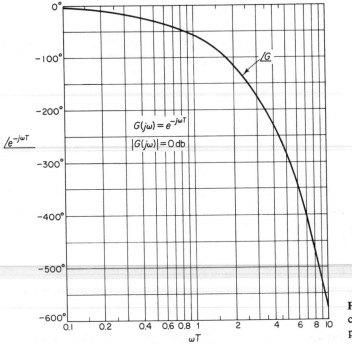

$\underline{/e^{-j\omega T}}$

Fig. 9-16. Phase angle characteristic of transportation lag.

Therefore, the log magnitude of the transportation lag $e^{-j\omega T}$ is equal to 0 db. The phase angle of the transportation lag is

$$\underline{/G(j\omega)} = -\omega T \quad \text{(radians)}$$
$$= -57.3\,\omega T \quad \text{(degrees)}$$

The phase angle varies linearly with the frequency $\omega$. The phase-angle characteristic of transportation lag is shown in Fig. 9-16.

---

*Example 9-3.* Draw the Bode diagram of the following transfer function:

$$G(j\omega) = \frac{e^{-j\omega L}}{1 + j\omega T}$$

The log magnitude is

$$20\log|G(j\omega)| = 20\log|e^{-j\omega L}| + 20\log\left|\frac{1}{1 + j\omega T}\right|$$
$$= 0 + 20\log\left|\frac{1}{1 + j\omega T}\right|$$

The phase angle of $G(j\omega)$ is

$$\underline{/G(j\omega)} = \underline{/e^{-j\omega L}} + \underline{/\frac{1}{1 + j\omega T}}$$
$$= -\omega L - \tan^{-1}\omega T$$

The log-magnitude and phase-angle curves for this transfer function with $L = 0.5$ and $T = 1$ are shown in Fig. 9-17.

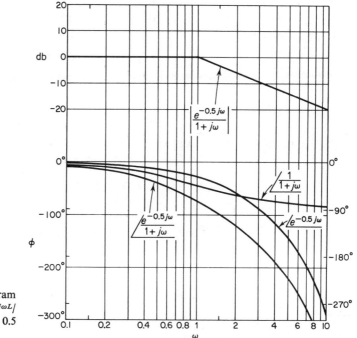

**Fig. 9-17.** Bode diagram for the system $e^{-j\omega L}/(1 + j\omega T)$ with $L = 0.5$ and $T = 1$.

**Relationship between system type and log-magnitude curve.** The static position, velocity, and acceleration error coefficients describe the low-frequency behavior of type 0, type 1, and type 2 systems, respectively. For a given system, only one of the static error coefficients is finite and significant. (The larger the value of the finite static error coefficient, the higher the loop gain is as $\omega$ approaches zero.)

The type of the system determines the slope of the log-magnitude curve at low frequencies. Thus, information concerning the existence and magnitude of the steady-state error of a control system to a given input can be determined from the observation of the low-frequency region of the log-magnitude curve.

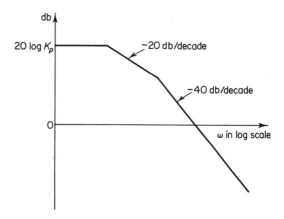

**Fig. 9-18.** Log-magnitude curve of a type 0 system.

**Determination of static position error coefficients.** Figure 9-18 shows an example of the log-magnitude plot of a type 0 system. In such a system, the magnitude of $G(j\omega)H(j\omega)$ equals $K_p$ at low frequencies, or

$$\lim_{\omega \to 0} G(j\omega)H(j\omega) = K_p$$

It follows that the low-frequency asymptote is a horizontal line at $20 \log K_p$ db.

**Determination of static velocity error coefficients.** Figure 9-19 shows an example of the log-magnitude plot of a type 1 system. The intersection of the initial $-20$-db/decade segment (or its extension) with the line $\omega = 1$ has the magnitude $20 \log K_v$. This may be seen as follows: In a type 1 system,

$$G(j\omega)H(j\omega) = \frac{K_v}{j\omega} \qquad \text{for } \omega \ll 1$$

Thus,

$$20 \log \left| \frac{K_v}{j\omega} \right|_{\omega=1} = 20 \log K_v$$

The intersection of the initial $-20$-db/decade segment (or its extension) with the 0-db line has a frequency numerically equal to $K_v$. To see this, define the frequency at this intersection to be $\omega_1$; then

$$\left| \frac{K_v}{j\omega_1} \right| = 1$$

or

$$K_v = \omega_1$$

As an example, consider the type 1 system with unity feedback whose open-loop transfer function is

$$G(s) = \frac{K}{s(Js + F)}$$

If we define the corner frequency to be $\omega_2$ and the frequency at the intersection of the $-40$-db/decade segment (or its extension) with 0-db line to be $\omega_3$, then

$$\omega_2 = \frac{F}{J}, \qquad \omega_3^2 = \frac{K}{J}$$

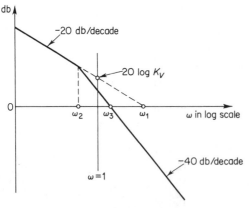

Fig. 9-19. Log-magnitude curve of a type 1 system.

Since

$$\omega_1 = K_v = \frac{K}{F}$$

it follows that

$$\omega_1 \omega_2 = \omega_3^2$$

or

$$\frac{\omega_1}{\omega_3} = \frac{\omega_3}{\omega_2}$$

On the logarithmic plot

$$\log \omega_1 - \log \omega_3 = \log \omega_3 - \log \omega_2$$

Thus, the $\omega_3$ point is just midway between the $\omega_2$ and $\omega_1$ points. The damping ratio $\zeta$ of the system is then

$$\zeta = \frac{F}{2\sqrt{KJ}} = \frac{\omega_2}{2\omega_3}$$

**Determination of static acceleration error coefficients.** Figure 9-20 shows an example of the log-magnitude plot of a type 2 system. The intersection of the initial $-40$-db/decade segmant (or its extension) with the $\omega = 1$ line has the magnitude of 20 log $K_a$. Since at low frequencies

$$G(j\omega)H(j\omega) = \frac{K_a}{(j\omega)^2}$$

it follows that

$$20 \log \left| \frac{K_a}{(j\omega)^2} \right|_{\omega=1} = 20 \log K_a$$

The frequency $\omega_a$ at the intersection of the initial $-40$-db/decade segment (or its extension) with the 0-db line gives the square root of $K_a$ numerically. This can be seen from the following:

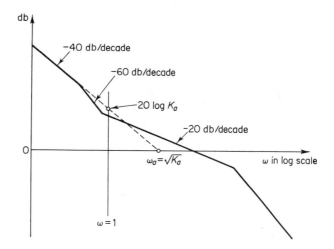

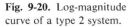

**Fig. 9-20.** Log-magnitude curve of a type 2 system.

$$20 \log \left| \frac{K_a}{(j\omega_a)^2} \right| = 20 \log 1 = 0$$

which yields

$$\omega_a = \sqrt{K_a}$$

### 9-3 POLAR PLOTS

The polar plot of a sinusoidal transfer function $G(j\omega)$ is a plot of the magnitude of $G(j\omega)$ versus the phase angle of $G(j\omega)$ on polar coordinates as $\omega$ is varied from zero to infinity. Thus, the polar plot is the locus of vectors $|G(j\omega)| \underline{/G(j\omega)}$ as $\omega$ is varied from zero to infinity. Note that in polar plots, a positive (negative) phase angle is measured counterclockwise (clockwise) from the positive real axis. The polar plot is often called the Nyquist plot. An example of such a plot is shown in Fig. 9-21. Each point on the polar plot of $G(j\omega)$ represents the terminal point of a vector at a particular value of $\omega$. In the polar plot, it is important to show the frequency graduation of the locus. The projections of $G(j\omega)$ on the real and imaginary axes are its real and imaginary components. Both the magnitude $|G(j\omega)|$ and phase angle $\underline{/G(j\omega)}$ must be calculated directly for each frequency $\omega$ in order to construct polar plots. Since the logarithmic plot is easy to construct, however, the data necessary for plotting the polar plot may be obtained directly from the logarithmic plot if the latter is drawn first and decibels are converted into ordinary magnitude by use of Fig. 9-6.

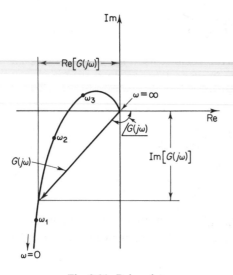

**Fig. 9-21.** Polar plot.

For two systems connected in cascade, the overall transfer function of the combination in the absence of loading effects is the product of the two individual transfer functions. If multiplication of two sinusoidal transfer functions is necessary, this can be done by multiplying together at each frequency the individual sinusoidal transfer functions by means of complex-algebra multiplication. That is, if $G(j\omega) = G_1(j\omega)G_2(j\omega)$, then

$$G(j\omega) = |G(j\omega)| \underline{/G(j\omega)}$$

where

$$|G(j\omega)| = |G_1(j\omega)| \cdot |G_2(j\omega)|$$

and

$$\underline{/G(j\omega)} = \underline{/G_1(j\omega)} + \underline{/G_2(j\omega)}$$

The product of $G_1(j\omega)$ and $G_2(j\omega)$ is shown in Fig. 9-22.

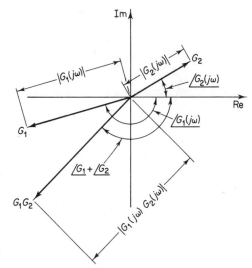

**Fig. 9-22.** Polar plots of $G_1(j\omega)$, $G_2(j\omega)$, and $G_1(j\omega)G_2(j\omega)$.

In general, if a polar plot of $G_1(j\omega)G_2(j\omega)$ is desired, it is convenient to draw a logarithmic plot of $G_1(j\omega)G_2(j\omega)$ first and then convert this into a polar plot rather than to draw the polar plots of $G_1(j\omega)$ and $G_2(j\omega)$ and multiply these two on the complex plane to obtain a polar plot of $G_1(j\omega)G_2(j\omega)$.

An advantage in using a polar plot is that it depicts the frequency-response characteristics of a system over the entire frequency range in a single plot. One disadvantage is that the plot does not clearly indicate the contributions of each of the individual factors of the open-loop transfer function.

**Integral and derivative factors** $(j\omega)^{\mp 1}$. The polar plot of $G(j\omega) = 1/j\omega$ is the negative imaginary axis since

$$G(j\omega) = \frac{1}{j\omega} = -j\frac{1}{\omega} = \frac{1}{\omega}\underline{/-90°}$$

The polar plot of $G(j\omega) = j\omega$ is the positive imaginary axis.

**First-order factors** $(1 + j\omega T)^{\mp 1}$. For the sinusoidal transfer function

$$G(j\omega) = \frac{1}{1 + j\omega T} = \frac{1}{\sqrt{1 + \omega^2 T^2}}\underline{/-\tan^{-1}\omega T}$$

the values of $G(j\omega)$ at $\omega = 0$ and $\omega = 1/T$ are, respectively,

$$G(j0) = 1\underline{/0°} \quad \text{and} \quad G\left(j\frac{1}{T}\right) = \frac{1}{\sqrt{2}}\underline{/-45°}$$

If $\omega$ approaches infinity, the magnitude of $G(j\omega)$ approaches zero and the phase angle approaches $-90°$. The polar plot of this transfer function is a semicircle as the frequency $\omega$ is varied from zero to infinity, as shown in Fig. 9-23 (a). The center is located at 0.5 on the real axis, and the radius is equal to 0.5.

To prove that the polar plot is a semicircle, define

$$G(j\omega) = X + jY$$

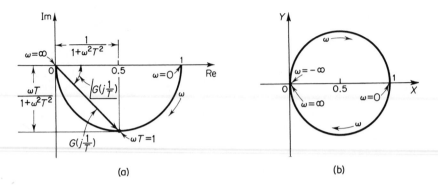

**Fig. 9-23.** (a) Polar plot of $1/(1 + j\omega T)$; (b) plot of $G(j\omega)$ in $X$-$Y$ plane.

where

$$X = \frac{1}{1 + \omega^2 T^2} = \text{real part of } G(j\omega)$$

$$Y = \frac{-\omega T}{1 + \omega^2 T^2} = \text{imaginary part of } G(j\omega)$$

Then we obtain

$$\left(X - \frac{1}{2}\right)^2 + Y^2 = \left(\frac{1}{2}\frac{1 - \omega^2 T^2}{1 + \omega^2 T^2}\right)^2 + \left(\frac{-\omega T}{1 + \omega^2 T^2}\right)^2 = \left(\frac{1}{2}\right)^2$$

Thus, in the $X$-$Y$ plane $G(j\omega)$ is a circle with center at $X = \frac{1}{2}$, $Y = 0$ and with radius $\frac{1}{2}$, as shown in Fig. 9-23 (b). The lower semicircle corresponds to $0 \leq \omega \leq \infty$ and the upper semicircle corresponds to $-\infty \leq \omega \leq 0$.

The polar plot of the transfer function $1 + j\omega T$ is simply the upper half of the straight line passing through point $(1, 0)$ in the complex plane and parallel to the imaginary axis, as shown in Fig. 9-24. The polar plot of $1 + j\omega T$ has an appearance completely different from that of $1/(1 + j\omega T)$.

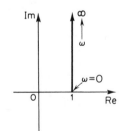

**Quadratic factors** $[1 + 2\zeta(j\omega/\omega_n) + (j\omega/\omega_n)^2]^{\mp 1}$. The low-frequency and high-frequency portions of the polar plot of the following sinusoidal transfer function

$$G(j\omega) = \frac{1}{1 + 2\zeta\left(j\dfrac{\omega}{\omega_n}\right) + \left(j\dfrac{\omega}{\omega_n}\right)^2} \qquad (\zeta > 0)$$

**Fig. 9-24.** Polar plot of $1 + j\omega T$.

are given, respectively, by

$$\lim_{\omega \to 0} G(j\omega) = 1\underline{/0^\circ} \qquad \text{and} \qquad \lim_{\omega \to \infty} G(j\omega) = 0\underline{/-180^\circ}$$

The polar plot of this sinusoidal transfer function starts at $1\underline{/0^\circ}$ and ends at $0\underline{/-180^\circ}$ as $\omega$ increases from zero to infinity. Thus, the high-frequency portion of $G(j\omega)$ is tangent to the negative real axis. The values of $G(j\omega)$ in the frequency range of interest can be calculated directly or by use of the logarithmic plot.

Examples of polar plots of the transfer function just considered are shown in Fig. 9-25. The exact shape of a polar plot depends on the value of the damping

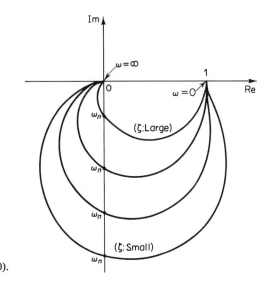

**Fig. 9-25.** Polar plots of

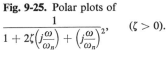
$$\frac{1}{1 + 2\zeta\left(j\dfrac{\omega}{\omega_n}\right) + \left(j\dfrac{\omega}{\omega_n}\right)^2}, \qquad (\zeta > 0).$$

ratio $\zeta$, but the general shape of the plot is the same for both the underdamped case $(1 > \zeta > 0)$ and overdamped case $(\zeta > 1)$.

For the underdamped case at $\omega = \omega_n$, we have $G(j\omega_n) = 1/(j2\zeta)$ and the phase angle at $\omega = \omega_n$ is $-90°$. Therefore, it can be seen that the frequency at which the $G(j\omega)$ locus intersects the imaginary axis is the undamped natural frequency, $\omega_n$. In the polar plot, the frequency point whose distance from the origin is maximum corresponds to the resonant frequency $\omega_r$. The peak value of $G(j\omega)$ is obtained as the ratio of the magnitude of the vector at the resonant frequency $\omega_r$ to the magnitude of the vector at $\omega = 0$. The resonant frequency $\omega_r$ is indicated in the polar plot shown in Fig. 9-26.

For the overdamped case, as $\zeta$ increases well beyond unity, the $G(j\omega)$ locus approaches a semicircle. This may be seen from the fact that for a heavily damped system the characteristic roots are real and one is much smaller than the other. Since for sufficiently large $\zeta$ the effect of the larger root on the response becomes very small, the system behaves like a first-order one.

For the sinusoidal transfer function

$$G(j\omega) = 1 + 2\zeta\left(j\frac{\omega}{\omega_n}\right) + \left(j\frac{\omega}{\omega_n}\right)^2$$

$$= \left(1 - \frac{\omega^2}{\omega_n^2}\right) + j\left(\frac{2\zeta\omega}{\omega_n}\right)$$

the low-frequency portion of the curve is

$$\lim_{\omega \to 0} G(j\omega) = 1\underline{/0°}$$

and the high-frequency portion is

$$\lim_{\omega \to \infty} G(j\omega) = \infty\underline{/180°}$$

Since the imaginary part of $G(j\omega)$ is positive for $\omega > 0$ and is monotonically increasing and the real part of $G(j\omega)$ is monotonically decreasing from unity, the general

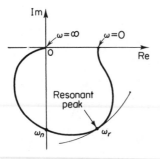

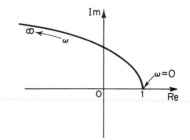

**Fig. 9-26.** Polar plot showing the resonant peak and resonant frequency $\omega_r$.

**Fig. 9-27.** Polar plot of

$$1 + 2\zeta\left(j\frac{\omega}{\omega_n}\right) + \left(j\frac{\omega}{\omega_n}\right)^2, \qquad (\zeta > 0).$$

shape of the polar plot of $G(j\omega)$ is as shown in Fig. 9-27. The phase angle is between $0°$ and $180°$.

---

*Example 9-4.* Consider the following second-order transfer function:

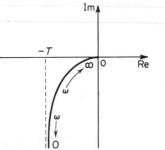

$$G(s) = \frac{1}{s(Ts + 1)}$$

Sketch a polar plot of this transfer function.

Since the sinusoidal transfer function can be written

$$G(j\omega) = \frac{1}{j\omega(1 + j\omega T)}$$

$$= -\frac{T}{1 + \omega^2 T^2} - j\frac{1}{\omega(1 + \omega^2 T^2)}$$

the low-frequency portion of the polar plot becomes

**Fig. 9-28.** Polar plot of $1/[j\omega\,(1 + j\omega T)]$.

$$\lim_{\omega \to 0} G(j\omega) = -T - j\infty = \infty\underline{/-90°}$$

and the high-frequency portion becomes

$$\lim_{\omega \to \infty} G(j\omega) = 0 - j0 = 0\underline{/-180°}$$

The general shape of the polar plot of $G(j\omega)$ is shown in Fig. 9-28. The $G(j\omega)$ plot is asymptotic to the vertical line passing through the point $(-T, 0)$. Since this transfer function involves integration $(1/s)$, the general shape of the polar plot differs substantially from those of second-order transfer functions which do not have the integration.

**Transportation lag.** The transportation lag

$$G(j\omega) = e^{-j\omega T}$$

can be written

$$G(j\omega) = 1\underline{/\cos \omega T - j \sin \omega T}$$

Since the magnitude of $G(j\omega)$ is always unity and the phase angle varies linearly with $\omega$, the polar plot of the transportation lag is a unit circle, as shown in Fig. 9-29.

At low frequencies, the transportation lag $e^{-j\omega T}$ and the first-order lag $1/(1 +$

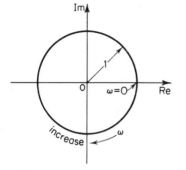

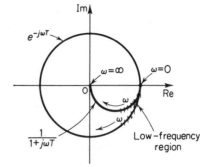

**Fig. 9-29.** Polar plot of transportation lag.

**Fig. 9-30.** Polar plots of $e^{-j\omega T}$
and $1/(1 + j\omega T)$.

$j\omega T$) behave similarly, as shown in Fig. 9-30. The polar plots of $e^{-j\omega T}$ and $1/(1 + j\omega T)$ are tangent to each other at $\omega = 0$. This may be seen from the fact that, for $\omega \ll 1/T$,

$$e^{-j\omega T} \doteq 1 - j\omega T \quad \text{and} \quad \frac{1}{1 + j\omega T} \doteq 1 - j\omega T$$

For $\omega \gg 1/T$, however, an essential difference exists between $e^{-j\omega T}$ and $1/(1 + j\omega T)$, as may also be seen from Fig. 9-30.

---

*Example 9-5.* Obtain the polar plot of the following transfer function:

$$G(j\omega) = \frac{e^{-j\omega L}}{1 + j\omega T}$$

Since $G(j\omega)$ can be written

$$G(j\omega) = (e^{-j\omega L})\left(\frac{1}{1 + j\omega T}\right)$$

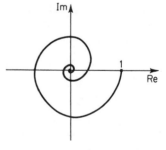

the magnitude and phase angle are, respectively,

$$|G(j\omega)| = |e^{-j\omega L}| \cdot \left|\frac{1}{1 + j\omega T}\right| = \frac{1}{\sqrt{1 + \omega^2 T^2}}$$

and

$$\underline{/G(j\omega)} = \underline{/e^{-j\omega L}} + \underline{/\frac{1}{1 + j\omega T}} = -\omega L - \tan^{-1}\omega T$$

**Fig. 9-31.** Polar plot of $e^{-j\omega L}/(1 + j\omega T)$.

Since the magnitude decreases from unity monoto-
nically and the phase angle also decreases monotonically and indefinitely, the polar plot of the given transfer function is a spiral, as shown in Fig. 9-31.

**General shapes of polar plots.** The polar plots of a transfer function of the form

$$G(j\omega) = \frac{K(1 + j\omega T_a)(1 + j\omega T_b)\cdots}{(j\omega)^\lambda(1 + j\omega T_1)(1 + j\omega T_2)\cdots}$$

$$= \frac{b_0(j\omega)^m + b_1(j\omega)^{m-1} + \cdots}{a_0(j\omega)^n + a_1(j\omega)^{n-1} + \cdots}$$

where the degree of the denominator polynomial is greater than that of the numerator, will have the following general shapes:

1. *For $\lambda = 0$ or type 0 systems:* The starting point of the polar plot (which corresponds to $\omega = 0$) is finite and on the positive real axis. The tangent to the polar plot at $\omega = 0$ is perpendicular to the real axis. The terminal point, which corresponds to $\omega = \infty$, is at the origin, and the curve is tangent to one of the axes.

2. *For $\lambda = 1$ or type 1 systems:* The $j\omega$ term in the denominator contributes $-90°$ to the total phase angle of $G(j\omega)$ for $0 \leq \omega \leq \infty$. At $\omega = 0$, the magnitude of $G(j\omega)$ is infinity, and the phase angle becomes $-90°$. At low frequencies, the polar plot is asymptotic to a line parallel to the negative imaginary axis. At $\omega = \infty$, the magnitude becomes zero, and the curve converges to the origin and is tangent to one of the axes.

3. *For $\lambda = 2$ or type 2 systems:* The $(j\omega)^2$ term in the denominator contributes $-180°$ to the total phase angle of $G(j\omega)$ for $0 \leq \omega \leq \infty$. At $\omega = 0$, the magnitude of $G(j\omega)$ is infinity, and the phase angle is equal to $-180°$. At low frequencies, the polar plot is asymptotic to a line parallel to the negative real axis. At $\omega = \infty$, the magnitude becomes zero, and the curve is tangent to one of the axes.

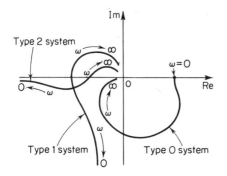

**Fig. 9-32.** Polar plots of type 0, type 1, and type 2 systems.

The general shapes of the low-frequency portions of the polar plots of type 0, type 1, and type 2 systems are shown in Fig. 9-32. It can be seen that if the degree of the denominator polynomial of $G(j\omega)$ is greater than that of the numerator, then the $G(j\omega)$ loci converge to the origin clockwise. At $\omega = \infty$, the loci are tangent to one or the other axes, as shown in Fig. 9-33.

For the case where the degrees of the denominator and numerator polynomials of $G(j\omega)$ are the same, the polar plot starts at a finite distance on the real axis and ends at a finite point on the real axis.

Note that any complicated shapes in the polar plot curves are caused by the numerator dynamics, namely, by the time constants in the numerator of the transfer function. Figure 9-34 shows examples of polar plots of transfer functions with numerator dynamics. In analyzing control systems, the polar plot of $G(j\omega)$ in the frequency range of interest must be accurately determined.

Table 9-1 shows sketches of polar plots of several transfer functions.

**Table 9-1.** POLAR PLOTS OF SIMPLE TRANSFER
FUNCTIONS

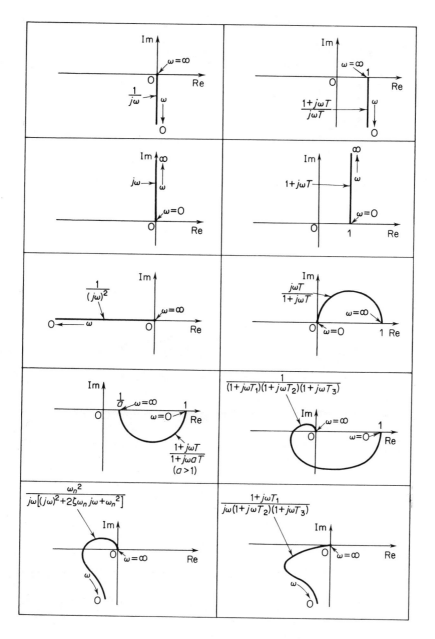

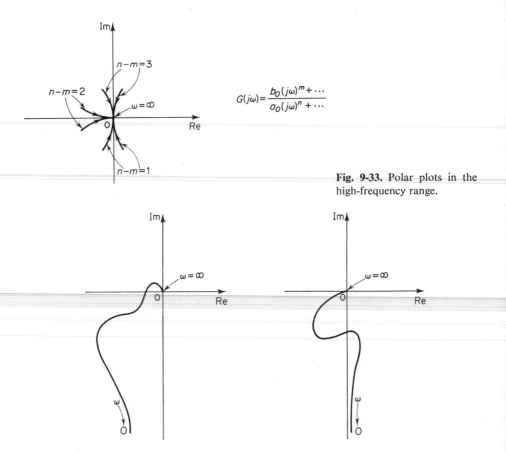

$$G(j\omega) = \frac{b_0(j\omega)^m + \cdots}{a_0(j\omega)^n + \cdots}$$

**Fig. 9-33.** Polar plots in the high-frequency range.

**Fig. 9-34.** Polar plots of transfer functions with numerator dynamics.

## 9-4 LOG-MAGNITUDE VERSUS PHASE PLOTS

Another approach to graphically portraying the frequency-response characteristics is to use the log-magnitude versus phase plot, which is a plot of the logarithmic magnitude in decibels versus the phase angle or phase margin for a frequency range of interest. [The phase margin is the difference between the actual phase angle $\phi$ and $-180°$; namely, $\phi - (-180°) = 180° + \phi$.] The curve is graduated in terms of the frequency $\omega$. Such log-magnitude versus phase plots are sometimes called Nichols plots.

In the Bode diagram, the frequency-response characteristics of $G(j\omega)$ are shown on semilog paper by two separate curves, the log-magnitude curve and the phase-angle curve, while in the log-magnitude versus phase plot, the two curves in the Bode diagram are combined into one. The log-magnitude versus phase plot can easily be constructed by reading values of the log magnitude and phase angle from the Bode diagram. Notice that in the log-magnitude versus phase plot, a change

**Table 9-2.** Log-Magnitude versus Phase Plots of
Simple Transfer Functions

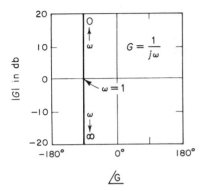

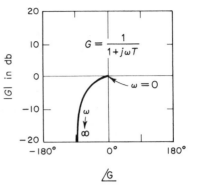

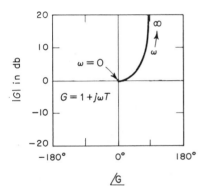

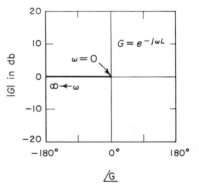

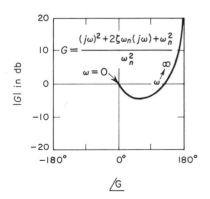

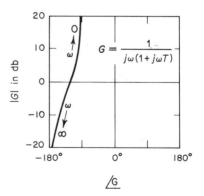

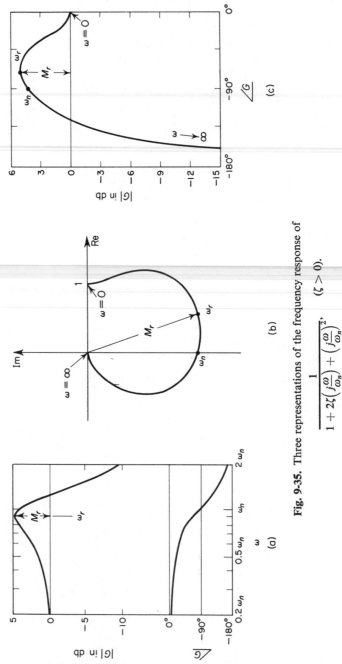

**Fig. 9-35.** Three representations of the frequency response of

$$\frac{1}{1 + 2\zeta\left(j\frac{\omega}{\omega_n}\right) + \left(j\frac{\omega}{\omega_n}\right)^2}, \qquad (\zeta > 0).$$

(a) Logarithmic plot; (b) polar plot; (c) log-magnitude versus phase plot.

in the gain constant of $G(j\omega)$ merely shifts the curve up (for increasing gain) or down (for decreasing gain), but the shape of the curve remains the same.

Advantages of the log-magnitude versus phase plot are that the relative stability of the closed-loop system can be determined quickly and that compensation can be worked out easily.

The log-magnitude versus phase plots for the sinusoidal transfer functions $G(j\omega)$ and $1/G(j\omega)$ are skew symmetrical about the origin since

$$\left|\frac{1}{G(j\omega)}\right| \text{ in db} = -|G(j\omega)| \text{ in db}$$

and

$$\underline{\Big/\frac{1}{G(j\omega)}} = -\underline{\Big/G(j\omega)}$$

Since log-magnitude and phase-angle characteristics of basic transfer functions have been discussed in detail in Sections 9-2 and 9-3, it will be sufficient here to give examples of some log-magnitude versus phase plots. Table 9-2 shows such examples.

Figure 9-35 compares frequency-response curves of

$$G(j\omega) = \frac{1}{1 + 2\zeta\left(j\dfrac{\omega}{\omega_n}\right) + \left(j\dfrac{\omega}{\omega_n}\right)^2}$$

in three different representations. In the log-magnitude versus phase plot, the vertical distance between the points $\omega = 0$ and $\omega = \omega_r$, where $\omega_r$ is the resonant frequency, is the peak value of $G(j\omega)$ in decibels.

## 9-5 NYQUIST STABILITY CRITERION

This section presents the Nyquist stability criterion and associated mathematical background.

Consider the closed-loop system shown in Fig. 9-36. The closed-loop transfer function is

$$\frac{C(s)}{R(s)} = \frac{G(s)}{1 + G(s)H(s)}$$

For stability, all roots of the characteristic equation

$$1 + G(s)H(s) = 0$$

must lie in the left-half $s$ plane. The Nyquist stability criterion is one which relates the open-loop frequency response $G(j\omega)H(j\omega)$ to the number of zeros and poles of $1 + G(s)H(s)$ that lie in the right-half $s$ plane. This criterion, due to H. Nyquist, is useful in control engineering because the absolute stability of the

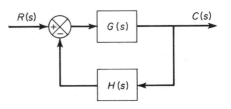

**Fig. 9-36.** Closed-loop system.

closed-loop system can be determined graphically from open-loop frequency-response curves and there is no need for actually determining the closed-loop poles. Analytically obtained open-loop frequency response curves, as well as experimentally obtained ones, can be used for the stability analysis. This is convenient because, in designing a control system, it often happens that mathematical expressions for some of the components are not known, only their frequency-response data are available.

The Nyquist stability criterion is based on a theorem from the theory of complex variables. To understand the criterion, we shall first discuss mappings of contours in the complex plane.

We shall assume that the open-loop transfer function $G(s)H(s)$ is representable as a ratio of polynomials in $s$. For a physically realizable system, the degree of the denominator polynomial of the closed-loop transfer function must be greater than or equal to that of the numerator polynomial. This means that the limit of $G(s)H(s)$ as $s$ approaches infinity is zero or a constant for any physically realizable system.

**Preliminary study.** The characteristic equation of the system shown in Fig. 9-36 is

$$F(s) = 1 + G(s)H(s) = 0$$

We shall show that for a given continuous closed path in the $s$ plane, which does not go through any singular points, there corresponds a closed curve in the $F(s)$ plane. The number and direction of encirclements of the origin of the $F(s)$ plane by the closed curve plays a particularly important role in what follows, for later we shall correlate the number and direction of encirclements with the stability of the system.

Consider, for example, the following open-loop transfer function:

$$G(s)H(s) = \frac{6}{(s+1)(s+2)}$$

The characteristic equation is

$$F(s) = 1 + G(s)H(s) = 1 + \frac{6}{(s+1)(s+2)}$$

$$= \frac{(s + 1.5 + j2.4)(s + 1.5 - j2.4)}{(s+1)(s+2)} = 0$$

The function $F(s)$ is analytic everywhere in the $s$ plane except at its singular points. For each point of analyticity in the $s$ plane, there corresponds a point in the $F(s)$ plane. For example, if $s = 1 + j2$, then $F(s)$ becomes

$$F(1 + j2) = 1 + \frac{6}{(2 + j2)(3 + j2)}$$

$$= 1.12 - j0.58$$

Thus, the point $s = 1 + j2$ in the $s$ plane maps into the point $1.12 - j0.58$ in the $F(s)$ plane.

Thus, as stated above, for a given continuous closed path in the $s$ plane, which does not go through any singular points, there corresponds a closed curve in the

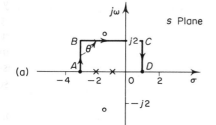
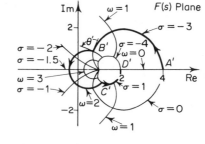

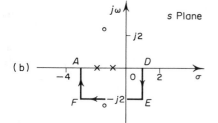
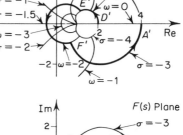

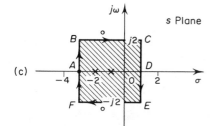
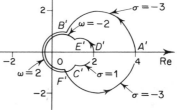

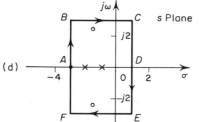

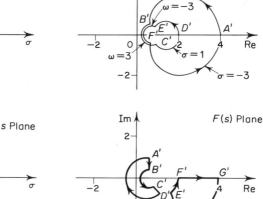

**Fig. 9-37.** Conformal mapping of $s$ plane grids into the $F(s)$ plane.

$F(s)$ plane. Figure 9-37 (a) shows conformal mappings of the lines $\omega = 0, 1, 2, 3$ and the lines $\sigma = 1, 0, -1, -2, -3, -4$ in the upper-half $s$ plane into the $F(s)$ plane. For example, the line $s = j\omega$ in the upper-half $s$ plane ($\omega \geq 0$) maps into the curve denoted by $\sigma = 0$ in the $F(s)$ plane. Figure 9-37 (b) shows conformal mappings of the lines $\omega = 0, -1, -2, -3$ and the lines $\sigma = 1, 0, -1, -2, -3, -4$ in the lower-half $s$ plane into the $F(s)$ plane. Notice that for a given $\sigma$ the curve for negative frequencies is symmetrical about the real axis with the curve for positive frequencies. Referring to Figs. 9-37 (a) and (b), we see that for the path $ABCD$ in the $s$ plane traversed in the clockwise direction, the corresponding curve in the $F(s)$ plane is $A'B'C'D'$. The arrows on the curves indicate directions of traversal. Similarly, the path $DEFA$ in the $s$ plane maps into the curve $D'E'F'A'$ in the $F(s)$ plane. Because of the property of conformal mapping, the corresponding angles in the $s$ plane and $F(s)$ plane are equal and have the same sense. (For example, since lines $AB$ and $BC$ intersect at right angles to each other in the $s$ plane, curves $A'B'$ and $B'C'$ also intersect at right angles at point $B'$ in the $F(s)$ plane.) Referring to Fig. 9-37 (c), we see that on the closed contour $ABCDEFA$ in the $s$ plane the variable $s$ starts at point $A$ and assumes values on this path in a clockwise direction until it returns to the starting point $A$. The corresponding curve in the $F(s)$ plane is denoted $A'B'C'D'E'F'A'$. If we define the area to the right of the contour when a representative point $s$ moves in the clockwise direction to be the inside of the contour and the area to the left to be the outside, then the shaded area in Fig. 9-37(c) is enclosed by the contour $ABCDEFA$ and is inside it. From Fig. 9-37(c), it can be seen that when the contour in the $s$ plane encloses two poles of $F(s)$, the locus of $F(s)$ encircles the origin of the $F(s)$ plane twice in the counterclockwise direction.

The number of encirclements of the origin of the $F(s)$ plane depends on the closed contour in the $s$ plane. If this contour encloses two zeros and two poles of $F(s)$, then the corresponding $F(s)$ locus does not encircle the origin, as shown in Fig. 9-37 (d). If this contour encloses only one zero, the corresponding locus of $F(s)$ encircles the origin once in the clockwise direction. This is shown in Fig. 9-37 (e). Finally, if the closed contour in the $s$ plane encloses neither zeros nor poles, then the locus of $F(s)$ does not encircle the origin of the $F(s)$ plane at all. This is also shown in Fig. 9-37 (e).

Note that for each point in the $s$ plane, except for the singular points, there is only one corresponding point in the $F(s)$ plane; that is, the mapping from the $s$ plane into the $F(s)$ plane is one-to-one. The mapping from the $F(s)$ plane into the $s$ plane may not be one-to-one, however, so that a given point in the $F(s)$ plane may correspond to more than one point in the $s$ plane. For example, point $B'$ in the $F(s)$ plane in Fig. 9-37 (d) corresponds to both point $(-3, 3)$ and point $(0, -3)$ in the $s$ plane.

From the foregoing analysis, we can see that the direction of encirclement of the origin of the $F(s)$ plane depends on whether the contour in the $s$ plane encloses a pole or a zero. Note that the location of a pole or zero in the $s$ plane, whether in the right-half or left-half $s$ plane, does not make any difference, but the enclosure of a pole or zero does. If the contour in the $s$ plane encloses $k$ zeros and $k$

poles ($k = 0, 1, 2, \ldots$), i.e., an equal number of each, then the corresponding closed curve in the $F(s)$ plane does not encircle the origin of the $F(s)$ plane. The foregoing discussion is a graphical explanation of the mapping theorem, which is the basis for the Nyquist stability criterion.

**Mapping theorem.** Let $F(s)$ be a ratio of two polynomials in $s$. Let $P$ be the number of poles and $Z$ be the number of zeros of $F(s)$ which lie inside some closed contour in the $s$ plane, with multiplicity of poles and zeros accounted for. Let this contour be such that it does not pass through any poles or zeros of $F(s)$. This closed contour in the $s$ plane is then mapped into the $F(s)$ plane as a closed curve. The total number $N$ of clockwise encirclements of the origin of the $F(s)$ plane, as a representative point $s$ traces out the entire contour in the clockwise direction, is equal to $Z - P$. (Note that by this mapping theorem the numbers of zeros and of poles cannot be found, only their difference.)

We shall not present a formal proof of this theorem here but leave the proof to Problem A-9-6. Note that a positive number $N$ indicates an excess of zeros over poles of the function $F(s)$ and a negative $N$ indicates an excess of poles over zeros. In control system applications, the number $P$ can be readily determined for $F(s) = 1 + G(s)H(s)$ from the function $G(s)H(s)$. Therefore, if $N$ is determined from the plot of $F(s)$, the number of zeros in the closed contour in the $s$ plane can be determined readily. Note that the exact shapes of the $s$-plane contour and $F(s)$ locus are immaterial as far as encirclements of the origin are concerned since encirclements depend only on the enclosure of poles and/or zeros of $F(s)$ by the $s$-plane contour.

**Application of the mapping theorem to the stability analysis of closed-loop systems.** For analyzing the stability of linear control systems, we let the closed contour in the $s$ plane enclose the entire right-half $s$ plane. The contour consists of the entire $j\omega$ axis from $\omega = -\infty$ to $+\infty$ and a semicircluar path of infinite radius in the right-half $s$ plane. Such a contour is called the Nyquist path. (The direction of the path is clockwise.) The Nyquist path encloses the entire right-half $s$ plane and encloses all the zeros and poles of $1 + G(s)H(s)$ that have positive real parts. [If there are no zeros of $1 + G(s)H(s)$ in the right-half $s$ plane, then there are no closed-loop poles there, and the system is stable.] It is necessary that the closed contour, or the Nyquist path, does not pass through any zeros and poles of $1 + G(s)H(s)$. If $G(s)H(s)$ has a pole or poles at the origin of the $s$ plane, mapping of the point $s = 0$ becomes indeterminate. In such cases, the origin is avoided by taking a detour around it. (A detailed discussion of this special case is given later.)

If the mapping theorem is applied to the special case in which $F(s)$ is equal to $1 + G(s)H(s)$, then we can make the following statement: If the closed contour in the $s$ plane encloses the entire right-half $s$ plane, as shown in Fig. 9-38, then the number of right-half plane zeros of the function $F(s) = 1 + G(s)H(s)$ is equal to the number of poles of the function $F(s) = 1 + G(s)H(s)$ in the right-half $s$ plane plus the number of clockwise encirclements of the origin of the $1 + G(s)H(s)$ plane by the corresponding closed curve in this latter plane.

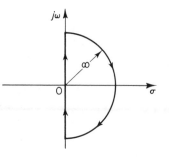

**Fig. 9-38.** Closed contour in the *s* plane.

Because of the assumed condition that

$$\lim_{s \to \infty} [1 + G(s)H(s)] = \text{constant}$$

the function $1 + G(s)H(s)$ remains constant as $s$ traverses the semicircle of infinite radius. Because of this, whether or not the locus of $1 + G(s)H(s)$ encircles the origin of the $1 + G(s)H(s)$ plane can be determined by considering only a part of the closed contour in the $s$ plane, namely, the $j\omega$ axis. Encirclements of the origin, if there are any, occur only while a representative point moves from $-j\infty$ to $+j\infty$ along the $j\omega$ axis, provided that no zeros or poles lie on the $j\omega$ axis.

Note that the portion of the $1 + G(s)H(s)$ contour from $\omega = -\infty$ to $\omega = \infty$ is simply $1 + G(j\omega)H(j\omega)$. Since $1 + G(j\omega)H(j\omega)$ is the vector sum of the unit vector and the vector $G(j\omega)H(j\omega)$, $1 + G(j\omega)H(j\omega)$ is identical to the vector drawn from the $-1 + j0$ point to the terminal point of the vector $G(j\omega)H(j\omega)$, as shown in Fig. 9-39. Encirclement of the origin by the graph of $1 + G(j\omega)H(j\omega)$ is equivalent to encirclement of the $-1 + j0$ point by just the $G(j\omega)H(j\omega)$ locus. Thus, stability of a closed-loop system can be investigated by examining encirclements of the $-1 + j0$ point by the locus of $G(j\omega)H(j\omega)$. The number of clockwise encirclements of the $-1 + j0$ point can be found by drawing a vector from the $-1 + j0$ point to the $G(j\omega)H(j\omega)$ locus, starting from $\omega = -\infty$, going through $\omega = 0$, and ending at $\omega = +\infty$ and by counting the number of clockwise rotations of the vector.

Plotting $G(j\omega)H(j\omega)$ for the Nyquist path is straightforward. The map of the negative $j\omega$ axis is the mirror image about the real axis of the map of the positive $j\omega$ axis. That is, the plot of $G(j\omega)H(j\omega)$ and the plot of $G(-j\omega)H(-j\omega)$ are symmetrical with each other about the real axis. The semicircle with infinite radius maps into either the origin of the $GH$ plane or a point on the real axis of the $GH$ plane.

In the preceding discussion, $G(s)H(s)$ has been assumed to be the ratio of two

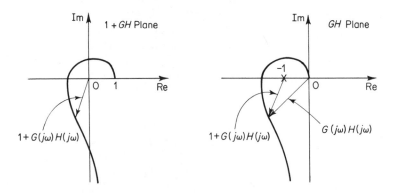

**Fig. 9-39.** Plots of $G(j\omega)H(j\omega)$ in the $1 + GH$ plane and $GH$ plane.

polynomials in $s$. Thus, the transportation lag $e^{-Ts}$ has been excluded from the discussion. Note, however, that a similar discussion applies to systems with transportation lag, although a proof of this is not given here. The stability of a system with transportation lag can be determined from the open-loop frequency-response curves by examining the number of encirclements of the $-1 + j0$ point, just as in the case of a system whose open-loop transfer function is a ratio of two polynomials in $s$.

**Nyquist stability criterion.** The foregoing analysis, utilizing the encirclement of the $-1 + j0$ point by the $G(j\omega)H(j\omega)$ locus, is summarized in the following Nyquist stability criterion:

*Nyquist stability criterion [For a special case where $G(s)H(s)$ has neither poles nor zeros on the $j\omega$ axis.]*: In the system shown in Fig. 9-36, if the open-loop transfer function $G(s)H(s)$ has $k$ poles in the right-half $s$ plane and $\lim_{s \to \infty} G(s)H(s) =$ constant, then for stability the $G(j\omega)H(j\omega)$ locus, as $\omega$ varies from $-\infty$ to $\infty$, must encircle the $-1 + j0$ point $k$ times in the counterclockwise direction.

**Remarks on the Nyquist stability criterion**

1. This criterion can be expressed as

$$Z = N + P$$

where

$Z$ = number of zeros of $1 + G(s)H(s)$ in the right-half $s$ plane
$N$ = number of clockwise encirclements of the $-1 + j0$ point
$P$ = number of poles of $G(s)H(s)$ in the right-half $s$ plane

If $P$ is not zero, for a stable control system, we must have $Z = 0$, or $N = -P$, which means that we must have $P$ counterclockwise encirclements of the $-1 + j0$ point.

If $G(s)H(s)$ does not have any poles in the right-half $s$ plane, then $Z = N$. Thus for stability there must be no encirclement of the $-1 + j0$ point by the $G(j\omega)H(j\omega)$ locus. In this case it is not necessary to consider the locus for the entire $j\omega$ axis, only for the positive frequency portion. The stability of such a system can be determined by seeing if the $-1 + j0$ point is enclosed by the Nyquist plot of $G(j\omega)H(j\omega)$. The region enclosed by the Nyquist plot is shown in Fig. 9-40. For stability, the $-1 + j0$ point must lie outside the shaded region.

2. We must be careful when testing the stability of multiple-loop systems since they may include poles in the right-half $s$ plane. (Note that although an inner loop may be unstable, the

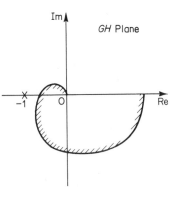

**Fig. 9-40.** Region enclosed by a Nyquist plot.

entire closed-loop system can be made stable by proper design.) Simple inspection of encirclements of the $-1 + j0$ point by the $G(j\omega)H(j\omega)$ locus is not sufficient to detect instability in multiple-loop systems. In such cases, however, whether or not any pole of $1 + G(s)H(s)$ is in the right-half $s$ plane can be determined easily by applying the Routh stability criterion to the denominator of $G(s)H(s)$.

If transcendental functions, such as transportation lag $e^{-Ts}$, are included in $G(s)H(s)$, they must be approximated by a series expansion before the Routh stability criterion can be applied. One form of a series expansion of $e^{-Ts}$ is given below:

$$e^{-Ts} = \frac{1 - \dfrac{Ts}{2} + \dfrac{(Ts)^2}{8} - \dfrac{(Ts)^3}{48} + \cdots}{1 + \dfrac{Ts}{2} + \dfrac{(Ts)^2}{8} + \dfrac{(Ts)^3}{48} + \cdots}$$

As a first approximation, we may take only the first two terms in the numerator and denominator, respectively, or

$$e^{-Ts} \doteq \frac{1 - \dfrac{Ts}{2}}{1 + \dfrac{Ts}{2}} = \frac{2 - Ts}{2 + Ts}$$

This gives a good approximation to transportation lag for the frequency range $0 \leq \omega \leq (0.5/T)$. [Note that the magnitude of $(2 - j\omega T)/(2 + j\omega T)$ is always unity and the phase lag of $(2 - j\omega T)/(2 + j\omega T)$ closely approximates that of transportation lag within the stated frequency range.]

3. If the locus of $G(j\omega)H(j\omega)$ passes through the $-1 + j0$ point, then zeros of the characteristic equation, or closed-loop poles, are located on the $j\omega$ axis. This is not desirable for practical control systems. For a well-designed closed-loop system, none of the roots of the characteristic equation should lie on the $j\omega$ axis.

**Special case where $G(s)H(s)$ involves poles and/or zeros on the $j\omega$ axis.** In the previous discussion, we assumed that the open-loop transfer function $G(s)H(s)$ has neither poles nor zeros at the origin. We now consider the case where $G(s)H(s)$ involves poles and/or zeros on the $j\omega$ axis.

Since the Nyquist path must not pass through poles or zeros of $G(s)H(s)$, if the function $G(s)H(s)$ has poles or zeros at the origin (or on the $j\omega$ axis at points other than the origin), the contour in the $s$ plane must be modified. The usual way of modifying the contour near the origin is to use a semicircle with infinitesimal radius $\epsilon$, as shown in Fig. 9-41. A representative point $s$ moves along the negative $j\omega$ axis from $-j\infty$ to $j0-$. From $s = j0-$ to $s = j0+$, the point moves along the semicircle of radius $\epsilon$ (where $\epsilon \ll 1$) and then moves along the positive $j\omega$ axis from $j0+$ to $j\infty$. From $s = j\infty$, the contour follows a semicircle with infinite radius, and the representative point moves back to the starting point. The area that the modified closed contour avoids is very small and approaches zero as the radius $\epsilon$ approaches zero. Therefore, all the poles and zeros, if any, in the right-half $s$ plane are enclosed by this contour.

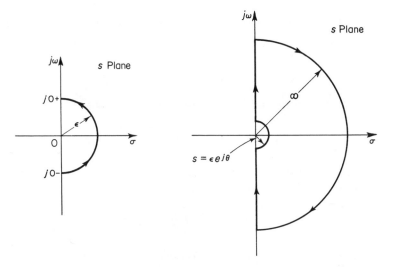

**Fig. 9-41.** Closed contours in the $s$ plane avoiding poles and zeros at the origin.

Consider, for example, a closed-loop system whose open-loop transfer function is given by

$$G(s)H(s) = \frac{K}{s(Ts + 1)}$$

The points corresponding to $s = j0+$ and $s = j0-$ on the locus of $G(s)H(s)$ in the $G(s)H(s)$ plane are $-j\infty$ and $j\infty$, respectively. On the semicircular path with radius $\epsilon$ (where $\epsilon \ll 1$), the complex variable $s$ can be written

$$s = \epsilon e^{j\theta}$$

where $\theta$ varies from $-90°$ to $+90°$. Then $G(s)H(s)$ becomes

$$G(\epsilon e^{j\theta})H(\epsilon e^{j\theta}) = \frac{K}{\epsilon e^{j\theta}} = \frac{K}{\epsilon}e^{-j\theta}$$

The value $K/\epsilon$ approaches infinity as $\epsilon$ approaches zero, and $-\theta$ varies from $90°$ to $-90°$ as a representative point $s$ moves along the semicircle. Thus, the points $G(j0-)H(j0-) = j\infty$ and $G(j0+)H(j0+) = -j\infty$ are joined by a semicircle of infinite radius in the right-half $GH$ plane. The infinitesimal semicircular detour around the origin maps into the $GH$ plane as a semicircle of infinite radius. Figure 9-42 shows the $s$-plane contour and the $G(s)H(s)$ locus in the $GH$ plane. Points $A$, $B$, and $C$ on the $s$-plane contour map into the respective points $A'$, $B'$, and $C'$ on the $G(s)H(s)$ locus. As seen from Fig. 9-42, points $D$, $E$, and $F$ on the semicircle of infinite radius in the $s$ plane map into the origin of the $GH$ plane. Since there is no pole in the right-half $s$ plane and the $G(s)H(s)$ locus does not encircle the $-1 + j0$ point, there are no zeros of the function $1 + G(s)H(s)$ in the right-half $s$ plane. Therefore, the system is stable.

For an open-loop transfer function $G(s)H(s)$ involving a $1/s^n$ factor (where $n = 2, 3, \ldots$), the plot of $G(s)H(s)$ has $n$ clockwise semicircles of infinite radius about the origin as a representative point $s$ moves along the semicircle of radius $\epsilon$ (where $\epsilon \ll 1$). For example, consider the following open-loop transfer function:

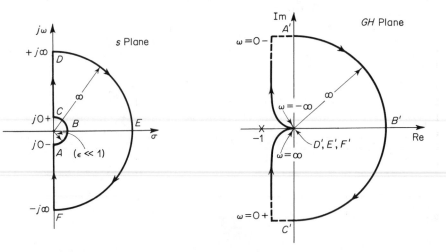

**Fig. 9-42.** *s*-plane contour and the $G(s)H(s)$ locus in the $GH$ plane where $G(s)H(s) = K/[s(Ts + 1)]$.

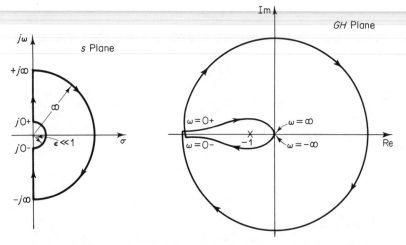

**Fig. 9-43.** *s*-plane contour and the $G(s)H(s)$ locus in the $GH$ plane where $G(s)H(s) = K/[s^2(Ts + 1)]$.

$$G(s)H(s) = \frac{K}{s^2(Ts + 1)}$$

Then

$$\lim_{s \to \epsilon e^{j\theta}} G(s)H(s) = \frac{K}{\epsilon^2 e^{2j\theta}} = \frac{K}{\epsilon^2} e^{-2j\theta}$$

As $\theta$ varies from $-90°$ to $90°$ in the $s$ plane, the angle of $G(s)H(s)$ varies from $180°$ to $-180°$, as shown in Fig. 9-43. Since there is no pole in the right-half $s$ plane and the locus encircles the $-1 + j0$ point twice clockwise for any positive value of $K$, there are two zeros of $1 + G(s)H(s)$ in the right-half $s$ plane. Therefore, this system is always unstable.

Note that a similar analysis can be made if $G(s)H(s)$ involves poles and/or zeros on the $j\omega$ axis. The Nyquist stability criterion can now be generalized as follows:

*Nyquist stability criterion [For a general case where $G(s)H(s)$ has poles and/or zeros on the $j\omega$ axis.]*: In the system shown in Fig. 9-36, if the open-loop transfer function $G(s)H(s)$ has $k$ poles in the right-half $s$ plane, then for stability the $G(s)H(s)$ locus, as a representative point $s$ traces out the modified Nyquist path in the clockwise direction, must encircle the $-1 + j0$ point $k$ times in the counter-clockwise direction.

## 9-6 STABILITY ANALYSIS

In this section, we shall present several illustrative examples of the stability analysis of control systems using the Nyquist stability criterion.

If the Nyquist path in the $s$ plane encircles $Z$ zeros and $P$ poles of $1 + G(s)H(s)$ and does not pass through any poles or zeros of $1 + G(s)H(s)$ as a representative point $s$ moves in the clockwise direction along the Nyquist path, then the corresponding contour in the $G(s)H(s)$ plane encircles the $-1 + j0$ point $N = Z - P$ times in the clockwise direction. (Negative values of $N$ imply counterclockwise encirclements.)

In examining stability of linear control systems using the Nyquist stability criterion, we see that three possibilities can occur.

1. There is no encirclement of the $-1 + j0$ point. This implies that the system is stable if there are no poles of $G(s)H(s)$ in the right-half $s$ plane; otherwise the system is unstable.

2. There is a counterclockwise encirclement or encirclements of the $-1 + j0$ point. In this case the system is stable if the number of counterclockwise encirclements is the same as the number of poles of $G(s)H(s)$ in the right-half $s$ plane, otherwise the system is unstable.

3. There is a clockwise encirclement or encirclements of the $-1 + j0$ point. In this case the system is unstable.

In the following examples, we assume that the values of the gain $K$ and the time constants (such as $T$, $T_1$, and $T_2$) are all positive.

---

*Example 9-6.* Consider a closed-loop system whose open-loop transfer function is given by

$$G(s)H(s) = \frac{K}{(T_1 s + 1)(T_2 s + 1)}$$

Examine the stability of the system.

A plot of $G(j\omega)H(j\omega)$ is shown in Fig. 9-44. Since $G(s)H(s)$ does not have any poles in the right-half $s$ plane and the $-1 + j0$ point is not encircled by the $G(j\omega)H(j\omega)$ locus, this system is stable for any positive values of $K$, $T_1$, and $T_2$.

*Example 9-7.* Consider the system with the following open-loop transfer function:

$$G(s)H(s) = \frac{K}{s(T_1s + 1)(T_2s + 1)}$$

Determine the stability of the system for the two cases, (1) the gain $K$ is small, (2) $K$ is large.

The Nyquist plots of the open-loop transfer function with a small value of $K$ and a large value of $K$ are shown in Fig. 9-45. The number of poles of $G(s)H(s)$ in the right-half $s$ plane is zero. Therefore, for this system to be stable, it is necessary that $N = Z = 0$ or that the $G(s)H(s)$ locus does not encircle the $-1 + j0$ point.

For small values of $K$, there is no encirclement of the $-1 + j0$ point. Hence the system is stable for small values of $K$. For large values of $K$, the locus of $G(s)H(s)$ encircles the $-1 + j0$

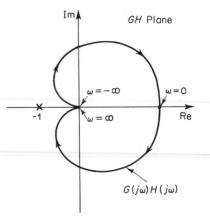

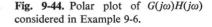

**Fig. 9-44.** Polar plot of $G(j\omega)H(j\omega)$ considered in Example 9-6.

point twice in the clockwise direction, indicating two closed-loop poles in the right-half $s$ plane and the system is unstable. (For good accuracy, $K$ should be large. From the stability viewpoint, however, a large value of $K$ causes poor stability or even instability. To compromise between accuracy and stability, it is necessary to insert a compensation network into the system. Compensating techniques in the frequency domain are discussed in Chapter 10.)

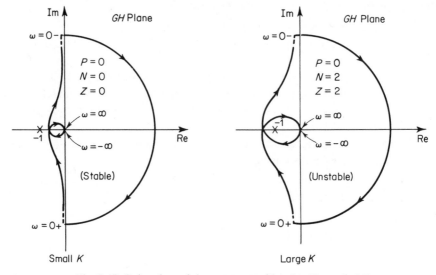

**Fig. 9-45.** Polar plots of the system considered in Example 9-7.

*Example 9-8.* The stability of a closed-loop system with the following open-loop transfer function

$$G(s)H(s) = \frac{K(T_2s + 1)}{s^2(T_1s + 1)}$$

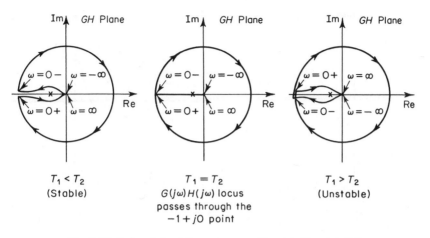

**Fig. 9-46.** Polar plots of the system considered in Example 9-8.

depends on the relative magnitudes of $T_1$ and $T_2$. Draw Nyquist plots and determine the stability of the system.

Plots of the locus of $G(s)H(s)$ for three cases, $T_1 < T_2$, $T_1 = T_2$, and $T_1 > T_2$, are shown in Fig. 9-46. For $T_1 < T_2$, the locus of $G(s)H(s)$ does not encircle the $-1 + j0$ point, and the closed-loop system is stable. For $T_1 = T_2$, the $G(s)H(s)$ locus passes through the $-1 + j0$ point, which indicates that the closed-loop poles are located on the $j\omega$ axis. For $T_1 > T_2$, the locus of $G(s)H(s)$ encircles the $-1 + j0$ point twice in the clockwise direction. Thus, the closed-loop system has two closed-loop poles in the right-half $s$ plane, and the system is unstable.

---

*Example 9-9.* Consider the closed-loop system having the following open-loop transfer function:

$$G(s)H(s) = \frac{K}{s(Ts - 1)}$$

Determine the stability of the system.

The function $G(s)H(s)$ has one pole ($s = 1/T$) in the right-half $s$ plane. Therefore, $P = 1$. The Nyquist plot shown in Fig. 9-47 indicates that the $G(s)H(s)$ plot encircles the $-1 + j0$ point once clockwise. Thus, $N = 1$. Since $Z = N + P$, we find that $Z = 2$. This means that the closed-loop system has two closed-loop poles in the right-half $s$ plane and is unstable.

---

*Example 9-10.* Investigate the stability of a closed-loop system with the following open-loop transfer function:

$$G(s)H(s) = \frac{K(s + 3)}{s(s - 1)}$$

The open-loop transfer function has one pole ($s = 1$) in the right-half $s$ plane, or $P = 1$. The open-loop system is unstable. The Nyquist plot shown in Fig. 9-48 indicates that the $-1 + j0$ point is encircled by the $G(s)H(s)$ locus once in the counterclockwise direction. Therefore, $N = -1$. Thus, $Z$ is found from $Z = N + P$ to be zero, which indicates that there is no zero of $1 + G(s)H(s)$ in the right-half $s$ plane, and the closed-loop

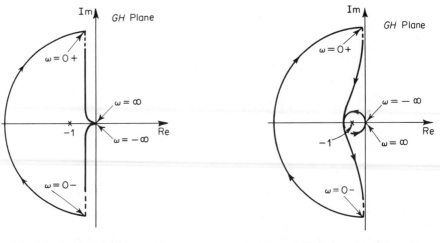

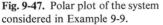

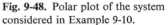

**Fig. 9-47.** Polar plot of the system considered in Example 9-9.

**Fig. 9-48.** Polar plot of the system considered in Example 9-10.

system is stable. This is one of the examples where an unstable open-loop system becomes stable when the loop is closed.

**Conditionally stable systems.** Figure 9-49 shows an example of a $G(j\omega)H(j\omega)$ locus for which the closed-loop system can be made unstable by varying the open-loop gain. If the open-loop gain is increased sufficiently, the $G(j\omega)H(j\omega)$ locus encloses the $-1 + j0$ point twice, and the system becomes unstable. If the open-loop gain is decreased sufficiently, again the $G(j\omega)H(j\omega)$ locus encloses the $-1 + j0$ point twice. The system is stable only for the limited range of values of the open-loop gain for which the $-1 + j0$ point is completely outside the $G(j\omega)H(j\omega)$ locus. Such a system is a conditionally stable one.

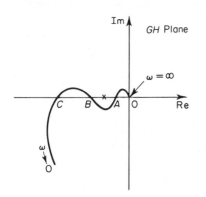

**Fig. 9-49.** Polar plot of a conditionally stable system.

A conditionally stable system is stable for the value of the open-loop gain lying between critical values, but it is unstable if the open-loop gain is either increased or decreased sufficiently. Such a system becomes unstable when large input signals are applied since a large signal may cause saturation, which in turn reduces the open-loop gain of the system. It is advisable to avoid such situations since the system may become unstable, should the open-loop gain drop beyond a critical value.

For stable operation of the conditionally stable system considered here, the critical point $-1 + j0$ must not be located in the regions between $OA$ and $BC$ shown in Fig. 9-49.

**Multiple-loop systems.** Consider the system shown in Fig. 9-50. This is a multiple-loop system. The inner loop has the transfer function

$$G(s) = \frac{G_2(s)}{1 + G_2(s)H_2(s)}$$

If $G(s)$ is unstable, the effects of instability are to produce a pole or poles in the right-half $s$ plane. Then the characteristic equation of the inner loop, $1 + G_2(s)H_2(s) = 0$, has a zero or zeros in this portion of the plane. If $G_2(s)$ and $H_2(s)$ have $P_1$ poles here, then the number $Z_1$ of right-half plane zeros of $1 + G_2(s)H_2(s)$ can be found from $Z_1 = N_1 + P_1$, where $N_1$ is the number of clockwise encirclements of the $-1 + j0$ point by the $G_2(s)H_2(s)$ locus. Since the open-loop transfer function of the entire system is given by $G_1(s)G(s)H_1(s)$, the stability of this closed-loop system can be found from the Nyquist plot of $G_1(s)G(s)H_1(s)$ and knowledge of the right-half plane poles of $G_1(s)G(s)H_1(s)$.

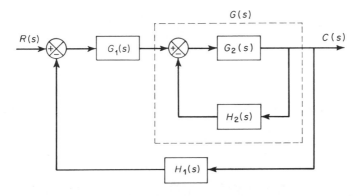

**Fig. 9-50.** Multiple-loop system.

Notice that if a feedback loop is eliminated by means of block diagram reductions, there is a possibility that unstable poles are introduced; if the feedforward branch is eliminated by means of block diagram reductions, there is a possibility that right-half plane zeros are introduced. Therefore, we must note all right-half plane poles and zeros as they appear from subsidiary loop reductions. This knowledge is necessary in determining the stability of multiple-loop systems.

In analyzing such a multiple-loop system, the inverse transfer function may sometimes be used in order to permit graphical analysis; this avoids much of the numerical calculation.

**Nyquist stability criterion applied to inverse polar plots.** In the previous analyses, the Nyquist stability criterion was applied to polar plots of the open-loop transfer function $G(s)H(s)$. The Nyquist stability criterion can equally well be applied to inverse polar plots. The mathematical derivation of the Nyquist stability criterion for inverse polar plots is the same as that for direct polar plots.

The inverse polar plot of $G(j\omega)H(j\omega)$ is a graph of $1/[G(j\omega)H(j\omega)]$ as a function of $\omega$. For example, if $G(j\omega)H(j\omega)$ is

$$G(j\omega)H(j\omega) = \frac{j\omega T}{1 + j\omega T}$$

then

$$\frac{1}{G(j\omega)H(j\omega)} = \frac{1}{j\omega T} + 1$$

The inverse polar plot for $\omega \geq 0$ is the lower half of the vertical line starting at the point $(1, 0)$ on the real axis.

The Nyquist stability criterion applied to inverse polar plots may be stated as follows: For a closed-loop system to be stable, the encirclement, if any, of the $-1 + j0$ point by the $1/[G(s)H(s)]$ locus (as $s$ moves along the Nyquist path) must be counterclockwise, and the number of such encirclements must be equal to the number of poles of $1/[G(s)G(s)]$ [i.e., the zeros of $G(s)H(s)$] that lie in the right-half $s$ plane. [The number of zeros of $G(s)H(s)$ in the right-half $s$ plane may be determined by use of the Routh stability criterion.] If the open-loop transfer function $G(s)H(s)$ has no zeros in the right-half $s$ plane, then for a closed-loop system to be stable the number of encirclements of the $-1 + j0$ point by the $1/[G(s)H(s)]$ locus must be zero.

Note that although the Nyquist stability criterion can be applied to inverse polar plots, if experimental frequency-response data are incorporated, counting the number of encirclements of the $1/[G(s)H(s)]$ locus may be difficult because the phase shift corresponding to the infinite semicircular path in the $s$ plane is difficult to measure. In general, if experimental frequency-response data cannot be put into analytical form, both the $G(j\omega)H(j\omega)$ and $1/[G(j\omega)H(j\omega)]$ loci must be plotted. In addition, the number of right-half plane zeros of $G(s)H(s)$ must be determined. It is more difficult to determine the right-half plane zeros of $G(s)H(s)$ (in other words, to determine whether or not a given component is a minimum phase one) than it is to determine the right-half plane poles of $G(s)H(s)$ (in other words, to determine whether or not the component is stable).

Depending on whether the data are graphical or analytical and whether or not nonminimum phase components are included, an appropriate stability test must be used for multiple loop systems. If the data are given in analytical form or if mathematical expressions for all the components are known, the application of the Nyquist stability criterion to inverse polar plots causes no difficulty and multiple-loop systems may be analyzed and designed in the inverse $GH$ plane.

---

*Example 9-11.* Consider the closed-loop system with the following open-loop transfer function:

$$G(s)H(s) = \frac{K}{s(Ts + 1)}$$

Determine the stability of the system by use of the inverse polar plot.

The inverse polar plot of $G(s)H(s)$, namely the graph of the locus of $1/[G(s)H(s)]$ as a representative point $s$ moves on the closed contour enclosing the entire right-half $s$ plane, is shown in Fig. 9-51, together with the $s$-plane contour. As $s$ traverses the closed contour in the clockwise direction, the $-1 + j0$ point is not encircled. Therefore, the

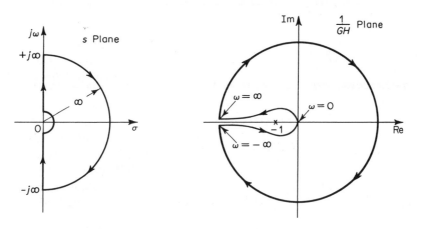

**Fig. 9-51.** $s$-plane contour and the $G(s)H(s)$ locus in the $1/GH$ plane where $G(s)H(s) = K/[s(Ts + 1)]$.

closed-loop system is stable. In this system, the origin of the $s$ plane is mapped into the origin of the $1/GH$ plane. The semicircle with infinite radius, or $s = \infty\, e^{j\theta}$, where $-90° \leq \theta \leq 90°$, corresponds to a circle of infinite radius in the $1/GH$ plane.

Note that if the open-loop transfer function $G(s)H(s)$ involves transportation lag, for example,

$$G(s)H(s) = \frac{Ke^{-j\omega L}}{s(Ts + 1)}$$

then the number of encirclements of the $-1 + j0$ point by the $1/[G(s)H(s)]$ locus becomes infinite, and the Nyquist stability criterion cannot be applied to the inverse polar plot of such an open-loop transfer function.

**Relative stability analysis through modified Nyquist paths.** The Nyquist path for stability tests can be modified in order that we may investigate the relative stability of closed-loop systems. For the following second-order characteristic equation,

$$s^2 + 2\zeta\omega_n s + \omega_n^2 = 0 \qquad (0 < \zeta < 1)$$

the roots are complex conjugates and are

$$s_1 = -\zeta\omega_n + j\omega_n\sqrt{1 - \zeta^2}$$
$$s_2 = -\zeta\omega_n - j\omega_n\sqrt{1 - \zeta^2}$$

If these roots are plotted in the $s$ plane, as shown in Fig. 9-52, then we see that $\sin\theta = \zeta$ or the angle $\theta$ is indicative of the damping ratio $\zeta$. As $\theta$ becomes smaller, so does the value of $\zeta$.

If we modify the Nyquist path and use radial lines with angle $\theta_x$, instead of the $j\omega$

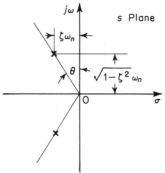

**Fig. 9-52.** Plot of complex-conjugate roots in the $s$ plane.

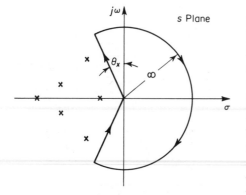

**Fig. 9-53.** Modified Nyquist path.

axis, as shown in Fig. 9-53, then it can be said, following the same reasoning as in the case of the Nyquist stability criterion, that if the $G(s)H(s)$ locus corresponding to the modified $s$-plane contour does not encircle the $-1 + j0$ point and none of the poles of $G(s)H(s)$ lie within the closed $s$-plane contour, then this contour does not enclose any zeros of $1 + G(s)H(s)$. The characteristic equation, $1 + G(s)H(s) = 0$, then does not have any roots within the modified $s$-plane contour. If no closed loop poles of a higher order system are enclosed by this contour, we can say that the damping ratio of each pair of complex-conjugate closed-loop poles of the system is greater than $\sin \theta_x$.

The locus of $G(s)H(s)$ can be plotted by replacing $s$ by

$$-\zeta_x \omega + j\omega\sqrt{1 - \zeta_x^2}$$

where $\zeta_x = \sin \theta_x$, letting $\omega$ take on all values from $-\infty$ to $\infty$, and sketching the resulting locus. The locus for

$$G(-\zeta_x \omega - j\omega\sqrt{1 - \zeta_x^2})H(-\zeta_x \omega - j\omega\sqrt{1 - \zeta_x^2})$$

is the mirror image with respect to the real axis of the locus for

$$G(-\zeta_x \omega + j\omega\sqrt{1 - \zeta_x^2})H(-\zeta_x \omega + j\omega\sqrt{1 - \zeta_x^2})$$

Such a diagram is called a Vazsonyi diagram. The Nyquist diagram is a special case of this latter one.

Similarly, if the $s$-plane contour consists of a line to the left of and parallel to the $j\omega$ axis at a distance $-\sigma_0$ (or the line $s = -\sigma_0 + j\omega$) and the semicircle of infinite radius enclosing the entire right-half $s$ plane and that part of the left-half $s$ plane between the lines $s = -\sigma_0 + j\omega$ and $s = j\omega$, as shown in Fig. 9-54 (a),

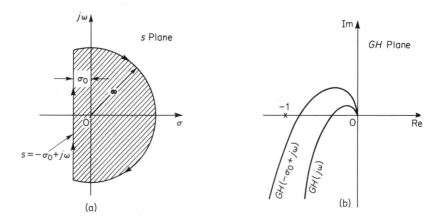

**Fig. 9-54.** (a) Modified Nyquist path; (b) polar plots of the $G(-\sigma_0 + j\omega)H(-\sigma_0 + j\omega)$ locus and $G(j\omega)H(j\omega)$ locus in the $GH$ plane.

then we can state the following: If the $G(s)H(s)$ locus corresponding to this $s$-plane contour does not encircle the $-1 + j0$ point and $G(s)H(s)$ has no poles within the enclosed $s$-plane contour, then the characteristic equation does not have any zeros in the region enclosed by the modified $s$-plane contour. All roots of the characteristic equation lie to the left of the line $s = -\sigma_0 + j\omega$. An example of a $G(-\sigma_0 + j\omega)H(-\sigma_0 + j\omega)$ locus, together with a $G(j\omega)H(j\omega)$ locus, is shown in Fig. 9-54 (b). The magnitude $1/\sigma_0$ is indicative of the time constant of the dominant closed-loop poles. If all roots lie outside the $s$-plane contour, all time constants of the closed-loop transfer function are less than $1/\sigma_0$. If the $s$-plane contour is chosen as shown in Fig. 9-55, then the test of encirclements of the $-1 + j0$ point reveals the existence or nonexistence of the roots of the characteristic equation of the closed-loop system within this $s$-plane contour. If the test reveals that no roots lie in the $s$-plane contour, then it is clear that all the closed-loop poles have damping ratios greater than $\zeta_x$ and time constants less than

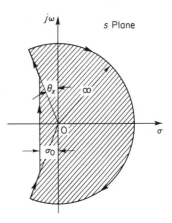

**Fig. 9-55.** Modified Nyquist path.

$1/\sigma_0$. Thus, by taking an appropriate $s$-plane contour, we can investigate time constants and damping ratios of closed-loop poles from given open-loop transfer functions.

### 9-7 RELATIVE STABILITY

In designing a control system, we require that the system be stable. Furthermore, it is necessary that the system have adequate relative stability.

In this section, we shall show that the Nyquist plot indicates not only whether or not a system is stable but also the degree of stability of a stable system. The Nyquist plot also gives information as to how stability may be improved, if this is necessary. (For details, see Chapter 10.)

In the following discussion, we shall assume that the systems considered have unity feedback. Note that it is always possible to reduce a system with feedback elements to a unity-feedback system, as shown in Fig. 9-56. Hence the extension of relative stability analysis for the unity-feedback system to nonunity-feedback systems is possible.

We shall also assume that, unless otherwise stated, the systems are minimum phase systems; that is, the open-loop transfer function $G(s)$ has neither poles nor zeros in the right-half $s$ plane.

**Relative stability analysis via conformal mapping.** One of the important problems in analyzing a control system is to find all closed-loop poles or at least those closest to the $j\omega$ axis (or the dominant pair of closed-loop poles.) If the open-loop frequency-response characteristics of a system are known, it may be possible to estimate the closed-loop poles closest to the $j\omega$ axis. The technique to be presented

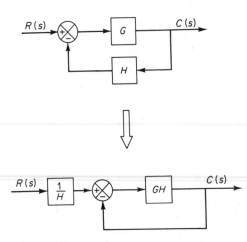

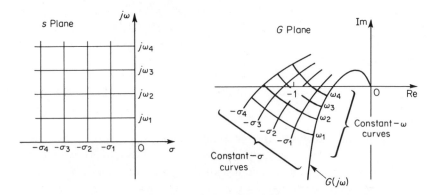

**Fig. 9-56.** Modification of a system with feedback elements to a unity-feedback system.

here is essentially graphical and is based on a conformal mapping of the $s$ plane into the $G(s)$ plane.

Consider the conformal mapping of constant-$\sigma$ lines (lines $s = \sigma + j\omega$ where $\sigma$ is constant and $\omega$ varies) and constant-$\omega$ lines (lines $s = \sigma + j\omega$, where $\omega$ is constant and $\sigma$ varies) in the $s$ plane into the $G(s)$ plane. The $\sigma = 0$ line (namely, the $j\omega$ axis) in the $s$ plane maps into the Nyquist plot in the $G(s)$ plane. The constant-$\sigma$ lines in the $s$ plane map into curves that are similar to the Nyquist plot and are in a sense parallel to the Nyquist plot, as shown in Fig. 9-57. The constant-$\omega$ lines in the $s$ plane map into curves, also shown in Fig. 9-57.

Although the shapes of constant-$\sigma$ and constant-$\omega$ loci in the $G(s)$ plane and the closeness of approach of the $G(j\omega)$ locus to the $-1 + j0$ point depend on a particular $G(s)$, the closeness of approach of the $G(j\omega)$ locus to the $-1 + j0$ point is an indication of the relative stability of a stable system. In general, we may expect that the closer the $G(j\omega)$ locus is to the $-1 + j0$ point, the larger the maximum overshoot is in the step transient response and the longer it takes to damp out.

Consider the two systems shown in Figs. 9-58 (a) and (b). (In Fig. 9-58, the crosses indicate closed-loop poles.) System (a) is obviously more stable than system (b) because the closed-loop poles of system (a) are located far to the left of those of system (b). Figures 9-59 (a) and (b) show the conformal mapping of $s$-

**Fig. 9-57.** Conformal mapping of $s$-plane grids into the $G(s)$ plane.

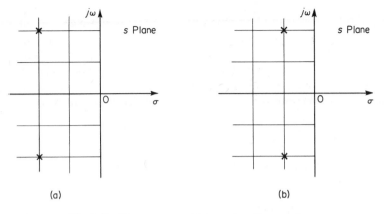

**Fig. 9-58.** Two systems with two closed-loop poles.

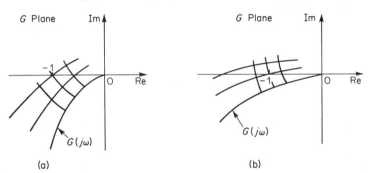

**Fig. 9-59.** Conformal mappings of *s*-plane grids for the systems shown in Fig. 9-58 into the *G(s)* plane.

plane grids into the *G(s)* plane. The closer the closed-loop poles are located to the *jω* axis, the closer the *G(jω)* locus is to the −1 + *j*0 point.

Suppose that the system possesses at least one pair of complex-conjugate closed-loop poles. If the −1 + *j*0 point is found at the intersection of a constant-*σ* curve and a constant-*ω* curve in the *G(s)* plane, then those particular values of *σ* and *ω*, which we define $\sigma_c$ and $\omega_c$, respectively, characterize the closed-loop pole closest to the *jω* axis in the upper-half *s* plane. (Note that $\sigma_c$ represents the exponential decay and $\omega_c$ represents the damped natural frequency of the step transient-response term due to the pair of the closed-loop poles closest to the *jω* axis.) If the analytical form of *G(s)* is given, the conformal mapping of the lines $s = j\omega_1$, $s = j\omega_2$, and $s = -\sigma_1$ into the *G(s)* plane can be made without difficulty. Then probable values of $\sigma_c$

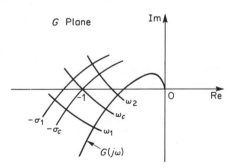

**Fig. 9-60.** Estimation of $\sigma_c$ and $\omega_c$.

and $\omega_c$ may be estimated from the plot, as shown in Fig. 9-60. Thus, the pair of complex-conjugate closed-loop poles which lie closest to the *jω* axis can be determined graphically. It should be noted that all closed-loop poles are mapped into the $-1 + j0$ point in the $G(s)$ plane. Although the complex-conjugate closed-loop poles closest to the $j\omega$ axis can be found easily by the technique above, finding other closed-loop poles, if any, by this technique is practically impossible.

If the data on $G(j\omega)$ are experimental, then a curvilinear square near the $-1 + j0$ point can be constructed by extrapolation. Referring to Fig. 9-61, we can find the location of the dominant closed-loop poles in the $s$ plane, or the damping ratio $\zeta$ and the damped natural frequency $\omega_d$, by drawing the line $AB$ which connects the $-1 + j0$ point (point $A$) and point $B$, the nearest approach to the $-1 + j0$ point, and then constructing a curvilinear square $CDEF$, as shown in Fig. 9-61. This

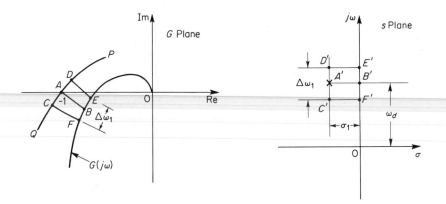

**Fig. 9-61.** Conformal mapping of a curvilinear square near the $-1 + j0$ point in the $G(s)$ plane into the $s$ plane.

curvilinear square $CDEF$ may be constructed by drawing the most-likely curve $PQ$ (where $PQ$ is the conformal mapping of a line parallel to the $j\omega$ axis in the $s$ plane) passing through the $-1 + j0$ point and "parallel" to the $G(j\omega)$ locus, and adjusting the points $C$, $D$, $E$, and $F$ such that $\widehat{FB} = \widehat{BE}$, $\widehat{CA} = \widehat{AD}$, and $\widehat{FE} + \widehat{CD} = \widehat{FC} + \widehat{ED}$. The corresponding $s$-plane contour $C'D'E'F'$, together with point $A'$, the closed-loop pole closest to the $j\omega$ axis, is shown in Fig. 9-61. The value of the frequency interval $\Delta\omega_1$ between points $E$ and $F$ is approximately equal to the value of $\sigma_1$ shown in Fig. 9-61. The frequency at point $B$ is approximately equal to the damped natural frequency $\omega_d$. The closed-loop poles closest to the $j\omega$ axis are then estimated as

$$s = -\sigma_1 \pm j\omega_d$$

Then the damping ratio $\zeta$ of these closed-loop poles can be obtained from

$$\frac{\zeta}{\sqrt{1 - \zeta^2}} = \frac{\sigma_1}{\omega_d} = \frac{\Delta\omega_1}{\omega_d}$$

It should be noted that the damped natural frequency $\omega_d$ of the step transient response actually is on the frequency contour that passes through the $-1 + j0$

point and is not necessarily the point of nearest approach to the $G(j\omega)$ locus. Therefore, the value $\omega_d$ obtained by the technique above is somewhat in error.

From the analysis above, we may conclude that it is possible to estimate the closed-loop poles closest to the $j\omega$ axis from the closeness of approach of the $G(j\omega)$ locus to the $-1 + j0$ point, the frequency at the point of nearest approach, and the frequency graduation near this point.

---

*Example 9-12.* Given the frequency-response plot of $G(j\omega)$ of a unity-feedback system, as shown in Fig. 9-62, find the closed-loop poles closest to the $j\omega$ axis.

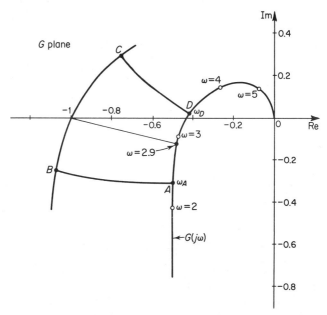

**Fig. 9-62.** Polar plot and a curvilinear square.

The line connecting the $-1 + j0$ point and the point of nearest approach of the $G(j\omega)$ locus to the $-1 + j0$ point is drawn first. Then the curvilinear square $ABCD$ is constructed. Since the frequency at the point of nearest approach is $\omega = 2.9$, the damped natural frequency is approximately 2.9, or $\omega_d = 2.9$. From the curvilinear square $ABCD$, it is found that

$$\Delta\omega = \omega_D - \omega_A = 3.4 - 2.4 = 1.0$$

The closed-loop poles closest to the $j\omega$ axis are then estimated as

$$s = -1 \pm j2.9$$

The $G(j\omega)$ locus shown in Fig. 9-62 is actually a plot of the following open-loop transfer function:

$$G(s) = \frac{5(s + 20)}{s(s + 4.59)(s^2 + 3.41s + 16.35)}$$

The exact closed-loop poles of this system are $s = -1 \pm j3$ and $s = -3 \pm j1$. The closed-loop poles closest to the $j\omega$ axis are $s = -1 \pm j3$. In this particular example, we see that the error involved is fairly small. In general, this error depends on a particular $G(j\omega)$ curve. The nearer the $G(j\omega)$ locus is to the $-1 + j0$ point, the smaller the error.

**Phase and gain margins.** Figure 9-63 shows the polar plots of $G(j\omega)$ for three different values of the open-loop gain $K$. For a large value of the gain $K$, the system is unstable. As the gain is decreased to a certain value, the $G(j\omega)$ locus passes through the $-1 + j0$ point. This means that with this gain value the system is on the verge of instability, and the system will exhibit sustained oscillations. For a small value of the gain $K$, the system is stable.

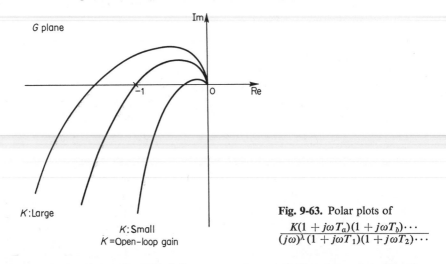

**Fig. 9-63.** Polar plots of
$$\frac{K(1 + j\omega T_a)(1 + j\omega T_b)\cdots}{(j\omega)^\lambda(1 + j\omega T_1)(1 + j\omega T_2)\cdots}$$

In general, the closer the $G(j\omega)$ locus comes to encircling the $-1 + j0$ point, the more oscillatory is the system response. The closeness of the $G(j\omega)$ locus to the $-1 + j0$ point can be used as a measure of the margin of stability. (This does not apply, however, to conditionally stable systems.) It is common practice to represent the closeness in terms of phase margin and gain margin.

*Phase margin:* The phase margin is that amount of additional phase lag at the gain crossover frequency required to bring the system to the verge of instability. The gain crossover frequency is the frequency at which $|G(j\omega)|$, the magnitude of the open-loop transfer function, is unity. The phase margin $\gamma$ is 180° plus the phase angle $\phi$ of the open-loop transfer function at the gain crossover frequency, or

$$\gamma = 180° + \phi$$

On the Nyquist diagram, a line may be drawn from the origin to the point at which the unit circle crosses the $G(j\omega)$ locus. The angle from the negative real axis to this line is the phase margin. The phase margin is positive for $\gamma > 0$ and negative for $\gamma < 0$. For a minimum phase system to be stable, the phase margin must be positive.

Figures 9-64 (a), (b), and (c) illustrate the phase margin of both a stable system and an unstable system in polar plots, logarithmic plots, and log-magnitude versus

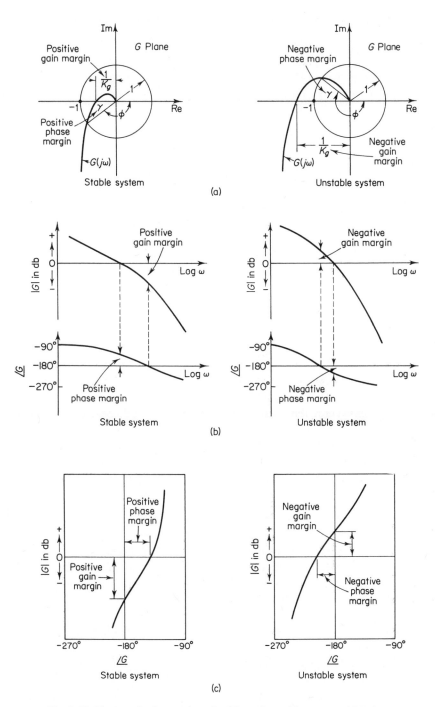

**Fig. 9-64.** Phase and gain margins of stable and unstable systems. (a) Polar plots; (b) logarithmic plots; (c) log-magnitude versus phase plots.

phase plots. In the logarithmic plots, the critical point in the complex plane corresponds to the 0 db and $-180°$ lines.

*Gain margin:* The gain margin is the reciprocal of the magnitude $|G(j\omega)|$ at the frequency where the phase angle is $-180°$. Defining the phase crossover frequency $\omega_1$ to be the frequency at which the phase angle of the open-loop transfer function equals $-180°$ gives the gain margin $K_g$:

$$K_g = \frac{1}{|G(j\omega_1)|}$$

In terms of decibels,

$$K_g \text{ db} = 20 \log K_g = -20 \log |G(j\omega_1)|$$

The gain margin expressed in decibels is positive if $K_g$ is greater than unity and negative if $K_g$ is smaller than unity. Thus, a positive gain margin (in decibels) means that the system is stable, and a negative gain margin (in decibels) means that the system is unstable. The gain margin is shown in Figs. 9-64 (a), (b), and (c).

For a stable minimum phase system, the gain margin indicates how much the gain can be increased before the system becomes unstable. For an unstable system, the gain margin is indicative of how much the gain must be decreased to make the system stable.

The gain margin of a first-order or second-order system is infinite since the polar plots for such systems do not cross the negative real axis. Thus, theoretically, first-order or second-order systems cannot be unstable. (Note, however, that so-called first-order or second-order systems are only approximations in the sense that small time lags are neglected in deriving the system equations. If these small lags are accounted for, the so-called first-order or second-order systems may become unstable.)

It is important to point out that for a nonminimum phase system the stability condition will not be satisfied unless the $G(j\omega)$ plot encircles the $-1 + j0$ point. Hence, a stable nonminimum phase system will have negative phase and gain margins.

**A few comments on phase and gain margins.** The phase and gain margins of a control system are a measure of the closeness of the polar plot to the $-1 + j0$ point. Therefore, these margins may be used as design criteria.

It should be noted that either the gain margin alone or the phase margin alone does not give a sufficient indication of the relative stability. Both should be given in the determination of relative stability.

For a minimum phase system, both the phase and gain margins must be positive for the system to be stable. Negative margins indicate instability.

Proper phase and gain margins ensure us against variations in the system components and are specified for definite values of frequency. The two values bound the behavior of the closed-loop system near the resonant frequency. For satisfactory performance, the phase margin should be between 30° and 60°, and the gain margin should be greater than 6 db. With these values, a minimum phase system has guaranteed stability, even if the open-loop gain and time constants of the components vary to a certain extent. Although the phase and gain margins give only rough estimates of the effective damping ratio of the closed-loop system, they do offer a convenient means for designing control systems or adjusting the gain constants of systems.

For minimum phase systems, the magnitude and phase characteristics of the open-loop transfer function are definitly related. The requirement that the phase margin be between 30° and 60° means that in a logarithmic plot the slope of the log-magnitude curve at the gain crossover frequency be more gradual than −40 db/decade. In most practical cases, a slope of −20 db/decade is desirable at the gain crossover frequency for stability. If it is −40 db/decade, the system could be either stable or unstable. (Even if the system is stable, however, the phase margin is small.) If the slope at the gain crossover frequency is −60 db/decade or steeper, the system is probably unstable.

---

*Example 9-13.* Obtain the phase and gain margins of the system shown in Fig. 9-65 for the two cases where $K = 10$ and $K = 100$.

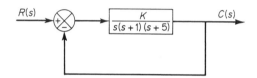

**Fig. 9-65.** Control system.

The phase and gain margins can easily be obtained from the logarithmic plot. A logarithmic plot of the given open-loop transfer function with $K = 10$ is shown in Fig. 9-66(a). The phase and gain margins for $K = 10$ are

$$\text{Phase margin} = 21°$$

$$\text{Gain margin} = 8 \text{ db}$$

Therefore, the system gain may be increased by 8 db before the instability occurs.

Increasing the gain from $K = 10$ to $K = 100$ shifts the 0-db axis down by 20 db, as shown in Fig. 9-66(b). The phase and gain margins are

$$\text{Phase margin} = -30°$$

$$\text{Gain margin} = -12 \text{ db}$$

Thus, the system is stable for $K = 10$ but unstable for $K = 100$.

Note that to obtain satisfactory performance, we muจt increase the phase margin to 30° ∼ 60°. This can be done by decreasing the gain $K$. Decreasing $K$ is not desirable, however, since a small value of $K$ will yield a large error for the ramp input. This suggests that reshaping of the open-loop frequency-response curve by adding compensation may be necessary. Such techniques are discussed in detail in Chapter 10.

**Correlation between step transient response and frequency response in second-order systems.** Exact mathematical relationships between the step transient response and frequency response can be derived for second-order systems. Consider the system shown in Fig. 9-67. The closed-loop transfer function is

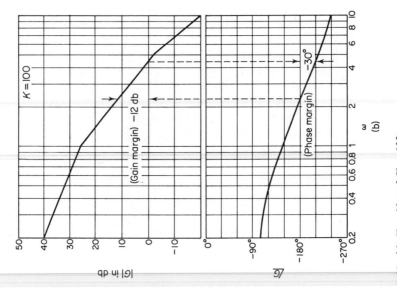

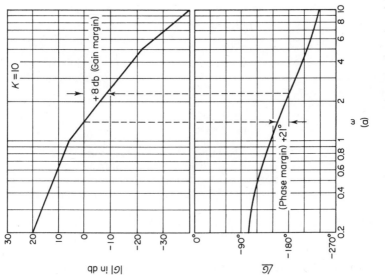

**Fig. 9-66.** Bode diagrams of the system shown in Fig. 9-65 with $K = 10$ and $K = 100$.

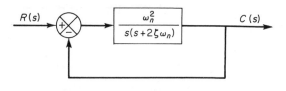

**Fig. 9-67.** Control system.

$$\frac{C(s)}{R(s)} = \frac{\omega_n^2}{s^2 + 2\zeta\omega_n s + \omega_n^2} \qquad (9\text{-}15)$$

where $\zeta$ and $\omega_n$ are the damping ratio and the undamped natural frequency, respectively. The closed-loop frequency response is

$$\frac{C(j\omega)}{R(j\omega)} = \frac{1}{\left(1 - \dfrac{\omega^2}{\omega_n^2}\right) + j2\zeta\dfrac{\omega}{\omega_n}} = Me^{j\alpha}$$

where

$$M = \frac{1}{\sqrt{\left(1 - \dfrac{\omega^2}{\omega_n^2}\right)^2 + \left(2\zeta\dfrac{\omega}{\omega_n}\right)^2}}, \qquad \alpha = -\tan^{-1}\frac{2\zeta\dfrac{\omega}{\omega_n}}{1 - \dfrac{\omega^2}{\omega_n^2}}$$

As given by Eq. (9-12), for $0 \leq \zeta \leq 0.707$ the maximum value of $M$ occurs at the resonant frequency $\omega_r$, where

$$\omega_r = \omega_n\sqrt{1 - 2\zeta^2} = \omega_n\sqrt{\cos 2\psi} \qquad (9\text{-}16)$$

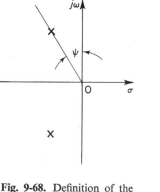

**Fig. 9-68.** Definition of the angle $\psi$.

The angle $\psi$ is defined in Fig. 9-68. At the resonant frequency, the value of $M$ is maximum and is given by Eq. (9-13), rewritten thus:

$$M_r = \frac{1}{2\zeta\sqrt{1 - \zeta^2}} = \frac{1}{\sin 2\psi} \qquad (9\text{-}17)$$

Remember that $\omega_r$ is real only if $\zeta < 0.707$. Thus, there is no closed-loop resonance if $\zeta > 0.707$. [The value of $M_r$ is unity for $\zeta > 0.707$. See Eq. (9-14).] Since the values of $M_r$ and $\omega_r$ can be easily measured in a physical system, they are quite useful for checking agreement between theoretical and experimental analyses.

For a unit-step input, the output of the system shown in Fig. 9-67 is given by Eq. (6-19), or

$$c(t) = 1 - e^{-\zeta\omega_n t}\left(\cos\omega_d t + \frac{\zeta}{\sqrt{1 - \zeta^2}}\sin\omega_d t\right) \qquad (t \geq 0)$$

where

$$\omega_d = \omega_n\sqrt{1 - \zeta^2} = \omega_n\cos\psi \qquad (9\text{-}18)$$

On the other hand, the maximum overshoot $M_p$ for the unit-step response is given by Eq. (6-28), or

$$M_p = e^{-(\zeta/\sqrt{1 - \zeta^2})\pi} \qquad (9\text{-}19)$$

This maximum overshoot occurs in the transient response which has the damped natural frequency $\omega_d = \omega_n\sqrt{1 - \zeta^2}$. The maximum overshoot becomes excessive for values of $\zeta < 0.4$.

Since the second-order system shown in Fig. 9-67 has the following open-loop transfer function

$$G(s) = \frac{\omega_n^2}{s(s + 2\zeta\omega_n)}$$

for sinusoidal operation, the magnitude of $G(j\omega)$ becomes unity when

$$\omega = \omega_n\sqrt{\sqrt{1 + 4\zeta^4} - 2\zeta^2}$$

At this frequency, the phase angle of $G(j\omega)$ is

$$\underline{/G(j\omega)} = \underline{\bigg/\frac{1}{j\omega}} + \underline{\bigg/\frac{1}{j\omega + 2\zeta\omega_n}}$$

$$= -90° - \tan^{-1}\frac{\sqrt{\sqrt{1 + 4\zeta^4} - 2\zeta^2}}{2\zeta}$$

Thus, the phase margin $\gamma$ is

$$\gamma = 180° + \underline{/G(j\omega)}$$

$$= 90° - \tan^{-1}\frac{\sqrt{\sqrt{1 + 4\zeta^4} - 2\zeta^2}}{2\zeta}$$

$$= \tan^{-1}\frac{2\zeta}{\sqrt{\sqrt{1 + 4\zeta^4} - 2\zeta^2}} \tag{9-20}$$

Equation (9-20) gives the relationship between the damping ratio $\zeta$ and the phase margin $\gamma$.

In the following, we shall summarize the correlation between the step transient response and frequency response of the second-order system given by Eq. (9-15):

1. The phase margin and the damping ratio are directly related. Figure 9-69 shows a plot of the phase margin $\gamma$ as a function of the damping ratio $\zeta$. Note that a phase margin of 60° corresponds to a damping ratio of 0.6.

2. Referring to Eqs. (9-16) and (9-18) we see that the values of $\omega_r$ and $\omega_d$ are almost the same for small values of $\zeta$. Thus, for small values of $\zeta$, the value of $\omega_r$ is indicative of the speed of the transient response of the system.

3. From Eqs. (9-17) and (9-19), we note that the smaller the value of $\zeta$ is, the larger the values of $M_r$ and $M_p$ are. The correlation between $M_r$ and $M_p$ as a function of $\zeta$ is shown in Fig. 9-70. A close relationship between $M_r$ and $M_p$ can be seen for $\zeta > 0.4$. For very small values of $\zeta$, $M_r$ becomes very large ($M_r \gg 1$), while the value of $M_p$ does not exceed 1.

**Correlation between step transient response and frequency response in higher-order systems.** The design of control systems is very often carried out on the basis of frequency response. The main reason for this is the relative simplicity of this approach as compared to others. Since in many applications it is the transient

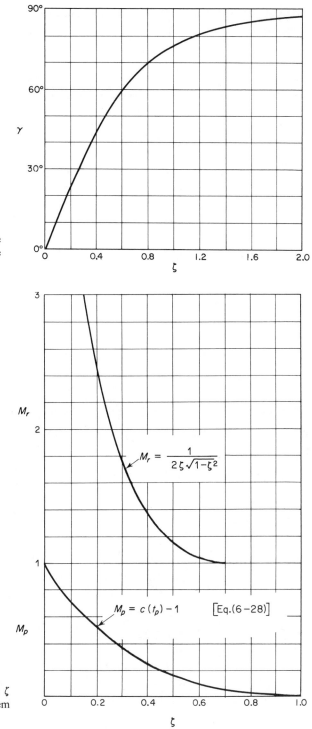

**Fig. 9-69.** Curve $\gamma$ (phase margin) versus $\zeta$ for the system shown in Fig. 9-67.

$$M_r = \frac{1}{2\zeta\sqrt{1-\zeta^2}}$$

$$M_p = c(t_p) - 1 \qquad [\text{Eq.}(6-28)]$$

**Fig. 9-70.** Curves $M_r$ versus $\zeta$ and $M_p$ versus $\zeta$ for the system shown in Fig. 9-67.

response of the system to aperiodic inputs rather than the steady-state response to sinusoidal inputs that is of primary concern, the question of correlation between transient response and frequency response arises.

For the second-order system shown in Fig. 9-67, mathematical relationships correlating the step transient response and frequency response can be obtained easily. The time response of a second-order system can be exactly predicted from a knowledge of the $M_r$ and $\omega_r$ of its closed-loop frequency response.

For higher-order systems, the correlation is more complex, and transient response may not be predicted easily from frequency response because additional poles may change the correlation between the step transient response and frequency response existing for a second-order system. Mathematical techniques for obtaining the exact correlation are available, but they are very laborious and of little practical value.

The applicability of the transient-response–frequency-response correlation existing for the second-order system shown in Fig. 9-67 to higher-order systems depends upon the presence of a dominant pair of complex-conjugate closed-loop poles in the latter systems. Clearly, if the frequency response of a higher-order system is dominated by a pair of complex-conjugate closed-loop poles, the transient-response–frequency-response correlation existing for the second-order system can be extended to the higher-order system.

For linear time-invariant higher-order systems having a dominant pair of complex-conjugate closed-loop poles, the following relationships generally exist between the step transient response and frequency response:

1. The value of $M_r$ is indicative of the relative stability. Satisfactory transient performance is usually obtained if the value of $M_r$ is in the range $1.0 < M_r < 1.4$ (0 db $< M_r < 3$ db), which corresponds to an effective damping ratio of $0.4 < \zeta < 0.7$. For values of $M_r$ greater than 1.5, the step transient response may exhibit several overshoots. (Note that in general a large value of $M_r$ corresponds to a large overshoot in the step transient response. If the system is subjected to noise signals whose frequencies are near the resonant frequency $\omega_r$, the noise will be amplified in the output and will cause serious trouble.)

2. The magnitude of the resonant frequency $\omega_r$ is indicative of the speed of the transient response. The larger the value of $\omega_r$, the faster the time response is. In other words, the rise time varies inversely with $\omega_r$. In terms of the open-loop frequency response, the damped natural frequency in the transient response is somewhere between the gain crossover frequency and phase crossover frequency.

3. The resonant frequency $\omega_r$ and the damped natural frequency $\omega_d$ for step transient response are very close to each other for lightly damped systems.

The three relationships just listed are useful for correlating the step transient response with the frequency response of higher-order systems, provided they can be approximated by a second-order system or a pair of complex-conjugate closed-loop poles. If a higher-order system satisfies this condition, a set of time-domain specifications may be translated into frequency-domain specifications. This simplifies greatly the design work or compensation work of higher-order systems.

In addition to the phase margin, gain margin, resonant peak $M_r$, and resonant frequency $\omega_r$, there are other frequency-domain quantities commonly used in performance specifications. They are the cutoff frequency, bandwidth, and the cutoff rate. These will be defined in what follows.

**Cutoff frequency and bandwidth.**
Referring to Fig. 9-71, the frequency $\omega_c$ at which the magnitude of the closed-loop frequency response is 3 db below its zero-frequency value is called the cutoff frequency. Thus

$$\left|\frac{C(j\omega)}{R(j\omega)}\right| < \left|\frac{C(j0)}{R(j0)}\right| - 3 \text{ db}$$

$$(\omega > \omega_c)$$

For systems where $|C(j0)/R(j0)| = 0$ db,

$$\left|\frac{C(j\omega)}{R(j\omega)}\right| < -3 \text{ db}$$

$$(\omega > \omega_c)$$

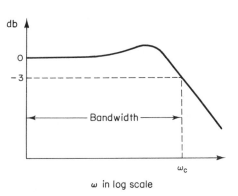

**Fig. 9-71.** Logarithmic plot showing cutoff frequency $\omega_c$ and bandwidth.

(Some authors use the −6-db point rather than the −3-db point. In this book, the −3-db point is used for defining the cutoff frequency.) The closed-loop system filters out the signal components whose frequencies are greater than the cutoff frequency and transmits those signal components with frequencies lower than the cutoff frequency.

The frequency range $0 \leq \omega \leq \omega_c$ in which the magnitude of the closed loop does not drop −3 decibels is called the bandwidth of the system. The bandwidth gives an indication of the speed of response of a control system.

The specification of the bandwidth may be determined by the following factors:

1. The ability to reproduce the input signal. (A large bandwidth corresponds to a small rise time, or fast response. Roughly speaking, we can say that the bandwidth is proportional to the speed of response.)

2. The necessary filtering characteristics for high-frequency noise.

In order that the system follow arbitrary inputs accurately, it is necessary that the system have a large bandwidth. From the viewpoint of noise, however, the bandwidth should not be too large. Thus, there are conflicting requirements on the bandwidth, and a compromise is usually necessary for good design. Note that a system with large bandwidth requires high performance components. (The cost of components usually increases with the bandwidth.)

**Cutoff rate.** The cutoff rate is the slope of the log-magnitude curve near the cutoff frequency. The cutoff rate indicates the ability of a system to distinguish the signal from noise.

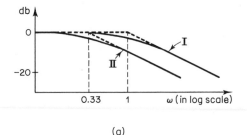

(a)

*Example 9-14.* Consider the following two systems:

System I:

$$\frac{C(s)}{R(s)} = \frac{1}{s+1}$$

System II:

$$\frac{C(s)}{R(s)} = \frac{1}{3s+1}$$

Compare the bandwidths of these two systems. Show that the system with the larger bandwidth has a faster speed of response and can follow the input much better than the one with a smaller bandwidth.

Figure 9-72(a) shows the closed-loop frequency-response curves for the two systems. (Asymptotic curves are shown by dotted lines.) We find that the bandwidth of system I is $0 \leq \omega \leq 1$ rad/sec and that of system II is $0 \leq \omega \leq 0.33$ rad/sec. Figures 9-72(b) and (c) show, respectively, the step-response and ramp-response curves for the two systems. Clearly, system I, whose bandwidth is three times wider than that of system II, has a faster speed of response and can follow the input much better.

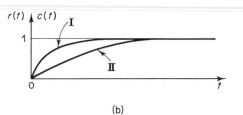

(b)

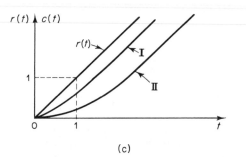

(c)

**Fig. 9-72.** Comparison of dynamic characteristics of the two systems considered in Example 9-14.

## 9-8 CLOSED-LOOP FREQUENCY RESPONSE

**Closed-loop frequency response of unity-feedback systems.** For a stable closed-loop system, the frequency response can be obtained easily from that of the open loop. Consider the unity-feedback system shown in Fig. 9-73 (a). The closed-loop transfer function is

$$\frac{C(s)}{R(s)} = \frac{G(s)}{1 + G(s)}$$

In the Nyquist or polar plot shown in Fig. 9-73 (b), the vector $\overrightarrow{OA}$ represents $G(j\omega_1)$, where $\omega_1$ is the frequency at point $A$. The length of the vector $\overrightarrow{OA}$ is $|G(j\omega_1)|$

and the angle of the vector $\overrightarrow{OA}$ is $\underline{/G(j\omega_1)}$.
The vector $\overrightarrow{PA}$, the vector from the
$-1+j0$ point to the Nyquist locus,
represents $1 + G(j\omega_1)$. Therefore, the
ratio of $\overrightarrow{OA}$ to $\overrightarrow{PA}$ represents the closed-
loop frequency response, or

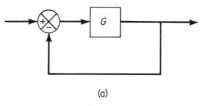

(a)

$$\frac{\overrightarrow{OA}}{\overrightarrow{PA}} = \frac{G(j\omega_1)}{1 + G(j\omega_1)} = \frac{C(j\omega_1)}{R(j\omega_1)}$$

The magnitude of the closed-loop transfer
function at $\omega = \omega_1$ is the ratio of the
magnitudes of $\overrightarrow{OA}$ to $\overrightarrow{PA}$. The phase angle
of the closed-loop transfer function at $\omega = \omega_1$ is the angle formed by the vectors
$\overrightarrow{OA}$ and $\overrightarrow{PA}$, namely $\phi - \theta$, shown in Fig.
9-73 (b). By measuring the magnitude and
phase angle at different frequency points,
the closed-loop frequency-response curve
can be obtained.

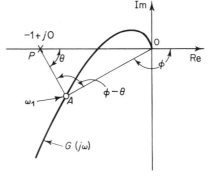

(b)

Let us define the magnitude of the
closed-loop frequency response as $M$ and
the phase angle as $\alpha$, or

**Fig. 9-73.** (a) Unity-feedback system;
(b) Determination of closed-loop frequency response from open-loop frequency response.

$$\frac{C(j\omega)}{R(j\omega)} = Me^{j\alpha}$$

In the following, we shall find the constant magnitude loci and constant phase-
angle loci. Such loci are convenient in determining the closed-loop frequency
response from the Nyquist plot.

**Constant magnitude loci ($M$ circles).** In order to obtain the constant magnitude loci, let us first note that $G(j\omega)$ is a complex quantity and can be written as
follows:

$$G(j\omega) = X + jY$$

where $X$ and $Y$ are real quantities. Then $M$ is given by

$$M = \frac{|X + jY|}{|1 + X + jY|}$$

and $M^2$ is

$$M^2 = \frac{X^2 + Y^2}{(1 + X)^2 + Y^2}$$

Hence

$$X^2(1 - M^2) - 2M^2X - M^2 + (1 - M^2)Y^2 = 0 \qquad (9\text{-}21)$$

If $M = 1$, then from Eq. (9-21) we obtain $X = -\frac{1}{2}$. This is the equation of a straight line parallel to the $Y$ axis and passing through the point $(-\frac{1}{2}, 0)$.

If $M \neq 1$, Eq. (9-21) can be written

$$X^2 + \frac{2M^2}{M^2 - 1} X + \frac{M^2}{M^2 - 1} + Y^2 = 0$$

If the term $M^2/(M^2 - 1)^2$ is added to both sides of this last equation, we obtain

$$\left(X + \frac{M^2}{M^2 - 1}\right)^2 + Y^2 = \frac{M^2}{(M^2 - 1)^2} \tag{9-22}$$

Equation (9-22) is the equation of a circle with center at $X = -M^2/(M^2 - 1)$, $Y = 0$ and with radius $|M/(M^2 - 1)|$.

The constant M loci on the $G(s)$ plane are thus a family of circles. The center and radius of the circle for a given value of $M$ can be easily calculated. For example, for $M = 1.3$, the center is at $(-2.45, 0)$ and the radius is 1.88. A family of constant $M$ circles is shown in Fig. 9-74. It is seen that as $M$ becomes larger and larger

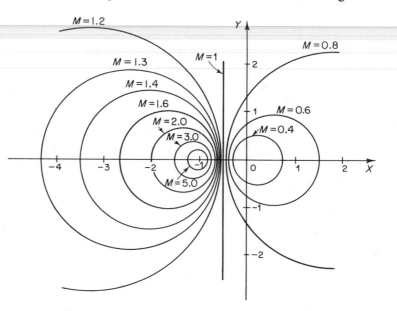

**Fig. 9-74.** A family of constant $M$ circles.

compared with 1, the $M$ circles become smaller and smaller and converge to the $-1 + j0$ point. For $M > 1$, the centers of the $M$ circles lie to the left of the $-1 + j0$ point. Similarly, as $M$ becomes smaller and smaller compared with 1, the $M$ circle becomes smaller and smaller and converges to the origin. For $M < 1$, the centers of the $M$ circles lie to the right of the origin. $M = 1$ corresponds to the locus of points equidistant from the origin and from the $-1 + j0$ point. As stated earlier, it is a straight line passing through the point $(-\frac{1}{2}, 0)$ and parallel to the imaginary axis. (The constant $M$ circles corresponding to $M > 1$ lie to the left of the $M = 1$

line and those corresponding to $M < 1$ lie to the right of the $M = 1$ line.) The $M$ circles are symmetrical with respect to the straight line corresponding to $M = 1$ and with respect to the real axis.

**Constant phase-angle loci** ($N$ **circles**). We shall first obtain the phase angle $\alpha$ in terms of $X$ and $Y$. Since

$$\underline{/e^{j\alpha}} = \left| \frac{X + jY}{1 + X + jY} \right|$$

the phase angle $\alpha$ is

$$\alpha = \tan^{-1}\left(\frac{Y}{X}\right) - \tan^{-1}\left(\frac{Y}{1 + X}\right)$$

If we define

$$\tan \alpha = N$$

then

$$N = \tan\left[\tan^{-1}\left(\frac{Y}{X}\right) - \tan^{-1}\left(\frac{Y}{1 + X}\right)\right]$$

Since

$$\tan(A - B) = \frac{\tan A - \tan B}{1 + \tan A \tan B}$$

we obtain

$$N = \frac{\dfrac{Y}{X} - \dfrac{Y}{1 + X}}{1 + \dfrac{Y}{X}\left(\dfrac{Y}{1 + X}\right)}$$

$$= \frac{Y}{X^2 + X + Y^2}$$

or

$$X^2 + X + Y^2 - \frac{1}{N}Y = 0$$

The addition of $(\frac{1}{4}) + 1/(2N)^2$ to both sides of this last equation yields

$$\left(X + \frac{1}{2}\right)^2 + \left(Y - \frac{1}{2N}\right)^2 = \frac{1}{4} + \left(\frac{1}{2N}\right)^2 \tag{9-23}$$

This is an equation of a circle with center at $X = -\frac{1}{2}$, $Y = 1/(2N)$ and with radius $\sqrt{(\frac{1}{4}) + 1/(2N)^2}$. For example, if $\alpha = 30°$, then $N = \tan \alpha = 0.577$, and the center and the radius of the circle corresponding to $\alpha = 30°$ are found to be $(-0.5, 0.866)$ and unity, respectively. Since Eq. (9-23) is satisfied when $X = Y = 0$ and $X = -1$, $Y = 0$ regardless of the value of $N$, each circle passes through the origin and the $-1 + j0$ point. The constant $\alpha$ loci can be drawn easily once the value of $N$ is given. A family of constant $N$ circles are shown in Fig. 9-75 with $\alpha$ as a parameter.

It should be noted that the constant $N$ locus for a given value of $\alpha$ is actually not the entire circle but only an arc. In other words, the $\alpha = 30°$ and $\alpha = -150°$ arcs are parts of the same circle. This is so because the tangent of an angle remains the same if $\pm 180°$ (or multiplies thereof) is added to the angle.

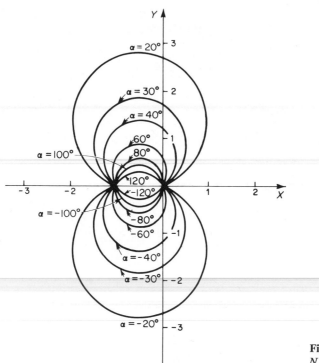

**Fig. 9-75.** A family of constant $N$ circles.

The use of the $M$ and $N$ circles enables us to find the entire closed-loop frequency response from the open-loop frequency response $G(j\omega)$ without calculating the magnitude and phase of the closed-loop transfer function at each frequency. The intersections of the $G(j\omega)$ locus and the $M$ circles and $N$ circles give the values of $M$ and $N$ at frequency points on the $G(j\omega)$ locus.

Since the $N$ circles are multivalued in the sense that the circle for $\alpha = \alpha_1$ and that for $\alpha = \alpha_1 \pm 180°n$ ($n = 1, 2, \ldots$) are the same, in using the $N$ circles for the determination of the phase angle of closed-loop systems, we must interpret the proper value of $\alpha$. To avoid any error, start at zero frequency, which corresponds to $\alpha = 0°$, and proceed to higher frequencies. The phase-angle curve must be continuous.

Figure 9-76 (a) shows the $G(j(\omega)$ locus superimposed on a family of $M$ circles. Figure 9-76 (b) shows the $G(j\omega)$ locus superimposed on a family of $N$ circles. From these plots, it is possible to obtain the closed-loop frequency response by inspection. It is seen that the $M = 1.1$ circle intersects the $G(j\omega)$ locus at frequency point $\omega = \omega_1$. This means that at this frequency the magnitude of the closed-loop transfer function is 1.1. In Fig. 9-76 (a), the $M = 2$ circle is just tangent to the $G(j\omega)$ locus. Thus, there is only one point on the $G(j\omega)$ locus for which $|C(j\omega)/R(j\omega)|$ is equal to 2. Figure 9-76 (c) shows the closed-loop frequency-response curve for the system. The upper curve is the $M$ versus frequency $\omega$ curve and the lower curve is the phase angle $\alpha$ versus frequency $\omega$ curve.

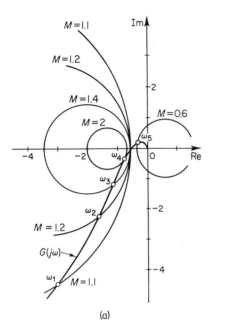

(a)

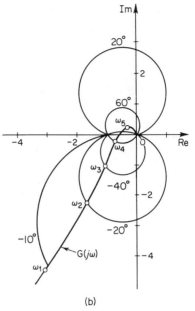

(b)

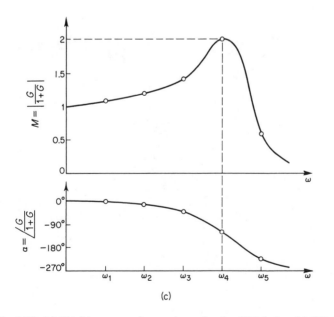

(c)

**Fig. 9-76.** (a) $G(j\omega)$ locus superimposed on a family of $M$ circles; (b) $G(j\omega)$ locus superimposed on a family of $N$ circles; (c) closed-loop frequency-response curves.

The resonant peak value is the value of $M$ corresponding to the $M$ circle of smallest radius that is tangent to the $G(j\omega)$ locus. Thus, in the Nyquist diagram, the resonant peak value $M_r$ and the resonant frequency $\omega_r$ can be found from the $M$-circle tangency to the $G(j\omega)$ locus. (In the present example, $M_r = 2$ and $\omega_r = \omega_4$.)

**Nichols chart.** In dealing with design problems, we find it convenient to construct the $M$ and $N$ loci in the log-magnitude versus phase plane. The chart consisting of the $M$ and $N$ loci in the log-magnitude versus phase diagram is called the Nichols chart. This chart is shown in Fig. 9-77 for phase angles between 0 and $-240°$. The Nichols chart is symmetric about the $-180°$ axis. The $M$ and $N$ loci repeat for every $360°$, and there is symmetry at every $180°$ interval. The $M$ loci are centered about the critical point (0 db, $-180°$). The Nichols chart is quite useful for determining the closed-loop frequency response from that of the open loop. If the open-loop frequency-response curve is superimposed on the Nichols chart, the intersections of the open-loop frequency-response curve $G(j\omega)$ and the $M$ and $N$ loci give the values of the magnitude $M$ and phase angle $\alpha$ of the closed-loop

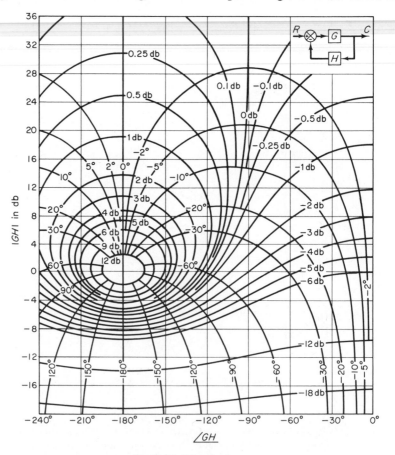

**Fig. 9-77.** Nichols chart.

frequency response at each frequency point. If the $G(j\omega)$ locus does not intersect the $M = M_r$ locus but is tangent to it, then the resonant peak value of $M$ of the closed-loop frequency response is given by $M_r$. The resonant frequency is given by the frequency at the point of tangency.

As an example, consider the unity-feedback system with the following open-loop transfer function:

$$G(s) = \frac{K}{s(s + 1)(0.5s + 1)}, \qquad K = 1$$

In order to find the closed-loop frequency response by use of the Nichols chart, the $G(j\omega)$ locus is constructed in the log-magnitude versus phase plane from the Bode diagram. The use of the Bode diagram eliminates the lengthy numerical calculation of $G(j\omega)$. Figure 9-78 (a) shows the $G(j\omega)$ locus together with the $M$ and $N$ loci. The closed-loop frequency-response curves may be constructed by reading the magnitudes and phase angles at various frequency points on the $G(j\omega)$ locus from the $M$ and $N$ loci, as shown in Fig. 9-78 (b). Since the $G(j\omega)$ locus is tangent to the $M = 5$ db locus, the peak value of the closed-loop frequency response is $M_r = 5$ db, and the resonant frequency is 0.8 rad/sec.

The bandwidth of the closed-loop system can easily be found from the $G(j\omega)$ locus in the Nichols diagram. The frequency at the intersection of the $G(j\omega)$ locus and the $M = -3$ db locus gives the bandwidth. The gain and phase margins can be read directly from the Nichols diagram.

If the open-loop gain $K$ is varied, the shape of the $G(j\omega)$ locus in the log-magnitude versus phase diagram remains the same, but it is shifted up (for increasing $K$) or down (for decreasing $K$) along the vertical axis. Therefore, the $G(j\omega)$ locus intersects the $M$ and $N$ loci differently, resulting in a different closed-loop frequency-response curve. For a small value of the gain $K$, the $G(j\omega)$ locus will not be tangent to any of the $M$ loci, which means that there is no resonance in the closed-loop frequency response.

**Closed-loop frequency response for nonunity-feedback systems.** If the closed-loop system involves a nonunity-feedback transfer function, then the closed-loop transfer function can be written

$$\frac{C(s)}{R(s)} = \frac{G(s)}{1 + G(s)H(s)}$$

where $G(s)$ is the feedforward transfer function and $H(s)$ is the feedback transfer function. Then $C(j\omega)/R(j\omega)$ can be written

$$\frac{C(j\omega)}{R(j\omega)} = \frac{1}{H(j\omega)} \frac{G(j\omega)H(j\omega)}{1 + G(j\omega)H(j\omega)}$$

The magnitude and phase angle of

$$\frac{G_1(j\omega)}{1 + G_1(j\omega)}$$

where $G_1(j\omega) = G(j\omega)H(j\omega)$, may be obtained easily by plotting the $G_1(j\omega)$ locus on the Nichols chart and reading the values of $M$ and $N$ at various frequency points.

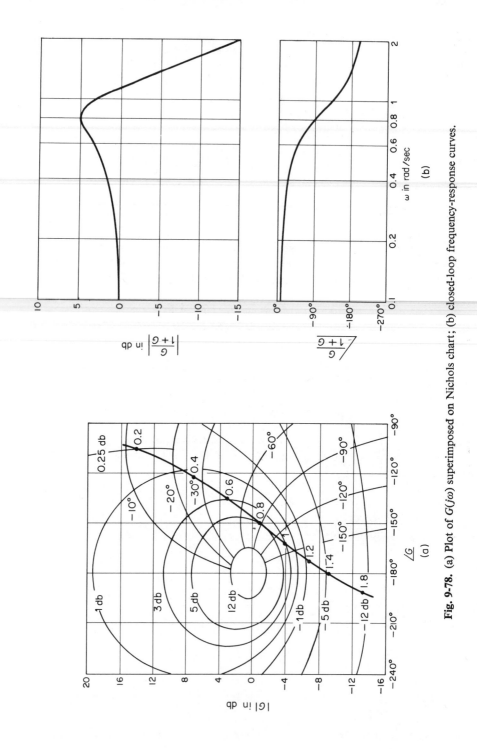

**Fig. 9-78.** (a) Plot of $G(j\omega)$ superimposed on Nichols chart; (b) closed-loop frequency-response curves.

The closed-loop frequency response $C(j\omega)/R(j\omega)$ may then be obtained by multiplying $G_1(j\omega)/[1 + G_1(j\omega)]$ by $1/H(j\omega)$. This multiplication can be made without difficulty if we draw the Bode diagrams for $G_1(j\omega)/[1 + G_1(j\omega)]$ and $H(j\omega)$ and then graphically subtract the magnitude of $H(j\omega)$ from that of $G_1(j\omega)/[1 + G_1(j\omega)]$ and also graphically subtract the phase angle of $H(j\omega)$ from that of $G_1(j\omega)/[1 + G_1(j\omega)]$. Then the resulting log-magnitude curve and phase-angle curve give the closed-loop frequency response $C(j\omega)/R(j\omega)$.

In order to obtain acceptable values of $M_r$, $\omega_r$, and $\omega_c$ for $|C(j\omega)/R(j\omega)|$, a trial-and-error process may be necessary. In each trial, the $G_1(j\omega)$ locus is varied in shape. Then the Bode diagrams for $G_1(j\omega)/[1 + G_1(j\omega)]$ and $H(j\omega)$ are drawn and the closed-loop frequency response $C(j\omega)/R(j\omega)$ is obtained. The values of $M_r$, $\omega_r$, and $\omega_c$ are checked until they are acceptable.

**Gain adjustments.** The concept of $M$ circles will now be applied to the design of control systems. In obtaining suitable performance, the adjustment of gain is usually the first consideration. The adjustment may be based on a desirable value for the resonant peak.

In the following, we shall demonstrate a method for determining the gain $K$ so that the system will have some maximum value $M_r$, not exceeded over the entire frequency range.

Referring to Fig. 9-79, we see that the tangent line drawn from the origin to the desired $M_r$ circle has an angle of $\psi$, as shown, if $M_r$ is greater than unity. The value of $\sin \psi$ is

$$\sin \psi = \frac{\left|\dfrac{M_r}{M_r^2 - 1}\right|}{\dfrac{M_r^2}{M_r^2 - 1}} = \frac{1}{M_r}$$

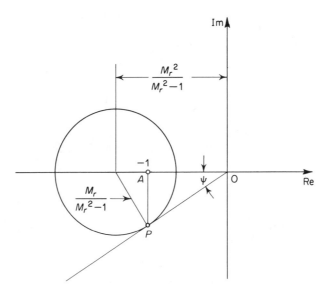

**Fig. 9-79.** $M$ circle.

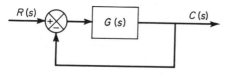

**Fig. 9-80.** Control system.

It can easily be proved that the line drawn from point $P$, perpendicular to the negative real axis, intersects this axis at the $-1 + j0$ point.

Consider the system shown in Fig. 9-80. The procedure for determining the gain $K$ so that $G(j\omega) = KG_1(j\omega)$ will have a desired value of $M_r$ (where $M_r > 1$) can be summarized as follows:

1. Draw the polar plot of the normalized open-loop transfer function $G_1(j\omega) = G(j\omega)/K$.

2. Draw from the origin the line which makes an angle of $\psi = \sin^{-1}(1/M_r)$ with the negative real axis.

3. Fit a circle with center on the negative real axis tangent to both the $G_1(j\omega)$ locus and the line $PO$.

4. Draw a perpendicular line to the negative real axis from point $P$, the point of tangency of this circle with the line $PO$. The perpendicular line $PA$ intersects the negative real axis at point $A$.

5. In order that the circle just drawn correspond to the desired $M_r$ circle, point $A$ should be the $-1 + j0$ point.

6. The desired value of the gain $K$ is that value which changes the scale so that point $A$ becomes the $-1 + j0$ point. Thus, $K = 1/\overline{OA}$.

Note that the resonant frequency $\omega_r$ is the frequency of the point at which the circle is tangent to the $G_1(j\omega)$ locus. The present procedure may not yield a satisfactory value for $\omega_r$. If this is the case, the system must be compensated in order to increase the value of $\omega_r$ without changing the value of $M_r$. (For the compensation of control systems, see Chapter 10.)

Note also that if the system has nonunity feedback, then the method requires some cut-and-try steps.

---

*Example 9-15.* Consider the unity-feedback control system whose open-loop transfer function is

$$G(j\omega) = \frac{K}{j\omega(1 + j\omega)}$$

Determine the value of the gain $K$ so that $M_r = 1.4$.

The first step in the determination of the gain $K$ is to sketch the polar plot of

$$\frac{G(j\omega)}{K} = \frac{1}{j\omega(1 + j\omega)}$$

as shown in Fig. 9-81. The value of $\psi$ corresponding to $M_r = 1.4$ is obtained from

$$\psi = \sin^{-1}\frac{1}{M_r} = \sin^{-1}\frac{1}{1.4} = 45.6°$$

The next step is to draw the line $OP$ which makes an angle $\psi = 45.6°$ with the negative real axis. Then draw the circle which is tangent to both the $G(j\omega)/K$ locus and the line $OP$. The perpendicular line drawn from the point $P$ interesects the negative real axis at $(-0.63, 0)$. Then the gain $K$ of the system is determined as follows:

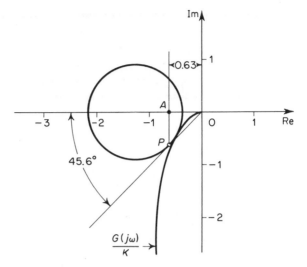

**Fig. 9-81.** Determination of the gain $K$ using an $M$ circle.

$$K = \frac{1}{0.63} = 1.58$$

It should be noted that such determination of the gain can also be done easily on the log-magnitude versus phase plot. In what follows, we shall demonstrate how the log-magnitude versus phase diagram can be used to determine the gain $K$ so that the system will have a desired value of $M_r$.

Figure 9-82 shows the $M_r = 1.4$ locus and the $G(j\omega)/K$ locus. Changing the gain has no effect on the phase angle but merely moves the curve vertically up for $K > 1$ and down for $K < 1$. In Fig. 9-82, the $G(j\omega)/K$ locus must be raised by 4 db in order that it be tangent to the desired $M_r$ locus and that the entire $G(j\omega)/K$ locus be outside the $M_r = 1.4$ locus. The amount of vertical shift of the $G(j\omega)/K$ locus determines the gain necessary to yield the desired value of $M_r$. Thus, by solving

$$20 \log K = 4$$

we obtain

$$K = 1.58$$

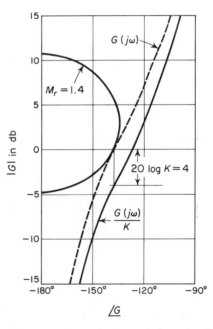

**Fig. 9-82.** Determination of the gain $K$ using the Nichols chart.

## 9-9 EXPERIMENTAL DETERMINATION OF TRANSFER FUNCTIONS

The first step in the analysis and design of a control system is to derive a mathematical model of the plant under consideration. Obtaining a model analytically may be quite difficult. We may have to obtain it by means of experimental analysis. The importance of the frequency-response methods is that the transfer function of the plant, or any other component of a system, may be determined by simple frequency-response measurements.

If the amplitude ratio and phase shift have been measured at a sufficient number of frequencies within the frequency range of interest, they may be plotted on the Bode diagram. Then the transfer function can be determined by asymptotic approximations. We build up asymptotic log-magnitude curves consisting of several segments. With some trial-and-error juggling of the corner frequencies, it is usually possible to find a very close fit to the curve. (Note that if the frequency is plotted in cycles per second rather than radians per second, the corner frequencies must be converted to radians per second before computing the time constants.)

**Sinusoidal-signal generators.** In performing a frequency-response test, suitable sinusoidal-signal generators must be available. The signal may have to be in mechanical, electrical, or pneumatic form. The frequency ranges needed for the test are approximately 0.001–10 cps for large time-constant systems and 0.1–1000 cps for small time-constant systems. The sinusoidal signal must be reasonably free from harmonics or distortion.

For very low frequency ranges (below 0.01 cps), a mechanical signal generator (together with a suitable pneumatic or electrical transducer if necessary) may be used. For the frequency range 0.01–1000 cps, a suitable electrical-signal generator (together with a suitable transducer if necessary) may be used.

**Determination of minimum phase transfer functions from Bode diagrams.** As stated previously, whether or not a system is a minimum phase one can be determined from the frequency-response curves by examining the high-frequency characteristics.

In order to determine the transfer function, we first draw asymptotes to the experimentally obtained log-magnitude curve. The asymptotes must have slopes of multiples of $\pm 20$ db/decade. If the slope of the experimentally obtained log-magnitude curve changes from $-20$ to $-40$ db/decade at $\omega = \omega_1$, it is clear that a factor $1/[1 + j(\omega/\omega_1)]$ exists in the transfer function. If the slope changes by $-40$ db/decade at $\omega = \omega_2$, there must be a quadratic factor of the form

$$\frac{1}{1 + 2\zeta\left(j\dfrac{\omega}{\omega_2}\right) + \left(j\dfrac{\omega}{\omega_2}\right)^2}$$

in the transfer function. The undamped natural frequency of this quadratic factor is equal to the corner frequency $\omega_2$. The damping ratio $\zeta$ can be determined from the experimentally obtained log-magnitude curve by measuring the amount of resonant peak near the corner frequency $\omega_2$ and comparing this with the curves shown in Fig. 9-11.

Once the factors of the transfer function $G(j\omega)$ have been determined, the gain can be determined from the low-frequency portion of the log-magnitude curve. Since such terms as $1 + j(\omega/\omega_1)$ and $1 + 2\zeta(j\omega/\omega_2) + (j\omega/\omega_2)^2$ become unity as

$\omega$ approaches zero, at very low frequencies, the sinusoidal transfer function $G(j\omega)$ can be written

$$\lim_{\omega \to 0} G(j\omega) = \frac{K}{(j\omega)^{\lambda}}$$

In many practical systems, $\lambda$ equals 0, 1, or 2.

1. For $\lambda = 0$, or type 0 systems,

$$G(j\omega) = K \quad \text{for } \omega \ll 1$$

or

$$20 \, |\log G(j\omega)| = 20 \log K \quad \text{for } \omega \ll 1$$

The low-frequency asymptote is a horizontal line at $20 \log K$ db. The value of $K$ can thus be found from this horizontal asymptote.

2. For $\lambda = 1$, or type 1 systems,

$$G(j\omega) = \frac{K}{j\omega} \quad \text{for } \omega \ll 1$$

or

$$20 \log |G(j\omega)| = 20 \log K - 20 \log \omega \quad \text{for } \omega \ll 1$$

which indicates that the low-frequency asymptote has the slope $-20$ db/decade. The frequency at which the low-frequency asymptote (or its extension) intersects the 0-db line is numerically equal to $K$.

3. For $\lambda = 2$, or type 2 systems,

$$G(j\omega) = \frac{K}{(j\omega)^2} \quad \text{for } \omega \ll 1$$

or

$$20 \log |G(j\omega)| = 20 \log K - 40 \log \omega \quad \text{for } \omega \ll 1$$

The slope of the low-frequency asymptote is $-40$ db/decade. The frequency at which this asymptote (or its extension) intersects the 0-db line is numerically equal to $\sqrt{K}$.

Examples of log-magnitude curves for type 0, type 1, and type 2 systems are shown in Fig. 9-83, together with the frequency to which the gain $K$ is related.

The experimentally obtained phase-angle curve provides a means of checking the transfer function obtained from the log-magnitude curve. For a minimum phase system, the experimental phase-angle curve should within reasonable accuracy agree with the theoretical phase-angle curve obtained from the transfer function just determined. These two phase-angle curves should agree exactly in both the very low frequency range and the very high frequency range. If the experimentally obtained phase angle at very high frequencies (compared with the corner frequencies) is not equal to $-90°(q-p)$, where $p$ and $q$ are the degrees of the numerator and denominator polynomials of the transfer function, respectively, then the transfer function must be a nonminimum phase transfer function.

**Nonminimum phase transfer functions.** If, at the high-frequency end, the computed phase lag is 180° less than the experimentally obtained phase lag, then one

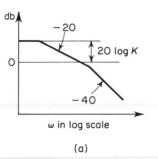

(a)

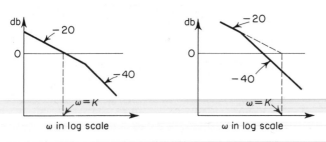

(b)

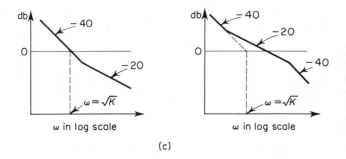

(c)

Fig. 9-83. (a) Log-magnitude curve of a type 0 system; (b) log-magnitude curves of type 1 systems; (c) log-magnitude curves of type 2 systems. (The slopes shown are in db/decade.)

of the zeros of the transfer function should have been in the right-half $s$ plane instead of the left-half $s$ plane.

If the computed phase lag differed from the experimentally obtained phase lag by a constant rate of change of phase, then transportation lag, or dead time, is present. If we assume the transfer function to be of the form

$$G(s)e^{-Ts}$$

where $G(s)$ is a ratio of two polynomials in $s$, then

$$\lim_{\omega \to \infty} \frac{d}{d\omega} \underline{/G(j\omega)e^{-j\omega T}} = -T$$

Thus from this last equation, we can evaluate the magnitude of the transportation lag $T$.

### A few remarks on the experimental determination of transfer functions

1. It is usually easier to make accurate amplitude measurements than accurate phase-shift measurements. Phase-shift measurements may involve errors which may be caused by instrumentation or by interpretation of the experimental records.

2. The frequency response of measuring equipment used to measure the system output must have nearly flat magnitude versus frequency curves. In addition, the phase angle must be nearly proportional to the frequency.

3. Physical systems have some kind of nonlinearities. Therefore, it is necessary to consider carefully the amplitude of input sinusoidal signals. If the amplitude of the input signal is too large, the system will saturate, and the frequency-response test will yield inaccurate results. On the other hand, a small signal will cause errors due to dead zone. Hence, a careful choice of the amplitude of the input sinusoidal signal must be made. It is necessary to sample the waveform of the system output in order to make sure that the waveform is sinusoidal and that the system is operating in the linear region during the test period. (The waveform of the system output is not sinusoidal when the system is operating in its nonlinear region.)

4. If the system under consideration is operating continuously for days and weeks, then normal operation need not be stopped for frequency-response tests. The sinusoidal test signal may be superimposed upon the normal inputs. Then, for linear systems, the output due to the test signal is superimposed upon the normal output. For the determination of the transfer function while the system is in normal operation, stochastic signals (white noise signals) also are often used. By use of correlation functions, the transfer function of the system can be determined without interrupting normal operation.

---

*Example 9-16.* Determine the transfer function of the system whose experimental frequency-response curves are as shown in Fig. 9-84.

The first step in determining the transfer function is to approximate the log-magnitude curve by asymptotes with slopes $\pm 20$ db/decade and multiplies thereof, as shown in Fig. 9-84. Then corner frequencies are found, and the following form of the transfer function is established:

$$G(j\omega) = \frac{K(1 + 0.5j\omega)}{j\omega(1 + j\omega)\left[1 + 2\zeta\left(j\dfrac{\omega}{8}\right) + \left(j\dfrac{\omega}{8}\right)^2\right]}$$

The value of the damping ratio $\zeta$ is found by examining the peak resonance near $\omega = 6$ rad/sec. Referring to Fig. 9-11, $\zeta$ is determined to be 0.5. The gain $K$ is numerically equal to the frequency at the intersection of the extension of the low-frequency asymptote and the 0-db line. The value of $K$ is thus found to be 10. Therefore, $G(j\omega)$ is tentatively determined as

$$G(j\omega) = \frac{10(1 + 0.5j\omega)}{j\omega(1 + j\omega)\left[1 + \left(j\dfrac{\omega}{8}\right) + \left(j\dfrac{\omega}{8}\right)^2\right]}$$

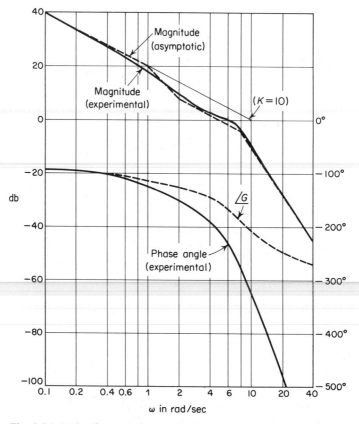

**Fig. 9-84.** Bode diagram of a system. (Solid curves are experimentally obtained curves.)

or

$$G(s) = \frac{320(s + 2)}{s(s + 1)(s^2 + 8s + 64)}$$

This transfer function is tentative because we have not examined the phase-angle curve yet.

Once the corner frequencies are noted on the log-magnitude curve, the corresponding phase-angle curve for each component factor of the transfer function can easily be drawn. The sum of these component phase-angle curves is that of the assumed transfer function. The phase-angle curve for $G(j\omega)$ is denoted by $\underline{/G}$ in Fig. 9-84. From Fig. 9-84 we clearly notice a discrepancy between the computed phase-angle curve and the experimentally obtained phase-angle curve. The difference between the two curves at very high frequencies appears to be a constant rate of change. Thus the discrepancy in the phase-angle curves must be caused by transportation lag.

Hence we assume the complete transfer function to be $G(s)e^{-Ts}$. Since the discrepancy between the computed and experimental phase angles is $-0.2\omega$ rad for very high frequencies, we can determine the value of $T$ as follows:

$$\lim_{\omega \to \infty} \frac{d}{d\omega} \underline{/G(j\omega)e^{-j\omega T}} = -T = -0.2$$

or

$$T = 0.2 \text{ sec}$$

The presence of transportation lag can thus be determined, and the complete transfer function determined from the experimental curves is

$$G(s)e^{-Ts} = \frac{320(s + 2)e^{-0.2s}}{s(s + 1)(s^2 + 8s + 64)}$$

## EXAMPLE PROBLEMS AND SOLUTIONS

**PROBLEM A-9-1.** Consider the circuit shown in Fig. 9-85. Assume that the voltage $e_i$ is applied to the input terminals and the voltage $e_0$ appears at the output terminals. Assume that the input to the system is

$$e_i(t) = E_i \sin \omega t$$

Obtain the steady-state current $i(t)$ flowing through the resistance $R$.

**Solution.** Applying Kirchhoff's law to the circuit, we obtain

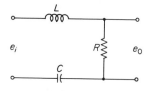

$$L\frac{di}{dt} + Ri + \frac{1}{C} \int i\, dt = e_i$$

Since $i = dq/dt$, we can rewrite the system equation as

**Fig. 9-85.** Electrical circuit.

$$L\frac{d^2q}{dt^2} + R\frac{dq}{dt} + \frac{1}{C}q = e_i$$

For the given input $e_i = E_i \sin \omega t$, the steady-state solution is

$$q(t) = \frac{E_i}{\sqrt{\left(\frac{1}{C} - L\omega^2\right)^2 + (R\omega)^2}} \sin(\omega t - \phi)$$

where

$$\phi = \frac{R\omega}{\frac{1}{C} - L\omega^2}$$

Then the current $i(t)$ is

$$i(t) = \frac{dq(t)}{dt} = \frac{E_i\omega}{\sqrt{\left(\frac{1}{C} - L\omega^2\right)^2 + (R\omega)^2}} \cos(\omega t - \phi)$$

**PROBLEM A-9-2.** Consider a system whose closed-loop transfer function is

$$\frac{C(s)}{R(s)} = \frac{10(s + 1)}{(s + 2)(s + 5)}$$

(This is the same system considered in Problem A-8-7.) Clearly, the closed-loop poles are located at $s = -2$ and $s = -5$, and the system is not oscillatory. (The unit-step response, however, exhibits overshoot due to the presence of a zero at $s = -1$. See Fig. 8-48.)

Show that the closed-loop frequency response of this system will exhibit a resonant peak, although the damping ratio of the closed-loop poles are greater than unity.

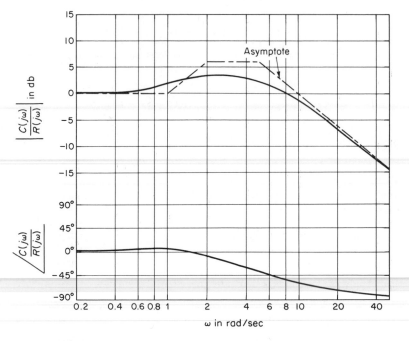

**Fig. 9-86.** Bode diagram for $10(1 + j\omega)/[(2 + j\omega)(5 + j\omega)]$.

**Solution.** Figure 9-86 shows the Bode diagram for the system. The resonant peak value is approximately 3.5 db. (Note that, in the absence of a zero, the second-order system with $\zeta > 0.7$ will not exhibit a resonant peak; however, the presence of a closed-loop zero will cause such a peak.)

**PROBLEM A-9-3.** Prove that the polar plot of the sinusoidal transfer function

$$G(j\omega) = \frac{j\omega T}{1 + j\omega T} \qquad (0 \leq \omega \leq \infty)$$

is a semicircle. Find the center and radius of the circle.

**Solution.** The given sinusoidal transfer function $G(j\omega)$ can be written as follows:

$$G(j\omega) = X + jY$$

where

$$X = \frac{\omega^2 T^2}{1 + \omega^2 T^2}, \qquad Y = \frac{\omega T}{1 + \omega^2 T^2}$$

Then

$$\left(X - \frac{1}{2}\right)^2 + Y^2 = \frac{(\omega^2 T^2 - 1)^2}{4(1 + \omega^2 T^2)^2} + \frac{\omega^2 T^2}{(1 + \omega^2 T^2)^2} = \frac{1}{4}$$

Hence we see that the plot of $G(j\omega)$ is a circle centered at $(0.5, 0)$ with radius equal to 0.5. The upper semicircle corresponds to $0 \leq \omega \leq \infty$, and the lower semicircle corresponds to $-\infty \leq \omega \leq 0$.

**PROBLEM A-9-4.** Draw the Bode diagram of the following nonminimum phase system:

$$\frac{C(s)}{R(s)} = 1 - Ts$$

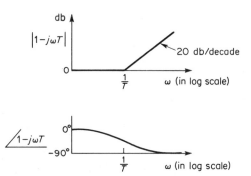

**Fig. 9-87.** Bode diagram of $1 - j\omega T$.

Obtain the unit-ramp response of the system and plot $c(t)$ versus $t$.

**Solution.** The Bode diagram of the system is shown in Fig. 9-87.

For a unit-ramp input, $R(s) = 1/s^2$, and

$$C(s) = \frac{1 - Ts}{s^2} = \frac{1}{s^2} - \frac{T}{s}$$

The inverse Laplace transform of $C(s)$ gives

$$c(t) = t - T \qquad (t \geq 0)$$

Figure 9-88 shows the response curve $c(t)$ versus $t$. (Note the faulty behavior at the start of the response.) A characteristic property of such a nonminimum phase system is that the transient response starts out in the opposite direction to the input but eventually comes back in the same direction.

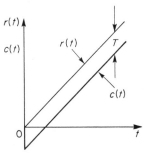

**PROBLEM A-9-5.** Consider the function

$$F(s) = \frac{s+1}{s-1}$$

**Fig. 9-88.** Unit-ramp response of the system considered in Problem A-9-4.

The conformal mapping of the lines $\omega = 0, \pm 1, \pm 2$ and the lines $\sigma = 0, \pm 1, \pm 2$ yields circles in the $F(s)$ plane, as shown in Fig. 9-89. Show that if the contour in the $s$ plane encloses the pole of $F(s)$, there is one encirclement of the origin of the $F(s)$ plane in the counterclockwise direction. If the contour in the $s$ plane encloses the zero of $F(s)$, there is one encirclement of the origin of the $F(s)$ plane in the clockwise direction. If the contour in the $s$ plane encloses both the zero and pole or if the contour encloses neither the zero nor pole, then there is no encirclement of the origin of the $F(s)$ plane by the locus of $F(s)$. (Note that in the $s$ plane a representative point $s$ traces out a contour in the clockwise direction.)

**Solution.** A graphical solution is given in Fig. 9-90; this shows closed contours in the $s$ plane and their corresponding closed curves in the $F(s)$ plane.

**PROBLEM A-9-6.** Prove the following mapping theorem: Let $F(s)$ be a ratio of polynomials in $s$. Let $P$ be the number of poles and $Z$ be the number of zeros of $F(s)$ which lie inside a closed contour in the $s$ plane, multiplicity accounted for. Let the closed contour be such that it does not pass through any poles or zeros of $F(s)$. The closed contour in the $s$ plane then maps into the $F(s)$ plane as a closed curve. The number $N$ of clockwise encirclements of the origin of the $F(s)$ plane, as a representative point $s$ traces out the entire contour in the $s$ plane in the clockwise direction, is equal to $Z - P$.

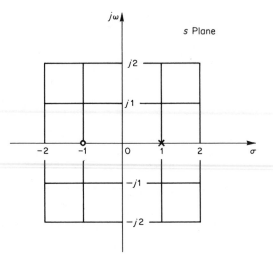

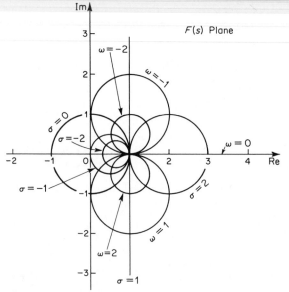

**Fig. 9-89.** Conformal mapping of the $s$-plane grids into the $F(s)$ plane where $F(s) = (s + 1)/(s - 1)$.

**Solution.** To prove this theorem, we use Cauchy's theorem and the residue theorem. Cauchy's theorem states that the integral of $F(s)$ around a closed contour in the $s$ plane is zero if $F(s)$ is analytic within and on the closed contour, or

$$\oint F(s)\, ds = 0$$

The residue theorem states that the integral of $F(s)$ taken in the clockwise direction around a closed contour in the $s$ plane is equal to $-2\pi j$ times the residues at the simple poles of $F(s)$, or

$$\oint F(s)\, ds = -2\pi j \left(\sum \text{residues}\right)$$

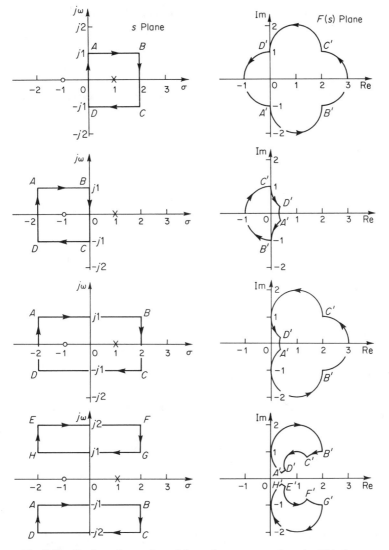

**Fig. 9-90.** Conformal mapping of the $s$-plane contours into the $F(s)$ plane where $F(s) = (s + 1)/(s - 1)$.

Suppose that $F(s)$ is given by

$$F(s) = \frac{(s + z_1)^{k_1}(s + z_2)^{k_2} \cdots}{(s + p_1)^{m_1}(s + p_2)^{m_2} \cdots} X(s)$$

where $X(s)$ is analytic in the closed contour in the $s$ plane and all the poles and zeros are located in the contour. Then the ratio $F'(s)/F(s)$ can be written

$$\frac{F'(s)}{F(s)} = \left(\frac{k_1}{s + z_1} + \frac{k_2}{s + z_2} + \cdots\right) - \left(\frac{m_1}{s + p_1} + \frac{m_2}{s + p_2} + \cdots\right) + \frac{X'(s)}{X(s)} \qquad (9\text{-}24)$$

This may be seen from the following consideration: If $F(s)$ is given by

$$F(s) = (s + z_1)^k X(s)$$

then $F(s)$ has a zero of $k$th order at $s = -z_1$. Differentiating $F(s)$ with respect to $s$ yields

$$F'(s) = k(s + z_1)^{k-1} X(s) + (s + z_1)^k X'(s)$$

Hence,

$$\frac{F'(s)}{F(s)} = \frac{k}{s + z_1} + \frac{X'(s)}{X(s)} \qquad (9\text{-}25)$$

We see that by taking the ratio $F'(s)/F(s)$ the $k$th-order zero of $F(s)$ becomes a simple pole of $F'(s)/F(s)$.

If the last term on the right-hand side of Eq. (9-25) does not contain any poles or zeros in the closed contour in the $s$ plane, $F'(s)/F(s)$ is analytic in this contour except at the zero $s = -z_1$. Then

$$\oint \frac{F'(s)}{F(s)} \, ds = \oint \frac{k}{s + z_1} + \oint \frac{X'(s)}{X(s)} = -2\pi jk$$

Referring to Eq. (9-24) and noting that $X'(s)/X(s)$ is analytic in the contour and that the factors in the parentheses are all simple poles located in the contour, we obtain the following relationship:

$$\oint \frac{F'(s)}{F(s)} \, ds = -2\pi j[(k_1 + k_2 + \cdots) - (m_1 + m_2 + \cdots)] = -2\pi j(Z - P)$$

where

$Z = k_1 + k_2 + \cdots = $ total number of zeros of $F(s)$ enclosed in the closed contour in the $s$ plane

$P = m_1 + m_2 + \cdots = $ total number of poles of $F(s)$ enclosed in the closed contour in the $s$ plane

(The $k$ multiple zeros or poles are considered $k$ zeros or poles located at the same point.) Since $F(s)$ is a complex number, $F(s)$ can be written

$$F(s) = |F|e^{j\theta}$$

and

$$\ln F(s) = \ln |F| + j\theta$$

Noting that $F'(s)/F(s)$ can be written

$$\frac{F'(s)}{F(s)} = \frac{d \ln F(s)}{ds}$$

we obtain

$$\frac{F'(s)}{F(s)} = \frac{d \ln |F|}{ds} + j\frac{d\theta}{ds}$$

If the closed contour in the $s$ plane is mapped into the closed contour $\Gamma$ in the $F(s)$ plane, then

$$\oint \frac{F'(s)}{F(s)} \, ds = \oint_\Gamma d \ln |F| + j \oint_\Gamma d\theta = j \int d\theta = 2\pi j(P - Z)$$

The integral $\int_\Gamma d \ln |F|$ is zero since the magnitude $\ln |F|$ is the same at the initial point and the final point of the contour $\Gamma$. Thus, we obtain

$$\frac{\theta_2 - \theta_1}{2\pi} = P - Z$$

The angular difference between the final and initial values of $\theta$ is equal to the total change in the phase angle of $F'(s)/F(s)$ as a representative point in the $s$ plane moves along the closed contour. Noting that $N$ is the number of clockwise encirclements of the origin of the $F(s)$ plane and $\theta_2 - \theta_1$ is zero or a multiple of $2\pi$ rad, we obtain

$$\frac{\theta_2 - \theta_1}{2\pi} = -N$$

Thus, we have the relationship

$$N = Z - P$$

This proves the theorem.

Note that by this mapping theorem, the exact numbers of zeros and of poles cannot be found, only their difference. Note also that from Figs. 9-91 (a) and (b) we see that if $\theta$ does not change through $2\pi$ rad, then the origin of the $F(s)$ plane cannot be encircled.

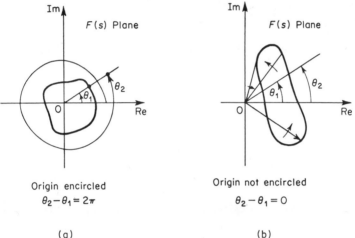

Fig. 9-91. Determination of encirclement of the origin of $F(s)$ plane.

**PROBLEM A-9-7.** Is the system with the following open-loop transfer function and with $K = 2$ stable?

$$G(s)H(s) = \frac{K}{s(s+1)(2s+1)}$$

Find the critical value of the gain $K$ for stability.

**Solution.** The open-loop transfer function is

$$G(j\omega)H(j\omega) = \frac{K}{j\omega(j\omega+1)(2j\omega+1)}$$

$$= \frac{K}{-3\omega^2 + j\omega(1 - 2\omega^2)}$$

This open-loop transfer function has no poles in the right half $s$ plane. Thus for stability, the $-1 + j0$ point should not be encircled by the Nyquist plot. Let us find the point where the Nyquist plot crosses the negative real axis. Let the imaginary part of $G(j\omega)H(j\omega)$ be zero, or

$$1 - 2\omega^2 = 0$$

from which

$$\omega = \pm \frac{1}{\sqrt{2}}$$

Substituting $\omega = 1/\sqrt{2}$ into $G(j\omega)H(j\omega)$, we obtain

$$G\left(j\frac{1}{\sqrt{2}}\right)H\left(j\frac{1}{\sqrt{2}}\right) = -\frac{2K}{3}$$

The critical value of the gain $K$ is obtained by equating $-2K/3$ to $-1$, or

$$-\frac{2}{3}K = -1$$

Hence

$$K = \tfrac{3}{2}$$

The system is stable if $0 < K < \tfrac{3}{2}$. Hence the system with $K = 2$ is unstable.

**PROBLEM A-9-8.** Consider the closed-loop system shown in Fig. 9-92. Determine the critical value of $K$ for stability by use of the Nyquist stability criterion.

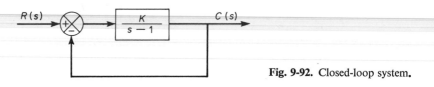

Fig. 9-92. Closed-loop system.

**Solution.** The polar plot of

$$G(j\omega) = \frac{K}{j\omega - 1}$$

is a circle with center at $-K/2$ on the negative real axis and radius $K/2$, as shown in Fig. 9-93 (a). As $\omega$ is increased from $-\infty$ to $\infty$, the $G(j\omega)$ locus makes a counterclockwise rotation. In this system, $P = 1$ because there is one pole of $G(s)$ in the right-half $s$ plane. For the closed-loop system to be stable, $Z$ must be equal to zero. Therefore, $N = Z - P$ must be equal to $-1$, or there must be one counterclockwise encirclement of the $-1 + j0$ point for stability. (If there is no encirclement of the $-1 + j0$ point, the system is unstable.) Thus, for stability, $K$ must be greater than unity, and $K = 1$ gives the stability limit. Fig. 9-93 (b) shows both stable and unstable cases of $G(j\omega)$ plots.

**PROBLEM A-9-9.** Consider a unity-feedback system whose open-loop transfer function is

$$G(s) = \frac{Ke^{-0.8s}}{s + 1}$$

Using the Nyquist plot, determine the critical value of $K$ for stability.

**Solution.** For this system,

$$\begin{aligned}
G(j\omega) &= \frac{Ke^{-0.8j\omega}}{j\omega + 1} \\
&= \frac{K(\cos 0.8\omega - j\sin 0.8\omega)(1 - j\omega)}{1 + \omega^2} \\
&= \frac{K}{1 + \omega^2}[(\cos 0.8\omega - \omega \sin 0.8\omega) - j(\sin 0.8\omega + \omega \cos 0.8\omega)]
\end{aligned}$$

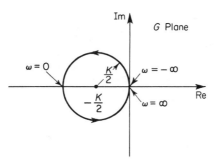

(a)

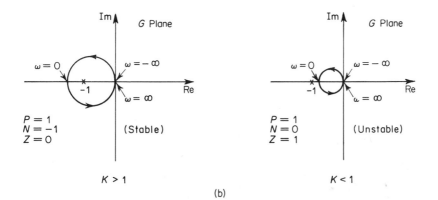

(b)

**Fig. 9-93.** (a) Polar plot of $K/(j\omega - 1)$; (b) polar plots of $K/(j\omega - 1)$ for stable and unstable cases.

The imaginary part of $G(j\omega)$ is equal to zero if

$$\sin 0.8\omega + \omega \cos 0.8\omega = 0$$

Hence

$$\omega = -\tan 0.8\omega$$

Solving this equation for the smallest positive value of $\omega$, we obtain

$$\omega = 2.45$$

Substituting $\omega = 2.45$ into $G(j\omega)$, we obtain

$$G(j2.45) = \frac{K}{1 + 2.45^2}(\cos 1.96 - 2.45 \sin 1.96)$$

$$= -0.378K$$

The critical value of $K$ for stability is obtained by letting $G(j2.45)$ equal $-1$. Hence

$$0.378K = 1$$

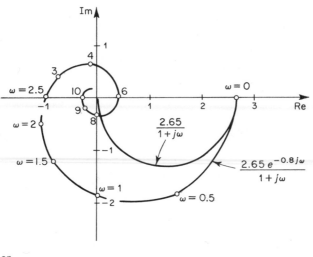

**Fig. 9-94.** Polar plots of $2.65e^{-0.8j\omega}/(1 + j\omega)$ and $2.65/(1 + j\omega)$.

or

$$K = 2.65$$

Figure 9-94 shows the Nyquist or polar plots of $2.65\,e^{-0.8j\omega}/(1 + j\omega)$ and $2.65/(1 + j\omega)$. The first-order system without transportation lag is stable for all values of $K$, but the one with transportation lag of 0.8 sec becomes unstable for $K > 2.65$.

**PROBLEM A-9-10\*.** We shall here consider a method for obtaining the step response directly from the closed-loop frequency-response data. In this method, a unit-step input is replaced by a square-wave input and the output to the square-wave input is approximated by a convergent sine series. The assumptions made in this method are summarized in the following:

1. The system is oscillatory. In other words, the method applies to oscillatory systems only.

2. The time of the first peak in the step response is $t_1 = \pi/\omega_0$, where $\omega_0$ is the frequency at which $\underline{/C(j\omega)/R(j\omega)} = -90°$.

3. The system will have a negligibly small error for a unit-step input for $t \geq 4.5t_1$.

For a second-order system of the form

$$\frac{\omega_n^2}{s^2 + 2\zeta\omega_n s + \omega_n^2}$$

the error in a unit-step response is less than 2% for $t > 4.5t_1$ if the damping ratio $\zeta$ is greater than 0.267. It can be seen that if the half period of the square wave is long enough compared with the system time constants, then replacement of the step input by a square-wave input places no additional restrictions on the system. For good accuracy, a proper choice of the frequency $\omega_1$ of the square wave is very important. In the present method, this period is chosen to be $18t_1$. The square-wave input and the corresponding system response are sketched in Fig. 9-95. The basis for this choice of the period of the square wave is given in the following: The ideal response of a system to a unit-step input may be as that shown in Fig. 9-96 (a), where $t_1$ is the time for the response to reach the maximum

\* Reference S–6

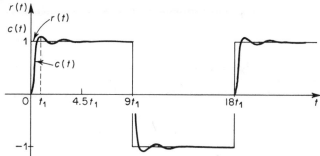

**Fig. 9-95.** Square-wave input and the corresponding response of a second-order system.

value for the first time. The corresponding ideal unit-impulse response is shown in Fig. 9-96 (b). The Laplace transform of the ideal impulse response is then

$$C(s) = \frac{1}{t_1 s}(1 - e^{-t_1 s})$$

Since the Laplace transform of the unit-impulse input is unity, $R(s) = 1$, and

$$\frac{C(s)}{R(s)} = \frac{1}{t_1 s}(1 - e^{-t_1 s})$$

from which

$$\frac{C(j\omega)}{R(j\omega)} = \frac{1}{j\omega t_1}(1 - e^{-j\omega t_1})$$

$$= \frac{\sqrt{2(1 - \cos \omega t_1)}}{\omega t_1} e^{j\theta}$$

where

$$\theta = -\frac{\pi}{2} + \tan^{-1}\frac{\sin \omega t_1}{1 - \cos \omega t_1}$$

At $\omega = \omega_0$, it has been assumed that

$$\left|\frac{C(j\omega_0)}{R(j\omega_0)}\right| = -\frac{\pi}{2}$$

Thus at $\omega = \omega_0$,

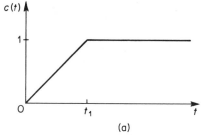

(a)

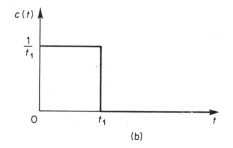

(b)

**Fig. 9-96.** (a) Ideal unit-step response; (b) ideal unit-impulse response.

$$\theta = -\frac{\pi}{2} = -\frac{\pi}{2} + \tan^{-1}\frac{\sin \omega_0 t_1}{1 - \cos \omega_0 t_1}$$

from which

$$\sin \omega_0 t_1 = 0$$

or

$$\omega_0 t_1 = 0, \pi, 2\pi, \ldots$$

The time for the first maximum to occur has been assumed to be

$$t_1 = \frac{\pi}{\omega_0}$$

Substituting this value of $t_1$ into $\omega_1 = \pi/(9t_1)$, we obtain

$$\omega_1 = \frac{\omega_0}{9} \qquad (9\text{-}26)$$

Thus, the square wave should have the frequency $\omega_0/9$ (or the period $18t_1$), where $\omega_0$ is the frequency at which the closed-loop frequency response exhibits a phase lag of $90°$. If the square-wave frequency is too fast, the accuracy will not be good, while too slow a frequency makes the convergence of the series slow.

The square-wave input $r(t)$ shown in Fig. 9-95 may be expanded into the following Fourier series:

$$r(t) = \frac{1}{2} + \frac{2}{\pi} \sum_{n=1,3,5,\ldots}^{\infty} \frac{\sin n\omega_1 t}{n}$$

where

$$\omega_1 = \frac{\pi}{9t_1}$$

The system response to this square-wave input is the sum of the responses to each element of the Fourier series. If we let

$$\frac{C(j\omega)}{R(j\omega)} = M(j\omega)$$

then

$$c(t) = \frac{1}{2} + \frac{2}{\pi} \sum_{n=1,3,5,\ldots}^{\infty} |M_n| \frac{\sin (n\omega_1 t + \underline{/M_n})}{n} \qquad (9\text{-}27)$$

where

$$M_n = M(jn\omega_1)$$

Since

$$M_n = |M_n| (\cos \underline{/M_n} + j \sin \underline{/M_n})$$
$$= \mathrm{Re}\,(M_n) + j \,\mathrm{Im}\,(M_n)$$

where

$$\mathrm{Re}\,(M_n) = |M_n| \cos \underline{/M_n}$$
$$\mathrm{Im}\,(M_n) = |M_n| \sin \underline{/M_n}$$

Eq. (9-27) may be written

$$c(t) = \frac{1}{2} + \frac{2}{\pi} \sum_{n=1,3,5,\ldots}^{\infty} \frac{\mathrm{Re}\,(M_n) \sin n\omega_1 t + \mathrm{Im}\,(M_n) \cos n\omega_1 t}{n} \qquad (9\text{-}28)$$

Since it has been assumed that the response is zero from time $-4.5t_1$ to time zero and unity from time $4.5t_1$ to time $9t_1$, substituting $-t$ into Eq. (9-28) results in

$$c(-t) = 0 = \frac{1}{2} + \frac{2}{\pi} \sum_{n=1,3,5,\ldots}^{\infty} \left[ \frac{-\mathrm{Re}\,(M_n) \sin n\omega_1 t}{n} + \frac{\mathrm{Im}\,(M_n) \cos n\omega_1 t}{n} \right] \qquad (9\text{-}29)$$
$$(0 \leq t \leq 4.5t_1)$$

Subtracting Eq. (9-29) from Eq. (9-28) yields

$$c(t) = \frac{4}{\pi} \sum_{n=1,3,5,\ldots}^{\infty} \frac{\mathrm{Re}\,(M_n) \sin n\omega_1 t}{n} \qquad (0 \leq t \leq 4.5t_1) \qquad (9\text{-}30)$$

From Eq. (9-30), we can compute the unit-step response directly from the closed-loop frequency-response data.

As a practical matter, the summation must be truncated. Generally, the first seven or eight terms in the infinite series in Eq. (9-30) will be sufficient to assure adequate accuracy. The convergence and accuracy may be checked easily by letting $\omega_1 t$ equal 90° and calculating the response as follows:

$$c\left(\frac{\pi}{2\omega_1}\right) = \frac{4}{\pi}\Big[\operatorname{Re}(M_1) - \frac{1}{3}\operatorname{Re}(M_3) + \frac{1}{5}\operatorname{Re}(M_5) - \frac{1}{7}\operatorname{Re}(M_7)$$

$$-\frac{1}{11}\operatorname{Re}(M_{11}) + \frac{1}{13}\operatorname{Re}(M_{13}) - \frac{1}{15}\operatorname{Re}(M_{15}) + \cdots\Big] \qquad (9\text{-}31)$$

The sum of this series must be equal to unity. In Eq. (9-31) the term at the ninth harmonic frequency vanishes because at this frequency the phase angle of $C(j\omega)/R(j\omega)$ is $-90°$, so the real part of $C(j\omega)/R(j\omega)$ is zero. Terms at higher frequencies are quite small, and the series converges rapidly.

Now consider the closed-loop frequency-response locus, as shown in Fig. 9-97. Obtain the maximum overshoot in the unit-step response of the system.

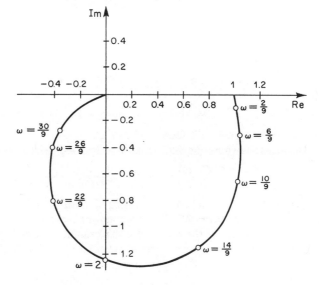

**Fig. 9-97.** Polar plot of a closed-loop frequency response.

**Solution.** From the given locus, $\omega_0$ is found to be 2. From Eq. (9-26), $\omega_1$ is

$$\omega_1 = \frac{2}{9}$$

From Eq. (9-30), the unit-step response can be computed and is

$$c(t) \doteq \frac{4}{\pi}\Big[\operatorname{Re}(M_1)\sin\frac{2}{9}t + \frac{\operatorname{Re}(M_3)}{3}\sin\frac{6}{9}t + \frac{\operatorname{Re}(M_5)}{5}\sin\frac{10}{9}t$$

$$+ \frac{\operatorname{Re}(M_7)}{7}\sin\frac{14}{9}t + \frac{\operatorname{Re}(M_{11})}{11}\sin\frac{22}{9}t + \frac{\operatorname{Re}(M_{13})}{13}\sin\frac{26}{9}t$$

$$+ \frac{\operatorname{Re}(M_{15})}{15}\sin\frac{30}{9}t\Big] \qquad (9\text{-}32)$$

From Fig. 9-97, Re $(M_i)$ can be found as follows;

$$Re\,(M_1) = 1.02$$
$$Re\,(M_3) = 1.04$$
$$Re\,(M_5) = 1.03$$
$$Re\,(M_7) = 0.72$$
$$Re\,(M_{11}) = -0.40$$
$$Re\,(M_{13}) = -0.42$$
$$Re\,(M_{15}) = -0.36$$

The accuracy of the solution may be checked by substituting these values of Re $(M_i)$ and $t = (\pi/2)(\frac{9}{2})$ into Eq. (9-32). The sum can be obtained as follows:

$$c\left(\frac{\pi}{2}\,\frac{9}{2}\right) = \frac{4}{\pi}\left[1.02 - \frac{1.04}{3} + \frac{1.03}{5} - \frac{0.72}{7} - \frac{(-0.40)}{11} + \frac{(-0.42)}{13} - \frac{(-0.36)}{15}\right] = 1.023$$

This value differs from unity, but this discrepancy may be well within the range of computational errors. The maximum overshoot in the unit-step response occurs at

$$t = \frac{\pi}{\omega_0} = \frac{\pi}{2}$$

Thus,

$$c\left(\frac{\pi}{2}\right) = \frac{4}{\pi}\left[1.02 \sin\frac{\pi}{9} + \frac{1.04}{3}\sin\frac{\pi}{3} + \frac{1.03}{5}\sin\frac{5\pi}{9} + \frac{0.72}{7}\sin\frac{7\pi}{9} - \frac{0.40}{11}\sin\frac{11\pi}{9}\right.$$
$$\left. - \frac{0.42}{13}\sin\frac{13\pi}{9} - \frac{0.36}{15}\sin\frac{15\pi}{9}\right] = 1.27$$

$$\text{Maximum overshoot} = 1.27 - 1 = 0.27$$

The maximum overshoot is found to be 27%.

The locus shown in Fig. 9-97 actually corresponds to that of the following system:

$$\frac{C(j\omega)}{R(j\omega)} = M(j\omega) = \frac{4}{(j\omega)^2 + 1.6(j\omega) + 4}$$

(This information is given here for checking the accuracy of the result and is not necessary for solving this problem.) The exact amount of maximum overshoot for this system is 25.4%. The accuracy of the solution of this example is usually adequate for engineering purposes. If the entire unit-step response curve is desired, it can be constructed from Eq. (9-32).

**PROBLEM A-9-11.** A closed-loop control system may include an unstable element within the loop. When the Nyquist stability criterion is to be applied to such a system, the frequency-response curves for the unstable element must be obtained.

How can we obtain experimentally the frequency-response curves for such an unstable element? Suggest a possible approach to the experimental determination of the frequency response of an unstable linear element.

**Solution.** One possible approach is to measure the frequency-response characteristics of the unstable element by using it as a part of a stable system.

Consider the system shown in Fig. 9-98. Suppose that the element $G_1(s)$ is unstable. The complete system may be made stable by choosing a suitable linear element $G_2(s)$. We apply a sinusoidal signal at the input. At steady state, all signals in the loop will be sinusoidal. We measure the signals $e(t)$, the input to the unstable element, and $x(t)$, the

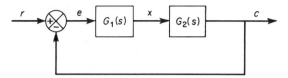

**Fig. 9-98.** Control system.

output of the unstable element. By changing the frequency [and possibly the amplitude for the convenience of measuring $e(t)$ and $x(t)$] of the input sinusoid and repeating this process, it is possible to obtain the frequency response of the unstable element.

## PROBLEMS

**PROBLEM B-9-1.** Consider the unity-feedback system with the open-loop transfer function

$$G(s) = \frac{10}{s+1}$$

Obtain the steady-state output of the system when it is subjected to each of the following inputs:

1. $\qquad r(t) = \sin (t + 30°)$
2. $\qquad r(t) = 2 \cos (2t - 45°)$
3. $\qquad r(t) = \sin (t + 30°) - 2 \cos (2t - 45°)$

**PROBLEM B-9-2.** Consider the system whose closed-loop transfer function is

$$\frac{C(s)}{R(s)} = \frac{K(T_2 s + 1)}{T_1 s + 1}$$

Obtain the steady-state output of the system when it is subjected to the input $r(t) = R \sin \omega t$.

**PROBLEM B-9-3.** Sketch the Bode diagrams of the following three transfer functions:

1. $\qquad G(s) = \dfrac{T_1 s + 1}{T_2 s + 1} \qquad (T_1 > T_2 > 0)$

2. $\qquad G(s) = \dfrac{T_1 s - 1}{T_2 s + 1} \qquad (T_1 > T_2 > 0)$

3. $\qquad G(s) = \dfrac{-T_1 s + 1}{T_2 s + 1} \qquad (T_1 > T_2 > 0)$

**PROBLEM B-9-4.** The pole-zero configurations of the complex functions $F_1(s)$ and $F_2(s)$ are shown in Figs. 9-99 (a) and (b), respectively, Assume that the closed contours in the $s$ plane are those shown in Figs. 9-99 (a) and (b). Sketch qualitatively the corresponding closed contours in the $F_1(s)$ plane and $F_2(s)$ plane.

**PROBLEM B-9-5.** Consider the unity-feedback control system whose open-loop transfer function is

$$G(s) = \frac{as + 1}{s^2}$$

Determine the value of $a$ so that the phase margin is equal to $45°$.

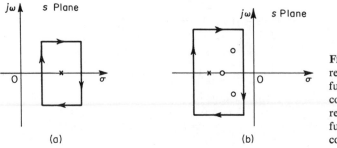

**Fig. 9-99.** (a) $s$ plane representation of complex function $F_1(s)$ and a closed contour; (b) $s$ plane representation of complex function $F_2(s)$ and a closed contour.

**PROBLEM B-9-6.** A system with the open-loop transfer function

$$G(s)H(s) = \frac{K}{s^2(T_1s + 1)}$$

is inherently unstable. This system can be stabilized by adding derivative control. Sketch the polar plots for the open-loop transfer function with and without derivative control.

**PROBLEM B-9-7.** Sketch the polar plots of the open-loop transfer function

$$G(s)H(s) = \frac{K(T_as + 1)(T_bs + 1)}{s^2(T_1s + 1)}$$

for the following two cases:

1. $\qquad\qquad T_a > T_1 > 0, \qquad T_b > T_1 > 0$
2. $\qquad\qquad T_1 > T_a > 0, \qquad T_1 > T_b > 0$

**PROBLEM B-9-8.** Consider the system with

$$G(s) = \frac{10}{s(s - 1)}, \qquad H(s) = 1 + K_h s$$

Determine the cirtical value of $K_h$ for stability of the closed-loop system.

**PROBLEM B-9-9.** Consider the closed-loop system with the following open-loop transfer function:

$$G(s)H(s) = \frac{10K(s + 0.5)}{s^2(s + 2)(s + 10)}$$

Plot both the direct and inverse polar plots of $G(s)H(s)$ with $K = 1$ and $K = 10$. Apply the Nyquist stability criterion to the plots and determine the stability of the system with these values of $K$.

**PROBLEM B-9-10.** Consider the closed-loop system whose open-loop transfer function is

$$G(s)H(s) = \frac{Ke^{-2s}}{s}$$

Find the maximum value of $K$ for which the system is stable.

**PROBLEM B-9-11.** Consider a closed-loop system whose open-loop transfer function is

$$G(s)H(s) = \frac{Ke^{-Ts}}{s(s + 1)}$$

Determine the maximum value of the gain $K$ for stability as a function of dead time $T$.

**PROBLEM B-9-12.** Sketch the polar plot of

$$G(s) = \frac{(Ts)^2 - 6(Ts) + 12}{(Ts)^2 + 6(Ts) + 12} \tag{9-33}$$

Show that for the frequency range $0 < \omega T < 2\sqrt{3}$, Eq. (9-33) gives a good approximation to transportation lag $e^{-Ts}$.

**PROBLEM B-9-13.** The experimentally determined Bode diagram of a system $G(j\omega)$ is shown in Fig. 9-100. Determine the transfer function $G(s)$.

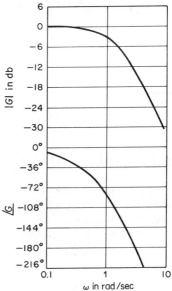

**Fig. 9-100.** Experimentally determined Bode diagram of a system.

# 10

# DESIGN AND COMPENSATION TECHNIQUES

## 10-1 INTRODUCTION

The primary objective of this chapter is to present procedures for the design and compensation of single-input–single-output linear time-invariant control systems. Compensation is the adjustment of a system in order to satisfy the given specifications. The approaches to the control system design and compensation used in this chapter are the root-locus approach and the frequency-response approach. (System design based on modern control theory will be presented in Chapter 16.)

**Performance specifications.** Control systems are designed to perform specific tasks. The requirements imposed upon the control system are usually spelled out as performance specifications. They generally relate to accuracy, relative stability, and speed of response.

For routine design problems, the performance specifications may be given in terms of precise numerical values. In other cases, they may be given partially in terms of precise numerical values and partially in terms of qualitative statements. In the latter case, the specifications may have to be modified during the course of design since the given specifications may never be satisfied (because of conflicting requirements) or may lead to a very expensive system.

Generally speaking, the performance specifications should not be more stringent than necessary to perform the given task. If the accuracy at steady-state operation is of prime importance in a given control system, then we should not require unnecessarily rigid performance specifications on the transient response since such specifications will require expensive components. Remember that the most important part of control system design is to state the performance specifications precisely so that they will yield an optimal control system for the given purpose.

**Trial-and-error approach to system design.** In most practical cases, the design method to be used may be determined by the performance specifications applicable to the particular case. In designing control systems, if the performance specifications are given in terms of time-domain performance measures, such as rise time, maximum overshoot, or settling time, or frequency-domain performance measures, such as phase margin, gain margin, resonant peak value, or bandwidth, then we have no choice but to use a trial-and-error approach based on the root-locus method and/or frequency-response methods.

The systems which may be designed by a trial-and-error approach are usually limited to single-input–single-output linear time-invariant systems. The designer seeks to satisfy all performance specifications by means of educated trial-and-error repetition. After a system is designed, the designer checks to see if the designed system satisfies all the performance specifications. If it does not, then he repeats the design process by adjusting parameter settings or by changing the system configuration until the given specifications are met. Although the design is based on a trial-and-error procedure, the ingenuity and know-how of the designer will play an important role in a successful design. An experienced designer may be able to design an acceptable system without using many trials.

**Modification of plant dynamics.** In building a control system, we know that proper modification of the plant dynamics may be a simple way to meet the performance specifications. This, however, may not be possible in many practical situations because the plant may be fixed and may not be modified. Then we must adjust parameters other than those in the fixed plant. In this chapter, we assume that the plant is given and unalterable.

**System compensation.** Setting the gain is the first step in adjusting the system for satisfactory performance. In many practical cases, however, the adjustment of the gain alone may not provide sufficient alteration of the system behavior to meet the given specifications. As is frequently the case, increasing the gain value will improve the steady-state behavior but will result in poor stability or even instability. It is then necessary to redesign the system (by modifying the structure or by incorporating additional devices or components) in order to alter the overall behavior so that the system will behave as desired.

An additional device inserted into the system for such purpose is called a *compensator*. This device compensates for deficient performance of the original system.

**Series compensation and feedback (or parallel) compensation.** If the compensator $G_c(s)$ is placed in series with the unalterable transfer function $G(s)$, as shown in Fig. 10-1 (a), then the compensation is called series compensation.

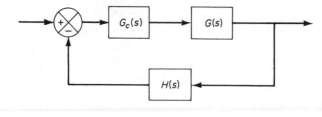

(a)

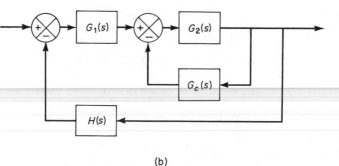

(b)

**Fig. 10-1.** (a) Series compensation; (b) feedback or parallel compensation.

An alternative to series compensation is to feed back the signal(s) from some element(s) and place a compensator in the resulting inner feedback path, as shown in Fig. 10-1 (b). Such compensation is called feedback compensation or parallel compensation.

In compensating control systems, we see that the problem usually boils down to a suitable design of a series or feedback compensator. The choice between series compensation and feedback compensation depends upon the nature of the signals in the system, the power levels at various points, available components, the designer's experience, economic considerations, etc.

In general, series compensation may be simpler than feedback compensation; however, series compensation frequently requires additional amplifiers to increase the gain and/or to provide isolation. (In order to avoid power dissipation, the series compensator is inserted at the lowest energy point in the feedforward path.) Note that, in general, the number of components required in feedback compensation will be less than the number of components in series compensation, provided a suitable signal is available, because the energy transfer is from a higher power level to a lower one. (This means that additional amplifiers may not be necessary.)

**Compensators.** If a compensator is needed to meet the performance specifications, the designer must realize a physical device which has the prescribed transfer function of the compensator.

Numerous physical devices have been used for such purposes. In fact, many noble and useful ideas for physically constructing compensators may be found in the literature.

Among the many compensators, the widely employed series compensators are the so-called lead compensators, lag compensators, and lag-lead compensators. In this chapter, we shall limit our discussions mostly to these three types. They are usually electrical, mechanical, pneumatic, hydraulic, or combinations thereof and consist of $RC$ networks (electrical, mechanical, pneumatic, or hydraulic) and amplifiers. (In many cases, a compensating $RC$ network is an integral part of the amplifier.)

In the actual design of a control system, whether or not to use an electrical, mechanical, pneumatic, or hydraulic compensator is a matter that must be decided partially upon the nature of the controlled plant. For example, if the controlled plant involves flammable fluid, then we have to choose a pneumatic compensator and actuator to avoid the possibility of sparks. If, however, no fire hazard exists, then electronic compensators are most commonly used. (In fact, we often transform nonelectrical signals into electrical signals because of the simplicity of transmission, increased accuracy, increased reliability, ease of compensation, etc.)

**Design procedures.** In the trial-and-error approach to system design, we set up a mathematical model of the control system and adjust the parameters of a compensator. The most time-consuming part of such work is the checking of the performance specifications by analysis with each adjustment of the parameters, and the designer should make use of an analog or digital computer to avoid much of the numerical drudgery necessary for this checking.

Once a satisfactory mathematical model has been obtained, the designer must construct a prototype and test the open-loop system. If absolute stability is assured, the designer closes the loop and tests the performance of the resulting closed-loop system. Because of the neglected loading effects among the components, nonlinearities, distributed parameters, etc., which were not taken into consideration in the original design work, the actual performance of the prototype system will probably differ from the theoretical predictions. Thus, the first design may not satisfy all the requirements on performance. By trial and error, the designer must make changes in the prototype until the system meets the specifications. In doing this, he must analyze each trial, and the results of the analysis must be incorporated into the next trial. The designer must see that the final system meets the performance specifications and, at the same time, is reliable and economical.

It is important to note that in design via a trial-and-error approach or in design through analysis the given specifications will not yield a unique system. In fact many (and possibly an infinite number of) systems may satisfy the given specifications. An optimal choice among the many possibilities may be made from such considerations as projected overall performance, cost, space, and weight.

**Design of complex systems.** The root-locus and frequency-response approaches to designs which essentially consist of gain adjustment and of the design of compensators are quite useful but are limited to idealized and relatively simple control systems, such as single-input–single-output linear time-invariant ones. Such approaches to design suffer from severe limitations and difficulties when applied to the design of multiple-input–multiple-output and time-varying systems.

While control system design via the root-locus and frequency-response approaches is an engineering endeavor, system design in the context of modern control theory (to be presented in Chapter 16) employs mathematical formulations of the problem and applies mathematical theory to design problems in which the system can have multiple inputs and outputs and can be time-varying. By applying modern control theory, the designer is able to start from a performance index, together with constraints imposed on the system, and to proceed to design a stable system by a completely analytical procedure. The advantage of design based on such control theory is that it enables the designer to produce a control system which is optimal with respect to the performance index considered.

It is important to note, however, that such a design technique cannot be applied if the performance specifications are given in terms of time-domain or frequency-domain quantities, in which case the root-locus or frequency-response techniques prove to be quite useful.

## 10-2 PRELIMINARY DESIGN CONSIDERATIONS

The design problems we consider in this chapter are those of improving system performance by insertion of a compensator. Compensation of a control system is reduced to the design of a filter whose characteristics tend to compensate for the undesirable and unalterable characteristics of the plant.

In Sections 10-3 through 10-5, we shall specifically consider the design of lead compensators, lag compensators, and lag-lead compensators. In such design problems, we place a compensator in series with the unalterable transfer function $G(s)$ in order to obtain desirable behavior. The main problem then involves the judicious choice of the pole(s) and zero(s) of the compensator $G_c(s)$ in order to alter the root locus or frequency response so that the performance specifications will be met.

**Root-locus approach to control system design.** The root-locus method is a graphical method for determining the locations of all closed-loop poles from knowledge of the locations of the open-loop poles and zeros as some parameter (usually the gain) is varied from zero to infinity. The method yields a clear indication of the effects of parameter adjustment. An advantage of the root-locus method is that we find that it is possible to obtain information on the transient response as well as on the frequency response from the pole-zero configuration of the system in the $s$ plane.

In practice, the root-locus plot of a system may indicate that the desired performance cannot be achieved just by the adjustment of gain. In fact, in some cases, the system may not be stable for all values of gain. Then it is necessary to reshape the root loci to meet the performance specifications.

In designing a control system, if other than a gain adjustment is required, we must modify the original root loci by inserting a suitable compensator. Once the effects on the root locus of the addition of poles and/or zeros are fully understood, we can readily determine the locations of the pole(s) and zero(s) of the compensator which will reshape the root locus as desired.

**Effects of the addition of poles.** The addition of a pole to the open-loop transfer function has the effect of pulling the root locus to the right, tending to lower the system's relative stability and to slow down the settling of the response. (Remember that the addition of integral control adds a pole at the origin, thus making the system less stable.) Figure 10-2 shows examples of root loci illustrating effects of

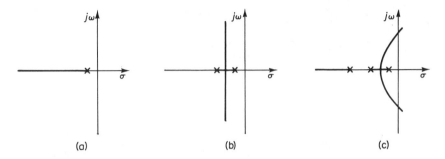

**Fig. 10-2.** (a) Root-locus plot of a single-pole system; (b) root-locus plot of a two-pole system; (c) root-locus plot of a three-pole system.

the addition of a pole to a single-pole system and the addition of two poles to a single-pole system.

**Effects of the addition of zeros.** The addition of a zero to the open-loop transfer function has the effect of pulling the root locus to the left, tending to make the system more stable and to speed up the settling of the response. (Physically, the addition of a zero in the feedforward transfer function means the addition of derivative control to the system. The effect of such control is to introduce a degree of anticipation into the system and speed up the transient response.) Figure 10-3 (a) shows the root loci for a system which is stable for small gain but unstable for large gain. Figures 10-3(b), (c), and (d) show root-locus plots for the system when a zero is added to the open-loop transfer function. Notice that when a zero is added to the system of Fig. 10-3 (a), it becomes stable for all values of gain.

**Frequency-response approach to control system design.** In dealing with the problem of compensating control systems via frequency-domain techniques, we secure control over the transient-response behavior in terms of such frequency-domain specifications as phase margin, gain margin, resonant peak value, and bandwidth. Design in the frequency domain is indirect because the system is designed to satisfy these frequency-domain specifications rather than time-domain

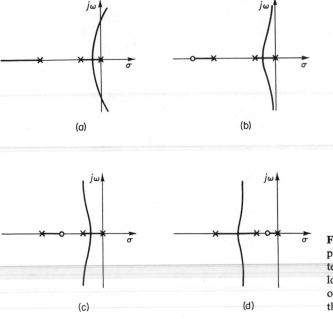

**Fig. 10-3.** (a) Root-locus plot of a three-pole system; (b), (c), and (d) root-locus plots showing effects of addition of a zero to the three-pole system.

specifications. After the open loop has been designed by the frequency-response method, the closed-loop poles and zeros can be obtained. Then transient-response characteristics must be checked to see whether or not the designed system satisfies the requirements in the time domain. If it does not, then the compensator must be modified and the analysis repeated until a satisfactory result is obtained.

Design in the frequency domain is simple and straightforward. The frequency-response plot indicates clearly the manner in which the system should be modified, although the exact quantitative prediction of the transient-response characteristics cannot be made. The frequency-response approach can be applied to systems or components whose dynamic characteristics are given in the form of frequency-response data. Note that because of difficulty in deriving the equations governing certain components, such as pneumatic and hydraulic components, the dynamic characteristics of such components are usually determined experimentally through frequency-response tests. The experimentally obtained frequency-response plots can be combined easily with other such plots. Note also that in dealing with high-frequency noises we find that the frequency-response approach is more convenient than other approaches.

In designing control systems in the frequency domain, if we desire a certain phase margin or gain margin, we note that Bode diagrams are more convenient than the polar plots. (In using Bode diagrams, unless the exact curves differ considerably from the straight-line asymptotes near the gain crossover frequency, we may employ the asymptotic plots for design purposes.) On the other hand, if we want a certain value of $M_r$, the polar plots or the log-magnitude versus phase plots are more convenient to use than the Bode diagram.

**Information obtainable from open-loop frequency response.** The low-frequency region (the region below the gain crossover frequency) of the locus indicates the steady-state behavior of the closed-loop system. The medium-frequency region (the region near the $-1 + j0$ point) of the locus indicates relative stability. The high-frequency region (the region above the gain crossover frequency) indicates the complexity of the system.

**Requirements on open-loop frequency response.** We might say that, in many practical cases, compensation is essentially a compromise between steady-state error and relative stability.

In order to have a high value of the velocity error coefficient and yet satisfactory relative stability, we find it necessary to reshape the open-loop frequency-response curve.

The gain in the low-frequency region should be large enough, and also, near

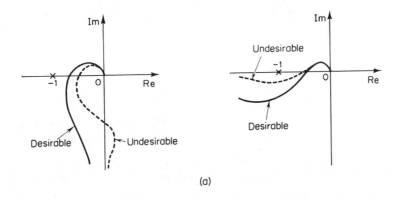

(a)

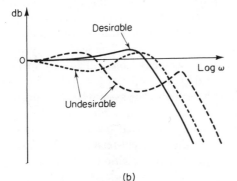

(b)

Fig. 10-4. (a) Examples of desirable and undesirable open-loop frequency-response curves; (b) examples of desirable and undesirable closed-loop frequency-response curves.

the gain crossover frequency, the slope of the log-magnitude curve in the Bode diagram should be $-20$ db/decade. This slope should extend over a sufficiently wide frequency band to assure a proper phase margin. For the high-frequency region, the gain should be attenuated as rapidly as possible in order to minimize the effects of noise.

Examples of generally desirable and undesirable open-loop and closed-loop frequency-response curves are shown in Fig. 10-4.

Referring to Fig. 10-5, we see that the reshaping of the open-loop frequency-

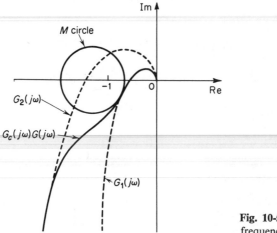

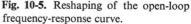

**Fig. 10-5.** Reshaping of the open-loop frequency-response curve.

response curve may be done if the high-frequency portion of the locus follows the $G_1(j\omega)$ locus, while the low-frequency portion of the locus follows the $G_2(j\omega)$ locus. The reshaped locus $G_c(j\omega)G(j\omega)$ should have reasonable phase and gain margins or should be tangent to a proper $M$ circle, as shown.

**Basic characteristics of lead, lag, and lag-lead compensation.** Lead compensation essentially yields an appreciable improvement in transient response and a small improvement in steady-state accuracy. Lag compensation, on the other hand, yields an appreciable improvement in steady-state accuracy at the expense of increasing the transient-response time. Lag-lead compensation combines the characteristics of both lead compensation and lag compensation. The use of a lead or lag compensator raises the order of the system by one. The use of a lag-lead compensator raises the order of the system by two (unless cancellation occurs between the zeros of the lag-lead network and the poles of the uncompensated open-loop transfer function), which means that the system becomes more complex and it is more difficult to control the transient-response behavior. The particular situation determines the type of the compensation to be used.

## 10-3  LEAD COMPENSATION

In this section, we shall first derive the transfer functions of an electrical lead network and of a mechanical lead network. Then we shall present procedures for designing lead compensators based on the root-locus and frequency-response approaches.

**Lead networks.** A schematic diagram of an electrical lead network is shown in Fig. 10-6 (a). The name "lead network" comes from the fact that for a sinusoidal

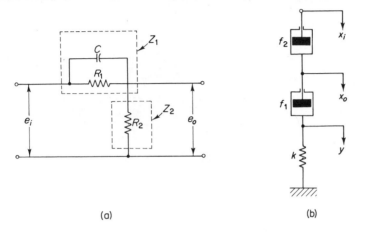

(a)                               (b)

**Fig. 10-6.** (a) Electrical lead network; (b) mechanical lead network.

input $e_i$ the output $e_0$ of the network is also sinusoidal with phase lead. The phase lead angle is a function of the input frequency. Let us derive the transfer function for this network. As usual in the derivation of the transfer function of any four-terminal network, we assume that the source impedance which the network sees is zero and that the output load impedance is infinite.

Using the sysmbols defined in Fig. 10-6 (a), we find that the complex imped-ances $Z_1$ and $Z_2$ are

$$Z_1 = \frac{R_1}{R_1 Cs + 1}, \qquad Z_2 = R_2$$

The transfer function between the output $E_0(s)$ and the input $E_i(s)$ is

$$\frac{E_0(s)}{E_i(s)} = \frac{Z_2}{Z_1 + Z_2} = \frac{R_2}{R_1 + R_2} \frac{R_1 Cs + 1}{\frac{R_1 R_2}{R_1 + R_2} Cs + 1}$$

Define

$$R_1 C = T, \qquad \frac{R_2}{R_1 + R_2} = \alpha < 1$$

Then the transfer function becomes

$$\frac{E_0(s)}{E_i(s)} = \alpha \frac{Ts + 1}{\alpha Ts + 1} = \frac{s + \dfrac{1}{T}}{s + \dfrac{1}{\alpha T}}$$

Figure 10-6 (b) shows a schematic diagram of a mechanical lead network. From the diagram, we obtain the following equations:

$$f_2(\dot{x}_i - \dot{x}_0) = f_1(\dot{x}_0 - \dot{y})$$
$$f_1(\dot{x}_0 - \dot{y}) = ky$$

Taking the Laplace transforms of these two equations, assuming zero initial conditions, and then eliminating $Y(s)$, we obtain

$$\frac{X_0(s)}{X_i(s)} = \frac{f_2}{f_1 + f_2} \frac{\dfrac{f_1}{k} s + 1}{\dfrac{f_2}{f_1 + f_2} \dfrac{f_1}{k} s + 1}$$

This is the transfer function between $X_0(s)$ and $X_i(s)$. By defining

$$\frac{f_1}{k} = T, \qquad \frac{f_2}{f_1 + f_2} = \alpha < 1$$

we obtain

$$\frac{X_0(s)}{X_i(s)} = \alpha \frac{Ts + 1}{\alpha Ts + 1} = \frac{s + \dfrac{1}{T}}{s + \dfrac{1}{\alpha T}}$$

**Characteristics of lead networks.** A lead network has the following transfer function:

$$\alpha \frac{Ts + 1}{\alpha Ts + 1} = \frac{s + \dfrac{1}{T}}{s + \dfrac{1}{\alpha T}} \qquad (\alpha < 1)$$

It has a zero at $s = -1/T$ and a pole at $s = -1/(\alpha T)$. Since $\alpha < 1$, we see that the zero is always located to the right of the pole in the complex plane. Note that for a small value of $\alpha$ the pole is located far to the left. The minimum value of $\alpha$ is limited by the physical construction of the lead network. The minimum value of $\alpha$ is usually taken to be about 0.07. If the value of $\alpha$ is small, it is necessary to cascade an amplifier in order to compensate for the attenuation of the lead network.

Figure 10-7 shows the polar plot of

$$\alpha \frac{j\omega T + 1}{j\omega \alpha T + 1} \qquad (0 < \alpha < 1)$$

For a given value of $\alpha$, the angle between the positive real axis and the tangent line drawn from the origin to the semicircle gives the maximum phase lead angle, $\phi_m$. We shall call the frequency at the tangent point $\omega_m$. From Fig. 10-7, the phase angle at $\omega = \omega_m$ is

$$\sin \phi_m = \frac{\dfrac{1 - \alpha}{2}}{\dfrac{1 + \alpha}{2}} = \frac{1 - \alpha}{1 + \alpha} \tag{10-1}$$

Equation (10-1) relates the maximum phase lead angle and the value of $\alpha$.

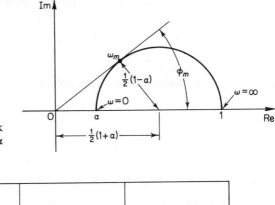

**Fig. 10-7.** Polar plot of a lead network $\alpha(j\omega T + 1)/(j\omega\alpha T + 1)$, where $0 < \alpha < 1$.

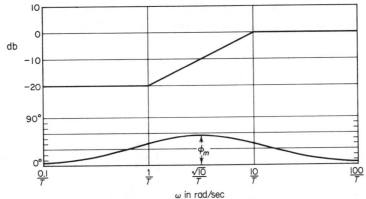

**Fig. 10-8.** Bode diagram of a lead network.

Figure 10-8 shows the Bode diagram of a lead network when $\alpha = 0.1$. The corner frequencies for the lead network are $\omega = 1/T$ and $\omega = 1/(\alpha T)$. By examining Fig. 10-8, we see that $\omega_m$ is the geometric mean of the two corner frequencies, or

$$\log \omega_m = \frac{1}{2}\left[\log \frac{1}{T} + \log \frac{1}{\alpha T}\right]$$

Hence

$$\omega_m = \frac{1}{\sqrt{\alpha}\, T} \tag{10-2}$$

As seen from Fig. 10-8, the lead network is basically a high-pass filter. (The high frequencies are passed but low frequencies are attenuated.) Therefore, an additional gain elsewhere is needed to increase the low-frequency gain.

**Lead compensation techniques based on the root-locus approach.** The root-locus approach to design is very powerful when the specifications are given in terms of time-domain quantities, such as maximum overshoot, rise time, settling time, damping ratio, and undamped natural frequency of the desired dominant closed-loop poles.

Consider a design problem in which the original system either is unstable for all values of gain or is stable but has undesirable transient-response characteristics. In such a case, the reshaping of the root locus is necessary in the broad neighborhood of the $j\omega$ axis and the origin in order that the dominant closed-loop poles be at desired locations in the complex plane. This problem may be solved by inserting an appropriate lead compensator in cascade with the feedforward transfer function.

The procedures for designing a lead compensator by the root-locus method may be stated as follows:

1. From the performance specifications, determine the desired location for the dominant closed-loop poles.

2. By drawing the root-locus plot, ascertain whether or not the gain adjustment alone can yield the desired closed-loop poles. If not, calculate the angle deficiency $\phi$. This angle must be contributed by the lead network if the new root locus is to pass through the desired locations for the dominant closed-loop poles.

3. If static error coefficients are not specified, determine the location of the pole and zero of the lead network so that the lead network will contribute the necessary angle $\phi$ and will require the minimum amount of additional gain. (If a particular static error coefficient is specified, it is generally simpler to use the frequency-response approach.)

4. Determine the open-loop gain of the compensated system from the magnitude condition.

Once a compensator has been designed, check to see whether or not all performance requirements have been met. (The use of analog or digital computers facilitates the check on the transient-response characteristics.) If the compensated system does not meet the performance specifications, then repeat the design procedure by adjusting the compensator pole and zero until all such specifications are met. If a large static error coefficient is required, cascade a lag network or alter the lead compensator to a lag-lead compensator.

Note that if the selected dominant closed-loop poles are not really dominant, it will be necessary to modify the location of the pair of closed-loop poles by a trial-and-error approach. (The closed-loop poles other than dominant ones modify the response obtained from the dominant closed-loop poles alone. The amount of modification depends on the location of these remaining closed-loop poles.)

---

*Example 10-1.* Consider the system shown in Fig. 10-9 (a). The feedforward transfer function is

$$G(s) = \frac{4}{s(s+2)}$$

The root-locus plot for this system is shown in Fig. 10-9 (b). The closed-loop transfer function becomes

$$\frac{C(s)}{R(s)} = \frac{4}{s^2 + 2s + 4}$$

$$= \frac{4}{(s + 1 + j\sqrt{3})(s + 1 - j\sqrt{3})}$$

The closed-loop poles are located at

$$s = -1 \pm j\sqrt{3}$$

The damping ratio of the closed-loop poles is 0.5. The undamped natural frequency of the closed-loop poles is 2 rad/sec. The static velocity error coefficient is 2 sec$^{-1}$.

It is desired to modify the closed-loop poles so that an undamped natural frequency $\omega_n = 4$ rad/sec is obtained, without changing the value of the damping ratio, $\zeta = 0.5$.

Recall that in the complex plane the damping ratio $\zeta$ of a pair of complex poles can be expressed in terms of the angle $\theta$ which is measured from the negative real axis, as shown in Fig. 10-10 (a), with

$$\zeta = \cos\theta$$

In other words, lines of constant damping ratio $\zeta$ are radial lines passing through the origin as shown in Fig. 10-10 (b). For example, a damping ratio of 0.5 requires that the complex poles lie on the lines drawn through the origin making angles of $\pm60°$ with the negative real axis. (If the real part of a pair of complex poles is positive, which means that the system is unstable, the corresponding $\zeta$ is negative.) The damping ratio determines the angular location of the poles, while the distance of the pole from the origin is determined by the undamped natural frequency $\omega_n$.

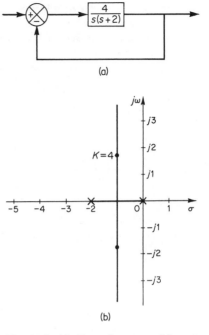

Fig. 10-9. (a) Control system; (b) root-locus plot.

In the present example, the desired locations of the closed-loop poles are

$$s = -2 \pm j2\sqrt{3}$$

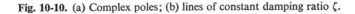

Fig. 10-10. (a) Complex poles; (b) lines of constant damping ratio $\zeta$.

In some cases, after the root loci of the original system have been obtained, the dominant closed-loop poles may be moved to the desired location by simple gain adjustment. This is, however, not the case for the present system. Therefore, we shall insert a lead compensator in the feedforward path.

A general procedure for determining the lead compensator is as follows: First, find the sum of the angles at the desired location of one of the dominant closed-loop poles with the open-loop poles and zeros of the original system, and determine the necessary angle $\phi$ to be added so that the total sum of the angles is equal to $\pm 180°(2k + 1)$. The lead network must contribute this angle. (If the angle is quite large, then two or more lead networks may be needed rather than a single one.)

If the original system has the open-loop transfer function $G(s)$, then the compensated system will have the open-loop transfer function

$$G_1(s) = \left(\alpha \frac{Ts + 1}{\alpha Ts + 1}\right) K_c G(s)$$

where the first term on the right-hand side corresponds to the lead network, the second term $K_c$ is the gain of the amplifier, and the last term $G(s)$ is the original open-loop transfer function. (Note that the amplifier provides the desired impedance matching as well as the desired gain $K_c$.) Note that there are many possible values for $T$ that will yield the necessary angle contribution at the desired closed-loop poles.

The next step is to determine the locations of the pole and zero of the lead network; in other words, the value of $T$. In choosing the value of $T$, we shall introduce a procedure to obtain the largest possible value for $\alpha$ so that the additional gain required of the amplifier is as small as possible. First, draw a horizontal line passing through point $P$, the desired location for one of the dominant closed-loop poles. This is shown as line $PA$ in Fig. 10-11. Draw also a line connecting point $P$ and the origin. Bisect the angle between the lines $PA$ and $PO$, as shown in Fig. 10-11. Draw two lines $PC$ and $PD$ which make angles $\pm \phi/2$

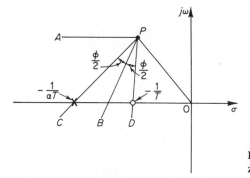

**Fig. 10-11.** Determination of the pole and zero of a lead network.

with the bisector $PB$. The intersections of $PC$ and $PD$ with the negative real axis give the necessary location for the pole and zero of the lead network. The compensator thus designed will make point $P$ a point on the root locus of the compensated system. The open-loop gain is determined by means of the magnitude condition.

In the present system, the angle of $G(s)$ at the desired closed-loop pole is

$$\left/ \frac{4}{s(s + 2)} \right|_{s = -2 + j2\sqrt{3}} = -210°$$

Thus for the root locus to go through the desired closed-loop pole, the lead network must

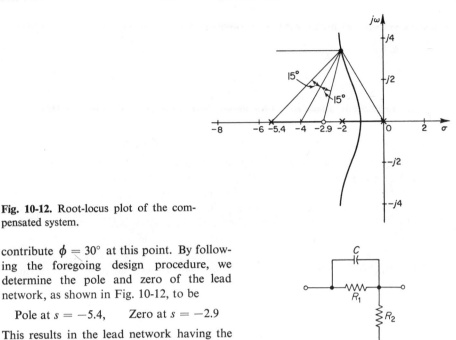

**Fig. 10-12.** Root-locus plot of the compensated system.

contribute $\phi = 30°$ at this point. By following the foregoing design procedure, we determine the pole and zero of the lead network, as shown in Fig. 10-12, to be

Pole at $s = -5.4$,     Zero at $s = -2.9$

This results in the lead network having the parameter values shown in Fig. 10-13. The lead compensator consisting of this lead network and an amplifier has the transfer function

$$G_c(s) = \frac{(s + 2.9)}{(s + 5.4)} K_c$$

Thus the open-loop transfer function of the compensated system becomes

$R_1 = 345 \text{ k}\Omega$
$R_2 = 400 \text{ k}\Omega$
$C = 1\mu\text{f}$

Transfer function $= \frac{400}{745} \left( \frac{0.345s + 1}{0.185s + 1} \right) = \frac{s + 2.9}{s + 5.4}$

**Fig. 10-13.** Lead network.

$$G_c(s)G(s) = \frac{(s + 2.9)}{(s + 5.4)} K_c \frac{4}{s(s + 2)}$$

$$= \frac{K(s + 2.9)}{s(s + 2)(s + 5.4)}$$

The root-locus plot for the compensated system is shown in Fig. 10-12. The gain $K$ is evaluated from the magnitude condition as follows:

$$\left| \frac{K(s + 2.9)}{s(s + 2)(s + 5.4)} \right|_{s = -2 + j2\sqrt{3}} = 1$$

or

$$K = 18.7$$

It follows that

$$G_c(s)G(s) = \frac{18.7(s + 2.9)}{s(s + 2)(s + 5.4)}$$

The gain constant $K_c$ of the amplifier is

$$K_c = \frac{K}{4} = 4.68$$

The static velocity error coefficient $K_v$ is obtained from the expression

$$K_v = \lim_{s \to 0} sG_c(s)G(s)$$
$$= \lim_{s \to 0} \frac{s18.7(s + 2.9)}{s(s + 2)(s + 5.4)}$$
$$= 5.02 \text{ sec}^{-1}$$

The third closed-loop pole is found by dividing the characteristic equation by the known factors as follows:

$$s(s + 2)(s + 5.4) + 18.7(s + 2.9) = (s + 2 + j2\sqrt{3})(s + 2 - j2\sqrt{3})(s + 3.4)$$

The foregoing compensation method enables us to place the dominant closed-loop poles at the desired points in the complex plane. The third pole at $s = -3.4$ is close to the added zero at $s = -2.9$. Therefore, the effect of this pole on the transient response is relatively small. Since no restriction has been imposed on the nondominant pole and no specification has been given concerning the value of the static velocity error coefficient, we conclude that the present design is satisfactory.

---

If the value of the static velocity error coefficient $K_v$ were specified, then we must meet the specification by modifying the pole-zero location of the lead network or by inserting a different network. Note that changing the pole-zero location of the lead network without changing the angle $\phi$ causes a change in the value of $K_v$. A certain change in the value of $K_v$ can thus be made by altering the pole-zero locations of the lead network. If a large increase in the value of $K_v$ is desired, then we must alter the lead compensator to a lag-lead compensator.

**Lead compensation techniques based on the frequency-response approach.** The primary function of the lead compensator is to reshape the frequency-response curve to provide sufficient phase lead angle to offset the excessive phase lag associated with the components of the fixed system.

Let us assume that a unity feedback system is given. We wish to satisfy the performance requirements, which are given in terms of phase margin, gain margin, error coefficients, etc. The procedures for designing a lead compensator by the frequency-response approach may be stated as follows:

1. Determine the open-loop gain $K$ to satisfy the requirement on the error coefficients.

2. Using the gain $K$ thus determined, evaluate the phase margin of the uncompensated system.

3. Determine the necessary phase lead angle $\phi$ to be added to the system.

4. Determine the attenuation factor $\alpha$ by use of Eq. (10-1). Determine the frequency where the magnitude of the uncompensated system is equal to $-20 \log(1/\sqrt{\alpha})$. Select this frequency as the new gain crossover frequency. This frequency corresponds to $\omega_m$ and the maximum phase shift $\phi_m$ occurs at this frequency.

5. Determine the corner frequencies of the lead network from

$$\omega = \frac{1}{T}, \qquad \omega = \frac{1}{\alpha T}$$

Finally, insert an amplifier with gain equal to $1/\alpha$, or increase the gain of the existing amplifier by a factor of $1/\alpha$.

---

*Example 10-2.* Consider the system shown in Fig. 10-14. The open-loop transfer function is

$$G(s) = \frac{4K}{s(s + 2)}$$

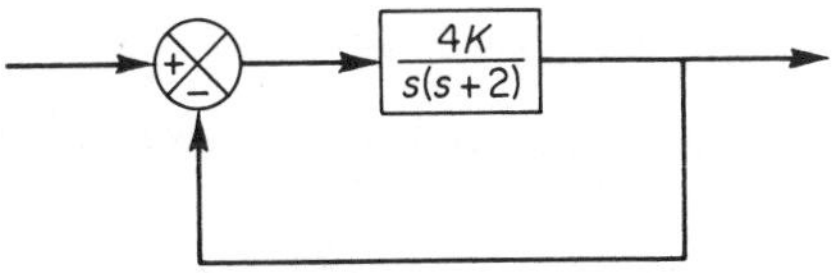

**Fig. 10-14.** Control system.

It is desired to find a compensator for the system so that the static velocity error coefficient $K_v$ is 20 sec$^{-1}$, the phase margin is at least 50°, and the gain margin is at least 10 db.

In the present example, the phase and gain margins have been specified. We shall therefore employ Bode diagrams.

The first step in the design is to adjust the gain $K$ to meet the steady-state performance specification or to provide the required static velocity error coefficient. Since this coefficient is given as 20 sec$^{-1}$, we obtain

$$K_v = \lim_{s \to 0} sG(s) = \lim_{s \to 0} \frac{s4K}{s(s + 2)} = 2K = 20$$

or

$$K = 10$$

With $K = 10$, the system of Fig. 10-14 satisfies the steady-state requirement.

We shall next plot the Bode diagram of

$$G(j\omega) = \frac{40}{j\omega(j\omega + 2)} = \frac{20}{j\omega(0.5j\omega + 1)}$$

Figure 10-15 shows the magnitude and phase angle curves of $G(j\omega)$. From this plot, the phase and gain margins of the system are found to be 17° and $+\infty$ db, respectively. (A phase margin of 17° implies that the system is quite oscillatory. Thus, satisfying the specification on the steady state yields a poor transient-response performance.) The specification calls for a phase margin of at least 50°. We thus find that the additional phase lead necessary to satisfy the relative stability requirement is 33°. In order to achieve a phase margin of 50° without decreasing the value of $K$, it is necessary to insert a suitable lead compensator into the system.

Noting that the addition of a lead compensator modifies the magnitude curve in the Bode diagram, we realize that the gain crossover frequency will be shifted to the right. We must offset the increased phase lag of $G(j\omega)$ due to this increase in the gain crossover frequency. Considering the shift of the gain crossover frequency, we may assume that $\phi_m$, the maximum phase lead required, is approximately 38°. (This means that 5° has been added to compensate for the shift in the gain crossover frequency.) Since

$$\sin \phi_m = \frac{1 - \alpha}{1 + \alpha}$$

$\phi_m = 38°$ corresponds to $\alpha = 0.24$. Once the attenuation factor $\alpha$ has been determined on the basis of the required phase lead angle, the next step is to determine the corner frequencies $\omega = 1/T$ and $\omega = 1/(\alpha T)$ of the lead network. To do so, we first note that the maximum phase lead angle $\phi_m$ occurs at the geometric mean of the two corner fre-

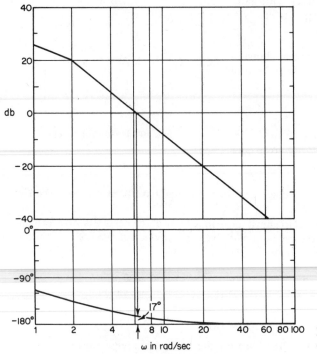

**Fig. 10-15.** Bode diagram for $G(s) = 40/[j\omega(j\omega + 2)]$.

quencies, or $\omega = 1/(\sqrt{\alpha}\,T)$. [See Eq. (10-2).] The amount of the modification in the magnitude curve at $\omega = 1/(\sqrt{\alpha}\,T)$ is

$$\left|\frac{1 + j\omega T}{1 + j\omega\alpha T}\right|_{\omega = 1/(\sqrt{\alpha}T)} = \left|\frac{1 + j\dfrac{1}{\sqrt{\alpha}}}{1 + j\alpha\dfrac{1}{\sqrt{\alpha}}}\right| = \frac{1}{\sqrt{\alpha}}$$

Note that

$$\frac{1}{\sqrt{\alpha}} = \frac{1}{\sqrt{0.24}} = \frac{1}{0.49} = 6.2 \text{ db}$$

and $|G(j\omega)| = -6.2$ db corresponds to $\omega = 9$ rad/sec. We shall select this frequency to be the new gain crossover frequency $\omega_c$. Noting that this frequency corresponds to $1/(\sqrt{\alpha}\,T)$, or $\omega_c = 1/(\sqrt{\alpha}\,T)$, we obtain

$$\frac{1}{T} = \sqrt{\alpha}\,\omega_c = 4.41$$

and

$$\frac{1}{\alpha T} = \frac{\omega_c}{\sqrt{\alpha}} = 18.4$$

The lead network thus determined is

$$\frac{s + 4.41}{s + 18.4} = \frac{0.24(0.227s + 1)}{0.054s + 1}$$

To compensate for the attenuation due to the lead network, we increase the amplifier gain by a factor of $1/0.24 = 4.17$. (If this were not done, the required static velocity error coefficient could not be realized.) Then the transfer function of the compensator which consists of the lead network and the amplifier becomes

$$G_c(s) = (4.17)\frac{s + 4.41}{s + 18.4} = \frac{0.227s + 1}{0.054s + 1}$$

The magnitude curve and phase-angle curve for $G_c(j\omega)$ are shown in Fig. 10-16. The compensated system has the following open-loop transfer function:

$$G_c(s)G(s) = \frac{0.227s + 1}{0.054s + 1}\frac{20}{s(0.5s + 1)} = (4.17)\frac{s + 4.41}{s + 18.4}\frac{40}{s(s + 2)}$$

The solid curves in Fig. 10-16 show the magnitude curve and phase-angle curve for the

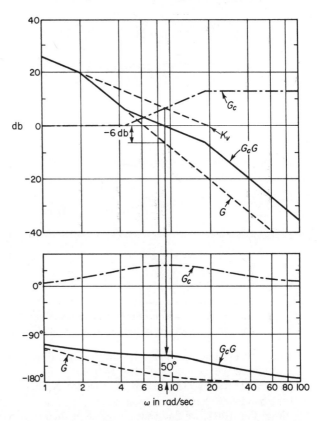

**Fig. 10-16.** Bode diagram for the compensated system.

compensated system. The lead compensator causes the gain crossover frequency to increase from 6.3 to 9 rad/sec. The increase in this frequency means an increase in bandwidth. This implies an increase in the speed of response. The phase and gain margins are seen to be approximately 50° and $+\infty$ db, respectively. The compensated system shown in Fig. 10-17 therefore meets both the steady-state and the relative-stability requirements.

Note that for type 1 systems, such as the system just considered (with the values considered here), the value of the static velocity error coefficient $K_v$ is merely the value of the frequency corresponding to the intersection of the initial $-20$ db/decade slope line and the 0-db line, as shown in Fig. 10-16.

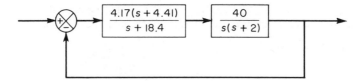

**Fig. 10-17.** Compensated system.

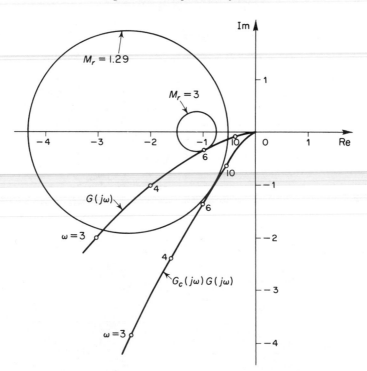

**Fig. 10-18.** Polar plots of the uncompensated and compensated open-loop transfer function. ($G$: uncompensated system, $G_cG$: compensated system.)

Figure 10-18 shows the polar plots of $G(j\omega)$ (with $K = 10$) and of $G_c(j\omega)G(j\omega)$. From Fig. 10-18, we see that the resonant frequency of the uncompensated system is about 6 rad/sec and that of the compensated system is about 7 rad/sec. (This also indicates that the bandwidth has been increased.)

From Fig. 10-18, we find that the value of the resonant peak $M_r$ for the uncompensated system with $K = 10$ is 3. The value of $M_r$ for the compensated system is found to be 1.29. This clearly shows that the compensated system has improved relative stability. (Note that the value of $M_r$ may be obtained easily by transferring the data from the Bode diagram to the Nichol's chart. See Example 10-4.)

Note that if the phase angle of $G(j\omega)$ decreases rapidly near the gain crossover frequency, lead compensation becomes ineffective because the shift in the gain crossover frequency to the right makes it difficult to provide enough phase lead at the new gain crossover frequency. This means that, in order to provide the desired phase margin, we must use a very small value for $\alpha$. The value of $\alpha$, however, should not be smaller than 0.07 nor should the maximum phase lead $\phi_m$ be more than 60° because such values will require an additional gain of excessive value. [If more than 60° is needed, two (or more) lead networks may be used in series with an isolating amplifier.]

## 10-4   LAG COMPENSATION

**Lag networks.** Figure 10-19 (a) shows an electrical lag network. The name "lag network" comes from the fact that when the input voltage $e_i$ is sinusoidal, the

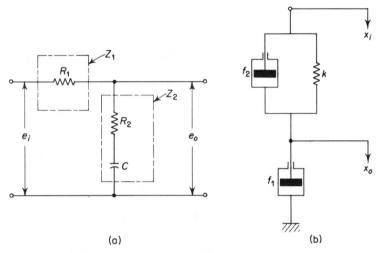

(a)                               (b)

**Fig. 10-19.** (a) Electrical lag network; (b) mechanical lag network.

output voltage $e_0$ is sinusoidal but lags the input by an angle which is a function of the frequency of the input sinusoid. The complex impedances $Z_1$ and $Z_2$ are

$$Z_1 = R_1, \qquad Z_2 = R_2 + \frac{1}{Cs}$$

The transfer function between the output voltage $E_0(s)$ and the input voltage $E_i(s)$ is given by

$$\frac{E_0(s)}{E_i(s)} = \frac{Z_2}{Z_1 + Z_2} = \frac{R_2 Cs + 1}{(R_1 + R_2)Cs + 1}$$

Define

$$R_2 C = T, \qquad \frac{R_1 + R_2}{R_2} = \beta > 1$$

Then the transfer function becomes

$$\frac{E_0(s)}{E_i(s)} = \frac{Ts + 1}{\beta Ts + 1} = \frac{1}{\beta}\left(\frac{s + \dfrac{1}{T}}{s + \dfrac{1}{\beta T}}\right)$$

Figure 10-19 (b) shows a mechanical lag network. It consists of a spring and two dashpots. The differential equation for this mechanical network is

$$f_1 \dot{x}_0 = k(x_i - x_0) + f_2(\dot{x}_i - \dot{x}_0)$$

Taking the Laplace transforms of both sides of this equation, assuming zero initial conditions and then rewriting, we obtain

$$\frac{X_0(s)}{X_i(s)} = \frac{f_2 s + k}{(f_1 + f_2)s + k}$$

$$= \frac{\dfrac{f_2}{k}s + 1}{\dfrac{f_1 + f_2}{k}s + 1}$$

If we define

$$\frac{f_2}{k} = T, \qquad \frac{f_1 + f_2}{f_2} = \beta > 1$$

then the transfer function $X_0(s)/X_i(s)$ becomes

$$\frac{X_0(s)}{X_i(s)} = \frac{Ts + 1}{\beta Ts + 1} = \frac{1}{\beta} \left( \frac{s + \dfrac{1}{T}}{s + \dfrac{1}{\beta T}} \right)$$

**Characteristics of lag networks.** A lag network has the following transfer function:

$$\frac{Ts + 1}{\beta Ts + 1} = \frac{1}{\beta} \left( \frac{s + \dfrac{1}{T}}{s + \dfrac{1}{\beta T}} \right) \qquad (\beta > 1)$$

In the complex plane, a lag network has a pole at $s = -1/(\beta T)$ and a zero at $s = -1/T$. (The pole is located to the right of the zero.)

Figure 10-20 shows a typical polar plot of a lag network. Figure 10-21 shows the Bode diagram of a lag network when $\beta = 10$. The corner frequencies of the lag network are $\omega = 1/T$ and $\omega = 1/(\beta T)$. As seen from Fig. 10-21, the lag network is essentially a low-pass filter.

**Lag compensation techniques based on the root-locus approach.** Consider the problem of finding a suitable compensation network for the case where the system exhibits satisfactory transient-response characteristics but unsatisfactory steady-state characteristics. Compensation in this case essentially consists of increasing the open-loop gain without appreciably changing the transient-response characteristics. This means that the root locus in the neighborhood of the dominant closed-loop poles should not be changed appreciably, but the open-loop gain should be increased as much as needed. This can be accomplished if a lag compensator is put in cascade with the given feedforward transfer function.

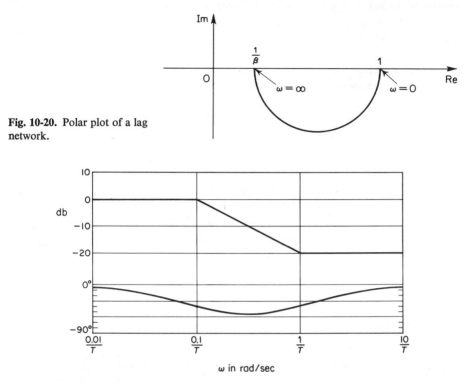

**Fig. 10-20.** Polar plot of a lag network.

**Fig. 10-21.** Bode diagram of a lag network $(j\omega T + 1)/(j\omega\beta T + 1)$ with $\beta = 10$.

To avoid an appreciable change in the root loci, the angle contribution of the lag network should be limited to a small amount, say 5°. To assure this, we place the pole and zero of the lag network relatively close together and near the origin of the $s$ plane. Then the closed-loop poles of the compensated system will be shifted only slightly from their original locations. Hence, the transient-response characteristics will be essentially unchanged.

Note that if we place the pole and zero of the lag network very close to each other, then $s_1 + (1/T)$ and $s_1 + (1/\beta T)$ are almost equal, where $s_1$ is the closed-loop pole. Thus

$$\left| \frac{1}{\beta} \left( \frac{s_1 + \dfrac{1}{T}}{s_1 + \dfrac{1}{\beta T}} \right) \right| \doteq \frac{1}{\beta}$$

This implies that the open-loop gain can be increased approximately by a factor of $\beta$ without altering the transient-response characteristics. If the pole and zero are placed very close to the origin, the value of $\beta$ can be made large. Usually, $1 < \beta < 15$, and $\beta = 10$ is a good choice.

An increase in the gain means an increase in the static error coefficients. If the open-loop transfer function of the uncompensated system is $G(s)$, then the static velocity error coefficient $K_v$ is

$$K_v = \lim_{s \to 0} sG(s)$$

If the compensator is chosen as

$$G_c(s) = K_c \frac{Ts + 1}{\beta Ts + 1} = \frac{K_c}{\beta} \left( \frac{s + \dfrac{1}{T}}{s + \dfrac{1}{\beta T}} \right) \tag{10-3}$$

then for the compensated system with the open-loop transfer function $G_c(s)G(s)$ the static velocity error coefficient $\hat{K}_v$ becomes

$$\begin{aligned} \hat{K}_v &= \lim_{s \to 0} sG_c(s)G(s) \\ &= \lim_{s \to 0} G_c(s)K_v \\ &= K_c K_v \end{aligned}$$

Thus if the compensator is given by Eq. (10-3), then the static velocity error coefficient is increased by a factor of $K_c$.

The procedures for designing a lag compensator by the root-locus method may be stated as follows: (We assume that the uncompensated system meets the transient-response specifications by simple gain adjustment. If this is not the case, refer to Section 10-5.)

1. Draw the root-locus plot for the uncompensated system. Based on the transient-response specifications, locate the dominant closed-loop poles on the root locus.

2. Determine the open-loop gain by use of the magnitude condition.

3. Evaluate the particular error coefficient specified in the problem.

4. Determine the amount of increase in the error coefficient necessary to satisfy the specifications.

5. Determine the pole and zero of the lag network which produce the necessary increase in the particular error coefficient without appreciably altering the original root loci.

6. Draw a new root-locus plot for the compensated system. Locate the desired dominant closed-loop poles on the root locus. (If the angle contribution of the lag network is very small, i.e., a few degrees, then the original and new root loci are almost identical. Otherwise there will be a slight discrepancy between them. Then locate, on the new root locus, the desired dominant closed-loop poles based on the transient-response specifications.)

7. Adjust the gain of the amplifier from the magnitude condition that the dominant closed-loop poles lie at the desired location.

---

*Example 10-3.* Consider the system shown in Fig. 10-22 (a). The feedforward transfer function is

$$G(s) = \frac{1.06}{s(s + 1)(s + 2)}$$

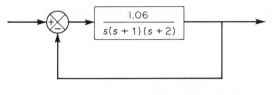

(a)

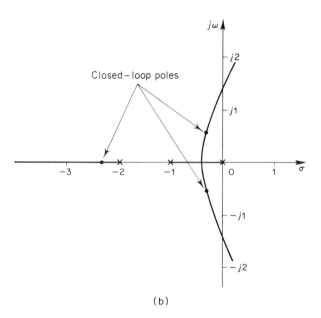

**Fig. 10-22.** (a) Control system;
(b) root-locus plot.                                                    (b)

The root-locus plot for the system is shown in Fig. 10-22 (b). The closed-loop transfer function becomes

$$\frac{C(s)}{R(s)} = \frac{1.06}{s(s+1)(s+2) + 1.06}$$

$$= \frac{1.06}{(s+0.33 - j0.58)(s+0.33 + j0.58)(s+2.33)}$$

The dominant closed-loop poles are

$$s = -0.33 \pm j0.58$$

The damping ratio of the dominant closed-loop poles is $\zeta = 0.5$. The undamped natural frequency of the dominant closed-loop poles is 0.67 rad/sec. The static velocity error coefficient is 0.53 sec$^{-1}$.

It is desired to increase the static velocity error coefficient $K_v$ to about 5 sec$^{-1}$ without appreciably changing the location of the dominant closed-loop poles.

To meet this specification, let us insert a lag compensator, which consists of a lag network and an amplifier, in cascade with the given feedforward transfer function. In order to increase the static velocity error coefficient by a factor of about 10, let us place the pole and zero of the lag network at $s = -0.01$ and $s = -0.1$, respectively. The transfer function of the lag network is then

$$\frac{1}{10}\left(\frac{s+0.1}{s+0.01}\right)$$

The angle contribution of this lag network near a dominant closed-loop pole is around seven degrees. (This is about the maximum we can allow.) Because the angle contribution of this lag network is not very small, there is a small change in the new root locus near the desired dominant closed-loop poles.

In order to account for the attenuation due to the lag network, we cascade an amplifier of gain $K_c$. The feedforward transfer function of the compensated system then becomes

$$G_1(s) = \frac{1}{10}\left(\frac{s+0.1}{s+0.01}\right)(K_c)\frac{1.06}{s(s+1)(s+2)}$$

$$= \frac{K(s+0.1)}{s(s+0.01)(s+1)(s+2)}$$

where

$$K = \frac{1.06K_c}{10}$$

The block diagram of the compensated system is shown in Fig. 10-23 (a). The root-locus

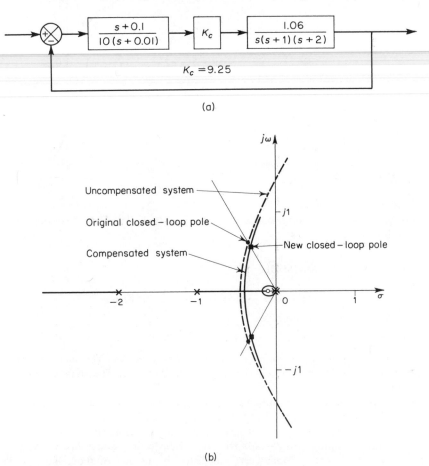

(a)

(b)

**Fig. 10-23.** (a) Compensated system; (b) root-locus plots for the compensated system and the uncompensated system.

plot for the compensated system near the dominant closed-loop poles is shown in Fig. 10-23 (b), together with the original root locus.

If the damping ratio of the new dominant closed-loop poles is kept the same, then the poles are obtained from the new root-locus plot as follows:

$$s_1 = -0.28 + j0.51, \qquad s_2 = -0.28 - j0.51$$

The open-loop gain $K$ is

$$K = \left| \frac{s(s + 0.01)(s + 1)(s + 2)}{s + 0.1} \right|_{s=-0.28+j0.51} = 0.98$$

Hence the amplifier gain $K_c$ is

$$K_c = \frac{10}{1.06} K = 9.25$$

Thus the compensated system has the following open-loop transfer function:

$$G_1(s) = \frac{0.98(s + 0.1)}{s(s + 0.01)(s + 1)(s + 2)} = \frac{4.9(10s + 1)}{s(100s + 1)(s + 1)(0.5s + 1)}$$

The static velocity error coefficient $K_v$ is

$$K_v = \lim_{s \to 0} sG_1(s) = 4.9 \text{ sec}^{-1}$$

In the compensated system, the static velocity error coefficient has increased to 4.9 sec$^{-1}$, or 4.9/0.53 = 9.25 times the original value. (The steady-state error with ramp inputs has decreased to about 11% of that of the original system.) We have essentially accomplished the design objective of increasing the static velocity error coefficient to about 5 sec$^{-1}$. (If we wish to increase the static velocity error coefficient to exactly 5 sec$^{-1}$, we can either modify the locations of the pole and zero of the lag network or use the present lag network and choose $K_c = 9.44$. In the latter case, however, the damping ratio of the dominant closed-loop poles is less than 0.5. In the present problem, we may consider the present design as quite acceptable.)

The two other closed-loop poles for the compensated system are found as follows:

$$s_3 = -2.31, \qquad s_4 = -0.137$$

The addition of the lag network increases the order of the system from three to four, adding one additional pole close to the zero of the lag network. Because the added pole at $s = -0.137$ is close to the zero at $s = -0.1$, the effect of this pole on the transient response is small. Since the pole at $s = -2.31$ is very far from the $j\omega$ axis compared with the dominant closed-loop poles, the effect of this pole on the transient response is also small. We may therefore discard, with little error, the closed-loop poles $s_3$ and $s_4$. The conclusion is that the two closed-loop poles $s_1$ and $s_2$ are truly dominant ones. We can predict a fairly accurate response by considering only the dominant closed-loop poles.

The undamped natural frequency of the compensated system is 0.6 rad/sec. This value is about 10% less than the original value, 0.67 rad/sec. This implies that the transient response of the compensated system is slower than that of the original. It will take a longer time to settle down. If this can be tolerated, the lag compensation as discussed here presents a satisfactory solution to the given design problem.

Finally, note that the crossing points of the root loci of the original and of the compensated systems with the $j\omega$ axis are approximately the same. This means that the compensated system is still stable, even if the open-loop gain is increased by a factor of about 10 over the original critical value.

**Lag compensation techniques based on the frequency-response approach.** The primary function of a lag network is to provide attenuation in the high-frequency range in order to give a system sufficient phase margin. The phase lag characteristic is of no consequence in lag compensation.

The procedures for designing a lag compensator by the frequency-response approach may be stated as follows: (We assume that the system has unity feedback.)

1. Determine the open-loop gain such that the requirement on the particular error coefficient is satisfied.

2. Using the gain thus determined, draw the Bode diagram of the uncompensated system and determine the phase and gain margins of the uncompensated system.

3. If the specifications on the phase and gain margins are not satisfied, then find the frequency point where the phase angle of the open-loop transfer function is equal to $-180°$ plus the required phase margin. The required phase margin is the specified phase margin plus $5°$ to $12°$. (The addition of $5°$ to $12°$ compensates for the phase lag of the lag network.) Choose this frequency as the new gain crossover frequency.

4. Choose the corner frequency $\omega = 1/T$ (corresponding to the zero of the lag network) one octave to one decade below the new gain crossover frequency. (If the time constants of the lag network do not become too large, the corner frequency $\omega = 1/T$ may be chosen one decade below the new gain crossover frequency.)

5. Determine the attenuation necessary to bring the magnitude curve down to 0 db at the new gain crossover frequency. Noting that this attenuation is $-20 \log \beta$, determine the value of $\beta$. Then the other corner frequency (corresponding to the pole of the lag network) is determined from $\omega = 1/(\beta T)$.

---

*Example 10-4.* Consider the system shown in Fig. 10-24. The open-loop transfer function is given by

$$G(s) = \frac{K}{s(s + 1)(0.5s + 1)}$$

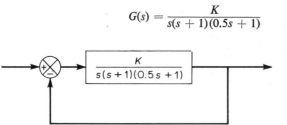

**Fig. 10-24.** Control system.

It is desired to compensate the system so that the static velocity error coefficient $K_v$ is 5 sec$^{-1}$, the phase margin is at least $40°$, and the gain margin is at least 10 db.

The first step in the design is to adjust the gain $K$ to meet the required static velocity error coefficient. Thus

$$K_v = \lim_{s \to 0} sG(s) = \lim_{s \to 0} \frac{sK}{s(s + 1)(0.5s + 1)} = K = 5$$

or

$$K = 5$$

With $K = 5$, the system of Fig. 10-24 satisfies the steady-state performance requirement. We shall next plot the Bode diagram of

$$G(j\omega) = \frac{5}{j\omega(j\omega + 1)(0.5j\omega + 1)}$$

The magnitude curve and phase-angle curve of $G(j\omega)$ are shown in Fig. 10-25. From this plot, the phase margin is found to be $-20°$, which means that the system is unstable.

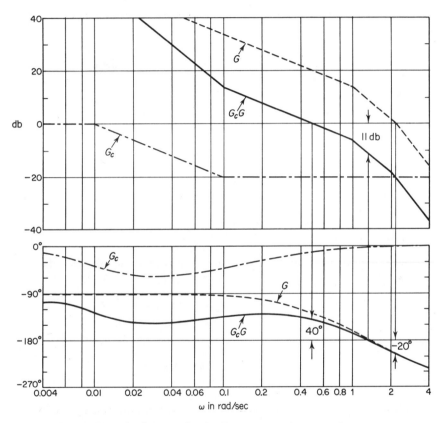

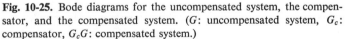

**Fig. 10-25.** Bode diagrams for the uncompensated system, the compensator, and the compensated system. ($G$: uncompensated system, $G_c$: compensator, $G_cG$: compensated system.)

Lead compensation is rather ineffective in this case. We shall demonstrate the use of lag compensation to meet the required specifications. Noting that the addition of a lag network modifies the phase curve of the Bode diagram, we must allow 5° to 12° to the specified phase margin to compensate for the modification of the phase curve. Since the frequency corresponding to a phase margin of 40° is 0.7 rad/sec, the new gain crossover frequency (of the compensated system) must be chosen near this value. In order to avoid overly large time constants for the lag network, we shall choose the corner frequency

$\omega = 1/T$ (which corresponds to the zero of the lag network) to be 0.1 rad/sec. Since this corner frequency is not too far below the new gain crossover frequency, the modification in the phase curve may not be small. Hence we add about 12° to the given phase margin as an allowance to account for the lag angle introduced by the lag network. The required phase margin is now 52°. The phase angle of the uncompensated open-loop transfer function is $-128°$ at about $\omega = 0.5$ rad/sec. So we choose the new gain crossover frequency to be 0.5 rad/sec. To bring the magnitude curve down to 0 db at this new gain crossover frequency, the lag network must give the necessary attenuation, which, in this case, is $-20$ db. Hence

$$20 \log \frac{1}{\beta} = -20$$

or

$$\beta = 10$$

The other corner frequency $\omega = 1/(\beta T)$ (which corresponds to the pole of the lag network) is then determined as

$$\frac{1}{\beta T} = 0.01 \text{ rad/sec}$$

Thus the transfer function of the necessary lag network is given by

$$\frac{1}{10} \frac{s + 0.1}{s + 0.01}$$

The open-loop transfer function of the compensated system is

$$G_c(s)G(s) = \frac{5(10s + 1)}{s(100s + 1)(s + 1)(0.5s + 1)}$$

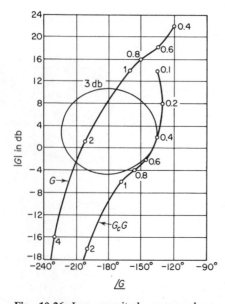

**Fig. 10-26.** Log magnitude versus phase ~ts of the uncompensated system and ~mpensated system. (*G*: uncompen-~n, *G_c G*: compensated system.)

The magnitude and phase-angle curves of $G_c(j\omega)G(j\omega)$ are also shown in Fig. 10-25. Note that the effect on the phase lag of the network is negligible at high frequencies.

The phase margin of the compensated system is about 40°, which is the required value. The gain margin is about 11 db, which is quite acceptable. The static velocity error coefficient is 5 $\text{sec}^{-1}$, as required. The compensated system, therefore, satisfies the requirements on both the steady state and the relative stability.

Note that the new gain crossover frequency is decreased from 2.1 to 0.5 rad/sec. This means that the bandwidth of the system is reduced. Therefore, the speed of transient response of the compensated system will be slower than that of the original system.

To further show the effects of lag compensation, the log-magnitude versus phase plots of the uncompensated system (with $K = 5$) and of the compensated system are shown in Fig. 10-26. The plot of $G(j\omega)$ clearly shows that the uncompensated system is unstable.

The addition of the lag network stabilizes the system. The plot of $G_c(j\omega)G(j\omega)$ is tangent to the $M = 3$ db locus. Thus the resonant peak value $M_r$ is 3 db, or 1.4, and this peak occurs at $\omega = 0.5$ rad/sec.

In comparing the compensated system designed here with that designed in Example 10-3, we note that they are quite close. This, however, may not necessarily be true in other cases. Compensators designed by different methods or by different designers (even using the same approach) may look sufficiently different. Any of the well-designed systems, however, will give similar transient and steady-state performance. The best among many alternatives may be chosen from the economic consideration that the time constants of the lag network should not be too large since the sizes of the resistors and capacitor required are directly related to the magnitudes of the time constants.

### A few comments on lag compensation

1. Lag networks are essentially low-pass filters. Therefore, lag compensation permits a high gain at low frequencies (which improves the steady-state error) and reduces gain in the higher critical range of frequencies so as to avoid system instability. Note that in lag compensation, we utilize the attenuation characteristic of the lag network at high frequencies rather than the phase lag characteristic. (The phase lag characteristic is of no use for compensation purposes.)

2. The attenuation due to the lag network will shift the gain crossover frequency to a lower frequency point where the phase margin is acceptable. Thus, the lag network will reduce the bandwidth of the system and will result in slower transient response. [The phase angle curve of $G_c(j\omega)G(j\omega)$ is relatively unchanged near and above the new gain crossover frequency.]

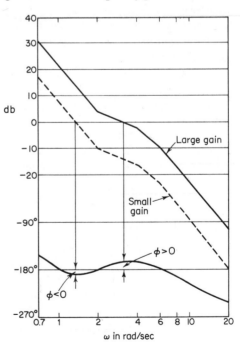

**Fig. 10-27.** Bode diagram of a conditionally stable system.

3. Since the lag compensator tends to integrate the input signal, it acts approximately as a proportional-plus-integral controller. Because of this, a lag compensated system tends to become less stable. In order to avoid this undesirable feature, the time constant $T$ should be made sufficiently larger than the largest time constant of the system.

4. Conditional stability may occur when a system having saturation or limiting is compensated by use of a lag compensator. When the saturation or limiting takes place in the system, it reduces the effective loop gain. Then the system becomes less stable, and even unstable operation may result, as shown in Fig. 10-27. To avoid this, the system must be designed so that the effect of lag compensation becomes significant only when the amplitude of the input to the saturating element is small. (This can be done by means of minor feedback-loop compensation.)

### 10-5 LAG-LEAD COMPENSATION

Lead compensation increases the bandwidth, which improves the speed of response, and also reduces the amount of overshoot. However, improvement in steady-state performance is rather small. Lag compensation results in a large improvement in steady-state performance but results in slower response due to the reduced bandwidth.

If improvements in both transient and steady-state response (namely, large increases in the gain and bandwidth) are desired, then both a lead network and a lag network may be used simultaneously. Rather than introducing both a lead network and lag network as separate elements, however, it is economical to use a single lag-lead network. The lag-lead network combines the advantages of the lag and lead networks.

The lag-lead network possesses two poles and two zeros. Therefore, such compensation increases the order of the system by two, unless cancellation of a pole and zero occurs in the compensated system.

**Lag-lead networks.** Figure 10-28 shows an electrical lag-lead network. For a sinusoidal input, the output is sinusoidal with a phase shift which is a function of the input frequency. This phase angle varies from lag to lead as the frequency is

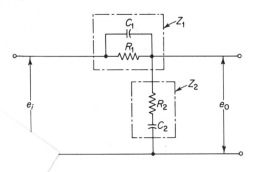

**Fig. 10-28.** Electrical lag-lead network.

increased from zero to infinity. Note that phase lead and lag occur in different frequency bands.

Let us obtain the transfer function of the lag-lead network. The complex impedances $Z_1$ and $Z_2$ are

$$Z_1 = \frac{R_1}{R_1 C_1 s + 1}, \qquad Z_2 = R_2 + \frac{1}{C_2 s}$$

The transfer function between $E_0(s)$ and $E_i(s)$ is

$$\frac{E_0(s)}{E_i(s)} = \frac{Z_2}{Z_1 + Z_2} = \frac{(R_1 C_1 s + 1)(R_2 C_2 s + 1)}{(R_1 C_1 s + 1)(R_2 C_2 s + 1) + R_1 C_2 s}$$

The denominator of this transfer function can be factored into two real terms. Let us define

$$R_1 C_1 = T_1, \qquad R_2 C_2 = T_2$$

$$R_1 C_1 + R_2 C_2 + R_1 C_2 = \frac{T_1}{\beta} + \beta T_2 \qquad (\beta > 1)$$

Then $E_0(s)/E_i(s)$ can be simplified to

$$\frac{E_0(s)}{E_i(s)} = \frac{(T_1 s + 1)(T_2 s + 1)}{\left(\frac{T_1}{\beta}s + 1\right)(\beta T_2 s + 1)}$$

$$= \frac{\left(s + \frac{1}{T_1}\right)\left(s + \frac{1}{T_2}\right)}{\left(s + \frac{\beta}{T_1}\right)\left(s + \frac{1}{\beta T_2}\right)}$$

**Characteristics of lag-lead networks.** Consider the transfer function of the lag-lead network

$$\frac{\left(s + \frac{1}{T_1}\right)\left(s + \frac{1}{T_2}\right)}{\left(s + \frac{\beta}{T_1}\right)\left(s + \frac{1}{\beta T_2}\right)}$$

The first term

$$\frac{s + \frac{1}{T_1}}{s + \frac{\beta}{T_1}} = \frac{1}{\beta}\left(\frac{T_1 s + 1}{\frac{T_1}{\beta}s + 1}\right) \qquad (\beta > 1)$$

produces the effect of the lead network and the second term

$$\frac{s + \frac{1}{T_2}}{s + \frac{1}{\beta T_2}} = \beta\left(\frac{T_2 s + 1}{\beta T_2 s + 1}\right) \qquad (\beta > 1)$$

produces the effect of the lag network.

Figure 10-29 shows the polar plot of the lag-lead network. It can be seen that for $0 < \omega < \omega_1$ the network acts as a lag network, while for $\omega_1 < \omega < \infty$ it acts as a lead network. The frequency $\omega_1$ at which the phase angle is zero is

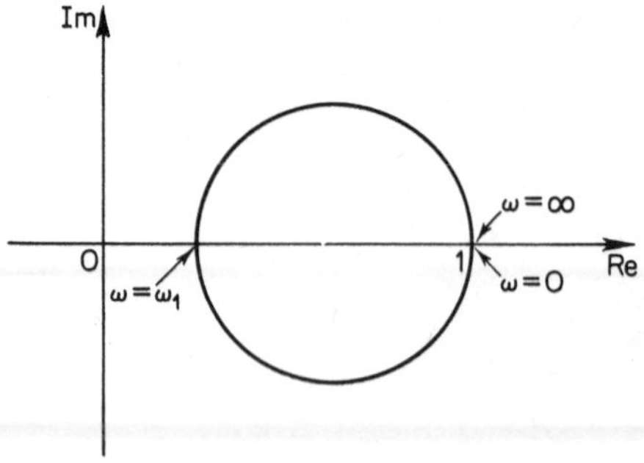

$$\omega_1 = \frac{1}{\sqrt{T_1 T_2}}$$

Figure 10-30 shows the Bode diagram of a lag-lead network when $\beta = 10$ and $T_2 = 10T_1$. Note that the magnitude curve has the value 0 db at the low-frequency and high-frequency regions. This is because the transfer function of the lag-lead network as a whole does not contain $\beta$ as a factor.

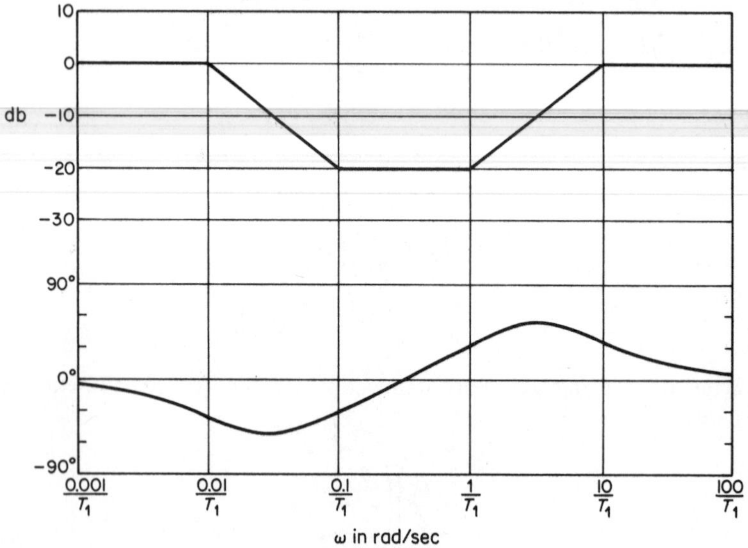

**Fig. 10-29.** Polar plot of a lag-lead network.

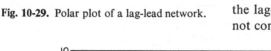

**Fig. 10-30.** Bode diagram of a lag-lead network.
$(j\omega T_1 + 1)(j\omega T_2 + 1)/\{[(j\omega T_1/\beta) + 1](j\omega\beta T_2 + 1)\}$ with $\beta = 10$ and $T_2 = 10T_1$.

**Lag-lead compensation techniques based on the root-locus approach.** The procedures for designing a lag-lead compensator may be stated as follows:

1. From the given performance specifications, determine the desired location for the dominant closed-loop poles.

2. In order to have the dominant closed-loop poles at the desired locations, calculate the angle contribution $\phi$ needed from the phase lead portion of the lag-lead network.

ing the following transfer function of the lag-lead compensator:

$$G_c(s) = \left( \frac{s + \dfrac{1}{T_1}}{s + \dfrac{\beta}{T_1}} \right) \left( \frac{s + \dfrac{1}{T_2}}{s + \dfrac{1}{\beta T_2}} \right) K_c$$

determine the constant $K_c$ from the requirement on the particular error coefficient specified in the design problem.

4. For the lag-lead compensator, we choose $T_2$ sufficiently large so that

$$\left| \frac{s_1 + \dfrac{1}{T_2}}{s_1 + \dfrac{1}{\beta T_2}} \right|$$

is approximately unity, where $s = s_1$ is one of the dominant closed-loop poles. Determine the values of $T_1$ and $\beta$ from the requirements that

$$\left| \frac{s_1 + \dfrac{1}{T_1}}{s_1 + \dfrac{\beta}{T_1}} \right| |K_c G(s_1)| = 1$$

$$\left/ \frac{s + \dfrac{1}{T_1}}{s + \dfrac{\beta}{T_1}} \right. = \phi$$

5. Using the value of $\beta$ just determined, choose $T_2$ so that

$$\left| \frac{s_1 + \dfrac{1}{T_2}}{s_1 + \dfrac{1}{\beta T_2}} \right| \doteqdot 1$$

$$0 < \left/ \frac{s_1 + \dfrac{1}{T_2}}{s_1 + \dfrac{1}{\beta T_2}} \right. < 3°$$

The value of $\beta T_2$, the largest time constant of the lag-lead network, should not be too large to be physically realized.

*Example 10-5.* Consider the control system shown in Fig. 10-31. The feedforward transfer function is

$$G(s) = \frac{4}{s(s + 0.5)}$$

This system has closed-loop poles at

$$s = -0.25 \pm j1.98$$

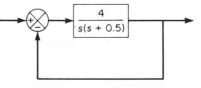

**Fig. 10-31.** Control system.

The damping ratio is 0.125, the undamped natural frequency is 2 rad/sec, and the static velocity error coefficient is 8 sec$^{-1}$.

It is desired to make the damping ratio of the dominant closed-loop poles equal to 0.5 and to increase the undamped natural frequency to 5 rad/sec and the static velocity error coefficient to 50 sec$^{-1}$. Design an appropriate compensator to meet all the performance specifications.

From the performance specifications, the dominant closed-loop poles must be

$$s = -2.50 \pm j4.33$$

Since

$$\left. \left/ \frac{4}{s(s + 0.5)} \right. \right|_{s=-2.50+j4.33} = -235°$$

the phase lead portion of the lag-lead network must contribute 55° so that the root locus passes through the desired location of the dominant closed-loop poles.

The lag-lead compensator has the transfer function

$$G_c(s) = \left( \frac{s + \dfrac{1}{T_1}}{s + \dfrac{\beta}{T_1}} \right) \left( \frac{s + \dfrac{1}{T_2}}{s + \dfrac{1}{\beta T_2}} \right) K_c$$

Thus the compensated system will have the transfer function

$$G_c(s)G(s) = \left( \frac{s + \dfrac{1}{T_1}}{s + \dfrac{\beta}{T_1}} \right) \left( \frac{s + \dfrac{1}{T_2}}{s + \dfrac{1}{\beta T_2}} \right) K_c G(s)$$

The static velocity error coefficient therefore becomes

$$K_v = \lim_{s \to 0} sG_c(s)G(s) = \lim_{s \to 0} sK_c G(s)$$

The requirement on the static velocity error coefficient is $K_v = 50$ sec$^{-1}$. Thus

$$K_v = \lim_{s \to 0} sK_c G(s)$$

$$= \lim_{s \to 0} \frac{s4K_c}{s(s + 0.5)}$$

$$= 8K_c = 50$$

Hence $K_c$ is

$$K_c = 6.25$$

Therefore, the compensated system will have the open-loop transfer function

$$G_c(s)G(s) = \left( \frac{s + \dfrac{1}{T_1}}{s + \dfrac{\beta}{T_1}} \right) \left( \frac{s + \dfrac{1}{T_2}}{s + \dfrac{1}{\beta T_2}} \right) \frac{25}{s(s + 0.5)}$$

Since we choose $T_2$ large enough that

$$\left. \left| \frac{s + \dfrac{1}{T_2}}{s + \dfrac{1}{\beta T_2}} \right| \right|_{s=-2.5+j4.33} \doteq 1$$

if we require that the closed-loop poles lie at $s = -2.5 \pm j4.33$, the magnitude condition becomes

$$|G_c(s)G(s)|_{s=-2.5+j4.33} = \left| \frac{s + \dfrac{1}{T_1}}{s + \dfrac{\beta}{T_1}} \right| \left| \frac{25}{s(s + 0.5)} \right|_{s=-2.5+j4.33} = \left| \frac{s + \dfrac{1}{T_1}}{s + \dfrac{\beta}{T_1}} \right| \frac{5}{4.77} = 1$$

and the angle condition becomes

$$\left| \frac{s + \dfrac{1}{T_1}}{s + \dfrac{\beta}{T_1}} \right|_{s=-2.5+j4.33} = 55°$$

It is a simple matter to graphically determine the values of $T_1$ and $\beta$ that satisfy these magnitude and angle conditions. Referring to Fig. 10-32, we can easily locate points $A$ and $B$ so that

$$\underline{/APB} = 55°, \qquad \frac{\overline{PA}}{\overline{PB}} = \frac{4.77}{5}$$

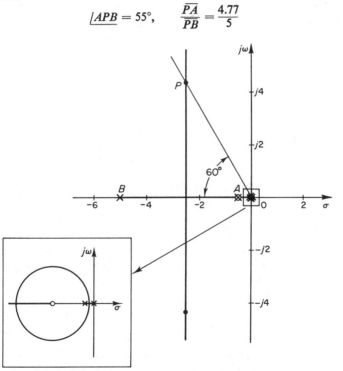

**Fig. 10-32.** Root-locus plot of the compensated system.

Graphically, from Fig. 10-32, we obtain

$$\overline{AO} = 0.5, \qquad \overline{BO} = 5$$

Hence

$$-\frac{1}{T_1} = -0.5, \qquad -\frac{\beta}{T_1} = -5$$

Thus

$$T_1 = 2, \quad \beta = 10$$

Therefore, the phase lead portion of the lag-lead network becomes

$$\frac{s + 0.5}{s + 5}$$

For the phase lag portion of the lag-lead network, it is required that

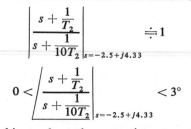

$$\left| \frac{s + \dfrac{1}{T_2}}{s + \dfrac{1}{10T_2}} \right|_{s=-2.5+j4.33} \doteq 1$$

$$0 < \left/ \frac{s + \dfrac{1}{T_2}}{s + \dfrac{1}{10T_2}} \right._{s=-2.5+j4.33} < 3°$$

To satisfy these relationships and, at the same time, to assure that the largest time constant $(10T_2)$ of the lag-lead network is not too large to be physically realized, we choose

$$T_2 = 10$$

Then the transfer function of the lag-lead compensator is

$$G_c(s) = \left(\frac{s + 0.5}{s + 5}\right)\left(\frac{s + 0.1}{s + 0.01}\right) \quad (6.25)$$

and the compensated system will have the open-loop transfer function

$$G_c(s)G(s) = \frac{(s + 0.5)(s + 0.1)25}{(s + 5)(s + 0.01)s(s + 0.5)}$$

$$= \frac{25(s + 0.1)}{s(s + 5)(s + 0.01)}$$

Because of the cancellation of the $(s + 0.5)$ terms, the compensated system is a third-order system. (Mathematically this cancellation is exact, but practically such cancellation will not be exact because some approximations are usually involved in deriving the mathematical model of the system and, as a result, the time constants are not precise.) The root-locus plot of the compensated system is shown in Fig. 10-32.

Because the angle contribution of the phase lag portion of the lag-lead network is quite small, there is no appreciable change in the location of the dominant closed-loop poles from the desired location, $s = -2.5 \pm j4.33$. Therefore, the compensated system meets all the required performance specifications. The third closed-loop pole of the compensated system is located at $s = -0.102$. Since this closed-loop pole is very close to the zero at $s = -0.01$, the effect of this pole on the response is relatively small. (Note that, in general, if a pole and a zero lie close to each other on the negative real axis near the origin, then such a pole-zero combination will yield a long tail of small amplitude in the transient response.)

The unit-step response curves and unit-ramp response curves before and after compensation are shown in Fig. 10-33.

---

*Example 10-6.* Consider the control system of Example 10-5. Suppose that the static velocity error coefficient is to be 80 sec$^{-1}$. The other specifications remain the same as

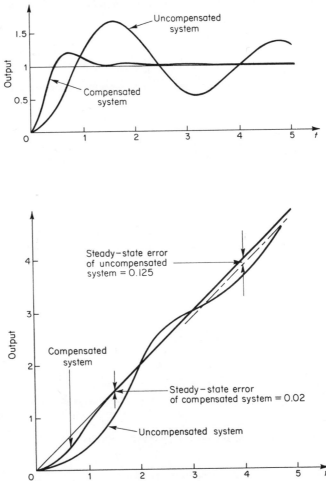

**Fig. 10-33.** Transient response curves for the uncompensated system and the compensated system. (a) Unit-step response curves; (b) unit-ramp response curves.

those given in Example 10-5. Namely, the damping ratio and the undamped natural frequency of the dominant closed-loop poles are given as 0.5 and 5, respectively. Design an appropriate lag-lead compensator.

The requirement on the static velocity error coefficient is that $K_v = 80$ sec$^{-1}$. This results in

$$K_c = 10$$

The time constant $T_1$ and the value of $\beta$ are determined from

$$\left|\frac{s + \dfrac{1}{T_1}}{s + \dfrac{\beta}{T_1}}\right|\left|\frac{40}{s(s + 0.5)}\right|_{s=-2.5+j4.33} = \left|\frac{s + \dfrac{1}{T_1}}{s + \dfrac{\beta}{T_1}}\right|\frac{5}{3} = 1$$

$$\left/\!\frac{s + \dfrac{1}{T_1}}{s + \dfrac{\beta}{T_1}}\right._{s=-2.5+j4.33} = 55°$$

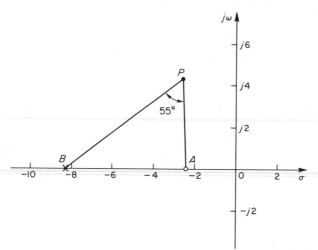

**Fig. 10-34.** Determination of the desired pole-zero location.

Referring to Fig. 10-34, we can easily locate points $A$ and $B$ so that

$$\underline{/APB} = 55°, \qquad \frac{\overline{PA}}{\overline{PB}} = \frac{3}{5}$$

The result is

$$\overline{AO} = 2.4, \qquad \overline{BO} = 8.3$$

or

$$T_1 = \frac{1}{2.4} = 0.416, \qquad \beta = 8.3 T_1 = 3.45$$

The phase lead portion of the lag-lead network thus becomes

$$\frac{s + 2.4}{s + 8.3}$$

For the phase lag portion, we may choose

$$T_2 = 10$$

Thus, the lag-lead compensator becomes

$$G_c(s) = \left(\frac{s + 2.4}{s + 8.3}\right)\left(\frac{s + 0.1}{s + 0.029}\right)(10)$$

The compensated system will have the open-loop transfer function

$$G_c(s)G(s) = \frac{(s + 2.4)(s + 0.1)40}{(s + 8.3)(s + 0.029)s(s + 0.5)}$$

No cancellation occurs in this case, and the compensated system is of fourth order. Because the angle contribution of the phase lag portion of the lag-lead network is quite small, the dominant closed-loop poles are located very near the desired location.

The two other closed-loop poles are obtained as

$$s = -0.09, \qquad s = -3.74$$

Since the closed-loop pole at $s = -0.09$ and zero at $s = -0.1$ almost cancel each other, the effect of this closed-loop pole is very small. The remaining closed-loop pole ($s = -3.74$) is relatively close to the zero at $s = -2.4$, and the effect of this closed-loop pole on the transient response will again be relatively small. The poles at $s = -2.5 \pm j4.33$ will serve as the dominant closed-loop poles.

**Lag-lead compensation based on the frequency-response approach.** The design of a lag-lead compensator by the frequency-response approach is based on the combination of the design techniques discussed under lead compensation and lag compensation. The value of $\alpha$ for the lead network must be equal to the reciprocal of the value of $\beta$ for the lag network. If $\alpha$ is chosen to be $1/\beta$, then we may simply combine the individually designed lead and lag compensators to produce the single lag-lead compensator.

The phase lead portion of the lag-lead network alters the frequency-response curve by adding phase lead angle and increasing the phase margin at the gain crossover frequency.

The phase lag portion of the lag-lead network provides attenuation near and above the gain crossover frequency and thereby allows an increase of gain at the low-frequency range to improve the steady-state performance.

We shall illustrate the details of the procedures for designing a lag-lead compensator by an example.

---

*Example 10-7.* Consider the unity-feedback system whose open-loop transfer function is

$$G(s) = \frac{K}{s(s + 1)(s + 2)}$$

It is desired that the static velocity error coefficient be 10 sec$^{-1}$, the phase margin be 50°, and the gain margin be 10 db or more.

From the requirement on the static velocity error coefficient, we obtain

$$K_v = \lim_{s \to 0} sG(s) = \lim_{s \to 0} \frac{sK}{s(s + 1)(s + 2)} = 10$$

Hence,

$$K = 20$$

We shall next draw the Bode diagram of the uncompensated system with $K = 20$, as shown in Fig. 10-35. The phase margin of the uncompensated system is found to be $-32°$, which indicates that the uncompensated system is unstable.

The next step in the design of a lag-lead compensator is to choose a new gain crossover frequency. From the phase angle curve for $G(j\omega)$, we notice that $\underline{/G(j\omega)} = -180°$ at $\omega = 1.5$ rad/sec. It is convenient to choose the new gain crossover frequency to be 1.5 rad/sec so that the phase lead angle required at $\omega = 1.5$ rad/sec is about 50°, which is quite possible by use of a single lag-lead network.

Once we choose the gain crossover frequency to be 1.5 rad/sec, we can determine the corner frequency of the phase lag portion of the lag-lead network. Let us choose the corner frequency $\omega = 1/T_2$ (which corresponds to the zero of the phase lag portion of the network) to be one decade below the new gain crossover frequency, or at $\omega = 0.15$ rad/sec. Let us choose

$$\beta = 10$$

Then the corner frequency $\omega = 1/\beta T_2$ (which corresponds to the pole of the phase lag portion of the network) becomes $\omega = 0.015$ rad/sec. The transfer function of the phase lag portion of the lag-lead network then becomes

$$\frac{s + 0.15}{s + 0.015} = 10\left(\frac{6.67s + 1}{66.7s + 1}\right)$$

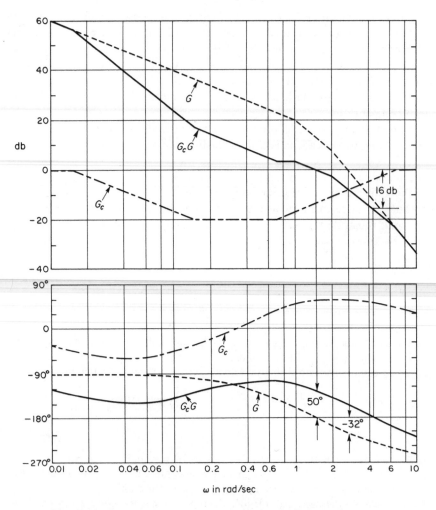

**Fig. 10-35.** Bode diagrams for the uncompensated system, the compensator, and the compensated system. ($G$: uncompensated system, $G_c$: compensator, $G_cG$: compensated system.)

The phase lead portion can be determined as follows: Since the new gain crossover frequency is $\omega = 1.5$ rad/sec, from Fig. 10.35, $G(j1.5)$ is found to be 13 db. Hence if the lag-lead network contributes $-13$ db at $\omega = 1.5$ rad/sec, then the new gain crossover frequency is as desired. From this requirement, it is possible to draw a straight line of slope 20 db/decade, passing through the point ($-13$ db, 1.5 rad/sec). The interesections of this line and the 0-db line and $-20$-db line determine the corner frequencies. Thus the corner frequencies for the lead potion are $\omega = 0.7$ rad/sec and $\omega = 7$ rad/sec. Thus, the transfer function of the lead portion of the lag-lead network becomes

$$\frac{s + 0.7}{s + 7} = \frac{1}{10}\left(\frac{1.43s + 1}{0.143s + 1}\right)$$

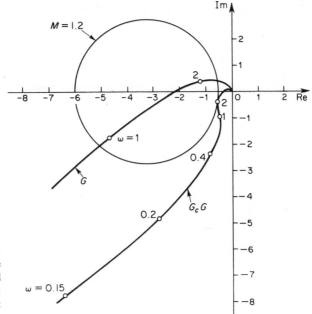

**Fig. 10-36.** Polar plots of the uncompensated system and the compensated system. ($G$: uncompensated system, $G_cG$: compensated system.)

Combining the transfer functions of the lag and lead portions of the network, we obtain the transfer function of the lag-lead compensator

$$\left(\frac{s + 0.7}{s + 7}\right)\left(\frac{s + 0.15}{s + 0.015}\right) = \left(\frac{1.43s + 1}{0.143s + 1}\right)\left(\frac{6.67s + 1}{66.7s + 1}\right)$$

The magnitude and phase-angle curves of the lag-lead compensator just designed are shown in Fig. 10-35. The open-loop transfer function of the compensated system is

$$G_c(s)G(s) = \frac{(s + 0.7)(s + 0.15)20}{(s + 7)(s + 0.015)s(s + 1)(s + 2)}$$

$$= \frac{10(1.43s + 1)(6.67s + 1)}{s(0.143s + 1)(66.7s + 1)(s + 1)(0.5s + 1)} \qquad (10\text{-}4)$$

The magnitude and phase-angle curves of the system of Eq. (10-4) are also shown in Fig. 10-35. The phase margin of the compensated system is 50°, the gain margin is 16 db, and the static velocity error coefficient is 10 sec$^{-1}$. All the requirements are therefore met, and the design has been completed.

Figure 10-36 shows the polar plots of the uncompensated system and compensated system. The $G_c(j\omega)G(j\omega)$ locus is tangent to the $M = 1.2$ circle at about $\omega = 2$ rad/sec. Clearly, this indicates that the compensated system has satisfactory relative stability. The bandwidth of the compensated system is slightly larger than 2 rad/sec.

## 10-6  SUMMARY OF CONTROL SYSTEM COMPENSATION METHODS

In Sections 10-3 through 10-5, we have presented detailed procedures for designing lead, lag, and lag-lead compensators by use of simple examples. A satisfactory design of a compensator for a given system will require the creative application of these basic design principles. (Remember that there is a wealth of literature on the design of complex control systems, and the reader may wish to refer to them for any specific problems he might face.)

**Comparison of lead, lag, and lag-lead compensation.**

1. Lead compensation achieves the desired result through the merits of its phase lead contribution; whereas lag compensation accomplishes the result through the merits of its attenuation property at high frequencies.

2. In the $s$ domain, lead compensation enables us to reshape the root locus and thus provide the desired closed-loop poles. In the frequency domain, lead compensation increases the phase margin and bandwidth. Large bandwidth means reduction in the settling time. The bandwidth of a system with lead compensation is always greater than that with lag compensation. Therefore, if a large bandwidth or fast response is desired, lead compensation should be employed. If, however, noise signals are present, then a large bandwidth may not be desirable since it makes the system more susceptible to noise signals because of increase in the high-frequency gain. In such a case, lag compensation should be used.

3. Lag compensation improves steady-state accuracy; however, it reduces the bandwidth. If the reduction of bandwidth is too excessive, the compensated system will exhibit sluggish response. If both fast response and good static accuracy are desired, a lag-lead compensator must be employed.

4. Lead compensation requires an additional increase in gain to offset the attenuation inherent in the lead network. This means that lead compensation will require a larger gain than that required by lag compensation. (A larger gain, in most cases, implies larger space, greater weight, and higher cost.)

5. Although a large number of practical compensation tasks can be accomplished with lead, lag, or lag-lead networks, for complicated systems, simple compensation by use of these networks may not yield satisfactory results. Then different compensators having different pole-zero configurations must be employed. Note that once the pole-zero configuration of a compensator has been specified, the necessary passive network may be realized by use of standard network synthesis techniques.

**Cancellation of undesirable poles.** Since the transfer function of elements in cascade is the product of their individual transfer functions, it is possible to cancel some undesirable poles or zeros by placing a compensating element in cascade, with its poles and zeros being adjusted to cancel the undesirable poles or zeros of the original system. For example, a large time constant $T_1$ may be canceled by use of the lead network $(T_1 s + 1)/(T_2 s + 1)$ as follows:

$$\left(\frac{1}{T_1 s + 1}\right)\left(\frac{T_1 s + 1}{T_2 s + 1}\right) = \frac{1}{T_2 s + 1}$$

If $T_2$ is much smaller than $T_1$, we can effectively eliminate the large time constant

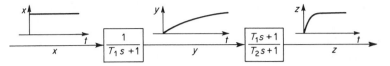

**Fig. 10-37.** Step-response curves showing the effect of canceling a large time constant.

$T_1$. Figure 10-37 shows the effect of canceling a large time constant in step transient response.

If an undesirable pole in the original system lies in the right-half $s$ plane, this compensation scheme should not be used since although mathematically it is possible to cancel the undesirable pole with an added zero, exact cancellation is physically impossible because of inaccuracies involved in the location of the poles and zeros. A pole in the right-half $s$ plane not exactly canceled by the compensator zero will eventually lead to unstable operation because the response will involve an exponential term which increases with time.

It should be noted that the ideal control system is not the one which has a transfer function of unity. Physically, such a control system cannot be built since it cannot instantaneously transfer energy from the input to the output. In addition, since noise is almost always present in one form or another, a system with a unity transfer function is not desirable. A desired control system, in many practical cases, may have one set of dominant complex-conjugate closed-loop poles with a reasonable damping ratio and undamped natural frequency. The determination of the significant part of the closed-loop pole-zero configuration, such as the location of the dominant closed-loop poles, is based on the specifications that give the required system performance.

**Cancellation of undesirable complex-conjugate poles.** If the transfer function of a plant contains one or more pairs of complex-conjugate poles, then a lead, lag, or lag-lead compensator may not give satisfactory results. In such a case, a network which has two zeros and two poles may prove to be useful. If the zeros are chosen so as to cancel the undesirable complex-conjugate poles of the plant, then we can essentially replace the undesirable poles by acceptable ones. That is, if the undesirable complex-conjugate poles are in the left-half $s$ plane and are in the form

$$\frac{1}{s^2 + 2\zeta_1 \omega_1 s + \omega_1^2}$$

then the insertion of a compensating network having the transfer function

$$\frac{s^2 + 2\zeta_1 \omega_1 s + \omega_1^2}{s^2 + 2\zeta_2 \omega_2 s + \omega_2^2}$$

will result in an effective change of the undesirable complex-conjugate poles to acceptable ones. Note that even though the cancellation may not be exact, the compensated system will exhibit better response characteristics. (As stated earlier, this approach cannot be used if the undesirable complex-conjugate poles are in the right-half $s$ plane.)

Familiar networks consisting only of $RC$ components whose transfer functions possess two zeros and two poles are the bridged-$T$ networks. Examples of bridged-$T$ networks and their transfer functions are shown in Fig. 10-38.

**Feedback compensation.** Tachometer feedback is very commonly used in positional servomechanisms. In Section 6-4, we discussed a simple design problem using tachometer feedback; therefore, we shall not repeat it here.

A tachometer is one of the so-called rate feedback devices. Another common rate feedback device is the rate gyro. Rate gyros are commonly used in aircraft autopilot systems.

**Eliminating the undesirable effects of disturbances by feedforward control.** If disturbances are measurable, feedforward control is a useful method of canceling their effects upon the system output. By feedforward control, we mean control of undesirable effects of measurable distur-

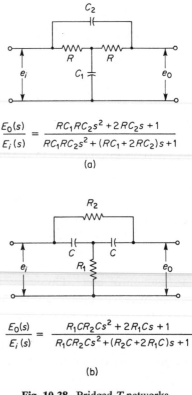

$$\frac{E_0(s)}{E_i(s)} = \frac{RC_1RC_2s^2 + 2RC_2s + 1}{RC_1RC_2s^2 + (RC_1 + 2RC_2)s + 1}$$

(a)

$$\frac{E_0(s)}{E_i(s)} = \frac{R_1CR_2Cs^2 + 2R_1Cs + 1}{R_1CR_2Cs^2 + (R_2C + 2R_1C)s + 1}$$

(b)

**Fig. 10-38.** Bridged-$T$ networks.

bances by approximately compensating for them before they materialize. This is advantageous because, in a usual feedback control system, the corrective action starts only after the output has been affected.

As an example, consider the temperature control system shown in Fig. 10-39 (a). In this system, it is desired to maintain the outlet temperature at some constant value. The disturbance in this system is a change in the inflow rate, which depends on the level in the tower. The effect of a change in this rate cannot be sensed immediately at the output due to the time lags involved in the system.

The temperature controller, which controls the heat input to the heat exchanger, will not act until an error has developed. If the system involves large time lags, it will take some time before any corrective action takes place. In fact, when the error shows up after a certain delay time and the corrective action starts, it may be too late to keep the outlet temperature within the desired limits.

If feedforward control is provided in such a system, then as soon as a change in the inflow occurs, a corrective measure will be taken simultaneously, by adjusting the heat input to the heat exchanger. This can be done by feeding both the signal from the flowmeter and the signal from the temperature-measuring element to the temperature controller.

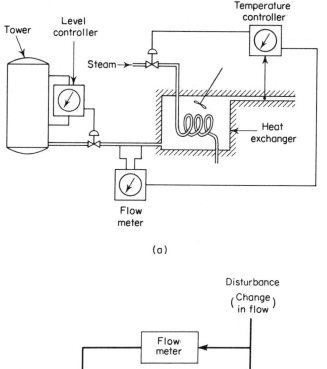

(a)

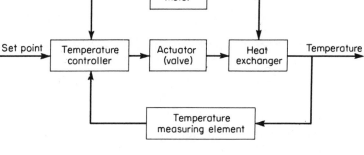

(b)

**Fig. 10-39.** (a) Temperature control system; (b) block diagram.

Feedforward control can minimize the transient error, but since feedforward control is open-loop control, there are limitations to its functional accuracy. Feedforward control will not cancel the effects of unmeasurable disturbances under normal operating conditions. It is, therefore, necessary that a feedforward control system include a feedback loop, as shown in Figs. 10-39 (a) and (b).

Essentially, feedforward control minimizes the transient error caused by measurable disturbances, while feedback control compensates for any imperfections in the functioning of the feedforward control and provides for corrections for unmeasurable disturbances.

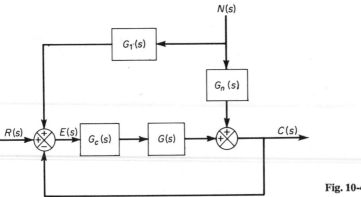

**Fig. 10-40.** Control system.

**Feedforward control of a plant.** Consider the system shown in Fig. 10-40. Suppose that the plant transfer function $G(s)$ and disturbance transfer function $G_n(s)$ are known. We shall illustrate a method for determining a suitable controller transfer function $G_c(s)$ and disturbance feedforward transfer function $G_1(s)$. Since the output $C(s)$ is given by

$$C(s) = G_c(s)G(s)E(s) + G_n(s)N(s)$$

where

$$E(s) = R(s) - C(s) + G_1(s)N(s)$$

we obtain

$$C(s) = G_c(s)G(s)[R(s) - C(s)] + [G_c(s)G(s)G_1(s) + G_n(s)]N(s)$$

The effects of $N(s)$ can be eliminated if $G_c(s)$ is chosen so that

$$G_c(s)G(s)G_1(s) + G_n(s) = 0$$

or

$$G_1(s) = -\frac{G_n(s)}{G_c(s)G(s)} \tag{10-5}$$

By properly designing the controller transfer function $G_c(s)$ (as discussed in Sections 10-3 through 10-5), the closed-loop control system can be assured desired performance. Once $G_c(s)$ is determined, then the disturbance feedforward transfer function $G_1(s)$ can be obtained from Eq. (10-5).

**Concluding comments.** In the design examples presented in this chapter, we have been primarily concerned only with the transfer functions of compensators. In actual design problems, we must choose the hardware. Thus we must satisfy additional design constraints such as cost, size, weight, and reliability.

The system designed may meet the specifications under normal operating conditions but may deviate considerably from the specifications when environmental changes are considerable. Since the changes in the environment affect the gain and time constants of the system, it is necessary to provide automatic or manual means to adjust the gain to compensate for such environmental changes, for nonlinear effects which were not taken into account in the design, and also to compensate for manufacturing tolerances from unit to unit in the production of system components. (The effects of manufacturing tolerances are suppressed in a closed-loop system; therefore, the effects may not be critical in closed-loop operation but critical in open-loop operation.) In addition to this, the designer must remember that any system is subject to small variations due mainly to the normal deterioration of the system.

### EXAMPLE PROBLEMS AND SOLUTIONS

**PROBLEM A-10-1.** Show that the lead network and lag network inserted in cascade in an open loop acts as proportional-plus-derivative control (in the region of small $\omega$) and proportional-plus-integral control (in the region of large $\omega$), respectively.

**Solution.** In the region of small $\omega$, the polar plot of the lead network is approximately the same as that of the proportional-plus-derivative controller. This is shown in Fig. 10-41 (a).

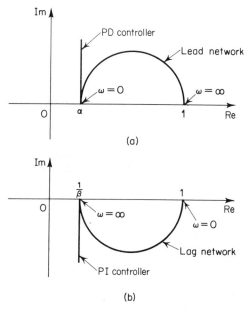

(a)

(b)

**Fig. 10-41.** (a) Polar plots of a lead network and a proportional-plus-derivative controller; (b) polar plots of a lag network and a proportional-plus-integral controller.

Similarly, in the region of large $\omega$, the polar plot of the lag network approximates the proportional-plus-integral controller, as shown in Fig. 10-41 (b).

**PROBLEM A-10-2.** If the open-loop transfer function $G(s)$ involves lightly damped complex-conjugate poles, then more than one $M$ locus may be tangent to the $G(j\omega)$ locus.

Consider the unity-feedback system whose open-loop transfer function is

$$G(s) = \frac{9}{s(s + 0.5)(s^2 + 0.6s + 10)} \tag{10-6}$$

Draw the Bode diagram for this open-loop transfer function. Draw also the log-magni-

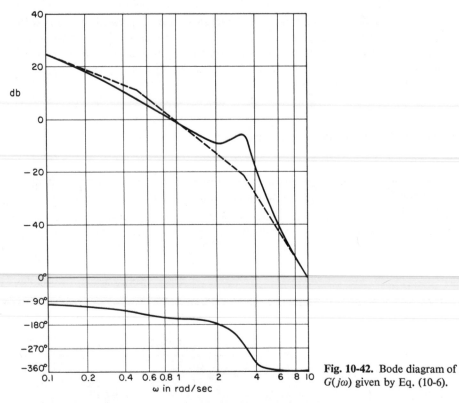

**Fig. 10-42.** Bode diagram of $G(j\omega)$ given by Eq. (10-6).

tude versus phase plot and show that two $M$ loci are tangent to the $G(j\omega)$ locus. Finally, plot the Bode diagram for the closed-loop transfer function.

**Solution.** Figure 10-42 shows the Bode diagram of $G(j\omega)$. Figure 10-43 shows the log-magnitude versus phase plot of $G(j\omega)$. It is seen that the $G(j\omega)$ locus is tangent to the $M = 8$ db locus at $\omega = 0.97$ rad/sec and it is tangent to $M = -4$ db locus at $\omega = 2.8$ rad/sec.

Figure 10-44 shows the Bode diagram of the closed-loop transfer function. The closed-loop frequency-response curves show two resonant peaks. Note that such a case occurs when the closed-loop transfer function involves the product of two lightly damped second-order terms and the two corresponding resonant frequencies are sufficiently separated from each other. As a matter of fact, the closed-loop transfer function of this system can be written

$$\frac{C(s)}{R(s)} = \frac{G(s)}{1 + G(s)}$$

$$= \frac{9}{(s^2 + 0.487s + 1)(s^2 + 0.613s + 9)}$$

Clearly, the closed-loop transfer function is a product of two lightly damped second-order terms (the damping ratios are 0.24 and 0.1), and the two resonant frequencies are sufficiently separated.

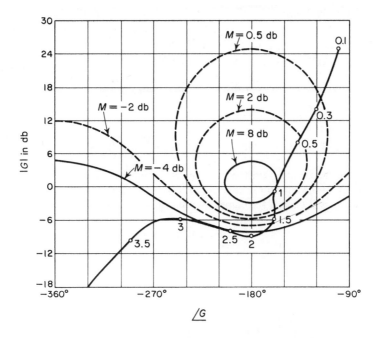

**Fig. 10-43.** Log-magnitude versus phase plot of $G(j\omega)$ given by Eq. (10-6).

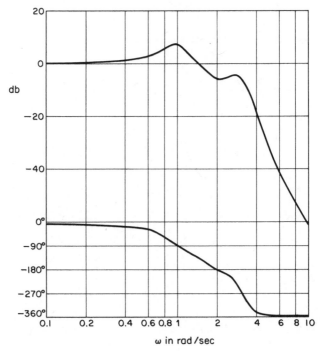

**Fig. 10-44.** Bode diagram of $G(j\omega)/[1 + G(j\omega)]$ where $G(j\omega)$ is given by Eq. (10-6).

**PROBLEM A-10-3.** Consider the system shown in Fig. 10-45. Determine the values of the amplifier gain $K$ and the velocity feedback gain $K_h$ so that the following specifications are satisfied:

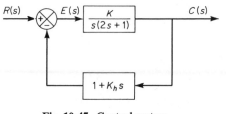

Fig. 10-45. Control system.

1. damping ratio of the closed-loop poles is 0.5
2. settling time $\leq 2$ sec
3. velocity error coefficient $K_v \geq 50$ sec$^{-1}$
4. $0 < K_h < 1$

**Solution.** The specification on the damping ratio requires that the closed-loop poles lie on lines at $60°$ in the left-half $s$ plane. The specification on the settling time can be written in terms of the real part of the complex-conjugate closed-loop poles as

$$t_s = \frac{4}{\sigma} \leq 2 \text{ sec}$$

or

$$\sigma \geq 2$$

Hence, the closed-loop poles must lie on the heavy lines $AB$ and $CD$ in the left-half $s$ plane, as shown in Fig. 10-46.

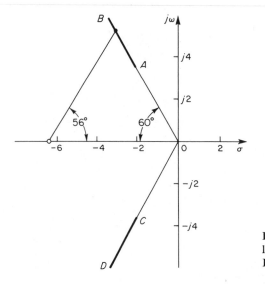

Fig. 10-46. Possible locations for the closed-loop poles in the $s$ plane for the system of Problem A-10-3.

Since the velocity error coefficient $K_v$ is defined by

$$K_v = \lim_{s \to 0} sG(s)H(s)$$

we obtain

$$K_v = \lim_{s \to 0} \frac{sK(1 + K_h s)}{s(2s + 1)} = K$$

From the given specification on the velocity error coefficient, we obtain

$$K \geq 50$$

For this system, the open-loop poles are at $s = 0$ and $s = -\frac{1}{2}$. The open-loop zero is at $s = -1/K_h$, where $K_h$ is an undetermined constant. First, let the closed-loop poles be $-2 \pm j3.4$ (points $A$ and $C$ in Fig. 10-46). The sum of the angles at the chosen closed-loop pole location with the open-loop poles is $120° + 115° = 235°$. Thus we need a $55°$ contribution from the zero, so that the total sum is $-180°$. To satisfy the angle condition, we choose the zero to be $s = -4.4$. Then $K_h$ is

$$K_h = \frac{1}{4.4} = 0.227$$

The magnitude condition states that

$$\left| \frac{K(1 + 0.227s)}{s(2s + 1)} \right|_{s=-2+j3.4} = 1$$

Hence

$$K = 31$$

Since $K < 50$, the choice of the closed-loop poles at $-2 \pm j3.4$ is not acceptable.

As a second trial, let the closed-loop poles be $-3 \pm j5.1$. The sum of the angle contributions from the open-loop poles is $236°$. We need a $56°$ contribution from the zero. This implies that the zero must be $s = -6.4$.

The application of the magnitude condition yields $K = 70$. This is quite satisfactory. Since $K_h = 0.156$, the requirement on $K_h$ is satisfied. Thus, all the given specifications are met. Hence a set of acceptable values of $K$ and $K_h$ is

$$K = 70, \qquad K_h = 0.156$$

**PROBLEM A-10-4.** A closed loop system has the characteristic that the closed-loop transfer function is nearly equal to the inverse of the feedback transfer function whenever the open-loop gain is much greater than unity.

The open-loop characteristic may be modified by adding an internal feedback loop with a characteristic equal to the inverse of the desired open-loop characteristic. Suppose that a unity-feedback system has the open-loop transfer function

$$G(s) = \frac{K}{(T_1 s + 1)(T_2 s + 1)}$$

Determine the transfer function $H(s)$ of the element in the internal feedback loop so that the inner loop becomes ineffective at both low and high frequencies.

**Solution.** Figure 10-47 (a) shows the original system. Figure 10-47 (b) shows the addition of the internal feedback loop around $G(s)$. Since

$$\frac{C(s)}{E(s)} = \frac{G(s)}{1 + G(s)H(s)} = \frac{1}{H(s)} \frac{G(s)H(s)}{1 + G(s)H(s)}$$

if the gain around the inner loop is large compared with unity, then $G(s)H(s)/[1 + G(s)H(s)]$ is approximately equal to unity, and the transfer function $C(s)/E(s)$ is approximately equal to $1/H(s)$.

On the other hand, if the gain $G(s)H(s)$ is much less than unity, the inner loop becomes ineffective and $C(s)/E(s)$ becomes approximately equal to $G(s)$.

To make the inner loop ineffective at both the low- and high-frequency ranges, we require that

$$G(j\omega)H(j\omega) \ll 1 \qquad \text{for } \omega \ll 1 \text{ and } \omega \gg 1$$

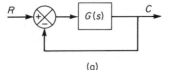

(a)

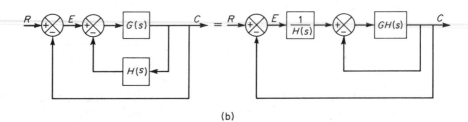

(b)

**Fig. 10-47.** (a) Control system; (b) addition of the internal feedback loop to modify the closed-loop characteristic.

Since in this problem

$$G(j\omega) = \frac{K}{(1 + j\omega T_1)(1 + j\omega T_2)}$$

the requirement can be satisfied if $H(s)$ is chosen to be

$$H(s) = ks$$

because

$$\lim_{\omega \to 0} G(j\omega)H(j\omega) = \lim_{\omega \to 0} \frac{Kkj\omega}{(1 + j\omega T_1)(1 + j\omega T_2)} = 0$$

$$\lim_{\omega \to \infty} G(j\omega)H(j\omega) = \lim_{\omega \to \infty} \frac{Kkj\omega}{(1 + j\omega T_1)(1 + j\omega T_2)} = 0$$

Thus, with $H(s) = ks$, the inner loop becomes ineffective at both the low- and high-frequency regions.

**PROBLEM A-10-5.** When a disturbance acts on a plant, it takes some time before any effect on the output can be detected. If we measure the disturbance itself (though this may not be possible or may be quite difficult) rather than the response to the disturbance, then a corrective action can be taken sooner, and we can expect a better result. Figure 10-48 is a block diagram showing feedforward compensation for the disturbance.

Discuss the limitations of the disturbance-feedforward scheme in general. Then discuss the advantages and limitations of the scheme shown in Fig. 10-48.

**Solution.** A disturbance-feedforward scheme is an open-loop scheme and thus depends on the constancy of the parameter values. Any drift in these values will result in imperfect compensation.

In the present system, both open-loop and closed-loop schemes are simultaneously in operation. Large errors due to the main disturbance source can be greatly reduced by the open-loop compensation without requiring a high loop gain. Smaller errors due to other disturbance sources can be taken care of by the closed-loop control scheme. Hence

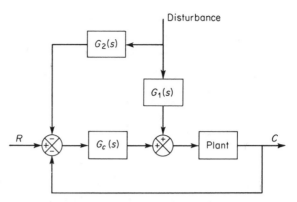

**Fig. 10-48.** Control system with feedforward compensation for the disturbance.

errors from all causes can be reduced without requiring a large loop gain. This is an advantage from the stability viewpoint.

Note that such a scheme cannot be used unless the main disturbance itself can be measured.

PROBLEMS

**PROBLEM B-10-1.** Draw Bode diagrams of the lead network and lag network shown in Figs. 10-49 (a) and (b), respectively.

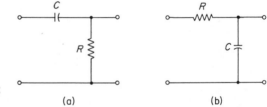

**Fig. 10-49.** (a) Lead network; (b) lag network.

(a)                    (b)

**PROBLEM B-10-2.** Consider a unity-feedback control system whose feedforward transfer function is given by

$$G(s) = \frac{K}{s(s + 1)(s + 2)(s + 3)}$$

Determine the value of $K$ so that the dominant closed-loop poles have a damping ratio of 0.5.

**PROBLEM B-10-3.** Consider a unity-feedback system whose feedforward transfer function is given by

$$G(s) = \frac{1}{s^2}$$

It is desired to insert a series compensator so that the open-loop frequency-response curve is tangent to the $M = 3$ db circle at $\omega = 3$ rad/sec. The system is subjected to high-frequency noises and sharp cutoff is desired. Design an appropriate series compensator.

**PROBLEM B-10-4.** Determine the values of $K$, $T_1$, and $T_2$ of the system shown in Fig. 10-50 so that the dominant closed-loop poles have $\zeta = 0.5$ and $\omega_n = 3$ rad/sec.

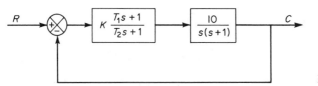

**Fig. 10-50.** Control system.

**PROBLEM B-10-5.** Consider a unity-feedback control system whose feedforward transfer function is given by

$$G(s) = \frac{10}{s(s + 2)(s + 8)}$$

Design a compensator so that the static velocity error coefficient $K_v$ is equal to 80 sec$^{-1}$ and the dominant closed-loop poles are located at $s = -2 \pm j2\sqrt{3}$.

**PROBLEM B-10-6.** Consider the system shown in Fig. 10-51. If the disturbance $N$ can be detected, it may be passed through a transfer function $G_3$ and added to the feedforward path between the amplifier and the plant, as shown in Fig. 10-51.

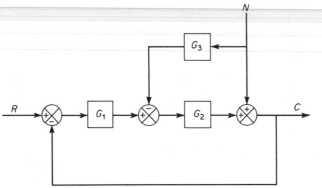

**Fig. 10-51.** Control system.

In order to reduce the effect of this disturbance $N$ on the steady-state error, determine an appropriate transfer function $G_3$. What will limit the present approach in reducing the effects of this disturbance?

**PROBLEM B-10-7.** Figure 10-52 is the block diagram of an attitude-rate control system. Design a compensator $G_c(s)$ so that (dominant) complex-conjugate poles are $s = -2 \pm j2$.

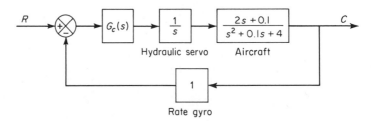

**Fig. 10-52.** Attitude-rate control system.

# 11

# DESCRIBING-FUNCTION ANALYSIS OF NONLINEAR CONTROL SYSTEMS

## 11-1 INTRODUCTION TO NONLINEAR SYSTEMS

It is a well-known fact that many relationships among physical quantities are not quite linear, although they are often approximated by linear equations mainly for mathematical simplicity. This simplification may be satisfactory as long as the resulting solutions are in agreement with experimental results. One of the most important characteristics of nonlinear systems is the dependence of the system response behavior upon the magnitude and type of the input. For example, a non-linear system may behave completely differently in response to step inputs of different magnitudes.

As pointed out in Chapter 4, nonlinear systems differ from linear systems greatly in that the principle of superposition does not hold for the former. Non-linear systems exhibit many phenomena that cannot be seen in linear systems, and in investigating such systems we must be familiar with these phenomena.

In this section, we shall present a brief discussion of several of them.

**Frequency-amplitude dependence.** Consider the free oscillation of the mechanical system shown in Fig. 11-1, which consists of a mass, dashpot, and nonlinear spring. The differential equation describing the dynamics of this system may be written

$$m\ddot{x} + f\dot{x} + kx + k'x^3 = 0 \tag{11-1}$$

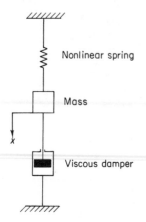

Nonlinear spring

Mass

$x$

Viscous damper

**Fig. 11-1.** Mechanical system.

where

$$x = \text{displacement of mass}$$
$$m = \text{mass}$$
$$f = \text{viscous-friction coefficient of damper}$$
$$kx + k'x^3 = \text{nonlinear spring force}$$

The parameters $m$, $f$, and $k$ are positive constants, while $k'$ may be either positive or negative. If $k'$ is positive, the spring is called a hard spring; if $k'$ is negative, a soft spring. The degree of nonlinearity of the system is characterized by the magnitude of $k'$. This nonlinear differential equation, Eq. (11-1), is known as Duffing's equation and has often been discussed in the field of nonlinear mechanics. The solution of Eq. (11-1) represents a damped oscillation if the system is subjected to a nonzero initial condition. In an experimental investigation, it is observed that as the amplitude decreases,

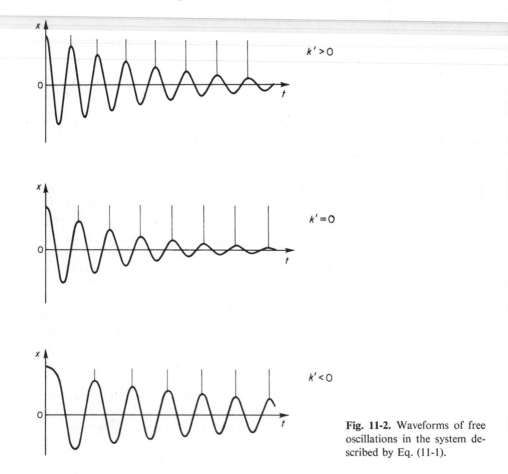

**Fig. 11-2.** Waveforms of free oscillations in the system described by Eq. (11-1).

the frequency of the free oscillation either decreases or increases, depending upon whether $k' > 0$ or $k' < 0$, respectively. When $k' = 0$, the frequency remains unchanged as the amplitude of the free oscillation decreases. (This corresponds to a linear system.) These characteristics are shown in Fig. 11-2, which displays the waveforms of free oscillations. Figure 11-3 depicts frequency-amplitude relationships for the three cases where $k'$ is greater than, equal to, and less than zero.

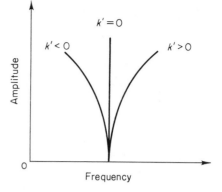

**Fig. 11-3.** Amplitude versus frequency curves for free oscillations in the system described by Eq. (11-1).

In an experimental study of nonlinear systems, the frequency-amplitude dependence can easily be detected. The frequency-amplitude dependence is one of the most fundamental characteristics of the oscillations of nonlinear systems. A graph of the form shown in Fig. 11-3 reveals whether or not a nonlinearity is present and also indicates the degree of the nonlinearity.

**Multivalued responses and jump resonances.** In carrying out experiments on the forced oscillations of the system shown in Fig. 11-1, the differential equation of which is

$$m\ddot{x} + f\dot{x} + kx + k'x^3 = P \cos \omega t$$

where

$P \cos \omega t$ = forcing function

we may observe a number of phenomena, such as multivalued responses, jump resonances, and a variety of periodic motions (such as subharmonic oscillations and superharmonic oscillations). These phenomena do not occur in the responses of linear systems.

In carrying out experiments in which the amplitude $P$ of the forcing function is held constant, while its frequency is varied slowly and the amplitude $X$ of the response is observed, we may obtain a frequency-response curve similar to those shown in Figs. 11-4 (a) and (b). Suppose that $k' > 0$ and that the forcing frequency $\omega$ is low at the start at point 1 on the curve of Fig. 11-4 (a). As the frequency $\omega$ is increased, the amplitude $X$ increases until point 2 is reached. A further increase in the frequency $\omega$ will cause a jump from point 2 to point 3, with accompanying changes in amplitude and phase. This phenomenon is called a jump resonance. As the frequency $\omega$ is increased further, the amplitude $X$ follows the curve from point 3 toward point 4. In performing the experiment in the other direction, i.e., starting from a high frequency, we observe that as $\omega$ is decreased, the amplitude $X$ slowly increases through point 3, until point 5 is reached. A further decrease in $\omega$ will cause another jump from point 5 to point 6, accompanied by changes in amplitude and phase. After this jump, the amplitude $X$ decreases with $\omega$ and follows the

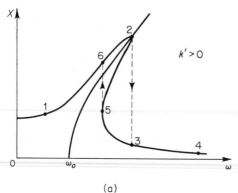

(a)

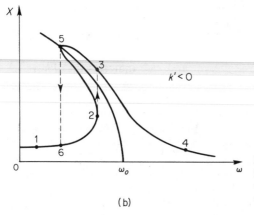

(b)

**Fig. 11-4.** Frequency response curves showing jump resonances. (a) Mechanical system with hard spring; (b) mechanical system with soft spring.

curve from point 6 toward point 1. Thus, the response curves are actually discontinuous, and a representative point on the response curve follows different paths for increasing and decreasing frequencies. The response oscillations corresponding to the curve between point 2 and point 5 correspond to unstable oscillations, and they cannot be observed experimentally. Similar jumps take place in the case of a system with a soft spring ($k' < 0$), as shown in Fig. 11-4(b). We thus see that for a given amplitude $P$ of the forcing function there is a range of frequencies in which either of the two stable oscillations can occur. It is noted that for jump resonance to take place, it is necessary that the damping term be small and that the amplitude of the forcing function be large enough to drive the system into a region of appreciably nonlinear operation.

**Subharmonic oscillations.** By the term *subharmonic oscillation*, we mean a nonlinear steady-state oscillation whose frequency is an integral submultiple of the forcing frequency. An example of an output waveform undergoing subharmonic oscillation is shown in Fig. 11-5, together with the input waveform. The generation of subharmonic oscillations is dependent upon system parameters and initial conditions, as well as upon the amplitude and frequency of the forcing function. By the

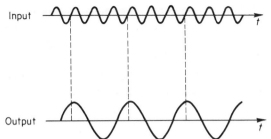

**Fig. 11-5.** Input and output waveforms under subharmonic oscillation.

phrase *dependence upon initial conditions*, we mean that subharmonic oscillations are not self-starting. It is necessary to give some sort of kick, say a sudden change in the amplitude or frequency of the forcing function, to initiate such oscillations. Once subharmonic oscillations are excited, they may be quite stable over certain frequency ranges. If the frequency of the forcing function is changed to a new value, either the subharmonic oscillation disappears or the frequency of the subharmonic oscillation changes also, to a value which is $\omega/n$, where $\omega$ is the forcing frequency and $n$ is the order of the subharmonic oscillation. (Note that an oscillation whose frequency is one-half that of the forcing function may occur in a linear system, if the system parameters are changed periodically with time. A linear conservative system may also exhibit oscillations that look like the subharmonic oscillations of nonlinear systems, but the oscillations in linear systems are essentially different from subharmonic oscillations.)

**Self-excited oscillations or limit cycles.** Another phenomenon which is observed in certain nonlinear systems is a self-excited oscillation or limit cycle. Consider a system described by the following equation:

$$m\ddot{x} - f(1 - x^2)\dot{x} + kx = 0$$

where $m$, $f$, and $k$ are positive quantities. This equation is called the Van der Pol equation. It is nonlinear in the damping term. Upon examining this term, we notice that for small values of $x$ the damping will be negative and will actually put energy into the system, while for large values of $x$ it is positive and removes energy from the system. Thus, it can be expected that such a system may exhibit a sustained oscillation. Since it is not a forced system, this oscillation is called a self-excited oscillation or a limit cycle. Note that if a system possesses only one limit cycle, as in the case of the present system, the amplitude of this limit cycle does not depend upon the initial condition.

**Frequency entrainment.** An example of an interesting phenomenon which may be observed in some nonlinear systems is frequency entrainment. If a periodic force of frequency $\omega$ is applied to a system capable of exhibiting a limit cycle of frequency $\omega_0$, the well-known phenomenon of beats is observed. As the difference between the two frequencies decreases, the beat frequency also decreases. In a linear system, it is found, both experimentally and theoretically, that the beat frequency decreases indefinitely as $\omega$ approaches $\omega_0$. In a self-excited nonlinear system, however, it is found experimentally that the frequency $\omega_0$ of the limit cycle falls in synchronistically with, or is entrained by, the forcing frequency $\omega$, within a certain band of frequencies. This phenomenon is usually called frequency entrainment, and the band of frequency in which entrainment occurs is called the zone of entrainment.

Figure 11-6 shows the relationship between $|\omega - \omega_0|$ and $\omega$. For a linear system, the relationship between $|\omega - \omega_0|$ and $\omega$ would follow the dotted lines, and $|\omega - \omega_0|$ would be zero for only one value of $\omega$; namely, $\omega = \omega_0$. For a self-excited nonlinear system, entrainment of frequency occurs, and in the zone of entrainment, which is indicated by the region $\Delta\omega$ in Fig. 11-6, frequencies $\omega$ and $\omega_0$ coalesce, and only one frequency $\omega$ exists. Such frequency entrainment is observed in the frequency response of nonlinear systems which exhibit limit cycles.

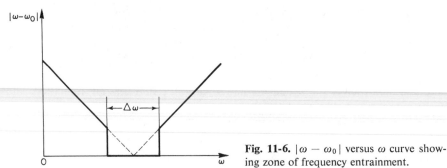

**Fig. 11-6.** $|\omega - \omega_0|$ versus $\omega$ curve showing zone of frequency entrainment.

**Asynchronous quenching.** In a nonlinear system which exhibits a limit cycle of frequency $\omega_0$, it is possible to quench the limit-cycle oscillation by forcing the system at a frequency $\omega_1$, where $\omega_1$ and $\omega_0$ are not related to each other. This phenomenon is called asynchronous quenching, or signal stabilization.

**Comment.** None of the phenomena mentioned above, as well as other nonlinear phenomena not mentioned here, occur in linear systems. These phenomena cannot be explained by linear theory; in order to explain them, we must solve the nonlinear differential equations describing the dynamics of the system analytically or graphically.

## 11-2 NONLINEAR CONTROL SYSTEMS

Many different types of nonlinearities may be found in practical control systems, and they may be divided into two classes, depending upon whether they are inherent in the system or intentionally inserted into the system.

In what follows, we shall first discuss inherent nonlinearities and then intentional nonlinearities. Finally we shall discuss approaches to the analysis and design of nonlinear control systems.

**Inherent nonlinearities.** Inherent nonlinearities are unavoidable in control systems. Examples of such nonlinearities are

1. saturation
2. dead zone
3. hysteresis
4. backlash
5. static friction, coulomb friction, and other nonlinear friction
6. nonlinear spring
7. compressibility of fluid

Generally speaking, the presence of such nonlinearities in the control system adversely affects system performance. For example, backlash may cause instability in the system, and dead zone will cause steady-state error.

**Intentional nonlinearities.** Some nonlinear elements are intentionally introduced into a system in order to improve the system performance or to simplify the construction of the system, or both. A nonlinear system properly designed to perform a certain function is often superior from economic, weight, space, and reliability points of view to linear systems designed to perform the same task. The simplest example of such an intentionally nonlinear system is a conventional relay-operated system. Other examples may be found in optimal control systems which often employ complicated nonlinear controllers. (For a discussion of optimal control systems, refer to Chapter 16.) It should be noted that although intentional nonlinear elements may improve the system performance under certain specified operating conditions, they will, in general, degrade system performance under other operating conditions.

**Effect of inherent nonlinearities on static accuracy.** One characteristic of control systems is that power is transmitted through the feedforward path, while the static accuracy of the system is determined by the elements in the feedback path. Thus, the measuring device determines the upper limit on the static accuracy; the static accuracy cannot be better than the accuracy of this measuring device. Therefore, any inherent nonlinearities in the feedback elements should be kept to a minimum.

If feedback elements suffer from friction, backlash, etc., then it is desirable to feed the error signal to an integrating device, because the system may not detect a very small error, unless the small error is integrated continuously, making the magnitude large enough to be detected.

**Approaches to the analysis and design of nonlinear control systems.** There is no general method for dealing with all nonlinear systems becuase nonlinear differential equations are virtually devoid of a general method of attack. (Exact solutions can be found only for certain simple nonlinear differential equations. For many nonlinear differential equations of practical importance, only approximate solutions are possible, and these solutions hold true only under the limited conditions for which they are obtained.) Because there is no general approach, we may take up each nonlinear equation, or group of similar equations, individually and attempt to develop a method of analysis which will satisfactorily apply to that particular group. (Note that although a very limited amount of generalization within the group of similar equations may be made, a broad generalization of a particular solution is impossible.)

One way to analyze and design a particular group of nonlinear control systems, in which the degree of nonlinearity is small, is to use equivalent linearization techniques and to solve the resulting linearized problem. The describing-function method to be discussed in this chapter is one of the equivalent linearization methods. In many practical cases, we are primarily concerned with the stability of nonlinear control systems, and analytical solutions of nonlinear differential equations may not be necessary. (Establishing stability criteria is very much simpler than obtaining analytical solutions.) The describing-function method enables us to study the stability of many simple nonlinear control systems from a frequency-domain point of view.

The describing-function method provides stability information for a system of any order, but it will not give exact information as to the time-response characteristics.

Other ways of analyzing and designing nonlinear control systems, which may be highly nonlinear, include solving the actual or simplified nonlinear differential equations by phase-plane techniques or applying techniques based on the second method of Liapunov. (For discussions of the phase-plane method and of the second method of Liapunov, refer to Chapters 12 and 15, respectively.)

The phase-plane method provides information on both stability and time-response behavior, but it is limited to first- and second-order systems.

The second method of Liapunov may be applied to stability analysis of any nonlinear system, but its application may be hampered because of difficulty in finding Liapunov functions for complicated nonlinear systems. (For Liapunov functions, see Chapter 15.)

**Computer solutions of nonlinear problems.** Modern computers have led to new methods for dealing with nonlinear problems. Computer simulation techniques by use of analog computers and/or digital computers are very powerful for analyzing and designing nonlinear control systems. It is now possible to handle rather complicated nonlinear systems by use of computers in a reasonable amount of time. When the complexity of a system precludes the use of any analytical approach, computer simulations may prove to be most advantageous in obtaining the necessary information for design purposes.

**Comment.** It is important to keep in mind that although the prediction of the behavior of nonlinear systems is usually difficult, in designing a control system we should not try to force the system to be as linear as possible, because the requirement of the system being linear may lead to the design of an expensive and less desirable system than a properly designed nonlinear one.

## 11-3 DESCRIBING FUNCTIONS

This section presents describing-function representations of commonly encountered nonlinear elements.

**Describing functions.** Suppose that the input to a nonlinear element is sinusoidal. The output of the nonlinear element is, in general, not sinusoidal. Suppose that the output is periodic with the same period as the input. (The output contains higher harmonics, in addition to the fundamental harmonic component.)

In the describing-function analysis, we assume that only the fundamental harmonic component of the output is significant. Such an assumption is often valid since the higher harmonics in the output of a nonlinear element are often of smaller amplitude than the amplitude of the fundamental harmonic component. In addition, most control systems are low-pass filters, with the result that the higher harmonics are very much attenuated compared with the fundamental harmonic component.

The describing function or sinusoidal describing function of a nonlinear element is defined to be the complex ratio of the fundamental harmonic component of the output to the input. That is,

$$N = \frac{Y_1}{X} \underline{/\phi_1}$$

where

$N =$ describing function

$X =$ amplitude of input sinusoid

$Y_1 =$ amplitude of the fundamental harmonic component of output

$\phi_1 =$ phase shift of the fundamental harmonic component of output

If no energy-storage element is included in the nonlinear element, then $N$ is a function only of the amplitude of the input to the element. On the other hand, if energy-storage elements are included, then $N$ is a function of both the amplitude and frequency of the input.

In calculating the describing function for a given nonlinear element, we need to find the fundamental harmonic component of the output. For the sinusoidal input $x(t) = X \sin \omega t$ to the nonlinear element, the output $y(t)$ may be expressed as a Fourier series as follows:

$$y(t) = A_0 + \sum_{n=1}^{\infty} (A_n \cos n\omega t + B_n \sin n\omega t)$$

$$= A_0 + \sum_{n=1}^{\infty} Y_n \sin (n\omega t + \phi_n)$$

where

$$A_n = \frac{1}{\pi} \int_0^{2\pi} y(t) \cos n\omega t \, d(\omega t)$$

$$B_n = \frac{1}{\pi} \int_0^{2\pi} y(t) \sin n\omega t \, d(\omega t)$$

$$Y_n = \sqrt{A_n^2 + B_n^2}$$

$$\phi_n = \tan^{-1} \left( \frac{A_n}{B_n} \right)$$

If the nonlinearity is symmetric, then $A_0 = 0$. The fundamental harmonic component of the output is

$$y_1(t) = A_1 \cos \omega t + B_1 \sin \omega t$$

$$= Y_1 \sin (\omega t + \phi_1)$$

The describing function is then given by

$$N = \frac{Y_1}{X} \underline{/\phi_1} = \frac{\sqrt{A_1^2 + B_1^2}}{X} \underline{/\tan^{-1}\left(\frac{A_1}{B_1}\right)}$$

Clearly, $N$ is a complex quantity when $\phi_1$ is nonzero.

Table 11-1 shows three nonlinearities and their describing functions. (In Table 11-1, $k_1$, $k_2$, and $k$ denote the slopes of the lines.) Illustrative calculations of describing functions for commonly encountered nonlinearities are given in what follows.

**Table 11-1.** THREE NONLINEARITIES AND THEIR DESCRIBING FUNCTIONS

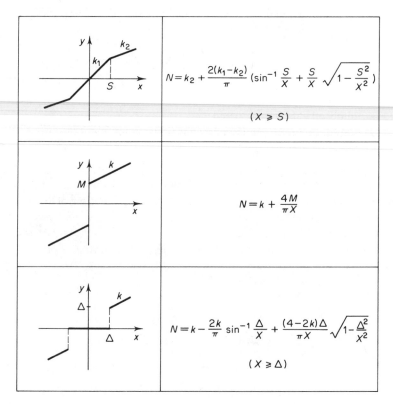

$$N = k_2 + \frac{2(k_1 - k_2)}{\pi}\left(\sin^{-1}\frac{S}{X} + \frac{S}{X}\sqrt{1 - \frac{S^2}{X^2}}\right)$$

$$(X \geqslant S)$$

$$N = k + \frac{4M}{\pi X}$$

$$N = k - \frac{2k}{\pi}\sin^{-1}\frac{\Delta}{X} + \frac{(4 - 2k)\Delta}{\pi X}\sqrt{1 - \frac{\Delta^2}{X^2}}$$

$$(X \geqslant \Delta)$$

**On-off nonlinearity.** The on-off nonlinearity is often called a two-position nonlinearity. Consider an on-off element whose input-output characteristic curve is shown in Fig. 11-7 (a). The output of this element is either a positive constant or a negative constant, and Fig. 11-7 (b) shows the input and output waveforms.

Let us obtain the Fourier series expansion of the output $y(t)$ of such an element.

$$y(t) = A_0 + \sum_{n=1}^{\infty}(A_n \cos n\omega t + B_n \sin n\omega t)$$

As shown in Fig. 11-7 (b), the output is an odd function. For any odd function,

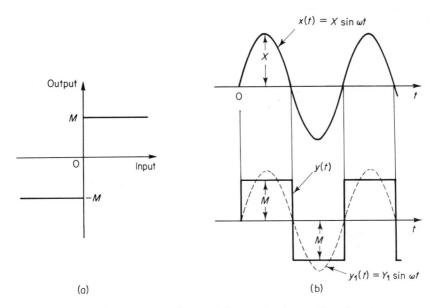

**Fig. 11-7.** (a) Input-output characteristic curve for the on-off nonlinearity; (b) input and output waveforms for the on-off nonlinearity.

we have $A_n = 0$ $(n = 0, 1, 2, \ldots)$. Hence

$$y(t) = \sum_{n=1}^{\infty} B_n \sin n\omega t$$

The fundamental harmonic component of $y(t)$ is

$$y_1(t) = B_1 \sin \omega t = Y_1 \sin \omega t$$

where

$$Y_1 = \frac{1}{\pi} \int_0^{2\pi} y(t) \sin \omega t \, d(\omega t) = \frac{2}{\pi} \int_0^{\pi} y(t) \sin \omega t \, d(\omega t)$$

Substituting $y(t) = M$ into this last equation gives

$$Y_1 = \frac{2M}{\pi} \int_0^{\pi} \sin \omega t \, d(\omega t) = \frac{4M}{\pi}$$

Thus

$$y_1(t) = \frac{4M}{\pi} \sin \omega t$$

The describing function $N$ is then given by

$$N = \frac{Y_1}{X} \underline{/0^\circ} = \frac{4M}{\pi X}$$

Clearly, the describing function for the on-off element is a real quantity and is a function only of the input amplitude $X$. A plot of this describing function versus $M/X$ is shown in Fig. 11-8.

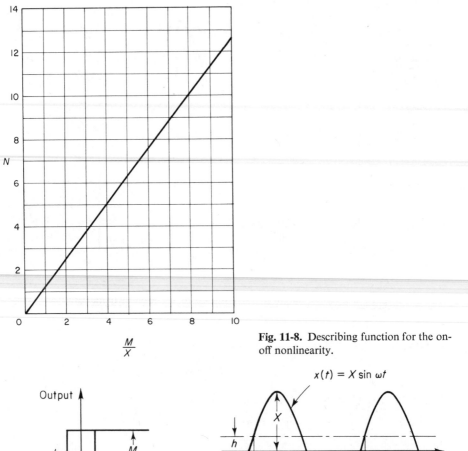

$$\frac{M}{X}$$

**Fig. 11-8.** Describing function for the on-off nonlinearity.

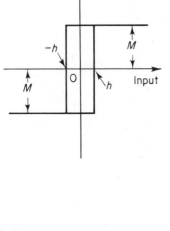

(a)

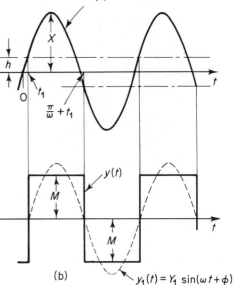

(b)

**Fig. 11-9.** (a) Input-output characteristic curve for the on-off nonlinearity with hysteresis; (b) input and output waveforms for the on-off nonlinearity with hysteresis.

**On-off nonlinearity with hysteresis.** Consider an on-off element with hysteresis whose input-output characteristic curve is shown in Fig. 11-9 (a). The input and output waveforms are shown in Fig. 11-9 (b). Clearly, the output is a square wave, but it lags behind the input by $\omega t_1 = \sin^{-1} (h/X)$. Hence the describing function for this nonlinear element is

$$N = \frac{4M}{\pi X} \left/ -\sin^{-1} \left( \frac{h}{X} \right) \right.$$

It is convenient to plot

$$\frac{h}{M} N = \frac{4h}{\pi X} \left/ -\sin^{-1} \left( \frac{h}{X} \right) \right.$$

versus $h/X$ rather than $N$ versus $h/X$ because $hN/M$ is a function only of $h/X$. A plot of $hN/M$ versus $h/X$ is shown in Fig. 11-10.

**Dead-zone nonlinearity.** The dead-zone nonlinearity is sometimes referred to as the threshold nonlinearity. A typical input-output characteristic curve is shown in Fig. 11-11 (a). Figure 11-11 (b) shows the input and output waveforms. In a dead-zone element, there is no output for inputs within the dead-zone amplitude.

For the element with dead-zone shown in Fig. 11-11 (a), the output $y(t)$ for $0 \leq \omega t \leq \pi$ is given by

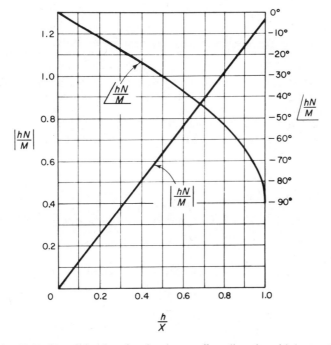

**Fig. 11-10.** Describing function for the on-off nonlinearity with hysteresis.

$$y(t) = 0 \qquad\qquad \text{for } 0 < t < t_1$$

$$= k(X \sin \omega t - \Delta) \qquad \text{for } t_1 < t < \frac{\pi}{\omega} - t_1$$

$$= 0 \qquad\qquad \text{for } \frac{\pi}{\omega} - t_1 < t < \frac{\pi}{\omega}$$

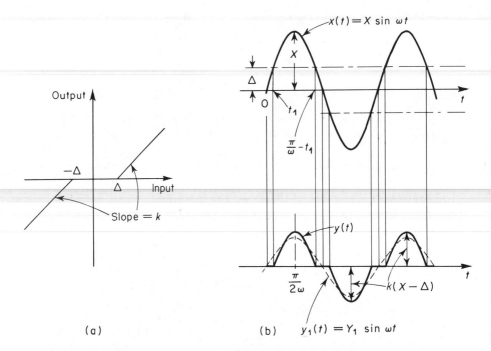

(a)        (b)     $y_1(t) = Y_1 \sin \omega t$

**Fig. 11-11.** (a) Input-output characteristic curve for the dead-zone non-linearity; (b) input and output waveforms for the dead-zone nonlinearity.

Since the output $y(t)$ is once again an odd function, its Fourier series expansion has only sine terms. The fundamental harmonic component of the output is given by

$$y_1(t) = Y_1 \sin \omega t$$

where

$$Y_1 = \frac{1}{\pi} \int_0^{2\pi} y(t) \sin \omega t \, d(\omega t)$$

$$= \frac{4}{\pi} \int_0^{\pi/2} y(t) \sin \omega t \, d(\omega t)$$

$$= \frac{4k}{\pi} \int_{\omega t_1}^{\pi/2} (X \sin \omega t - \Delta) \sin \omega t \, d(\omega t)$$

Note that

$$\Delta = X \sin \omega t_1$$

or

$$\omega t_1 = \sin^{-1}\left(\frac{\Delta}{X}\right)$$

Hence

$$Y_1 = \frac{4Xk}{\pi}\left[\int_{\omega t_1}^{\pi/2} \sin^2 \omega t \, d(\omega t) - \sin \omega t_1 \int_{\omega t_1}^{\pi/2} \sin \omega t \, d(\omega t)\right]$$

$$= \frac{2Xk}{\pi}\left[\frac{\pi}{2} - \sin^{-1}\left(\frac{\Delta}{X}\right) - \frac{\Delta}{X}\sqrt{1 - \left(\frac{\Delta}{X}\right)^2}\right]$$

The describing function for an element with dead zone can be obtained as

$$N = \frac{Y_1}{X}\underline{/0°}$$

$$= k - \frac{2k}{\pi}\left[\sin^{-1}\frac{\Delta}{X} + \frac{\Delta}{X}\sqrt{1 - \left(\frac{\Delta}{X}\right)^2}\right]$$

Figure 11-12 shows a plot of $N/k$ as a function of $\Delta/X$. Note that for $(\Delta/X) > 1$ the output is zero and the value of the describing function is also zero.

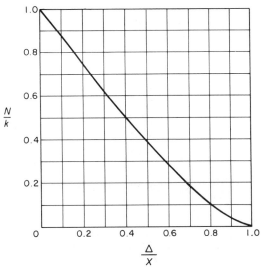

**Fig. 11-12.** Describing function for the dead-zone nonlinearity.

**Saturation nonlinearity.** A typical input-output characteristic curve for the saturation nonlinearity is shown in Fig. 11-13 (a). For small input signals, the output of a saturation element is proportional to the input. For larger input signals the output will not increase proportionally, and finally for very large input signals the output is constant. Figure 11-13 (b) depicts the input and output waveforms for the saturation nonlinearity.

The describing function for such an element can be obtained as

$$N = \frac{2k}{\pi}\left[\sin^{-1}\left(\frac{S}{X}\right) + \frac{S}{X}\sqrt{1 - \left(\frac{S}{X}\right)^2}\right]$$

Figure 11-14 shows a plot of $N/k$ as a function of $S/X$. For $(S/X) > 1$, the value of the describing function is unity.

Note that the describing function for the dead-zone nonlinearity and that for the saturation nonlinearity are related as follows:

$$N_{\text{dead zone}} = k - N_{\text{saturation}} \qquad \text{for } \Delta = S$$

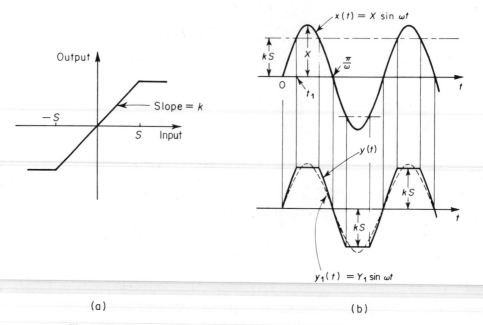

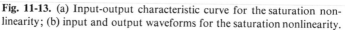

(a)

(b)

**Fig. 11-13.** (a) Input-output characteristic curve for the saturation non-linearity; (b) input and output waveforms for the saturation nonlinearity.

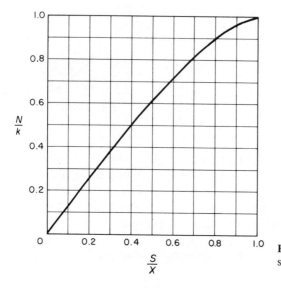

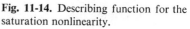

**Fig. 11-14.** Describing function for the saturation nonlinearity.

## 11-4 DESCRIBING-FUNCTION ANALYSIS OF NONLINEAR CONTROL SYSTEMS

Many control systems containing nonlinear elements may be represented by the block diagram shown in Fig. 11-15. If the higher harmonics generated by the

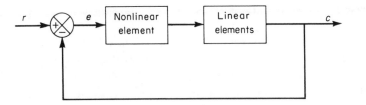

**Fig. 11-15.** Nonlinear control system.

nonlinear element are sufficiently attenuated by the linear elements so that only the fundamental harmonic component of the output of the nonlinear element is significant, then the stability of the system can be predicted by a describing-function analysis.

**Describing-function analysis.** We shall first discuss how the describing functions of nonlinear elements can be used for stability analysis of nonlinear control systems. We shall show that if a sustained oscillation exists in the system output, then the amplitude and frequency of the oscillation may be determined from a graphical study in the frequency domain.

Consider the system shown in Fig. 11-16. The block $N$ denotes the describing function of the nonlinear element. If the higher harmonics are sufficiently at-

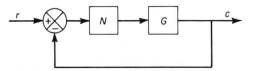

**Fig. 11-16.** Nonlinear control system.

tenuated, the describing function $N$ can be treated as a real variable or complex variable gain. Then the closed-loop frequency response becomes

$$\frac{C(j\omega)}{R(j\omega)} = \frac{NG(j\omega)}{1 + NG(j\omega)}$$

The characteristic equation is

$$1 + NG(j\omega) = 0$$

or

$$G(j\omega) = -\frac{1}{N} \qquad (11\text{-}2)$$

If Eq. (11-2) is satisfied, then the system output will exhibit a limit cycle. This situation corresponds to the case where the $G(j\omega)$ locus passes through the critical point. (In the conventional frequency-response analysis of linear control systems, the critical point is the $-1 + j0$ point.)

In the describing-function analysis, the conventional frequency-response analysis is modified so that the entire $-1/N$ locus becomes a locus of critical points. Thus, the relative location of the $-1/N$ locus and $G(j\omega)$ locus will provide the stability information.

To determine the stability of the system, we plot the $-1/N$ locus and the $G(j\omega)$ locus. In the present analysis, we assume that the linear part of the system is of minimum phase or that all poles and zeros of $G(s)$ lie in the left half of the $s$ plane, including the $j\omega$ axis. The criterion for stability is that if the $-1/N$ locus is not enclosed by the $G(j\omega)$ locus, then the system is stable, or there is no limit cycle at steady state.

On the other hand, if the $-1/N$ locus is enclosed by the $G(j\omega)$ locus, then the system is unstable, and the system output when subjected to any disturbance will increase until breakdown occurs or increase to some limiting value determined by a mechanical stop or other safety device.

If the $-1/N$ locus and the $G(j\omega)$ locus intersect, then the system output may exhibit a sustained oscillation, or a limit cycle. Such a sustained oscillation is not sinusoidal, but it may be approximated by a sinusoidal one. The sustained oscillation is characterized by the value of $X$ on the $-1/N$ locus and the value of $\omega$ on the $G(j\omega)$ locus at the intersection.

In general, a control system should not exhibit limit-cycle behavior, although a limit cycle of small amplitude may be acceptable in certain applications.

**Stability of sustained oscillations, or limit cycles.** The stability of the limit cycle can be predicted as follows: Consider the system shown in Fig. 11-17. Assume that point $A$ on the $-1/N$ locus corresponds to a small value of $X$, where $X$ is the ampli-

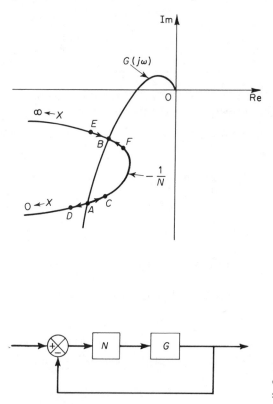

**Fig. 11-17.** Stability analysis of limit-cycle operations of nonlinear control system.

tude of the sinusoidal input signal to the nonlinear element, and that point $B$ on the $-1/N$ locus corresponds to a large value of $X$. The value of $X$ on the $-1/N$ locus increases in the direction from point $A$ to point $B$.

Let us assume that the system is originally operated at point $A$. The oscillation has amplitude $X_A$ and frequency $\omega_A$, determined from the $-1/N$ locus and the $G(j\omega)$ locus, respectively. Assume that a slight disturbance is given to the system operating at point $A$ so that the amplitude of the input to the nonlinear element is increased slightly. (For example, assume that the operating point moves from point $A$ to point $C$ on the $-1/N$ locus.) Then the operating point $C$ corresponds to the critical point or to the $-1 + j0$ point in the complex plane for linear control systems. Therefore, as seen from Fig. 11-17, the $G(j\omega)$ locus encloses point $C$ in the Nyquist sense. Since this is similar to the case where the open-loop locus of a linear system encloses the $-1 + j0$ point, the amplitude will increase and the operating point moves toward point $B$.

Next suppose that a slight disturbance decreases the amplitude of the sinusoidal input to the nonlinear element. Assume that the operating point is moved from point $A$ to point $D$ on the $-1/N$ locus. Point $D$ then corresponds to the critical point. In this case, the $G(j\omega)$ locus does not enclose the critical point and, therefore, the amplitude of the input to the nonlinear element decreases, and the operating point moves further from point $D$ to the left. Thus, point $A$ possesses divergent characteristics and corresponds to an unstable limit cycle.

Consider next the case where a slight disturbance is given to the system operating at point $B$. Assume that the operating point is moved to point $E$ on the $-1/N$ locus. Then the $G(j\omega)$ locus in this case does not enclose the critical point (point $E$). The amplitude of the sinusoidal input to the nonlinear element decreases, and the operating point moves toward point $B$.

Similarly, assume that a slight disturbance causes the system operating point to move from point $B$ to point $F$. Then the $G(j\omega)$ locus will enclose the critical point (point $F$). Therefore, the amplitude of oscillation will increase, and the operating point moves from point $F$ toward point $B$. Thus, point $B$ possesses convergent characteristics, and the system operation at point $B$ is stable; in other words, the limit cycle at this point is stable.

For the system shown in Fig. 11-17, the stable limit cycle corresponding to point $B$ can be experimentally observed, but the unstable limit cycle corresponding to point $A$ cannot be.

**Accuracy of describing-function analysis.** Note that the amplitude and frequency of the limit cycle indicated by the intersection of the $-1/N$ locus and $G(j\omega)$ locus are approximate values.

If the $-1/N$ locus and the $G(j\omega)$ locus intersect almost perpendicularly, then the accuracy of the describing-function analysis is generally good. (If the higher harmonics are all attenuated, the accuracy is excellent. Otherwise the accuracy is good to fair.)

If the $G(j\omega)$ locus is tangent, or almost tangent, to the $-1/N$ locus, then the accuracy of the information obtained from the describing-function analysis depends on how well $G(j\omega)$ will attenuate the higher harmonics. In some cases, there is a sustained oscillation; in other cases, there is no such oscillation. It depends on the nature of $G(j\omega)$. One can say, however, that the system is almost on the verge of exhibiting a limit cycle when the $-1/N$ locus and $G(j\omega)$ locus are tangent to each other.

*Example 11-1.* Figure 11-18 shows a control system with saturation nonlinearity. We assume that $G(s)$ is a minimum phase transfer function. Figure 11-19 shows a plot of the

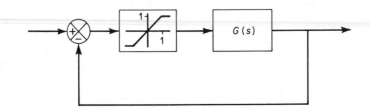

**Fig. 11-18.** Control system with saturation nonlinearity.

$-1/N$ locus and the $G(j\omega)$ locus. The $-1/N$ locus starts from the $-1$ point on the negative real axis and extends to $-\infty$. Clearly $N$ is a function only of the amplitude of the input signal $x(t) = X \sin \omega t$. The $G(j\omega)$ locus is a function only of $\omega$.

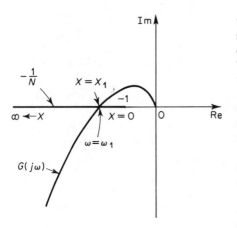

**Fig. 11-19.** Plot of $-1/N$ and $G(j\omega)$ for stability analysis.

Figure 11-19 shows that the two loci intersect. This intersection corresponds to a stable limit cycle. The amplitude of the limit cycle is read from the $-1/N$ locus as $X = X_1$. The frequency of the limit cycle is read from the $G(j\omega)$ locus as $\omega = \omega_1$.

In the absence of any reference input, the output of this system at steady state exhibits a sustained oscillation with amplitude equal to $X_1$ and frequency equal to $\omega_1$.

If the gain of the transfer function $G(s)$ is decreased so that the $-1/N$ locus and the $G(j\omega)$ locus do not intersect, as shown in Fig. 11-20, then the system becomes stable, and any oscillations which may occur in the system output as a result of disturbances will die out and no sustained oscillation will exist at steady state. This is because the $-1/N$ locus is to the left of the $G(j\omega)$ locus, or the $G(j\omega)$ locus does not enclose the $-1/N$ locus.

*Example 11-2.* Figure 11-21 shows a plot of the $-1/N$ locus for the dead-zone nonlinearity and the $G(j\omega)$ locus. In this system, the $-1/N$ locus and the $G(j\omega)$ locus intersect each other. The limit cycle in this case is unstable. The oscillation either dies out or increases in amplitude indefinitely. This indicates an undesirable situation and must be avoided.

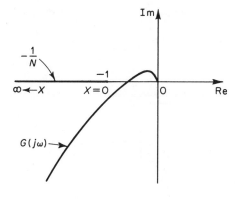

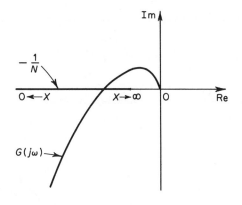

**Fig. 11-20.** Plot of $-1/N$ and $G(j\omega)$ for stability analysis.

**Fig. 11-21.** Plot of $-1/N$ and $G(j\omega)$ for stability analysis.

*Example 11-3.* Consider the system shown in Fig. 11-22. Determine the effect of hysteresis on the amplitude and frequency of the limit-cycle operation of the system.

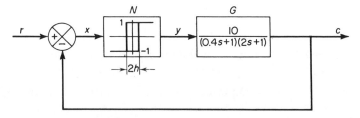

**Fig. 11-22.** Nonlinear control system.

The $-1/N$ loci for three different values of $h$, namely, $h = 0.1, 0.2$, and $0.3$, are shown in Fig. 11-23, together with the $G(j\omega)$ locus. The $-1/N$ loci are straight lines parallel to the real axis. The values of $N$ are obtained from Fig. 11-10.

From Fig. 11-23, one can see that the amplitude and frequency of the limit cycles are

$$X = 0.27, \qquad \omega = 7 \qquad \text{if } h = 0.1$$
$$X = 0.42, \qquad \omega = 5.9 \qquad \text{if } h = 0.2$$
$$X = 0.57, \qquad \omega = 5.1 \qquad \text{if } h = 0.3$$

Inspection of these values reveals that increasing the hysteresis amplitude decreases the frequency but increases the amplitude of the limit cycle, as expected.

## 11-5  CONCLUDING COMMENTS

The describing-function analysis is an extension of linear techniques to the study of nonlinear systems. Therefore, typical applications are to systems with a low degree of nonlinearity. The use of describing functions in the analysis of high-degree nonlinear systems may lead to seriously erroneous results; this limits the applicability of the describing function to the analysis and synthesis of low-degree nonlinear systems.

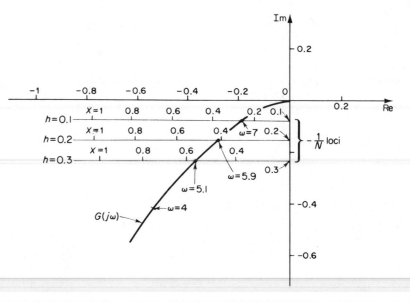

**Fig. 11-23.** Plot of $-1/N$ and $G(j\omega)$ for the system shown in Fig. 11-22.

In concluding this chapter, we shall summarize the describing-function approach to the analysis and design of nonlinear control systems.

1. The describing-function method is an approximate method for determining the stability of unforced nonlinear control systems. In applying this method, one must keep in mind the basic assumptions and limitations. Although many practical control systems satisfy the basic assumptions of the describing function method, there are some systems which do not. Therefore, it is always necessary to examine the validity of the method in each case.

2. In the describing-function analysis, the nature of the nonlinearity present in the system determines the complexity of the analysis. In other words, the linear elements of whatever order do not affect the complexity. It is an advantage of this method that the analysis is not materially complicated for systems with complex dynamics in their linear parts. The accuracy of the analysis is better for higher-order systems than for lower-order ones because higher-order systems generally have better low-pass filtering characteristics.

3. Although the describing-function method is quite useful in predicting the stability of unforced systems, it presents little information concerning transient-response characteristics.

4. The describing function method is convenient to apply to design problems. The use of describing functions enables one to apply frequency-response methods to the reshaping of the $G(j\omega)$ locus. The describing-function analysis is particularly useful when the designer wants a rough idea of the effects of certain nonlinearities or of the effects of modifying linear or nonlinear components within the loop. The analysis yields graphically information concerning the stability and suggests ways to improve the response characteristics, if

necessary. When the $-1/N$ locus and the $G(j\omega)$ locus are plotted in the complex plane, the system performance can be estimated quickly from the plot. If any performance improvements are necessary, they may be accomplished by modifying the loci. This modification of the loci suggests the type of a suitable compensating network. [Design through describing-function analysis may suggest a particular $-1/N$ locus rather than suggesting the reshaping of the $G(j\omega)$ locus. However, the realization of a nonlinear element with a specified describing function may be difficult.] It is interesting to note that although the describing-function method may enable one to predict limit cycles with good engineering accuracy, in design problems the method is used as a negative criterion, in that system parameters are adjusted until limit-cycle conditions are eliminated and a proper relative stability is assured.

5. A physical system may possess two or more significant nonlinear elements. When only one nonlinear element becomes significant for a particular operating condition, effects of other nonlinear elements may be neglected in the analysis. For example, if the system has both small-signal nonlinearities and large-signal nonlinearities, the former may be neglected when the signal amplitude is large and vice versa. It is important to keep in mind that the describing function of two nonlinearities in series is, in general, not the same as the product of the individual describing functions. Therefore, if two or more nonlinear elements, which are not separated from each other by effective low-pass filters, become significant simultaneously under certain operating conditions, they may be combined in one block, and the equivalent describing function of this block may be obtained. In this case, the describing function may become both amplitude and frequency dependent.

6. In the usual describing-function analysis, the input to the nonlinearity is assumed to be sinusoidal, but this assumption may be extended. The input to the nonlinearity may be a sinusoidal input, plus an additional signal, although this additional complication may make the analysis very tedious. Describing functions corresponding to this case are called dual-input describing functions.

7. In some cases the stability analysis of control systems may be of primary concern, but in other cases an optimal response (in some sense) may be desired. Optimal design problems may involve the determination of a nonlinear controller (or computer) for insertion into the sytem. The performance of the nonlinear system depends markedly upon the input signals. This means that precise descriptions of the input and desired output are necessary. Because of a weak correlation between the frequency response and time response of nonlinear systems, the describing-function approach ceases to be useful in the design of optimal control systems with aperiodic inputs. (Optimal control systems with aperiodic inputs are discussed in Chapter 16.)

EXAMPLE PROBLEMS AND SOLUTIONS

**PROBLEM A-11-1.** Figure 11-24 (a) is a schematic diagram of a liquid-level control system. The movement of the float positions the electric mercury switch which energizes

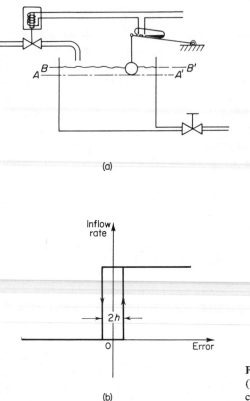

(a)

(b)

**Fig. 11-24.** (a) Liquid-level control system; (b) inflow-rate versus error curve of the controller.

or deenergizes the electric solenoid-operated valve. When the valve is open, liquid is admitted to the tank. The control action is on-off with hysteresis. The inflow rate versus error curve is shown in Fig. 11-24 (b). (The hysteresis width $2h$ is called a differential gap.)

Plot the head versus time curves at steady-state operation under the following two conditions:

1. The rate of rise in head when the inflow valve is open is considerably smaller than the rate of fall in head when the valve is closed.
2. The rate of rise in head when the inflow valve is opened is considerably greater than the rate of fall in head when the valve is closed.

**Solution.** Assume that the liquid level is falling and that the inflow valve is closed. When the head falls to level $AA'$ in Fig. 11-24 (a), the bottom of the differential gap, the mercury switch contacts will be closed and the inflow valve will be opened, admitting liquid to the tank. The head will start to rise as the inflow valve is opened. At the start there will be a high rate of rise. As the level in the tank rises, the outflow will increase due to the greater head. The result is a smaller net inflow to the tank. When the head rises to level $BB'$ in Fig. 11-24 (a), the mercury switch contacts will be opened and the inflow valve will be closed. The head will then start to fall.

Figure 11-25 (a) shows a head versus time curve under Condition 1, when the rate of rise in head with the inflow valve open is considerably smaller than the rate of fall in head with the valve closed. Here the on time is considerably longer than the off time. Figure 11-25(b) shows a head versus time curve under Condition 2. The on time is considerably shorter than the off time.

Under either condition, the level oscillates about the desired value. Thus the system exhibits limit-cycle behavior. In any case, the average inflow is equal to the average outflow. The inflow-outflow rates determine the shape of the level versus time curve. If the head rises at a rate which is equal to the fall with the valve closed, then the on time and off time will be equal.

Note that widening the gap will cause less frequent operation of the mercury switch and solenoid-operated valve. This normally means longer life for the equipment. The disadvantage of operating less frequently is that the variation in the head of the tank becomes larger.

Finally, it should be noted that on-off type control offers the best economy, the highest sensitivity, and ease of maintenance. Hence if a limit cycle of small amplitude is permissible in a particular application, then the use of any other type of control would be a mistake.

**PROBLEM A-11-2.** Obtain the describing function for the on-off nonlinearity with dead zone shown in Fig. 11-26.

**Solution.** Figure 11-27 shows the input and output waveforms for an element with the given nonlinearity.

The output of the nonlinear element for $0 \leq \omega t \leq \pi$ is given by

$$y(t) = 0 \qquad \text{for } 0 < t < t_1$$

$$= M \qquad \text{for } t_1 < t < \frac{\pi}{\omega} - t_1$$

$$= 0 \qquad \text{for } \frac{\pi}{\omega} - t_1 < t < \pi$$

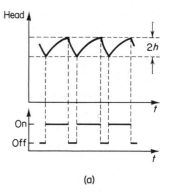

(a)

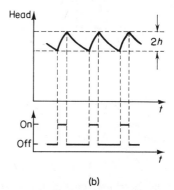

(b)

**Fig. 11-25.** (a) Head versus time curve under Condition 1; (b) head versus time curve under Condition 2.

The output waveform is an odd function. Hence the fundamental harmonic component of the output $y(t)$ is given by

$$y_1(t) = Y_1 \sin \omega t$$

where

$$Y_1 = \frac{1}{\pi} \int_0^{2\pi} y(t) \sin \omega t \, d(\omega t)$$

$$= \frac{4}{\pi} \int_0^{\pi/2} y(t) \sin \omega t \, d(\omega t)$$

$$= \frac{4}{\pi} \int_{\omega t_1}^{\pi/2} M \sin \omega t \, d(\omega t)$$

$$= \frac{4M}{\pi} \cos \omega t_1$$

Since $\sin \omega t_1 = \Delta/X$, we obtain

$$\cos \omega t_1 = \sqrt{1 - \left(\frac{\Delta}{X}\right)^2}$$

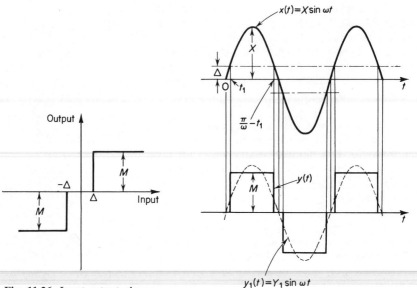

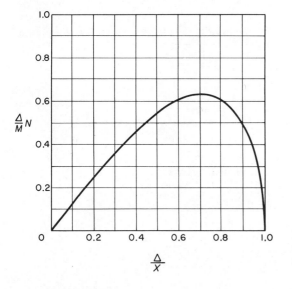

**Fig. 11-26.** Input-output characteristic curve for the on-off nonlinearity with dead zone.

**Fig. 11-27.** Input and output waveforms for the on-off nonlinearity with dead zone.

Thus

$$y_1(t) = \frac{4M}{\pi} \sqrt{1 - \left(\frac{\Delta}{X}\right)^2} \sin \omega t$$

The describing function for an on-off element with dead zone is

$$N = \frac{4M}{\pi X} \sqrt{1 - \left(\frac{\Delta}{X}\right)^2}$$

The describing function in this case is a function only of the input amplitude $X$. Figure 11-28 shows a plot of $\Delta N/M$ versus $\Delta/X$.

**Fig. 11-28.** Describing function for the on-off nonlinearity with dead zone.

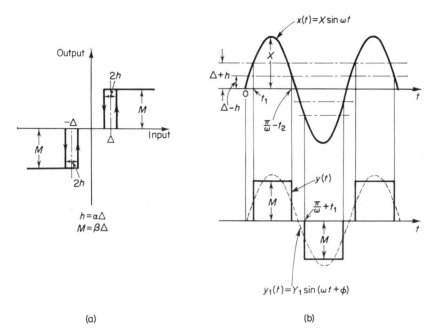

$$h = \alpha \Delta$$
$$M = \beta \Delta$$

(a)                                                      (b)

**Fig. 11-29.** (a) Input-output characteristic curve for the on-off nonlinearity with dead zone and hysteresis; (b) input and output waveforms for the on-off nonlinearity with dead zone and hysteresis.

**PROBLEM A-11-3.** Figure 11-29 (a) shows the input-output characteristic curve for an on-off nonlinearity with dead zone and hysteresis. Figure 11-29 (b) shows the input and output waveforms of a contactor with this nonlinearity.

The contactor does not close until the input exceeds the value $\Delta + h$. The contactor remains closed until the input becomes less than $\Delta - h$. In the region between $\Delta - h$ and $\Delta + h$, the output depends on the past history of the input. The contactor opens or closes in a similar fashion for a negative input.

Obtain the describing function for the contactor with this nonlinearity.

**Solution.** From Fig. 11-29 (b) we obtain the following equation:

$$y(t) = M \qquad \text{for } t_1 < t < \frac{\pi}{\omega} - t_2$$

$$= 0 \qquad \text{for } \frac{\pi}{\omega} - t_2 < t < \frac{\pi}{\omega} + t_1$$

where $t_1$ and $t_2$ are defined by

$$\sin \omega t_1 = \frac{\Delta + h}{X}$$

$$\sin \omega t_2 = \frac{\Delta - h}{X}$$

As seen from Fig. 11-29 (b), the output lags the input. The fundamental harmonic component of the output can be obtained as follows:

$$y_1(t) = A_1 \cos \omega t + B_1 \sin \omega t$$

where

$$A_1 = \frac{2}{\pi} \int_0^\pi y(t) \cos \omega t \, d(\omega t)$$

$$= \frac{2}{\pi} \int_{\omega t_1}^{\pi - \omega t_2} M \cos \omega t \, d(\omega t)$$

$$= -\frac{4hM}{\pi X}$$

$$B_1 = \frac{2}{\pi} \int_0^\pi y(t) \sin \omega t \, d(\omega t)$$

$$= \frac{2}{\pi} \int_{\omega t_1}^{\pi - \omega t_2} M \sin \omega t \, d(\omega t)$$

$$= \frac{2M}{\pi} \left[ \sqrt{1 - \left(\frac{\Delta - h}{X}\right)^2} + \sqrt{1 - \left(\frac{\Delta + h}{X}\right)^2} \right]$$

Let us define

$$h = \alpha \Delta, \qquad M = \beta \Delta$$

The values of $\alpha$ and $\beta$ are constant for a given on-off nonlinearity with dead zone and hysteresis. Then

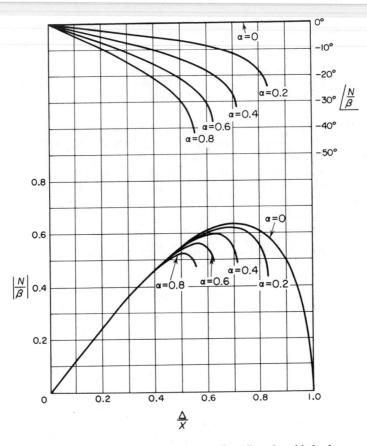

**Fig. 11-30.** Describing function for the on-off nonlinearity with dead zone and hysteresis.

$$\frac{A_1}{X} = -\frac{4\alpha\beta}{\pi}\left(\frac{\Delta}{X}\right)^2$$

$$\frac{B_1}{X} = \frac{2\beta}{\pi}\frac{\Delta}{X}\left[\sqrt{1 - \left(\frac{\Delta}{X}\right)^2(1 - \alpha)^2} + \sqrt{1 - \left(\frac{\Delta}{X}\right)^2(1 + \alpha)^2}\right]$$

Thus the describing function for this nonlinearity is

$$N = \sqrt{\left(\frac{A_1}{X}\right)^2 + \left(\frac{B_1}{X}\right)^2}\ \bigg/ \tan^{-1}\left(\frac{A_1}{B_1}\right)$$

This describing function is a complex quantity. Figure 11-30 shows plots of $|N/\beta|$ versus $\Delta/X$ and $\underline{/N/\beta}$ versus $\Delta/X$.

**PROBLEM A-11-4.** The system shown in Fig. 11-31 exhibits a limit cycle with the frequency of oscillation at 5.9 rad/sec, as shown in Fig. 11-32. It is desired to decrease the frequency of the limit cycle to 4 rad/sec. Determine the necessary change in the gain of $G(s)$, assuming that the nonlinear element is fixed.

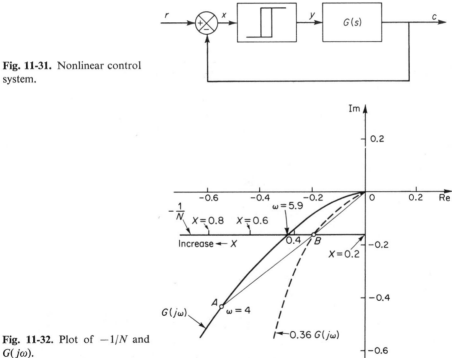

**Fig. 11-31.** Nonlinear control system.

**Fig. 11-32.** Plot of $-1/N$ and $G(j\omega)$.

**Solution.** From Fig. 11-32, $\overline{OB}/\overline{OA}$ is found to be 0.36. Hence, if the value of the gain of $G(s)$ is decreased to 36% of the original value, the frequency of the new limit cycle will become 4 rad/sec. The amplitude also decreases, from 0.42 to 0.35.

**PROBLEM A-11-5.** Figure 11-33(a) shows the block diagram for a servomechanism consisting of an amplifier, a motor, and a gear train. The position of the output is fed back to the input to generate the error signal. It is assumed that the inertia of the gears is negligible compared with that of the motor and that the gear ratio is unity.

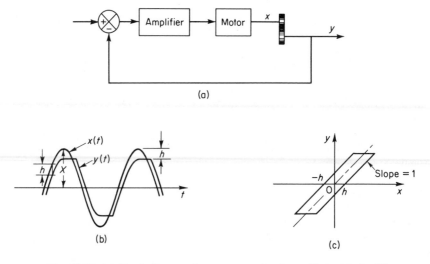

(a)

(b)

(c)

**Fig. 11-33.** (a) Block diagram for a servomechanism with backlash; (b) characteristic curves for a backlash nonlinearity; (c) input-output characteristic curve for a backlash (or hysteresis) nonlinearity.

Because of backlash, the signals $x(t)$ and $y(t)$ are related as shown in Fig. 11-33 (b). Figure 11-33 (c) shows the input-output characteristic curve for the backlash nonlinearity considered here. The describing function for this nonlinearity is plotted in Fig. 11-34.

Using this describing function, determine the amplitude and frequency of the limit cycle when the transfer function of the amplifier-motor combination is given by

$$\frac{5}{s(s+1)}$$

and the backlash amplitude is given as unity, or $h = 1$,

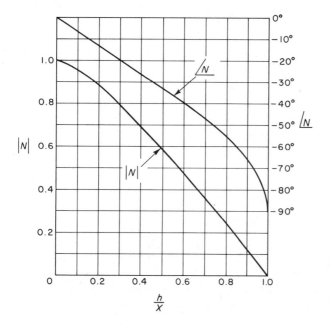

**Fig. 11-34.** Describing function for a backlash nonlinearity, or a hysteresis nonlinearity shown in Fig. 11-33(c).

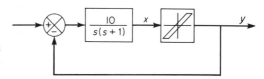

**Fig. 11-35.** Block diagram representation for the servomechanism shown in Fig. 11-33 (a).

**Solution.** From the problem statement, the block diagram for the system may be drawn as shown in Fig. 11-35. Information about the limit-cycle operation of the system may easily be obtained if a frequency-domain diagram is plotted.

Figure 11-36 shows a plot of the $-1/N$ locus and the $G(j\omega)$ locus on the log-magnitude versus phase diagram. As seen from the plot, there are two intersections of the two loci. Applying the stability test for the limit cycle discussed in Section 11-4 reveals that point $A$ corresponds to a stable limit cycle and point $B$ corresponds to an unstable limit cycle. The stable limit cycle has a frequency of 1.6 rad/sec and an amplitude of 2. (The unstable limit cycle cannot physically occur.)

To avoid limit-cycle behavior, the gain of the amplifier must be decreased sufficiently so that the $G(j\omega)$ locus lies well below the $-1/N$ locus in the log-magnitude versus phase diagram. If the two loci are tangent, or almost tangent, the accuracy of the describing-function analysis is poor, and we may expect a limit cycle or a slowly damped oscillation.

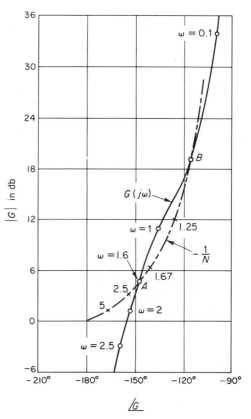

**Fig. 11-36.** Plot of $-1/N$ and $G(j\omega)$ of the servomechanism shown in Fig. 11-35.

## Problems

**PROBLEM B-11-1.** For the system shown in Fig. 11-37, determine the amplitude and frequency of the limit cycle.

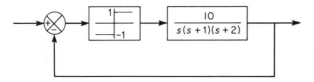

**Fig. 11-37.** Nonlinear control system.

**PROBLEM B-11-2.** Determine the describing function for the nonlinear element described by

$$y = x^3$$

where

    $x$ = input to the nonlinear element (sinusoidal signal)
    $y$ = output of the nonlinear element

**PROBLEM B-11-3.** Determine the stability of the system shown in Fig. 11-38.

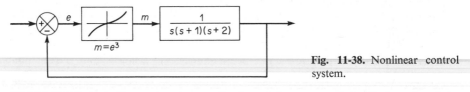

**Fig. 11-38.** Nonlinear control system.

**PROBLEM B-11-4.** Determine the amplitude and frequency of the limit cycle of the system shown in Fig. 11-39.

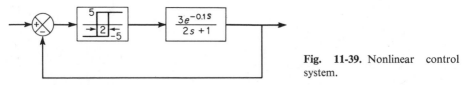

**Fig. 11-39.** Nonlinear control system.

**PROBLEM B-11-5.** Derive the equation for the describing function $N$ for the hysteresis nonlinearity shown in Fig. 11-33 (c).

**PROBLEM B-11-6.** Referring to the input and output waveforms for the saturation nonlinearity shown in Fig. 11-13 (b), obtain the amplitude of the third harmonic component of the output and plot $Y_3/Y_1$ as a function of $S/X$ where $Y_1$ is the amplitude of the fundamental harmonic component and $Y_3$ is the amplitude of the third harmonic component.

**PROBLEM B-11-7.*** Consider a nonlinear element whose input-output characteristic is defined by

$$y = b_1 x + b_3 x^3 + b_5 x^5 + b_7 x^7 + \cdots$$

where

    $x$ = input to the nonlinear element (sinusoidal signal)
    $y$ = output of the nonlinear element

Show that the describing function for this nonlinearity can be given by

$$N = b_1 + \tfrac{3}{4}b_3 X^2 + \tfrac{5}{8}b_5 X^4 + \tfrac{35}{64}b_7 X^6 + \cdots$$

where $X$ is the amplitude of the input sinusoid $x = X \sin \omega t$.

    *Reference W-1.

# 12

# PHASE-PLANE ANALYSIS

## 12-1 INTRODUCTION

Consider a second-order system described by the following ordinary differential equation:

$$\ddot{x} + f(x, \dot{x}) = 0$$

where $f(x, \dot{x})$ is either a linear or nonlinear function of $x$ and $\dot{x}$. The time solution of this system may be illustrated by a plot of $x(t)$ versus $t$. It also can be illustrated by plotting $\dot{x}(t)$ versus $x(t)$ using time $t$ as a parameter.

If we take $x$ and $\dot{x}$ as the coordinates of a plane, to each state of the system there corresponds a point in this plane. As $t$ varies, this point describes a curve in the $x$-$\dot{x}$ plane, indicating the history of the system. Such a curve is called a *trajectory*.

The geometrical representation of the system behavior in terms of trajectories is called a phase-plane representation of the system dynamics. Although the phase-plane diagram will give a clear picture of the trajectories for second-order systems, it is usually difficult to visualize or construct trajectories for systems of third order. For systems of order higher than third, it is impossible to visualize trajectories; however, notions of the motion of a representative point in two-dimensional space can be conceptually extended to $n$-dimensional space.

This chapter presents the phase-plane method of analysis of second-order systems. The method is especially useful for dealing with systems with gross non-linearities. Different from the describing-function method, the phase-plane method is not restricted to small nonlinearities. Familiarity with this method may enable one to solve specific nonlinear control problems, as well as to understand the fundamental characteristics of certain nonlinear systems. Although systems amenable to this method are practically limited to first- and second-order ones, the results obtained from studying the effects of a nonlinearity on the response of a second-order system give considerable insight into the behavior of higher-order systems possessing the same nonlinearity. In this sense, the phase-plane method is quite useful for analyzing nonlinear control systems. It is also useful for synthesizing such systems. In particular, the time-optimal control of second-order systems can be studied conveniently in the phase plane (see Chapter 16). Note, however, that the limitations on the order of the system hamper the usefulness of this approach in the study of time-optimal higher-order control systems.

**Phase-plane method.** The phase-plane method introduced by Poincare is a method for obtaining graphically the solution of the following two simultaneous first-order differential equations:

$$\frac{dx_1}{dt} = f_1(x_1, x_2) \tag{12-1}$$

$$\frac{dx_2}{dt} = f_2(x_1, x_2) \tag{12-2}$$

where $f_1(x_1, x_2)$ and $f_2(x_1, x_2)$ are either linear or nonlinear functions of the variables $x_1$ and $x_2$, respectively. Equations (12-1) and (12-2) are called autonomous, which means that the independent variable $t$ appears only in the form of derivatives. (Thus, in an autonomous system, neither the forces nor the constraints vary with time.)

The plane with rectangular coordinates $x_1$ and $x_2$ is called the phase plane, or state plane. (The phase plane or state plane is a two-dimensional state space. We shall postpone detailed discussions of the state space to Chapter 14.)

Very often Eqs. (12-1) and (12-2) assume the following simpler form:

$$\frac{dx_1}{dt} = x_2$$

$$\frac{dx_2}{dt} = f(x_1, x_2)$$

If we define $x_1 = x$, then $x_2 = \dot{x}$. The most common phase plane is the $x$-$\dot{x}$ plane. In this chapter, unless otherwise stated, we assume the phase plane to be the $x$-$\dot{x}$ plane.

The phase-plane analysis of systems of Eqs. (12-1) and (12-2) yields a bird's-eye view of the solutions for any possible initial conditions. In the field of control systems, the phase-plane method is particularly well suited to the analysis and synthesis of second-order systems when they are subjected to initial conditions and/or aperiodic inputs, such as step inputs, ramp inputs, pulse inputs, and impulse inputs.

**Phase-plane portraits.** From the fundamental theorem on the uniqueness of the solution of simultaneous differential equations, we know that the solution of Eqs. (12-1) and (12-2) with a given initial condition is unique, provided that $f_1(x_1, x_2)$ and $f_2(x_1, x_2)$ in Eqs. (12-1) and (12-2) are analytic. (A function is analytic at a given point if it is possible to obtain a Taylor series expansion of the function about the given point.) This uniqueness result does not apply to the points where simultaneously $f_1(x_1, x_2) = 0$ and $f_2(x_1, x_2) = 0$. Such points are called singular points. Singular points are equilibrium points. Any other point in the phase plane is called an ordinary point.

If there are no other equilibrium points in the neighborhood of a given equilibrium point, then this equilibrium point is called an isolated one. Although many practical systems involve only isolated equilibrium points, there are some cases where this is not true. For example, for the system

$$\ddot{x} + \dot{x} = 0$$

all points on the $x$ axis are equilibrium points and, therefore, the equilibrium points are not isolated.

**Obtaining first-order differential equations for second-order systems.** Elimination of the independent variable $t$ from Eqs. (12-1) and (12-2) gives

$$\frac{dx_2}{dx_1} = \frac{f_2(x_1, x_2)}{f_1(x_1, x_2)} \tag{12-3}$$

Equation (12-3) is a first-order differential equation relating $x_1$ to $x_2$, and, in fact, this equation gives the slope of the tangent to the trajectory passing through point $(x_1, x_2)$.

The state of the system given by Eqs. (12-1) and (12-2) [or Eq. (12-3)] can be determined at any time $t$ by the values of $x_1$ and $x_2$.

The solution to Eq. (12-3) may be written

$$x_2 = \phi(x_1) \tag{12-4}$$

Equation (12-4) represents a curve in the phase plane and indicates a motion of a representative point on the curve. The solution curve, or trajectory, which is a plot of $x_2$ as a function of $x_1$, is an integral curve of the system represented by Eq. (12-3). The trajectory does not show time information explicitly. If needed, however, the trajectory can be graduated in time units.

**Phase-plane portraits.** A family of trajectories is called a phase-plane portrait. The initial condition determines the initial location of a representative point on the trajectory. As time increases, the representative point moves along the trajectory. The phase-plane representation of an autonomous system depicts the totality of all possible states of the system, and thus the nature of the response of the system is shown directly in the phase-plane portrait. Since there is one and only one trajectory passing through any given ordinary point in the phase plane, the trajectories generated by all possible initial conditions do not cross each other, except at singular points. At singular points, $dx_2/dx_1$ is indeterminate since it is of the form zero over zero. An infinity of trajectories may approach or leave a singular point.

For a system having a double-valued nonlinearity, such as the hysteresis nonlinearity, the system loses its analyticity. In such a case, however, it may be possible to divide the region into subregions within which the system is analytic so that the phase-plane method may be applied. Then the complete solution may be obtained by connecting piecewise analytic solutions. It should be noted that when the system response is defined by two or more second-order differential equations, trajectories may cross each other.

*Example 12-1.* Consider the first-order system described by

$$\dot{x} = -x \qquad (12\text{-}5)$$

In the phase plane, or $x$-$\dot{x}$ plane, Eq. (12-5) represents a straight line, as shown in Fig. 12-1. For any initial condition $x(0)$, the system returns to its singular point, the origin,

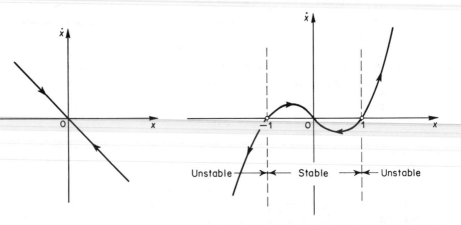

**Fig. 12-1.** Phase-plane plot of a system described by $\dot{x} = -x$.

**Fig. 12-2.** Phase-plane plot of a system described by $\dot{x} = -x + x^3$.

after an infinite time. The initial condition $x(0)$ determines the starting point of the trajectory.

Consider next the following first-order system:

$$\dot{x} = -x + x^3$$

The trajectory is shown in Fig. 12-2. It can be seen that the trajectory is divided into three parts, one stable part and two unstable parts. That is, if $x(0) > 1$, then $x(\infty) \longrightarrow \infty$. If $1 > x(0) > -1$, then $x(\infty) \longrightarrow 0$. If $x(0) < -1$, then $x(\infty) \longrightarrow -\infty$.

*Example 12-2.* Consider the system represented by the following equation:

$$\ddot{x} + \dot{x} + x = 0 \qquad (12\text{-}6)$$

Figure 12-3 shows a phase-plane portrait for this system. Equation (12-6) gives

$$\frac{\ddot{x}}{\dot{x}} = \frac{\dfrac{d\dot{x}}{dt}}{\dfrac{dx}{dt}} = \frac{d\dot{x}}{dx} = \frac{-\dot{x} - x}{\dot{x}}$$

At the origin

$$\frac{d\dot{x}}{dx} = \frac{0}{0}$$

The origin is thus a singular point; i.e., the system is in equilibrium at the origin. When the initial conditions of Eq. (12-6) are $x(0) = 0$ and $\dot{x}(0) = 10$, the corresponding trajectory is the curve $ABCDO$ in Fig. 12-3. Similarly, for $x(0) = 2$ and $\dot{x}(0) = -7$, the trajectory is the curve $EFO$.

If a representative point is in the upper half of the $x$-$\dot{x}$ plane (for example, point $B$ in Fig. 12-3), the point moves to the right on a trajectory as time increases since a positive velocity ($\dot{x} > 0$) corresponds to an increase in the value of $x$ with time. Similarly, if a representative point is in the lower half of the $x$-$\dot{x}$ plane (for example, point $D$ in Fig. 12-3), then the point moves to the left on a trajectory as time increases since a negative velocity ($\dot{x} < 0$) corresponds to a decrease in $x$ with time. Thus, motion along a trajectory in the $x$-$\dot{x}$ plane is in the clockwise direction. When the trajectory crosses the $x$ axis, the velocity $\dot{x}$ is zero. Therefore, the trajectory will cross the $x$ axis perpendicularly.

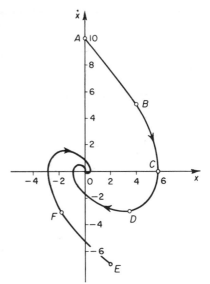

Fig. 12-3. Phase-plane plot of a system described by $\ddot{x} + \dot{x} + x = 0$.

## 12-2 METHODS FOR CONSTRUCTING TRAJECTORIES

In the phase-plane analysis of second-order systems, trajectories may be constructed analytically, graphically, or experimentally. In this section we shall present analytical and graphical methods for constructing trajectories.

Analytical methods are useful for systems whose differential equations are simple and piecewise linear. Obviously, if the analysis of the system is of primary interest and if the time solution of the system differential equation can be obtained easily by analytical methods, there is little justification for making the phase-plane plot of the system. However, analytical methods for obtaining the equations of trajectories prove to be useful in cases where analytical expressions of trajectories of simple linear differential equations are needed for the purpose of synthesizing nonlinear systems or for the purpose of aiding the proof of existence of a closed curve in the phase plane, etc.

Graphical methods are useful when it is tedious, difficult, or impossible to solve the given differential equation analytically. Graphical methods can be applied to both linear and nonlinear equations. These methods yield directly plots of trajectories in the phase plane. Since most nonlinear second-order differential equations cannot be solved analytically, it is for such equations that the graphical methods have much to offer.

All the graphical methods are essentially based on step-by-step procedures. The number of steps needed to obtain a trajectory depends on the size of the increments used. These increments should be of proper size so that only relatively small changes occur within each increment. They must be neither too small nor too large. If the increments are too small, the calculations may be too time-consuming.

The accuracy of graphical methods depends on the way the construction is performed. Obviously, the degree of accuracy may be increased up to a point, as the size of the figure is increased. It should be clear that if the increments are too large, the accuracy is likely to be poor. We should note, however, that even if the increments are small, if the number of increments is too large, accuracy is also likely to be poor. This is because unavoidable errors, even though they may be small at each step, tend to accumulate and may cause the latter portion of the solution to be inaccurate.

The two graphical methods discussed in this chapter are the isocline method and the delta method. Any parameters in the given equation must, of course, be assigned numerical values before graphical methods can be applied.

Experimental techniques are also available for obtaining trajectories. Once the equations for the phase-plane analysis can be set up, an analog computer and a cathode-ray oscilloscope or an $xy$ recorder may be used. For example, trajectories can be visualized on the screen by applying, to the deflector plates of an oscilloscope, voltages proportional to $x$ for the horizontal deflection and to $\dot{x}$ for the vertical deflection. Such an experimental technique for obtaining trajectories makes it possible to carry out a general phase-plane analysis of the problem and yet to save tedious hand calculations otherwise necessary for constructing trajectories.

**Symmetry in phase-plane portraits.** In certain cases, the phase-plane portrait may be symmetrical about the $x$ axis, the $\dot{x}$ axis, or both. Symmetry in the phase-plane portrait can easily be established from the original differential equation or the equation representing the trajectory.

*Symmetry about the x axis:* Consider the following equation:

$$\ddot{x} + f(x, \dot{x}) = 0$$

or

$$\frac{d\dot{x}}{dx} = -\frac{f(x, \dot{x})}{\dot{x}}$$

For the trajectories to be symmetrical about the $x$ axis, the slope $d\dot{x}/dx$ must be equal but of opposite sign for $\dot{x} > 0$ and $\dot{x} < 0$ for all $x$. From this requirement it can easily be shown that if the phase-plane portrait is to be symmetrical about the $x$ axis, then

$$f(x, \dot{x}) = f(x, -\dot{x})$$

namely, $f(x, \dot{x})$ must be an even function of $\dot{x}$. Thus the phase-plane portrait of a differential equation which does not involve $\dot{x}$ is always symmetrical about the $x$ axis.

If the phase-plane portrait is symmetrical about the $x$ axis, the only difference between the portrait in the upper-half plane and that in the lower-half plane is the direction of motion. That is, the motion of a representative point in the upper-half plane is to the right; in the lower-half plane, it is to the left. Thus, for this case, it may not be necessary to plot the phase-plane portrait in the lower-half plane.

*Symmetry about the $\dot{x}$ axis:* For symmetry about the $\dot{x}$ axis, the slope $d\dot{x}/dx$ must be equal but of opposite sign for $x > 0$ and $x < 0$ for all $\dot{x}$. This requires that

$$f(x, \dot{x}) = -f(-x, \dot{x})$$

or $f(x, \dot{x})$ must be an odd function of $x$.

*Symmetry about both the $x$ axis and the $\dot{x}$ axis:* For the phase-plane portrait to be symmetrical about both the $x$ and $\dot{x}$ axes, the requirements for the previous two cases must be met simultaneously. This requires that

$$f(-x, \dot{x}) = -f(x, -\dot{x})$$

**Analytical methods for constructing trajectories.** There are two methods available for analytical determination of the equations of the trajectories. One method is to integrate Eq. (12-3) to obtain the following equation of the trajectory

$$x_2 = \phi(x_1) \tag{12-7}$$

This method is applicable only if Eq. (12-3) can be integrated. Once Eq. (12-7) is obtained, the relationship between $x_1$ and $x_2$ can be plotted directly. The other method is to obtain $x_1$ and $x_2$ as functions of $t$ and then to eliminate $t$ from these two equations. This method can be applied if elimination of $t$ from Eqs. (12-1) and (12-2) can be made without too much difficulty.

---

*Example 12-3.* Consider the following simple second-order differential equation:

$$\ddot{x} + \omega^2 x = 0$$

By use of the relationship $\ddot{x} = d\dot{x}/dt = \dot{x}\, d\dot{x}/dx$ we can change the equation to the following form:

$$\dot{x}\frac{d\dot{x}}{dx} + \omega^2 x = 0 \tag{12-8}$$

The first analytic method is to integrate Eq. (12-8). This can be done quite easily, and the equation for the trajectory is obtained as

$$\frac{\dot{x}^2}{\omega^2} + x^2 = A^2 \tag{12-9}$$

where $A$ is a constant, determined by the initial conditions.

For a simple system such as this, the second analytic method can also be applied. The solution $x(t)$ can be found as

$$x(t) = A\cos(\omega t + \alpha) \tag{12-10}$$

where $A$ and $\alpha$ are constants to be determined by the initial conditions. Differentiating Eq. (12-10) with respect to $t$ gives

$$\dot{x}(t) = -A\omega\sin(\omega t + \alpha) \tag{12-11}$$

Eliminating $t$ from Eqs. (12-10) and (12-11), we can obtain Eq. (12-9).

---

*Example 12-4.* Consider the following equation:

$$\ddot{x} = -M \qquad (M = \text{constant}) \tag{12-12}$$

with the initial conditions

$$x(0) = x_0, \qquad \dot{x}(0) = 0$$

Equation (12-12) can be written as

$$\dot{x}\frac{d\dot{x}}{dx} = -M \tag{12-13}$$

Separating variables and integrating Eq. (12-13) gives

$$\dot{x}^2 = 2(x_0 - x)M \tag{12-14}$$

where the integration constant has been determined by the given initial conditions. Equation (12-14) represents a trajectory passing through the point $(x_0, 0)$.

Equation (12-14) can also be obtained by finding $x = \phi_1(t)$, $\dot{x} = \phi_2(t)$ and eliminating $t$ from these two equations. Using the given initial conditions, $\dot{x}(t)$ and $x(t)$ are found to be

$$\dot{x}(t) = -Mt \tag{12-15}$$

$$x(t) = -\tfrac{1}{2}Mt^2 + x_0 \tag{12-16}$$

Eliminating $t$ from Eqs. (12-15) and (12-16) yields Eq. (12-14). The phase-plane portraits of the system corresponding to $M = 1$ and $M = -1$ are shown in Fig. 12-4. The par-

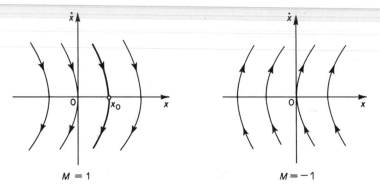

**Fig. 12-4.** Phase-plane portraits of a system described by $\ddot{x} = -M$ for $M = 1$ and $M = -1$.

ticular trajectory corresponding to $M = 1$ and passing through the point $(x_0, 0)$ is shown by a heavy curve.

**Graphical method for constructing trajectories—the isocline method.** The isocline method is one used to obtain the field of trajectories graphically, without solving a given differential equation. The isocline method is applicable to the following first-order differential equation:

$$\frac{dx_2}{dx_1} = \frac{f_2(x_1, x_2)}{f_1(x_1, x_2)} \tag{12-17}$$

where $f_1(x_1, x_2)$ and $f_2(x_1, x_2)$ are analytic. In Eq. (12-17), $x_1$ is considered the independent variable and $x_2$ the dependent variable.

The locus of a constant slope of the trajectory, that is, the locus of

$$\frac{dx_2}{dx_1} = \alpha = \text{constant} \tag{12-18}$$

in the phase plane may be obtained from Eqs. (12-17) and (12-18). Namely, the following equation

$$f_2(x_1, x_2) = \alpha f_1(x_1, x_2)$$

gives the locus of constant slope $\alpha$. The locus of points where trajectories have a given slope is called an isocline. If we plot isoclines corresponding to various values of $\alpha$, the field of directions of tangents to trajectories can be obtained.

When we construct a phase-plane portrait by the isocline method, the entire phase plane may be filled with short line segments fixing directions of the field, as shown in Fig. 12-5. In many cases, the isoclines for zero and infinite slopes may

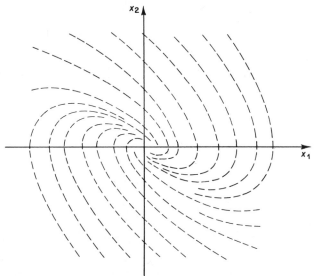

**Fig. 12-5.** An example of a phase-plane diagram filled with short line segments fixing the directions of the field.

easily be constructed. Since the slope is indeterminate at a singular point, no line segment may be drawn at this point. The trajectory passing through any given ordinary point in the phase plane can be constructed by drawing a continuous curve following the directions of the field.

As an example, consider the differential equation

$$\ddot{x} + 2\zeta\omega\dot{x} + \omega^2 x = 0 \tag{12-19}$$

Equation (12-19) may be rewritten as follows:

$$\dot{x}\frac{d\dot{x}}{dx} + 2\zeta\omega\dot{x} + \omega^2 x = 0 \tag{12-20}$$

Letting $d\dot{x}/dx = \alpha$, we may write Eq. (12-20) as

$$\frac{\dot{x}}{x} = \frac{-\omega^2}{2\zeta\omega + \alpha} \tag{12-21}$$

Equation (12-21) represents an isocline, the locus of constant tangent $\alpha$, and is an equation of a line. The isoclines for linear second-order differential equations are straight lines and pass through the origin of the phase plane.

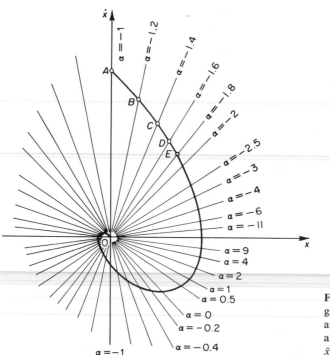

**Fig. 12-6.** Phase-plane diagram showing isoclines and a typical trajectory for a system described by $\ddot{x} + \dot{x} + x = 0$.

Figure 12-6 shows isoclines and one particular trajectory for the system given by Eq. (12-19) when $\zeta = 0.5$ and $\omega = 1$. The trajectory starting at point $A$ in Fig. 12-6 may be constructed as follows: When the trajectory is in the region bounded by the isoclines corresponding to $\alpha = -1$ and $\alpha = -1.2$, the slope of the trajectory is approximately $(-1 - 1.2)/2 = -1.1$. Therefore, if the line with slope $-1.1$ is drawn from point $A$, intersecting the isocline corresponding to $\alpha = -1.2$ at point $B$, line $AB$ is approximately a part of the trajectory. Similarly, the trajectory passing through point $B$ has the slope of approximately $(-1.2 - 1.4)/2 = -1.3$ in the region bounded by the isoclines corresponding to $\alpha = -1.2$ and $\alpha = -1.4$. Then line $BC$ with slope $-1.3$, intersecting the isocline $\alpha = -1.4$ at point $C$, is a part of the trajectory. The slope of the trajectory between points $C$ and $D$ is $-1.5$. Constructing the trajectory in this way, we may obtain the curve $ABCDEO$.

For a linear system, there is only one trajectory passing through any ordinary point in the phase plane. Therefore, the trajectory starting from point $D$ in Fig. 12-6 is the curve $DEO$, a part of the curve $ABCDEO$. Obviously, the trajectory can be constructed in either a clockwise direction or counterclockwise direction from any point in the phase plane.

It should be noted that constructing the trajectory in this way may accumulate errors that may arise during the course of the construction, and the resulting trajectory may not be accurate. For good accuracy, it is advantageous to first fill a certain region of the phase plane by short line segments which indicate directions of the field. Then construct a trajectory by sketching it carefully, following the directions of the field. In this way, it may be possible to avoid the possibility of accumulating certain errors.

Remember that the accuracy of the result may not be good if, as in the case of certain nonlinear differential equations, the slopes of trajectories vary rapidly in certain regions in the phase plane. The accuracy of this method, in general, depends on the number of isoclines used. For reasonable accuracy, the isoclines may be plotted every 5°–10°.

For the system given by Eq. (12-19), the equation of the isocline, Eq. (12-21), may be written

$$\alpha = -\left(2\zeta\omega + \frac{\omega^2}{\tan\theta}\right) \tag{12-22}$$

where we have used the relationship

$$\frac{\dot{x}}{x} = \tan\theta$$

and angle $\theta$ is that shown in Fig. 12-7. By use of Eq. (12-22), we can draw isoclines for any value of $\theta$.

The isocline method discussed here is convenient when the isoclines are straight lines. If they are not, the delta method, which will be discussed subsequently, may be less tedious for constructing the trajectory passing through a given point.

It should be noted that the slope $\alpha = d\dot{x}/dx$ is a quantity with a dimension and its value is dependent on the unit of time. It is most convenient to use the same scales along the $x$ and $\dot{x}$ axes so that the value of $\alpha$ and the corresponding geometrical slope of the trajectory are the same.

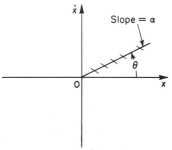

**Fig. 12-7.** Diagram showing the angle $\theta$ in Eq. (12-22).

---

*Example 12-5.* Obtain a phase-plane portrait of the following equation by using the isocline method:

$$\ddot{x} + a|\dot{x}| + x = 0 \qquad (a > 0) \tag{12-23}$$

In general, when the system equation involves absolute values of $x$ or $\dot{x}$ or both, replacing the nonlinear differential equation by several linear differential equations simplifies the construction of isoclines.

Let us replace Eq. (12-23) by

$$\begin{aligned}
\ddot{x} + a\dot{x} + x &= 0 \qquad \text{for } \dot{x} > 0 \\
\ddot{x} - a\dot{x} + x &= 0 \qquad \text{for } \dot{x} < 0
\end{aligned} \tag{12-24}$$

In the upper half of the $x$-$\dot{x}$ plane, the equation of isoclines is

$$\frac{\dot{x}}{x} = \frac{-1}{a + \alpha}$$

and in the lower half it is

$$\frac{\dot{x}}{x} = \frac{-1}{-a + \alpha}$$

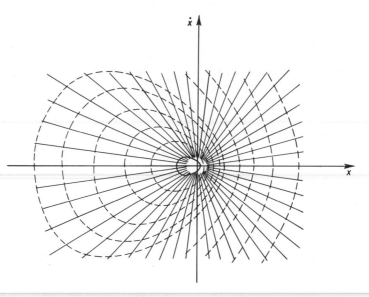

**Fig. 12-8.** Phase-plane portrait for a system described by
$\ddot{x} + |\dot{x}| + x = 0$.

Figure 12-8 shows a phase-plane portrait, together with isoclines for the system of Eq. (12-23) when $a = 1$. The straight lines are isoclines and short line segments on each line indicate the field of directions of tangents to the trajectories. (Short line segments are placed so that the trajectories can easily be visualized.) For this system, the trajectories are concentric ovals, and the motion is periodic for any given initial conditions, except at the origin. This may be clear also from Eq. (12-24) since the energy dissipated during the half cycle (corresponding to a period for which $\dot{x} > 0$) is the same as the energy put into the system during the other half cycle (corresponding to a period for which $\dot{x} < 0$), and thus the net energy dissipation is zero for one cycle. In this example, the symmetric nature of the phase-plane portrait can be predicted directly from Eq. (12-23). Since replacement of $\dot{x}$ by $-\dot{x}$ does not change the equation, it is symmetrical about the $x$ axis.

**Graphical method for constructing trajectories—the delta method.** In the delta method, the trajectory is obtained as a sequence of circular arcs whose centers slide along the $x$ axis. The delta method can be applied to equations of the form

$$\ddot{x} = -f(\dot{x}, x, t) \tag{12-25}$$

where $f(\dot{x}, x, t)$ may be either linear or nonlinear and may be time-varying but must be continuous and single-valued. In applying this method, we modify Eq. (12-25) to the following form:

$$\ddot{x} + \omega^2 x = -f(\dot{x}, x, t) + \omega^2 x \tag{12-26}$$

A term $\omega^2 x$ is added to both sides of Eq. (12-25). This term should be chosen properly so that values of the function $\delta$ defined below are neither too small nor too large for the range of values of $\dot{x}$, $x$, and $t$ considered.

$$\delta(\dot{x}, x, t) = \frac{-f(\dot{x}, x, t) + \omega^2 x}{\omega^2} \tag{12-27}$$

Using Eq. (12-27), we can write Eq. (12-26) as

$$\ddot{x} + \omega^2 x = \omega^2\, \delta(\dot{x}, x, t) \tag{12-28}$$

The function $\delta(\dot{x}, x, t)$ depends on the variables $\dot{x}$, $x$, and $t$. For small changes in these variables, however, $\delta(\dot{x}, x, t)$ may be considered constant. Then in the neighborhood of the state $x = x_1$, $\dot{x} = \dot{x}_1$, $t = t_1$; namely, $x = x \pm \Delta x_1$, $\dot{x} = \dot{x}_1 \pm \Delta \dot{x}_1$, $t = t_1 \pm \Delta t_1$, where $\Delta x_1$, $\Delta \dot{x}_1$, $\Delta t_1$ are assumed to be small, the value of $\delta$ can be assumed constant, $\delta_1$, and Eq. (12-28) can be changed to

$$\ddot{x} + \omega^2(x - \delta_1) = 0 \tag{12-29}$$

Equation (12-29) shows a simple harmonic motion. The trajectories for this system are circles centered at $x = \delta_1$, $\dot{x}/\omega = 0$ in the normalized phase plane [$x$-($\dot{x}/\omega$) plane]. Thus, we can see that for a small increment in the neighborhood of the state $x = x_1$, $\dot{x}/\omega = \dot{x}_1/\omega$, $t = t_1$ the trajectory is an arc of the circle centered at $x = \delta_1$, $\dot{x}/\omega = 0$ with radius equal to

$$\sqrt{\left(\frac{\dot{x}_1}{\omega}\right)^2 + (x_1 - \delta_1)^2}$$

It is noted that in the delta method it is essential to use a normalized phase plane so that the trajectory of Eq. (12-29) becomes a circle. (Also, it is necessary to use equal numerical scales along the $x$ and $\dot{x}/\omega$ axes.) The construction of a trajectory is shown in Fig. 12-9. Point $P$ represents the state of the system on the trajectory at time $t = t_1$. The value of $\delta_1$ is determined from Eq. (12-27) as

$$\delta_1 = -\frac{f(\dot{x}_1, x_1, t_1)}{\omega^2} + x_1$$

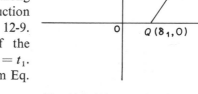

**Fig. 12-9.** Diagram showing the construction of a trajectory by the delta method.

Once point $Q$, the center of the circular arc, is located on the $x$ axis, the radius is fixed as $\overline{PQ}$. The true trajectory in the neighborhood of point $P$ is then approximated by a short circular arc. The arc must be short enough to ensure that the changes in the variables are small.

Consider next the following equation:

$$\frac{d^2x}{dt^2} + 2\zeta\omega\frac{dx}{dt} + \omega^2 x = 0 \tag{12-30}$$

Equation (12-30) contains a term in $x$ with a positive coefficient. Therefore, Eq. (12-30) can be changed into the form of Eq. (12-26) by simply bringing the second term of Eq. (12-30) to the right-hand side of the equation. Namely,

$$\frac{d^2x}{dt^2} + \omega^2 x = -2\zeta\omega\frac{dx}{dt} \tag{12-31}$$

Equation (12-31) can be normalized by substituting

$$t = \frac{1}{\omega}\tau$$

to give

$$\frac{d^2x}{d\tau^2} + x = -2\zeta \frac{dx}{d\tau} \qquad (12\text{-}32)$$

If we let

$$\frac{\dot{x}}{\omega} = \frac{dx}{dt}\frac{1}{\omega} = \frac{dx}{d\tau} = y$$

then Eq. (12-32) becomes

$$\frac{d^2x}{d\tau^2} + x = \delta \qquad (12\text{-}33)$$

where $\delta = -2\zeta y$.

For $y = y_1 = $ constant, Eq. (12-33) indicates that the trajectory in the $x$-$(dx/d\tau)$ plane is a circle, whose center is located at $x = -2\zeta y_1$, $y = 0$. Since the location of the center depends on $y_1$, the center of the circle moves along the $x$ axis as $y$ changes. In constructing the trajectory of Eq. (12-32), we find it convenient to draw a line $x = \delta$, namely,

$$x = -2\zeta y$$

in the $x$-$y$ plane. Figures 12-10 (a) and (b) display the construction of trajectories.

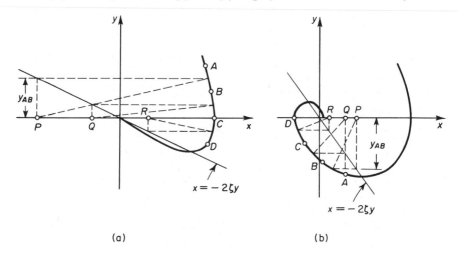

(a)                               (b)

Fig. 12-10. Diagrams showing the construction of trajectories by the delta method. (a) Overdamped system; (b) underdamped system.

In both Figs. 12-10 (a) and (b), the center of arc $AB$ is located at point $P$, $(x = -2\zeta y_{AB}, y = 0)$. Here $y_{AB}$ is the mean value of $y$ between point $A$ and point $B$. The location of the center of the circle can be obtained directly and simply by use of the line $x = -2\zeta y$ drawn in the $x$-$y$ plane. In Figs. 12-10 (a) and (b), segments of arcs from point $B$ to point $C$ and from point $C$ to point $D$ have their centers at point $Q$ and point $R$, respectively.

The delta method is a general method in that trajectories can be constructed in the phase plane regardless of whether the phase-plane equation represents a linear or a nonlinear physical system, and moreover regardless of whether or not the phase-plane equation includes time-varying elements or a time-varying forcing function.

---

*Example 12-6.* Consider the system described by the following equation:

$$\ddot{x} + \dot{x} + x^3 = 0 \tag{12-34}$$

Given the initial conditions $x(0) = 1$, $\dot{x}(0) = 0$, construct the trajectory starting at the initial point. Use the delta method.

We see that Eq. (12-34) does not contain a term in $x$ with a positive coefficient. Therefore, it is necessary to add a term $\omega^2 x$ to both sides of Eq. (12-34). The value of $\omega$ must be chosen so that values of $\delta$ are neither too small nor too large. If we choose $\omega = 1$, Eq. (12-34) can be written as

$$\ddot{x} + x = -\dot{x} - x^3 + x$$
$$\delta = -\dot{x} - x^3 + x$$

Since the value of $\delta$ depends on both $x$ and $\dot{x}$, it will be necessary to make successive trials in the construction process.

The trajectory starts at point $A$, $(x = 1, \dot{x} = 0)$, in Fig. 12-11. In the neighborhood of point $A$,

$$\delta = -0 - 1 + 1 = 0$$

Therefore, the initial arc is centered at point $(0, 0)$ with radius equal to unity. Then a short arc is drawn. The mean value of $x$ and the mean value of $\dot{x}$ for this arc are used for finding a more accurate value of $\delta$. A few successive trials will be sufficient to obtain a reasonably accurate value of $\delta$. In this example, the first arc $AB$ of Fig. 12-11 is centered at point $P_1$, $(x = 0.12, \dot{x} = 0)$. For the second arc, similar successive trials are made, showing that the second arc $BC$ of Fig. 12-11 is centered at point $P_2$, $(x = 0.37, \dot{x} = 0)$. By continuing in this way, we can construct the trajectory as far as desired.

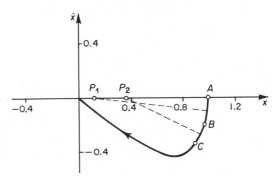

**Fig. 12-11.** Phase-plane diagram showing the trajectory, starting from point $x(0) = 1$, $\dot{x}(0) = 0$, of a system described by $\ddot{x} + \dot{x} + x^3 = 0$.

From the preceding discussion, it may be seen that a system with $\delta$, which depends on three variables $\dot{x}$, $x$, and $t$, can be handled in much the same way. In constructing trajectories, however, we must keep track of $t$. (Time information can be obtained from the trajectory. A few methods for obtaining time information from a phase-plane plot are given in the following section.)

## 12-3 OBTAINING TIME SOLUTIONS FROM PHASE-PLANE PLOTS

The trajectory plotted in the $x$-$\dot{x}$ plane is a plot of $\dot{x}$ as a function of $x$. Time does not appear explicitly in this plot. For the analysis of a system, we may need only trajectories; however, it is sometimes desirable to obtain a plot of the variable $x$ as a function of time $t$. In such a case, it is possible to obtain the time solution from the phase-plane plot, even though the original differential equation cannot be solved for $x$ and $\dot{x}$ as functions of $t$. The process for obtaining a time solution is essentially a step-by-step one and may be performed in several different ways. Once the trajectory is graduated in $t$, the behavior of the system response with respect to time may be visualized.

**Time information based on $\Delta t = \Delta x/\dot{x}$.** It can be seen that for the small increments $\Delta x$ and $\Delta t$, the average velocity is given by $\dot{x}_{av} = \Delta x/\Delta t$. The incremental time $\Delta t$ is then

$$\Delta t = \frac{\Delta x}{\dot{x}_{av}} \tag{12-35}$$

Figure 12-12 (a) shows a trajectory in the $x$-$\dot{x}$ plane. The incremental time $\Delta t_{AB}$ needed for the representative point to traverse the incremental displacement

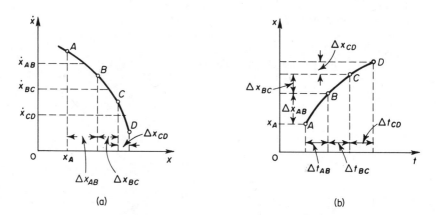

(a)                                    (b)

**Fig. 12-12.** (a) A trajectory in the $x$-$\dot{x}$ plane; (b) a plot of the time solution $x(t)$ versus $t$.

$\Delta x_{AB}$ is $\Delta t_{AB} = \Delta x_{AB}/\dot{x}_{AB}$. Similarly, $\Delta t_{BC} = \Delta x_{BC}/\dot{x}_{BC}$. Then the time solution $x = x(t)$ can be easily plotted as shown in Fig. 12-12 (b). For good accuracy, the incremental displacement $\Delta x$ must be chosen small enough so that the corresponding incremental changes in $\dot{x}$ and $t$ are reasonably small. However, the value of $\Delta x$ need not be constant. It can be changed according to the shapes of portions of the trajectory to minimize the labor needed to obtain reasonable accuracy.

For trajectories in the normalized phase plane [$x$-($\dot{x}/\omega$) plane], Eq. (12-35) is changed into

$$\Delta t = \frac{\Delta x}{\dot{x}_{av}} = \frac{1}{\omega} \frac{\Delta x}{\left(\dfrac{\dot{x}}{\omega}\right)_{av}}$$

In a general case where the original differential equations are given by Eqs. (12-1) and (12-2), $\Delta t$ can be obtained from the following relationship:

$$\Delta t = \frac{\Delta x_1}{f_1(x_{1av}, x_{2av})} = \frac{\Delta x_2}{f_2(x_{1av}, x_{2av})}$$

where $x_{1av}$ and $x_{2av}$ are the average values of $x_1$ and $x_2$ during a given time interval $\Delta t$. The increments in $\Delta x_1$, $\Delta x_2$, and $\Delta t$ must be reasonably small for good accuracy.

**Time information based on** $t = \int (1/\dot{x}) \, dx$. Since $\dot{x} = dx/dt$, it follows that the time interval $t_2 - t_1$ may be expressed as

$$t_2 - t_1 = \int_{x_1}^{x_2} \frac{1}{\dot{x}} \, dx \tag{12-36}$$

Equation (12-36) shows that if the trajectory is replotted with $1/\dot{x}$ as the ordinate and $x$ the abscissa, the area under the resulting curve represents the corresponding time interval. The time required for the representative point on the trajectory to traverse from point $A$ to point $B$ shown in Fig. 12-13 (a) may be obtained from the following equation:

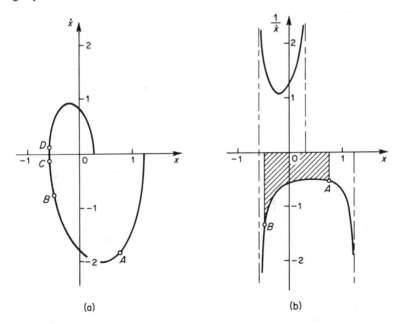

(a)                                                  (b)

**Fig. 12-13.** (a) A trajectory in the $x$-$\dot{x}$ plane; (b) graphical computation of $t_{AB}$. (The hatched area equals $t_{AB}$.)

$$t_{AB} = \int_A^B \frac{1}{\dot{x}} \, dx \tag{12-37}$$

The right-hand side of Eq. (12-37) is the hatched area in Fig. 12-13 (b). This area may be obtained analytically or graphically by conventional means.

When the value of $\dot{x}$ becomes zero between the limits of integration, the value of $1/\dot{x}$ becomes infinity, resulting in a difficulty in evaluating the integral given by Eq. (12-36). For example, the value of $\dot{x}$ becomes zero between point $C$ and point $D$ in Fig. 12-13 (a). Therefore, the time interval required for the representative point to move from point $C$ to point $D$ must be evaluated by some other means. (See the method based on circular arc approximations presented in the following.)

For trajectories in the normalized phase plane $[x\text{-}(\dot{x}/\omega)$ plane], Eq. (12-36) is modified into

$$t_2 - t_1 = \frac{1}{\omega} \int_{x_1}^{x_2} \frac{1}{\left(\dfrac{\dot{x}}{\omega}\right)} \, dx$$

**Time information based on circular arc approximations.** In this method, the trajectory is approximated by a sequence of arcs of circles centered on the $x$ axis. In the $x\text{-}\dot{x}$ plane, if the variable $x$ exhibits simple harmonic motion about the point $(x_0, 0)$ with an angular velocity of 1 rad/sec, namely,

$$\ddot{x} + (x - x_0) = 0$$

then the trajectory can be written

$$\dot{x}^2 + (x - x_0)^2 = K^2 \tag{12-38}$$

where $K$ is an integration constant. Equation (12-38) indicates that the trajectory is a circle with radius $K$, centered at $x = x_0$, $\dot{x} = 0$. The velocity of a representative point along the trajectory is given by

$$\sqrt{\left(\frac{dx}{dt}\right)^2 + \left(\frac{d\dot{x}}{dt}\right)^2}$$

(This velocity must not be confused with the velocity $\dot{x}$, the time rate of change of displacement $x$.) In general, the velocity of a representative point varies as it moves along the trajectory. The velocity is nonzero and finite, except at the singular points, where this velocity becomes zero, because both $dx/dt$ and $d\dot{x}/dt$ become zero simultaneously. The speed of the representative point on the circular trajectory is given by

$$\sqrt{\dot{x}^2 + \ddot{x}^2} = \sqrt{\dot{x}^2 + (x - x_0)^2} = K$$

This indicates that the velocity of a representative point along the trajectory is constant. Since the circumference of the circle is $2\pi K$ and the velocity of a representative point is a constant $K$, this point requires $2\pi$ seconds to make a complete cycle, so that the period is $2\pi$ seconds.

Figure 12-14 shows a circular trajectory. The time required for the representative point on the trajectory to move from point $A$ to point $B$ may be obtained as

$$t_{AB} = 2\pi \frac{\theta_{AB}}{2\pi} = \theta_{AB} \text{ sec}$$

where the angle $\theta_{AB}$ is the magnitude of the angle subtending the circular arc $AB$ and is measured in radians. Thus, time $t_{AB}$ sec is numerically equal to $\theta_{AB}$ rad if the coordinates of the phase plane are $x$ and $\dot{x}$. That is, if $\theta_{AB}$ equals 0.5 rad, then $t_{AB}$ equals 0.5 sec.

In general, trajectories are not circular; however, trajectories may be well approximated by a series of arcs of circles centered on the $x$ axis. For example, consider the trajectory shown in Fig. 12-15. A part of the trajectory, $ABCD$, may

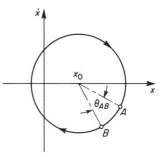

**Fig. 12-14.** Circular trajectory.

be approximated by three circular arcs centered on the $x$ axis. $AB$ may be approximated by a part of the circle centered at point $P$. Similarly, $BC$ and $CD$ are parts of circles centered at point $Q$ and point $R$, respectively. Then $t_{AD}$, the time necessary

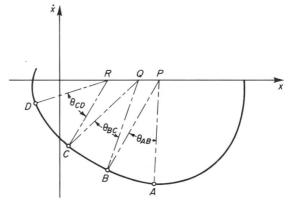

**Fig. 12-15.** Circular-arc approximation of a trajectory in the $x$-$\dot{x}$ plane.

for a representative point to move from point $A$ to point $D$ on the trajectory, may be obtained as

$$t_{AD} = t_{AB} + t_{BC} + t_{CD}$$

where $t_{AB}$, $t_{BC}$, and $t_{CD}$ are numerically equal to $\theta_{AB}$, $\theta_{BC}$, and $\theta_{CD}$ respectively; and $\theta_{AB} = \underline{/APB}$, $\theta_{BC} = \underline{/BQC}$, and $\theta_{CD} = \underline{/CRD}$, all measured in radians.

If the phase plane is normalized with $x$ and $\dot{x}/\omega$ as coordinates, then $t_{AB}$ and $\theta_{AB}$ are not numerically equal. Let us consider this case further. If the variable $x$ exhibits simple harmonic motion with angular velocity $\omega$ rad/sec; namely,

$$\ddot{x} + \omega^2(x - x_0) = 0$$

the equation of the trajectory can be written

$$\left(\frac{\dot{x}}{\omega}\right)^2 + (x - x_0)^2 = K^2 \qquad (12\text{-}39)$$

where $K$ is a constant. The trajectory given by Eq. (12-39) is a circle centered at the point $(x_0, 0)$ in the $x$-$(\dot{x}/\omega)$ plane with radius equal to $K$. The speed of a repre-

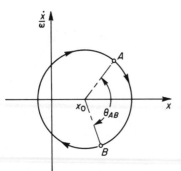

**Fig. 12-16.** Graphical computation of $t_{AB}$ in the $x$-$(\dot{x}/\omega)$ plane.

sentative point on the trajectory is constant, and this point requires $2\pi/\omega$ seconds to make a complete cycle. Therefore, in Fig. 12-16, $t_{AB}$, the time interval required for a representative point to move from point $A$ to point $B$, may be obtained as

$$t_{AB} = \frac{2\pi}{\omega} \times \frac{\theta_{AB}}{2\pi} = \frac{\theta_{AB}}{\omega} \text{ sec}$$

Thus, in the $x$-$(\dot{x}/\omega)$ plane, the time interval $t_{AB}$ sec is equal to $\theta_{AB}/\omega$, where $\theta_{AB}$ is measured in radians and $\omega$ is measured in radians per second.

## 12-4  SINGULAR POINTS

In this section, we shall examine the behavior of trajectories near singular points. Consider the system

$$\frac{dx_1}{dt} = f_1(x_1, x_2)$$

$$\frac{dx_2}{dt} = f_2(x_1, x_2)$$

where $f_1(x_1, x_2)$ and $f_2(x_1, x_2)$ are analytic functions of the variables $x_1$ and $x_2$ in the neighborhood of the origin. Suppose that the origin is a singular point, or an equilibrium point; so that

$$f_1(0, 0) = 0, \qquad f_2(0, 0) = 0$$

Let us expand $f_1(x_1, x_2)$ and $f_2(x_1, x_2)$ in Taylor series in the neighborhood of the origin. The system equations then become

$$\frac{dx_1}{dt} = a_1 x_1 + b_1 x_2 + a_{11}x_1^2 + a_{12}x_1 x_2 + a_{22}x_2^2 + g_1(x_1, x_2) \qquad (12\text{-}40)$$

$$\frac{dx_2}{dt} = a_2 x_1 + b_2 x_2 + b_{11}x_1^2 + b_{12}x_1 x_2 + b_{22}x_2^2 + g_2(x_1, x_2) \qquad (12\text{-}41)$$

where $g_1(x_1, x_2)$ and $g_2(x_1, x_2)$ involve only third- and higher-order powers of $x_1$ and $x_2$.

In the neighborhood of the origin, where $x_1$ and $x_2$ are very small, Eqs. (12-40) and (12-41) may be approximated by linear terms only, provided they are dominant in the neighborhood of the origin. Then

$$\frac{dx_1}{dt} = a_1 x_1 + b_1 x_2 \qquad (12\text{-}42)$$

$$\frac{dx_2}{dt} = a_2 x_1 + b_2 x_2 \qquad (12\text{-}43)$$

An examination of the solutions of Eqs. (12-42) and (12-43) is useful in drawing trajectories near the origin.

Equations (12-42) and (12-43) may be modified to

$$\ddot{x} + a\dot{x} + bx = 0 \tag{12-44}$$

where

$$x = x_1$$
$$a = -a_1 - b_2$$
$$b = a_1 b_2 - a_2 b_1$$

Note that Eq. (12-44) is equivalent to Eqs. (12-42) and (12-43). If the roots of the characteristic equation of Eq. (12-44) have negative real parts, all trajectories near the origin will approach it as $t$ increases indefinitely. If at least one root is zero, then the stability cannot be determined from the linearized equation, Eq. (12-44). In this case the behavior of trajectories near the origin depends on the higher-order terms in Eqs. (12-40) and (12-41).

**Classification of singular points.** The phase-plane portrait of a linear autonomous system is a family of noncrossing trajectories which describe the response of the system to all possible initial conditions. Consider the linearized second-order differential equation, Eq. (12-44). The location of the singular point in the $x$-$\dot{x}$ plane is the origin. The nature of the solution of Eq. (12-44) is determined by the two roots $\lambda_1$ and $\lambda_2$ of the following characteristic equation:

$$\lambda^2 + a\lambda + b = 0$$

Here we assume that $a$ and $b$ are constants with $b \neq 0$. The locations of $\lambda_1$ and $\lambda_2$ in the complex plane determine the characteristics of the singular point. It can be seen that the following six cases exist:

1. $\lambda_1$ and $\lambda_2$ are complex conjugates and lie in the left-half plane.
2. $\lambda_1$ and $\lambda_2$ are complex conjugates and lie in the right-half plane.
3. $\lambda_1$ and $\lambda_2$ are real and lie in the left-half plane.
4. $\lambda_1$ and $\lambda_2$ are real and lie in the right-half plane.
5. $\lambda_1$ and $\lambda_2$ are complex conjugates and lie on the $j\omega$ axis.
6. $\lambda_1$ and $\lambda_2$ are real; $\lambda_1$ lies in the left-half plane, and $\lambda_2$ lies in the right-half plane.

According to the nature of the responses corresponding to each case, the singular points are classified as stable focus, unstable focus, stable node, unstable node, center, and saddle point, respectively. The phase-plane portraits of each of the six cases are illustrated in Fig. 12-17, and the associated singular points are indicated. If the singular point is a saddle point, there are particular trajectories which enter the saddle point and separate the phase plane into regions of distinct motion. Such trajectories are called separatrices. The trajectories shown in Fig. 12-17 can be constructed easily by the isocline method. The phase-plane portraits clearly show the type of response of the system, once the initial condition is given.

The classification of singular points just given applies only to second-order

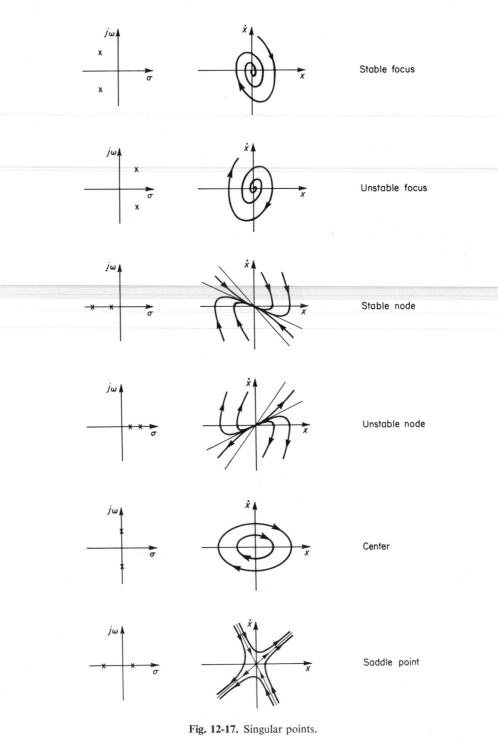

**Fig. 12-17.** Singular points.

systems. For third- or higher-order systems, singular points are usually classified as being stable, asymptotically stable, or unstable. For details, refer to Chapter 15.

**Limit cycles.** Limit cycles occur frequently in such physical systems as electronic oscillators. Limit cycles have a distinct geometric configuration in the phase-plane portrait, namely, that of an isolated closed path in the phase plane. A given system may have more than one limit cycle. A limit cycle represents a steady-state oscillation, to which or from which all trajectories nearby will converge or diverge. Thus it divides the plane into an inside and an outside. No trajectories inside (outside) a particular limit cycle ever cross the limit cycle to enter the outside (inside) region.

A limit cycle in a nonlinear system describes the amplitude and period of a self-sustained oscillation. It should be pointed out that not all closed curves in the phase plane are limit cycles. A phase-plane portrait of a conservative system, in which there is no damping to dissipate energy, is a continuous family of closed curves. Closed curves of this kind are not limit cycles because none of these curves are isolated from one another. Such trajectories always occur as a continuous family, so that there are closed curves in any neighborhood of any particular closed curve. On the other hand, limit cycles are periodic motions exhibited only by non-linear nonconservative systems. (If the system is everywhere dissipative, then the net energy loss along any trajectory in the phase plane is positive, and there cannot be any limit cycles. Thus if a nonlinear nonconservative system exhibits limit-cycle behavior, the equivalent damping must be equal to zero.) Since limit cycles are isolated from each other, there can be no other limit cycles in the neighborhood of any limit cycle.

A limit cycle is called stable if trajectories near the limit cycle, originating from outside or inside, converge to that limit cycle. In this case, the system exhibits a sustained oscillation with constant amplitude. This is shown in Fig. 12-18 (a). The inside of the limit cycle is an unstable region in the sense that trajectories diverge to the limit cycle, and the outside is a stable region in the sense that trajectories converge to the limit cycle. In the case of control-systems having this type of limit cycle, the design criterion is often to make the magnitude of the limit cycle small enough to satisfy accuracy requirements.

A limit cycle is called an unstable one if trajectories near it diverge from this limit cycle. In this case, an unstable region surrounds a stable region. If a trajectory starts within the stable region, it converges to a singular point within the limit cycle. If a trajectory starts in the unstable region, however, it diverges and increases with time to infinity. This is shown in Fig. 12-18 (b). The inside of an unstable limit cycle is the stable region, and the outside the unstable region. The design criterion for a control system having such an unstable limit cycle is to make the stable region as large as possible. Unstable limit-cycle operation can be theoretically realized if the initial conditions can be set exactly. However, any small disturbances will cause instability in this limit-cycle operation.

A limit cycle is called semistable if the trajectories originating at points outside the limit cycle diverge from it, while the trajectories originating at points inside

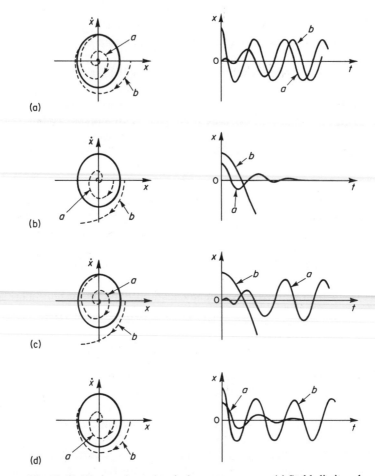

**Fig. 12-18.** Limit cycles and typical $x$ versus $t$ curves. (a) Stable limit cycle; (b) unstable limit cycle; (c) semistable limit cycle; (d) semistable limit cycle.

the limit cycle converge to it, as shown in Fig. 12-18 (c), or vice versa, as shown in Fig. 12-18 (d). A control system may have no limit cycle, or it may have one or more.

Consider a system possessing two limit cycles and assume that these two limit cycles are located close to each other. If the larger limit cycle is unstable, while the smaller limit cycle is stable as shown in Fig. 12-19, then the characteristics of the system become similar to that having the semistable limit cycle shown in Fig. 12-18 (c). Similarly, if the larger limit cycle is stable, while the smaller limit cycle is unstable, the characteristics of the system become similar to that having the semistable limit cycle shown in Fig. 12-18 (d).

Stable limit cycles can be observed experimentally, but because of "noise" unstable limit cycles and semistable limit cycles cannot be.

With regard to limit-cycle operation, we should note that energy must, of course, be provided to the system in some manner, for example, as a dc voltage, constant

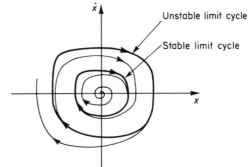

**Fig. 12-19.** Phase-plane portrait of a system with two limit cycles.

wind, etc. In a self-oscillatory system, the work accomplished by the source of energy, which may be constant, is periodic. In other words, a self-oscillatory system can generate a periodic motion from a nonperiodic energy source.

Finally, note that except in simple cases it is difficult or impossible to determine the exact location of limit cycles in the phase plane by analytical techniques. The exact location of limit cycles may be found only by graphical, experimental, or computational techniques.

*Example 12-7.* Obtain a phase-plane portrait of the system given by

$$\ddot{x} + 0.5\dot{x} + 2x + x^2 = 0 \tag{12-45}$$

The singular points for this system are

$$x = 0, \dot{x} = 0 \quad\text{and}\quad x = -2, \dot{x} = 0$$

The nature of these points can be determined as follows. In a neighborhood of the origin, Eq. (12-45) can be linearized to

$$\ddot{x} + 0.5\dot{x} + 2x = 0$$

The two roots of the characteristic equation

$$\lambda^2 + 0.5\lambda + 2 = 0$$

are

$$\lambda_1 = -0.25 + j1.39, \quad \lambda_2 = -0.25 - j1.39$$

Thus this singular point is a stable focus.

In a neighborhood of the singular point $(-2, 0)$, Eq. (12-45) can be written as follows: By letting

$$y = x + 2$$

we obtain

$$\ddot{y} + 0.5\dot{y} - 2y + y^2 = 0 \tag{12-46}$$

Near the point $y = 0, \dot{y} = 0$, Eq. (12-46) becomes

$$\ddot{y} + 0.5\dot{y} - 2y = 0$$

The two roots of the characteristic equation

$$\mu^2 + 0.5\mu - 2 = 0$$

are

$$\mu_1 = 1.19, \qquad \mu_2 = -1.69$$

Thus the singular point $(-2, 0)$ is a saddle point.

By use of the isocline method, the phase-plane portrait, as shown in Fig. 12-20, may be obtained. The two trajectories which enter the saddle point $(-2, 0)$ are separatrices. In

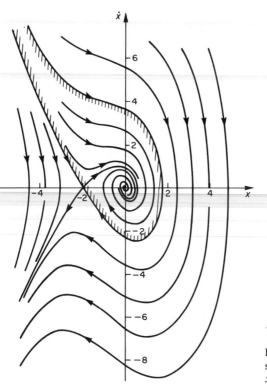

**Fig. 12-20.** Phase-plane portrait of a system described by $\ddot{x} + 0.5\dot{x} + 2x + x^2 = 0$.

this system one of the separatrices divides the phase plane into two regions of distinct motion. In other words, the solution of Eq. (12-45) has two distinctively different types of trajectories. The type of trajectory depends on the initial condition. If the initial point falls within the hatched region (the region bounded by one of the two separatrices), the trajectories converge to the origin. If the initial point is outside the hatched region, the trajectories approach infinity. (The displacement will grow in the negative direction without bound as time increases indefinitely.)

## 12-5  PHASE-PLANE ANALYSIS OF LINEAR CONTROL SYSTEMS

The phase-plane method is quite useful for analyzing nonlinear second-order systems. Before we present the analysis of nonlinear systems, however, it seems desirable to consider an application of the phase-plane method to the analysis of linear second-order systems because many control systems with signal-dependent nonlinearities may be approximated by piecewise-linear systems.

In this section, we shall illustrate the application of this method to the transient-response analysis of the second-order control system shown in Fig. 12-21.

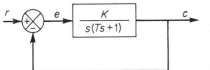

**Fig. 12-21.** Second-order control system.

**Step response.** Assume that the system is initially at rest. For this system, we have

$$T\ddot{c} + \dot{c} = Ke \qquad (12\text{-}47)$$

Since $e = r - c$, Eq. (12-47) may be written

$$T\ddot{e} + \dot{e} + Ke = T\ddot{r} + \dot{r} \qquad (12\text{-}48)$$

For the step input $r(t) = R$, $\ddot{r} = \dot{r} = 0$ for $t > 0$. Therefore, Eq. (12-48) may be written

$$T\ddot{e} + \dot{e} + Ke = 0 \qquad \text{for } t > 0 \qquad (12\text{-}49)$$

Since the system is assumed to be initially at rest, the initial conditions for the error signal are $e(0) = R$ and $\dot{e}(0) = 0$. The trajectory in the $e$-$\dot{e}$ plane starts at point $(R, 0)$ and converges to the origin, the singular point of the system. Figure 12-22 (a) shows the trajectory when the roots of the characteristic equation of the system are a complex-conjugate pair and lie in the left-half plane. Figure 12-22 (b) shows

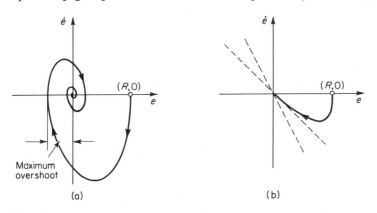

**Fig. 12-22.** Trajectories corresponding to step responses of the system of Fig. 12-21. (a) Underdamped case; (b) overdamped case.

the trajectory when the roots are real and lie in the left-half plane. In either case, the error of the system is zero at steady state. The nature of the response can clearly be seen in the phase-plane plots. For example, the maximum overshoot in the underdamped case can be obtained from the phase-plane plot of Fig. 12-22 (a).

**Ramp response.** For the ramp input $r(t) = Vt$ or ramp-plus-step input $r(t) = Vt + R$, derivatives of $r(t)$ become $\ddot{r} = 0$ and $\dot{r} = V$ for $t > 0$. Therefore, Eq. (12-48) may be written

$$T\ddot{e} + \dot{e} + Ke = V \qquad (12\text{-}50)$$

Let

$$e - \frac{V}{K} = x$$

Then, Eq. (12-50) becomes

$$T\ddot{x} + \dot{x} + Kx = 0 \tag{12-51}$$

The phase-plane portrait of the differential equation given by Eq. (12-51) in the $x$-$\dot{x}$ plane is the same as that of Eq. (12-49) in the $e$-$\dot{e}$ plane.

It is noted that the nature of the singular point is determined by the roots of the characteristic equation. The location of the singular point in the $e$-$\dot{e}$ plane, in general, depends on both the step and ramp components of the input. [In this particular example, the location of the singular point in the phase plane depends on the ramp input only because the right-hand side of Eq. (12-48) does not contain a term in $r$.] Since the system is assumed to be at rest initially, the initial conditions for the error signal may be written

$$e(0) = R, \qquad \dot{e}(0) = V$$

where $R$ may be zero. Figures 12-23 (a) and (b) show trajectories in the $e$-$\dot{e}$ plane. [Figure 12-23 (a) corresponds to the case where the roots of the characteristic equa-

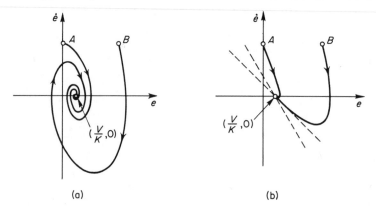

Fig. 12-23. Trajectories corresponding to ramp responses of the system of Fig. 12-21. (a) Underdamped case; (b) overdamped case.

tion corresponding to Eq. (12-49) are a complex-conjugate pair and lie in the left-half plane. Figure 12-23 (b) corresponds to the case where the roots are real and lie in the left-half plane.] In Figs. 12-23 (a) and (b), for a ramp input the trajectory starts, for example, at point $A$. For a ramp-plus-step input the trajectory starts, for example, at point $B$. In either case, the trajectories converge to the singular point $(V/K, 0)$.

From the foregoing analysis, it can be seen that the phase-plane portrait of the system shown in Fig. 12-21 for a ramp or a ramp-plus-step input is the same as that for a step input, with the exception that the entire portrait is shifted to the right by $V/K$. The steady-state error of the system output when the system is subjected to the input $r(t) = Vt + R$ is $V/K$.

Note that for a system with one singular point it is possible to shift the location of the singular point to the origin of a new phase plane by introducing a new variable. Thus, for such a system, step-response and ramp-response characteristics can be shown on one phase-plane plot. For a nonlinear system having two or more singular points, however, it is impossible to shift all these singular points to the origin at the same time. Thus, for such a system, it is imperative to use a separate phase-plane plot for a step response and for a ramp response.

**Impulse response.** For the unit-impulse input, the system equation is

$$T\ddot{c} + \dot{c} + Kc = 0 \qquad \text{for } t > 0$$

with initial conditions $c(0-) = \dot{c}(0-) = 0$ and

$$c(0+) = \lim_{s \to \infty} \frac{sK}{Ts^2 + s + K} = 0, \qquad \dot{c}(0+) = \lim_{s \to \infty} \frac{s^2K}{Ts^2 + s + K} = \frac{K}{T}$$

The starting point of the trajectory in the $c$-$\dot{c}$ plane is $(0, K/T)$. In terms of the error signal,

$$T\ddot{e} + \dot{e} + Ke = 0 \qquad \text{for } t > 0$$

with initial conditions

$$e(0+) = 0, \qquad \dot{e}(0+) = -\frac{K}{T}$$

In the $e$-$\dot{e}$ plane, the starting point (corresponding to $t = 0+$) of the trajectory is $(0, -K/T)$. The trajectories indicating unit-impulse responses for $t > 0$ are shown in Figs. 12-24 and 12-25. Figure 12-24 corresponds to an underdamped system, and Fig. 12-25 corresponds to an overdamped one. In these figures, the starting points (at $t = 0+$) of the trajectories are shown by point $P$.

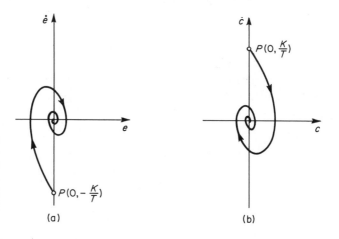

**Fig. 12-24.** Trajectories corresponding to unit-impulse responses of the system of Fig. 12-21 (underdamped case). (a) Plot in the $e$-$\dot{e}$ plane; (b) plot in the $c$-$\dot{c}$ plane.

(a)

(b)

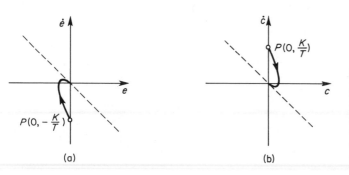

**Fig. 12-25.** Trajectories corresponding to unit-impulse responses of the system of Fig. 12-21 (overdamped case). (a) Plot in the $e$-$\dot{e}$ plane; (b) plot in the $c$-$\dot{c}$ plane.

## 12-6 PHASE-PLANE ANALYSIS OF NONLINEAR CONTROL SYSTEMS*

In second-order systems with signal-dependent nonlinearities, it is possible to approximate the system by several piecewise-linear systems. The entire phase plane is divided into several subregions, each corresponding to an individual linear operation. There is one singular point for each region, although it may be located outside of that particular region.

If the singular point lies within its subregion, then it is called an actual singular point. But, if the singular point lies outside the subregion to which it belongs and therefore can never be reached by its own trajectories, it is called a *virtual* singular point. A second order system with signal-dependent nonlinearity can have only one actual singular point.

All regions adjacent to a region with an actual singular point will have virtual singular points. The location and nature of each singular point are determined from the differential equation governing the given subregion. The location of a singular point may depend on the input. The phase-plane portrait of each subregion is that of a linear system. The composite trajectory, which is obtained by joining trajectories at the boundaries of each operating region, gives the transient response of the nonlinear system.

**Control systems with nonlinear gains.** Consider the nonlinear system shown in Fig. 12-26 (a). The block denoted $G_N$ is the nonlinear gain element. The input-output characteristic curve of this element is shown in Fig. 12-26 (b). The gain of the element is unity or $k$, whenever the magnitude of the error signal $e$ is greater than or less than $e_0$, respectively. Namely,

$$m = e \qquad \text{for } |e| > e_0 \tag{12-52}$$

$$m = ke \qquad \text{for } |e| < e_0 \tag{12-53}$$

The system has a large gain for large error signals and has a small gain for small error signals. Switching the gain may be accomplished by use of a switching device which changes the gain of the amplifier abruptly from one value to another. For

*This section closely follows Reference K-1.

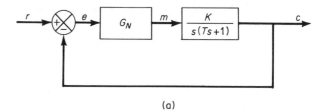

(a)

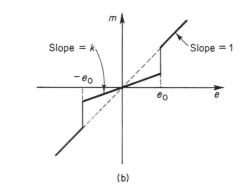

**Fig. 12-26.** (a) Nonlinear system; (b) input-output characteristic curve of the nonlinear gain element.

(b)

small values of $e$, the system exhibits slow response, and for large errors fast response. This characteristic may be desirable for systems subjected to low-amplitude high-frequency noises since unwanted noise signals will be suppressed substantially, while command signals can be transmitted satisfactorily.

Since this system is piecewise linear, it is possible to obtain trajectories for this system by combining the trajectories of two linear systems.

In this example, we assume that the system is at rest initially. The differential equation relating the variables $c$ and $m$ is

$$T\ddot{c} + \dot{c} = Km$$

Since $e = r - c$, this last equation can be rewritten

$$T\ddot{e} + \dot{e} + Km = T\ddot{r} + \dot{r} \tag{12-54}$$

Two types of input will be considered here, namely, the step input and the ramp (or ramp-plus-step) input.

*For step inputs:* Since for step inputs $\ddot{r} = \dot{r} = 0$ for $t > 0$, Eq. (12-54) becomes

$$T\ddot{e} + \dot{e} + Km = 0 \qquad \text{for } t > 0 \tag{12-55}$$

From Eqs. (12-52), (12-53), and (12-55), we obtain the following two equations:

$$T\ddot{e} + \dot{e} + Ke = 0 \qquad \text{for } |e| > e_0 \tag{12-56}$$

$$T\ddot{e} + \dot{e} + kKe = 0 \qquad \text{for } |e| < e_0 \tag{12-57}$$

In order to obtain the trajectory of the error signal, the singular points of the system must be determined first. The singular points are where $\dot{e} = \ddot{e} = 0$. It is clear from Eqs. (12-56) and (12-57) that the origin $(0, 0)$ is the singular point for the system represented by these two equations. In the following analysis, it will be assumed that the roots of the characteristic equation corresponding to Eq. (12-56) are complex conjugates and lie in the left half of the complex plane. Thus, the nature of the singular point $(0, 0)$ in this case is that of a stable focus.

The value of $k$ is assumed to be such that the damping ratio of the characteristic equation corresponding to Eq. (12-57) equals unity. Then the singular point $(0, 0)$ corresponding to Eq. (12-57) is a stable node. Thus, the response is underdamped for large errors and critically damped for small errors. If the relationship between $m$ and $e$ were given by either $m = ke$ or $m = e$ for all magnitudes of the error signal, then the phase-plane plot would look like the one shown in Fig. 12-27 (a) or the one shown in Fig. 12-27 (b).

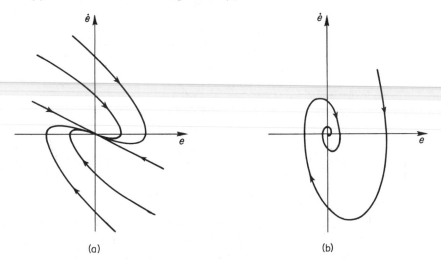

(a)                                        (b)

**Fig. 12-27.** Phase-plane plots of linear systems. (a) Overdamped case $(m = ke)$; (b) underdamped case $(m = e)$.

Figure 12-28 shows the trajectory of the error signal of the system shown in Fig. 12-26 (a), with the system constants being $T = 1$, $K = 4$, $k = 0.0625$, and $e_0 = 0.2$. The input to the system is assumed to be a unit step. As shown in Fig. 12-28, the phase plane is divided into three regions. In the region bounded by the lines $e = e_0$ and $e = -e_0$, a linear operation represented by Eq. (12-57) takes place. Outside this region, the other linear operation, corresponding to Eq. (12-56), will occur. In Fig. 12-28, the trajectory starting at point $A$, which is determined by the initial conditions $e(0) = 1$, $\dot{e}(0) = 0$, tends to converge to the stable focus $(0, 0)$. The operation of the system switches at point $B$, however, where the trajectory intersects the boundary line $e = e_0$. At this point, the trajectory is connected to a trajectory belonging to the singular point of the adjacent region and going through this point. From point $B$ on, until the system operation switches again, the trajectory tends to converge to the stable node $(0, 0)$. The operation of the system changes again at point $C$, and the trajectory tends to converge to the stable focus

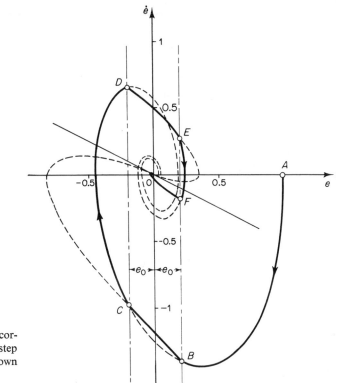

**Fig. 12-28.** Trajectory corresponding to a unit-step response of the system shown in Fig. 12-26 (a).

$(0, 0)$ until point $D$ is reached. At this point, the operation of the system switches once again. Repeating the same process, the trajectory finally converges to the stable node $(0, 0)$. At steady state, there is no error.

It can be seen that the trajectory shown in Fig. 12-28 represents generally more desirable step-response characteristics than those trajectories shown in Figs. 12-27 (a) and (b), in the sense that the former response is faster than the latter. For small step inputs, the response exhibits no overshoots. For medium-sized step inputs, the system response exhibits a single overshoot. For larger step inputs, there can be overshoots and undershoots in the response curves. Figure 12-29 shows typical step-response curves.

*For ramp inputs (or ramp-plus-step inputs):* For the input signal

$$r(t) = R + Vt$$

Eq. (12-54) becomes

$$T\ddot{e} + \dot{e} + Km = V \qquad \text{for } t > 0 \qquad (12\text{-}58)$$

From Eqs. (12-52), (12-53), and (12-58), we obtain for $t > 0$

$$T\ddot{e} + \dot{e} + Ke = V \qquad \text{for } |e| > e_0 \qquad (12\text{-}59)$$

$$T\ddot{e} + \dot{e} + Kke = V \qquad \text{for } |e| < e_0 \qquad (12\text{-}60)$$

The singular point corresponding to Eq. (12-59) is $(V/K, 0)$, which is assumed to

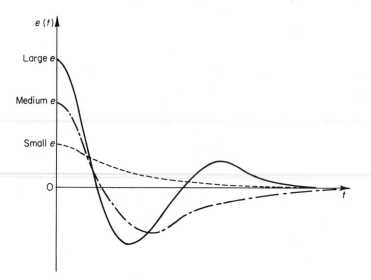

**Fig. 12-29.** Typical step-response curves for the system shown in Fig. 12-26 (a).

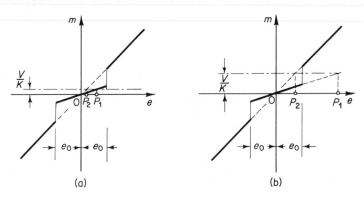

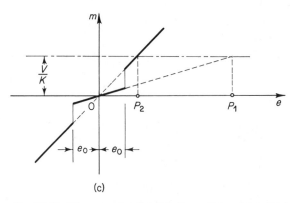

**Fig. 12-30.** Diagrams showing locations of singular points.

be a stable focus, and that corresponding to Eq. (12-60) is $(V/Kk, 0)$, which is assumed to be a stable node.

The locations of the singular points can also be obtained graphically. The values of $e$ corresponding to the intersections of the horizontal line $m = V/K$ with the line $m = ke$ and with the line $m = e$ give the values of $e$ of the singular points. This is shown in Figs. 12-30 (a), (b), and (c). For small values of $V$, namely for $|V| < kKe_0$, the singular points are located inside the strip bounded by the lines $e = \pm e_0$ in the $e$-$\dot{e}$ plane. For medium values of $V$, namely $kK_0e_0 < |V| < Ke_0$, the singular point corresponding to Eq. (12-59) is inside the strip bounded by the lines $e = \pm e_0$ in the $e$-$\dot{e}$ plane, and the other singular point corresponding to Eq. (12-60) lies outside this strip. For larger values of $V$, namely $|V| > Ke_0$, the singular points lie outside this strip.

If the relationship between $m$ and $e$ were such that $m = ke$ throughout the response time, then the phase-plane plot of the system subjected to the input $r(t) = R + Vt$ would be the same as that shown in Fig. 12-27 (a), with the exception that the former is shifted to the right by the amount $V/Kk$, as shown in Fig. 12-31 (a). Similarly, if $m$ were equal to $e$ throughout the response time, then the

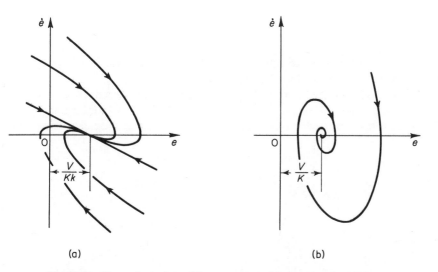

(a)                                      (b)

**Fig. 12-31.** Phase-plane plots of linear systems with ramp inputs. (a) Overdamped case ($m = ke$); (b) underdamped case ($m = e$).

phase-plane plot would be the same as that shown in Fig. 12-27 (b), except that the former is shifted to the right by the amount $V/K$, as shown in Fig. 12-31 (b).

Figure 12-32 shows the trajectory of the error signal for the case $V < kKe_0$. (The numerical values used for this example are $T = 1$, $K = 4$, $k = 0.0625$, $e_0 = 0.2$, $R = 0.3$, and $V = 0.04$.) The starting point $A$ of the trajectory is determined by the initial conditions $e(0) = R = 0.3$ and $\dot{e}(0) = V = 0.04$. The trajectory starting at point $A$ tends to converge to the stable focus $P_2$. As soon as the trajectory reaches point $B$, however, the operation of the system switches, and the tra-

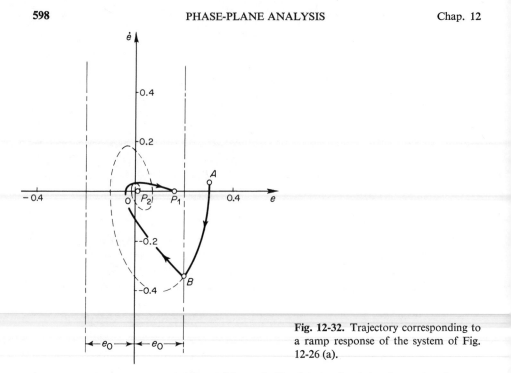

**Fig. 12-32.** Trajectory corresponding to a ramp response of the system of Fig. 12-26 (a).

jectory starts to converge to the stable node $P_1$. At steady state, the system has an error, the magnitude of which is $\overline{OP_1}$.

Figure 12-33 shows the trajectory of the error signal when $kKe_0 < V < Ke_0$. [In Fig. 12-33, the system constants are $T = 1$, $K = 4$, $k = 0.0625$, and $e_0 = 0.2$.

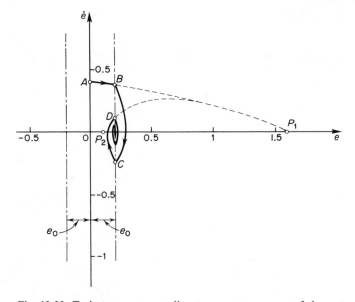

**Fig. 12-33.** Trajectory corresponding to a ramp response of the system of Fig. 12-26 (a).

The input is $r(t) = Vt \doteq 0.4t$.] The trajectory starts at point $A$, which corresponds
to the initial conditions $e(0) = 0$, $\dot{e}(0) = 0.4$, and tends to converge to the stable
node $P_1$, $(1.6, 0)$. The switch in the operation occurs at point $B$, and then the trajectory tends to converge to the stable focus $P_2$, $(0.1, 0)$. The system again switches
at point $C$, however, and the trajectory tends to converge to the stable node $P_1$.
Another switch in the system operation occurs at point $D$. The same process continues until the trajectory converges to the point where $e = e_0$, $\dot{e} = 0$. It can be
seen from Fig. 12-33 that as the trajectory approaches the singular point $(e_0, 0)$,
the error signal exhibits small oscillations, and the magnitude of the steady-state
error becomes $e_0$.

Steady-state operation of the system depends on the physical construction of
the nonlinear gain element. If the switch from one linear operation to the other
involves some delay, the system response would exhibit a limit cycle around the
point $(e_0, 0)$. If the switch occurs instantaneously, the system, at steady state,
exhibits so-called "chattering."

Figure 12-34 shows the trajectory of the error signal for the case where $V >
Ke_0$. [The input is $r(t) = Vt = 1.2t$.] The trajectory starting at point $A$, the point

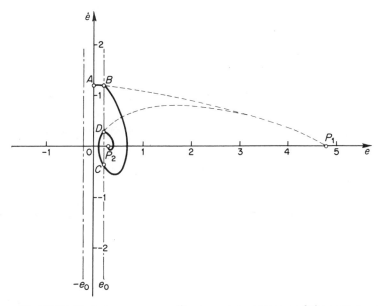

**Fig. 12-34.** Trajectory corresponding to a ramp response of the system
of Fig. 12-26 (a).

corresponding to the initial condition, tends to converge to the stable focus $P_1$,
$(4.8, 0)$. Operation of the system switches at point $B$, however, and the trajectory
tends to converge to the stable focus $P_2$, $(0.3, 0)$. Operation is again switched at
point $C$, then at point $D$, and finally the trajectory converges to the stable focus
$P_2$. As the trajectory approaches the neighborhood of point $P_2$, the error signal
exhibits small oscillations, which eventually die out. At steady state, the magnitude
of the error is $\overline{OP_2}$.

The preceding analysis illustrates that the types of responses of nonlinear systems are dependent on the input. For example, for a small step input, the response does not overshoot. For a large step input, the response is oscillatory. For a ramp input of small magnitude, the response is aperiodic, while for a ramp input of large magnitude the response is oscillatory.

---

*Example 12-8.* Consider the system shown in Fig. 12-35 (a). The input-output characteristic curve of the nonlinear element is shown in Fig. 12-35 (b). It is assumed that there is

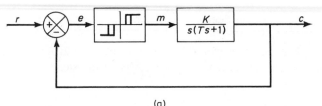

(a)

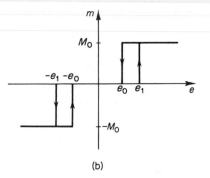

(b)

**Fig. 12-35.** (a) Nonlinear control system; (b) input-output characteristic curve of the nonlinear element.

no time delay in switching from the on to the off condition. Determine the step-response behavior and ramp-response behavior of the system. The constants of the system are assumed to be $T = 1$, $K = 4$, $e_0 = 0.1$, $e_1 = 0.2$, and $M_0 = 0.2$.

The equation for this system is

$$T\ddot{e} + \dot{e} + Km = T\ddot{r} + \dot{r}$$

For $\dot{e} > 0$, we have

$$
\begin{aligned}
m &= M_0 && \text{for } e > e_1 \\
m &= 0 && \text{for } e_1 > e > -e_0 \\
m &= -M_0 && \text{for } e < -e_0
\end{aligned}
$$

For $\dot{e} < 0$, we have

$$
\begin{aligned}
m &= M_0 && \text{for } e > e_0 \\
m &= 0 && \text{for } e_0 > e > -e_1 \\
m &= -M && \text{for } e < -e_1
\end{aligned}
$$

*For step inputs:* The trajectory of this system with a unit-step input is shown in Fig.

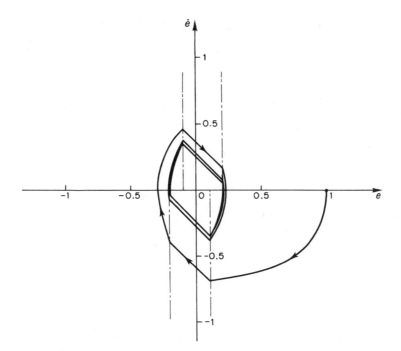

**Fig. 12-36.** Trajectory corresponding to a unit-step response of the system shown in Fig. 12-35 (a).

12-36. The response shows a limit cycle at steady state. Thus, system output oscillation continues indefinitely.

*For ramp inputs:* For a ramp input $r(t) = Vt$, the system equation becomes

$$T\ddot{e} + \dot{e} + Km = V$$

We now investigate the response to the ramp input for three different cases.

*Case 1 ($V > KM_0$):* A phase-plane portrait for the case where $V = 1.2$ is shown in Fig. 12-37. The trajectory starting at point $A$, for example, will follow the path $ABCD$. The trajectory is asymptotic to the line $\dot{e} = 0.4$. The error approaches infinity as time increases indefinitely.

*Case 2 ($V < KM_0$):* A phase-plane portrait for the case where $V = 0.4$ is shown in Fig. 12.38. The trajectory starting at point $A$ follows the path $ABCDEF$ and converges to a limit cycle. Output oscillation continues indefinitely.

*Case 3 ($V = KM_0$):* A phase-plane portrait for the case where $V = 0.8$ is shown in Fig. 12-39. The trajectory starting at point $A$ follows the path $ABCD$. The trajectory converges to point $D$. At steady state, the system will have a steady-state error equal to $\overline{OD}$.

**Summary.** The material presented in this section may be summarized as follows: Consider the system shown in Fig. 12-40. The block $N$ in the feedforward path represents a nonlinear gain element which is amplitude dependent. If $n(e)$ is a single-valued, continuous, differentiable function of $e$, then its slope $dn(e)/de = n'(e)$ defines an incremental loop gain. As long as the trajectory stays in a vertical

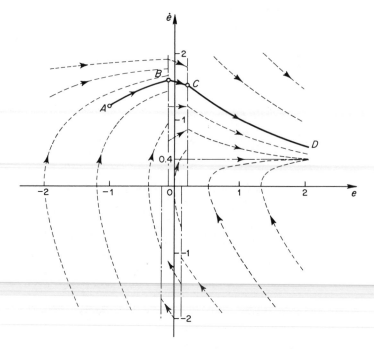

**Fig. 12-37.** Phase-plane portrait of the system shown in Fig. 12-35 (a).
$[r(t) = 1.2t]$.

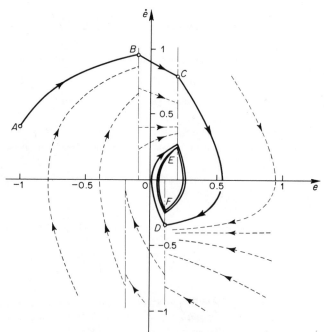

**Fig. 12-38.** Phase-plane portrait of the system shown in Fig. 12-35 (a).
$[r(t) = 0.4t]$.

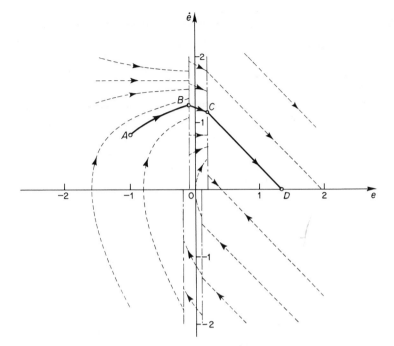

**Fig. 12-39.** Phase-plane portrait of the system shown in Fig. 12-35 (a). $[r(t) = 0.8t]$.

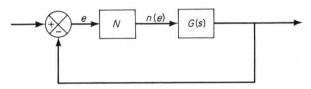

**Fig. 12-40.** Nonlinear control system.

strip about $e = e_1$ in the phase plane, the nonlinearity may be replaced by a constant $n'(e_1) = K$. The behavior of the trajectories in a strip about $e = e_1$ will then be governed by a linear time-invariant differential equation. That is, to each straight-line segment of the curve $n(e)$ versus $e$, there corresponds a region in the phase plane within which the transient solutions obey a linear differential equation. Thus, the entire phase plane is divided into a number of regions.

1. If $n'(e)$ is replaced by a variable gain $K$ corresponding to all possible values of $n'(e)$, and if it is found that the closed-loop system is stable for all values of $K$, then it is intuitively clear that the system is stable, and all trajectories will converge to a unique, stable singular point.

2. If the closed-loop system is unstable for some values of $K$ and stable for others, the system may possess a limit cycle, or limit cycles. Consider the case where the entire phase plane is divided into three regions, as shown in Fig. 12-41. If the trajectories in the given region have a singular point within this region, then it is an actual singular point. If the singular point is located outside the given region, it is a virtual singular point. Referring to Fig. 12-41, assume

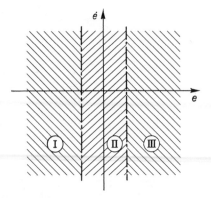

**Fig. 12-41.** Phase plane divided into three regions.

that region II has an actual singular point and regions I and III have virtual singular points. Assume also that the actual singular point is unstable and that the virtual singular points are stable. It can be seen that all trajectories will ultimately enter region II. Since the singular point in this region is unstable, the trajectories cannot remain in region II but ultimately must leave it. Therefore, it can be seen that the trajectories cannot end up at a singular point since the only actual singular point is unstable. They cannot tend to infinity either since the virtual singular points are stable. Then the only possible behavior for the trajectories is to tend to a stable limit cycle. The concept of actual and virtual singular points is very easy to handle, and it can be extended readily to systems of higher than second order.

An $n$th-order system of ordinary differential equations containing an arbitrary number of single-valued continuous nonlinearities, where each nonlinearity may be approximated by a sufficiently large number of straight-line segments (in other words, where the topological aspects of the solution of a differential equation are preserved whenever a nonlinearity is replaced by a suitably chosen straight-line approximation), is called *monostable* if the incremental linear differential equation at each point in the phase space is stable. In other words, the differential equation just referred to fails to be monostable only if the incremental linear differential equation is unstable in some region in the phase space. The dynamic behavior of monostable systems is essentially similar to that of linear systems. If the trajectory in the phase space is trapped between two concentric spherical surfaces with no singular point available, then we cannot assert that the trajectory will tend to a periodic orbit or limit cycle. It is true, however, that the trajectory will tend to a nearly periodic limit trajectory. In a control system, such a trajectory generally represents undersirable performance, just as a periodic orbit does.

## 12-7 CONCLUDING COMMENTS

In this chapter, we have presented the graphical analysis associated with second-order systems. Most of the discussion was given in the phase plane with $x$ and $\dot{x}$ as the coordinates. Other variables may be used as coordinates, but the phase-plane portrait will change accordingly.

We have shown that the limiting behavior of the trajectories of a second-order system as time $t$ approaches infinity will be one of the following three possibilities:

1. Trajectories tend to one or more stable equilibrium points.

2. Trajectories tend to infinity.
3. Trajectories tend to a limit cycle or limit cycles.

We have shown that it is possible to establish all the qualitative features of the trajectories by examining the field of directions defined by the differential equations at a sufficiently large number of points in the phase plane. The accuracy of such a phase-plane plot can be made as good as desired, within limits.

Although the graphical phase-plane analysis is limited to second-order systems, the concepts of phase-plane analysis can be extended to higher-order systems. Because of the difficulties of working graphically in three-dimensional space and of the impossibility of visualizing trajectories in $n$-dimensional spaces if $n > 3$, however, it has become necessary to resort to other methods of analysis. The state-space methods of analysis, which are indispensable to modern control theory and which are presented in Chapters 14 through 16, are extensions of the phase-plane method and are applicable to the analysis and synthesis of dynamic systems in $n$-dimensional space.

EXAMPLE PROBLEMS AND SOLUTIONS

**PROBLEM A-12-1.** Consider the simple pendulum shown in Fig. 12-42. The equation for this system is

$$\ddot{\theta} = -\frac{g}{l}\sin\theta$$

Obtain the equation for the trajectory. Then construct a phase-plane portrait.

**Solution.** Rewriting the system equation, we obtain

$$\dot{\theta}\frac{d\dot{\theta}}{d\theta} = -\frac{g}{l}\sin\theta$$

or

$$\dot{\theta}\,d\dot{\theta} = -\frac{g}{l}\sin\theta\,d\theta \tag{12-61}$$

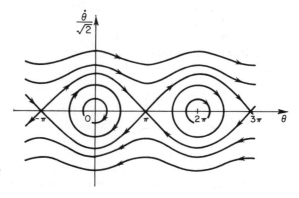

**Figure 12-42.** Simple pendulum.

Integrating both sides of Eq. (12-61), we obtain the equation for the trajectory:

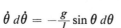

**Fig. 12-43.** Phase-plane portrait for the simple pendulum system.

$$\frac{1}{2}\dot{\theta}^2 - \frac{g}{l}\cos\theta = k$$

where $k$ is a constant. Let us put

$$\frac{d\dot{\theta}}{d\theta} = \alpha, \qquad \sqrt{\frac{g}{l}} = \omega$$

Then Eq. (12-61) becomes

$$-\omega^2\frac{\sin\theta}{\dot{\theta}} = \alpha$$

The isoclines are sinusoidal curves. Figure 12-43 shows the phase-plane portrait for the system.

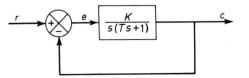

**Fig. 12-44.** Control system.

**PROBLEM A-12-2.** Consider the system shown in Fig. 12-44. The system is assumed to be initially at rest. The input to the system is the pulse function shown in Fig. 12-45 (a). The height and width of the pulse input are $h$ and $\alpha$, respectively. This pulse input may be considered to be the sum of two step inputs, as shown in Fig. 12-45 (b). The positive step input $h1(t)$ is given at $t = 0$ and then the negative step input $-h1(t - \alpha)$ at $t = \alpha$.

Construct the trajectory of the system when subjected to this pulse input.

**Solution.** The equations for the system are

$$T\ddot{e} + \dot{e} + Ke = 0 \qquad \text{for } 0 < t < \alpha$$
$$T\ddot{e} + \dot{e} + Ke = -\infty \qquad \text{for } t = \alpha$$
$$T\ddot{e} + \dot{e} + Ke = 0 \qquad \text{for } \alpha < t$$

We shall assume that the poles of the closed-loop transfer function are complex conjugates and lie in the left-half plane. (Other cases can be treated similarly.)

Since the system is assumed to be initially at rest, the initial conditions for the error signal are $e(0) = h$ and $\dot{e}(0) = 0$. Then, as shown in Fig. 12-46 (a), the trajectory on the $e$-$\dot{e}$ plane starts at point $A$ and follows the path which converges to the stable focus $(0, 0)$ until $t = \alpha$. At $t = \alpha$, the trajectory reaches point $B$. Since the negative step input $-h1(t - \alpha)$ is given at this moment, the trajectory jumps from point $B$ to point $C$, where $\dot{e} =$

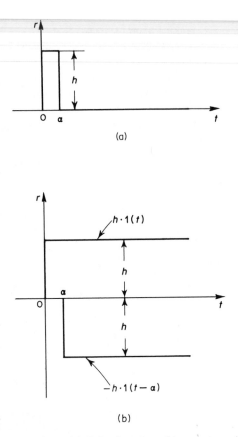

**Fig. 12-45.** (a) Pulse function; (b) two step functions to obtain the pulse function.

$-\infty$. From point $C$, the trajectory jumps to point $D$. These jumps from point $B$ to point $C$ to point $D$ occur in the instant $t = \alpha$. From point $D$, the trajectory converges to the origin, as shown in Fig. 12-46 (a). In the phase plane, such jumps may be indicated by a dashed line, as shown in Fig. 12-46 (b).

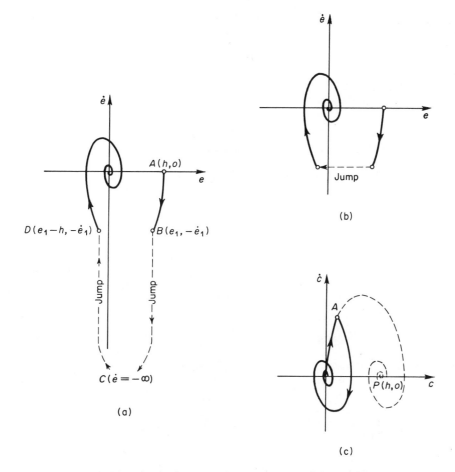

**Fig. 12-46.** (a) Trajectory in the $e$-$\dot{e}$ plane; (b) trajectory indicating a jump; (c) trajectory in the $c$-$\dot{c}$ plane.

In terms of the output signal, the system equations become

$$T\ddot{c} + \dot{c} + Kc = Kh \qquad \text{for } 0 < t < \alpha$$
$$T\ddot{c} + \dot{c} + Kc = 0 \qquad \text{for } \alpha < t$$

In the $c$-$\dot{c}$ plane, the trajectory starts from the origin as shown in Fig. 12-46 (c). This is because the initial conditions are $c(0) = \dot{c}(0) = 0$. The trajectory tends to converge to a stable focus $P$. At $t = \alpha$, the trajectory reaches point $A$. At this moment, the singular point of the system is switched from point $P$ to the origin. Therefore, for $t > \alpha$, the trajectory converges to the origin, a stable focus, as shown in Fig. 12-46 (c). The trajectory in the $c$-$\dot{c}$ plane does not exhibit any jumps. As seen from Fig. 12-46 (c), the pulse response of the system output $c(t)$ is oscillatory since the poles of the closed-loop transfer function are assumed to be complex conjugates and to lie in the left-half plane. If $c(t)$ were plotted as a function of time, point $A$ in Fig. 12-46 (c) would correspond to the first inflection point of the time-response curve.

**PROBLEM A-12-3.** Figure 12-47 (a) shows a control system with saturation nonlinearity. The input-output characteristic curve of the saturation nonlinearity is shown in Fig. 12-47 (b). Assuming that the system is initially at rest, construct the trajectories in the

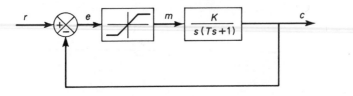

(a)

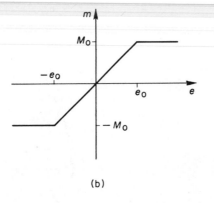

(b)

**Fig. 12-47.** (a) Control system with saturation nonlinearity; (b) input-output characteristic curve of the saturation nonlinearity.

phase plane when the system is subjected to a step input $r(t) = R$ and a ramp input $r(t) = Vt$, $(V > 0)$. The constants of the system are assumed to be $T = 1$, $K = 4$, $e_0 = 0.2$, and $M_0 = 0.2$.

**Solution.** From the saturation characteristic curve shown in Fig. 12-47 (b), we obtain

$$m = e \qquad \text{for } |e| \leq e_0$$
$$m = M_0 \qquad \text{for } e > e_0$$
$$m = -M_0 \qquad \text{for } e < -e_0$$

The equation for the system is

$$T\ddot{e} + \dot{e} + Km = T\ddot{r} + \dot{r}$$

*For step input:* For the step input, $\ddot{r} = \dot{r} = 0$ for $t > 0$. Hence

$$T\ddot{e} + \dot{e} + Km = 0$$

For the linear operation of the system,

$$T\ddot{e} + \dot{e} + Ke = 0$$

The singular point $(0, 0)$ is either a stable node or a stable focus. For the nonlinear operations of the system,

$$T\ddot{e} + \dot{e} + KM_0 = 0 \qquad \text{for } e > e_0$$
$$T\ddot{e} + \dot{e} - KM_0 = 0 \qquad \text{for } e < -e_0$$

Let us define

$$\frac{d\dot{e}}{de} = \alpha$$

then we obtain

$$\dot{e} = \frac{-\dfrac{KM_0}{T}}{\alpha + \dfrac{1}{T}} \qquad \text{for } e > e_0 \tag{12-62}$$

$$\dot{e} = \frac{\dfrac{KM_0}{T}}{\alpha + \dfrac{1}{T}} \qquad \text{for } e < -e_0 \tag{12-63}$$

From Eq. (12-62), it can be seen that for $e > e_0$ all trajectories are asymptotic to the line

$$\dot{e} = -KM_0$$

which corresponds to $\alpha = 0$. Similarly, from Eq. (12-63), for $e < -e_0$ all trajectories are asymptotic to the line

$$\dot{e} = KM_0$$

Figure 12-48 shows a phase-plane portrait for the region $|e| > e_0$. Figure 12-49 shows the trajectory when the system is subjected to a step input of magnitude 2.

*For ramp inputs:* For the ramp input, $r = Vt$ and $\dot{r} = V = $ constant. Hence the system equation becomes

$$T\ddot{e} + \dot{e} + Km = V$$

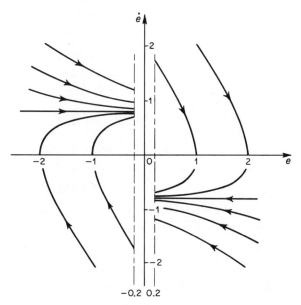

**Fig. 12-48.** Phase-plane portrait for the region $|e| > e_0$ for the system shown in Fig. 12-47 (a).

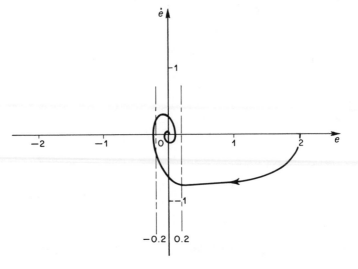

**Fig. 12-49.** Trajectory corresponding to a step response of the system shown in Fig. 12-47 (a).

or

$$T\ddot{e} + \dot{e} + Ke = V \qquad \text{for } |e| < e_0$$
$$T\ddot{e} + \dot{e} + KM_0 = V \qquad \text{for } e > e_0$$
$$T\ddot{e} + \dot{e} - KM_0 = V \qquad \text{for } e < -e_0$$

The singular point for the linear operation is located at $(V/K, 0)$. It is either a stable focus or a stable node. For the nonlinear operations

$$\dot{e} = \frac{\frac{V}{T} - \frac{KM_0}{T}}{\alpha + \frac{1}{T}} \qquad \text{for } e > e_0$$

$$\dot{e} = \frac{\frac{V}{T} + \frac{KM_0}{T}}{\alpha + \frac{1}{T}} \qquad \text{for } e < -e_0$$

For $e > e_0$, except for the special case where $V = KM_0$ the trajectories are asymptotic to the line

$$\dot{e} = V - KM_0 \qquad (12\text{-}64)$$

and, for $e < -e_0$, the trajectories are asymptotic to the line

$$\dot{e} = V + KM_0 \qquad (12\text{-}65)$$

From Eqs. (12-64) and (12-65), it can be seen that the line to which the trajectories asymptotically approach is either above the $e$ axis or below the $e$ axis, depending on whether $V > KM_0$ or $V < KM_0$.

*Case 1 ($V > KM_0$):* The phase-plane portrait for the case where $V = 1.2$ is shown in Fig. 12-50. If the initial condition is given by point $A$, then the trajectory follows path $ABCD$. From point $B$ to point $C$, the trajectory tends to converge to a stable focus, $(V/K,$

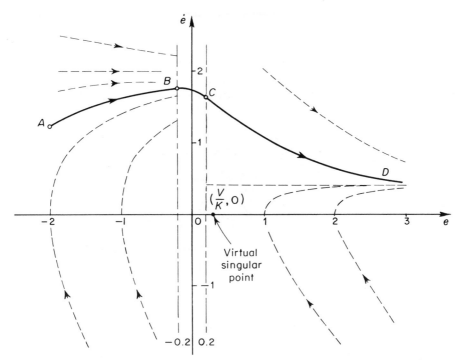

**Fig. 12-50.** Phase-plane portrait of the system shown in Fig. 12-47 (a). $[r(t) = 1.2t]$.

0); however, the trajectory cannot converge to this stable focus. Instead, it becomes asymptotic to the horizontal line $\dot{e} = 0.4$. At steady state, the error becomes infinity.

*Case 2 ($V < KM_0$):* The phase-plane portrait for the case where $V = 0.4$ is shown in Fig. 12-51. If the initial condition is given by point $A$, then the trajectory converges to a stable focus, $(V/K, 0)$, the actual singular point.

*Case 3 ($V = KM_0$):* In this case, for $e > e_0$,

$$T\ddot{e} + \dot{e} = 0$$

or

$$\dot{e}\left(T\frac{d\dot{e}}{de} + 1\right) = 0$$

This means that for $e > e_0$ the trajectories are either straight lines with slope $-1/T$ or the straight line $\dot{e} = 0$. The phase-plane portrait for the case where $V = 0.8$ is shown in Fig. 12-52. If the initial condition is given by point $A$, the trajectory follows path $ABCD$. From point $B$ to point $C$ the trajectory tends to converge to the stable focus at $(V/K, 0)$. From point $C$ the trajectory converges to point $D$. The response terminates at point $D$. The magnitude of the steady-state error is $\overline{OD}$.

**PROBLEM A-12-4.*** Figure 12-53 (a) shows a block diagram of a control system with nonlinear damping. The input-output characteristic curve of the nonlinear element $G_N$

*Reference L-8.

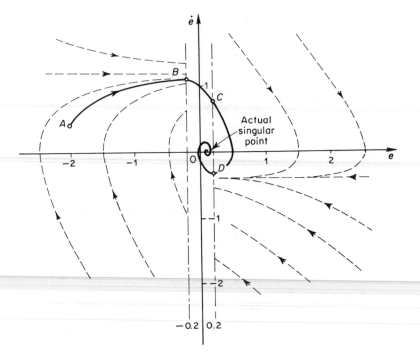

**Fig. 12-51.** Phase-plane portrait of the system shown in Fig. 12-47 (a). $[r(t) = 0.4t]$.

is shown in Fig. 12-53 (b). The characteristic of the nonlinear element is such that the signal $m$ is zero if the magnitude of the error signal is greater than $e_0$. This means that for large error signals the system has zero damping or an infinite velocity constant. For smaller error signals ($|e| < e_0$), the system has the damping term $K_0\dot{c}$. Thus, the amount of tachometric feedback signal is controlled nonlinearly. The transient-response characteristics of this system are, in general, superior to those of linear systems, in the sense that the former has characteristics of faster response with smaller overshoot than the latter. (For a large error signal, the system exhibits faster response; for a small error signal, the system is well damped.)

Assuming that the system is initially at rest and that the system constants are $K = 4$, $K_0 = 1$, and $e_0 = 0.2$, construct the trajectory in the normalized phase plane when the system is subjected to a unit-step input. Then construct trajectories in the normalized phase plane when the system is subjected to an input $r(t) = 0.5 + 0.1t$ and an input $r(t) = t$.

**Solution.** From Fig. 12-53 (a), we obtain

$$\ddot{c} = Kb$$
$$b = e - m\dot{c}$$
$$c = r - e$$

The equation for the error signal can be written

$$\ddot{e} + mK\dot{e} + Ke = \ddot{r} + mK\dot{r} \tag{12-66}$$

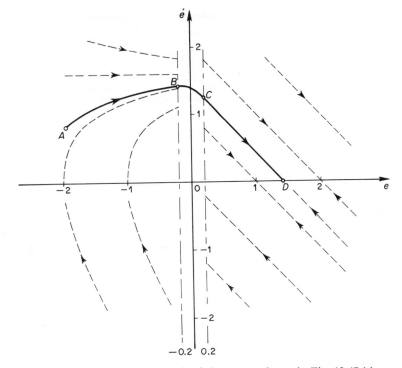

**Fig. 12-52.** Phase-plane portrait of the system shown in Fig. 12-47 (a). [$r(t) = 0.8t$].

From Fig. 12-53 (b)

$$m = K_0 \quad \text{for } |e| < e_0$$
$$m = 0 \quad \text{for } |e| > e_0$$

*For step inputs:* Since for the step input $\ddot{r} = \dot{r} = 0$ for $t > 0$, Eq. (12-66) can be simplified to

$$\ddot{e} + KK_0\dot{e} + Ke = 0 \quad \text{for } |e| < e_0 \tag{12-67}$$
$$\ddot{e} + Ke = 0 \quad \text{for } |e| > e_0 \tag{12-68}$$

Thus, system operation consists of two linear modes. From Eqs. (12-67) and (12-68), it can be seen that for both modes of linear operation the origin is the singular point. For $|e| < e_0$, the singular point is either a stable node or a stable focus. For $|e| > e_0$, the singular point is a center.

From Eq. (12-68), it can be seen that the trajectory for $|e| > e_0$ is a circle in the normalized phase plane [$e$-($\dot{e}/\sqrt{K}$) plane]. Figure 12-54 shows the trajectory of the error signal in the normalized phase plane when the system is subjected to a unit-step input.

*For ramp or ramp-plus-step inputs:* Let us denote

$$r(t) = R + V(t)$$

For $t > 0$, Eq. (12-66) can be rewritten

$$\ddot{e} + mK\dot{e} + Ke = mKV$$

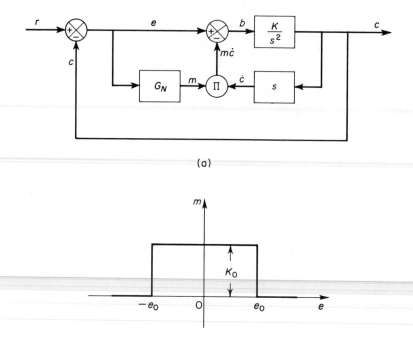

(a)

(b)

**Fig. 12-53.** (a) Control system with nonlinear damping; (b) input-output characteristic curve of the nonlinear element $G_N$.

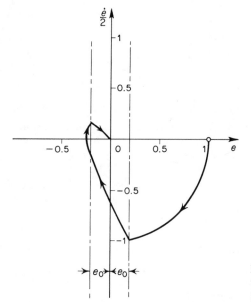

**Fig. 12-54.** Trajectory corresponding to a unit-step response of the system shown in Fig. 12-53 (a).

or

$$\ddot{e} + KK_0\dot{e} + Ke = KK_0 V \qquad \text{for } |e| < e_0 \qquad (12\text{-}69)$$

$$\ddot{e} + Ke = 0 \qquad \text{for } |e| > e_0 \qquad (12\text{-}70)$$

From Eqs. (12-69) and (12-70), we find that the singular point for $|e| < e_0$ is $(K_0V, 0)$ and that for $|e| > e_0$ it is the origin $(0, 0)$. Since the system has positive damping for $|e| < e_0$, the singular point $(K_0V, 0)$ is either a stable node or a stable focus. For smaller values of $V$, namely, $V < e_0/K_0$ the singular point lies on the $e$ axis between 0 and $e_0$; for larger values of $V$, namely, $V > e_0/K_0$, it lies on the $e$ axis to the right of point $e = e_0$.

Figure 12-55 shows the trajectory of the error signal in the normalized phase plane when $R = 0.5$ and $V = 0.1$. The trajectory starts at point $A$, which is specified by the initial

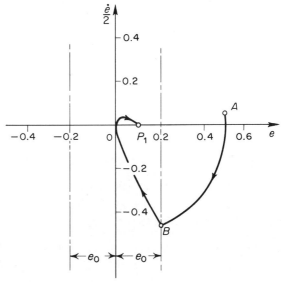

**Fig. 12-55.** Trajectory corresponding to a ramp response of the system shown in Fig. 12-53 (a). $[r(t) = 0.5 + 0.1t]$.

conditions $e(0) = 0.5$, $\dot{e}(0)/2 = 0.05$. The representative point moves along the arc centered at the origin until the trajectory hits the line $e = e_0$ at point $B$, where system operation switches to the other linear operation. From this point on, the trajectory converges to the singular point $P_1$, $(K_0V, 0)$. At steady state, the magnitude of the error is $\overline{OP_1}$.

Figure 12-56 shows the trajectory of the error signal in the normalized phase plane when $R = 0$ and $V = 1$. The trajectory starts at point $A$, which corresponds to the initial conditions $e(0) = 0$, $\dot{e}(0)/2 = 0.5$, and begins to converge to the singular point $P_1$, $(1, 0)$. At point $B$ the switch in system operation takes place, and the trajectory begins to converge to a center $(0, 0)$. At point $C$, another switch in operation takes place. Repeating the same process, the trajectory finally converges to the point $(e_0, 0)$. In the neighborhood of point $(e_0, 0)$, the error signal exhibits small oscillations.

The steady-state behavior of the system depends on the physical construction of the system. If the switch between two linear operations takes place instantaneously, the system will exhibit chattering at steady state. If there is considerable time delay in switching, however, the system will exhibit a limit cycle about the point $(e_0, 0)$.

From the preceding analysis, it can be seen that for the ramp input $r(t) = Vt$ the

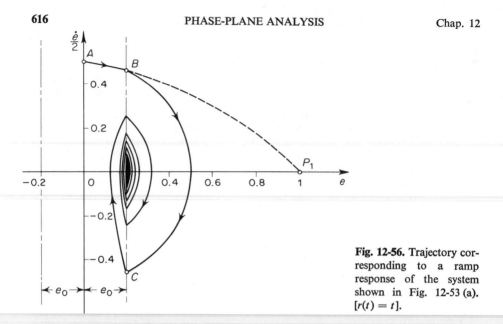

**Fig. 12-56.** Trajectory corresponding to a ramp response of the system shown in Fig. 12-53 (a). $[r(t) = t]$.

present nonlinear system has a steady-state error of magnitude $e_0$ or less, depending on the magnitude of $V$. This can be considered an advantage of the present nonlinear system over its linear counterpart because for a ramp input $r(t) = Vt$ the linesr system whose differential equation is given by Eq. (12-69) has a steady-state error of $K_0 V$, which is proportional to $V$, so it can be a large value for a large value of $V$; in the case of the present nonlinear system, the magnitude of the steady-state error for ramp inputs is at most $e_0$, which can be made small.

**PROBLEM A-12-5.** Figure 12-57 shows a second-order servomechanism with coulomb friction. Coulomb friction is a frictional force independent of the magnitude of the veloc-

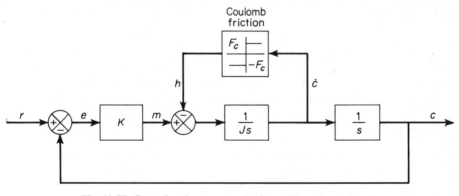

**Fig. 12-57.** Second-order servomechanism with coulomb function.

ity but always opposing it. Draw the trajectories in the $e$-$\dot{e}$ plane when the system is subjected to the initial conditions (i) $e(0) = 2.2$, $\dot{e}(0) = 0$; (ii) $e(0) = 3.5$, $\dot{e}(0) = 0$. Assume that the system constants are $K = 2, J = 1$, and $F_c = 1$.

**Solution.** The equations describing the system dynamics are

$$J\ddot{c} = m - h = Ke - h$$

$$c = r - e$$

where

$$h = F_c \qquad \text{for } \dot{c} > 0$$

$$= -F_c \qquad \text{for } \dot{c} < 0$$

In terms of the error signal, the system equations can be simplified to give

$$J\ddot{e} + Ke - h = J\ddot{r} \tag{12-71}$$

In the present analysis, $r = 0$. Thus Eq. (12-71) becomes

$$J\ddot{e} + Ke - h = 0$$

or

$$J\dot{e}\frac{d\dot{e}}{de} + Ke + F_c = 0 \qquad \text{for } \dot{e} > 0 \tag{12-72}$$

$$J\dot{e}\frac{d\dot{e}}{de} + Ke - F_c = 0 \qquad \text{for } \dot{e} < 0 \tag{12-73}$$

Let us define

$$\frac{K}{J} = \omega^2$$

Then Eq. (12-72) can be written

$$\frac{1}{\omega^2}\dot{e}\,d\dot{e} + \left(e + \frac{F_c}{K}\right) de = 0$$

Integrating both sides of this last equation gives

$$\left(\frac{\dot{e}}{\omega}\right)^2 + \left(e + \frac{F_c}{K}\right)^2 = A^2 \tag{12-74}$$

where $A$ is a constant. This is the equation of a circle in the normalized phase plane [$e$-($\dot{e}/\omega$) plane]. The center of the circle is at $e = -F_c/K$, $\dot{e}/\omega = 0$, and the radius is equal to $A$. Since Eq. (12-74) holds for $\dot{e} > 0$, the trajectories in the upper half of the normalized phase plane are semicircles.

Similarly, Eq. (12-73) can be rewritten as

$$\left(\frac{\dot{e}}{\omega}\right)^2 + \left(e - \frac{F_c}{K}\right)^2 = B^2$$

where $B$ is a constant. The trajectories in the lower half of the normalized phase plane are also semicircles.

It can be seen that the effect of coulomb friction on the system response is to shift the phase-plane portrait of the corresponding linear system, without coulomb friction, the distance $F_c/K$ to the left in the upper half plane and the same distance to the right in the lower half plane.

Figure 12-58 shows two trajectories corresponding to the given initial conditions. The trajectory denoted by (i) indicates that there is a steady-state error of magnitude 0.2. The trajectory denoted by (ii) terminates at $e(0) = -0.5$, $\dot{e}(0) = 0$; thus the system exhibits a steady-state error of magnitude 0.5.

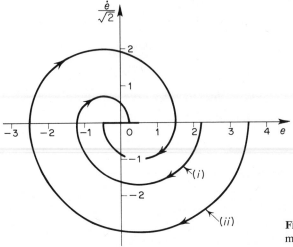

**Fig. 12-58.** Trajectories for the servo-mechanism shown in Fig. 12-57.

---

## PROBLEMS

**PROBLEM B-12-1.** Draw a phase-plane portrait of the system defined by

$$\dot{x}_1 = x_1 + x_2$$
$$\dot{x}_2 = 2x_1 + x_2$$

**PROBLEM B-12-2.** Draw a phase-plane portrait of the following system:

$$\ddot{x} + \dot{x} + |x| = 0$$

**PROBLEM B-12-3.** Draw a phase-plane portrait of the following system:

$$\ddot{\theta} + \dot{\theta} + \sin\theta = 0$$

**PROBLEM B-12-4.** Determine the locations and types of singular points of the nonlinear system described by

$$\dot{x}_1 = 0.3 - 0.1x_1 + x_2 - 0.188x_1^2 x_2 - 0.75x_2^3$$
$$\dot{x}_2 = -0.25x_1 - 0.1x_2 + 0.047x_1^3 + 0.188x_1 x_2^2$$

**PROBLEM B-12-5.** The following equation is called the Van der Pol equation.

$$\ddot{x} - (1 - x^2)\dot{x} + x = 0$$

Determine the type of the singular point. Draw a phase-plane portrait.

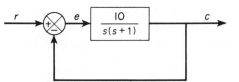

**Fig. 12-59.** Control system.

**PROBLEM B-12-6.** Obtain the trajectory representing the response of the system shown in Fig. 12-59 when it is subjected to the input

$$r(t) = R_1\, 1(t) + R_2\, 1(t - 2) + R_3\, 1(t - 3)$$

where $1(t - t_i)$ is a unit-step function occurring at $t = t_i$. Assume that the system is at rest initially.

**PROBLEM B-12-7.** Consider the system shown in Fig. 12-60. Draw phase-plane portraits for it in the $e$-$\dot{e}$ plane when $K = 0$ and $K = 1$. Assume that the input $r(t)$ is zero for $t > 0$, and the system is subjected only to the initial condition.

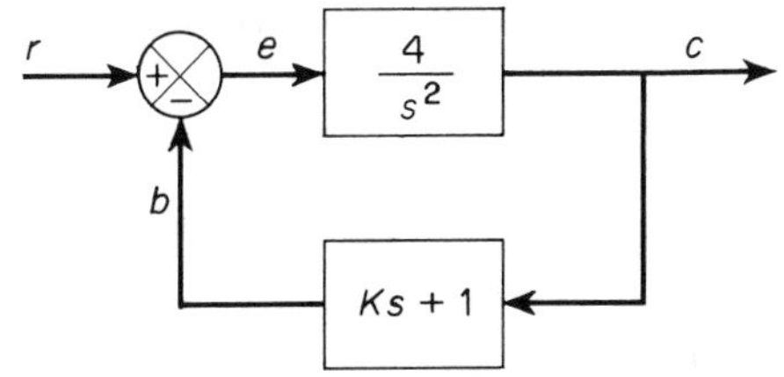

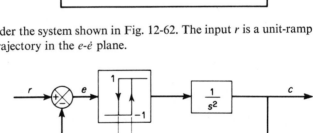

**Fig. 12-60.** Control system.

**PROBLEM B-12-8.** Draw phase-plane diagrams for the system shown in Fig. 12-61 when $\Delta = 0$ and $\Delta = 0.1$. The input $r$ is a unit-step function. Take $e$ and $\dot{e}$ as the coordinates.

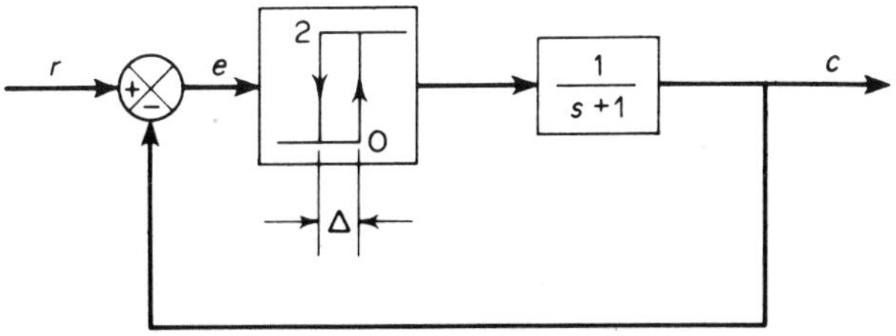

**Fig. 12-61.** Nonlinear control system.

**PROBLEM B-12-9.** Consider the system shown in Fig. 12-62. The input $r$ is a unit-ramp function. Draw a typical trajectory in the $e$-$\dot{e}$ plane.

**Fig. 12-62.** Nonlinear control system.

**PROBLEM B-12-10.** Consider the system shown in Fig. 12-63. Assume that it is subjected only to the initial condition. Draw a typical trajectory in the $c$-$\dot{c}$ plane.

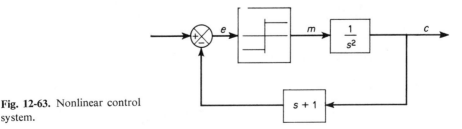

**Fig. 12-63.** Nonlinear control system.

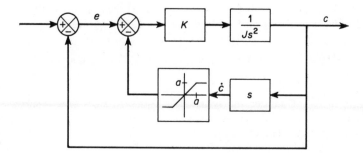

**Fig. 12-64.** Nonlinear control system.

**PROBLEM B-12-11.** Figure 12-64 shows a second-order system with a nonlinear feedback gain. The block $K$ indicates a proportional control gain, and the block $1/Js^2$ indicates a pure inertia load. The characteristic of the nonlinear element in the tachometric feedback loop is that of saturation.

Plot typical trajectories in the $e$-$\dot{e}$ plane, showing the responses to various initial conditions. Assume that $K = 5$, $J = 1$, and $a = 1$.

# 13

# DISCRETE-TIME SYSTEMS AND THE $z$-TRANSFORM METHOD

### 13-1 INTRODUCTION TO DISCRETE-TIME SYSTEMS

Discrete-time systems, or sampled-data systems, are dynamic systems in which one or more variables can change only at discrete instants of time. These instants, which we shall denote by $kT$ or $t_k$ ($k = 0, 1, 2, \ldots$), may specify the time at which some physical measurement is performed or the time at which the memory of a digital computer is read out, etc. The time interval between two discrete instants is taken to be sufficiently short so that the data for the time between these discrete instants can be approximated by simple interpolation.

Discrete-time systems differ from continuous-time ones in that the signals for a discrete-time system are in sampled-data form.

Discrete-time systems arise in practice whenever the measurements necessary for control are obtained in an intermittent fashion, or a large scale controller or computer is time-shared by several plants so that a control signal is sent out to each plant only periodically or whenever a digital computer is used to perform computations necessary for control. Many modern industrial control systems are discrete-time systems since they invariably include some elements whose inputs and/or outputs are discrete in time. Sometimes, however, the sampling operation, or discretization, may be entirely fictitious and introduced only to simplify the analysis of a control system which actually contains only continuous elements.

In this chapter, we shall be concerned with discrete-time systems where the signal representing the control effort is piecewise constant and changes only at discrete points in time. Since there are several different types of sampling operations of practical importance, we shall list them as follows:

1. Periodic (conventional) sampling: In this case, the sampling instants are equally spaced, or $t_k = kT$ ($k = 0, 1, 2, \ldots$).

2. Multiple-order sampling: The pattern of the $t_k$ is repeated periodically, or $t_{k+r} - t_k = $ constant for all $k$.

3. Multiple-rate sampling: In this case, two concurrent sampling operations occur at $t_k = pT_1$ and $qT_2$, where $T_1, T_2$ are constants and $p, q$ are integers.

4. Random sampling: In this case, the sampling instants are random, or $t_k$ is a random variable.

In this book we shall treat only the case where the sampling is periodic.

**Quantization.** The inclusion of a digital computer in an otherwise analog system produces signals in digital form (usually as binary numbers) in part of the system. The system then takes the form of a mixed digital-analog combination. The introduction of a digital computer in a control system requires the use of digital-to-analog and analog-to-digital converters. The conversion of an analog signal to the corresponding digital signal (binary number) is an approximation because the analog signal can take an infinite number of values, whereas the variety of different numbers which can be formed by a finite set of digits is limited. This approximation process is called *quantization*.

The process of quantizing (converting a signal in analog form to digital form) may be illustrated by means of the characteristic curve of Fig. 13-1. The range of

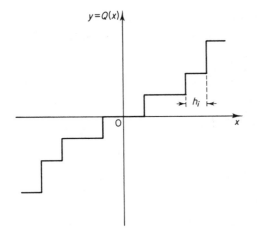

**Fig. 13-1.** A curve showing quantization.

input magnitudes is divided into a finite number of disjoint intervals $h_i$ which are not necessarily equal. All magnitudes falling within each interval are equated to a single value within the interval. This single value is the digital approximation to the magnitudes of the analog input signal. Thus, if $x$ is the analog input, the digital output is given by $y = Q(x)$ where $Q$ is the quantizing function.

The function $x(t)$ illustrated in Fig. 13-2 (a) is a discrete-time function; the one shown in Fig. 13-2 (b) is a quantized function; and the one shown in Fig. 13-2 (c) is quantized both in amplitude and in time. The operation of digital control systems involves quantization both in amplitude and in time.

**Definitions.** We shall next present the definitions of several terms.

*Transducer.* A transducer is a device which converts an input signal into an output signal of another form. (The output signal, in general, depends on the past history of the input.)

*Analog transducer.* An analog transducer is a transducer in which the input and output signals are continuous functions of time. The magnitudes of these signals may be any value within the physical limitations of the system.

*Sampled-data transducer.* This is a transducer in which the input and output signals occur only at discrete instants of time (usually periodic), but the magnitudes of the signal, as in the case of the analog transducer, are unquantized.

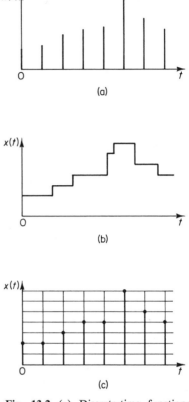

Fig. 13-2. (a) Discrete-time function; (b) quantized function; (c) discrete-time quantized function.

*Digital transducer.* A digital transducer is one in which the input and output signals occur only at discrete instants of time, and the signal magnitudes are quantized; i.e., they can assume only certain discrete levels.

*Analog-to-digital transducer.* This is a transducer in which the input signal is a continuous function of time and the output signal is a quantized signal which can assume only certain discrete levels.

*Digital-to-analog transducer.* A digital-to-analog transducer is one in which the input signal is a quantized signal and the output signal is a smoothed continuous function of time.

**Analog controllers and digital controllers.** In considering the types of controllers which are used in industrial control systems, we may divide them into the following three categories:

*Analog controllers or computers:* Analog controllers or computers represent the variables in the equations by continuous physical quantities. Analog controllers can be designed which will satisfactorily serve as nondecision making controllers.

*Digital controllers or computers:* These operate only on numbers. Decision making is an important function in digital controllers, and they are currently being used for the solution of problems involving the optimal overall operation of industrial plants.

*Analog-digital controllers or computers:* These are often called hybrid controllers. They are combinations of analog controllers and digital controllers. Some of the high performance controllers are of this type.

**Advantages of digital controllers over analog controllers.** Some of the advantages of digital controllers over analog controllers may be summarized as follows:

1. Digital controllers are capable of performing complex computations with constant accuracy at high speed. Digital computers can have almost any desired degree of accuracy in computations at relatively little increase in cost. On the other hand, the cost of analog computers increases rapidly as the complexity of the computations increases if constant accuracy is to be maintained.

2. Digital controllers are extremely versatile. By merely issuing a new program, one can completely change the operations being performed. This feature is particularly important if the control system is to receive operating information or instructions from some computing center, where economic analysis and optimization studies are being made.

Because of the inability of conventional techniques to adequately handle complex control problems, it has been customary to subdivide a process into smaller units and handle each of these as a separate control problem. Human operators are normally used to coordinate the operation of units. Recent advances in computer control systems have caused changes in this use of industrial process controls. Recent developments in large-scale computers and mathematical methods provide a basis for use of all available information in the control system. In conventional control, this part of the control loop is being done directly by humans.

**Computer control of complex systems.** Current trends in the control of large-scale systems are to consolidate the multiplicity of independently controlled units into single optimally controlled processes. In industrial process control systems, it is, in general, not practical to operate for a very long time at steady state because certain changes in production requirements, raw materials, economic factors, and processing equipment and techniques may occur. Thus, the transient behavior of industrial processes must be taken into consideration. Since there are interactions among process variables, using only one process variable for each control agent is not suitable for really complete control. By use of computer control, it is possible to take into account all process variables together with economic factors, production requirements, equipment performance, etc., and to thereby accomplish optimal control of industrial processes.

Note that a system capable of controlling a process as completely as possible will have to solve complex equations. The more complete the control, the more important it is that the correct relations between operating variables be known and be used. The system must be capable of accepting instructions from such varied sources as computers and human operators and must also be capable of changing its control subsystem completely in a short time.

**z-transform approach and state-space approach to the analysis of discrete-time systems.** The analysis of discrete-time systems may be carried out easily in either of two different approaches. One is the z-transform approach and the other is the state-space approach.

The z-transform approach has the same relationship to linear time-invariant discrete-time systems as the Laplace transform approach bears to linear time-invariant continuous-time systems. This chapter presents only the z-transform approach to the analysis of linear time-invariant discrete-time systems. The state-space approach to the analysis of linear discrete-time systems is given in Sections 14-6 and 14-7.

### 13-2 THE z TRANSFORMATION

This section presents the z-transform method for treating discrete-time functions. As stated previsouly, the role played by the z transformation in discrete-time systems is quite similar to that of the Laplace transformation in continuous-time systems. Since discrete-time functions arise when continuous signals are sampled, we shall first discuss samplers and holding devices.

**Samplers and holding devices.** The essential element of a discrete-time system is the sampler. In a conventional sampler, a switch closes to admit an input signal every $T$ seconds. In practice, the sampling duration is very short in comparison with the most significant time constant of the plant. A sampler converts a continuous signal into a train of pulses occurring at the sampling instants $0, T, 2T, \ldots$, where $T$ is the sampling period. (Between sampling instants, the sampler transmits no information.) Two signals whose respective values at the sampling instants are equal will give rise to the same sampled signal.

A holding device converts the sampled signal into a continuous signal, which approximately reproduces the signal applied to a sampler. The simplest holding device converts the sampled signal into one which is constant between two consecutive sampling instants, as shown in Fig. 13-3. Such a device is called a zero-order holding device. The transfer function $G_h$ of a zero-order holding device is

$$G_h = \frac{1 - e^{-Ts}}{s}$$

When the input signal $x(t)$ is sampled at discrete instants, the sampled signal is passed through the holding device. This device, which is a low-pass filter, smoothes the sampled signal $x^*(t)$ to produce the signal $x_h(t)$, which is constant from the last sampled value until the next sample is available. That is,

$$x_h(kT + t) = x(kT) \qquad \text{for } 0 \leq t < T$$

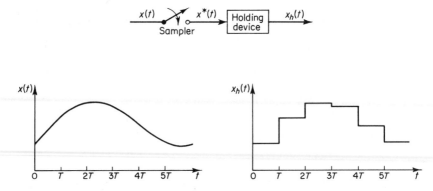

**Fig. 13-3.** Signals before and after the sampler and holding device.

In the following analysis, we assume that the holding device is of zero order. Essentially, a zero-order holding device integrates the signal $x^*(t)$ between two consecutive sampling instants. Noting that the integral of an impulse function is a constant, we see that the input to a zero-order holding device is a train of impulse functions.

Considering the sampler output to be a train of weighted impulses, we can relate the continuous signal $x(t)$ to the sampler output $x^*(t)$ by

$$x^*(t) = \delta_T(t)\, x(t)$$

where $\delta_T(t)$ represents a train of unit impulses, as shown in Fig. 13-4 (a). The sampler output is equal to the product of the continuous input $x(t)$ and the train of unit impulses. In other words, the sampler may be considered a modulator

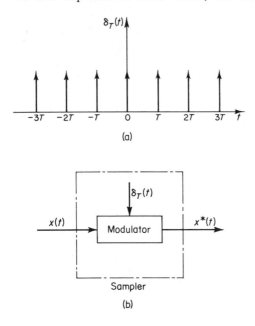

**Fig. 13-4.** (a) Train of unit impulses; (b) sampler as a modulator.

with the input $x(t)$ as the modulating signal and the train of unit impulses as the carrier, as shown in Fig. 13-4 (b). Note that $\delta_T(t)$ may be written

$$\delta_T(t) = \sum_{k=-\infty}^{\infty} \delta(t - kT)$$

where $\delta(t - kT)$ is the unit-impulse function occurring at $t = kT$. If the continuous signal $x(t)$ is sampled in a periodic manner, the sampled signal may be represented by

$$x^*(t) = \sum_{k=-\infty}^{\infty} x(t)\,\delta(t - kT) \tag{13-1}$$

or

$$x^*(t) = \sum_{k=-\infty}^{\infty} x(kT)\,\delta(t - kT) \tag{13-2}$$

Figure 13-5 shows $\delta_T(t)$, $x(t)$, and $x^*(t)$. Since the amplitude of any impulse function

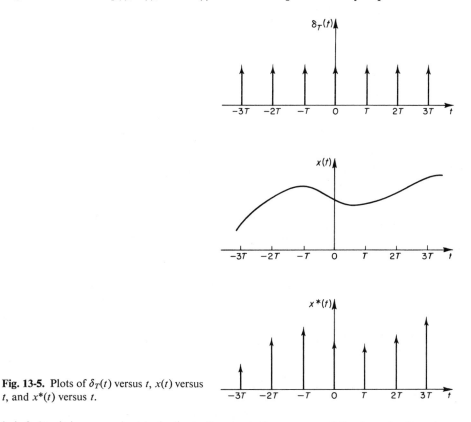

**Fig. 13-5.** Plots of $\delta_T(t)$ versus $t$, $x(t)$ versus $t$, and $x^*(t)$ versus $t$.

is infinite, it is convenient to indicate the strength, or area, of the impulse function by the length of an arrow. In Fig. 13-5, the length of each arrow in the plot of the sampled signal $x^*(t)$ indicates the strength of each sampled value $x(kT)$.

Most of the time functions we consider in this book are zero for $t < 0$. Therefore, unless otherwise stated, we shall assume that this is the case. Then, for the signal $x(t)$, Eqs. (13-1) and (13-2) become, respectively

$$x^*(t) = \sum_{k=0}^{\infty} x(t)\, \delta(t - kT) \tag{13-3}$$

and

$$x^*(t) = \sum_{k=0}^{\infty} x(kT)\, \delta(t - kT) \tag{13-4}$$

In the following analyses, we take Eq. (13-2) or (13-4) [alternatively Eq. (13-1) or (13-3)] to be the definition of the sampled signal $x^*(t)$. That is, we consider the sampler output to be a train of impulses, the strengths of which are equal to the sampled values at the respective sampling instants.

$z$ **transformation.** We shall now define the $z$ transform. Taking the Laplace transform of Eq. (13-4), we obtain

$$X^*(s) = \mathcal{L}[x^*(t)] = \sum_{k=0}^{\infty} x(kT)e^{-kTs}$$

Define

$$e^{Ts} = z$$

and write $X^*(s)$ as $X(z)$. Then

$$X(z) = X^*(s) = X^*\!\left(\frac{1}{T} \ln z\right) = \sum_{k=0}^{\infty} x(kT)z^{-k}$$

$X(z)$ is called the $z$ transform of $x^*(t)$, and the notation for the $z$ transform of $x^*(t)$ is $\mathcal{Z}[x^*(t)]$.

In the $z$ transformation, we consider only the values of the signal at the sampling instants. Therefore, the $z$ transform of $x(t)$ and that of $x^*(t)$ yield the same result, or

$$\mathcal{Z}[x(t)] = \mathcal{Z}[x^*(t)] = X(z) = \sum_{k=0}^{\infty} x(kT)z^{-k} \tag{13-5}$$

Because $X(z)$ depends only upon the values of $x(t)$ at $t = kT$ ($k = 0, 1, 2, \ldots$), the inverse $z$ transform of $X(z)$ gives information on $x(t)$ only at the sampling instants. (For the inverse $z$ transformation, refer to Section 13-4.)

Note that Eq. (13-5) is not the only form giving the $z$ transform of $x(t)$. There are two other expressions for the $z$ transform. (See Problems A-13-5 through A-13-8.)

A table of $z$ transforms of common time functions is given in Table 13-1. Table 13-2 lists useful properties of the $z$ transform.

---

*Example 13-1.* Find the $z$ transform of the unit-step function $1(t)$ by using Eq. (13-5).

$$\mathcal{Z}[1(t)] = \sum_{k=0}^{\infty} 1(kT)z^{-k}$$

$$= 1 + z^{-1} + z^{-2} + \cdots$$

$$= \frac{z}{z - 1}$$

**Table 13-1.** A TABLE OF z TRANSFORMS

| | $X(s)$ | $x(t)$ or $x(k)$ | $X(z)$ |
|---|---|---|---|
| 1 | 1 | $\delta(t)$ | 1 |
| 2 | $e^{-kTs}$ | $\delta(t - kT)$ | $z^{-k}$ |
| 3 | $\dfrac{1}{s}$ | $1(t)$ | $\dfrac{z}{z - 1}$ |
| 4 | $\dfrac{1}{s^2}$ | $t$ | $\dfrac{Tz}{(z - 1)^2}$ |
| 5 | $\dfrac{1}{s + a}$ | $e^{-at}$ | $\dfrac{z}{z - e^{-aT}}$ |
| 6 | $\dfrac{a}{s(s + a)}$ | $1 - e^{-at}$ | $\dfrac{(1 - e^{-aT})z}{(z - 1)(z - e^{-aT})}$ |
| 7 | $\dfrac{\omega}{s^2 + \omega^2}$ | $\sin \omega t$ | $\dfrac{z \sin \omega T}{z^2 - 2z \cos \omega T + 1}$ |
| 8 | $\dfrac{s}{s^2 + \omega^2}$ | $\cos \omega t$ | $\dfrac{z(z - \cos \omega T)}{z^2 - 2z \cos \omega T + 1}$ |
| 9 | $\dfrac{1}{(s + a)^2}$ | $te^{-at}$ | $\dfrac{Tze^{-aT}}{(z - e^{-aT})^2}$ |
| 10 | $\dfrac{\omega}{(s + a)^2 + \omega^2}$ | $e^{-at} \sin \omega t$ | $\dfrac{ze^{-aT} \sin \omega T}{z^2 - 2ze^{-aT} \cos \omega T + e^{-2aT}}$ |
| 11 | $\dfrac{s + a}{(s + a)^2 + \omega^2}$ | $e^{-at} \cos \omega t$ | $\dfrac{z^2 - ze^{-aT} \cos \omega T}{z^2 - 2ze^{-aT} \cos \omega T + e^{-2aT}}$ |
| 12 | $\dfrac{2}{s^3}$ | $t^2$ | $\dfrac{T^2z(z + 1)}{(z - 1)^3}$ |
| 13 | | $a^k$ | $\dfrac{z}{z - a}$ |
| 14 | | $a^k \cos k\pi$ | $\dfrac{z}{z + a}$ |

Note that whenever the infinite series $X(z)$ representing the z transform of a function converges in a region of the z plane, in using the z-transform method for solving discrete-time problems, it is not necessary to specify the values of z over which $X(z)$ is convergent.

*Example 13-2.* Obtain the z transform of the signal $x(t)$ where

$$x(t) = 0 \qquad \text{for } t < 0$$
$$= e^{-at} \qquad \text{for } t \geq 0$$

$$\mathcal{Z}[e^{-at}] = \sum_{k=0}^{\infty} e^{-akT}z^{-k}$$

$$= 1 + e^{-aT}z^{-1} + e^{-2aT}z^{-2} + \cdots$$

$$= \frac{z}{z - e^{-aT}}$$

**Table 13-2.** PROPERTIES OF THE z TRANFORM

| | $x(t)$ or $x(k)$ | $\mathcal{Z}[x(t)$ or $\mathcal{Z}[x(k)]$ |
|---|---|---|
| 1 | $ax(t)$ | $aX(z)$ |
| 2 | $x_1(t) + x_2(t)$ | $X_1(z) + X_2(z)$ |
| 3 | $x(t + T)$ or $x(k + 1)$ | $zX(z) - zx(0)$ |
| 4 | $x(t + 2T)$ | $z^2 X(z) - z^2x(0) - zx(T)$ |
| 5 | $x(k + 2)$ | $z^2 X(z) - z^2x(0) - zx(1)$ |
| 6 | $x(t + kT)$ | $z^k X(z) - z^kx(0) - z^{k-1}x(T) - \cdots - zx(kT - T)$ |
| 7 | $x(k + m)$ | $z^m X(z) - z^mx(0) - z^{m-1}x(1) - \cdots - zx(m - 1)$ |
| 8 | $tx(t)$ | $-Tz \dfrac{d}{dz} [X(z)]$ |
| 9 | $kx(k)$ | $-z \dfrac{d}{dz} [X(z)]$ |
| 10 | $e^{-at}x(t)$ | $X(ze^{aT})$ |
| 11 | $e^{-ak}x(k)$ | $X(ze^a)$ |
| 12 | $a^kx(k)$ | $X\left(\dfrac{z}{a}\right)$ |
| 13 | $ka^kx(k)$ | $-z \dfrac{d}{dz}\left[X\left(\dfrac{z}{a}\right)\right]$ |
| 14 | $x(0)$ | $\displaystyle\lim_{z \to \infty} X(z)$ if the limit exists |
| 15 | $x(\infty)$ | $\displaystyle\lim_{z \to 1} [(z - 1)X(z)]$ if $\dfrac{z - 1}{z} X(z)$ is analytic on and outside the unit circle |
| 16 | $\displaystyle\sum_{k=0}^{\infty} x(k)$ | $X(1)$ |
| 17 | $\displaystyle\sum_{k=0}^{n} x(kT)y(nT - kT)$ | $X(z)Y(z)$ |

*Example 13-3.* Obtain the z transform of $x(t)$ where

$$x(t) = 0 \qquad \text{for } t < 0$$
$$= \sin \omega t \qquad \text{for } t \geq 0$$

Since

$$\mathcal{Z}[e^{-at}] = \frac{z}{z - e^{-aT}}$$

we obtain

$$\mathcal{Z}[\sin \omega t] = \mathcal{Z}\left[\frac{e^{j\omega t} - e^{-j\omega t}}{2j}\right]$$

$$= \frac{1}{2j}\left(\frac{z}{z - e^{j\omega T}} - \frac{z}{z - e^{-j\omega T}}\right)$$

$$= \frac{1}{2j}\frac{z(e^{j\omega T} - e^{-j\omega T})}{z^2 - z(e^{j\omega T} + e^{-j\omega T}) + 1}$$

$$= \frac{z \sin \omega T}{z^2 - 2z \cos \omega T + 1}$$

*Example 13-4.* Obtain the $z$ transform of

$$X(s) = \mathscr{L}[x(t)] = \frac{1}{s(s + 1)}$$

Whenever a function in $s$ is given, the corresponding $z$ transform may be obtained by first expanding the given function in $s$ into partial fractions and then combining the $z$ transform of each partial fraction term. Let us expand $X(s)$ into partial fractions.

$$X(s) = \frac{1}{s(s + 1)} = \frac{1}{s} - \frac{1}{s + 1}$$

The $z$ transform corresponding to $1/s$ [or $1(t)$] is $z/(z - 1)$ and that corresponding to $1/(s + 1)$ [or $e^{-t}$] is $z/(z - e^{-T})$. Hence

$$\mathscr{Z}[x(t)] = X(z) = \frac{z}{z - 1} - \frac{z}{z - e^{-T}}$$

$$= \frac{z(1 - e^{-T})}{(z - 1)(z - e^{-T})}$$

## 13-3  SOLVING DIFFERENCE EQUATIONS BY THE $z$-TRANSFORM METHOD

The solution of difference equations by the $z$-transform method is very useful, as is the solution of the differential equations by Laplace transforms. Essentially, by using the $z$-transform method, we can transform difference equations into algebraic equations in $z$. In the following we shall use the simplified notation $x(k)$ to denote $x(kT)$.

**$z$ transform of $x(k + 1)$.** The $z$ transform of $x(k + 1)$ is given by

$$\mathscr{Z}[x(k + 1)] = zX(z) - zx(0) \tag{13-6}$$

where $X(z) = \mathscr{Z}[x(k)]$. This can be proved as follows:

$$\mathscr{Z}[x(k + 1)] = \sum_{k=0}^{\infty} x(k + 1)z^{-k}$$

$$= \sum_{k=1}^{\infty} x(k)z^{-k+1}$$

$$= z[\sum_{k=0}^{\infty} x(k)z^{-k} - x(0)]$$

$$= zX(z) - zx(0)$$

Note that if $x(0) = 0$, then

$$\mathscr{Z}[x(k + 1)] = z\mathscr{Z}[x(k)]     \text{if } x(0) = 0$$

Thus, if $x(0) = 0$, then multiplication of the $z$ transform of a function $x(k)$ by $z$ corresponds to a forward time shift of one period.

Equation (13-6) can be easily modified to obtain the following relationship:

$$\mathscr{Z}[x(k + 2)] = z\mathscr{Z}[x(k + 1)] - zx(1)$$
$$= z^2 X(z) - z^2 x(0) - zx(1)$$

Similarly,

$$\mathscr{Z}[x(k + m)] = z^m X(z) - z^m x(0) - z^{m-1}x(1) - z^{m-2}x(2) - \cdots - zx(m - 1)$$

where $m$ is a positive integer. Note that when the difference equation is transformed into an algebraic equation in $z$ by the $z$-transform method, the initial data are automatically included in the algebraic representation.

---

*Example 13-5.* Solve the following difference equation by using the $z$-transform method:

$$x(k + 2) + 3x(k + 1) + 2x(k) = 0, \qquad x(0) = 0, \qquad x(1) = 1$$

Taking the $z$ transforms of both sides of this difference equation, we obtain

$$z^2 X(z) - z^2 x(0) - zx(1) + 3zX(z) - 3zx(0) + 2X(z) = 0$$

Substituting the initial data and simplifying gives

$$X(z) = \frac{z}{z^2 + 3z + 2}$$

$$= \frac{z}{(z + 1)(z + 2)}$$

$$= \frac{z}{z + 1} - \frac{z}{z + 2}$$

Noting that

$$\mathscr{Z}[a^k] = \frac{z}{z - a}$$

we have

$$\mathscr{Z}[(-1)^k] = \frac{z}{z + 1}, \qquad \mathscr{Z}[(-2)^k] = \frac{z}{z + 2}$$

Hence

$$x(k) = (-1)^k - (-2)^k \qquad (k = 0, 1, 2, \ldots)$$

---

*Example 13-6.* Find the response $x(k)$ of the following system:

$$x(k + 2) - 3x(k + 1) + 2x(k) = u(k) \tag{13-7}$$

where

$$x(k) = 0 \qquad \text{for } k \leq 0$$
$$u(0) = 1$$
$$u(k) = 0 \qquad \text{for } k < 0, k > 0$$

By substituting $k = -1$ into Eq. (13-7), we obtain

$$x(1) = 0$$

Taking the $z$ transform of Eq. (13-7) with the initial data $x(0) = x(1) = 0$, we obtain

$$(z^2 - 3z + 2)X(z) = U(z)$$

Note that the $z$ transform of the forcing function $u(k)$ is

$$U(z) = \sum_{k=0}^{\infty} u(k)z^{-k} = 1$$

Hence

$$X(z) = \frac{1}{z^2 - 3z + 2} = \frac{-1}{z - 1} + \frac{1}{z - 2}$$

Using the relationship

$$\mathcal{Z}[x(k + 1)] = zX(z) - zx(0)$$

and noting that $x(0) = 0$, we obtain

$$\mathcal{Z}[x(k + 1)] = zX(z)$$

$$= -\frac{z}{z - 1} + \frac{z}{z - 2}$$

Since

$$\mathcal{Z}[1^k] = \frac{z}{z - 1}, \qquad \mathcal{Z}[2^k] = \frac{z}{z - 2}$$

we obtain

$$x(k + 1) = -1 + 2^k \qquad (k = 0, 1, 2, \ldots)$$

or

$$x(k) = -1 + 2^{k-1} \qquad (k = 1, 2, 3, \ldots)$$

**Initial value theorem.** If $x(t)$ has the $z$ transform $X(z)$ and $\lim_{z \to \infty} X(z)$ exists, then the initial value $x(0)$ of $x(t)$ or $x(k)$ is given by

$$x(0) = \lim_{z \to \infty} X(z) \qquad (13\text{-}8)$$

To prove this, note that

$$X(z) = \sum_{k=0}^{\infty} x(k)z^{-k}$$
$$= x(0) + x(1)z^{-1} + x(2)z^{-2} + \cdots$$

Letting $z \to \infty$, we obtain Eq. (13-8).

**Final value theorem.** If $x(t)$ has the $z$ transform $X(z)$, $X(z)$ has no double- or higher-order poles on the unit circle centered at the origin of the $z$ plane and no poles outside the unit circle [this is the condition for the stablility of $X(z)$, or the condition that the $x(k)$ $(k = 0, 1, 2, \ldots)$ remain finite (see Section 13-6)], then the final value of $x(t)$ or $x(k)$ is given by

$$\lim_{k \to \infty} x(k) = \lim_{t \to \infty} x(t) = \lim_{z \to 1} [(z - 1)X(z)] \qquad (13\text{-}9)$$

To prove this, note that

$$\mathcal{Z}[x(k)] = X(z) = \sum_{k=0}^{\infty} x(k)z^{-k}$$
$$\mathcal{Z}[x(k + 1)] = zX(z) - zx(0) = \sum_{k=0}^{\infty} x(k + 1)z^{-k}$$

Hence

$$zX(z) - zx(0) - X(z) = (z - 1)X(z) - zx(0)$$
$$= \sum_{k=0}^{\infty} x(k+1)z^{-k} - \sum_{k=0}^{\infty} x(k)z^{-k}$$

from which we obtain

$$(z - 1)X(z) = zx(0) + \sum_{k=0}^{\infty} [x(k+1) - x(k)]z^{-k}$$

Because of the assumed stability condition, we obtain, as $z \rightarrow 1$,

$$\lim_{z \to 1} [(z - 1)X(z)] = x(0) + x(\infty) - x(0) = x(\infty)$$

which is Eq. (13-9).

## 13-4 THE INVERSE $z$ TRANSFORMATION

Given $X(z)$, there are three methods of obtaining the inverse $z$ transform, $x(kT)$ or $x(k)$. The three methods are based on expansion by infinite power series, partial fraction expansion, and the inversion integral. In obtaining the inverse $z$ transform, we assume as usual that the time series $x(kT)$ or $x(k)$ is zero for $k < 0$.

**Obtaining the inverse $z$ transform by expanding $X(z)$ into an infinite power series.** If $X(z)$ is expanded into a convergent power series in $z^{-1}$; namely,

$$X(z) = \sum_{k=0}^{\infty} x(kT)z^{-k}$$
$$= x(0) + x(T)z^{-1} + x(2T)z^{-2} + \cdots + x(kT)z^{-k} + \cdots$$

then the values of $x(kT)$ can be determined by inspection.

If $X(z)$ is given in the form of a rational function, the expansion into an infinite power series can be accomplished by simply dividing the denominator into the numerator. If the resulting series is convergent, the coefficients of the $z^{-k}$ in the series are the values $x(kT)$ of the time sequence. In obtaining the coefficients by long division, both the numerator and denominator of $X(z)$ must be written in ascending powers of $z^{-1}$.

Although the present method gives the values of $x(0)$, $x(T)$, $x(2T)$, ... in a sequential manner, it is usually difficult to obtain an expression for the general term from a set of values of $x(kT)$.

---

*Example 13-7.* Find $x(kT)$ for $k = 0, 1, 2, 3, 4$ when $X(z)$ is given by

$$X(z) = \frac{10z}{(z - 1)(z - 2)}$$

$X(z)$ can be written

$$X(z) = \frac{10z^{-1}}{1 - 3z^{-1} + 2z^{-2}}$$

By long division,

$$X(z) = 10z^{-1} + 30z^{-2} + 70z^{-3} + 150z^{-4} + \cdots$$

This infinite series converges. Therefore, we obtain by inspection

$$x(0) = 0$$
$$x(T) = 10$$
$$x(2T) = 30$$
$$x(3T) = 70$$
$$x(4T) = 150$$

**Obtaining the inverse $z$ transform by expanding $X(z)$ into partial fractions.** An alternate method for obtaining the $x(kT)$ is based upon the partial fraction expansion of $X(z)/z$ and the identification of each of the terms by the use of a table of $z$ transforms. (Tables of complicated $z$ transforms may not be readily available. Hence polynomials in $z$ may have to be expanded into partial fractions before the inverse $z$ transforms can be obtained.) Note that the reason we expand $X(z)/z$ into partial fractions is that the functions of $z$ appearing in tables of $z$ transforms usually have the factor $z$ in their numerators.

Consider $X(z)$ given by

$$X(z) = \frac{b_0 z^m + b_1 z^{m-1} + \cdots + b_{m-1} z + b_m}{a_0 z^n + a_1 z^{n-1} + \cdots + a_{n-1} z + a_n} \qquad (m \le n)$$

We first factor the denominator polynomial of $X(z)$ and find the poles of $X(z)$. Then we expand $X(z)/z$ into partial fractions so that each of the terms is easily recognizable in a table of $z$ transforms. The inverse $z$ transform of $X(z)$ is obtained as the sum of the inverse $z$ transforms of the partial fractions.

---

*Example 13-8.* Find $x(kT)$ if $X(z)$ is given by

$$X(z) = \frac{10z}{(z-1)(z-2)}$$

We first expand $X(z)/z$ into partial fractions as follows:

$$\frac{X(z)}{z} = \frac{10}{(z-1)(z-2)}$$

$$= \frac{-10}{z-1} + \frac{10}{z-2}$$

Then we obtain

$$X(z) = -\frac{10z}{z-1} + \frac{10z}{z-2}$$

From Table 13-1 we obtain

$$\mathcal{Z}^{-1}\left[\frac{z}{z-1}\right] = 1, \qquad \mathcal{Z}^{-1}\left[\frac{z}{z-2}\right] = 2^k$$

Hence

$$x(kT) = 10(-1 + 2^k) \qquad (k = 0, 1, 2, \ldots)$$

or

$$x(0) = 0$$
$$x(T) = 10$$
$$x(2T) = 30$$
$$x(3T) = 70$$
$$x(4T) = 150$$
$$\cdot \ \cdot \ \cdot$$

Clearly, the result checks with the $x(kT)$ obtained in Example 13-7.

**Obtaining the inverse $z$ transform by the inversion integral.** The third method of finding the inverse $z$ transform is to use the inversion integral. From Eq. (13-5), we have

$$X(z) = \sum_{k=0}^{\infty} x(kT)z^{-k}$$
$$= x(0) + x(T)z^{-1} + x(2T)z^{-2} + \cdots + x(kT)z^{-k} + \cdots$$

By multiplying both sides of this last equation by $z^{k-1}$, we obtain

$$X(z)z^{k-1} = x(0)z^{k-1} + x(T)z^{k-2} + x(2T)z^{k-3} + \cdots + x(kT)z^{-1} + \cdots \quad (13\text{-}10)$$

Note that because

$$z = e^{Ts}$$

if $s = \sigma + j\omega$ is substituted in this last equation, we obtain $z = e^{T(\sigma + j\omega)}$ or

$$|z| = e^{T\sigma}, \qquad \underline{/z} = \omega T$$

If the poles of $\mathscr{L}[x]$ lie to the left of the line $s = \sigma_1$ in the $s$ plane, the poles of $\mathscr{Z}[x]$ will lie inside the circle with its center at the origin and radius equal to $e^{T\sigma_1}$ in the $z$ plane.

Suppose we integrate both sides of Eq. (13-10) along this circle in the counter-clockwise direction:

$$\oint X(z)z^{k-1}\,dz = \oint x(0)z^{k-1}\,dz + \oint x(T)z^{k-2}\,dz + \cdots + \oint x(kT)z^{-1}\,dz + \cdots$$

Applying Cauchy's theorem, we see that all terms on the right-hand side of this last equation are zero except one term

$$\oint x(kT)z^{-1}\,dz$$

Hence

$$\oint X(z)z^{k-1}\,dz = \oint x(kT)z^{-1}\,dz$$

from which we obtain

$$x(kT) = \frac{1}{2\pi j}\oint X(z)z^{k-1}\,dz \quad (13\text{-}11)$$

Equation (13-11) is the inversion integral for the $z$ transform. Equation (13-11) is equivalent to stating that

$$x(kT) = \sum \, [\text{residues of } X(z)z^{k-1} \text{ at the poles of } X(z)] \quad (13\text{-}12)$$

*Example 13-9.* Obtain $x(kT)$ by use of the inversion integral when $X(z)$ is given by

$$X(z) = \frac{10z}{(z-1)(z-2)}$$

From Eqs. (13-11) and (13-12),

$$x(kT) = \frac{1}{2\pi j} \oint \left[ \frac{10z}{(z-1)(z-2)} z^{k-1} \right] dz$$

$$= \frac{1}{2\pi j} \oint \left[ -\frac{10z^k}{z-1} + \frac{10z^k}{z-2} \right] dz$$

$$= \left( \text{residue of } -\frac{10z^k}{z-1} \text{ at pole } z = 1 \right) + \left( \text{residue of } \frac{10z^k}{z-2} \text{ at pole } z = 2 \right)$$

$$= 10(-1 + 2^k) \qquad (k = 0, 1, 2, \ldots)$$

## 13-5 PULSE TRANSFER FUNCTIONS

This section presents the basic material necessary for analyzing discrete-time systems, or sampled-data systems, by the z-transform method.

In analyzing discrete-time systems by this method, it is important to note that although the values of the system response at the sampling instants are correct, the system response obtained by the z-transform method may not portray the correct time-response behavior of the actual system unless the transfer function $G(s)$ of the continuous part of the system has at least two more poles than zeros, so that $\lim_{s \to \infty} sG(s) = 0$.

**Sampling theorem.** Shannon's sampling theorem presented here is important in designing discrete-time systems since it gives the minimum sampling frequency necessary to reconstruct the original signal from a sampled signal.

We shall assume that a continuous signal $x(t)$ has the frequency spectrum as shown in Fig. 13-6. This signal $x(t)$ does not contain any frequency components above $\omega_1$ rad/sec.

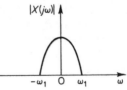

**Fig. 13-6.** A frequency spectrum.

*Sampling theorem.* If $\omega_s = 2\pi/T$, where $T$ is the sampling period, is greater than $2\omega_1$, or

$$\omega_s > 2\omega_1$$

where $2\omega_1$ corresponds to the frequency spectrum of the continuous signal $x(t)$, then the signal $x(t)$ can be reconstructed completely from the sampled signal $x^*(t)$.

We shall demonstrate this when $\lim_{s \to \infty} sX(s) = x(0+) = 0$. Let us define

$$X(s) = \mathscr{L}[x(t)]$$

Referring to Problem A-13-5 the Laplace transform of the sampled signal $x^*(t)$ is given by

$$X^*(s) = \frac{1}{T} \sum_{k=-\infty}^{\infty} X(s + j\omega_s k)$$

By substituting $s = j\omega$ into this last equation, we obtain the frequency spectrum for $X^*(s)$ as follows:

$$|X^*(j\omega)| = \frac{1}{T}\left|\sum_{k=-\infty}^{\infty} X[j(\omega + \omega_s k)]\right|$$

$$= \cdots + \frac{1}{T}|X[j(\omega - \omega_s)]| + \frac{1}{T}|X(j\omega)|$$

$$+ \frac{1}{T}|X[j(\omega + \omega_s)]| + \cdots$$

Figure 13-7 shows plots of $|X^*(j\omega)|$ versus $\omega$ for two values of $T$. Each plot of

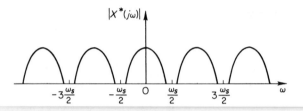

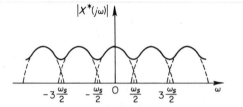

Fig. 13-7. Plots of $|X^*(j\omega)|$ versus $\omega$.

$|X^*(j\omega)|$ versus $\omega$ consists of $|X(j\omega)|$ repeated every $\omega_s = 2\pi/T$ rad/sec. In the frequency spectrum, the component $|X(j\omega)|/T$ is called the primary component and the other components $|X[j(\omega \pm \omega_s k)]|/T$ are called complementary components.

If $\omega_s > 2\omega_1$ or $T < (\pi/\omega_1)$, no two components of $|X^*(j\omega)|$ will overlap. Thus the original shape of $|X(j\omega)|$ is preserved by the sampling process.

If $\omega_s < 2\omega_1$ or $(\pi/\omega_1) < T$, then the original shape of $|X(j\omega)|$ no longer appears in the plot of $|X^*(j\omega)|$ versus $\omega$.

Therefore, we see that the continuous signal $x(t)$ can be reproduced from the sampled signal $x^*(t)$ by filtering if and only if $\omega_s > 2\omega_1$ or $T < (\pi/\omega_1)$.

Hence, if the sampling period $T$ is smaller than $\pi/\omega_1$, then the continuous signal can be reconstructed by use of a low-pass filter, after the signal has been sampled. If the low-pass filter has a characteristic such that it passes only signals having frequencies lower than $\omega_1$, then we can obtain the frequency spectrum at the output of the filter exactly as $1/T$ times $|X(j\omega)|$.

Note that in most sampled-data or discrete-time systems, the circuit following the sampler has the characteristic of a low-pass filter. This reduces the high-frequency components, smoothes out the sampled signal, and roughly reproduces the original continuous signal.

**Frequency response characteristics of zero-order holding devices.** In order to attenuate the complementary components introduced by the sampler, the sampled signal is usually fed to a hold circuit, a low-pass filter.

If the signal between two consecutive sampling instants is approximated by an $n$th-degree polynomial, then the holding device is called an $n$th-order holding device. In this book, we consider only zero-order holding devices. In such devices, the output signal $x(t)$ is approximated by a zeroth-degree polynomial, or constant. Hence,

$$x(t) = x(kT) \qquad \text{for } kT \le t < (k+1)T$$

where $k = 0, 1, 2, \ldots$. The transfer function of a zero-order holding device is

$$G_h(s) = \frac{1 - e^{-Ts}}{s}$$

By substituting $s = j\omega$ in this transfer function, we obtain

$$G_h(j\omega) = \frac{1 - e^{-Tj\omega}}{j\omega}$$

$$= \frac{2e^{-Tj\omega/2}(e^{Tj\omega/2} - e^{-Tj\omega/2})}{2j\omega}$$

$$= T\frac{\sin\left(\frac{\omega T}{2}\right)}{\frac{\omega T}{2}}e^{-Tj\omega/2}$$

In terms of the sampling frequency $\omega_s = 2\pi/T$,

$$G_h(j\omega) = \frac{2\pi}{\omega_s}\frac{\sin\left(\frac{\pi\omega}{\omega_s}\right)}{\frac{\pi\omega}{\omega_s}}e^{-j\pi(\omega/\omega_s)}$$

Figure 13-8 shows the frequency-response curve of the zero-order holding device. Clearly, it has the characteristics of a low-pass filter.

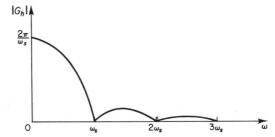

**Fig. 13-8.** Frequency-response curve of the zero-order holding device.

**Convolution summation.** Consider the discrete-time system shown in Fig. 13-9. Here the sequence of impulses $x^*(t)$ is the input to the continuous-time plant whose transfer function is $G(s)$. The output of the plant is a continuous signal $y(t)$. If at the output there is another sampler, which is synchronized in phase with the

**Fig. 13-9.** Discrete-time system.

input sampler and operates at the same sampling period, then the output is a train of impulses. If $y(t) = 0$ for $t < 0$, then the $z$ transform of $y(t)$ is

$$\mathcal{Z}[y(t)] = Y(z) = \sum_{k=0}^{\infty} y(kT)z^{-k} \tag{13-13}$$

In the absence of the output sampler, if we consider a fictitious sampler (which is synchronized in phase with the input sampler and operates at the same sampling period) at the output and observe the sequence of values taken by $y$ only at the instants $t = kT$, then the $z$ transform of the output can also be given by Eq. (13-13).

It should be noted that, in feeding the output signal back to the input, i.e., closing the loop, the existence or nonexistence of the output sampler within the loop makes a difference as to the system behavior. (If the output sampler is outside the loop, it will make no difference to the closed-loop operation.)

For a linear time-invariant stable plant, it is a well-known fact that the output $y(t)$ of the plant is related to the input $x(t)$ by the convolution integral, or

$$y(t) = \int_0^t g(t - \tau)x(\tau)\, d\tau = \int_0^t x(t - \tau)g(\tau)\, d\tau$$

where $g(t)$ is the impulse-response function of the plant. For discrete-time systems we have a convolution summation, which is similar to the convolution integral. Since $x^*(t)$ is a train of impulses, the response of the plant to the input $x^*(t)$ is the sum of the individual impulse responses. Hence for $0 \le t \le kT$,

$$y(t) = g(t)x(0) + g(t - T)x(T) + g(t - 2T)x(2T) + \cdots + g(t - kT)x(kT)$$
$$= \sum_{h=0}^{k} g(t - hT)x(hT)$$

The values of the output $y(t)$ at the sampling instants $t = kT$ $(k = 0, 1, 2, \ldots)$ are given by

$$y(kT) = \sum_{h=0}^{k} g(kT - hT)x(hT) \tag{13-14}$$

$$= \sum_{h=0}^{k} x(kT - hT)g(hT) \tag{13-15}$$

The summation in Eq. (13-14) or (13-15) is called convolution summation. Note that the simplified notation

$$y(kT) = x(kT) * g(kT)$$

is often used for convolution summation.

**Pulse transfer functions.** We shall first note that $y(kT)$ can also be written

$$y(kT) = \sum_{h=0}^{\infty} g(kT - hT)x(hT) \qquad (k = 0, 1, 2, \ldots)$$

since $g(kT - hT) = 0$ for $h > k$. Hence

$$Y(z) = \sum_{k=0}^{\infty} y(kT)z^{-k}$$

$$= \sum_{k=0}^{\infty} \sum_{h=0}^{\infty} g(kT - hT)x(hT)z^{-k}$$

$$= \sum_{m=0}^{\infty} \sum_{h=0}^{\infty} g(mT)x(hT)z^{-(m+h)}$$

$$= \sum_{m=0}^{\infty} g(mT)z^{-m} \sum_{h=0}^{\infty} x(hT)z^{-h}$$

$$= \sum_{m=0}^{\infty} g(mT)z^{-m} X(z)$$

$$= G(z)X(z) \tag{13-16}$$

where

$$G(z) = \sum_{k=0}^{\infty} g(kT)z^{-k}$$

$$= g(0) + g(T)z^{-1} + g(2T)z^{-2} + \cdots$$

$$= z \text{ transform of } g(t)$$

Equation (13-16) relates the pulsed output of the system to its pulse input. The function $G(z)$ where

$$G(z) = \frac{Y(z)}{X(z)}$$

is called the $z$-transfer function, or pulse transfer function, of the discrete-time system. Figure 13-10 shows a block diagram for a pulse transfer function $G(z)$, together with the input $X(z)$ and the output $Y(z)$.

**Fig. 13-10.** Block diagram for a pulse transfer function system.

**General procedure for obtaining pulse transfer functions.** The pulse transfer function of a system may be obtained by the following procedure:

1. Obtain the transfer function $G(s)$ of the system.
2. Obtain the impulse-response function $g(t)$, where $g(t) = \mathscr{L}^{-1}[G(s)]$.
3. Evaluate

$$G(z) = \sum_{k=0}^{\infty} g(kT)z^{-k}$$

where $g(kT)$ is obtained from $g(t)$ by substituting $kT$ for $t$. For stable systems, the infinite series converges.

It is important to remember that in the $z$-transform approach, or pulse transfer function approach, to the analysis of discrete-time systems, the sampled signal is assumed to be a train of impulses whose strengths, or areas, are equal to the continuous-time signal at the sampling instants. Such assumption is valid only if the sampling duration of the sampler is small compared with the largest time constant of the system.

*Example 13-10.* Obtain the pulse transfer function of the system shown in Fig. 13-11.
The transfer function $G(s)$ can be expanded in partial fractions as follows:

$$G(s) = \frac{K}{(s + a)(s + b)}$$

$$= \frac{K}{b - a}\left(\frac{1}{s + a} - \frac{1}{s + b}\right)$$

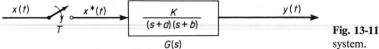

**Fig. 13-11.** Discrete-time system.

The impulse-response function is then obtained as

$$g(t) = \frac{K}{b - a}(e^{-at} - e^{-bt})$$

Hence

$$g(kT) = \frac{K}{b - a}(e^{-akT} - e^{-bkT})$$

Then $G(z)$ is obtained as

$$G(z) = \sum_{k=0}^{\infty} \frac{K}{b - a}(e^{-akT} - e^{-bkT})z^{-k} \qquad (13\text{-}17)$$

Since

$$\sum_{k=0}^{\infty} e^{-akT}z^{-k} = 1 + \sum_{k=1}^{\infty} e^{-akT}z^{-k}$$

$$= 1 + \sum_{k=1}^{\infty} e^{-aT}z^{-1}(e^{-akT+aT}z^{-k+1})$$

$$= 1 + e^{-aT}z^{-1} \sum_{k=0}^{\infty} e^{-akT}z^{-k}$$

we obtain

$$\sum_{k=0}^{\infty} e^{-akT}z^{-k} = \frac{1}{1 - e^{-aT}z^{-1}} \qquad (13\text{-}18)$$

Substituting Eq. (13-18) into Eq. (13-17), we obtain the following pulse transfer function $G(z)$:

$$G(z) = \frac{K}{b - a}\left(\frac{1}{1 - e^{-aT}z^{-1}} - \frac{1}{1 - e^{-bT}z^{-1}}\right)$$

$$= \frac{K}{b - a}\frac{z(e^{-aT} - e^{-bT})}{(z - e^{-aT})(z - e^{-bT})}$$

*Example 13-11.* Obtain the pulse transfer function of the system shown in Fig. 13-12.
From the block diagram, we obtain

$$G(s) = \frac{C(s)}{M^*(s)} = \frac{1 - e^{-s}}{s^2(s + 1)}$$

$$= (1 - e^{-s})\left(\frac{1}{s^2} - \frac{1}{s} + \frac{1}{s + 1}\right)$$

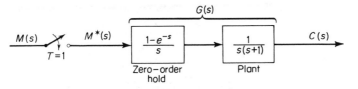

**Fig. 13-12.** Discrete-time system.

Hence

$$g(t) = (t - 1 + e^{-t})1(t) - (t - 1 - 1 + e^{-(t-1)})1(t - 1)$$

Since $T = 1$, we have $kT = k$ and

$$g(k) = (k - 1 + e^{-k}) - (k - 2 + e^{-(k-1)})$$
$$= e^{-k} + 1 - e^{-(k-1)} \qquad (k = 1, 2, 3, \ldots)$$
$$g(0) = 0$$

Thus $G(z)$ is obtained as

$$G(z) = \sum_{k=0}^{\infty} g(k)z^{-k}$$

$$= \sum_{k=0}^{\infty} (e^{-k} + 1 - e^{-(k-1)})z^{-k} + e - 2$$

$$= \frac{1 - e}{1 - e^{-1}z^{-1}} + \frac{1}{1 - z^{-1}} + e - 2$$

$$= \frac{e^{-1}z^{-1} + (1 - 2e^{-1})z^{-2}}{1 - (1 + e^{-1})z^{-1} + e^{-1}z^{-2}}$$

$$= \frac{e^{-1}z + 1 - 2e^{-1}}{z^2 - (1 + e^{-1})z + e^{-1}}$$

**Pulse transfer functions of cascaded elements.** It is important to note that the pulse transfer functions of the systems shown in Figs. 13-13 (a) and (b) are different.

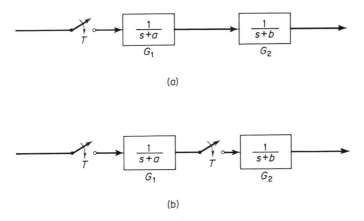

(a)

(b)

**Fig. 13-13.** (a) Discrete-time system with one sampler; (b) discrete-time system with two synchronized samplers.

For the system shown in Fig. 13-13 (a), the pulse transfer function is, from Example 13-10 (with $K = 1$),

$$G_1G_2(z) = \frac{z(e^{-aT} - e^{-bT})}{(b - a)(z - e^{-aT})(z - e^{-bT})}$$

For the system shown in Fig. 13-13 (b) (assuming that the two samplers are synchronized and have the same sampling period), we obtain

$$G_1(z)G_2(z) = \frac{z^2}{(z - e^{-aT})(z - e^{-bT})}$$

Thus

$$G_1G_2(z) \neq G_1(z)G_2(z)$$

Therefore, we must be careful and observe whether or not there is a sampler between cascaded elements.

**Pulse transfer functions of closed-loop systems.** Consider the closed-loop system shown in Fig. 13-14. In this system, the actuating error is sampled. From the block

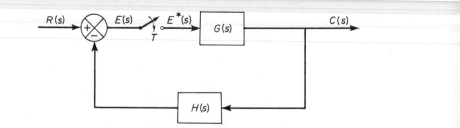

**Fig. 13-14.** Closed-loop discrete-time system.

diagram,

$$E(s) = R(s) - H(s)C(s)$$
$$C(s) = G(s)E^*(s)$$

Hence

$$E(s) = R(s) - G(s)H(s)E^*(s)$$

Then we obtain

$$E^*(s) = R^*(s) - GH^*(s)E^*(s)$$

or

$$E^*(s) = \frac{R^*(s)}{1 + GH^*(s)}$$

Since

$$C^*(s) = G^*(s)E^*(s)$$

we obtain

$$C^*(s) = \frac{G^*(s)R^*(s)}{1 + GH^*(s)}$$

In terms of the $z$ transform, $C(z)$ is given by

$$C(z) = \frac{G(z)R(z)}{1 + GH(z)} \tag{13-19}$$

The inverse $z$ transform of Eq. (13-19) gives the values of the output at the sampling instants. The pulse transfer function of the present closed-loop system is

$$\frac{C(z)}{R(z)} = \frac{G(z)}{1 + GH(z)} \tag{13-20}$$

Table 13-3 shows five typical configurations of closed-loop discrete-time systems. For each configuration, the corresponding output $C(z)$ is shown.

**Table 13-3.** TYPICAL CONFIGURATIONS OF CLOSED-LOOP DISCRETE-TIME SYSTEMS AND THE CORRESPONDING OUTPUTS $C(z)$

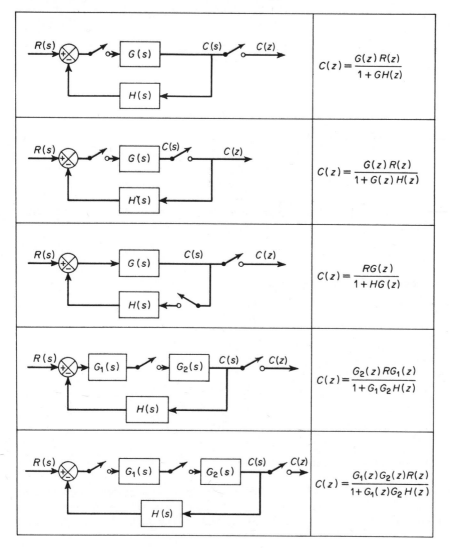

*Example 13-12.* Obtain the unit-step response of the system shown in Fig. 13-15.

The pulse transfer function of this closed-loop system is

$$\frac{C(z)}{R(z)} = \frac{G(z)}{1 + G(z)}$$

Since

$$G(s) = \frac{1 - e^{-s}}{s^2(s + 1)}$$

referring to Example 13-11, we obtain

$$G(z) = \frac{e^{-1}z + 1 - 2e^{-1}}{z^2 - (1 + e^{-1})z + e^{-1}}$$

$$= \frac{0.368z + 0.264}{z^2 - 1.368z + 0.368}$$

Thus

$$\frac{C(z)}{R(z)} = \frac{0.368z + 0.264}{z^2 - z + 0.632}$$

For a unit-step input,

$$R(z) = \frac{z}{z - 1}$$

The output $C(z)$ is then obtained as follows:

$$C(z) = \frac{(0.368z + 0.264)z}{(z^2 - z + 0.632)(z - 1)}$$

$$= \frac{0.368z^2 + 0.264z}{z^3 - 2z^2 + 1.632z - 0.632}$$

$$= \frac{0.368z^{-1} + 0.264z^{-2}}{1 - 2z^{-1} + 1.632z^{-2} - 0.632z^{-3}}$$

$$= 0.368z^{-1} + z^{-2} + 1.4z^{-3} + 1.4z^{-4}$$
$$+ 1.147z^{-5} + 0.895z^{-6} + 0.802z^{-7} + \cdots$$

The inverse $z$ transform of $C(z)$ gives

$$c(0) = 0$$
$$c(1) = 0.368$$
$$c(2) = 1$$
$$c(3) = 1.4$$
$$c(4) = 1.4$$
$$c(5) = 1.147$$
$$c(6) = 0.895$$
$$c(7) = 0.802$$
$$\cdot \ \cdot \ \cdot$$

The output $c(k)$ is plotted in Fig. 13-16.

The $z$-transform analysis will not give information on the response between sampling instants. Thus the smooth dotted curve connecting the data points is only approximate.

If we wish to obtain information on the response between sampling instants by the $z$-transform approach, we need to modify the method. The modified $z$-transform method,

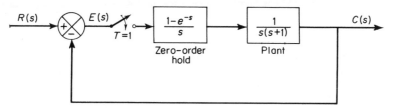

**Fig. 13-15.** Closed-loop discrete-time system.

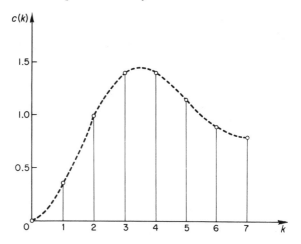

**Fig. 13-16.** Plot of $c(k)$ versus $k$.

the submultiple sampling method, etc., enable us to obtain the system response between sampling instants. (The reader interested in such modified methods should refer to specialized books on sampled-data systems.)

## 13-6　STABILITY ANALYSIS IN THE z PLANE

This section presents the stability analysis of discrete-time systems in the $z$ plane. Stability analysis of such systems by the second method of Liapunov will be discussed in Section 15-4.

**Mapping of the left-half $s$ plane into the $z$ plane.** A linear dynamic system is stable if all poles of the transfer function lie in the left-half $s$ plane. In the $z$ plane, the left-half $s$ plane corresponds to the unit circle centered at the origin, or the left-half $s$ plane maps into the inside of the unit circle in the $z$ plane. This can be proved easily. Since

$$z = e^{Ts}, \qquad s = \sigma + j\omega$$

we obtain

$$|z| = e^{T\sigma}, \qquad \underline{/z} = \omega T$$

In the left-half $s$ plane, $\sigma < 0$. Therefore, the magnitude of $z$ varies between 0 and 1. The imaginary axis, or $\sigma = 0$, corresponds to the unit circle in the $z$ plane. The inside of the circle corresponds to the left-half $s$ plane.

Note that since $\underline{/z} = \omega T$, the angle of $z$ varies from $-\infty$ to $\infty$ as $\omega$ varies from $-\infty$ to $\infty$. Consider a representative point on the $j\omega$ axis in the $s$ plane. As this point moves from $-\pi/T$ to $\pi/T$ on the $j\omega$ axis, we have $|z| = 1$, and $\underline{/z}$ varies from $-\pi$ to $\pi$ in the counterclockwise direction in the $z$ plane. As the representative point moves from $\pi/T$ to $3\pi/T$ on the $j\omega$ axis, the corresponding point in the $z$ plane traces out the unit circle once in the counterclockwise direction. Thus, as the point in the $s$ plane moves from $-\infty$ to $\infty$ on the $j\omega$ axis, we trace the unit circle in the $z$ plane an infinite number of times.

From this analysis, it is clear that each strip of width $2\pi/T$ in the left-half $s$ plane maps into the inside of the unit circle in the $z$ plane, as shown in Fig. 13-17 (a). Figures 13-17 (b) and (c) show corresponding regions in the $s$ and $z$ planes.

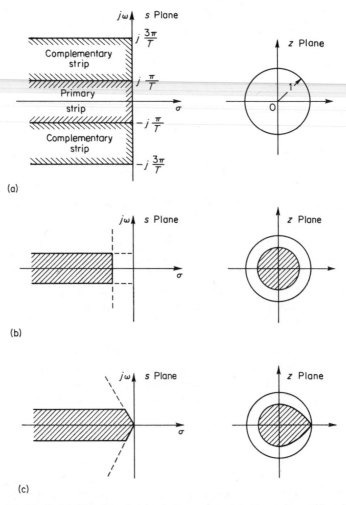

**Fig. 13-17.** (a) Mapping of strips in the $s$ plane into the $z$ plane; (b) and (c) regions in the $s$ plane and the corresponding regions in the $z$ plane.

**Stability analysis.** We shall now discuss the stability of the closed-loop system shown in Fig. 13-14 in the $z$ plane. The output-input ratio of the error-sampled system is given by Eq. (13-20), rewritten

$$\frac{C(z)}{R(z)} = \frac{G(z)}{1 + GH(z)}$$

The stability of such a system can be determined from the location of the roots of the characteristic equation

$$1 + GH(z) = 0 \tag{13-21}$$

Since this polynomial in $z$ can be converted into a ratio of two polynomials in $z$, for stability all the roots $z_i$ of the characteristic equation, Eq. (13-21), must lie inside the unit circle, or

$$|z_i| < 1$$

The closed-loop system $C(z)/R(z)$ becomes unstable if any of the closed-loop poles lie outside the unit circle and/or any multiple poles lie on the unit circle.

A few methods are available for determining whether or not a polynomial in $z$ contains a root or roots on or outside the unit circle. One method is to modify the Routh stability criterion. The Routh stability criterion tells us whether or not any of the roots of a polynomial lie in the right half of the complex plane. Since the following transformation

$$z = \frac{r + 1}{r - 1}$$

maps the interior of the unit circle in the $z$ plane to the left-half $r$ plane, with this transformation, the Routh stability criterion may be applied to the polynomial in $r$ in the same manner as continuous-time systems. Example 13-13 illustrates this approach. Another approach is to apply the Schur-Cohn stability criterion. (For a discussion of this criterion, see a specialized book on sampled-data systems.)

Alternatively, by writing a state-space equation for the discrete-time system, we may apply state space methods to determine the stability. We shall discuss this matter in Chapters 14 and 15.

---

*Example 13-13.* Consider the discrete-time system shown in Fig. 13-18. The open-loop transfer function of the system is

$$G(s) = \frac{10}{s(s + 1)}$$

The $z$ transform of $G(s)$ is

$$G(z) = \frac{10(1 - e^{-1})z}{(z - 1)(z - e^{-1})}$$

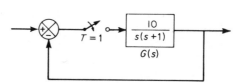

**Fig. 13-18.** Closed-loop discrete-time system.

The characteristic equation

$$1 + G(z) = 0$$

becomes

$$(z - 1)(z - e^{-1}) + 10(1 - e^{-1})z = 0$$

Noting that $e^{-1} = 0.368$, we can simplify the characteristic equation to

$$z^2 + 4.952z + 0.368 = 0 \tag{13-22}$$

from which the roots are found to be

$$z = -0.076, \qquad z = -4.876$$

Thus, one root of the characteristic equation has a magnitude greater than unity, and the system is unstable.

It is important to note that in the absence of a sampler, a second-order system is always stable. In the presence of a sampler, however, a second-order system such as this can become unstable for large values of the gain. In fact, it can be shown that the second-order system shown in Fig. 13-18 with

$$G(s) = \frac{K}{s(s + 1)}$$

is stable only for $0 < K < 4.32$. (See Problem A-13-14.)

In the present system, the characteristic equation is of quadratic form and can be solved easily. For a characteristic equation of higher degree, it is not simple to solve for $z$. In such a case, it is convenient to transform the characteristic equation in $z$ to a polynomial in $r$ by use of the transformation

$$z = \frac{r + 1}{r - 1} \tag{13-23}$$

and apply the Routh stability criterion to the resulting polynomial.

We shall illustrate this approach using the present system. By use of the transformation given by Eq. (13-23) we can write the characteristic equation, Eq. (13-22), as

$$\left(\frac{r + 1}{r - 1}\right)^2 + 4.952\left(\frac{r + 1}{r - 1}\right) + 0.368 = 0$$

or

$$6.32r^2 + 1.264r - 3.584 = 0$$

The Routh array becomes

$$
\begin{array}{lll}
r^2 & 6.32 & -3.584 \\
r^1 & 1.264 & 0 \\
r^0 & -3.584 &
\end{array}
$$

There is one change of sign in the first column of the Routh array. Thus, there is one root in the right-half $r$ plane, which implies that there is one root outside the unit circle in the $z$ plane. This corresponds to the result obtained previously.

Note that, in general, the stability of a discrete-time system is improved as the sampling period is shortened because increasing the sampling rate tends to make the system behave more like the corresponding continuous-time system. (In the present example, the corresponding continuous-time system—a second-order system—is stable for all positive values of $K$.)

## Example Problems and Solutions

**PROBLEM A-13-1.** Obtain the $z$ transforms of $a^k$ and $\mathbf{A}^k$, where $\mathbf{A}$ is an $n \times n$ matrix.

**Solution.** By definition, the $z$ transform of $a^k$ is

$$\mathcal{Z}[a^k] = \sum_{k=0}^{\infty} a^k z^{-k}$$
$$= 1 + az^{-1} + a^2 z^{-2} + \cdots$$
$$= \frac{1}{1 - az^{-1}}$$
$$= \frac{z}{z - a}$$

Similarly,

$$\mathcal{Z}[\mathbf{A}^k] = \sum_{k=0}^{\infty} \mathbf{A}^k z^{-k}$$
$$= \mathbf{I} + \mathbf{A}z^{-1} + \mathbf{A}^2 z^{-2} + \cdots$$
$$= (\mathbf{I} - \mathbf{A}z^{-1})^{-1}$$
$$= (z\mathbf{I} - \mathbf{A})^{-1}z$$

Note that $\mathbf{A}^k$ can be obtained by taking the inverse $z$ transform of $(z\mathbf{I} - \mathbf{A})^{-1}z$, or

$$\mathbf{A}^k = \mathcal{Z}^{-1}[(z\mathbf{I} - \mathbf{A})^{-1}z]$$

**PROBLEM A-13-2.** Consider the system shown in Fig. 13-19. Derive the difference equation describing the system dynamics when the input voltage applied is piecewise constant, or

$$e(t) = e(kT) \quad \text{for } kT \le t < (k + 1)T$$

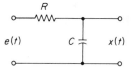

**Fig. 13-19.** Network system.

**Solution.** From Fig. 13-19, we obtain

$$RC\dot{x}(t) + x(t) = e(kT) \quad \text{for } kT \le t < (k + 1)T$$

Taking the Laplace transform of this last equation, considering $t = kT$ to be the initial time, we obtain

$$RC[sX(s) - x(kT)] + X(s) = \frac{e(kT)}{s}$$

or

$$X(s) = \frac{e(kT)}{s} + \frac{x(kT) - e(kT)}{s + \dfrac{1}{RC}}$$

The inverse Lapalce transform of $X(s)$ gives

$$x(t) = e(kT) + [x(kT) - e(kT)]e^{-(t-kT)/RC} \quad \text{for } kT \le t < (k + 1)T$$

Substituting $t = (k + 1)T - 0$ into this last equation and noting that $x(t)$ is a continuous function of time and therefore $x((k + 1)T - 0) = x((k + 1)T + 0) = x((k + 1)T)$, we obtain the desired difference equation, or

$$x((k + 1)T) = e^{-T/RC} x(kT) + (1 - e^{-T/RC})e(kT)$$

**PROBLEM A-13-3.** Given

$$X(z) = \frac{2z(z^2 - 1)}{(z^2 + 1)^2}$$

obtain $x(kT)$.

**Solution.** Expanding $X(z)/z$ into partial fractions, we obtain

$$\frac{X(z)}{z} = \frac{1}{(z + j)^2} + \frac{0}{z + j} + \frac{1}{(z - j)^2} + \frac{0}{z - j}$$

Hence

$$X(z) = \frac{z}{(z + j)^2} + \frac{z}{(z - j)^2}$$

$$= (e^{j(\pi/2)}) \frac{ze^{-j(\pi/2)}}{(z - e^{-j(\pi/2)})^2} + (e^{-j(\pi/2)}) \frac{ze^{j(\pi/2)}}{(z - e^{j(\pi/2)})^2}$$

Using Table 13-1, we obtain the inverse $z$ transform of $X(z)$ as follows:

$$x(kT) = e^{j(\pi/2)}ke^{-j(\pi/2)k} + e^{-j(\pi/2)}ke^{j(\pi/2)k}$$

$$= jk\left[\cos\frac{\pi k}{2} - j\sin\frac{\pi k}{2}\right] - jk\left[\cos\frac{\pi k}{2} + j\sin\frac{\pi k}{2}\right]$$

$$= 2k\sin\frac{\pi k}{2} \qquad (k = 0, 1, 2, \ldots)$$

**PROBLEM A-13-4.** Show that the Laplace transform of

$$x^*(t) = \sum_{k=0}^{\infty} x(t)\,\delta(t - kT)$$

is

$$X^*(s) = \sum_{k=0}^{\infty} x(kT)e^{-kTs}$$

$$= X(s) * \frac{1}{1 - e^{-Ts}}$$

$$= \frac{1}{2\pi j} \int_{c-j\infty}^{c+j\infty} X(p)\frac{1}{1 - e^{-T(s-p)}}\,dp$$

where $c$ is the abscissa of convergence for $X(s) = \mathscr{L}[x(t)]$.

**Solution.** First, we note that

$$\mathscr{L}[\delta(t - kT)] = e^{-kTs}$$

Thus

$$X^*(s) = \mathscr{L}[x^*(t)] = \mathscr{L}\left[\sum_{k=0}^{\infty} x(t)\,\delta(t - kT)\right]$$

$$= \mathscr{L}\left[\sum_{k=0}^{\infty} x(kT)\delta(t - kT)\right]$$

$$= \sum_{k=0}^{\infty} x(kT)e^{-kTs}$$

Since

$$\mathcal{L}[\sum_{k=0}^{\infty} \delta(t - kT)] = 1 + e^{-Ts} + e^{-2Ts} + \cdots$$

$$= \frac{1}{1 - e^{-Ts}}$$

we obtain

$$X^*(s) = \mathcal{L}[x^*(t)] = \mathcal{L}[\sum_{k=0}^{\infty} x(t) \, \delta(t - kT)]$$

$$= \mathcal{L}[x(t) \sum_{k=0}^{\infty} \delta(t - kT)]$$

$$= X(s) * \frac{1}{1 - e^{-Ts}}$$

The Laplace transform of the product of two functions $f(t)$ and $g(t)$ is

$$\mathcal{L}[f(t)g(t)] = \int_0^{\infty} f(t)g(t) \, e^{-st} \, dt$$

The inversion integral defined by Eq. (2-8) is

$$f(t) = \frac{1}{2\pi j} \int_{c-j\infty}^{c+j\infty} F(s) \, e^{st} \, ds \qquad (t > 0)$$

where $c$ is the abscissa of convergence for $F(s)$. Thus,

$$\mathcal{L}[f(t)g(t)] = \frac{1}{2\pi j} \int_0^{\infty} \int_{c-j\infty}^{c+j\infty} F(p) \, e^{pt} \, dp \, g(t) \, e^{-st} \, dt$$

Because of the uniform convergence of the integrals considered, we may invert the order of integration, or

$$\mathcal{L}[f(t)g(t)] = \frac{1}{2\pi j} \int_{c-j\infty}^{c+j\infty} F(p) \, dp \int_0^{\infty} g(t) \, e^{-(s-p)t} \, dt$$

Noting that

$$\int_0^{\infty} g(t) \, e^{-(s-p)t} \, dt = G(s - p)$$

we obtain

$$\mathcal{L}[f(t)g(t)] = \frac{1}{2\pi j} \int_{c-j\infty}^{c+j\infty} F(p)G(s - p) \, dp$$

Hence

$$\mathcal{L}[x^*(t)] = \mathcal{L}[x(t) \sum_{k=0}^{\infty} \delta(t - kT)]$$

$$= \frac{1}{2\pi j} \int_{c-j\infty}^{c+j\infty} X(p) \frac{1}{1 - e^{-T(s-p)}} \, dp$$

**PROBLEM A-13-5.** Assuming that the poles of $X(s)$ lie in the left-half $s$ plane and that $X(s)$ can be expressed as a ratio of the polynomials with a denominator of degree at least 2 greater than the degree of the numerator [which means that $\lim_{s \to \infty} sX(s) = x(0+) = 0$], evaluate

$$X^*(s) = \frac{1}{2\pi j} \int_{c-j\infty}^{c+j\infty} \frac{X(p)}{1 - e^{-T(s-p)}} \, dp$$

where the contour is a line parallel to the imaginary axis in the $p$ plane which separates the poles of $X(p)$ from those of $1/[1 - e^{-T(s-p)}]$, and show that

$$X(z) = \frac{1}{T} \sum_{k=-\infty}^{\infty} X\left(s + j\frac{2\pi}{T}k\right)\bigg|_{s=(1/T)\ln z}$$

**Solution.** The poles of $1/[1 - e^{-T(s-p)}]$ may be obtained by solving

$$1 - e^{-T(s-p)} = 0$$

so that

$$p = s \pm j\frac{2\pi}{T}k \qquad (k = 0, 1, 2, \ldots)$$

Thus there is an infinite number of poles.

In order to evaluate the given integral, let us choose the contour which consists of the line from $c - j\infty$ to $c + j\infty$ and the semicircle $\Gamma$ of an infinite radius in the right-half $p$ plane as shown in Fig. 13-20. The closed contour encloses all poles of $1/[1 - e^{-T(s-p)}]$, but it does not enclose any poles of $X(p)$.

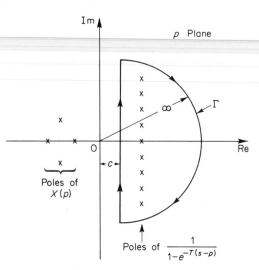

Fig. 13-20. $p$-plane contour.

Now $X^*(s)$ can be written

$$X^*(s) = \frac{1}{2\pi j} \int_{c-j\infty}^{c+j\infty} \frac{X(p)}{1 - e^{-T(s-p)}} \, dp$$

$$= \frac{1}{2\pi j} \oint \frac{X(p)}{1 - e^{-T(s-p)}} \, dp - \frac{1}{2\pi j} \int_{\Gamma} \frac{X(p)}{1 - e^{-T(s-p)}} \, dp$$

Let us consider the integral along the semicircle $\Gamma$. Since the degree of the denominator of $X(s)$ is at least 2 greater than the degree of the numerator, it can be shown that the value of this integral is zero, or

$$\frac{1}{2\pi j} \int_{\Gamma} \frac{X(p)}{1 - e^{-T(s-p)}} \, dp = 0$$

Therefore,

$$X^*(s) = \frac{1}{2\pi j} \oint \frac{X(p)}{1 - e^{-T(s-p)}} \, dp$$

The integral along the closed contour can be obtained by evaluating the residues. Hence

$$X^*(s) = -\sum_{k=-\infty}^{\infty} \frac{X(p)}{\frac{d}{dp}[1 - e^{-T(s-p)}]}\Bigg|_{p=s+j(2\pi/T)k}$$

$$= \frac{1}{T}\sum_{k=-\infty}^{\infty} X\left(s + j\frac{2\pi}{T}k\right)$$

Thus,

$$X(z) = \frac{1}{T}\sum_{k=-\infty}^{\infty} X\left(s + j\frac{2\pi}{T}k\right)\Bigg|_{s=(1/T)\ln z}$$

**Problem A-13-6.** Assuming that the poles of $X(s)$, a ratio of two polynomials in $s$, lie in the left-half $s$ plane and $\lim_{s\to\infty} X(s) = 0$, evaluate

$$X^*(s) = \frac{1}{2\pi j}\int_{c-j\infty}^{c+j\infty} \frac{X(p)}{1 - e^{-T(s-p)}}\, dp$$

and show that

$$X(z) = \sum_{i=1}^{m}\left\{\frac{1}{(n_i - 1)!}\frac{d^{n_i-1}}{ds^{n_i-1}}\left[\frac{(s - s_i)^{n_i}X(s)z}{z - e^{Ts}}\right]\right\}_{s=s_i}$$

where the $s_i$ are the poles of $X(s)$ and $n_i$ is the order of the pole at $s = s_i$, or

$$X(z) = \sum\left[\text{residues of } \frac{X(s)z}{z - e^{Ts}} \text{ at the poles of } X(s)\right]$$

**Solution.** As obtained in Problem A-13-5, the poles of $1/[1 - e^{-T(s-p)}]$ in the $p$ plane are

$$p = s \pm j\frac{2\pi}{T}k \qquad (k = 0, 1, 2, \ldots)$$

In order to evaluate the given integral, let us write

$$X^*(s) = \frac{1}{2\pi j}\int_{c-j\infty}^{c+j\infty} \frac{X(p)}{1 - e^{-T(s-p)}}\, dp$$

$$= \frac{1}{2\pi j}\oint \frac{X(p)}{1 - e^{-T(s-p)}}\, dp - \frac{1}{2\pi j}\int_{\Gamma} \frac{X(p)}{1 - e^{-T(s-p)}}\, dp$$

The closed contour consists of the line from $c - j\infty$ to $c + j\infty$ and the semicircle $\Gamma$ of infinite radius in the left half $p$ plane as shown in Fig. 13-21. In the present case we find that the integral along the infinite semicircle $\Gamma$ is not always zero. If it is not zero, consider $X(p)$ as $\lim_{\varepsilon\to 0} e^{\varepsilon p}X(p)$, $(\varepsilon > 0)$. Then we can show that the integral along $\Gamma$ is zero. Thus

$$\frac{1}{2\pi j}\int_{\Gamma} \frac{X(p)}{1 - e^{-T(s-p)}}\, dp = 0$$

Hence

$$X^*(s) = \frac{1}{2\pi j}\oint \frac{X(p)}{1 - e^{-T(s-p)}}\, dp$$

Since the poles of $X(p)$ are $p_1, p_2, \ldots p_m$,

$$X^*(s) = \sum\left[\text{residues of } \frac{X(p)}{1 - e^{-T(s-p)}} \text{ at the poles of } X(p)\right]$$

$$= \sum_{i=1}^{m}\frac{1}{(n_i - 1)!}\frac{d^{n_i-1}}{dp^{n_i-1}}\left[(p - p_i)^{n_i}\frac{X(p)}{1 - e^{-T(s-p)}}\right]_{p=p_i}$$

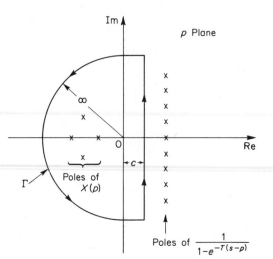

Poles of $\dfrac{1}{1-e^{-T(s-p)}}$      **Fig. 13-21.** $p$-plane contour.

By substituting $z = e^{Ts}$ into this last equation, we obtain

$$X(z) = \sum_{i=1}^{m} \frac{1}{(n_i - 1)!} \frac{d^{n_i-1}}{dp^{n_i-1}} \left[ (p - p_i)^{n_i} X(p) \frac{z}{z - e^{Tp}} \right]_{p=p_i}$$

By changing the complex variable from $p$ to $s$, we obtain

$$X(z) = \sum_{i=1}^{m} \frac{1}{(n_i - 1)!} \frac{d^{n_i-1}}{ds^{n_i-1}} \left[ (s - s_i)^{n_i} X(s) \frac{z}{z - e^{Ts}} \right]_{s=s_i}$$

$$= \sum \left[ \text{residues of } \frac{X(s)z}{z - e^{Ts}} \text{ at the poles of } X(s) \right]$$

**PROBLEM A-13-7.** Obtain the $z$ transform of $te^{-at}$ where $t \geq 0$. Note that $\mathscr{L}[te^{-at}] = 1/(s + a)^2$.

**Solution**

$$F(z) = \left[ \text{residue of } \frac{X(s)z}{z - e^{Ts}} \text{ at } s = -a \right]$$

$$= \left[ \text{residue of } \frac{z}{(s + a)^2(z - e^{Ts})} \text{ at } s = -a \right]$$

$$= \frac{d}{ds} \left[ \frac{z}{z - e^{Ts}} \right]_{s=-a}$$

$$= \frac{Tze^{Ts}}{(z - e^{Ts})^2} \bigg|_{s=-a}$$

$$= \frac{Tze^{-aT}}{(z - e^{-aT})^2}$$

**PROBLEM A-13-8.** Given

$$X(s) = \frac{s + 3}{(s + 1)(s + 2)}$$

obtain $X(z)$.

**Solution**

$$X(z) = \left[ \text{residue of } \frac{(s+3)z}{(s+1)(s+2)(z-e^{Ts})} \text{ at } s = -1 \right]$$
$$+ \left[ \text{residue of } \frac{(s+3)z}{(s+1)(s+2)(z-e^{Ts})} \text{ at } s = -2 \right]$$
$$= \frac{(s+3)z}{(s+2)(z-e^{Ts})}\bigg|_{s=-1} + \frac{(s+3)z}{(s+1)(z-e^{Ts})}\bigg|_{s=-2}$$
$$= \frac{2z}{z-e^{-T}} - \frac{z}{z-e^{-2T}}$$

**PROBLEM A-13-9.** Prove that

$$\mathcal{Z}[\sum_{k=0}^{n} x(k)] = \frac{z}{z-1} X(z)$$

$$\mathcal{Z}[\sum_{k=0}^{n-1} x(k)] = \frac{1}{z-1} X(z)$$

Then prove that

$$\sum_{k=0}^{\infty} x(k) = \lim_{z \to 1} X(z)$$

**Solution.** The $z$ transform of the first forward difference is

$$\mathcal{Z}[\Delta f(k)] = \mathcal{Z}[f(k+1)] - \mathcal{Z}[f(k)]$$
$$= [z F(z) - z f(0)] - F(z)$$
$$= (z-1) F(z) - z f(0)$$

Let us write

$$x(n) = \sum_{k=0}^{n} x(k) - \sum_{k=0}^{n-1} x(k) = \Delta \sum_{k=0}^{n-1} x(k)$$

Then

$$\mathcal{Z}[\Delta \sum_{k=0}^{n-1} x(k)] = \mathcal{Z}[x(n)] = X(z) = (z-1) \mathcal{Z}[\sum_{k=0}^{n-1} x(k)]$$

since the value of the sum $\sum_{k=0}^{n-1} x(k)$ for $n = 0$ equals zero. Hence

$$\mathcal{Z}[\sum_{k=0}^{n-1} x(k)] = \frac{X(z)}{z-1}$$

and

$$\mathcal{Z}[\sum_{k=0}^{n} x(k)] = \mathcal{Z}[\sum_{k=0}^{n-1} x(k)] + \mathcal{Z}[x(n)] = \frac{X(z)}{z-1} + X(z)$$

or

$$\mathcal{Z}[\sum_{k=0}^{n} x(k)] = \frac{z}{z-1} X(z)$$

By use of the final value theorem, we find

$$\lim_{n \to \infty} \left[ \sum_{k=0}^{n} x(k) \right] = \lim_{z \to 1} \left[ (z-1) \frac{z}{z-1} X(z) \right]$$

or

$$\sum_{k=0}^{\infty} x(k) = \lim_{z \to 1} X(z)$$

**PROBLEM A-13-10.** Prove the following relationships. Note that $X(z) = \mathcal{Z}[x(t)]$.

1. $$\mathcal{Z}[a^k x(t)] = X\left(\frac{z}{a}\right)$$

2. $$\mathcal{Z}[e^{-at}x(t)] = X(ze^{aT})$$

3. $$\mathcal{Z}[tx(t)] = -Tz\frac{d}{dz}X(z)$$

4. $$\mathcal{Z}[t] = \frac{Tz}{(z-1)^2}$$

**Solution**

1. $$\mathcal{Z}[a^k x(t)] = \sum_{k=0}^{\infty} a^k x(kT)z^{-k}$$
$$= \sum_{k=0}^{\infty} x(kT)\left(\frac{z}{a}\right)^{-k}$$
$$= X\left(\frac{z}{a}\right)$$

2. $$\mathcal{Z}[e^{-at}x(t)] = \sum_{k=0}^{\infty} e^{-akT}x(kT)z^{-k}$$
$$= \sum_{k=0}^{\infty} x(kT)(ze^{aT})^{-k}$$
$$= X(ze^{aT})$$

3. $$\mathcal{Z}[tx(t)] = \sum_{k=0}^{\infty} kTx(kT)z^{-k}$$
$$= -T\sum_{k=0}^{\infty} x(kT)z\frac{d}{dz}(z^{-k})$$
$$= -Tz\frac{d}{dz}[\sum_{k=0}^{\infty} x(kT)z^{-k}]$$
$$= -Tz\frac{d}{dz}X(z)$$

4. Since $\mathcal{Z}[1(t)] = z/(z-1)$, referring to the relationship just obtained, we see that
$$\mathcal{Z}[t] = -Tz\frac{d}{dz}\{\mathcal{Z}[1(t)]\}$$
$$= -Tz\frac{d}{dz}\left(\frac{z}{z-1}\right)$$
$$= \frac{Tz}{(z-1)^2}$$

**PROBLEM A-13-11.** Obtain the solution of the following difference equation in terms of $x(0)$ and $x(1)$:
$$x(k+2) + ax(k+1) + bx(k) = 0$$
where $a$ and $b$ are constants and $k \geq 0$.

**Solution.** The $z$ transforms of $x(k+2)$, $x(k+1)$, and $x(k)$ are given, respectively, by
$$\mathcal{Z}[x(k+2)] = z^2X(z) - z^2x(0) - zx(1)$$
$$\mathcal{Z}[x(k+1)] = zX(z) - zx(0)$$
$$\mathcal{Z}[x(k)] = X(z)$$

Therefore, the $z$ transform of the given difference equation becomes

$$[z^2 X(z) - z^2 x(0) - z x(1)] + a[z X(z) - z x(0)] + b X(z) = 0$$

or

$$(z^2 + az + b)X(z) = (z^2 + az)x(0) + z x(1)$$

Solving for $X(z)$ gives

$$X(z) = \frac{(z^2 + az)x(0) + z x(1)}{z^2 + az + b}$$

By denoting the roots of the denominator polynomial as $m_1$ and $m_2$, or

$$z^2 + az + b = (z - m_1)(z - m_2)$$

and expanding $X(z)/z$ into partial fractions, we obtain

$$\frac{X(z)}{z} = \frac{(m_1 + a)x(0) + x(1)}{m_1 - m_2} \frac{1}{z - m_1} + \frac{(m_2 + a)x(0) + x(1)}{m_2 - m_1} \frac{1}{z - m_2}$$

Noting that $a = -m_1 - m_2$, we obtain

$$X(z) = \frac{m_2 x(0) - x(1)}{m_2 - m_1} \frac{z}{z - m_1} + \frac{m_1 x(0) - x(1)}{m_1 - m_2} \frac{z}{z - m_2}$$

The inverse $z$ transform of $X(z)$ gives the solution of the given difference equation. Namely,

$$x(k) = \frac{m_2 x(0) - x(1)}{m_2 - m_1} m_1{}^k + \frac{m_1 x(0) - x(1)}{m_1 - m_2} m_2{}^k \qquad (k = 0, 1, 2, \ldots)$$

**PROBLEM A-13-12.** Obtain the pulse transfer function of the closed-loop system shown in Fig. 13-22.

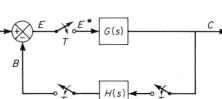

**Fig. 13-22.** Closed-loop discrete-time system.

**Solution.** From the block diagram of Fig. 13-22, we obtain

$$E(s) = R(s) - B(s)$$

or

$$E(z) = R(z) - B(z)$$

Also

$$C(z) = G(z)E(z)$$
$$B(z) = H(z)C(z)$$

Hence

$$C(z) = [R(z) - H(z)C(z)]G(z)$$

or

$$C(z)[1 + H(z)G(z)] = R(z)G(z)$$

The pulse transfer function becomes

$$\frac{C(z)}{R(z)} = \frac{G(z)}{1 + H(z)G(z)}$$

**PROBLEM A-13-13.** Obtain the pulse transfer function of the closed-loop system shown in Fig. 13-23.

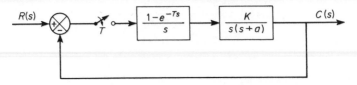

**Fig. 13-23.** Closed-loop discrete-time system.

**Solution.** The $z$ transform of the open-loop transfer function is

$$\mathcal{Z}\left[\frac{K(1 - e^{-Ts})}{s^2(s + a)}\right] = K(1 - z^{-1})\mathcal{Z}\left[\frac{1}{s^2(s + a)}\right]$$

Note that

$$\frac{1}{s^2(s + a)} = \frac{1}{as^2} - \frac{1}{a^2 s} + \frac{1}{a^2(s + a)}$$

From Table 13-1, we obtain the $z$ transform of this last equation as follows:

$$\mathcal{Z}\left[\frac{1}{s^2(s + a)}\right] = \frac{Tz}{a(z - 1)^2} - \frac{z}{a^2(z - 1)} + \frac{z}{a^2(z - e^{-aT})}$$

Hence

$$\mathcal{Z}\left[\frac{K(1 - e^{-Ts})}{s^2(s + a)}\right] = \frac{KT}{a(z - 1)} - \frac{K}{a^2} + \frac{K(z - 1)}{a^2(z - e^{-aT})}$$

$$= \frac{K[(aT - 1 + e^{-aT})z + (1 - e^{-aT} - aTe^{-aT})]}{a^2(z - 1)(z - e^{-aT})}$$

The closed-loop pulse transfer function is then

$$\frac{C(z)}{R(z)} = \frac{K[(aT - 1 + e^{-aT})z + (1 - e^{-aT} - aTe^{-aT})]}{a^2 z^2 + [K(aT - 1 + e^{-aT}) - a^2(1 + e^{-aT})]z + (1 - e^{-aT} - aTe^{-aT} + a^2 e^{-aT})}$$

**PROBLEM A-13-14.** Consider the system shown in Fig. 13-24. Show that the discrete-time system is stable if and only if

$$0 < K < 2 \coth\left(\frac{T}{2T_1}\right)$$

**Solution.** Since

$$G(s) = \frac{K}{s(T_1 s + 1)}$$

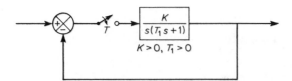

$K > 0, \ T_1 > 0$

**Fig. 13-24.** Closed-loop discrete-time system.

we obtain

$$G(z) = \frac{K(1 - e^{-T/T_1})z}{(z - 1)(z - e^{-T/T_1})}$$

The characteristic equation is

$$z^2 + [K(1 - e^{-T/T_1}) - (1 + e^{-T/T_1})]z + e^{-T/T_1} = 0 \qquad (13\text{-}24)$$

By use of the following transformation,

$$z = \frac{r + 1}{r - 1}$$

Eq. (13-24) becomes

$$[K(1 - e^{-T/T_1})]r^2 + 2(1 - e^{-T/T_1})r + 2(1 + e^{-T/T_1}) - K(1 - e^{-T/T_1}) = 0$$

The Routh array is

$$\begin{array}{lll}
r^2 & K(1 - e^{-T/T_1}) & 2(1 + e^{-T/T_1}) - K(1 - e^{-T/T_1}) \\
r^1 & 2(1 - e^{-T/T_1}) & 0 \\
r^0 & 2(1 + e^{-T/T_1}) - K(1 - e^{-T/T_1}) &
\end{array}$$

Since $1 - e^{-T/T_1} > 0$, the system is stable if and only if

$$K > 0$$
$$\frac{2(1 + e^{-T/T_1})}{1 - e^{-T/T_1}} > K$$

Hence the system is stable if and only if

$$0 < K < \frac{2(1 + e^{-T/T_1})}{1 - e^{-T/T_1}} = 2\coth\left(\frac{T}{2T_1}\right)$$

## PROBLEMS

**PROBLEM B-13-1.** Obtain the $z$ transform of $g(t - kT)$, where $g(t)$ is the impulse-response function of a linear system.

**PROBLEM B-13-2.** If $X(z) = X_1(z)X_2(z)$, show that

$$x(kT) = \sum_{h=0}^{k} x_1(kT - hT)x_2(hT)$$

**PROBLEM B-13-3.** Given

$$X(z) = \frac{z}{(z - 1)^2(z - 2)}$$

find $x(kT)$.

**PROBLEM B-13-4.** Given

$$X(z) = \frac{z^2}{(ze - 1)^3}$$

find $x(kT)$ by computing the residue of $X(z)z^{k-1}$ at the pole $z = 1/e$.

**PROBLEM B-13-5.** Given

$$X(z) = \frac{z(1 - e^{-T})}{(z - 1)(z - e^{-T})}$$

find $x(kT)$ by expanding $X(z)$ into a convergent power series in $z^{-1}$.

**PROBLEM B-13-6.** Given

$$G(s) = \frac{K}{s(s + a)}$$

find the pulse transfer function $G(z)$.

**PROBLEM B-13-7.** Given

$$G(s) = \frac{\omega_0}{s^2 + \omega_0^2}$$

find the pulse transfer function $G(z)$.

**PROBLEM B-13-8.** Consider the $RC$ circuit shown in Fig. 13-25. Find $y(kT)$ when the input voltage is given by $x(t) = 100e^{-t}$ volts.

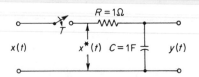

Fig. 13-25. *RC* circuit with sampler.

**PROBLEM B-13-9.** Find the solution of the following difference equation:

$$x(k + 2) + 2x(k + 1) + x(k) = u(k), \qquad x(0) = 0, \qquad x(1) = 0$$

where

$$u(k) = k \qquad (k = 0, 1, 2, \ldots)$$

# 14

# STATE-SPACE ANALYSIS OF CONTROL SYSTEMS

## 14-1 INTRODUCTION

**Limitations of conventional control theory.** In Chapters 8 through 10, we showed that the root-locus method and frequency-response methods are quite useful for dealing with single-input–single-output systems. For example, by means of open-loop frequency-response tests, we can predict the dynamic behavior of the closed-loop system. If necessary, the dynamic behavior of a complex system may be improved by inserting a simple lead or lag compensator. The techniques of conventional control theory are conceptually simple and require only a reasonable amount of computation.

In conventional control theory, only the input, output, and error signals are considered important; the analysis and design of control systems are carried out using transfer functions, together with a variety of graphical techniques such as root-locus plots and Nyquist plots. The unique characteristic of conventional control theory is that it is based on the input-output relation of the system, or the transfer function.

The main disadvantage of conventional control theory is that, generally speaking, it is applicable only to linear time-invariant systems having a single input and a single output. It is powerless for time-varying systems, nonlinear systems (except simple ones), and multiple-input–multiple-output systems. Thus conventional techniques (the root-locus and frequency-response methods) do not apply to the design of optimal and adaptive control systems, which are mostly time-varying and/or nonlinear.

**A new approach to control system analysis and design—modern control theory.** The modern trend in engineering systems is toward greater complexity, due mainly to the requirements of complex tasks and good accuracy. Complex systems may have multiple inputs and multiple outputs and may be time-varying. Because of the necessity of meeting increasingly stringent requirements on the performance of control systems, the increase in system complexity, and easy access to large-scale computers, modern control theory, which is a new approach to the analysis and design of complex control systems, has been developed since around 1960. This new approach is based on the concept of state. The concept of state by itself is not new since it has been in existence for a long time in the field of classical dynamics and other fields. (In fact, the phase-plane method, discussed in Chapter 13, is a two-dimensional state-space method.)

**Modern control theory versus conventional control theory.** Modern control theory is contrasted with conventional control theory in that the former is applicable to multiple-input–multiple-output systems, which may be linear or nonlinear, time-invariant or time-varying, while the latter is applicable only to linear time-invariant single-input–single-output systems. Also, modern control theory is essentially a time-domain approach, while conventional control theory is a complex frequency-domain approach.

System design in classical control theory is based on trial-and-error procedures which, in general, will not yield optimal control systems. System design in modern control theory, on the other hand, enables the engineer to design optimal control systems with respect to given performance indexes. In addition, design in modern control theory can be carried out for a class of inputs, instead of a specific input function, such as the impulse function, step function, or sinusoidal function. Also, modern control theory enables the engineer to include initial conditions in the design.

Before we proceed further, we must define state, state variables, state vector, and state space.

**State.** The state of a dynamic system is the smallest set of variables (called state variables) such that the knowledge of these variables at $t = t_0$, together with the input for $t \geq t_0$, completely determines the behavior of the system for any time $t \geq t_0$.

Thus, the state of a dynamic system at time $t$ is uniquely determined by the state at time $t_0$ and the input for $t \geq t_0$, and it is independent of the state and input before $t_0$. Note that, in dealing with linear time-invariant systems, we usually choose the reference time $t_0$ to be zero.

**State variables.** The state variables of a dynamic system are the smallest set of variables which determine the state of the dynamic system. If at least $n$ variables $x_1(t), x_2(t), \ldots, x_n(t)$ are needed to completely describe the behavior of a dynamic system (such that once the input is given for $t \geq t_0$ and the initial state at $t = t_0$ is specified, the future state of the system is completely determined), then such $n$

variables $x_1(t)$, $x_2(t)$. . . . , $x_n(t)$ are a set of state variables. Note that the state variables need not be physically measurable or observable quantities. Practically, however, it is convenient to choose easily measurable quantities for the state variables because optimal control laws will require the feedback of all state variables with suitable weighting.

**State vector.** If $n$ state variables are needed to completely describe the behavior of a given system, then these $n$ state variables can be considered to be the $n$ components of a vector $\mathbf{x}(t)$. Such a vector is called a state vector. A state vector is thus a vector which determines uniquely the system state $\mathbf{x}(t)$ for any $t \geq t_0$, once the input $\mathbf{u}(t)$ for $t \geq t_0$, is specified.

**State space.** The $n$-dimensional space whose coordinate axes consist of the $x_1$ axis, $x_2$ axis, . . . , $x_n$ axis is called a state space. Any state can be represented by a point in the state space.

---

*Example 14-1.* Consider the *RLC* network system shown in Fig. 14-1. The dynamic behavior of the system is completely defined for $t \geq t_0$ if the initial values of the current $i(t_0)$, the capacitor voltage $v_c(t_0)$, and the input voltage $v(t)$ for $t \geq t_0$ are known. Thus, the state of the network for $t \geq t_0$ is completely determined by $i(t)$, $v_c(t)$, and the input voltage $v(t)$ for $t \geq t_0$. Hence $i(t)$ and $v_c(t)$ are a set of the state variables for this system. [Note, however, that the choice of the state variables for a given system is not unique. For example, in this system $x_1(t) = v_c(t) + Ri(t)$ and $x_2(t) = v_c(t)$ can be chosen as a set of state variables.]

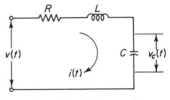

**Fig. 14-1.** *RLC* network.

Suppose that we choose $i(t)$ and $v_c(t)$ as the state variables. Then the equations describing the system dynamics are

$$L\frac{di}{dt} + Ri + v_c = v$$

$$C\frac{dv_c}{dt} = i$$

In vector-matrix notation, we have

$$\begin{bmatrix} \dot{i} \\ \dot{v}_c \end{bmatrix} = \begin{bmatrix} -\dfrac{R}{L} & -\dfrac{1}{L} \\ \dfrac{1}{C} & 0 \end{bmatrix} \begin{bmatrix} i \\ v_c \end{bmatrix} + \begin{bmatrix} \dfrac{1}{L} \\ 0 \end{bmatrix} [v]$$

This is a state-space representation for the given network system.

**Analysis of complex systems.** A modern complex system may have many inputs and many outputs, and these may be interrelated in a complicated manner. To analyze such a system, it is essential to reduce the complexity of the mathematical expressions, as well as to resort to computers for most of the tedious computations necessary in the analysis. The state-space approach to system analysis is best suited from this viewpoint.

While conventional control theory is based on the input-output relationship, or transfer function, modern control theory is based on the description of system equations in terms of $n$ first-order differential equations, which may be combined into a first-order vector-matrix differential equation. The use of vector-matrix notation greatly simplifies the mathematical representation of systems of equations. The increase in the number of state variables, the number of inputs, or the number of outputs does not increase the complexity of the equations. In fact, the analysis of complicated multiple-input–multiple-output systems can be carried out by the procedures that are only slightly more complicated than those required for the analysis of systems of first-order scalar differential equations.

From the computational viewpoint, the state-space methods are particularly suited for digital-computer computations because of their time-domain approach. This relieves the engineer of the burden of tedious computations otherwise necessary and enables him to devote his efforts solely to the analytical aspects of the problem. This is one of the advantages of the state-space methods.

Finally, it is important to note that it is not necessary that the state variables represent physical quantities of the system. Variables which do not represent physical quantities, and those which are neither measurable nor observable, may be chosen as state variables. Such freedom in choosing state variables is another advantage of the state-space methods.

## 14-2  STATE-SPACE REPRESENTATION OF SYSTEMS

A dynamic system consisting of a finite number of lumped elements may be described by ordinary differential equations in which time is the independent variable. By use of vector-matrix notation, an $n$th-order differential equation may be expressed by a first-order vector-matrix differential equation. If $n$ elements of the vector are a set of state variables, then the vector-matrix differential equation is called a *state* equation. In this section we shall present methods for obtaining state-space representations of continuous-time systems.

**State-space representation of $n$th-order systems of linear differential equations in which the forcing function does not involve derivative terms.** Consider the following $n$th-order system:

$$\overset{(n)}{y} + a_1\overset{(n-1)}{y} + \cdots + a_{n-1}\dot{y} + a_n y = u \tag{14-1}$$

Noting that the knowledge of $y(0), \dot{y}(0), \ldots, \overset{(n-1)}{y}(0)$, together with the input $u(t)$ for $t \geq 0$, determines completely the future behavior of the system, we may take $y(t), \dot{y}(t), \ldots, \overset{(n-1)}{y}(t)$ as a set of $n$ state variables. (Mathematically, such a choice of state variables is quite convenient. Practically, however, because higher-order derivative terms are inaccurate, due to the noise effects inherent in any practical situations, such a choice of the state variables may not be desirable.)

Let us define

$$x_1 = y$$
$$x_2 = \dot{y}$$
$$\cdot \quad \cdot \quad \cdot$$
$$x_n = \overset{(n-1)}{y}$$

Then Eq. (14-1) can be written as

$$\dot{x}_1 = x_2$$
$$\dot{x}_2 = x_3$$
$$\cdot / \cdot \quad \cdot$$
$$\dot{x}_{n-1} = x_n$$
$$\dot{x}_n = -a_n x_1 - \cdots - a_1 x_n + u$$

or

$$\dot{\mathbf{x}} = \mathbf{Ax} + \mathbf{B}u \tag{14-2}$$

where

$$\mathbf{x} = \begin{bmatrix} x_1 \\ x_2 \\ \cdot \\ \cdot \\ \cdot \\ x_n \end{bmatrix}, \quad \mathbf{A} = \begin{bmatrix} 0 & 1 & 0 & \cdots & 0 \\ 0 & 0 & 1 & \cdots & 0 \\ \cdot & & \cdot & & \cdot \\ \cdot & & \cdot & & \cdot \\ 0 & 0 & 0 & \cdots & 1 \\ -a_n & -a_{n-1} & -a_{n-2} & \cdots & -a_1 \end{bmatrix}, \quad \mathbf{B} = \begin{bmatrix} 0 \\ 0 \\ \cdot \\ \cdot \\ 0 \\ 1 \end{bmatrix}$$

The output equation becomes

$$y = [1 \quad 0 \quad \cdots \quad 0] \begin{bmatrix} x_1 \\ x_2 \\ \cdot \\ \cdot \\ x_n \end{bmatrix}$$

or

$$y = \mathbf{Cx} \tag{14-3}$$

where

$$\mathbf{C} = [1 \quad 0 \quad \cdots \quad 0]$$

The first-order differential equation, Eq. (14-2), is the state equation, and the algebraic equation, Eq. (14-3), is the output equation.

---

*Example 14-2.* Consider the system defined by

$$\dddot{y} + 6\ddot{y} + 11\dot{y} + 6y = 6u \tag{14-4}$$

where $y$ is the output and $u$ is the input of the system. Obtain a state-space representation of the system.

Let us choose the state variables as

$$x_1 = y$$
$$x_2 = \dot{y}$$
$$x_3 = \ddot{y}$$

Then we obtain

$$\dot{x}_1 = x_2$$
$$\dot{x}_2 = x_3$$
$$\dot{x}_3 = -6x_1 - 11x_2 - 6x_3 + 6u$$

The last of these three equations was obtained by solving the original differential equation for the highest derivative term $\dddot{y}$ and then substituting $y = x_1$, $\dot{y} = x_2$, $\ddot{y} = x_3$ into the resulting equation. By use of vector-matrix notation, these three first-order differential equations can be combined into one as follows:

$$\begin{bmatrix} \dot{x}_1 \\ \dot{x}_2 \\ \dot{x}_3 \end{bmatrix} = \begin{bmatrix} 0 & 1 & 0 \\ 0 & 0 & 1 \\ -6 & -11 & -6 \end{bmatrix} \begin{bmatrix} x_1 \\ x_2 \\ x_3 \end{bmatrix} + \begin{bmatrix} 0 \\ 0 \\ 6 \end{bmatrix} [u] \tag{14-5}$$

The output equation is given by

$$y = \begin{bmatrix} 1 & 0 & 0 \end{bmatrix} \begin{bmatrix} x_1 \\ x_2 \\ x_3 \end{bmatrix} \tag{14-6}$$

Equations (14-5) and (14-6) can be put in a standard form as

$$\dot{\mathbf{x}} = \mathbf{A}\mathbf{x} + \mathbf{B}u \tag{14-7}$$
$$y = \mathbf{C}\mathbf{x} \tag{14-8}$$

where

$$\mathbf{A} = \begin{bmatrix} 0 & 1 & 0 \\ 0 & 0 & 1 \\ -6 & -11 & -6 \end{bmatrix}, \quad \mathbf{B} = \begin{bmatrix} 0 \\ 0 \\ 6 \end{bmatrix}, \quad \mathbf{C} = \begin{bmatrix} 1 & 0 & 0 \end{bmatrix}$$

Figure 14-2 shows the block-diagram representation of the present state equation and output equation. Notice that the transfer functions of the feedback blocks are identical with the negatives of the coefficients of the original differential equation, Eq. (14-4).

**Nonuniqueness of the set of state variables.** It has been stated that a set of state variables is not unique for a given system. Suppose that $x_1, x_2, \ldots, x_n$ are a set of state variables. Then we may take as another set of state variables any set of functions

$$\hat{x}_1 = X_1(x_1, x_2, \ldots, x_n)$$
$$\hat{x}_2 = X_2(x_1, x_2, \ldots, x_n)$$
$$\cdot \; \cdot \; \cdot$$
$$\hat{x}_n = X_n(x_1, x_2, \ldots, x_n)$$

provided that, for every set of values $\hat{x}_1, \hat{x}_2, \ldots, \hat{x}_n$, there corresponds a unique set of values $x_1, x_2, \ldots, x_n$ and vice versa. Thus, if $\mathbf{x}$ is a state vector, then $\hat{\mathbf{x}}$ where

$$\hat{\mathbf{x}} = \mathbf{P}\mathbf{x}$$

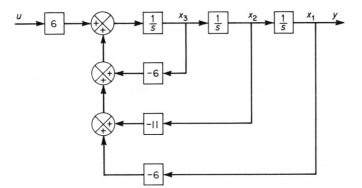

**Fig. 14-2.** Block-diagram representation of the system described by Eqs. (14-7) and (14-8).

is also a state vector, provided the matrix $\mathbf{P}$ is nonsingular. Different state vectors convey the same information about the system behavior.

**Eigenvalues of an $n \times n$ matrix A.** The eigenvalues of an $n \times n$ matrix $\mathbf{A}$ are the roots of the characteristic equation

$$|\lambda\mathbf{I} - \mathbf{A}| = 0$$

The eigenvalues are sometimes called the characteristic roots.

Consider, for example, the following matrix $\mathbf{A}$:

$$\mathbf{A} = \begin{bmatrix} 0 & 1 & 0 \\ 0 & 0 & 1 \\ -6 & -11 & -6 \end{bmatrix}$$

The characteristic equation is

$$\begin{aligned} |\lambda\mathbf{I} - \mathbf{A}| &= \begin{vmatrix} \lambda & -1 & 0 \\ 0 & \lambda & -1 \\ 6 & 11 & \lambda+6 \end{vmatrix} \\ &= \lambda^3 + 6\lambda^2 + 11\lambda + 6 \\ &= (\lambda + 1)(\lambda + 2)(\lambda + 3) = 0 \end{aligned}$$

The eigenvalues of $\mathbf{A}$ are the roots of the characteristic equation, or $-1$, $-2$, and $-3$.

---

*Example 14-3.* Consider the same system as discussed in Example 14-2. We shall show that Eq. (14-5) is not the only state equation possible for this system. Suppose we define a set of new state variables $z_1, z_2, z_3$ by the transformation

$$\begin{bmatrix} x_1 \\ x_2 \\ x_3 \end{bmatrix} = \begin{bmatrix} 1 & 1 & 1 \\ -1 & -2 & -3 \\ 1 & 4 & 9 \end{bmatrix} \begin{bmatrix} z_1 \\ z_2 \\ z_3 \end{bmatrix}$$

or

$$\mathbf{x} = \mathbf{Pz} \tag{14-9}$$

where

$$\mathbf{P} = \begin{bmatrix} 1 & 1 & 1 \\ -1 & -2 & -3 \\ 1 & 4 & 9 \end{bmatrix} \tag{14-10}$$

Then by substituting Eq. (14-9) into Eq. (14-7), we obtain

$$\mathbf{P\dot{z}} = \mathbf{APz} + \mathbf{B}u$$

By premultiplying both sides of this last equation by $\mathbf{P}^{-1}$, we get

$$\dot{\mathbf{z}} = \mathbf{P}^{-1}\mathbf{APz} + \mathbf{P}^{-1}\mathbf{B}u \tag{14-11}$$

or

$$\begin{bmatrix} \dot{z}_1 \\ \dot{z}_2 \\ \dot{z}_3 \end{bmatrix} = \begin{bmatrix} 3 & 2.5 & 0.5 \\ -3 & -4 & -1 \\ 1 & 1.5 & 0.5 \end{bmatrix} \begin{bmatrix} 0 & 1 & 0 \\ 0 & 0 & 1 \\ -6 & -11 & -6 \end{bmatrix} \begin{bmatrix} 1 & 1 & 1 \\ -1 & -2 & -3 \\ 1 & 4 & 9 \end{bmatrix} \begin{bmatrix} z_1 \\ z_2 \\ z_3 \end{bmatrix}$$

$$+ \begin{bmatrix} 3 & 2.5 & 0.5 \\ -3 & -4 & -1 \\ 1 & 1.5 & 0.5 \end{bmatrix} \begin{bmatrix} 0 \\ 0 \\ 6 \end{bmatrix} [u]$$

Simplifying gives

$$\begin{bmatrix} \dot{z}_1 \\ \dot{z}_2 \\ \dot{z}_3 \end{bmatrix} = \begin{bmatrix} -1 & 0 & 0 \\ 0 & -2 & 0 \\ 0 & 0 & -3 \end{bmatrix} \begin{bmatrix} z_1 \\ z_2 \\ z_3 \end{bmatrix} + \begin{bmatrix} 3 \\ -6 \\ 3 \end{bmatrix} [u] \tag{14-12}$$

Equation (14-12) is also a state equation which describes the same system as defined by Eq. (14-5).

The output equation, Eq. (14-8), is modified to

$$y = \mathbf{CPz}$$

or

$$y = [1 \quad 0 \quad 0] \begin{bmatrix} 1 & 1 & 1 \\ -1 & -2 & -3 \\ 1 & 4 & 9 \end{bmatrix} \begin{bmatrix} z_1 \\ z_2 \\ z_3 \end{bmatrix}$$

$$= [1 \quad 1 \quad 1] \begin{bmatrix} z_1 \\ z_2 \\ z_3 \end{bmatrix} \tag{14-13}$$

Notice that the transformation matrix $\mathbf{P}$, defined by Eq. (14-10), modifies the coefficient matrix of $\mathbf{z}$ into the diagonal matrix. As clearly seen from Eq. (14-12), the three separate state equations are uncoupled. Notice also that the diagonal elements of the matrix $\mathbf{P}^{-1}\mathbf{AP}$ in Eq. (14-11) are identical with the three eigenvalues of $\mathbf{A}$. It is very important to note that the eigenvalues of $\mathbf{A}$ and those of $\mathbf{P}^{-1}\mathbf{AP}$ are identical. We shall prove this for a general case in what follows.

**Invariance of eigenvalues.** To prove the invariance of the eigenvalues under a linear transformation, we must show that the characteristic polynomials $|\lambda\mathbf{I} - \mathbf{A}|$ and $|\lambda\mathbf{I} - \mathbf{P}^{-1}\mathbf{A}\mathbf{P}|$ are identical.

Since the determinant of a product is the product of the determinants, we obtain

$$
\begin{aligned}
|\lambda\mathbf{I} - \mathbf{P}^{-1}\mathbf{A}\mathbf{P}| &= |\lambda\mathbf{P}^{-1}\mathbf{P} - \mathbf{P}^{-1}\mathbf{A}\mathbf{P}| \\
&= |\mathbf{P}^{-1}(\lambda\mathbf{I} - \mathbf{A})\mathbf{P}| \\
&= |\mathbf{P}^{-1}||\lambda\mathbf{I} - \mathbf{A}||\mathbf{P}| \\
&= |\mathbf{P}^{-1}||\mathbf{P}||\lambda\mathbf{I} - \mathbf{A}|
\end{aligned}
$$

Noting that the product of the determinants $|\mathbf{P}^{-1}|$ and $|\mathbf{P}|$ is the determinant of the product $|\mathbf{P}^{-1}\mathbf{P}|$, we obtain

$$
\begin{aligned}
|\lambda\mathbf{I} - \mathbf{P}^{-1}\mathbf{A}\mathbf{P}| &= |\mathbf{P}^{-1}\mathbf{P}||\lambda\mathbf{I} - \mathbf{A}| \\
&= |\lambda\mathbf{I} - \mathbf{A}|
\end{aligned}
$$

Thus we have proved that the eigenvalues of $\mathbf{A}$ are invariant under a linear transformation.

**Diagonalization of $n \times n$ matrix.** Note that if an $n \times n$ matrix $\mathbf{A}$ with distinct eigenvalues is given by

$$
\mathbf{A} = \begin{bmatrix}
0 & 1 & 0 & \cdots & 0 \\
0 & 0 & 1 & \cdots & 0 \\
\cdot & \cdot & \cdot & & \cdot \\
\cdot & \cdot & \cdot & & \cdot \\
\cdot & \cdot & \cdot & & \cdot \\
0 & 0 & 0 & \cdots & 1 \\
-a_n & -a_{n-1} & -a_{n-2} & \cdots & -a_1
\end{bmatrix} \tag{14-14}
$$

the transformation $\mathbf{x} = \mathbf{P}\mathbf{z}$ where

$$
\mathbf{P} = \begin{bmatrix}
1 & 1 & \cdots & 1 \\
\lambda_1 & \lambda_2 & \cdots & \lambda_n \\
\lambda_1^2 & \lambda_2^2 & \cdots & \lambda_n^2 \\
\cdot & \cdot & & \cdot \\
\cdot & \cdot & & \cdot \\
\cdot & \cdot & & \cdot \\
\lambda_1^{n-1} & \lambda_2^{n-1} & \cdots & \lambda_n^{n-1}
\end{bmatrix}
$$

$\lambda_1, \lambda_2, \ldots, \lambda_n = n$ distinct eigenvalues of $\mathbf{A}$

will transform $\mathbf{P}^{-1}\mathbf{A}\mathbf{P}$ into the diagonal matrix, or

$$
\mathbf{P}^{-1}\mathbf{A}\mathbf{P} = \begin{bmatrix}
\lambda_1 & & & & 0 \\
& \lambda_2 & & & \\
& & \cdot & & \\
& & & \cdot & \\
& & & & \cdot \\
0 & & & & \lambda_n
\end{bmatrix}
$$

If the matrix $\mathbf{A}$ defined by Eq. (14-14) involves multiple eigenvalues, then diagonalization is impossible. For example, if the $3 \times 3$ matrix $\mathbf{A}$ where

$$\mathbf{A} = \begin{bmatrix} 0 & 1 & 0 \\ 0 & 0 & 1 \\ -a_3 & -a_2 & -a_1 \end{bmatrix}$$

has the eigenvalues $\lambda_1, \lambda_1, \lambda_3$, then the transformation $\mathbf{x} = \mathbf{S}\mathbf{z}$ where

$$\mathbf{S} = \begin{bmatrix} 1 & 0 & 1 \\ \lambda_1 & 1 & \lambda_3 \\ \lambda_1^2 & 2\lambda_1 & \lambda_3^2 \end{bmatrix}$$

will yield

$$\mathbf{S}^{-1}\mathbf{A}\mathbf{S} = \begin{bmatrix} \lambda_1 & 1 & 0 \\ 0 & \lambda_1 & 0 \\ 0 & 0 & \lambda_3 \end{bmatrix}$$

Such a form is called the Jordan canonical form.

---

*Example 14-4.* Consider the same system as discussed in Examples 14-2 and 14-3, rewritten thus

$$\dddot{y} + 6\ddot{y} + 11\dot{y} + 6y = 6u \tag{14-15}$$

We shall demonstrate that the state-space representation as given by Eqs. (14-12) and (14-13) can also be obtained by use of the so-called partial fraction expansion technique.

Let us rewrite Eq. (14-15) in the form of a transfer function:

$$\frac{Y(s)}{U(s)} = \frac{6}{s^3 + 6s^2 + 11s + 6} = \frac{6}{(s+1)(s+2)(s+3)}$$

By expanding this transfer function into partial fractions, we obtain

$$\frac{Y(s)}{U(s)} = \frac{3}{s+1} + \frac{-6}{s+2} + \frac{3}{s+3}$$

Hence

$$Y(s) = \frac{3}{s+1}U(s) + \frac{-6}{s+2}U(s) + \frac{3}{s+3}U(s) \tag{14-16}$$

Let us define

$$X_1(s) = \frac{3}{s+1}U(s) \tag{14-17}$$

$$X_2(s) = \frac{-6}{s+2}U(s) \tag{14-18}$$

$$X_3(s) = \frac{3}{s+3}U(s) \tag{14-19}$$

The inverse Laplace transforms of Eqs. (14-17), (14-18), and (14-19) give

$$\dot{x}_1 = -x_1 + 3u$$
$$\dot{x}_2 = -2x_2 - 6u$$
$$\dot{x}_3 = -3x_3 + 3u$$

Since Eq. (14-16) can be written as

$$Y(s) = X_1(s) + X_2(s) + X_3(s)$$

we obtain

$$y = x_1 + x_2 + x_3$$

In terms of vector-matrix notation, we obtain

$$\begin{bmatrix} \dot{x}_1 \\ \dot{x}_2 \\ \dot{x}_3 \end{bmatrix} = \begin{bmatrix} -1 & 0 & 0 \\ 0 & -2 & 0 \\ 0 & 0 & -3 \end{bmatrix} \begin{bmatrix} x_1 \\ x_2 \\ x_3 \end{bmatrix} + \begin{bmatrix} 3 \\ -6 \\ 3 \end{bmatrix} [u] \qquad (14\text{-}20)$$

$$y = \begin{bmatrix} 1 & 1 & 1 \end{bmatrix} \begin{bmatrix} x_1 \\ x_2 \\ x_3 \end{bmatrix} \qquad (14\text{-}21)$$

Equations (14-20) and (14-21) are clearly identical with Eqs. (14-12) and (14-13), respectively.

Figure 14-3 shows a block-diagram representation of Eqs. (14-20) and (14-21). Notice that the transfer functions in the feedback blocks are identical with the eigenvalues

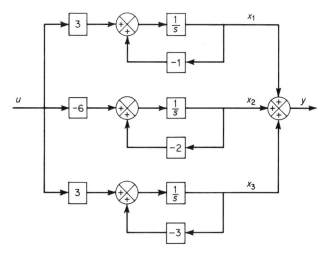

**Fig. 14-3.** Block-diagram representation of the system described by Eqs. (14-20) and (14-21).

of the system. Notice also that the residues of the poles of the transfer function, or the coefficients in the partial fraction expansions of $Y(s)/U(s)$, appear in the feedforward blocks.

**State-space representation of $n$th-order systems of linear differential equations with $r$ forcing functions.** Consider the multiple-input–multiple-output system shown in Fig. 14-4. In this system, $x_1, x_2, \ldots, x_n$ represent the state variables; $u_1, u_2, \ldots, u_r$ denote the input variables; and $y_1, y_2, \ldots, y_m$ are the output variables. From Fig. 14-4, we obtain the system equations as follows:

$$\dot{x}_1 = a_{11}(t)x_1 + a_{12}(t)x_2 + \cdots + a_{1n}(t)x_n + b_{11}(t)u_1 + b_{12}(t)u_2 + \cdots + b_{1r}(t)u_r$$
$$\dot{x}_2 = a_{21}(t)x_1 + a_{22}(t)x_2 + \cdots + a_{2n}(t)x_n + b_{21}(t)u_1 + b_{22}(t)u_2 + \cdots + b_{2r}(t)u_r$$

$$\cdots$$

$$\dot{x}_n = a_{n1}(t)x_1 + a_{n2}(t)x_2 + \cdots + a_{nn}(t)x_n + b_{n1}(t)u_1 + b_{n2}(t)u_2 + \cdots + b_{nr}(t)u_r$$

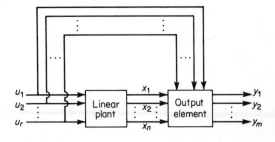

**Fig. 14-4.** Multiple-input–multiple-output system.

where the $a(t)$'s and $b(t)$'s are constants or functions of $t$. In terms of vector-matrix notation, these $n$ equations can be written compactly as

$$\dot{\mathbf{x}} = \mathbf{A}(t)\mathbf{x} + \mathbf{B}(t)\mathbf{u} \tag{14-22}$$

where

$$\mathbf{x} = \begin{bmatrix} x_1 \\ x_2 \\ \cdot \\ \cdot \\ \cdot \\ x_n \end{bmatrix} = \text{state vector}$$

$$\mathbf{u} = \begin{bmatrix} u_1 \\ u_2 \\ \cdot \\ \cdot \\ \cdot \\ u_r \end{bmatrix} = \text{input (or control) vector}$$

$$\mathbf{A}(t) = \begin{bmatrix} a_{11}(t) & a_{12}(t) & \cdots & a_{1n}(t) \\ a_{21}(t) & a_{22}(t) & \cdots & a_{2n}(t) \\ \cdot & \cdot & & \cdot \\ \cdot & \cdot & & \cdot \\ \cdot & \cdot & & \cdot \\ a_{n1}(t) & a_{n2}(t) & \cdots & a_{nn}(t) \end{bmatrix}$$

$$\mathbf{B}(t) = \begin{bmatrix} b_{11}(t) & b_{12}(t) & \cdots & b_{1r}(t) \\ b_{21}(t) & b_{22}(t) & \cdots & b_{2r}(t) \\ \cdot & \cdot & & \cdot \\ \cdot & \cdot & & \cdot \\ \cdot & \cdot & & \cdot \\ b_{n1}(t) & b_{n2}(t) & \cdots & b_{nr}(t) \end{bmatrix}$$

Equation (14-22) is the state equation for the system. [Note that a vector-matrix differential equation such as Eq. (14-22) (or the equivalent $n$ first-order differential equations) describing the dynamics of a system is a state equation if and only if the set of dependent variables in the vector-matrix differential equation satisfies the definition of state variables.]

For the output signals, we obtain

$$y_1 = c_{11}(t)x_1 + c_{12}(t)x_2 + \cdots + c_{1n}(t)x_n + d_{11}(t)u_1 + d_{12}(t)u_2 + \cdots + d_{1r}(t)u_r$$
$$y_2 = c_{21}(t)x_1 + c_{22}(t)x_2 + \cdots + c_{2n}(t)x_n + d_{21}(t)u_1 + d_{22}(t)u_2 + \cdots + d_{2r}(t)u_r$$
$$\cdot \quad \cdot \quad \cdot$$
$$y_m = c_{m1}(t)x_1 + c_{m2}(t)x_2 + \cdots + c_{mn}(t)x_n + d_{m1}(t)u_1 + d_{m2}(t)u_2 + \cdots + d_{mr}(t)u_r$$

In terms of vector-matrix notation, these $m$ equations can be written compactly as

$$\mathbf{y} = \mathbf{C}(t)\mathbf{x} + \mathbf{D}(t)\mathbf{u} \tag{14-23}$$

where

$$\mathbf{y} = \begin{bmatrix} y_1 \\ y_2 \\ \cdot \\ \cdot \\ \cdot \\ y_m \end{bmatrix} = \text{output vector}$$

$$\mathbf{C}(t) = \begin{bmatrix} c_{11}(t) & c_{12}(t) & \cdots & c_{1n}(t) \\ c_{21}(t) & c_{22}(t) & \cdots & c_{2n}(t) \\ \cdot & \cdot & & \cdot \\ \cdot & \cdot & & \cdot \\ \cdot & \cdot & & \cdot \\ c_{m1}(t) & c_{m2}(t) & \cdots & c_{mn}(t) \end{bmatrix}$$

$$\mathbf{D}(t) = \begin{bmatrix} d_{11}(t) & d_{12}(t) & \cdots & d_{1r}(t) \\ d_{21}(t) & d_{22}(t) & \cdots & d_{2r}(t) \\ \cdot & \cdot & & \cdot \\ \cdot & \cdot & & \cdot \\ \cdot & \cdot & & \cdot \\ d_{m1}(t) & d_{m2}(t) & \cdots & d_{mr}(t) \end{bmatrix}$$

Equation (14-23) is the output equation for the system. The matrices $\mathbf{A}(t)$, $\mathbf{B}(t)$, $\mathbf{C}(t)$, and $\mathbf{D}(t)$ completely characterize the system dynamics.

A block-diagram representation and a signal flow graph representation of the system defined by Eqs. (14-22) and (14-23) are shown in Figs. 14-5 (a) and (b), respectively. To indicate vector quantities, we have used double arrows in the diagrams.

**State-space representation of $n$th-order systems of linear differential equations in which the forcing function involves derivative terms.** If the differential equation of the system involves derivatives of the forcing function, such as

$$\overset{(n)}{y} + a_1 \overset{(n-1)}{y} + \cdots + a_{n-1}\dot{y} + a_n y = b_0 \overset{(n)}{u} + b_1 \overset{(n-1)}{u} + \cdots + b_{n-1}\dot{u} + b_n u \tag{14-24}$$

then the set of $n$ variables $y, \dot{y}, \ddot{y}, \ldots, \overset{(n-1)}{y}$ do not qualify as a set of state variables,

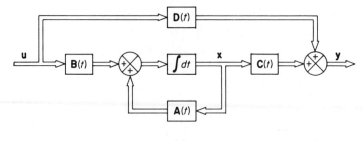

(a)

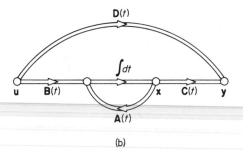

(b)

**Fig. 14-5.** (a) Block diagram representation of the system described by Eqs. (14-22) and (14-23); (b) signal flow graph representation of the system of (a).

and the straightforward method previously employed cannot be used. This is because $n$ first-order differential equations

$$\dot{x}_1 = x_2$$
$$\dot{x}_2 = x_3$$
$$\cdots$$
$$\dot{x}_n = -a_n x_1 - a_{n-1} x_2 - \cdots - a_1 x_n + b_0 \overset{(n)}{u} + b_1 \overset{(n-1)}{u} + \cdots + b_n u$$

where $x_1 = y$ may not yield a unique solution.

The main problem in defining the state variables for this case lies in the derivative terms on the right-hand side of the last of the preceding $n$ equations. The state variables must be such that they will eliminate the derivatives of $u$ in the state equation.

It is a well-known fact in modern control theory that if we define the following $n$ variables as a set of $n$ state variables

$$x_1 = y - \beta_0 u$$
$$x_2 = \dot{y} - \beta_0 \dot{u} - \beta_1 u = \dot{x}_1 - \beta_1 u$$
$$x_3 = \ddot{y} - \beta_0 \ddot{u} - \beta_1 \dot{u} - \beta_2 u = \dot{x}_2 - \beta_2 u \qquad (14\text{-}25)$$
$$\cdots$$
$$x_n = \overset{(n-1)}{y} - \beta_0 \overset{(n-1)}{u} - \beta_1 \overset{(n-2)}{u} - \cdots - \beta_{n-2} \dot{u} - \beta_{n-1} u = \dot{x}_{n-1} - \beta_{n-1} u$$

where $\beta_0, \beta_1, \beta_2, \ldots, \beta_n$ are determined from

$$
\begin{aligned}
\beta_0 &= b_0 \\
\beta_1 &= b_1 - a_1\beta_0 \\
\beta_2 &= b_2 - a_1\beta_1 - a_2\beta_0 \\
\beta_3 &= b_3 - a_1\beta_2 - a_2\beta_1 - a_3\beta_0
\end{aligned}
\tag{14-26}
$$

$$
\cdots
$$

$$
\beta_n = b_n - a_1\beta_{n-1} - \cdots - a_{n-1}\beta_1 - a_n\beta_0
$$

then the existence and uniqueness of the solution of the state equation is guaranteed. (Note that this is not the only choice of a set of state variables.) With the present choice of state variables, we obtain the following state equation and output equation for the system of Eq. (14-24):

$$
\begin{bmatrix} \dot{x}_1 \\ \dot{x}_2 \\ \vdots \\ \dot{x}_{n-1} \\ \dot{x}_n \end{bmatrix} =
\begin{bmatrix}
0 & 1 & 0 & \cdots & 0 \\
0 & 0 & 1 & \cdots & 0 \\
\vdots & \vdots & \vdots & & \vdots \\
0 & 0 & 0 & \cdots & 1 \\
-a_n & -a_{n-1} & -a_{n-2} & \cdots & -a_1
\end{bmatrix}
\begin{bmatrix} x_1 \\ x_2 \\ \vdots \\ x_{n-1} \\ x_n \end{bmatrix} +
\begin{bmatrix} \beta_1 \\ \beta_2 \\ \vdots \\ \beta_{n-1} \\ \beta_n \end{bmatrix} [u]
$$

$$
y = \begin{bmatrix} 1 & 0 & \cdots & 0 \end{bmatrix}
\begin{bmatrix} x_1 \\ x_2 \\ \vdots \\ x_n \end{bmatrix} + \beta_0 u
$$

or

$$
\dot{\mathbf{x}} = \mathbf{A}\mathbf{x} + \mathbf{B}u
\tag{14-27}
$$

$$
y = \mathbf{C}\mathbf{x} + Du
\tag{14-28}
$$

where

$$
\mathbf{x} = \begin{bmatrix} x_1 \\ x_2 \\ \vdots \\ x_{n-1} \\ x_n \end{bmatrix}, \qquad
\mathbf{A} = \begin{bmatrix}
0 & 1 & 0 & \cdots & 0 \\
0 & 0 & 1 & \cdots & 0 \\
\vdots & \vdots & \vdots & & \vdots \\
0 & 0 & 0 & \cdots & 1 \\
-a_n & -a_{n-1} & -a_{n-2} & \cdots & -a_1
\end{bmatrix}
$$

$$
\mathbf{B} = \begin{bmatrix} \beta_1 \\ \beta_2 \\ \vdots \\ \beta_{n-1} \\ \beta_n \end{bmatrix}, \qquad
\mathbf{C} = \begin{bmatrix} 1 & 0 & \cdots & 0 \end{bmatrix}, \qquad
D = \beta_0 = b_0
$$

The initial condition $\mathbf{x}(0)$ may be determined by use of Eq. (14-25).

In this state-space representation, the matrix $\mathbf{A}$ is essentially the same as in the system of Eq. (14-1). The derivatives on the right-hand side of Eq. (14-24) affect only the elements of the $\mathbf{B}$ matrix.

Note that the state-space representation for the following transfer function

$$\frac{Y(s)}{U(s)} = \frac{b_0 s^n + b_1 s^{n-1} + \cdots + b_{n-1} s + b_n}{s^n + a_1 s^{n-1} + \cdots + a_{n-1} s + a_n}$$

is given also by Eqs. (14-27) and (14-28).

---

*Example 14-5.* Consider the control system shown in Fig. 14-6. The closed-loop transfer function is

$$\frac{Y(s)}{U(s)} = \frac{160(s + 4)}{s^3 + 18s^2 + 192s + 640}$$

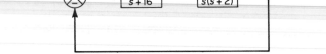

**Fig. 14-6.** Control system.

The corresponding differential equation is

$$\dddot{y} + 18\ddot{y} + 192\dot{y} + 640y = 160\dot{u} + 640u$$

Obtain a state-space representation of the system.

Referring to Eq. (14-25), let us define

$$x_1 = y - \beta_0 u$$
$$x_2 = \dot{y} - \beta_0 \dot{u} - \beta_1 u = \dot{x}_1 - \beta_1 u$$
$$x_3 = \ddot{y} - \beta_0 \ddot{u} - \beta_1 \dot{u} - \beta_2 u = \dot{x}_2 - \beta_2 u$$

where $\beta_0$, $\beta_1$, and $\beta_2$ are determined from Eq. (14-26) as follows:

$$\beta_0 = b_0 = 0$$
$$\beta_1 = b_1 - a_1 \beta_0 = 0$$
$$\beta_2 = b_2 - a_1 \beta_1 - a_2 \beta_0 = 160$$
$$\beta_3 = b_3 - a_1 \beta_2 - a_2 \beta_1 - a_3 \beta_0 = -2240$$

Then the state equation for the system becomes

$$\begin{bmatrix} \dot{x}_1 \\ \dot{x}_2 \\ \dot{x}_3 \end{bmatrix} = \begin{bmatrix} 0 & 1 & 0 \\ 0 & 0 & 1 \\ -640 & -192 & -18 \end{bmatrix} \begin{bmatrix} x_1 \\ x_2 \\ x_3 \end{bmatrix} + \begin{bmatrix} 0 \\ 160 \\ -2240 \end{bmatrix} [u]$$

The output equation becomes

$$y = \begin{bmatrix} 1 & 0 & 0 \end{bmatrix} \begin{bmatrix} x_1 \\ x_2 \\ x_3 \end{bmatrix}$$

## 14-3　SOLVING THE TIME-INVARIANT STATE EQUATION

In this section, we shall obtain the general solution of the linear time-invariant state equation. We shall first consider the homogeneous case and then consider the nonhomogeneous case.

**Solution of homogeneous state equations.** Before we solve vector-matrix differential equations, let us review the solution of the scalar differential equation

$$\dot{x} = ax \tag{14-29}$$

In solving this equation, we may assume a solution $x(t)$ of the form

$$x(t) = b_0 + b_1 t + b_2 t^2 + \cdots + b_k t^k + \cdots \tag{14-30}$$

By substituting this assumed solution into Eq. (14-29), we obtain

$$b_1 + 2b_2 t + 3b_3 t^2 + \cdots + kb_k t^{k-1} + \cdots$$
$$= a(b_0 + b_1 t + b_2 t^2 + \cdots + b_k t^k + \cdots) \tag{14-31}$$

If the assumed solution is to be the true solution, Eq. (14-31) must hold for any $t$. Hence, equating the coefficients of the equal powers of $t$, we obtain

$$b_1 = ab_0$$
$$b_2 = \frac{1}{2} ab_1 = \frac{1}{2} a^2 b_0$$
$$b_3 = \frac{1}{3} ab_2 = \frac{1}{3 \times 2} a^3 b_0$$
$$\cdot \quad \cdot \quad \cdot$$
$$b_k = \frac{1}{k!} a^k b_0$$

The value of $b_0$ is determined by substituting $t = 0$ into Eq. (14-30), or

$$x(0) = b_0$$

Hence the solution $x(t)$ can be written as

$$x(t) = \left(1 + at + \frac{1}{2!} a^2 t^2 + \cdots + \frac{1}{k!} a^k t^k + \cdots\right) x(0)$$
$$= e^{at} x(0)$$

We shall now solve the vector-matrix differential equation

$$\dot{\mathbf{x}} = \mathbf{A}\mathbf{x} \tag{14-32}$$

where

$\mathbf{x} = n$-dimensional vector

$\mathbf{A} = n \times n$ constant matrix

By analogy with the scalar case we assume that the solution is in the form of a vector power series in $t$, or

$$\mathbf{x}(t) = \mathbf{b}_0 + \mathbf{b}_1 t + \mathbf{b}_2 t^2 + \cdots + \mathbf{b}_k t^k + \cdots \tag{14-33}$$

By substituting this assumed solution into Eq. (14-32), we obtain

$$\mathbf{b}_1 + 2\mathbf{b}_2 t + 3\mathbf{b}_3 t^2 + \cdots + k\mathbf{b}_k t^{k-1} + \cdots$$
$$= \mathbf{A}(\mathbf{b}_0 + \mathbf{b}_1 t + \mathbf{b}_2 t^2 + \cdots + k\mathbf{b}_k t^k + \cdots) \quad (14\text{-}34)$$

If the assumed solution is to be the true solution, Eq. (14-34) must hold for all $t$. Thus we require that the coefficients of the equal powers of $t$ be identical, or

$$\mathbf{b}_1 = \mathbf{A}\mathbf{b}_0$$

$$\mathbf{b}_2 = \frac{1}{2}\mathbf{A}\mathbf{b}_1 = \frac{1}{2}\mathbf{A}^2\mathbf{b}_0$$

$$\mathbf{b}_3 = \frac{1}{3}\mathbf{A}\mathbf{b}_2 = \frac{1}{3 \times 2}\mathbf{A}^3\mathbf{b}_0$$

$$\cdots$$

$$\mathbf{b}_k = \frac{1}{k!}\mathbf{A}^k\mathbf{b}_0$$

By substituting $t = 0$ into Eq. (14-33), we obtain

$$\mathbf{x}(0) = \mathbf{b}_0$$

Thus the solution $\mathbf{x}(t)$ can be written as

$$\mathbf{x}(t) = \left(\mathbf{I} + \mathbf{A}t + \frac{1}{2!}\mathbf{A}^2 t^2 + \cdots + \frac{1}{k!}\mathbf{A}^k t^k + \cdots\right)\mathbf{x}(0)$$

The expression in the parentheses in the right-hand side of this last equation is an $n \times n$ matrix. Because of its similarity to the infinite power series for a scaler exponential, we call it the matrix exponential and write

$$\mathbf{I} + \mathbf{A}t + \frac{1}{2!}\mathbf{A}^2 t^2 + \cdots + \frac{1}{k!}\mathbf{A}^k t^k + \cdots = e^{\mathbf{A}t}$$

In terms of the matrix exponential, the solution to Eq. (14-32) can be written as

$$\mathbf{x}(t) = e^{\mathbf{A}t}\mathbf{x}(0) \quad (14\text{-}35)$$

Since the matrix exponential is very important in the state-space analysis of linear systems, we shall next examine the properties of the matrix exponential.

**Matrix exponential.** It can be proved that the matrix exponential of an $n \times n$ matrix $\mathbf{A}$

$$e^{\mathbf{A}t} = \sum_{k=0}^{\infty} \frac{\mathbf{A}^k t^k}{k!}$$

converges absolutely for all finite $t$. (Hence computer calculations for evaluating the elements of $e^{\mathbf{A}t}$ by using the series expansion can easily be carried out.)

Because of the convergence of the infinite series $\sum_{k=0}^{\infty} \mathbf{A}^k t^k / k!$, the series can be differentiated term by term to give

$$\frac{d}{dt}e^{\mathbf{A}t} = \mathbf{A} + \mathbf{A}^2 t + \frac{\mathbf{A}^3 t^2}{2!} + \cdots + \frac{\mathbf{A}^k t^{k-1}}{(k-1)!} + \cdots$$

$$= \mathbf{A}\left[\mathbf{I} + \mathbf{A}t + \frac{\mathbf{A}^2 t^2}{2!} + \cdots + \frac{\mathbf{A}^{k-1} t^{k-1}}{(k-1)!} + \cdots\right] = \mathbf{A}e^{\mathbf{A}t}$$

$$= \left[\mathbf{I} + \mathbf{A}t + \frac{\mathbf{A}^2 t^2}{2!} + \cdots + \frac{\mathbf{A}^{k-1} t^{k-1}}{(k-1)!} + \cdots\right]\mathbf{A} = e^{\mathbf{A}t}\mathbf{A}$$

The matrix exponential has the property that

$$e^{A(t+s)} = e^{At}e^{As}$$

This can be proved as follows:

$$e^{At}e^{As} = \left(\sum_{k=0}^{\infty} \frac{A^k t^k}{k!}\right)\left(\sum_{k=0}^{\infty} \frac{A^k s^k}{k!}\right)$$

$$= \sum_{k=0}^{\infty} A^k \left(\sum_{i=0}^{k} \frac{t^i s^{k-i}}{i!(k-i)!}\right)$$

$$= \sum_{k=0}^{\infty} A^k \frac{(t+s)^k}{k!}$$

$$= e^{A(t+s)}$$

In particular, if $s = -t$, then

$$e^{At}e^{-At} = e^{-At}e^{At} = e^{A(t-t)} = I$$

Thus the inverse of $e^{At}$ is $e^{-At}$. Since the inverse of $e^{At}$ always exists, $e^{At}$ is non-singular.

It is very important to remember that

$$e^{(A+B)t} = e^{At}e^{Bt} \qquad \text{if } AB = BA$$

$$e^{(A+B)t} \neq e^{At}e^{Bt} \qquad \text{if } AB \neq BA$$

To prove this, note that

$$e^{(A+B)t} = I + (A+B)t + \frac{(A+B)^2}{2!}t^2 + \frac{(A+B)^3}{3!}t^3 + \cdots$$

$$e^{At}e^{Bt} = \left(I + At + \frac{A^2 t^2}{2!} + \frac{A^3 t^3}{3!} + \cdots\right)\left(I + Bt + \frac{B^2 t^2}{2!} + \frac{B^3 t^3}{3!} + \cdots\right)$$

$$= I + (A+B)t + \frac{A^2 t^2}{2!} + ABt^2 + \frac{B^2 t^2}{2!} + \frac{A^3 t^3}{3!}$$

$$+ \frac{A^2 B t^3}{2!} + \frac{AB^2 t^3}{2!} + \frac{B^3 t^3}{3!} + \cdots$$

Hence

$$e^{(A+B)t} - e^{At}e^{Bt} = \frac{BA - AB}{2!}t^2$$

$$+ \frac{BA^2 + ABA + B^2 A + BAB - 2A^2 B - 2AB^2}{3!}t^3 + \cdots$$

The difference between $e^{(A+B)t}$ and $e^{At}e^{Bt}$ vanishes if $A$ and $B$ commute.

**Laplace transform approach to the solution of homogeneous state equations.** Let us first consider the scalar case:

$$\dot{x} = ax \tag{14-36}$$

Taking the Laplace transform of Eq. (14-36), we obtain

$$sX(s) - x(0) = aX(s) \tag{14-37}$$

where $X(s) = \mathscr{L}[x]$. Solving Eq. (14-37) for $X(s)$ gives

$$X(s) = \frac{x(0)}{s-a} = (s-a)^{-1}x(0)$$

The inverse Laplace transform of this last equation gives the solution

$$x(t) = e^{at}x(0)$$

The foregoing approach to the solution of the homogeneous scalar differential equation can be extended to the homogeneous state equation:

$$\dot{\mathbf{x}}(t) = \mathbf{A}\mathbf{x}(t) \tag{14-38}$$

Taking the Laplace transform of both sides of Eq. (14-38), we obtain

$$s\mathbf{X}(s) - \mathbf{x}(0) = \mathbf{A}\mathbf{X}(s)$$

where $\mathbf{X}(s) = \mathscr{L}[\mathbf{x}]$. Hence

$$(s\mathbf{I} - \mathbf{A})\mathbf{X}(s) = \mathbf{x}(0)$$

Premultiplying both sides of this last equation by $(s\mathbf{I} - \mathbf{A})^{-1}$, we obtain

$$\mathbf{X}(s) = (s\mathbf{I} - \mathbf{A})^{-1}\mathbf{x}(0)$$

The inverse Laplace transform of $\mathbf{X}(s)$ gives the solution $\mathbf{x}(t)$. Thus

$$\mathbf{x}(t) = \mathscr{L}^{-1}[(s\mathbf{I} - \mathbf{A})^{-1}]\mathbf{x}(0) \tag{14-39}$$

Note that

$$(s\mathbf{I} - \mathbf{A})^{-1} = \frac{\mathbf{I}}{s} + \frac{\mathbf{A}}{s^2} + \frac{\mathbf{A}^2}{s^3} + \cdots$$

Hence, the inverse Laplace transform of $(s\mathbf{I} - \mathbf{A})^{-1}$ gives

$$\mathscr{L}^{-1}[(s\mathbf{I} - \mathbf{A})^{-1}] = \mathbf{I} + \mathbf{A}t + \frac{\mathbf{A}^2 t^2}{2!} + \frac{\mathbf{A}^3 t^3}{3!} + \cdots = e^{\mathbf{A}t} \tag{14-40}$$

(The inverse Laplace transform of a matrix is the matrix consisting of the inverse Laplace transforms of all elements.) From Eqs. (14-39) and (14-40), the solution of Eq. (14-38) is obtained as

$$\mathbf{x}(t) = e^{\mathbf{A}t}\mathbf{x}(0)$$

The importance of Eq. (14-40) lies in the fact that it provides a convenient means for finding the closed solution for the matrix exponential.

**State transition matrix.** We can write the solution of the homogeneous state equation

$$\dot{\mathbf{x}} = \mathbf{A}\mathbf{x} \tag{14-41}$$

as

$$\mathbf{x}(t) = \boldsymbol{\Phi}(t)\mathbf{x}(0) \tag{14-42}$$

where $\boldsymbol{\Phi}(t)$ is an $n \times n$ matrix and is the unique solution of

$$\dot{\boldsymbol{\Phi}}(t) = \mathbf{A}\boldsymbol{\Phi}(t), \qquad \boldsymbol{\Phi}(0) = \mathbf{I}$$

To verify this, note that

$$\mathbf{x}(0) = \boldsymbol{\Phi}(0)\mathbf{x}(0) = \mathbf{I}\mathbf{x}(0)$$

and

$$\dot{\mathbf{x}}(t) = \dot{\boldsymbol{\Phi}}(t)\mathbf{x}(0) = \mathbf{A}\boldsymbol{\Phi}(t)\mathbf{x}(0) = \mathbf{A}\mathbf{x}(t)$$

We thus confirm that Eq. (14-42) is the solution of Eq. (14-41).

From Eqs. (14-35), (14-39), and (14-42), we obtain

$$\mathbf{\Phi}(t) = e^{\mathbf{A}t} = \mathscr{L}^{-1}[(s\mathbf{I} - \mathbf{A})^{-1}]$$

Note that

$$\mathbf{\Phi}^{-1}(t) = e^{-\mathbf{A}t} = \mathbf{\Phi}(-t)$$

From Eq. (14-42), we see that the solution of Eq. (14-41) is simply a transformation of the initial condition. Hence the unique matrix $\mathbf{\Phi}(t)$ is called the state-transition matrix. The state-transition matrix contains all the information about the free motions of the system defined by Eq. (14-41).

If the eigenvalues $\lambda_1, \lambda_2, \ldots, \lambda_n$ of the matrix $\mathbf{A}$ are distinct, then $\mathbf{\Phi}(t)$ will contain the $n$ exponentials

$$e^{\lambda_1 t}, e^{\lambda_2 t}, \ldots, e^{\lambda_n t}$$

In particular, if the matrix $\mathbf{A}$ is diagonal, then

$$\mathbf{\Phi}(t) = e^{\mathbf{A}t} = \begin{bmatrix} e^{\lambda_1 t} & & & & 0 \\ & e^{\lambda_2 t} & & & \\ & & \cdot & & \\ & & & \cdot & \\ 0 & & & & e^{\lambda_n t} \end{bmatrix} \qquad (\mathbf{A}: \text{diagonal})$$

If there is a multiplicity in the eigenvalues, for example, if the eigenvalues of $\mathbf{A}$ are

$$\lambda_1, \lambda_1, \lambda_1, \lambda_4, \lambda_5, \ldots, \lambda_n$$

then $\mathbf{\Phi}(t)$ will contain, in addition to the exponentials $e^{\lambda_1 t}, e^{\lambda_4 t}, e^{\lambda_5 t}, \ldots, e^{\lambda_n t}$, terms like $te^{\lambda_1 t}$ and $t^2 e^{\lambda_1 t}$.

**Properties of state-transition matrices.** We shall now summarize the important properties of the state-transition matrix $\mathbf{\Phi}(t)$. For the time-invariant system

$$\dot{\mathbf{x}} = \mathbf{A}\mathbf{x}$$

for which

$$\mathbf{\Phi}(t) = e^{\mathbf{A}t}$$

we have

1. $\mathbf{\Phi}(0) = e^{\mathbf{A}0} = \mathbf{I}$
2. $\mathbf{\Phi}(t) = e^{\mathbf{A}t} = (e^{-\mathbf{A}t})^{-1} = [\mathbf{\Phi}(-t)]^{-1}$
   or $\mathbf{\Phi}^{-1}(t) = \mathbf{\Phi}(-t)$
3. $\mathbf{\Phi}(t_1 + t_2) = e^{\mathbf{A}(t_1+t_2)} = e^{\mathbf{A}t_1} e^{\mathbf{A}t_2} = \mathbf{\Phi}(t_1)\mathbf{\Phi}(t_2) = \mathbf{\Phi}(t_2)\mathbf{\Phi}(t_1)$
4. $[\mathbf{\Phi}(t)]^n = \mathbf{\Phi}(nt)$
5. $\mathbf{\Phi}(t_2 - t_1)\mathbf{\Phi}(t_1 - t_0) = \mathbf{\Phi}(t_2 - t_0) = \mathbf{\Phi}(t_1 - t_0)\mathbf{\Phi}(t_2 - t_1)$

---

*Example 14-6.* Obtain the state-transition matrix $\mathbf{\Phi}(t)$ of the following system:

$$\begin{bmatrix} \dot{x}_1 \\ \dot{x}_2 \end{bmatrix} = \begin{bmatrix} 0 & 1 \\ -2 & -3 \end{bmatrix} \begin{bmatrix} x_1 \\ x_2 \end{bmatrix}$$

Obtain also the inverse of the state-transition matrix, $\mathbf{\Phi}^{-1}(t)$.

For this system,

$$\mathbf{A} = \begin{bmatrix} 0 & 1 \\ -2 & -3 \end{bmatrix}$$

The state-transition matrix $\mathbf{\Phi}(t)$ is given by

$$\mathbf{\Phi}(t) = e^{\mathbf{A}t} = \mathscr{L}^{-1}[(s\mathbf{I} - \mathbf{A})^{-1}]$$

Since

$$s\mathbf{I} - \mathbf{A} = \begin{bmatrix} s & 0 \\ 0 & s \end{bmatrix} - \begin{bmatrix} 0 & 1 \\ -2 & -3 \end{bmatrix} = \begin{bmatrix} s & -1 \\ 2 & s+3 \end{bmatrix}$$

The inverse of $(s\mathbf{I} - \mathbf{A})$ is given by

$$(s\mathbf{I} - \mathbf{A})^{-1} = \frac{1}{(s+1)(s+2)} \begin{bmatrix} s+3 & 1 \\ -2 & s \end{bmatrix}$$

$$= \begin{bmatrix} \dfrac{s+3}{(s+1)(s+2)} & \dfrac{1}{(s+1)(s+2)} \\ \dfrac{-2}{(s+1)(s+2)} & \dfrac{s}{(s+1)(s+2)} \end{bmatrix}$$

Hence

$$\mathbf{\Phi}(t) = e^{\mathbf{A}t} = \mathscr{L}^{-1}[(s\mathbf{I} - \mathbf{A})^{-1}]$$

$$= \begin{bmatrix} 2e^{-t} - e^{-2t} & e^{-t} - e^{-2t} \\ -2e^{-t} + 2e^{-2t} & -e^{-t} + 2e^{-2t} \end{bmatrix}$$

Noting that $\mathbf{\Phi}^{-1}(t) = \mathbf{\Phi}(-t)$, we obtain the inverse of the state-transition matrix as follows:

$$\mathbf{\Phi}^{-1}(t) = e^{-\mathbf{A}t} = \begin{bmatrix} 2e^{t} - e^{2t} & e^{t} - e^{2t} \\ -2e^{t} + 2e^{2t} & -e^{t} + 2e^{2t} \end{bmatrix}$$

**Solution of nonhomogeneous state equations.** We shall begin by considering the scalar case

$$\dot{x} = ax + bu \qquad (14\text{-}43)$$

Let us rewrite Eq. (14-43) as

$$\dot{x} - ax = bu$$

Multiplying both sides of this equation by $e^{-at}$, we obtain

$$e^{-at}[\dot{x}(t) - ax(t)] = \frac{d}{dt}[e^{-at}x(t)] = e^{-at}bu(t)$$

Integrating this equation between 0 and $t$ gives

$$e^{-at}x(t) = x(0) + \int_0^t e^{-a\tau}bu(\tau)\,d\tau$$

or

$$x(t) = e^{at}x(0) + e^{at}\int_0^t e^{-a\tau}bu(\tau)\,d\tau$$

The first term on the right-hand side is the response to the initial condition and the second term is the response to the input $u(t)$.

Let us now consider the nonhomogeneous state equation described by

$$\dot{\mathbf{x}} = \mathbf{A}\mathbf{x} + \mathbf{B}\mathbf{u} \tag{14-44}$$

where

$\mathbf{x} = n$-dimensional vector
$\mathbf{u} = r$-dimensional vector
$\mathbf{A} = n \times n$ constant matrix
$\mathbf{B} = n \times r$ constant matrix

By writing Eq. (14-44) as

$$\dot{\mathbf{x}}(t) - \mathbf{A}\mathbf{x}(t) = \mathbf{B}\mathbf{u}(t)$$

and premultiplying both sides of this equation by $e^{-\mathbf{A}t}$, we obtain

$$e^{-\mathbf{A}t}[\dot{\mathbf{x}}(t) - \mathbf{A}\mathbf{x}(t)] = \frac{d}{dt}[e^{-\mathbf{A}t}\mathbf{x}(t)] = e^{-\mathbf{A}t}\mathbf{B}\mathbf{u}(t)$$

Integrating the preceding equation between 0 and $t$ gives

$$e^{-\mathbf{A}t}\mathbf{x}(t) = \mathbf{x}(0) + \int_0^t e^{-\mathbf{A}\tau}\mathbf{B}\mathbf{u}(\tau)\,d\tau$$

or

$$\mathbf{x}(t) = e^{\mathbf{A}t}\mathbf{x}(0) + \int_0^t e^{\mathbf{A}(t-\tau)}\mathbf{B}\mathbf{u}(\tau)\,d\tau \tag{14-45}$$

Equation (14-45) can also be written as

$$\mathbf{x}(t) = \mathbf{\Phi}(t)\mathbf{x}(0) + \int_0^t \mathbf{\Phi}(t-\tau)\mathbf{B}\mathbf{u}(\tau)\,d\tau \tag{14-46}$$

where

$$\mathbf{\Phi}(t) = e^{\mathbf{A}t}$$

Equation (14-45) or (14-46) is the solution of Eq. (14-44). The solution $\mathbf{x}(t)$ is clearly the sum of a term consisting of the transition of the initial state and a term arising from the input vector.

**Laplace transform approach to the solution of nonhomogeneous state equations.** The solution of the nonhomogeneous state equation

$$\dot{\mathbf{x}} = \mathbf{A}\mathbf{x} + \mathbf{B}\mathbf{u}$$

can also be obtained by the Laplace transform approach. The Laplace transform of Eq. (14-44) yields

$$s\mathbf{X}(s) - \mathbf{x}(0) = \mathbf{A}\mathbf{X}(s) + \mathbf{B}\mathbf{U}(s)$$

or

$$(s\mathbf{I} - \mathbf{A})\mathbf{X}(s) = \mathbf{x}(0) + \mathbf{B}\mathbf{U}(s)$$

Premultiplying both sides of this last equation by $(s\mathbf{I} - \mathbf{A})^{-1}$, we obtain

$$\mathbf{X}(s) = (s\mathbf{I} - \mathbf{A})^{-1}\mathbf{x}(0) + (s\mathbf{I} - \mathbf{A})^{-1}\mathbf{B}\mathbf{U}(s)$$

Using the relationship given by Eq. (14-40) gives

$$\mathbf{X}(s) = \mathscr{L}[e^{\mathbf{A}t}]\mathbf{x}(0) + \mathscr{L}[e^{\mathbf{A}t}]\mathbf{B}\mathbf{U}(s)$$

The inverse Laplace transform of this last equation can be obtained by use of the convolution integral as follows:

$$\mathbf{x}(t) = e^{\mathbf{A}t}\mathbf{x}(0) + \int_0^t e^{\mathbf{A}(t-\tau)}\mathbf{B}\mathbf{u}(\tau)\,d\tau$$

**Solution in terms of $\mathbf{x}(t_0)$.** Thus far we have assumed the initial time is zero. If, however, the initial time is given by $t_0$ instead of 0, then the solution to Eq. (14-44) must be modified to

$$\mathbf{x}(t) = e^{\mathbf{A}(t-t_0)}\mathbf{x}(t_0) + \int_{t_0}^t e^{\mathbf{A}(t-\tau)}\mathbf{B}\mathbf{u}(\tau)\,d\tau \qquad (14\text{-}47)$$

---

*Example 14-7.* Obtain the time response of the following system:

$$\begin{bmatrix} \dot{x}_1 \\ \dot{x}_2 \end{bmatrix} = \begin{bmatrix} 0 & 1 \\ -2 & -3 \end{bmatrix}\begin{bmatrix} x_1 \\ x_2 \end{bmatrix} + \begin{bmatrix} 0 \\ 1 \end{bmatrix}[u]$$

where $u(t)$ is the unit-step function occurring at $t = 0$, or

$$u(t) = 1(t)$$

For this system

$$\mathbf{A} = \begin{bmatrix} 0 & 1 \\ -2 & -3 \end{bmatrix}, \quad \mathbf{B} = \begin{bmatrix} 0 \\ 1 \end{bmatrix}$$

The state-transition matrix $\boldsymbol{\Phi}(t) = e^{\mathbf{A}t}$ was obtained in Example 14-6 as

$$\boldsymbol{\Phi}(t) = e^{\mathbf{A}t} = \begin{bmatrix} 2e^{-t} - e^{-2t} & e^{-t} - e^{-2t} \\ -2e^{-t} + 2e^{-2t} & -e^{-t} + 2e^{-2t} \end{bmatrix}$$

The response to the unit-step input is then obtained as

$$\mathbf{x}(t) = e^{\mathbf{A}t}\mathbf{x}(0) + \int_0^t \begin{bmatrix} 2e^{-(t-\tau)} - e^{-2(t-\tau)} & e^{-(t-\tau)} - e^{-2(t-\tau)} \\ -2e^{-(t-\tau)} + 2e^{-2(t-\tau)} & -e^{-(t-\tau)} + 2e^{-2(t-\tau)} \end{bmatrix}\begin{bmatrix} 0 \\ 1 \end{bmatrix}[1]\,d\tau$$

or

$$\begin{bmatrix} x_1(t) \\ x_2(t) \end{bmatrix} = \begin{bmatrix} 2e^{-t} - e^{-2t} & e^{-t} - e^{-2t} \\ -2e^{-t} + 2e^{-2t} & -e^{-t} + 2e^{-2t} \end{bmatrix}\begin{bmatrix} x_1(0) \\ x_2(0) \end{bmatrix} + \begin{bmatrix} \frac{1}{2} - e^{-t} + \frac{1}{2}e^{-2t} \\ e^{-t} - e^{-2t} \end{bmatrix}$$

If the initial state is zero, or $\mathbf{x}(0) = \mathbf{0}$, then $\mathbf{x}(t)$ can be simplified to

$$\begin{bmatrix} x_1(t) \\ x_2(t) \end{bmatrix} = \begin{bmatrix} \frac{1}{2} - e^{-t} + \frac{1}{2}e^{-2t} \\ e^{-t} - e^{-2t} \end{bmatrix}$$

## 14-4  TRANSFER MATRIX

In Section 4-6 we defined the transfer matrix. Since the concept of the transfer matrix is an extension of that of the transfer function, here we shall first obtain transfer functions of single-input–single-output systems and then transfer matrices of multiple-input–multiple-output systems from state and output equations.

**Transfer functions.** We shall derive the transfer function of a single-input–single-output system from the Laplace-transformed version of the state and output equations.

Let us consider the system whose transfer function is given by

$$\frac{Y(s)}{U(s)} = G(s) \tag{14-48}$$

In Section 14-3, we showed that the state-space representation for this system is given by

$$\dot{\mathbf{x}} = \mathbf{A}\mathbf{x} + \mathbf{B}u \tag{14-49}$$

$$y = \mathbf{C}\mathbf{x} + Du \tag{14-50}$$

where $\mathbf{x}$ is the state vector, $u$ is the input, and $y$ is the output. The Laplace transforms of Eqs. (14-49) and (14-50) are given by

$$s\mathbf{X}(s) - \mathbf{x}(0) = \mathbf{A}\mathbf{X}(s) + \mathbf{B}U(s) \tag{14-51}$$

$$Y(s) = \mathbf{C}\mathbf{X}(s) + DU(s) \tag{14-52}$$

Since the transfer function was previously defined as the ratio of the Laplace transform of the output to the Laplace transform of the input when the initial conditions were zero, we assume that $\mathbf{x}(0)$ in Eq. (14-51) is zero.

By substituting $\mathbf{X}(s) = (s\mathbf{I} - \mathbf{A})^{-1}\mathbf{B}U(s)$ into Eq. (14-52), we obtain

$$Y(s) = [\mathbf{C}(s\mathbf{I} - \mathbf{A})^{-1}\mathbf{B} + D]U(s) \tag{14-53}$$

Upon comparing Eq. (14-53) with Eq. (14-48), we see that

$$G(s) = \mathbf{C}(s\mathbf{I} - \mathbf{A})^{-1}\mathbf{B} + D \tag{14-54}$$

This is the transfer-function expression in terms of $\mathbf{A}$, $\mathbf{B}$, $\mathbf{C}$, and $D$.

Note that the right-hand side of Eq. (14-54) involves $(s\mathbf{I} - \mathbf{A})^{-1}$. Hence $G(s)$ can be written as

$$G(s) = \frac{Q(s)}{|s\mathbf{I} - \mathbf{A}|}$$

where $Q(s)$ is a polynomial in $s$. Therefore, $|s\mathbf{I} - \mathbf{A}|$ is equal to the characteristic polynomial of $G(s)$. In other words, the eigenvalues of $\mathbf{A}$ are identical to the poles of $G(s)$.

---

*Example 14-8.* Obtain the transfer function of the system shown in Fig. 14-7.

From the diagram, we obtain the following state and output equations:

$$\dot{x}_1 = -5x_1 - x_2 + 2u$$
$$\dot{x}_2 = 3x_1 - x_2 + 5u$$
$$y = x_1 + 2x_2$$

In vector-matrix form, we have

$$\begin{bmatrix} \dot{x}_1 \\ \dot{x}_2 \end{bmatrix} = \begin{bmatrix} -5 & -1 \\ 3 & -1 \end{bmatrix} \begin{bmatrix} x_1 \\ x_2 \end{bmatrix} + \begin{bmatrix} 2 \\ 5 \end{bmatrix} [u]$$

$$y = [1 \quad 2] \begin{bmatrix} x_1 \\ x_2 \end{bmatrix}$$

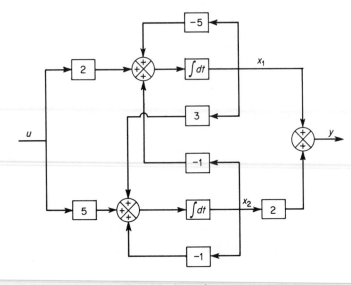

**Fig. 14-7.** Control system.

The transfer function for the system is then

$$G(s) = C(sI - A)^{-1}B$$

$$= [1 \quad 2] \begin{bmatrix} s+5 & 1 \\ -3 & s+1 \end{bmatrix}^{-1} \begin{bmatrix} 2 \\ 5 \end{bmatrix}$$

$$= [1 \quad 2] \begin{bmatrix} \dfrac{s+1}{(s+2)(s+4)} & \dfrac{-1}{(s+2)(s+4)} \\ \dfrac{3}{(s+2)(s+4)} & \dfrac{s+5}{(s+2)(s+4)} \end{bmatrix} \begin{bmatrix} 2 \\ 5 \end{bmatrix}$$

$$= \dfrac{12s+59}{(s+2)(s+4)}$$

**Transfer matrix.** The transfer matrix $G(s)$ relates the output $Y(s)$ to the input $U(s)$, or

$$Y(s) = G(s)U(s) \tag{14-55}$$

If the input vector $u$ is $r$ dimensional and the output vector $y$ is $m$ dimensional, then the transfer matrix is an $m \times r$ matrix. In an expanded form, Eq. (14-55) can be written as

$$\begin{bmatrix} y_1 \\ y_2 \\ \cdot \\ \cdot \\ \cdot \\ y_m \end{bmatrix} = \begin{bmatrix} G_{11}(s) & G_{12}(s) & \cdots & G_{1r}(s) \\ G_{21}(s) & G_{22}(s) & \cdots & G_{2r}(s) \\ \cdot & \cdot & & \cdot \\ \cdot & \cdot & & \cdot \\ \cdot & \cdot & & \cdot \\ G_{m1}(s) & G_{m2}(s) & \cdots & G_{mr}(s) \end{bmatrix} \begin{bmatrix} u_1 \\ u_2 \\ \cdot \\ \cdot \\ \cdot \\ u_r \end{bmatrix}$$

The $(i, j)$th element $G_{ij}(s)$ of $G(s)$ is the transfer function relating the $i$th output to the $j$th input.

By following the same steps as used in deriving Eq. (14-54) we obtain the transfer matrix for multiple-input–multiple-output systems as follows:

$$\mathbf{G}(s) = \mathbf{C}(s\mathbf{I} - \mathbf{A})^{-1}\mathbf{B} + \mathbf{D}$$

Clearly, the transfer function expression given by Eq. (14-54) is a special case of this transfer matrix expression.

**Transfer matrix of closed-loop systems.** Consider the system shown in Fig. 14-8. The system has multiple inputs and multiple outputs. The transfer matrix of the feedforward path is $\mathbf{G}_0(s)$, and that of the feedback path is $\mathbf{H}(s)$. The transfer matrix between the feedback signal vector $\mathbf{B}(s)$ and the error vector $\mathbf{E}(s)$ is obtained as follows: Since

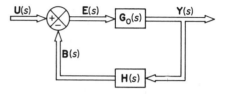

**Fig. 14-8.** Block diagram of a multiple-input–multiple-output system.

$$\mathbf{B}(s) = \mathbf{H}(s)\mathbf{Y}(s)$$
$$= \mathbf{H}(s)\mathbf{G}_0(s)\mathbf{E}(s)$$

we obtain the transfer matrix between $\mathbf{B}(s)$ and $\mathbf{E}(s)$ to be $\mathbf{H}(s)\mathbf{G}_0(s)$. Thus the transfer matrix of the cascaded elements is the product of the transfer matrices of the individual elements. (Note that the order of the matrix multiplication is very important since matrix multiplication is, in general, not commutative.)

The transfer matrix of the closed-loop system is obtained as follows: Since

$$\mathbf{Y}(s) = \mathbf{G}_0(s)[\mathbf{U}(s) - \mathbf{B}(s)]$$
$$= \mathbf{G}_0(s)[\mathbf{U}(s) - \mathbf{H}(s)\mathbf{Y}(s)]$$

we obtain

$$[\mathbf{I} + \mathbf{G}_0(s)\mathbf{H}(s)]\mathbf{Y}(s) = \mathbf{G}_0(s)\mathbf{U}(s)$$

Premultiplying both sides of this last equation by $[\mathbf{I} + \mathbf{G}_0(s)\mathbf{H}(s)]^{-1}$, we obtain

$$\mathbf{Y}(s) = [\mathbf{I} + \mathbf{G}_0(s)\mathbf{H}(s)]^{-1}\mathbf{G}_0(s)\mathbf{U}(s)$$

The closed-loop transfer matrix $\mathbf{G}(s)$ is then given by

$$\mathbf{G}(s) = [\mathbf{I} + \mathbf{G}_0(s)\mathbf{H}(s)]^{-1}\mathbf{G}_0(s) \qquad (14\text{-}56)$$

**Noninteraction in multiple-input–multiple-output systems.** Many process control systems have multiple inputs and multiple outputs, and it is often desired that changes in one reference input affect only one output. (If such noninteraction can be achieved, it is easier to maintain each output value at a desired constant value in the absence of external disturbances.)

Let us consider the transfer matrix $\mathbf{G}_p(s)$ (an $n \times n$ matrix) of a plant and design a series compensator $\mathbf{G}_c(s)$ (also an $n \times n$ matrix) such that the $n$ inputs and $n$ outputs are uncoupled. If noninteraction, or uncoupling, between the $n$ inputs and

$n$ outputs is desired, the closed-loop transfer matrix must be diagonal, or

$$\mathbf{G}(s) = \begin{bmatrix} G_{11}(s) & & & 0 \\ & G_{22}(s) & & \\ & & \cdot & \\ & & & \cdot \\ 0 & & & G_{nn}(s) \end{bmatrix}$$

We shall consider the case where the feedback matrix $\mathbf{H}(s)$ is the identity matrix. Then from Eq. (14-56), we obtain

$$\mathbf{G}(s) = [\mathbf{I} + \mathbf{G}_0(s)]^{-1}\mathbf{G}_0(s) \tag{14-57}$$

where

$$\mathbf{G}_0(s) = \mathbf{G}_p(s)\mathbf{G}_c(s)$$

From Eq. (14-57) we obtain

$$[\mathbf{I} + \mathbf{G}_0(s)]\mathbf{G}(s) = \mathbf{G}_0(s)$$

or

$$\mathbf{G}_0(s)[\mathbf{I} - \mathbf{G}(s)] = \mathbf{G}(s)$$

Postmultiplying both sides of this last equation by $[\mathbf{I} - \mathbf{G}(s)]^{-1}$, we obtain

$$\mathbf{G}_0(s) = \mathbf{G}(s)[\mathbf{I} - \mathbf{G}(s)]^{-1}$$

Since $\mathbf{G}(s)$ is a diagonal matrix, $\mathbf{I} - \mathbf{G}(s)$ is also a diagonal matrix. Then $\mathbf{G}_0(s)$, a product of two diagonal matrices, is also a diagonal matrix. This means that, to achieve noninteraction, we must make $\mathbf{G}_0(s)$ a diagonal matrix, provided the feedback matrix $\mathbf{H}(s)$ is the identity matrix.

---

*Example 14-9.* Consider the system shown in Fig. 14-9. Determine the transfer matrix of the series compensator such that the closed-loop transfer matrix is

$$\mathbf{G}(s) = \begin{bmatrix} \dfrac{1}{s+1} & 0 \\ 0 & \dfrac{1}{5s+1} \end{bmatrix}$$

Since

$$\mathbf{G}_0 = \mathbf{G}(\mathbf{I} - \mathbf{G})^{-1}$$

$$= \begin{bmatrix} \dfrac{1}{s+1} & 0 \\ 0 & \dfrac{1}{5s+1} \end{bmatrix} \begin{bmatrix} \dfrac{s+1}{s} & 0 \\ 0 & \dfrac{5s+1}{5s} \end{bmatrix} = \begin{bmatrix} \dfrac{1}{s} & 0 \\ 0 & \dfrac{1}{5s} \end{bmatrix}$$

$$\begin{bmatrix} Y_1(s) \\ Y_2(s) \end{bmatrix} = \begin{bmatrix} \dfrac{1}{2s+1} & 0 \\ 1 & \dfrac{1}{s+1} \end{bmatrix} \begin{bmatrix} U_1(s) \\ U_2(s) \end{bmatrix}$$

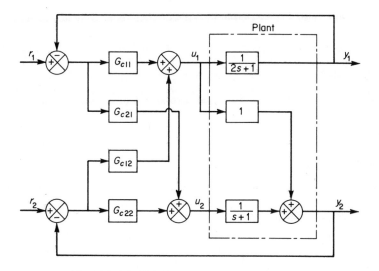

**Fig. 14-9.** Multiple-input–multiple-output system.

and

$$
\begin{bmatrix} U_1(s) \\ U_2(s) \end{bmatrix} = \begin{bmatrix} G_{c11}(s) & G_{c12}(s) \\ G_{c21}(s) & G_{c22}(s) \end{bmatrix} \begin{bmatrix} R_1(s) - Y_1(s) \\ R_2(s) - Y_2(s) \end{bmatrix}
$$

we obtain

$$
\begin{bmatrix} Y_1(s) \\ Y_2(s) \end{bmatrix} = \begin{bmatrix} \dfrac{1}{2s+1} & 0 \\ 1 & \dfrac{1}{s+1} \end{bmatrix} \begin{bmatrix} G_{c11}(s) & G_{c12}(s) \\ G_{c21}(s) & G_{c22}(s) \end{bmatrix} \begin{bmatrix} R_1(s) - Y_1(s) \\ R_2(s) - Y_2(s) \end{bmatrix}
$$

$$
= \begin{bmatrix} \dfrac{1}{s} & 0 \\ 0 & \dfrac{1}{5s} \end{bmatrix} \begin{bmatrix} R_1(s) - Y_1(s) \\ R_2(s) - Y_2(s) \end{bmatrix}
$$

Hence

$$
\mathbf{G}_c(s) = \begin{bmatrix} G_{c11}(s) & G_{c12}(s) \\ G_{c21}(s) & G_{c22}(s) \end{bmatrix}
$$

$$
= \begin{bmatrix} \dfrac{1}{2s+1} & 0 \\ 1 & \dfrac{1}{s+1} \end{bmatrix}^{-1} \begin{bmatrix} \dfrac{1}{s} & 0 \\ 0 & \dfrac{1}{5s} \end{bmatrix}
$$

$$
= \begin{bmatrix} \dfrac{2s+1}{s} & 0 \\ -\dfrac{(s+1)(2s+1)}{s} & \dfrac{s+1}{5s} \end{bmatrix} \tag{14-58}
$$

Equation (14-58) gives the transfer matrix of the series compensator. Note that $G_{c11}(s)$ and $G_{c22}(s)$ are proportional-plus-integral controllers and $G_{c21}(s)$ is a proportional-plus-integral-plus-derivative controller.

It is very important to note that in the present analysis we have not considered external disturbances. In general, in the present approach, cancellations take place in the numerator and denominator. Therefore, some of the eigenvalues will be lost in $\mathbf{G}_p(s)\mathbf{G}_c(s)$. This means that although the present approach will yield the desired result of noninteraction in the responses to reference inputs in the absence of external disturbances, if the system is disturbed by external forces, then the system may become uncontrollable because any motion caused by the canceled eigenvalue cannot be controlled. (We shall discuss details of system controllability in Section 16-2.)

### 14-5  LINEAR TIME-VARYING SYSTEMS

An advantage of the state-space approach to the control system analysis is that it can be extended to linear time-varying systems.

Most of the results obtained in Section 14-4 carry over to linear time-varying systems by changing the transition matrix $\mathbf{\Phi}(t)$ to $\mathbf{\Phi}(t, t_0)$. (For time-varying systems, the transition matrix depends upon both $t$ and $t_0$ and not the difference $t - t_0$. Thus, we cannot always set the initial time equal to zero. There are, of course, cases where $t_0$ is zero.) It is important to realize, however, that the transition matrix for a time-varying system cannot, in general, be given as a matrix exponential.

**Solution of time-varying state equations.** For a scalar differential equation

$$\dot{x} = a(t)x$$

the solution can be given by

$$x(t) = e^{\int_{t_0}^{t} a(\tau)\, d\tau} x(t_0)$$

and the state-transition function is given by

$$\phi(t, t_0) = \exp\left[\int_{t_0}^{t} a(\tau)\, d\tau\right]$$

The same result, however, does not carry over to the vector-matrix differential equation.

Consider the state equation

$$\dot{\mathbf{x}} = \mathbf{A}(t)\mathbf{x} \tag{14-59}$$

where

$\mathbf{x}(t) = n$-dimensional vector
$\mathbf{A}(t) = n \times n$ matrix whose elements are piecewise continuous functions
of $t$ in the interval $t_0 \leq t \leq t_1$

The solution to Eq. (14-59) is given by

$$\mathbf{x}(t) = \mathbf{\Phi}(t, t_0)\mathbf{x}(t_0) \tag{14-60}$$

where $\mathbf{\Phi}(t, t_0)$ is the $n \times n$ nonsingular matrix satisfying the following matrix differential equation:

$$\dot{\mathbf{\Phi}}(t, t_0) = \mathbf{A}(t)\mathbf{\Phi}(t, t_0), \qquad \mathbf{\Phi}(t_0, t_0) = \mathbf{I} \tag{14-61}$$

The fact that Eq. (14-60) is the solution of Eq. (14-61) can be verified easily since

$$\mathbf{x}(t_0) = \mathbf{\Phi}(t_0, t_0)\mathbf{x}(t_0) = \mathbf{I}\mathbf{x}(t_0)$$

and

$$\dot{\mathbf{x}}(t) = \frac{d}{dt}[\mathbf{\Phi}(t, t_0)\mathbf{x}(t_0)]$$

$$= \dot{\mathbf{\Phi}}(t, t_0)\mathbf{x}(t_0)$$

$$= \mathbf{A}(t)\mathbf{\Phi}(t, t_0)\mathbf{x}(t_0) = \mathbf{A}(t)\mathbf{x}(t)$$

We see that the solution of Eq. (14-59) is simply a transformation of the initial state. The matrix $\mathbf{\Phi}(t, t_0)$ is the state-transition matrix for the time-varying system described by Eq. (14-59).

**State-transition matrix for the time-varying case.** It is important to note that the state-transition matrix $\mathbf{\Phi}(t, t_0)$ is given by a matrix exponential if and only if $\mathbf{A}(t)$ and $\int_{t_0}^{t} \mathbf{A}(\tau)\, d\tau$ commute. That is,

$$\mathbf{\Phi}(t, t_0) = \exp\left[\int_{t_0}^{t} \mathbf{A}(\tau)\, d\tau\right] \qquad \text{(If and only if } \mathbf{A}(t) \text{ and } \int_{t_0}^{t} \mathbf{A}(\tau)\, d\tau \text{ commute.)}$$

Note that if $\mathbf{A}(t)$ is a constant matrix or diagonal matrix, $\mathbf{A}(t)$ and $\int_{t_0}^{t} \mathbf{A}(\tau)\, d\tau$ commute. If $\mathbf{A}(t)$ and $\int_{t_0}^{t} \mathbf{A}(\tau)\, d\tau$ do not commute, there is no simple way to compute the state-transition matrix.

To compute $\mathbf{\Phi}(t, t_0)$ numerically, we may use the following series expansion for $\mathbf{\Phi}(t, t_0)$:

$$\mathbf{\Phi}(t, t_0) = \mathbf{I} + \int_{t_0}^{t} \mathbf{A}(\tau)\, d\tau + \int_{t_0}^{t} \mathbf{A}(\tau_1)\left[\int_{t_0}^{\tau_1} \mathbf{A}(\tau_2)\, d\tau_2\right] d\tau_1 + \cdots \qquad (14\text{-}62)$$

This, in general, will not give $\mathbf{\Phi}(t, t_0)$ in a closed form.

---

*Example 14-10.* Obtain $\mathbf{\Phi}(t, 0)$ for the time-varying system

$$\begin{bmatrix} \dot{x}_1 \\ \dot{x}_2 \end{bmatrix} = \begin{bmatrix} 0 & 1 \\ 0 & t \end{bmatrix}\begin{bmatrix} x_1 \\ x_2 \end{bmatrix}$$

To compute $\mathbf{\Phi}(t, 0)$, let us use Eq. (14-62). Since

$$\int_0^t \mathbf{A}(\tau)\, d\tau = \int_0^t \begin{bmatrix} 0 & 1 \\ 0 & \tau \end{bmatrix} d\tau = \begin{bmatrix} 0 & t \\ 0 & \dfrac{t^2}{2} \end{bmatrix}$$

$$\int_0^t \begin{bmatrix} 0 & 1 \\ 0 & \tau_1 \end{bmatrix}\left\{\int_0^{\tau_1} \begin{bmatrix} 0 & 1 \\ 0 & \tau_2 \end{bmatrix} d\tau_2\right\} d\tau_1 = \int_0^t \begin{bmatrix} 0 & 1 \\ 0 & \tau_1 \end{bmatrix}\begin{bmatrix} 0 & \tau_1 \\ 0 & \dfrac{\tau_1^2}{2} \end{bmatrix} d\tau_1 = \begin{bmatrix} 0 & \dfrac{t^3}{6} \\ 0 & \dfrac{t^4}{8} \end{bmatrix}$$

we obtain

$$\mathbf{\Phi}(t, 0) = \begin{bmatrix} 1 & 0 \\ 0 & 1 \end{bmatrix} + \begin{bmatrix} 0 & t \\ 0 & \dfrac{t^2}{2} \end{bmatrix} + \begin{bmatrix} 0 & \dfrac{t^3}{6} \\ 0 & \dfrac{t^4}{8} \end{bmatrix} + \cdots$$

$$= \begin{bmatrix} 1 & t + \dfrac{t^3}{6} + \cdots \\ 0 & 1 + \dfrac{t^2}{2} + \dfrac{t^4}{8} + \cdots \end{bmatrix}$$

**Properties of the state-transition matrix $\Phi(t, t_0)$.** We shall list the properties of the state-transition matrix $\Phi(t, t_0)$ in what follows.

1.
$$\Phi(t_2, t_1)\Phi(t_1, t_0) = \Phi(t_2, t_0)$$

To prove this, note that

$$\mathbf{x}(t_1) = \Phi(t_1, t_0)\mathbf{x}(t_0)$$
$$\mathbf{x}(t_2) = \Phi(t_2, t_0)\mathbf{x}(t_0)$$

Also

$$\mathbf{x}(t_2) = \Phi(t_2, t_1)\mathbf{x}(t_1)$$

Hence

$$\mathbf{x}(t_2) = \Phi(t_2, t_1)\Phi(t_1, t_0)\mathbf{x}(t_0) = \Phi(t_2, t_0)\mathbf{x}(t_0)$$

Thus

$$\Phi(t_2, t_1)\Phi(t_1, t_0) = \Phi(t_2, t_0)$$

2.
$$\Phi(t_1, t_0) = \Phi^{-1}(t_0, t_1)$$

To prove this, note that

$$\Phi(t_1, t_0) = \Phi^{-1}(t_2, t_1)\Phi(t_2, t_0)$$

If we let $t_2 = t_0$ in this last equation, then

$$\Phi(t_1, t_0) = \Phi^{-1}(t_0, t_1)\Phi(t_0, t_0) = \Phi^{-1}(t_0, t_1)$$

**Solution of linear time-varying state equations.** Consider the following state equation:

$$\dot{\mathbf{x}} = \mathbf{A}(t)\mathbf{x} + \mathbf{B}(t)\mathbf{u} \qquad (14\text{-}63)$$

where

$\mathbf{x} = n$-dimentional vector

$\mathbf{u} = r$-dimensional vector

$\mathbf{A}(t) = n \times n$ matrix

$\mathbf{B}(t) = n \times r$ matrix

The elements of $\mathbf{A}(t)$ and $\mathbf{B}(t)$ are assumed to be piecewise continuous functions of $t$ in the interval $t_0 \leq t \leq t_1$.

In order to obtain the solution of Eq. (14-63), let us put

$$\mathbf{x}(t) = \Phi(t, t_0)\xi(t)$$

where $\Phi(t, t_0)$ is the unique matrix satisfying the following equation:

$$\dot{\Phi}(t, t_0) = \mathbf{A}(t)\Phi(t, t_0), \qquad \Phi(t_0, t_0) = \mathbf{I}$$

Then

$$\dot{\mathbf{x}}(t) = \frac{d}{dt}[\Phi(t, t_0)\xi(t)]$$

$$= \dot{\Phi}(t, t_0)\xi(t) + \Phi(t, t_0)\dot{\xi}(t)$$

$$= \mathbf{A}(t)\Phi(t, t_0)\xi(t) + \Phi(t, t_0)\dot{\xi}(t)$$

$$= \mathbf{A}(t)\Phi(t, t_0)\xi(t) + \mathbf{B}(t)\mathbf{u}(t)$$

Therefore

$$\boldsymbol{\Phi}(t, t_0)\dot{\boldsymbol{\xi}}(t) = \mathbf{B}(t)\mathbf{u}(t)$$

or

$$\dot{\boldsymbol{\xi}}(t) = \boldsymbol{\Phi}^{-1}(t, t_0)\mathbf{B}(t)\mathbf{u}(t)$$

Hence

$$\boldsymbol{\xi}(t) = \boldsymbol{\xi}(t_0) + \int_{t_0}^{t} \boldsymbol{\Phi}^{-1}(\tau, t_0)\mathbf{B}(\tau)\mathbf{u}(\tau)\, d\tau$$

Since

$$\boldsymbol{\xi}(t_0) = \boldsymbol{\Phi}^{-1}(t_0, t_0)\mathbf{x}(t_0) = \mathbf{x}(t_0)$$

the solution of Eq. (14-63) is obtained as

$$\mathbf{x}(t) = \boldsymbol{\Phi}(t, t_0)\mathbf{x}(t_0) + \boldsymbol{\Phi}(t, t_0)\int_{t_0}^{t} \boldsymbol{\Phi}^{-1}(\tau, t_0)\mathbf{B}(\tau)\mathbf{u}(\tau)\, d\tau$$

$$= \boldsymbol{\Phi}(t, t_0)\mathbf{x}(t_0) + \int_{t_0}^{t} \boldsymbol{\Phi}(t, \tau)\mathbf{B}(\tau)\mathbf{u}(\tau)\, d\tau \tag{14-64}$$

Computing the right-hand side of Eq. (14-64) for practical cases will require a digital computer.

## 14-6　STATE-SPACE REPRESENTATION OF DISCRETE-TIME SYSTEMS

The state-space approach to the analysis of dynamic systems can be extended to the discrete-time case. The discrete form of the state-space representation is quite analogous to the continuous form.

The most general state-space representation of linear discrete-time systems is

$$\mathbf{x}(k + 1) = \mathbf{G}(k)\mathbf{x}(k) + \mathbf{H}(k)\mathbf{u}(k) \tag{14-65}$$

$$\mathbf{y}(k) = \mathbf{C}(k)\mathbf{x}(k) + \mathbf{D}(k)\mathbf{u}(k) \tag{14-66}$$

where $\mathbf{x}(k)$ is the state vector, $\mathbf{u}(k)$ is the input vector, and $\mathbf{y}(k)$ is the output vector, each specified at $t = kT$, $k = 0, 1, 2, \ldots$, and $T$ is the sampling period. [Note that unless otherwise stated we shall use the simplified notation $\mathbf{x}(k)$ to indicate $\mathbf{x}(kT)$. That is, $\mathbf{x}(k)$ implies the vector $\mathbf{x}(t)$ at $t = kT$. Similarly, we use the simplified notation $\mathbf{u}(k)$, $\mathbf{y}(k)$, $\mathbf{G}(k)$, $\mathbf{H}(k)$, $\mathbf{C}(k)$, and $\mathbf{D}(k)$.] Equations (14-65) and (14-66) correspond to the time-varying case. Figure 14-10 shows the block diagram of the discrete-time system described by Eqs. (14-65) and (14-66). The unit-delay element has a delay time of $T$ seconds.

If the linear discrete-time system is time-invariant, then Eqs. (14-65) and (14-66) are modified to

$$\mathbf{x}(k + 1) = \mathbf{G}\mathbf{x}(k) + \mathbf{H}\mathbf{u}(k) \tag{14-67}$$

$$\mathbf{y}(k) = \mathbf{C}\mathbf{x}(k) + \mathbf{D}\mathbf{u}(k) \tag{14-68}$$

In this section, we shall be concerned mostly with systems described by Eqs. (14-67) and (14-68).

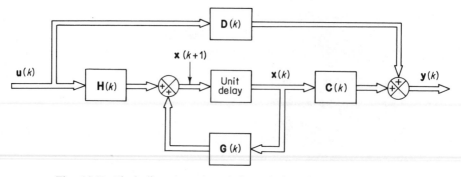

**Fig. 14-10.** Block-diagram representation of the discrete-time system described by Eqs. (14-65) and (14-66).

**State-space representation of time-invariant scalar difference equations where the forcing function is** $bu(k)$**.** Consider the following scalar difference equation:

$$y(k + n) + a_1y(k + n - 1) + a_2y(k + n - 2) + \cdots$$
$$+ a_{n-1}y(k + 1) + a_ny(k) = bu(k) \qquad (14\text{-}69)$$

where $k$ denotes the $k$th sampling instant, $y(k)$ is the system output at the $k$th sampling instant, and $u(k)$ is the input at the $k$th sampling instant. Let us define

$$x_1(k) = y(k)$$
$$x_1(k + 1) = x_2(k)$$
$$x_2(k + 1) = x_3(k)$$
$$\cdots$$
$$x_{n-1}(k + 1) = x_n(k)$$
$$x_n(k + 1) = -a_1x_n(k) - a_2x_{n-1}(k) - \cdots - a_nx_1(k) + bu(k)$$

Then Eq. (14-69) can be written in the following form:

$$\begin{bmatrix} x_1(k+1) \\ x_2(k+1) \\ \vdots \\ \vdots \\ x_{n-1}(k+1) \\ x_n(k+1) \end{bmatrix} = \begin{bmatrix} 0 & 1 & \cdots & 0 & 0 \\ 0 & 0 & \cdots & 0 & 0 \\ \vdots & \vdots & & \vdots & \vdots \\ \vdots & \vdots & & \vdots & \vdots \\ 0 & 0 & \cdots & 0 & 1 \\ -a_n & -a_{n-1} & \cdots & -a_2 & -a_1 \end{bmatrix} \begin{bmatrix} x_1(k) \\ x_2(k) \\ \vdots \\ \vdots \\ x_{n-1}(k) \\ x_n(k) \end{bmatrix} + \begin{bmatrix} 0 \\ 0 \\ \vdots \\ \vdots \\ 0 \\ b \end{bmatrix} [u(k)]$$

$$y(k) = [1 \quad 0 \quad \cdots \quad 0] \begin{bmatrix} x_1(k) \\ x_2(k) \\ \vdots \\ \vdots \\ x_n(k) \end{bmatrix}$$

or

$$\mathbf{x}(k + 1) = \mathbf{G}\mathbf{x}(k) + \mathbf{H}u(k)$$
$$y(k) = \mathbf{C}\mathbf{x}(k)$$

where

$$\mathbf{x}(k) = \begin{bmatrix} x_1(k) \\ x_2(k) \\ \vdots \\ x_{n-1}(k) \\ x_n(k) \end{bmatrix}, \quad \mathbf{G} = \begin{bmatrix} 0 & 1 & \cdots & 0 & 0 \\ 0 & 0 & \cdots & 0 & 0 \\ \vdots & \vdots & & \vdots & \vdots \\ 0 & 0 & \cdots & 0 & 1 \\ -a_n & -a_{n-1} & \cdots & -a_2 & -a_1 \end{bmatrix}, \quad \mathbf{H} = \begin{bmatrix} 0 \\ 0 \\ \vdots \\ 0 \\ b \end{bmatrix}$$

$$\mathbf{C} = [1 \quad 0 \quad \cdots \quad 0 \quad 0]$$

**State-space representation of time-invariant scalar difference equations where the forcing function involves** $u(k), u(k+1), \ldots, u(k+n)$. Consider next the following scalar difference equation:

$$y(k+n) + a_1 y(k+n-1) + a_2 y(k+n-2) + \cdots + a_{n-1} y(k+1) + a_n y(k)$$
$$= b_0 u(k+n) + b_1 u(k+n-1) + b_2 u(k+n-2) + \cdots + b_{n-1} u(k+1)$$
$$+ b_n u(k) \tag{14-70}$$

where $k$ denotes the $k$th sampling instant, $y(k)$ is the system output at the $k$th sampling instant, and $u(k)$ is the input at the $k$th sampling instant.

Similar to the case of the system of scalar differential equations given by Eq. (14-24), define the state variables as follows:

$$x_1(k) = y(k) - h_0 u(k)$$
$$x_2(k) = x_1(k+1) - h_1 u(k)$$
$$x_3(k) = x_2(k+1) - h_2 u(k)$$
$$\cdots$$
$$x_n(k) = x_{n-1}(k+1) - h_{n-1} u(k)$$

where $h_0, h_1, h_2, \ldots, h_n$ are determined from

$$h_0 = b_0$$
$$h_1 = b_1 - a_1 h_0$$
$$h_2 = b_2 - a_1 h_1 - a_2 h_0$$
$$\cdots$$
$$h_n = b_n - a_1 h_{n-1} - \cdots - a_{n-1} h_1 - a_n h_0$$

With this choice of the state variables, we obtain the following discrete-time state equation and output equation for the system of Eq. (14-70):

$$\begin{bmatrix} x_1(k+1) \\ x_2(k+1) \\ \vdots \\ x_{n-1}(k+1) \\ x_n(k+1) \end{bmatrix} = \begin{bmatrix} 0 & 1 & \cdots & 0 & 0 \\ 0 & 0 & \cdots & 0 & 0 \\ \vdots & \vdots & & \vdots & \vdots \\ 0 & 0 & \cdots & 0 & 1 \\ -a_n & -a_{n-1} & \cdots & -a_2 & -a_1 \end{bmatrix} \begin{bmatrix} x_1(k) \\ x_2(k) \\ \vdots \\ x_{n-1}(k) \\ x_n(k) \end{bmatrix} + \begin{bmatrix} h_1 \\ h_2 \\ \vdots \\ h_{n-1} \\ h_n \end{bmatrix} [u(k)]$$

$$y(k) = [1 \quad 0 \quad \cdots \quad 0] \begin{bmatrix} x_1(k) \\ x_2(k) \\ \vdots \\ \vdots \\ x_n(k) \end{bmatrix} + h_0 u(k)$$

or

$$\mathbf{x}(k+1) = \mathbf{G}\mathbf{x}(k) + \mathbf{H}u(k)$$
$$y(k) = \mathbf{C}\mathbf{x}(k) + Du(k)$$

where

$$\mathbf{x}(k) = \begin{bmatrix} x_1(k) \\ x_2(k) \\ \vdots \\ \vdots \\ x_{n-1}(k) \\ x_n(k) \end{bmatrix}, \quad \mathbf{G} = \begin{bmatrix} 0 & 1 & \cdots & 0 & 0 \\ 0 & 0 & \cdots & 0 & 0 \\ \vdots & \vdots & & \vdots & \vdots \\ \vdots & \vdots & & \vdots & \vdots \\ 0 & 0 & \cdots & 0 & 1 \\ -a_n & -a_{n-1} & \cdots & -a_2 & -a_1 \end{bmatrix}, \quad \mathbf{H} = \begin{bmatrix} h_1 \\ h_2 \\ \vdots \\ h_{n-1} \\ h_n \end{bmatrix}$$

$$\mathbf{C} = [1 \quad 0 \quad \cdots \quad 0], \quad D = h_0 = b_0$$

The initial conditions $x_1(0), x_2(0), \ldots, x_n(0)$ are determined from

$$\bar{x}_1(0) = y(0) - h_0 u(0)$$
$$x_2(0) = y(1) - h_0 u(1) - h_1 u(0)$$
$$x_3(0) = y(2) - h_0 u(2) - h_1 u(1) - h_2 u(0)$$
$$\cdots$$
$$x_n(0) = y(n-1) - h_0 u(n-1) - h_1 u(n-2) - \cdots - h_{n-2} u(1) - h_{n-1} u(0)$$

*Example 14-11.* Obtain a state-space representation of the system described by

$$y(k+2) + y(k+1) + 0.16y(k) = u(k+1) + 2u(k)$$

By defining the state variables as follows,

$$x_1(k) = y(k)$$
$$x_2(k) = x_1(k+1) - u(k)$$

the difference equation can be put in the standard state-space representation:

$$x_1(k+1) = x_2(k) + u(k)$$
$$x_2(k+1) = -0.16x_1(k) - x_2(k) + u(k)$$
$$y(k) = x_1(k)$$

Rewriting,

$$\begin{bmatrix} x_1(k+1) \\ x_2(k+1) \end{bmatrix} = \begin{bmatrix} 0 & 1 \\ -0.16 & -1 \end{bmatrix} \begin{bmatrix} x_1(k) \\ x_2(k) \end{bmatrix} + \begin{bmatrix} 1 \\ 1 \end{bmatrix} [u(k)]$$

$$y(k) = [1 \quad 0] \begin{bmatrix} x_1(k) \\ x_2(k) \end{bmatrix}$$

The initial conditions are given by

$$\begin{bmatrix} x_1(0) \\ x_2(0) \end{bmatrix} = \begin{bmatrix} y(0) \\ y(1) - u(0) \end{bmatrix}$$

## 14-7  SOLVING THE DISCRETE-TIME STATE EQUATION

In this section, we first present the solution of the discrete-time state equation:

$$\mathbf{x}(k + 1) = \mathbf{Gx}(k) + \mathbf{Hu}(k) \qquad (14\text{-}71)$$

by use of a recursion procedure and then by use of the $z$-transform method. Then we discuss the discretization of the continuous-time state equation:

$$\dot{\mathbf{x}} = \mathbf{Ax} + \mathbf{Bu} \qquad (14\text{-}72)$$

That is, we derive the discrete-time state equation of the form given by Eq. (14-71) from the continuous-time state equation, Eq. (14-72).

**Solution of difference equations.** In general, difference equations are easier to solve than differential equations because the former can be solved simply by means of a recursion procedure.

As an example, consider the following difference equation:

$$x(k + 1) + 0.2x(k) = 2u(k)$$

where $x(0) = 0$ and $u(k) = 1$ for $k = 0, 1, 2, \ldots$. The solution $x(1)$ can be found by recursion.

$$x(1) = -0.2x(0) + 2u(0) = 2$$

Similarly,

$$x(2) = -0.2x(1) + 2u(1) = 1.6$$
$$x(3) = -0.2x(2) + 2u(2) = 1.68$$
$$x(4) = -0.2x(3) + 2u(3) = 1.664$$

$\cdot\ \cdot\ \cdot$

[Note that in this method $x(k + 1)$ cannot be computed unless $x(k)$ is known.] This procedure is quite simple and convenient for digital computations.

**Solution of discrete-time state equations.** The foregoing procedure for computing the solution of scalar difference equations by recursion can be carried over to the vector-matrix difference equation, or the discrete-time state equation.

Consider the following state equation and output equation:

$$\mathbf{x}(k + 1) = \mathbf{Gx}(k) + \mathbf{Hu}(k) \qquad (14\text{-}73)$$

$$\mathbf{y}(k) = \mathbf{Cx}(k) + \mathbf{Du}(k) \qquad (14\text{-}74)$$

The solution of Eq. (14-73) for any $k > 0$ may be obtained directly by recursion as follows:

$$\mathbf{x}(1) = \mathbf{Gx}(0) + \mathbf{Hu}(0)$$
$$\mathbf{x}(2) = \mathbf{Gx}(1) + \mathbf{Hu}(1) = \mathbf{G}^2\mathbf{x}(0) + \mathbf{GHu}(0) + \mathbf{Hu}(1)$$
$$\mathbf{x}(3) = \mathbf{Gx}(2) + \mathbf{Hu}(2) = \mathbf{G}^3\mathbf{x}(0) + \mathbf{G}^2\mathbf{Hu}(0) + \mathbf{GHu}(1) + \mathbf{Hu}(2)$$
$$\cdot \quad \cdot \quad \cdot$$

By repeating this procedure, we obtain

$$\mathbf{x}(k) = \mathbf{G}^k\mathbf{x}(0) + \sum_{j=0}^{k-1} \mathbf{G}^{k-j-1}\mathbf{Hu}(j) \qquad (k = 1, 2, 3, \ldots) \qquad (14\text{-}75)$$

Clearly $\mathbf{x}(k)$ consists of two parts, one representing the contribution of the initial state $\mathbf{x}(0)$, and the other the contribution of the input $\mathbf{u}(j), j = 0, 1, 2, \ldots, k-1$.

From Eq. (14-75), we see that the state-transition matrix of the system of Eq. (14-73) is

$$\mathbf{\Phi}(k) = \mathbf{G}^k \qquad (14\text{-}76)$$

It is a unique matrix satisfying

$$\mathbf{\Phi}(k+1) = \mathbf{G}\mathbf{\Phi}(k), \qquad \mathbf{\Phi}(0) = \mathbf{I}$$

In terms of the state transition matrix $\mathbf{\Phi}(k)$, Eq. (14-75) can be written

$$\mathbf{x}(k) = \mathbf{\Phi}(k)\mathbf{x}(0) + \sum_{j=0}^{k-1} \mathbf{\Phi}(k-j-1)\mathbf{Hu}(j) \qquad (14\text{-}77)$$

$$= \mathbf{\Phi}(k)\mathbf{x}(0) + \sum_{j=0}^{k-1} \mathbf{\Phi}(j)\mathbf{Hu}(k-j-1) \qquad (14\text{-}78)$$

Substituting Eq. (14-77) [or Eq. (14-78)] into Eq. (14-74), the output equation can be written

$$\mathbf{y}(k) = \mathbf{C}\mathbf{\Phi}(k)\mathbf{x}(0) + \mathbf{C}\sum_{j=0}^{k-1} \mathbf{\Phi}(k-j-1)\mathbf{Hu}(j) + \mathbf{Du}(k)$$

$$= \mathbf{C}\mathbf{\Phi}(k)\mathbf{x}(0) + \mathbf{C}\sum_{j=0}^{k-1} \mathbf{\Phi}(j)\mathbf{Hu}(k-j-1) + \mathbf{Du}(k)$$

**$z$-transform approach to the solution of discrete-time state equations.** We next present the solution of discrete-time state equations by use of the $z$-transform method. Consider the discrete-time system described by Eq. (14-73) rewritten as

$$\mathbf{x}(k+1) = \mathbf{Gx}(k) + \mathbf{Hu}(k) \qquad (14\text{-}79)$$

Taking the $z$ transform of both sides of Eq. (14-79) gives

$$z\mathbf{X}(z) - z\mathbf{x}(0) = \mathbf{GX}(z) + \mathbf{HU}(z)$$

where $\mathbf{X}(z) = \mathcal{Z}[\mathbf{x}(k)]$ and $\mathbf{U}(z) = \mathcal{Z}[\mathbf{u}(k)]$. Then

$$(z\mathbf{I} - \mathbf{G})\mathbf{X}(z) = z\mathbf{x}(0) + \mathbf{HU}(z) \qquad (14\text{-}80)$$

Premultiplying both sides of Eq. (14-80) by $(z\mathbf{I} - \mathbf{G})^{-1}$, we obtain

$$\mathbf{X}(z) = (z\mathbf{I} - \mathbf{G})^{-1}z\mathbf{x}(0) + (z\mathbf{I} - \mathbf{G})^{-1}\mathbf{HU}(z) \qquad (14\text{-}81)$$

Taking the inverse $z$ transform of both sides of Eq. (14-81) gives

$$\mathbf{x}(k) = \mathcal{Z}^{-1}[(z\mathbf{I} - \mathbf{G})^{-1}z]\mathbf{x}(0) + \mathcal{Z}^{-1}[(z\mathbf{I} - \mathbf{G})^{-1}\mathbf{HU}(z)] \qquad (14\text{-}82)$$

Comparing Eqs. (14-75) and (14-82), we obtain

$$\mathbf{G}^k = \mathcal{Z}^{-1}[(z\mathbf{I} - \mathbf{G})^{-1}z] \qquad (14\text{-}83)$$

$$\sum_{j=0}^{k-1} \mathbf{G}^{k-j-1}\mathbf{H}u(j) = \mathcal{Z}^{-1}[(z\mathbf{I} - \mathbf{G})^{-1}\mathbf{H}U(z)] \qquad (14\text{-}84)$$

where $k = 1, 2, 3, \ldots$ [For Eq. (14-83), see Problem A-13-1.]

Equation (14-84) can also be obtained directly. Taking the $z$ transform of $\sum_{j=0}^{k-1} \mathbf{G}^{k-j-1}\mathbf{H}u(j)$ where $k = 1, 2, 3, \ldots$, we obtain

$$\mathcal{Z}[\sum_{j=0}^{k-1} \mathbf{G}^{k-j-1}\mathbf{H}u(j)] = \sum_{k=0}^{\infty} \sum_{j=0}^{k-1} \mathbf{G}^{k-j-1}\mathbf{H}u(j)z^{-k}$$

$$= \sum_{k=0}^{\infty} z^{-k+j} \sum_{j=0}^{k-1} \mathbf{G}^{k-j-1}\mathbf{H}u(j)z^{-j}$$

$$= \sum_{k=1}^{\infty} z^{-k+j} \sum_{j=0}^{k-1} \mathbf{G}^{k-j-1}\mathbf{H}u(j)z^{-j}$$

$$= \sum_{k=1}^{\infty} [\mathbf{G}^{k-1}\mathbf{H}u(0)z^{-k} + \mathbf{G}^{k-2}\mathbf{H}u(1)z^{-k} + \mathbf{G}^{k-3}\mathbf{H}u(2)z^{-k} + \cdots]$$

$$= (\mathbf{H}z^{-1} + \mathbf{G}\mathbf{H}z^{-2} + \mathbf{G}^2\mathbf{H}z^{-3} + \cdots)$$
$$\times [u(0) + u(1)z^{-1} + u(2)z^{-2} + \cdots]$$

$$= (\mathbf{I} + \mathbf{G}z^{-1} + \mathbf{G}^2z^{-2} + \cdots)\mathbf{H}z^{-1}$$
$$\times [u(0) + u(1)z^{-1} + u(2)z^{-2} + \cdots]$$

$$= (\mathbf{I} - \mathbf{G}z^{-1})^{-1}\mathbf{H}z^{-1} \sum_{k=0}^{\infty} u(k)z^{-k}$$

$$= (z\mathbf{I} - \mathbf{G})^{-1}\mathbf{H}U(z)$$

Finally, note that from Eq. (14-81) we can see that the characteristic equation of the discrete-time system is

$$|z\mathbf{I} - \mathbf{G}| = 0 \qquad (14\text{-}85)$$

Referring to Section 13-6, we know that the discrete-time system is stable if and only if all the roots of the characteristic equation, Eq. (14-85), are in the unit circle centered at the origin of the $z$ plane.

---

*Example 14-12.* Obtain the state-transition matrix of the following discrete-time system:

$$\mathbf{x}(k + 1) = \mathbf{G}\mathbf{x}(k) + \mathbf{H}u(k)$$

where

$$\mathbf{G} = \begin{bmatrix} 0 & 1 \\ -0.16 & -1 \end{bmatrix}, \qquad \mathbf{H} = \begin{bmatrix} 1 \\ 1 \end{bmatrix}$$

Then obtain $\mathbf{x}(k)$ when $u(k) = 1$ for $k = 0, 1, 2, \ldots$. Assume that the initial condition is given by

$$\mathbf{x}(0) = \begin{bmatrix} x_1(0) \\ x_2(0) \end{bmatrix} = \begin{bmatrix} 1 \\ -1 \end{bmatrix}$$

From Eqs. (14-76) and (14-83), we see that the state-transition matrix $\boldsymbol{\Phi}(k)$ is

$$\boldsymbol{\Phi}(k) = \mathbf{G}^k = \mathcal{Z}^{-1}[(z\mathbf{I} - \mathbf{G})^{-1}z]$$

Therefore, we first obtain $(z\mathbf{I} - \mathbf{G})^{-1}$.

$$(z\mathbf{I} - \mathbf{G})^{-1} = \begin{bmatrix} z & -1 \\ 0.16 & z+1 \end{bmatrix}^{-1}$$

$$= \begin{bmatrix} \dfrac{z+1}{(z+0.2)(z+0.8)} & \dfrac{1}{(z+0.2)(z+0.8)} \\ \dfrac{-0.16}{(z+0.2)(z+0.8)} & \dfrac{z}{(z+0.2)(z+0.8)} \end{bmatrix}$$

$$= \begin{bmatrix} \dfrac{4}{3} \\ \dfrac{z+0.2} + \dfrac{-\dfrac{1}{3}}{z+0.8} & \dfrac{\dfrac{5}{3}}{z+0.2} + \dfrac{-\dfrac{5}{3}}{z+0.8} \\ -\dfrac{0.8}{3} \\ \dfrac{z+0.2} + \dfrac{\dfrac{0.8}{3}}{z+0.8} & \dfrac{-\dfrac{1}{3}}{z+0.2} + \dfrac{\dfrac{4}{3}}{z+0.8} \end{bmatrix}$$

Thus $\mathbf{\Phi}(k)$ is obtained as

$$\mathbf{\Phi}(k) = \mathbf{G}^k = \mathcal{Z}^{-1}[(z\mathbf{I} - \mathbf{G})^{-1}z]$$

$$= \mathcal{Z}^{-1} \begin{bmatrix} \dfrac{4}{3}\left(\dfrac{z}{z+0.2}\right) - \dfrac{1}{3}\left(\dfrac{z}{z+0.8}\right) & \dfrac{5}{3}\left(\dfrac{z}{z+0.2}\right) - \dfrac{5}{3}\left(\dfrac{z}{z+0.8}\right) \\ -\dfrac{0.8}{3}\left(\dfrac{z}{z+0.2}\right) + \dfrac{0.8}{3}\left(\dfrac{z}{z+0.8}\right) & -\dfrac{1}{3}\left(\dfrac{z}{z+0.2}\right) + \dfrac{4}{3}\left(\dfrac{z}{z+0.8}\right) \end{bmatrix}$$

$$= \begin{bmatrix} \dfrac{4}{3}(-0.2)^k - \dfrac{1}{3}(-0.8)^k & \dfrac{5}{3}(-0.2)^k - \dfrac{5}{3}(-0.8)^k \\ -\dfrac{0.8}{3}(-0.2)^k + \dfrac{0.8}{3}(-0.8)^k & -\dfrac{1}{3}(-0.2)^k + \dfrac{4}{3}(-0.8)^k \end{bmatrix}$$

Next, compute $\mathbf{x}(k)$. The $z$ transform of $\mathbf{x}(k)$ is given by

$$\mathcal{Z}[\mathbf{x}(k)] = \mathbf{X}(z) = (z\mathbf{I} - \mathbf{G})^{-1}z\mathbf{x}(0) + (z\mathbf{I} - \mathbf{G})^{-1}\mathbf{H}U(z)$$
$$= (z\mathbf{I} - \mathbf{G})^{-1}[z\mathbf{x}(0) + \mathbf{H}U(z)]$$

Since

$$U(z) = \dfrac{z}{z-1}$$

we obtain

$$z\mathbf{x}(0) + \mathbf{H}U(z) = \begin{bmatrix} z \\ -z \end{bmatrix} + \begin{bmatrix} \dfrac{z}{z-1} \\ \dfrac{z}{z-1} \end{bmatrix} = \begin{bmatrix} \dfrac{z^2}{z-1} \\ \dfrac{-z^2+2z}{z-1} \end{bmatrix}$$

Hence

$$\mathbf{X}(z) = (z\mathbf{I} - \mathbf{G})^{-1}[z\mathbf{x}(0) + \mathbf{H}U(z)]$$

$$= \begin{bmatrix} \dfrac{(z^2+2)z}{(z+0.2)(z+0.8)(z-1)} \\ \dfrac{(-z^2+1.84z)z}{(z+0.2)(z+0.8)(z-1)} \end{bmatrix}$$

$$= \begin{bmatrix} \dfrac{-\frac{17}{6}z}{z+0.2} + \dfrac{\frac{22}{9}z}{z+0.8} + \dfrac{\frac{25}{18}z}{z-1} \\ \dfrac{\frac{3.4}{6}z}{z+0.2} + \dfrac{-\frac{17.6}{9}z}{z+0.8} + \dfrac{\frac{7}{18}z}{z-1} \end{bmatrix}$$

Thus

$$\mathbf{x}(k) = \mathcal{Z}^{-1}[\mathbf{X}(z)] = \begin{bmatrix} -\dfrac{17}{6}(-0.2)^k + \dfrac{22}{9}(-0.8)^k + \dfrac{25}{18} \\[2ex] \dfrac{3.4}{6}(-0.2)^k - \dfrac{17.6}{9}(-0.8)^k + \dfrac{7}{18} \end{bmatrix}$$

**Discretization of continuous-time state equations.** If we wish to compute the state $\mathbf{x}(t)$ by use of a digital computer, we must convert a continuous-time state equation to a discrete-time state equation. In the following, we shall present such a procedure. We assume that the input vector $\mathbf{u}(t)$ changes only at the equally spaced sampling instants. (The sampling operation here is fictitious. We shall derive the discrete-time state equation which yields the exact values at $t = kT$, $k = 0, 1, 2, \ldots$.)

Consider the continuous-time state equation

$$\dot{\mathbf{x}} = \mathbf{Ax} + \mathbf{Bu} \qquad (14\text{-}86)$$

In the following, in order to clarify the analysis, we shall use the notation $kT$ and $(k + 1)T$ instead of $k$ and $k + 1$. The discrete-time representation of Eq. (14-86) will take the form

$$\mathbf{x}((k + 1)T) = \mathbf{G}(T)\mathbf{x}(kT) + \mathbf{H}(T)\mathbf{u}(kT) \qquad (14\text{-}87)$$

Note that the matrices $\mathbf{G}$ and $\mathbf{H}$ depend on the sampling period $T$. (Once this period is fixed, $\mathbf{G}$ and $\mathbf{H}$ are constant matrices.)

In order to determine $\mathbf{G}(T)$ and $\mathbf{H}(T)$, we use the solution of Eq. (14-86), or

$$\mathbf{x}(t) = e^{\mathbf{A}t}\mathbf{x}(0) + e^{\mathbf{A}t} \int_0^t e^{-\mathbf{A}\tau}\mathbf{Bu}(\tau)\, d\tau$$

We assume that all the components of $\mathbf{u}(t)$ are constant over the interval between any two consecutive sampling instants, or $\mathbf{u}(t) = \mathbf{u}(kT)$ for the $k$th sampling period. Since

$$\mathbf{x}((k + 1)T) = e^{\mathbf{A}(k+1)T}\mathbf{x}(0) + e^{\mathbf{A}(k+1)T} \int_0^{(k+1)T} e^{-\mathbf{A}\tau}\mathbf{Bu}(\tau)\, d\tau \qquad (14\text{-}88)$$

and

$$\mathbf{x}(kT) = e^{\mathbf{A}kT}\mathbf{x}(0) + e^{\mathbf{A}kT} \int_0^{kT} e^{-\mathbf{A}\tau}\mathbf{Bu}(\tau)\, d\tau \qquad (14\text{-}89)$$

multiplying Eq. (14-89) by $e^{\mathbf{A}T}$ and subtracting from Eq. (14-88), we obtain

$$\mathbf{x}((k + 1)T) = e^{\mathbf{A}T}\mathbf{x}(kT) + e^{\mathbf{A}(k+1)T} \int_{kT}^{(k+1)T} e^{-\mathbf{A}\tau}\mathbf{Bu}(\tau)\, d\tau$$

$$= e^{\mathbf{A}T}\mathbf{x}(kT) + e^{\mathbf{A}T} \int_0^T e^{-\mathbf{A}t}\mathbf{Bu}(kT)\, dt$$

$$= e^{\mathbf{A}T}\mathbf{x}(kT) + \int_0^T e^{\mathbf{A}\lambda}\mathbf{Bu}(kT)\, d\lambda \qquad (14\text{-}90)$$

where $\lambda = T - t$. If we define

$$\mathbf{G}(T) = e^{\mathbf{A}T} \qquad (14\text{-}91)$$

$$\mathbf{H}(T) = \left( \int_0^T e^{\mathbf{A}t}\, dt \right)\mathbf{B} \qquad (14\text{-}92)$$

then Eq. (14-90) becomes

$$\mathbf{x}((k + 1)T) = \mathbf{G}(T)\mathbf{x}(kT) + \mathbf{H}(T)\,\mathbf{u}(kT)$$

which is Eq. (14-87). Thus Eqs. (14-91) and (14-92) give the desired matrices $\mathbf{G}(T)$ and $\mathbf{H}(T)$.

---

*Example 14-13.* Obtain a discrete-time state-space representation of the following continuous-time system:

$$\begin{bmatrix} \dot{x}_1 \\ \dot{x}_2 \end{bmatrix} = \begin{bmatrix} 0 & 1 \\ 0 & -2 \end{bmatrix} \begin{bmatrix} x_1 \\ x_2 \end{bmatrix} + \begin{bmatrix} 0 \\ 1 \end{bmatrix} [u]$$

The desired discrete-time state equation will be of the following form:

$$\mathbf{x}((k + 1)T) = \mathbf{G}(T)\mathbf{x}(kT) + \mathbf{H}(T)u(kT)$$

The matrices $\mathbf{G}(T)$ and $\mathbf{H}(T)$ may be obtained from Eqs. (14-91) and (14-92) as

$$\mathbf{G}(T) = e^{\mathbf{A}T} = \begin{bmatrix} 1 & \frac{1}{2}(1 - e^{-2T}) \\ 0 & e^{-2T} \end{bmatrix}$$

$$\mathbf{H}(T) = \left( \int_0^T e^{\mathbf{A}t}\, dt \right)\mathbf{B}$$

$$= \left\{ \int_0^T \begin{bmatrix} 1 & \frac{1}{2}(1 - e^{-2t}) \\ 0 & e^{-2t} \end{bmatrix} dt \right\} \begin{bmatrix} 0 \\ 1 \end{bmatrix}$$

$$= \begin{bmatrix} \frac{1}{2}\left(T + \dfrac{e^{-2T} - 1}{2}\right) \\ \frac{1}{2}(1 - e^{-2T}) \end{bmatrix}$$

Thus

$$\begin{bmatrix} x_1((k + 1)T) \\ x_2((k + 1)T) \end{bmatrix} = \begin{bmatrix} 1 & \frac{1}{2}(1 - e^{-2T}) \\ 0 & e^{-2T} \end{bmatrix} \begin{bmatrix} x_1(kT) \\ x_2(kT) \end{bmatrix} + \begin{bmatrix} \frac{1}{2}\left(T + \dfrac{e^{-2T} - 1}{2}\right) \\ \frac{1}{2}(1 - e^{-2T}) \end{bmatrix} u(kT)$$

If, for example, the sampling period is 1 second, or $T = 1$, then the discrete-time state equation becomes

$$\begin{bmatrix} x_1(k + 1) \\ x_2(k + 1) \end{bmatrix} = \begin{bmatrix} 1 & 0.432 \\ 0 & 0.135 \end{bmatrix} \begin{bmatrix} x_1(k) \\ x_2(k) \end{bmatrix} + \begin{bmatrix} 0.284 \\ 0.432 \end{bmatrix} [u(k)]$$

## EXAMPLE PROBLEMS AND SOLUTIONS

**PROBLEM A-14-1.** Obtain a state-space representation of the system shown in Fig. 14-11.

**Solution.** In this problem, we shall illustrate a method for deriving a state-space representation of systems given in block-diagram form.

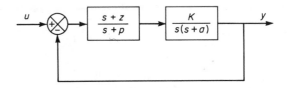

**Fig. 14-11.** Control system.

In the present problem, first expand $(s + z)/(s + p)$ into partial fractions.

$$\frac{s + z}{s + p} = 1 + \frac{z - p}{s + p}$$

Next convert $K/[s(s + a)]$ into the product of $K/s$ and $1/(s + a)$. Then redraw the block diagram, as shown in Fig. 14-12. Defining a set of state variables, as shown in Fig. 14-12, we obtain the following equations:

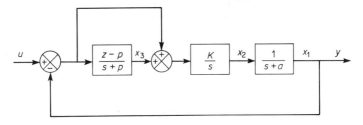

**Fig. 14-12.** Block diagram defining the state variables for the system shown in Fig. 14-11.

$$\dot{x}_1 = -ax_1 + x_2$$
$$\dot{x}_2 = -Kx_1 + Kx_3 + Ku$$
$$\dot{x}_3 = -(z - p)x_1 - px_3 + (z - p)u$$
$$y = x_1$$

Rewriting gives

$$\begin{bmatrix} \dot{x}_1 \\ \dot{x}_2 \\ \dot{x}_3 \end{bmatrix} = \begin{bmatrix} -a & 1 & 0 \\ -K & 0 & K \\ -(z - p) & 0 & -p \end{bmatrix} \begin{bmatrix} x_1 \\ x_2 \\ x_3 \end{bmatrix} + \begin{bmatrix} 0 \\ K \\ z - p \end{bmatrix} [u]$$

$$y = [1 \quad 0 \quad 0] \begin{bmatrix} x_1 \\ x_2 \\ x_3 \end{bmatrix}$$

**PROBLEM A-14-2.** Derive a state equation for the system shown in Fig. 14-13. Let the current through the inductance $L$ be $x_1$, namely, $x_1 = i_1 - i_2$ and the voltage across the capacitor be $x_2$. Assume that $e(t)$ is the input to the system. Take $x_1$ and $x_2$ as the state variables for the system.

**Solution.** From Fig. 14-13, we obtain the following equations:

$$L\left(\frac{di_1}{dt} - \frac{di_2}{dt}\right) + R_1 i_1 = e(t)$$

$$\frac{1}{C}\int i_2 dt + R_2 i_2 + L\left(\frac{di_2}{dt} - \frac{di_1}{dt}\right) = 0$$

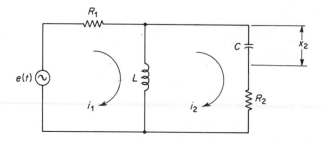

**Fig. 14-13.** Network system.

In terms of $x_1$ and $x_2$ we can rewrite these equations as

$$L\frac{dx_1}{dt} + R_1(x_1 + i_2) = e(t)$$

$$x_2 + R_2i_2 + R_1(x_1 + i_2) = e(t)$$

$$x_2 = \frac{1}{C}\int i_2\,dt$$

To obtain the state equation, we must eliminate $i_2$. To do so, let us take the Laplace transforms of these equations:

$$L[sX_1(s) - x_1(0)] + R_1[X_1(s) + I_2(s)] = E(s)$$
$$X_2(s) + R_2I_2(s) + R_1[X_1(s) + I_2(s)] = E(s)$$

$$sX_2(s) - x_2(0) = \frac{1}{C}I_2(s)$$

Eliminating $I_2(s)$ from the last three equations gives

$$L[sX_1(s) - x_1(0)] = \frac{-R_1R_2}{R_1 + R_2}X_1(s) + \frac{R_1}{R_1 + R_2}X_2(s) + \frac{R_2}{R_1 + R_2}E(s)$$

$$sX_2(s) - x_2(0) = \frac{-R_1}{C(R_1 + R_2)}X_1(s) + \frac{-1}{C(R_1 + R_2)}X_2(s) + \frac{1}{C(R_1 + R_2)}E(s)$$

Inverse Laplace transforming gives

$$\dot{x}_1(t) = \frac{-R_1R_2}{L(R_1 + R_2)}x_1(t) + \frac{R_1}{L(R_1 + R_2)}x_2(t) + \frac{R_2}{L(R_1 + R_2)}e(t)$$

$$\dot{x}_2(t) = \frac{-R_1}{C(R_1 + R_2)}x_1(t) + \frac{-1}{C(R_1 + R_2)}x_2(t) + \frac{1}{C(R_1 + R_2)}e(t)$$

The state equation for the system is then given by

$$\begin{bmatrix} \dot{x}_1(t) \\ \dot{x}_2(t) \end{bmatrix} = \begin{bmatrix} \dfrac{-R_1R_2}{L(R_1 + R_2)} & \dfrac{R_1}{L(R_1 + R_2)} \\ \dfrac{-R_1}{C(R_1 + R_2)} & \dfrac{-1}{C(R_1 + R_2)} \end{bmatrix} \begin{bmatrix} x_1(t) \\ x_2(t) \end{bmatrix} + \begin{bmatrix} \dfrac{R_2}{L(R_1 + R_2)} \\ \dfrac{1}{C(R_1 + R_2)} \end{bmatrix} [e(t)]$$

**PROBLEM A-14-3.** Gyros for sensing angular motion are commonly used in inertial guidance systems, autopilot systems, etc.

Figure 14-14 (a) shows a single degree of freedom gyro. The spinning wheel is mounted in a movable gimbal which is, in turn, mounted in a gyro case. The gimbal is free to move relative to the case, about the output axis OB. Note that the output axis is perpendicular to the wheel spin axis. The input axis around which a turning rate, or angle, is measured is perpendicular to both the output and spin axes. Information on the input signal (the

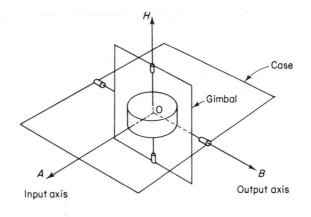

(a)

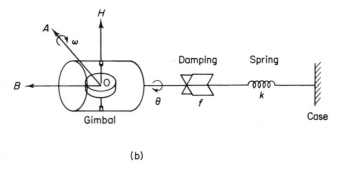

**Fig. 14-14.** (a) Schematic diagram of a single degree of freedom gyro; (b) functional diagram of the gyro shown in (a).

(b)

turning rate or angle around the input axis) is obtained from the resulting motion of the gimbal about the output axis, relative to the case.

Figure 14-14 (b) shows a functional diagram of the gyro system. The equation of motion about the output axis can be obtained by equating the rate of change of angular momentum to the sum of the external torques.

The change in angular momentum about axis $OB$ consists of two parts: $I\ddot{\theta}$, the change due to acceleration of the gimbal around axis $OB$, and $-H\omega\cos\theta$, the change due to the turning of the wheel angular-momentum vector around axis $OA$. The external torques consist of $-f\dot{\theta}$, the damping torque, and $-k\theta$, the spring torque. Thus the equation of the gyro system is

$$I\ddot{\theta} - H\omega\cos\theta = -f\dot{\theta} - k\theta$$

or

$$I\ddot{\theta} + f\dot{\theta} + k\theta = H\omega\cos\theta \tag{14-93}$$

In practice, $\theta$ is a very small angle, usually not more than $\pm 2.5$ degrees.

Obtain a state-space representation of the gyro system.

**Solution.** In this system, $\theta$ and $\dot{\theta}$ may be chosen as state variables. The input variable is $\omega$. Let us define

$$\mathbf{x} = \begin{bmatrix} x_1 \\ x_2 \end{bmatrix} = \begin{bmatrix} \theta \\ \dot{\theta} \end{bmatrix}, \quad u = \omega$$

Then Eq. (14-93) can be written as follows:

$$\dot{x}_1 = x_2$$

$$\dot{x}_2 = -\frac{k}{I} x_1 - \frac{f}{I} x_2 + \frac{H}{I} u \cos x_1$$

or

$$\dot{\mathbf{x}} = \mathbf{f}(\mathbf{x}, u)$$

where

$$\mathbf{x} = \begin{bmatrix} x_1 \\ x_2 \end{bmatrix}, \quad \mathbf{f}(\mathbf{x}, u) = \begin{bmatrix} f_1(\mathbf{x}, u) \\ f_2(\mathbf{x}, u) \end{bmatrix} = \begin{bmatrix} x_2 \\ -\dfrac{k}{I} x_1 - \dfrac{f}{I} x_2 + \dfrac{H}{I} u \cos x_1 \end{bmatrix}$$

Clearly $f_2(\mathbf{x}, u)$ involves a nonlinear term in $x_1$ and $u$. By expanding $\cos x_1$ into its series representation,

$$\cos x_1 = 1 - \frac{1}{2} x_1^2 + \cdots$$

and noting that $x_1$ is a very small angle, we may approximate $\cos x_1$ by unity to obtain the following linearized state equation:

$$\begin{bmatrix} \dot{x}_1 \\ \dot{x}_2 \end{bmatrix} = \begin{bmatrix} 0 & 1 \\ -\dfrac{k}{I} & -\dfrac{f}{I} \end{bmatrix} \begin{bmatrix} x_1 \\ x_2 \end{bmatrix} + \begin{bmatrix} 0 \\ \dfrac{H}{I} \end{bmatrix} [u]$$

If $\theta$ is considered the output $y$, then

$$y = \theta$$

or

$$y = \begin{bmatrix} 1 & 0 \end{bmatrix} \begin{bmatrix} x_1 \\ x_2 \end{bmatrix}$$

**PROBLEM A-14-4.** Consider the system described by

$$\dot{\mathbf{x}} = \mathbf{A}\mathbf{x} + \mathbf{B}\mathbf{u}$$

where

$\mathbf{x} = n$-dimensional vector
$\mathbf{u} = r$-dimensional vector
$\mathbf{A} = n \times n$ constant matrix
$\mathbf{B} = n \times r$ constant matrix

Obtain the response to each of the following inputs:

1. The $r$ components of $\mathbf{u}$ are impulse functions of various magnitudes.
2. The $r$ components of $\mathbf{u}$ are step functions of various magnitudes.
3. The $r$ components of $\mathbf{u}$ are ramp functions of various magnitudes.

Assume that each input is given at $t = 0$.

**Solution.** Substituting $t_0 = 0-$ into Eq. (14-47), we can write the solution $\mathbf{x}(t)$ as follows:

$$\mathbf{x}(t) = e^{\mathbf{A}t}\mathbf{x}(0-) + \int_{0-}^{t} e^{\mathbf{A}(t-\tau)}\mathbf{B}\mathbf{u}(\tau)\, d\tau$$

1. *Impulse response:* Let us write the impulse input $\mathbf{u}(t)$ as

$$\mathbf{u}(t) = \delta(t)\mathbf{w}$$

where $\mathbf{w}$ is a vector whose components are the magnitudes of $r$ impulse functions applied at $t = 0$. The solution to the impulse $\delta(t)\mathbf{w}$ given at $t = 0$ is

$$\mathbf{x}(t) = e^{\mathbf{A}t}\mathbf{x}(0-) + \int_{0-}^{t} e^{\mathbf{A}(t-\tau)}\mathbf{B}\delta(\tau)\mathbf{w}\, d\tau$$

$$= e^{\mathbf{A}t}\mathbf{x}(0-) + e^{\mathbf{A}t}\mathbf{B}\mathbf{w}$$

2. *Step response:* Let us write the step input $\mathbf{u}(t)$ as

$$\mathbf{u}(t) = \mathbf{k}$$

where $\mathbf{k}$ is a vector whose components are the magnitudes of $r$ step functions applied at $t = 0$. The solution to the step input at $t = 0$ is given by

$$\mathbf{x}(t) = e^{\mathbf{A}t}\mathbf{x}(0) + \int_{0}^{t} e^{\mathbf{A}(t-\tau)}\mathbf{B}\mathbf{k}\, d\tau$$

$$= e^{\mathbf{A}t}\mathbf{x}(0) + e^{\mathbf{A}t}\left[\int_{0}^{t}\left(\mathbf{I} - \mathbf{A}\tau + \frac{\mathbf{A}^2\tau^2}{2!} - \cdots\right)d\tau\right]\mathbf{B}\mathbf{k}$$

$$= e^{\mathbf{A}t}\mathbf{x}(0) + e^{\mathbf{A}t}\left(\mathbf{I}t - \frac{\mathbf{A}t^2}{2!} + \frac{\mathbf{A}^2 t^3}{3!} - \cdots\right)\mathbf{B}\mathbf{k}$$

If $\mathbf{A}$ is nonsingular, then this last equation can be simplified to give

$$\mathbf{x}(t) = e^{\mathbf{A}t}\mathbf{x}(0) + e^{\mathbf{A}t}[-(\mathbf{A}^{-1})(e^{-\mathbf{A}t} - \mathbf{I})]\mathbf{B}\mathbf{k}$$

$$= e^{\mathbf{A}t}\mathbf{x}(0) + \mathbf{A}^{-1}(e^{\mathbf{A}t} - \mathbf{I})\mathbf{B}\mathbf{k}$$

3. *Ramp response:* Let us write the ramp input $\mathbf{u}(t)$ as

$$\mathbf{u}(t) = t\mathbf{v}$$

where $\mathbf{v}$ is a vector whose components are magnitudes of ramp functions applied at $t = 0$. The solution to the ramp input $t\mathbf{v}$ given at $t = 0$ is

$$\mathbf{x}(t) = e^{\mathbf{A}t}\mathbf{x}(0) + \int_{0}^{t} e^{\mathbf{A}(t-\tau)}\mathbf{B}\tau\mathbf{v}\, d\tau$$

$$= e^{\mathbf{A}t}\mathbf{x}(0) + e^{\mathbf{A}t}\int_{0}^{t} e^{-\mathbf{A}\tau}\tau\, d\tau\, \mathbf{B}\mathbf{v}$$

$$= e^{\mathbf{A}t}\mathbf{x}(0) + e^{\mathbf{A}t}\left(\frac{\mathbf{I}}{2}t^2 - \frac{2\mathbf{A}}{3!}t^3 + \frac{3\mathbf{A}^2}{4!}t^4 - \frac{4\mathbf{A}^3}{5!}t^5 + \cdots\right)\mathbf{B}\mathbf{v}$$

If $\mathbf{A}$ is nonsingular, then this last equation can be simplified to give

$$\mathbf{x}(t) = e^{\mathbf{A}t}\mathbf{x}(0) + (\mathbf{A}^{-2})(e^{\mathbf{A}t} - \mathbf{I} - \mathbf{A}t)\mathbf{B}\mathbf{v}$$

$$= e^{\mathbf{A}t}\mathbf{x}(0) + [\mathbf{A}^{-2}(e^{\mathbf{A}t} - \mathbf{I}) - \mathbf{A}^{-1}t]\mathbf{B}\mathbf{v}$$

**PROBLEM A-14-5.** A useful method for evaluating $e^{\mathbf{A}t}$ is based on the Cayley-Hamilton theorem, which states that an $n \times n$ matrix $\mathbf{A}$ satisfies its own characteristic equation. This enables us to write all powers of $\mathbf{A}$ in terms of a polynomial in $\mathbf{A}$ of degree equal to or less than $n - 1$, i.e.,

$$e^{\mathbf{A}t} = \alpha_0(t)\mathbf{I} + \alpha_1(t)\mathbf{A} + \cdots + \alpha_{n-1}(t)\mathbf{A}^{n-1} \tag{14-94}$$

If $\mathbf{A}$ has distinct eigenvalues, then to determine the coefficients $\alpha_0(t)$, $\alpha_1(t)$, ..., $\alpha_{n-1}(t)$, we substitute each of the eigenvalues of $\mathbf{A}$ into the scalar version of Eq. (14-94) and obtain $n$ simultaneous equations. (Note that if the eigenvalues of $\mathbf{A}$ are multiple, then the procedure must be modified slightly. For details, refer to Chapter 6 of Reference O-1.)

By use of the present approach, obtain $e^{\mathbf{A}t}$ where

$$\mathbf{A} = \begin{bmatrix} 0 & 1 \\ -2 & -3 \end{bmatrix}$$

**Solution.** The eigenvalues of $\mathbf{A}$ are obtained from

$$|\lambda \mathbf{I} - \mathbf{A}| = 0$$

or

$$\begin{vmatrix} \lambda & -1 \\ 2 & \lambda + 3 \end{vmatrix} = 0$$

Thus

$$\lambda_1 = -1, \qquad \lambda_2 = -2$$

$e^{\mathbf{A}t}$ can then be expressed as

$$e^{\mathbf{A}t} = \alpha_0(t)\mathbf{I} + \alpha_1(t)\mathbf{A} \tag{14-95}$$

The scalar version of this equation is

$$e^{\lambda t} = \alpha_0(t) + \alpha_1(t)\lambda \tag{14-96}$$

Substituting $\lambda = \lambda_1 = -1$ into Eq. (14-96) gives

$$e^{-t} = \alpha_0(t) - \alpha_1(t) \tag{14-97}$$

Similarly, substituting $\lambda = \lambda_2 = -2$ into Eq. (14-96) gives

$$e^{-2t} = \alpha_0(t) - 2\alpha_1(t) \tag{14-98}$$

Solving Eqs. (14-97) and (14-98) for $\alpha_0(t)$ and $\alpha_1(t)$ yields

$$\alpha_0(t) = 2e^{-t} - e^{-2t}$$
$$\alpha_1(t) = e^{-t} - e^{-2t}$$

Substituting these values into Eq. (14-95) gives

$$e^{\mathbf{A}t} = (2e^{-t} - e^{-2t})\mathbf{I} + (e^{-t} - e^{-2t})\mathbf{A}$$
$$= \begin{bmatrix} 2e^{-t} - e^{-2t} & e^{-t} - e^{-2t} \\ -2e^{-t} + 2e^{-2t} & -e^{-t} + 2e^{-2t} \end{bmatrix}$$

**PROBLEM A-14-6.** Consider the system described by

$$\begin{bmatrix} \dot{x}_1 \\ \dot{x}_2 \end{bmatrix} = \begin{bmatrix} a_{11} & a_{12} \\ a_{21} & a_{22} \end{bmatrix} \begin{bmatrix} x_1 \\ x_2 \end{bmatrix}, \qquad a_{21} \neq 0$$

Assuming that the two eigenvalues $\lambda_1$ and $\lambda_2$ of the coefficient matrix are distinct, find a diagonalizing transformation matrix $\mathbf{P}$ in terms of $a_{ij}$, $\lambda_1$, and $\lambda_2$.

**Solution**

$$\mathbf{P} = \begin{bmatrix} a_{22} - \lambda_1 & a_{22} - \lambda_2 \\ -a_{21} & -a_{21} \end{bmatrix}$$

or

$$\mathbf{P} = \begin{bmatrix} \lambda_2 - a_{11} & \lambda_1 - a_{11} \\ -a_{21} & -a_{21} \end{bmatrix}$$

**PROBLEM A-14-7.** Obtain a discrete-time state-space representation of the system shown in Fig. 14-15.

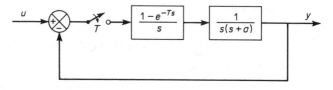

**Fig. 14-15.** Discrete-time system.

**Solution.** The open-loop transfer function of the system is

$$G(s) = \frac{1 - e^{-Ts}}{s} \frac{1}{s(s + a)}$$

The $z$ transform of $G(s)$ is given by

$$G(z) = \frac{T}{a(z - 1)} - \frac{1 - e^{-aT}}{a^2(z - e^{-aT})}$$

In terms of the pulse transfer function, Fig. 14-15 may be modified as shown in Fig. 14-16.

Let us define the state variables as shown in Fig. 14-16. Noting that $z\mathcal{Z}[\mathbf{x}(k)] = \mathcal{Z}[\mathbf{x}(k + 1)]$, we obtain the following discrete-time state equation:

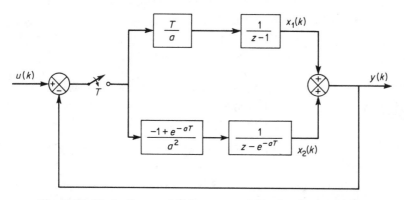

**Fig. 14-16.** Block diagram defining state variables for the system shown in Fig. 14-15.

$$\begin{bmatrix} x_1(k + 1) \\ x_2(k + 1) \end{bmatrix} = \begin{bmatrix} 1 - \dfrac{T}{a} & -\dfrac{T}{a} \\ \dfrac{1 - e^{-aT}}{a^2} & \dfrac{1 + (a^2 - 1)e^{-aT}}{a^2} \end{bmatrix} \begin{bmatrix} x_1(k) \\ x_2(k) \end{bmatrix} + \begin{bmatrix} \dfrac{T}{a} \\ \dfrac{-1 + a^{-aT}}{a^2} \end{bmatrix} [u(k)]$$

**PROBLEM A-14-8.** Country $A$ has a population of 100 million in 1970. The population of City $B$ in Country $A$ is 10 million in 1970.

Suppose that every year 4% of the previous year's population of City $B$ leaves this city, and 2% of the previous year's population of the country outside City $B$ moves into City $B$.

Find the population of City $B$ in the year 1980. Assume that the natural increase of the population is 1% per year.

**Solution.** Let us call year 1970 the 0th year. Let the population of City $B$ at the $k$th year be $x_1(k)$ and the population of Country $A$ excluding City $B$ be $x_2(k)$. Then we obtain the following equations:

$$x_1(k + 1) = 1.01[(1 - 0.04)x_1(k) + 0.02x_2(k)]$$
$$x_2(k + 1) = 1.01[0.04x_1(k) + (1 - 0.02)x_2(k)]$$

with initial conditions

$$x_1(0) = 10 \times 10^6, \qquad x_2(0) = 90 \times 10^6$$

In terms of vector-matrix notation

$$\mathbf{x}(k + 1) = \mathbf{G}\mathbf{x}(k) \tag{14-99}$$

where

$$\mathbf{G} = \begin{bmatrix} (1.01)(1 - 0.04) & (1.01)(0.02) \\ (1.01)(0.04) & (1.01)(1 - 0.02) \end{bmatrix}$$

The solution of Eq. (14-99) is given by

$$\mathbf{x}(k) = \mathbf{G}^k\mathbf{x}(0)$$

Referring to Eq. (14-83), we have

$$\mathbf{G}^k = \mathcal{Z}^{-1}[(z\mathbf{I} - \mathbf{G})^{-1}z]$$

Since

$$(z\mathbf{I} - \mathbf{G})^{-1} = \begin{bmatrix} z - (1.01)(0.96) & -(1.01)(0.02) \\ -(1.01)(0.04) & z - (1.01)(0.98) \end{bmatrix}^{-1}$$

$$= \begin{bmatrix} \dfrac{z - (1.01)(0.98)}{(z - 1.01)(z - 0.9494)} & \dfrac{(1.01)(0.02)}{(z - 1.01)(z - 0.9494)} \\ \dfrac{(1.01)(0.04)}{(z - 1.01)(z - 0.9494)} & \dfrac{z - (1.01)(0.96)}{(z - 1.01)(z - 0.9494)} \end{bmatrix}$$

$$= \begin{bmatrix} \dfrac{\frac{1}{3}}{z - 1.01} + \dfrac{\frac{2}{3}}{z - 0.9494} & \dfrac{\frac{1}{3}}{z - 1.01} + \dfrac{-\frac{1}{3}}{z - 0.9494} \\ \dfrac{\frac{2}{3}}{z - 1.01} + \dfrac{-\frac{2}{3}}{z - 0.9494} & \dfrac{\frac{2}{3}}{z - 1.01} + \dfrac{\frac{1}{3}}{z - 0.9494} \end{bmatrix}$$

$\mathbf{G}^k$ can be obtained as

$$\mathbf{G}^k = \mathcal{Z}^{-1}[(z\mathbf{I} - \mathbf{G})^{-1}z]$$

$$= \mathcal{Z}^{-1}\begin{bmatrix} \dfrac{1}{3}\dfrac{z}{z - 1.01} + \dfrac{2}{3}\dfrac{z}{z - 0.9494} & \dfrac{1}{3}\dfrac{z}{z - 1.01} - \dfrac{1}{3}\dfrac{z}{z - 0.9494} \\ \dfrac{2}{3}\dfrac{z}{z - 1.01} - \dfrac{2}{3}\dfrac{z}{z - 0.9494} & \dfrac{2}{3}\dfrac{z}{z - 1.01} + \dfrac{1}{3}\dfrac{z}{z - 0.9494} \end{bmatrix}$$

$$= \begin{bmatrix} \frac{1}{3}(1.01)^k + \frac{2}{3}(0.9494)^k & \frac{1}{3}(1.01)^k - \frac{1}{3}(0.9494)^k \\ \frac{2}{3}(1.01)^k - \frac{2}{3}(0.9494)^k & \frac{2}{3}(1.01)^k + \frac{1}{3}(0.9494)^k \end{bmatrix}$$

Hence $\mathbf{x}(k)$ is obtained as

$$\mathbf{x}(k) = \mathbf{G}^k \mathbf{x}(0)$$

$$= \begin{bmatrix} \frac{1}{3}(1.01)^k(10^8) - \frac{0.7}{3}(0.9494)^k(10^8) \\ \frac{2}{3}(1.01)^k(10^8) + \frac{0.7}{3}(0.9494)^k(10^8) \end{bmatrix}$$

For $k = 10$, we obtain

$$\mathbf{x}(10) = \begin{bmatrix} \frac{1}{3}(1.01)^{10}[1 - 0.7(0.94)^{10}](10^8) \\ \frac{2}{3}(1.01)^{10}[1 + \frac{0.7}{2}(0.94)^{10}](10^8) \end{bmatrix}$$

$$= \begin{bmatrix} 22.94 \times 10^6 \\ 87.46 \times 10^6 \end{bmatrix}$$

Thus, City $B$ will have population of 22.94 million at year $k = 10$.

Note that as $k$ approaches infinity, $\lim\limits_{k \to \infty} (0.9494)^k \to 0$. Hence

$$\lim_{k \to \infty} \begin{bmatrix} x_1(k) \\ x_2(k) \end{bmatrix} = \lim_{k \to \infty} \begin{bmatrix} \frac{1}{3}(1.01)^k(10^8) \\ \frac{2}{3}(1.01)^k(10^8) \end{bmatrix}$$

The ratio of the population of City $B$ to that of the rest of Country $A$ becomes 1 to 2, or one-third of the population of Country $A$ will live in City $B$.

## PROBLEMS

**PROBLEM B-14-1.** Consider the network system shown in Fig. 14-17. Choosing $v_c$ and $i_L$ as the state variables, obtain the state equation of the system.

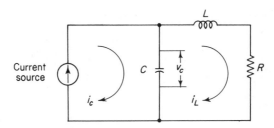

**Fig. 14-17.** Network system.

**PROBLEM B-14-2.** Consider the system described by

$$\dddot{y} + 3\ddot{y} + 2\dot{y} = u$$

Derive a state-space representation of the system. Choose the state variables such that the coefficient matrix of the state vector is diagonal.

**PROBLEM B-14-3.** Obtain a state-space representation of the following system:

$$\frac{Y(s)}{U(s)} = \frac{2(s + 3)}{(s + 1)(s + 2)}$$

**PROBLEM B-14-4.** Derive a state-space representation of the thermal system shown in Fig. 14-18. In order to simplify the derivation, assume the following: The mass of metal of the inner vessel is small and its thermal capacitance is negligible. The heat loss from the outer vessel is negligible. Also, the heat loss from the free surface is negligible. The

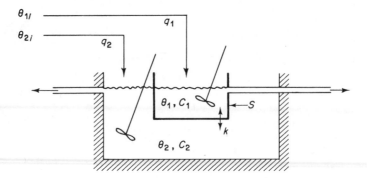

**Fig. 14-18.** Thermal system.

fluid in the inner vessel is assumed to be thoroughly mixed at all times. The same is true for the fluid in the outer vessel. In the diagram

$\theta_{1i}$ = temperature of the inflow liquid to the inner vessel, °F
$\theta_{2i}$ = temperature of the inflow liquid to the outer vessel, °F
$\theta_1$ = temperature of the liquid in the inner vessel, °F
$\theta_2$ = temperature of the liquid in the outer vessel, °F
$C_1$ = heat capacitance of the liquid in the inner vessel, Btu/°F
$C_2$ = heat capacitance of the liquid in the outer vessel, Btu/°F
$q_1$ = heat inflow to the inner vessel, Btu/min °F
$q_2$ = heat inflow to the outer vessel, Btu/min °F
$S$ = heat-transfer surface area, in.$^2$
$k$ = coefficient of heat transfer, Btu/in.$^2$ min °F

Assume that $\theta_1$ and $\theta_2$ are the state variables and $\theta_{1i}$ and $\theta_{2i}$ are the input variables.

**PROBLEM B-14-5.** Find $x_1(t)$ and $x_2(t)$ of the system described by

$$\begin{bmatrix} \dot{x}_1 \\ \dot{x}_2 \end{bmatrix} = \begin{bmatrix} 0 & 1 \\ -3 & -2 \end{bmatrix} \begin{bmatrix} x_1 \\ x_2 \end{bmatrix}$$

where the initial conditions are

$$\begin{bmatrix} x_1(0) \\ x_2(0) \end{bmatrix} = \begin{bmatrix} 1 \\ -1 \end{bmatrix}$$

**PROBLEM B-14-6.** Given the system equation

$$\begin{bmatrix} \dot{x}_1 \\ \dot{x}_2 \\ \dot{x}_3 \end{bmatrix} = \begin{bmatrix} 2 & 1 & 0 \\ 0 & 2 & 1 \\ 0 & 0 & 2 \end{bmatrix} \begin{bmatrix} x_1 \\ x_2 \\ x_3 \end{bmatrix}$$

Find the solution in terms of the initial conditions $x_1(0)$, $x_2(0)$, and $x_3(0)$.

**PROBLEM B-14-7.** Show that the solution of the following matrix differential equation:

$$\frac{d\mathbf{X}}{dt} = \mathbf{AX} + \mathbf{XB}, \qquad \mathbf{X}(0) = \mathbf{C}$$

is given by

$$\mathbf{X} = e^{\mathbf{A}t}\mathbf{C}e^{\mathbf{B}t}$$

**PROBLEM B-14-8.** Obtain the response of the following system:

$$\begin{bmatrix} \dot{x}_1 \\ \dot{x}_2 \end{bmatrix} = \begin{bmatrix} 0 & 1 \\ -2 & -3 \end{bmatrix} \begin{bmatrix} x_1 \\ x_2 \end{bmatrix} + \begin{bmatrix} 2 \\ 0 \end{bmatrix} [u], \qquad \begin{bmatrix} x_1(0) \\ x_2(0) \end{bmatrix} = \begin{bmatrix} 0 \\ 1 \end{bmatrix}$$

where

$$\begin{aligned} u(t) &= 0 &\quad \text{for } t < 0 \\ &= e^{-t} &\quad \text{for } t \geq 0 \end{aligned}$$

**PROBLEM B-14-9.** Consider the system described by

$$\begin{bmatrix} \dot{x}_1 \\ \dot{x}_2 \end{bmatrix} = \begin{bmatrix} -5 & -1 \\ 3 & -1 \end{bmatrix} \begin{bmatrix} x_1 \\ x_2 \end{bmatrix} + \begin{bmatrix} 1 \\ 1 \end{bmatrix} [u]$$

$$y = \begin{bmatrix} 1 & 2 \end{bmatrix} \begin{bmatrix} x_1 \\ x_2 \end{bmatrix}$$

Obtain the transfer function of the system.

**PROBLEM B-14-10.** Consider a continuous-time system described by

$$\begin{bmatrix} \dot{x}_1 \\ \dot{x}_2 \end{bmatrix} = \begin{bmatrix} 0 & 1 \\ 0 & 0 \end{bmatrix} \begin{bmatrix} x_1 \\ x_2 \end{bmatrix} + \begin{bmatrix} 0 \\ 1 \end{bmatrix} [u]$$

Discretize the system equation and derive a discrete-time state equation. Assume the sampling period is 2 sec.

**PROBLEM B-14-11.** Obtain a state-space representation of the following system:

$$\frac{Y(z)}{U(z)} = \frac{z + 2}{z^2 + z + 0.16}$$

**PROBLEM B-14-12.** Consider the difference equation

$$x(k + 2) = x(k + 1) + x(k)$$

where $x(0) = 0$, $x(1) = 1$. Note that $x(2) = 1$, $x(3) = 2$, $x(4) = 3, \ldots$. The series 0, 1, 1, 2, 3, 5, 8, 13, $\ldots$ is known as the Fibonacci series.

Obtain the general solution $x(k)$. Show that the limiting value of $x(k + 1)/x(k)$ as $k$ approaches infinity is $(\sqrt{5} + 1)/2$, or approximately 1.62.

# 15

## LIAPUNOV STABILITY ANALYSIS

### 15-1   INTRODUCTION

For a given control system, stability is usually the most important thing to be determined. If the system is linear and time-invariant, many stability criteria are available. Among them are the Nyquist stability criterion, Routh's stability criterion, etc. If the system is nonlinear, or linear but time-varying, however, then such stability criteria do not apply. Although a technique employing the Nyquist stability plot may be applied to a special group of nonlinear systems, as discussed in Chapter 11, the describing-function approach for the determination of stability is only approximate. Stability analysis based on the phase-plane method presented in Chapter 12 applies only to first- and second-order systems.

The second method of Liapunov (which is also called the direct method of Liapunov) to be presented in this chapter is the most general method for the determination of stability of nonlinear and/or time-varying systems. The method applies to systems of any order.

By using the second method of Liapunov, we can determine the stability of a system without solving the state equations. This is quite advantageous because solving nonlinear and/or time-varying state equations is usually very difficult.

Although the second method of Liapunov requires considerable experience and ingenuity, it can answer the question of stability of nonlinear systems when other methods fail.

The purpose of this chapter is to present the second method of Liapunov

**716**

and to illustrate its application to the stability analysis of both linear and nonlinear systems.

**Outline of the chapter.** In Section 15-2, we present preliminary materials such as definitions of various types of system stability and the concepts of the definiteness of scalar functions. Section 15-3 presents the main stability theorem of the second method and introduces the Liapunov function. Section 15-4 discusses the application of the second method of Liapunov to the stability analysis of linear time-invariant systems. Then in Section 15-5 a useful technique based on the second method for estimating transient behavior of systems is given. Section 15-6 discusses stability analysis of nonlinear systems. We present two methods for constructing Liapunov functions for such systems. Finally, Section 15-7 summarizes concluding comments for the chapter.

## 15-2  DEFINITIONS

In this section, we first give definitions of a system, an equilibrium state, stability, asymptotic stability, and instability. Then we define the definiteness, semidefiniteness, and indefiniteness of scalar functions.

**System.** The system we consider in this chapter is defined by

$$\dot{\mathbf{x}} = \mathbf{f}(\mathbf{x}, t) \tag{15-1}$$

where $\mathbf{x}$ is a state vector ($n$-dimensional vector) and $\mathbf{f}(\mathbf{x}, t)$ is an $n$-dimensional vector whose elements are functions of $x_1, x_2, \ldots, x_n$, and $t$. We assume that the system of Eq. (15-1) has a unique solution starting at the given initial condition.

We shall denote the solution of Eq. (15-1) as $\boldsymbol{\phi}(t; \mathbf{x}_0, t_0)$, where $\mathbf{x} = \mathbf{x}_0$ at $t = t_0$ and $t$ is the observed time. Thus,

$$\boldsymbol{\phi}(t_0; \mathbf{x}_0, t_0) = \mathbf{x}_0$$

**Equilibrium state.** In the system of Eq. (15-1), a state $\mathbf{x}_e$ where

$$\mathbf{f}(\mathbf{x}_e, t) = \mathbf{0} \qquad \text{for all } t \tag{15-2}$$

is called an equilibrium state of the system. If the system is linear time-invariant, namely, if $\mathbf{f}(\mathbf{x}, t) = \mathbf{A}\mathbf{x}$, then there exists only one equilibrium state if $\mathbf{A}$ is nonsingular, and there exist infinitely many equilibrium states if $\mathbf{A}$ is singular. For nonlinear systems, there may be one or more equilibrium states. These states correspond to the constant solutions of the system ($\mathbf{x} = \mathbf{x}_e$ for all $t$). Determination of the equilibrium states does not involve the solution of the differential equations of the system, Eq. (15-1), but only the solution of Eq. (15-2).

Any isolated equilibrium state (i.e., isolated from each other) can be shifted to the origin of the coordinates, or $\mathbf{f}(\mathbf{0}, t) = \mathbf{0}$, by a translation of coordinates. In this chapter, we shall treat stability analysis of only such states.

**Stability in the sense of Liapunov.** In the following, we shall denote a spherical region of radius $k$ about an equilibrium state $\mathbf{x}_e$ as

$$\| \mathbf{x} - \mathbf{x}_e \| \leq k$$

where $\| \mathbf{x} - \mathbf{x}_e \|$ is called the Euclidean norm and is defined by

$$\| \mathbf{x} - \mathbf{x}_e \| = [(x_1 - x_{1e})^2 + (x_2 - x_{2e})^2 + \cdots + (x_n - x_{ne})^2]^{1/2}$$

Let $S(\delta)$ consists of all points such that

$$\| \mathbf{x}_0 - \mathbf{x}_e \| \leq \delta$$

and let $S(\epsilon)$ consists of all points such that

$$\| \boldsymbol{\phi}(t; \mathbf{x}_0, t_0) - \mathbf{x}_e \| \leq \epsilon \qquad \text{for all } t \geq t_0$$

An equilibrium state $\mathbf{x}_e$ of the system of Eq. (15-1) is said to be stable in the sense of Liapunov if, corresponding to each $S(\epsilon)$, there is an $S(\delta)$ such that trajectories starting in $S(\delta)$ do not leave $S(\epsilon)$ as $t$ increases indefinitely. The real number $\delta$ depends on $\epsilon$ and, in general, also depends on $t_0$. If $\delta$ does not depend on $t_0$, the equilibrium state is said to be uniformly stable.

What we have stated here is that we first choose the region $S(\epsilon)$, and for each $S(\epsilon)$, there must be a region $S(\delta)$ such that trajectories starting within $S(\delta)$ do not leave $S(\epsilon)$ as $t$ increases indefinitely.

**Asymptotic stability.** An equilibrium state $\mathbf{x}_e$ of the system of Eq. (15-1) is said to be asymptotically stable if it is stable in the sense of Liapunov and if every solution starting within $S(\delta)$ converges, without leaving $S(\epsilon)$, to $\mathbf{x}_e$ as $t$ increases indefinitely.

In practice, asymptotic stability is more important than mere stability. Also since asymptotic stability is a local concept, simply to establish asymptotic stability may not mean that the system will operate properly. Some knowledge of the size of the largest region of asymptotic stability is usually necessary. This region is called the domain of attraction. It is that part of the state space in which asymptotically stable trajectories originate. In other words, every trajectory originating in the domain of attraction is asymptotically stable.

**Asymptotic stability in the large.** If asymptotic stability holds for all states (all points in the state space) from which trajectories originate, the equilibrium state is said to be asymptotically stable in the large. That is, the equilibrium state $\mathbf{x}_e$ of the system given by Eq. (15-1) is said to be asymptotically stable in the large if it is stable and if every solution converges to $\mathbf{x}_e$ as $t$ increases indefinitely. Obviously a necessary condition for asymptotic stability in the large is that there be only one equilibrium state in the whole state space.

In control engineering problems, a desirable feature is asymptotic stability in the large. If the equilibrium state is not asymptotically stable in the large, then the problem becomes that of determining the largest region of asymptotic stability. This is usually very difficult. For all practical purposes, however, it is sufficient to determine a region of asymptotic stability large enough so that no disturbance will exceed it.

**Instability.** An equilibrium state $\mathbf{x}_e$ is said to be unstable if for some real number $\epsilon > 0$ and any real number $\delta > 0$, no matter how small, there is always a state $\mathbf{x}_0$ in $S(\delta)$ such that the trajectory starting at this state leaves $S(\epsilon)$.

**Graphical representation of stability, asymptotic stability, and instability.** A graphical representation of the foregoing definitions will clarify their notions.

Let us consider the two-dimensional case. Figures 15-1 (a), (b), and (c) show equilibrium states and typical trajectories corresponding to stability, asymptotic stability, and instability, respectively. In Fig. 15-1 (a), (b), or (c), the region $S(\delta)$ bounds the initial state $\mathbf{x}_0$, and the region $S(\epsilon)$ corresponds to the boundary for the trajectory starting at $\mathbf{x}_0$.

Note that the foregoing definitions do not specify the exact region of allowable initial conditions. Thus the definitions apply to the neighborhood of the equilibrium state, unless $S(\epsilon)$ corresponds to the entire state plane.

Note that in Fig. 15-1 (c), the trajectory leaves $S(\epsilon)$ and implies that the equilibrium state is unstable. We cannot, however, say that the trajectory will go to infinity since it may approach a limit cycle outside the region $S(\epsilon)$. (If a linear time-invariant system is unstable, trajectories starting near the unstable equilibrium state go to infinity. But in the case of nonlinear systems this is not necessarily true.)

Knowledge of the foregoing definitions is the minimum requirement for understanding the stability analysis of linear and nonlinear systems as presented in this chapter. Note that these definitions are not the only ones defining concepts of stability of an equilibrium state. In fact, various other ways are available in the literature. For example, in classical control theory, only systems that are asymptotically stable are called stable systems, and those systems that are stable in the sense of Liapunov, but are not asymptotically stable, are called unstable.

**Positive definiteness of scalar functions.** A scalar function $V(\mathbf{x})$ is said to be *positive definite* in a region $\Omega$ (which includes the origin of the state space) if $V(\mathbf{x}) > 0$ for all nonzero states $\mathbf{x}$ in the region $\Omega$ and $V(\mathbf{0}) = 0$.

A time-varying function $V(\mathbf{x}, t)$ is said to be positive definite in a region $\Omega$ (which includes the origin of the state space) if it is bounded from below by a time-invariant positive-definite function, that is, if there exists a positive definite function $V(\mathbf{x})$ such that

$$V(\mathbf{x}, t) > V(\mathbf{x}) \quad \text{for all } t \geq t_0$$
$$V(\mathbf{0}, t) = 0 \quad \text{for all } t \geq t_0$$

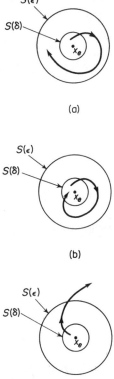

(a)

(b)

(c)

**Fig. 15-1.** (a) Stable equilibrium state and a representative trajectory; (b) asymptotically stable equilibrium state and a representative trajectory; (c) unstable equilibrium state and a representative trajectory.

**Negative definiteness of scalar functions.** A scalar function $V(\mathbf{x})$ is said to be *negative definite* if $-V(\mathbf{x})$ is positive definite.

**Positive semidefiniteness of scalar functions.** A scalar function $V(\mathbf{x})$ is said to be *positive semidefinite* if it is positive at all states in the region $\Omega$ except at the origin and at certain other states, where it is zero.

**Negative semidefiniteness of scalar functions.** A scalar function $V(\mathbf{x})$ is said to be *negative semidefinite* if $-V(\mathbf{x})$ is positive semidefinite.

**Indefiniteness of scalar functions.** A scalar function $V(\mathbf{x})$ is said to be *indefinite* if in the region $\Omega$ it assumes both positive and negative values, no matter how small the region $\Omega$ is.

---

*Example 15-1.* In this example, we give several scalar functions and their classifications according to the foregoing definitions. Here we assume $\mathbf{x}$ to be a two-dimensional vector.

1. $\qquad V(\mathbf{x}) = x_1^2 + 2x_2^2$ $\qquad\qquad$ positive definite
2. $\qquad V(\mathbf{x}) = (x_1 + x_2)^2$ $\qquad\qquad$ positive semidefinite
3. $\qquad V(\mathbf{x}) = -x_1^2 - (3x_1 + 2x_2)^2$ $\qquad$ negative definite
4. $\qquad V(\mathbf{x}) = x_1 x_2 + x_2^2$ $\qquad\qquad$ indefinite
5. $\qquad V(\mathbf{x}) = x_1^2 + \dfrac{2x_2^2}{1 + x_2^2}$ $\qquad\qquad$ positive definite

**Quadratic forms.** A class of scalar functions which plays an important role in the stability analysis based on the second method of Liapunov is the quadratic form. An example is

$$V(\mathbf{x}) = \mathbf{x}'\mathbf{P}\mathbf{x}$$

$$= [x_1 \quad x_2 \quad \cdots \quad x_n]
\begin{bmatrix}
p_{11} & p_{12} & \cdots & p_{1n} \\
p_{12} & p_{22} & \cdots & p_{2n} \\
\cdot & \cdot & & \cdot \\
\cdot & \cdot & & \cdot \\
\cdot & \cdot & & \cdot \\
p_{1n} & p_{2n} & \cdots & p_{nn}
\end{bmatrix}
\begin{bmatrix}
x_1 \\
x_2 \\
\cdot \\
\cdot \\
\cdot \\
x_n
\end{bmatrix}$$

Note that $\mathbf{P}$ is real and symmetric.

The positive definiteness of the quadratic form $V(\mathbf{x})$ can be determined by Sylvester's criterion, which states that the necessary and sufficient conditions that the quadratic form $V(\mathbf{x})$ be positive definite are that all the successive principal minors of $\mathbf{P}$ be positive; i.e.,

$$p_{11} > 0, \qquad \begin{vmatrix} p_{11} & p_{12} \\ p_{12} & p_{22} \end{vmatrix} > 0, \ldots,$$

$$\begin{vmatrix}
p_{11} & p_{12} & \cdots & p_{1n} \\
p_{12} & p_{22} & \cdots & p_{2n} \\
\cdot & \cdot & & \cdot \\
\cdot & \cdot & & \cdot \\
\cdot & \cdot & & \cdot \\
p_{1n} & p_{2n} & \cdots & p_{nn}
\end{vmatrix} > 0$$

$V(\mathbf{x}) = \mathbf{x}'\mathbf{P}\mathbf{x}$ is positive semidefinite if $\mathbf{P}$ is singular and all the principal minors are nonnegative.

$V(\mathbf{x})$ is negative definite if $-V(\mathbf{x})$ is positive definite. Similarly, $V(\mathbf{x})$ is negative semidefinite if $-V(\mathbf{x})$ is positive semidefinite.

---

*Example 15-2.* Show that the following quadratic form is positive definite:

$$V(\mathbf{x}) = 10x_1^2 + 4x_2^2 + x_3^2 + 2x_1x_2 - 2x_2x_3 - 4x_1x_3$$

The quadratic form $V(\mathbf{x})$ can be written

$$V(\mathbf{x}) = \mathbf{x}'\mathbf{P}\mathbf{x} = [x_1 \quad x_2 \quad x_3] \begin{bmatrix} 10 & 1 & -2 \\ 1 & 4 & -1 \\ -2 & -1 & 1 \end{bmatrix} \begin{bmatrix} x_1 \\ x_2 \\ x_3 \end{bmatrix}$$

Applying Sylvester's criterion, we obtain

$$10 > 0, \qquad \begin{vmatrix} 10 & 1 \\ 1 & 4 \end{vmatrix} > 0, \qquad \begin{vmatrix} 10 & 1 & -2 \\ 1 & 4 & -1 \\ -2 & -1 & 1 \end{vmatrix} > 0$$

Since all the successive principal minors of the matrix $\mathbf{P}$ are positive, $V(\mathbf{x})$ is positive definite.

## 15-3  SECOND METHOD OF LIAPUNOV

In 1892, A. M. Liapunov presented two methods (called the first and the second methods) for determining the stability of dynamic systems described by ordinary differential equations.

The first method consists of all procedures in which the explicit form of the solutions of the differential equations are used for the analysis.

The second method, on the other hand, does not require the solutions of the differential equations. Therefore, the second method is quite convenient for the stability analysis of nonlinear systems, for which exact solutions may not be obtained.

**Second method of Liapunov.** From the classical theory of mechanics, we know that a vibratory system is stable if its total energy (a positive definite function) is continually decreasing (which means that the time derivative of the total energy must be negative definite) until an equilibrium state is reached.

The second method of Liapunov is based on a generalization of this fact: if the system has an asymptotically stable equilibrium state, then the stored energy of the system displaced within the domain of attraction decays with increasing time until it finally assumes its minimum value at the equilibrium state. For purely mathematical systems, however, there is no simple way of defining an "energy function." In order to circumvent this difficulty, Liapunov introduced the so-called Liapunov function, a fictitious energy function. This idea is, however, more general than that of energy and is more widely applicable. In fact, any scalar function satisfying the hypotheses of Liapunov's stability theorems (see Theorems 15-1 and 15-2) can serve as Liapunov functions. (For simple systems, we may be able to guess suitable Liapunov functions; but, for a complicated system, finding a Liapunov function may be quite difficult.)

Liapunov functions depend on $x_1, x_2, \ldots, x_n$, and $t$. We denote them by $V(x_1, x_2, \ldots, x_n, t)$, or simply by $V(\mathbf{x}, t)$, If Liapunov functions do not include $t$ explicitly, then we denote them by $V(x_1, x_2, \ldots, x_n)$, or $V(\mathbf{x})$. In the second method of Liapunov, the sign behavior of $V(\mathbf{x}, t)$ and that of its time derivative $\dot{V}(\mathbf{x}, t) = dV(\mathbf{x}, t)/dt$ give us information as to the stability, asymptotic stability, or instability of an equilibrium state without requiring us to solve directly for the solution. (This applies to both linear and nonlinear systems.)

**Liapunov's main stability theorem.** It can be shown that if a scalar function $V(\mathbf{x})$, where $\mathbf{x}$ is an $n$-dimensional vector, is positive definite, then the states $\mathbf{x}$ which satisfy

$$V(\mathbf{x}) = C$$

where $C$ is a positive constant, lie on a closed hypersurface in the $n$-dimensional state space, at least in the neighborhood of the origin. If $V(\mathbf{x}) \rightarrow \infty$ as $\| \mathbf{x} \| \rightarrow \infty$, then such closed surfaces extend over the entire state space. The hypersurface $V(\mathbf{x}) = C_1$ lies entirely inside the hypersurface $V(\mathbf{x}) = C_2$ if $C_1 < C_2$.

For a given system, if a positive-definite scalar function $V(\mathbf{x})$ can be found such that its time derivative taken along a trajectory is always negative, then as time increases, $V(\mathbf{x})$ takes smaller and smaller values of $C$. As time increases, $V(\mathbf{x})$ finally shrinks to zero, and therefore $\mathbf{x}$ also shrinks to zero. This implies the asymptotic stability of the origin of the state space. Liapunov's main stability theorem, which is a generalization of the foregoing fact, provides a sufficient condition for asymptotic stability. This theorem may be stated as follows:

**Theorem 15-1.** Suppose that a system is described by

$$\dot{\mathbf{x}} = \mathbf{f}(\mathbf{x}, t)$$

where

$$\mathbf{f}(0, t) = 0 \qquad \text{for all } t$$

If there exists a scalar function $V(\mathbf{x}, t)$ having continuous, first partial derivatives and satisfying the following conditions,

1. $V(\mathbf{x}, t)$ is positive definite
2. $\dot{V}(\mathbf{x}, t)$ is negative definite

then the equilibrium state at the origin is uniformly asymptotically stable.

If, in addition, $V(\mathbf{x}, t) \rightarrow \infty$ as $\| \mathbf{x} \| \rightarrow \infty$, then the equilibrium state at the origin is uniformly asymptotically stable in the large.

We shall not give details of the proof of this theorem here. (The proof follows directly from the definition of asymptotic stability. For details, see, for example, Chapter 8 of Reference O-1.)

---

*Example 15-3.* Consider the system described by

$$\dot{x}_1 = x_2 - x_1(x_1^2 + x_2^2)$$
$$\dot{x}_2 = -x_1 - x_2(x_1^2 + x_2^2)$$

Clearly, the origin ($x_1 = 0$, $x_2 = 0$) is the only equilibrium state. Determine its stability.

If we define a scalar function $V(\mathbf{x})$ by

$$V(\mathbf{x}) = x_1^2 + x_2^2$$

which is positive definite, then the time derivative of $V(\mathbf{x})$ along any trajectory is

$$\dot{V}(\mathbf{x}) = 2x_1\dot{x}_1 + 2x_2\dot{x}_2$$
$$= -2(x_1^2 + x_2^2)^2$$

which is negative definite. This shows that $V(\mathbf{x})$ is continually decreasing along any trajectory; hence $V(\mathbf{x})$ is a Liapunov function. Since $V(\mathbf{x})$ becomes infinite with infinite deviation from the equilibrium state, by Theorem 15-1, the equilibrium state at the origin of the system is asymptotically stable in the large.

Note that if we let $V(\mathbf{x})$ take constant values, $0, C_1, C_2, \ldots, (0 < C_1 < C_2 < \cdots)$, then $V(\mathbf{x}) = 0$ corresponds to the origin of the state plane and $V(\mathbf{x}) = C_1$, $V(\mathbf{x}) = C_2, \ldots$ describe nonintersecting circles enclosing the origin of the state plane, as shown in Fig. 15-2. Note also that since $V(\mathbf{x})$ is radially unbounded, or $V(\mathbf{x}) \to \infty$ as $\|\mathbf{x}\| \to \infty$, the circles extend over the entire state plane.

Since circle $V(\mathbf{x}) = C_k$ lies completely inside circle $V(\mathbf{x}) = C_{k+1}$, a representative trajectory crosses the boundary of $V$ contours from outside toward the inside. From this, the geometric interpretation of a Liapunov function may be stated as follows: $V(\mathbf{x})$ is a measure of the distance of the state $\mathbf{x}$ from the origin of the state space. If the distance between the origin and the instantaneous state $\mathbf{x}(t)$ is continually decreasing as $t$ increases [that is, $\dot{V}(\mathbf{x}(t)) < 0$], then $\mathbf{x}(t) \to \mathbf{0}$. Since a Liapunov function $V(\mathbf{x})$ is regarded as defining a distance from the origin in the state space, its derivative can be used to give a quantitative estimate of the speed with which the origin is approached. (For details, see Section 15-5.)

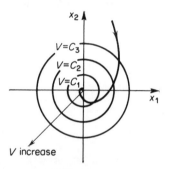

**Fig. 15-2.** Constant-$V$ contours and representative trajectory.

**Comment.** Although Theorem 15-1 is a basic theorem of the second method, it is somewhat restrictive because $\dot{V}(\mathbf{x}, t)$ must be negative definite. If, however, an additional restriction is imposed on $\dot{V}(\mathbf{x}, t)$ that it does not vanish identically along any trajectory except at the origin, then it is possible to replace the requirement of $\dot{V}(\mathbf{x}, t)$ being negative definite by stating that $\dot{V}(\mathbf{x}, t)$ be negative semidefinite.

**Theorem 15-2.** Suppose that a system is described by

$$\dot{\mathbf{x}} = \mathbf{f}(\mathbf{x}, t)$$

where

$$\mathbf{f}(\mathbf{0}, t) = \mathbf{0} \qquad \text{for all } t \geq t_0$$

If there exists a scalar function $V(\mathbf{x}, t)$ having continuous, first partial derivatives and satisfying the following conditions,

1. $V(\mathbf{x}, t)$ is positive definite
2. $\dot{V}(\mathbf{x}, t)$ is negative semidefinite
3. $\dot{V}(\boldsymbol{\phi}(t; \mathbf{x}_0, t_0), t)$ does not vanish identically in $t \geq t_0$ for any $t_0$ and any

$\mathbf{x}_0 \neq \mathbf{0}$, where $\boldsymbol{\phi}(t; \mathbf{x}_0, t_0)$ denotes the trajectory or solution starting from $\mathbf{x}_0$ at $t_0$

then the equilibrium state at the origin of the system is uniformly asymptotically stable in the large.

Note that if $\dot{V}(\mathbf{x}, t)$ is not negative definite, but only negative semidefinite, then the trajectory of a representative point can become tangent to some particular surface $V(\mathbf{x}, t) = C$. However, since $\dot{V}(\boldsymbol{\phi}(t; \mathbf{x}_0, t_0), t)$ does not vanish identically in $t \geq t_0$ for any $t_0$ and any $\mathbf{x}_0 \neq \mathbf{0}$, the representative point cannot remain at the tangent point [the point which corresponds to $\dot{V}(\mathbf{x}, t) = 0$] and therefore must move toward the origin.

If, however, there exists a positive-definite scalar function $V(\mathbf{x}, t)$ such that $\dot{V}(\mathbf{x}, t)$ is identically zero, then the system can remain in a limit cycle. The equilibrium state at the origin, in this case, is said to be stable in the sense of Liapunov.

---

*Example 15-4.* Consider the following system:

$$\begin{bmatrix} \dot{x}_1 \\ \dot{x}_2 \end{bmatrix} = \begin{bmatrix} 0 & 1 \\ -1 & -1 \end{bmatrix} \begin{bmatrix} x_1 \\ x_2 \end{bmatrix}$$

Clearly the only equilibrium state is the origin, $\mathbf{x} = \mathbf{0}$. Determine the stability of this state.

Let us choose the following scalar function as a possible Liapunov function:

$$V(\mathbf{x}) = 2x_1^2 + x_2^2 = \text{positive definite}$$

Then $\dot{V}(\mathbf{x})$ becomes

$$\dot{V}(\mathbf{x}) = 4x_1\dot{x}_1 + 2x_2\dot{x}_2 = 2x_1x_2 - 2x_2^2$$

$\dot{V}(\mathbf{x})$ is indefinite. This implies that this particular $V(\mathbf{x})$ is not a Liapunov function, and therefore stability cannot be determined by its use. [Since the eigenvalues of the coefficient matrix are $(-1 + j\sqrt{3})/2$ and $(-1 - j\sqrt{3})/2$, clearly the origin of the system is stable. This means that we have not chosen a suitable Liapunov function.]

If we choose the following scalar function as a possible Liapunov function,

$$V(\mathbf{x}) = x_1^2 + x_2^2 = \text{positive definite}$$

then

$$\dot{V}(\mathbf{x}) = 2x_1\dot{x}_1 + 2x_2\dot{x}_2 = -2x_2^2$$

which is negative semidefinite. If $\dot{V}(\mathbf{x})$ is to vanish identically for $t \geq t_1$, then $x_2$ must be zero for all $t \geq t_1$. This requires that $\dot{x}_2 = 0$ for $t \geq t_1$. Since

$$\dot{x}_2 = -x_1 - x_2$$

$x_1$ must also be equal to zero for $t \geq t_1$. This means that $\dot{V}(x)$ vanishes identically only at the origin. Hence by Theorem 15-2 the equilibrium state at the origin is asymptotically stable in the large.

To show that a different choice of a Liapunov function yields the same stability information, let us choose the following scalar function as another possible Liapunov function:

$$V(\mathbf{x}) = \tfrac{1}{2}[(x_1 + x_2)^2 + 2x_1^2 + x_2^2] = \text{positive definite}$$

then $\dot{V}(\mathbf{x})$ becomes

$$\dot{V}(\mathbf{x}) = (x_1 + x_2)(\dot{x}_1 + \dot{x}_2) + 2x_1\dot{x}_1 + x_2\dot{x}_2 = -(x_1^2 + x_2^2)$$

which is negative definite. Since $V(\mathbf{x}) \longrightarrow \infty$ as $\|\mathbf{x}\| \longrightarrow \infty$, by Theorem 15-1, the equilibrium state at the origin is asymptotically stable in the large.

Note that since the stability theorems of the second method require positive definiteness of $V(\mathbf{x})$, we often (but not always) choose $V(\mathbf{x})$ to be a quadratic form in $\mathbf{x}$. (Note that the simplest positive-definite function is a quadratic form.) Then we examine if $\dot{V}(\mathbf{x})$ is at least negative semidefinite.

**Instability.** If an equilibrium state $\mathbf{x} = \mathbf{0}$ of a system is unstable, then there exists a scalar function $W(\mathbf{x}, t)$ which determines the instability of the equilibrium state. We shall present a theorem on instability without a proof.

**Theorem 15-3.** Suppose a system is described by

$$\dot{\mathbf{x}} = \mathbf{f}(\mathbf{x}, t)$$

where

$$\mathbf{f}(\mathbf{0}, t) = \mathbf{0} \qquad \text{for all } t \geq t_0$$

If there exists a scalar function $W(\mathbf{x}, t)$ having continuous, first partial derivatives and satisfying the following conditions,

1. $W(\mathbf{x}, t)$ is positive definite in some region about the origin
2. $\dot{W}(\mathbf{x}, t)$ is positive definite in the same region

then the equilibrium state at the origin is unstable.

## 15-4  STABILITY ANALYSIS OF LINEAR SYSTEMS

There are many approaches to the investigation of the asymptotic stability of linear time-invariant systems. For example, for a continuous-time system

$$\dot{\mathbf{x}} = \mathbf{A}\mathbf{x}$$

the necessary and sufficient condition for the asymptotic stability of the origin of the system can be stated that all eigenvalues of $\mathbf{A}$ have negative real parts, or the zeros of the characteristic polynomial

$$|s\mathbf{I} - \mathbf{A}| = s^n + a_1 s^{n-1} + \cdots + a_{n-1} s + a_n$$

have negative real parts. Similarly, for a discrete-time system

$$\mathbf{x}(k + 1) = \mathbf{G}\mathbf{x}(k)$$

the necessary and sufficient condition for the asymptotic stability of the origin can be stated that all eigenvalues of $\mathbf{G}$ are less than unity in their magnitudes, or the zeros of the characteristic polynomial

$$|z\mathbf{I} - \mathbf{G}| = z^n + a_1 z^{n-1} + \cdots + a_{n-1} z + a_n$$

lie within the unit circle centered at the origin of the $z$ plane.

Finding the eigenvalues becomes difficult or impossible in the case of higher order systems or if some of the coefficients of the characteristic polynomial are nonnumerical. In such a case, Routh's stability criterion may conveniently be applied. One alternative to this approach is available; it is based on the second method of Liapunov. The Liapunov approach is algebraic and does not require factoring of the characteristic polynomial. In addition, this approach can be used to estimate the transient response behavior of the system (see Section 15-5) and can be applied to the system design (see Section 16-6). The purpose of this section is to present the Liapunov approach to the stability analysis of linear time-invariant systems.

**Liapunov stability analysis of linear time-invariant systems.** Consider the following linear time-invariant system,

$$\dot{\mathbf{x}} = \mathbf{A}\mathbf{x} \qquad (15\text{-}3)$$

where $\mathbf{x}$ is a state vector ($n$-dimensional vector) and $\mathbf{A}$ is an $n \times n$ constant matrix. We assume that $\mathbf{A}$ is nonsingular. Then the only equilibrium state is the origin $\mathbf{x} = \mathbf{0}$. The stability of the equilibrium state of the linear time-invariant system can be investigated easily by use of the second method of Liapunov.

For the system defined by Eq. (15-3), let us choose a possible Liapunov function as

$$V(\mathbf{x}) = \mathbf{x}^*\mathbf{P}\mathbf{x}$$

where $\mathbf{P}$ is a positive-definite Hermitian matrix. (If $\mathbf{x}$ is a real vector, then $\mathbf{P}$ can be chosen to be a positive-definite real symmetric matrix.) The time derivative of $V(\mathbf{x})$ along any trajectory is

$$\begin{aligned} \dot{V}(\mathbf{x}) &= \dot{\mathbf{x}}^*\mathbf{P}\mathbf{x} + \mathbf{x}^*\mathbf{P}\dot{\mathbf{x}} \\ &= (\mathbf{A}\mathbf{x})^*\mathbf{P}\mathbf{x} + \mathbf{x}^*\mathbf{P}\mathbf{A}\mathbf{x} \\ &= \mathbf{x}^*\mathbf{A}^*\mathbf{P}\mathbf{x} + \mathbf{x}^*\mathbf{P}\mathbf{A}\mathbf{x} \\ &= \mathbf{x}^*(\mathbf{A}^*\mathbf{P} + \mathbf{P}\mathbf{A})\mathbf{x} \end{aligned}$$

Since $V(\mathbf{x})$ was chosen to be positive definite, we require, for asymptotic stability, that $\dot{V}(\mathbf{x})$ be negative definite. Therefore, we require that

$$\dot{V} = -\mathbf{x}^*\mathbf{Q}\mathbf{x}$$

where

$$\mathbf{Q} = -(\mathbf{A}^*\mathbf{P} + \mathbf{P}\mathbf{A}) = \text{positive definite}$$

Hence, for the asymptotic stability of the system of Eq. (15-3), it is sufficient that $\mathbf{Q}$ be positive definite. For a test of positive definiteness of an $n \times n$ matrix, we apply Sylvester's criterion, which states that a necessary and sufficient condition that the matrix be positive definite is that the determinants of all the successive principal minors of the matrix be positive.

Instead of first specifying a positive-definite matrix $\mathbf{P}$ and examining whether or not $\mathbf{Q}$ is positive definite, it is convenient to specify a positive-definite matrix

**Q** first and then examine whether or not **P** determined from

$$A^*P + PA = -Q$$

is positive definite. Note that **P** being positive definite is a necessary and sufficient condition. We shall summarize what we have just stated in the form of a theorem.

**Theorem 15-4.** Consider the system described by

$$\dot{x} = Ax$$

where **x** is a state vector ($n$-dimensional vector) and **A** is an $n \times n$ constant non-singular matrix. A necessary and sufficient condition that the equilibrium state **x** $= 0$ be asymptotically stable in the large is that, given any positive-definite Hermitian (or real symmetric) matrix **Q**, there exists a positive-definite Hermitian (or real symmetric) matrix **P** such that

$$A^*P + PA = -Q$$

The scalar function **x**\***Px** is a Liapunov function for this system. [Note that in the linear system considered, if the equilibrium state (the origin) is asymptotically stable, then it is asymptotically stable in the large.]

In applying this theorem several important remarks are in order.

1. If $\dot{V}(x) = -x^*Qx$ does not vanish identically along any trajectory, then **Q** may be chosen to be positive semidefinite.

2. If we choose an arbitrary positive-definite matrix as **Q** [or an arbitrary positive-semidefinite matrix as **Q** if $\dot{V}(x)$ does not vanish identically along any trajectory] and solve the matrix equation

$$A^*P + PA = -Q$$

to determine **P**, then the positive definiteness of **P** is a necessary and sufficient condition for the asymptotic stability of the equilibrium state **x** $= 0$.

3. The final result does not depend on a particular **Q** matrix chosen so long as it is positive definite (or positive semidefinite, as the case may be).

4. To determine the elements of the **P** matrix, we equate the matrices $A^*P + PA$ and $-Q$ element by element. This results in $n(n + 1)/2$ linear equations for the determination of the elements $p_{ij} = \bar{p}_{ji}$ of **P**. If we denote the eigenvalues of **A** by $\lambda_1, \lambda_2, \ldots, \lambda_n$, each repeated as often as its multiplicity as a root of the characteristic equation, and if for every sum of two roots

$$\lambda_j + \lambda_k \neq 0$$

then the elements of **P** is uniquely determined. Note that if the matrix **A** represents a stable system, then the sums $\lambda_j + \lambda_k$ are always nonzero.

5. In determining whether or not there exists a positive-definite Hermitian or a real symmetric matrix **P**, it is convenient to choose **Q** $= I$, where **I** is the identity matrix. Then the elements of **P** are determined from

$$A^*P + PA = -I$$

and the matrix **P** is tested for positive definiteness.

*Example 15-5.* Consider the second-order system described by

$$\begin{bmatrix} \dot{x}_1 \\ \dot{x}_2 \end{bmatrix} = \begin{bmatrix} 0 & 1 \\ -1 & -1 \end{bmatrix} \begin{bmatrix} x_1 \\ x_2 \end{bmatrix}$$

Clearly, the equilibrium state is the origin. Determine the stability of this state.
   Let us assume a tentative Liapunov function

$$V(\mathbf{x}) = \mathbf{x'Px}$$

where $\mathbf{P}$ is to be determined from

$$\mathbf{A'P + PA = -I}$$

or

$$\begin{bmatrix} 0 & -1 \\ 1 & -1 \end{bmatrix} \begin{bmatrix} p_{11} & p_{12} \\ p_{12} & p_{22} \end{bmatrix} + \begin{bmatrix} p_{11} & p_{12} \\ p_{12} & p_{22} \end{bmatrix} \begin{bmatrix} 0 & 1 \\ -1 & -1 \end{bmatrix} = \begin{bmatrix} -1 & 0 \\ 0 & -1 \end{bmatrix}$$

By expanding this matrix equation, we obtain three simultaneous equations as follows:

$$-2p_{12} = -1$$
$$p_{11} - p_{12} - p_{22} = 0$$
$$2p_{12} - 2p_{22} = -1$$

Solving for $p_{11}, p_{12}, p_{13}$, we obtain

$$\begin{bmatrix} p_{11} & p_{12} \\ p_{12} & p_{22} \end{bmatrix} = \begin{bmatrix} \dfrac{3}{2} & \dfrac{1}{2} \\ \dfrac{1}{2} & 1 \end{bmatrix}$$

To test the positive definiteness of $\mathbf{P}$, we check the determinants of the successive principal minors:

$$p_{11} = \frac{3}{2} > 0, \qquad \begin{vmatrix} p_{11} & p_{12} \\ p_{12} & p_{22} \end{vmatrix} = \begin{vmatrix} \dfrac{3}{2} & \dfrac{1}{2} \\ \dfrac{1}{2} & 1 \end{vmatrix} > 0$$

Clearly, $\mathbf{P}$ is positive definite. Hence the equilibrium state at the origin is asymptotically stable in the large, and a Liapunov function is

$$V = \mathbf{x'Px} = \tfrac{1}{2}(3x_1^2 + 2x_1 x_2 + 2x_2^2)$$

and

$$\dot{V} = -(x_1^2 + x_2^2)$$

*Example 15-6.* Determine the stability range for the gain $K$ of the system shown in Fig. 15-3.

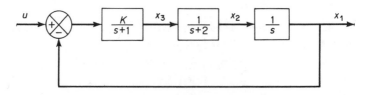

**Fig. 15-3.** Control system.

The state equation of the system is

$$\begin{bmatrix} \dot{x}_1 \\ \dot{x}_2 \\ \dot{x}_3 \end{bmatrix} = \begin{bmatrix} 0 & 1 & 0 \\ 0 & -2 & 1 \\ -K & 0 & -1 \end{bmatrix} \begin{bmatrix} x_1 \\ x_2 \\ x_3 \end{bmatrix} + \begin{bmatrix} 0 \\ 0 \\ K \end{bmatrix} [u] \qquad (15\text{-}4)$$

In determining the stability range for $K$, we assume the input $u$ to be zero. Then Eq. (15-4) may be written

$$\dot{x}_1 = x_2 \qquad (15\text{-}5)$$
$$\dot{x}_2 = -2x_2 + x_3 \qquad (15\text{-}6)$$
$$\dot{x}_3 = -Kx_1 - x_3 \qquad (15\text{-}7)$$

From Eqs. (15-5) through (15-7) we find that the origin is the equilibrium state. Let us choose the positive-semidefinite real symmetric matrix $\mathbf{Q}$ to be

$$\mathbf{Q} = \begin{bmatrix} 0 & 0 & 0 \\ 0 & 0 & 0 \\ 0 & 0 & 1 \end{bmatrix} \qquad (15\text{-}8)$$

This choice of $\mathbf{Q}$ is permissible since $\dot{V}(\mathbf{x}) = -\mathbf{x}'\mathbf{Q}\mathbf{x}$ cannot be identically equal to zero except at the origin. To verify this, note that

$$\dot{V}(\mathbf{x}) = -\mathbf{x}'\mathbf{Q}\mathbf{x} = -x_3^2$$

$\dot{V}(\mathbf{x})$ being identically zero implies that $x_3$ is identically zero. If $x_3$ is identically zero, then $x_1$ must be identically zero since from Eq. (15-7) we obtain

$$0 = -Kx_1 - 0$$

If $x_1$ is identically zero, then $x_2$ must also be identically zero since from Eq. (15-5)

$$0 = x_2$$

Thus, $\dot{V}(\mathbf{x})$ is identically zero only at the origin. Hence we may use the $\mathbf{Q}$ matrix defined by Eq. (15-8) for stability analysis.

Let us solve

$$\mathbf{A}'\mathbf{P} + \mathbf{P}\mathbf{A} = -\mathbf{Q}$$

or

$$\begin{bmatrix} 0 & 0 & -K \\ 1 & -2 & 0 \\ 0 & 1 & -1 \end{bmatrix} \begin{bmatrix} p_{11} & p_{12} & p_{13} \\ p_{12} & p_{22} & p_{23} \\ p_{13} & p_{23} & p_{33} \end{bmatrix} + \begin{bmatrix} p_{11} & p_{12} & p_{13} \\ p_{12} & p_{22} & p_{23} \\ p_{13} & p_{23} & p_{33} \end{bmatrix} \begin{bmatrix} 0 & 1 & 0 \\ 0 & -2 & 1 \\ -K & 0 & -1 \end{bmatrix} = \begin{bmatrix} 0 & 0 & 0 \\ 0 & 0 & 0 \\ 0 & 0 & -1 \end{bmatrix}$$

for the elements of $\mathbf{P}$. The result is

$$\mathbf{P} = \begin{bmatrix} \dfrac{K^2 + 12K}{12 - 2K} & \dfrac{6K}{12 - 2K} & 0 \\[3mm] \dfrac{6K}{12 - 2K} & \dfrac{3K}{12 - 2K} & \dfrac{K}{12 - 2K} \\[3mm] 0 & \dfrac{K}{12 - 2K} & \dfrac{6}{12 - 2K} \end{bmatrix}$$

In order for $\mathbf{P}$ to be positive definite, it is necessary and sufficient that

$$12 - 2K > 0 \qquad \text{and} \qquad K > 0$$

or

$$0 < K < 6$$

Thus, for $0 < K < 6$, the system is stable in the conventional sense: that is, the origin is asymptotically stable in the large.

**Liapunov stability analysis of discrete-time systems.** Consider the discrete-time system described by

$$\mathbf{x}(k + 1) = \mathbf{Gx}(k) \tag{15-9}$$

where $\mathbf{x}$ is a state vector ($n$-dimensional vector) and $\mathbf{G}$ is an $n \times n$ constant nonsingular matrix. The origin $\mathbf{x} = \mathbf{0}$ is the equilibrium state. We shall investigate the stability of this state by use of the second method of Liapunov.

Let us choose a possible Liapunov function as

$$V(\mathbf{x}(k)) = \mathbf{x}^*(k)\mathbf{Px}(k)$$

where $\mathbf{P}$ is a positive-definite Hermitian (or real symmetric) matrix. Noting that for discrete-time systems, instead of $\dot{V}(\mathbf{x})$, we use the difference between $V(\mathbf{x}(k + 1))$ and $V(\mathbf{x}(k))$, or

$$\Delta V(\mathbf{x}(k)) = V(\mathbf{x}(k + 1)) - V(\mathbf{x}(k))$$

which is analogous to the derivative of $V(\mathbf{x})$. Then

$$\begin{aligned}
\Delta V(\mathbf{x}(k)) &= V(\mathbf{x}(k + 1)) - V(\mathbf{x}(k)) \\
&= \mathbf{x}^*(k + 1)\mathbf{Px}(k + 1) - \mathbf{x}^*(k)\mathbf{Px}(k) \\
&= [\mathbf{Gx}(k)]^*\mathbf{P}[\mathbf{Gx}(k)] - \mathbf{x}^*(k)\mathbf{Px}(k) \\
&= \mathbf{x}^*(k)\mathbf{G}^*\mathbf{PGx}(k) - \mathbf{x}^*(k)\mathbf{Px}(k) \\
&= \mathbf{x}^*(k)[\mathbf{G}^*\mathbf{PG} - \mathbf{P}]\mathbf{x}(k)
\end{aligned}$$

Since $V(\mathbf{x}(k))$ was chosen to be positive definite, we require, for asymptotic stability, that $\Delta V(\mathbf{x}(k))$ be negative definite. Therefore,

$$\Delta V(\mathbf{x}(k)) = -\mathbf{x}^*(k)\mathbf{Qx}(k)$$

where

$$\mathbf{Q} = -(\mathbf{G}^*\mathbf{PG} - \mathbf{P}) = \text{positive definite}$$

Hence, for the asymptotic stability of the discrete-time system of Eq. (15-9), it is sufficient that $\mathbf{Q}$ be positive definite.

Similar to the case of linear continuous-time systems, it is convenient to specify a positive-definite Hermitian (or real symmetric) matrix $\mathbf{Q}$ and then see whether or not the $\mathbf{P}$ matrix determined from

$$\mathbf{G}^*\mathbf{PG} - \mathbf{P} = -\mathbf{Q}$$

is positive definite. Note that $\mathbf{P}$ being positive definite is a necessary and sufficient condition. We shall summarize what we have stated here as a theorem.

**Theorem 15-5.** Consider the discrete-time system

$$\mathbf{x}(k + 1) = \mathbf{Gx}(k)$$

where $\mathbf{x}$ is a state vector ($n$-dimensional vector) and $\mathbf{G}$ is an $n \times n$ constant nonsingular matrix. A necessary and sufficient condition for the equilibrium state $\mathbf{x} = \mathbf{0}$ to be asymptotically stable is that, given any positive-definite Hermitian (or real symmetric) matrix $\mathbf{Q}$, there exists a positive definite Hermitian (or real symmetric) matrix $\mathbf{P}$ such that

$$\mathbf{G*PG} - \mathbf{P} = -\mathbf{Q}$$

The scalar function $\mathbf{x*Px}$ is a Liapunov function for this system.

If $\Delta V(\mathbf{x}(k)) = -\mathbf{x}^*(k)\mathbf{Q}\mathbf{x}(k)$ does not vanish identically along any solution series, then $\mathbf{Q}$ may be chosen to be positive semidefinite.

## 15-5  ESTIMATING THE TRANSIENT-RESPONSE BEHAVIOR OF DYNAMIC SYSTEMS

The applicability of the second method of Liapunov to control theory is not limited to stability analysis. It can be applied to the study of the transient-response behavior of linear and nonlinear systems. In this section, we shall discuss the relationship between Liapunov functions and the transient-response behavior of dynamic systems.

As stated earlier in this chapter, Liapunov functions give a measure of distance in state space. Therefore, for an asymptotically stable system for which a Liapunov function has been obtained, the latter can be used to estimate the rapidity of the transient response.

Let us assume that the equilibrium state is the origin of the state space and define

$$\eta = -\frac{\dot{V}(\mathbf{x}, t)}{V(\mathbf{x}, t)} \tag{15-10}$$

in some region of the state space excluding the origin.

For uniformly asymptotically stable systems, a Liapunov function with negative definite $\dot{V}$ can be found, at least in principle. In the present analysis, we therefore exclude the case where $\dot{V}(\mathbf{x}, t) = 0$ for $\mathbf{x} \neq \mathbf{0}$. Then the value of $\eta$ is always positive since $V$ is positive definite and $\dot{V}$ is negative definite. Equation (15-10) gives an indication of how fast the system approaches its equilibrium state.

If we let the minimum value of the ratio $-\dot{V}(\mathbf{x}, t)/V(\mathbf{x}, t)$ be $\eta_{\min}$, or

$$\eta_{\min} = \min\left[-\frac{\dot{V}(\mathbf{x}, t)}{V(\mathbf{x}, t)}\right] \tag{15-11}$$

then

$$\dot{V}(\mathbf{x}, t) \leq -\eta_{\min} V(\mathbf{x}, t)$$

and

$$V(\mathbf{x}, t) \leq V(\mathbf{x}_0, t_0)e^{-\eta_{\min}(t-t_0)} \tag{15-12}$$

where $V(\mathbf{x}_0, t_0)$ corresponds to the initial value of $V$ starting at state $\mathbf{x} = \mathbf{x}_0$ at time $t = t_0$.

Interpreting $V$ as the distance from the origin, it is clear that Eq. (15-12) gives an estimate of how fast equilibrium is approached. In other words, given the initial state $\mathbf{x}_0$ and the initial value $V(\mathbf{x}_0, t_0)$, the system reaches the region

$$V(\mathbf{x}, t) = V(\mathbf{x}_0, t_0)e^{-\eta_{\min}(t-t_0)}$$

in $t - t_0$ seconds.

From Eq. (15-11), we see that $1/\eta_{\min}$ corresponds to the largest time constant relating to changes in the Liapunov function $V$. Since $V(\mathbf{x})$ may be expanded in a series starting with a quadratic term in $\mathbf{x}$, this time constant $1/\eta_{\min}$ is about half the conventional time constant defined for the system. Larger values of $\eta_{\min}$ correspond to faster response.

Since there may be a number of Liapunov functions for a given system, a number of different values of $\eta_{\min}$ may be obtained. The largest value of $\eta_{\min}$ represents the figure of merit. It is, however, usually not known how to construct the Liapunov function which yields the largest value of $\eta_{\min}$.

For a nonlinear system, $\eta_{\min}$ may depend on the state of the system. This approach is useful only if the Liapunov function available is of a sufficiently simple form.

**Calculation of $\eta$ for linear time-invariant systems.** Consider the linear time-invariant system

$$\dot{\mathbf{x}} = \mathbf{A}\mathbf{x}$$

where

$\mathbf{x} = n$-dimensional vector

$\mathbf{A} = n \times n$ nonsingular matrix

Suppose that the eigenvalues of $\mathbf{A}$ have negative real parts. Then a Liapunov function and its time derivative are

$$V(\mathbf{x}) = \mathbf{x}^*\mathbf{P}\mathbf{x}, \qquad \dot{V}(\mathbf{x}) = -\mathbf{x}^*\mathbf{Q}\mathbf{x}$$

where

$\mathbf{P} = $ positive-definite Hermitian (or real symmetric) matrix

$\mathbf{Q} = -(\mathbf{A}^*\mathbf{P} + \mathbf{P}\mathbf{A})$

A definition of $\eta_{\min}$ equivalent to Eq. (15-11) is

$$\eta_{\min} = \min_{\mathbf{x}} \{\mathbf{x}^*\mathbf{Q}\mathbf{x}; \mathbf{x}^*\mathbf{P}\mathbf{x} = 1\}$$

The right-hand side of this equation means that we take the minimum value of $\mathbf{x}^*\mathbf{Q}\mathbf{x}$ as $\mathbf{x}$ varies, satisfying the condition that $\mathbf{x}^*\mathbf{P}\mathbf{x} = 1$. This is the same as considering the minimum value of the ratio

$$-\frac{\dot{V}(\mathbf{x}, t)}{V(\mathbf{x}, t)} = \frac{\mathbf{x}^*\mathbf{Q}\mathbf{x}}{\mathbf{x}^*\mathbf{P}\mathbf{x}}$$

along one particular surface $V(\mathbf{x}) = 1$, or $\mathbf{x}^*\mathbf{P}\mathbf{x} = 1$.

The required minimization can be performed by use of the Lagrange multiplier technique. Let $\mu$ be the Lagrange multiplier. Then minimizing $\mathbf{x}^*\mathbf{Q}\mathbf{x}$ with the constraint that $\mathbf{x}^*\mathbf{Px} = 1$ is the same as minimizing $\mathbf{x}^*\mathbf{Qx} - \mu\mathbf{x}^*\mathbf{Px}$ with respect to $\mathbf{x}$. The minimum value of $\mathbf{x}^*\mathbf{Qx} - \mu\mathbf{x}^*\mathbf{Px}$ occurs at a value $\mathbf{x}_{\min}$ of $\mathbf{x}$ such that

$$(\mathbf{Q} - \mu\mathbf{P})\mathbf{x}_{\min} = 0 \qquad (15\text{-}13)$$

Hence, we obtain

$$\mathbf{x}_{\min}^*(\mathbf{Q} - \mu\mathbf{P})\mathbf{x}_{\min} = 0$$

or

$$\mathbf{x}_{\min}^*\mathbf{Qx}_{\min} = \mu\mathbf{x}_{\min}^*\mathbf{Px}_{\min} = \mu > 0$$

which is a minimum if $\mu$ is. But, by Eq. (15-13), $\mu$ is an eigenvalue of the matrix $\mathbf{QP}^{-1}$. Hence $\eta_{\min}$ is equal to the minimum eigenvalue $\lambda_{\min}$ of $\mathbf{QP}^{-1}$.

---

*Example 15-7.* Consider the following system:

$$\begin{bmatrix} \dot{x}_1 \\ \dot{x}_2 \end{bmatrix} = \begin{bmatrix} 0 & 1 \\ -1 & -1 \end{bmatrix} \begin{bmatrix} x_1 \\ x_2 \end{bmatrix}$$

Find a Liapunov function for the system. Then obtain an upper bound on the response time that it takes the system to go from a point on the boundary of the closed curve

$$V(\mathbf{x}) = 150$$

to a point within the closed curve

$$V(\mathbf{x}) = 0.06$$

Let us assume $\mathbf{Q} = \mathbf{I}$ and determine $\mathbf{P}$ by solving

$$\begin{bmatrix} 0 & -1 \\ 1 & -1 \end{bmatrix} \begin{bmatrix} p_{11} & p_{12} \\ p_{12} & p_{22} \end{bmatrix} + \begin{bmatrix} p_{11} & p_{12} \\ p_{12} & p_{22} \end{bmatrix} \begin{bmatrix} 0 & 1 \\ -1 & -1 \end{bmatrix} = \begin{bmatrix} -1 & 0 \\ 0 & -1 \end{bmatrix}$$

The result is

$$\begin{bmatrix} p_{11} & p_{12} \\ \\ p_{12} & p_{22} \end{bmatrix} = \begin{bmatrix} \dfrac{3}{2} & \dfrac{1}{2} \\ \\ \dfrac{1}{2} & 1 \end{bmatrix}$$

Hence we obtain a Liapunov function as follows:

$$V = \mathbf{x}'\mathbf{Px} = \tfrac{1}{2}(3x_1^2 + 2x_1x_2 + 2x_2^2)$$

Noting that

$$\dot{V} = -\mathbf{x}'\mathbf{x} = -(x_1^2 + x_2^2)$$

we obtain the value of $\eta$ as

$$\eta = -\frac{\dot{V}}{V} = \frac{2(x_1^2 + x_2^2)}{3x_1^2 + 2x_1x_2 + 2x_2^2}$$

The minimum value of $\eta$ can be obtained as the minimum eigenvalue of $\mathbf{QP}^{-1}$. Since we assumed $\mathbf{Q} = \mathbf{I}$, we obtain

$$|\mathbf{QP}^{-1} - \lambda\mathbf{I}| = |\mathbf{P}^{-1} - \lambda\mathbf{I}| = 0$$

Finding the eigenvalues of $\mathbf{P}^{-1}$ is equivalent to solving

$$|\mathbf{I} - \mathbf{P}\lambda| = 0$$

or

$$\begin{vmatrix} 1 - \frac{3}{2}\lambda & -\frac{1}{2}\lambda \\ -\frac{1}{2}\lambda & 1 - \lambda \end{vmatrix} = 0$$

The two eigenvalues are

$$\lambda_1 = 1.447, \qquad \lambda_2 = 0.533$$

Hence

$$\eta_{\min} = \lambda_2 = 0.553$$

From Eq. (15-12)

$$V(\mathbf{x}, t) \le V(\mathbf{x}_0, t_0)e^{-\eta_{\min}(t-t_0)}$$

or

$$-\frac{1}{\eta_{\min}} \ln \left[ \frac{V(\mathbf{x}, t)}{V(\mathbf{x}_0, t_0)} \right] \ge t - t_0 \qquad (15\text{-}14)$$

To compute $t - t_0$, we use the value of $\eta_{\min}$ just obtained. Substitution of $\eta_{\min} = 0.553$, $V(\mathbf{x}, t) = 0.06$, and $V(\mathbf{x}_0, t_0) = 150$ into Eq. (15-14) yields

$$-\frac{1}{0.553} \ln \left( \frac{0.06}{150} \right) \ge t - t_0$$

or

$$14.1 \ge t - t_0$$

Any trajectory starting on the curve $V(\mathbf{x}_0, t_0) = 150$ will be within the region surrounded by the curve $V(\mathbf{x}, t) = 0.06$ within 14.1 units of time. (Note that this time interval depends on the choice of the $\mathbf{P}$ and $\mathbf{Q}$ matrices. If we choose appropriate $\mathbf{P}$ and $\mathbf{Q}$ matrices, we may have a smaller number than 14.1 as an upper limit on $t - t_0$.)

### 15-6 STABILITY ANALYSIS OF NONLINEAR SYSTEMS

In a linear system, if the equilibrium state is locally asymptotically stable, then it is asymptotically stable in the large. In a nonlinear system, however, an equilibrium state can be locally asymptotically stable without being asymptotically stable in the large. Hence, the implications of asymptotic stability of equilibrium states of linear systems and those of nonlinear systems are quite different. For example, consider the nonlinear system discussed in Example 12-7, rewritten

$$\ddot{x} + 0.5\dot{x} + 2x + x^2 = 0$$

Figure 15-4 shows the phase-plane diagram for this system. In this case, the origin is asymptotically stable. But trajectories starting at points outside the hatched region go to infinity, as shown. The fact that asymptotic stability is a local property can be clearly seen in this example.

Because the stability of nonlinear systems is of local nature, we are usually concerned with finding a Liapunov function which satisfies stability conditions in the largest region about the origin.

Several methods based on the second method of Liapunov are available for

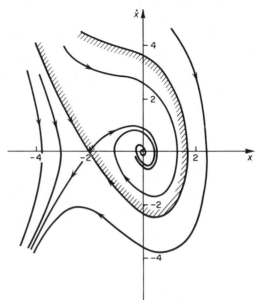

**Fig. 15-4.** Phase-plane diagram of the system described by $\ddot{x} + 0.5\dot{x} + 2x + x^2 = 0$

testing the stability of nonlinear systems. This section presents two such methods, namely Krasovskii's method for testing sufficient conditions for asymptotic stability and Schultz-Gibson's variable-gradient method for generating Liapunov functions.

**Krasovskii's method.** For certain nonlinear systems, a possible Liapunov function may be chosen in terms of $\dot{\mathbf{x}}$ rather than in terms of the state variables. Krasovskii suggested a possible Liapunov function to be the Euclidean norm of $\dot{\mathbf{x}}$, or $V = \|\dot{\mathbf{x}}\|^2$.

Krasovskii's theorem, to be presented in the following, is not limited to small departures from the equilibrium state. It essentially differs from usual linearization approaches. Krosovskii's theorem for asymptotic stability in the large gives sufficient conditions for nonlinear systems and necessary and sufficient conditions for linear systems. (An equilibrium state of a nonlinear system may be stable even if the conditions specified in this theorem are not satisfied. Hence, in using Krasovskii's theorem, we should be careful not to make erroneous conclusions about the stability of the equilibrium state of the given nonlinear system.)

In a nonlinear system, there may be more than one equilibrium state. It is, however, possible to transfer any isolated equilibrium state to the origin of the state space by an appropriate transformation of coordinates. We shall, therefore, consider the equilibrium state under consideration to be the origin.

We shall now present Krasovskii's theorem.

**Theorem 15-6** (Krasovskii's theorem). Consider the system defined by

$$\dot{\mathbf{x}} = \mathbf{f}(\mathbf{x})$$

where $\mathbf{x}$ is an $n$-dimensional vector. Assume that $\mathbf{f}(0) = 0$ and that $\mathbf{f}(\mathbf{x})$ is differentiable with respect to $x_i$, $i = 1, 2, \ldots, n$. The Jacobian matrix $\mathbf{F}(\mathbf{x})$ for the system is

$$\mathbf{F}(\mathbf{x}) = \begin{bmatrix} \dfrac{\partial f_1}{\partial x_1} & \dfrac{\partial f_1}{\partial x_2} & \cdots & \dfrac{\partial f_1}{\partial x_n} \\[2mm] \dfrac{\partial f_2}{\partial x_1} & \dfrac{\partial f_2}{\partial x_2} & \cdots & \dfrac{\partial f_2}{\partial x_n} \\[2mm] \cdot & \cdot & & \cdot \\ \cdot & \cdot & & \cdot \\ \cdot & \cdot & & \cdot \\[2mm] \dfrac{\partial f_n}{\partial x_1} & \dfrac{\partial f_n}{\partial x_2} & \cdots & \dfrac{\partial f_n}{\partial x_n} \end{bmatrix}$$

Define

$$\hat{\mathbf{F}}(\mathbf{x}) = \mathbf{F}^*(\mathbf{x}) + \mathbf{F}(\mathbf{x})$$

where $\mathbf{F}^*(\mathbf{x})$ is the conjugate transpose of $\mathbf{F}(\mathbf{x})$. If the Hermitian matrix $\hat{\mathbf{F}}(\mathbf{x})$ is negative definite, then the equilibirum state $\mathbf{x} = 0$ is asymptotically stable. A Liapunov function for this system is

$$V(\mathbf{x}) = \mathbf{f}^*(\mathbf{x})\mathbf{f}(\mathbf{x})$$

If, in addition, $\mathbf{f}^*(\mathbf{x})\mathbf{f}(\mathbf{x}) \longrightarrow \infty$ as $\|\mathbf{x}\| \longrightarrow \infty$, then the equilibrium state is asymptotically stable in the large.

**PROOF:** If $\hat{\mathbf{F}}(\mathbf{x})$ is negative definite for all $\mathbf{x} \neq 0$, the determinant of $\mathbf{F}$ is nonzero everywhere except at $\mathbf{x} = 0$. (Refer to Problems A-15-5 and A-15-7.) There is no other equilibirum state than $\mathbf{x} = 0$ in the entire state space. Since $\mathbf{f}(0) = 0$, $\mathbf{f}(\mathbf{x}) \neq 0$ for $\mathbf{x} \neq 0$, and $V(\mathbf{x}) = \mathbf{f}^*(\mathbf{x})\mathbf{f}(\mathbf{x})$, $V(\mathbf{x})$ is positive definite. Noting that

$$\dot{\mathbf{f}}(\mathbf{x}) = \mathbf{F}(\mathbf{x})\dot{\mathbf{x}} = \mathbf{F}(\mathbf{x})\mathbf{f}(\mathbf{x})$$

we can obtain $\dot{V}$ as

$$\begin{aligned} \dot{V}(\mathbf{x}) &= \dot{\mathbf{f}}^*(\mathbf{x})\mathbf{f}(\mathbf{x}) + \mathbf{f}^*(\mathbf{x})\dot{\mathbf{f}}(\mathbf{x}) \\ &= [\mathbf{F}(\mathbf{x})\mathbf{f}(\mathbf{x})]^*\mathbf{f}(\mathbf{x}) + \mathbf{f}^*(\mathbf{x})\mathbf{F}(\mathbf{x})\mathbf{f}(\mathbf{x}) \\ &= \mathbf{f}^*(\mathbf{x})[\mathbf{F}^*(\mathbf{x}) + \mathbf{F}(\mathbf{x})]\mathbf{f}(\mathbf{x}) \\ &= \mathbf{f}^*(\mathbf{x})\hat{\mathbf{F}}(\mathbf{x})\mathbf{f}(\mathbf{x}) \end{aligned}$$

If $\hat{\mathbf{F}}(\mathbf{x})$ is negative definite, we see that $\dot{V}(\mathbf{x})$ is negative definite. Hence $V(\mathbf{x})$ is a Liapunov function. Therefore, the origin is asymptotically stable. If $V(\mathbf{x}) = \mathbf{f}^*(\mathbf{x})\mathbf{f}(\mathbf{x})$ tends to infinity as $\|\mathbf{x}\| \longrightarrow \infty$, then by Theorem 15-1 the equilibrium state is asymptotically stable in the large.

Note that $\hat{\mathbf{F}}(\mathbf{x})$ being negative definite requires that $\mathbf{F}(\mathbf{x})$ must have nonzero elements on its main diagonal. Hence if $f_i(\mathbf{x})$ does not involve $x_i$, then $\hat{\mathbf{F}}(\mathbf{x})$ cannot be negative definite.

---

*Example 15-8.* By use of Krasovskii's theorem, examine the stability of the equilibrium state $\mathbf{x} = 0$ of the following system:

$$\begin{aligned} \dot{x}_1 &= -x_1 \\ \dot{x}_2 &= x_1 - x_2 - x_2^3 \end{aligned}$$

For this system

$$\mathbf{f(x)} = \begin{bmatrix} -x_1 \\ x_1 - x_2 - x_2^3 \end{bmatrix}$$

and

$$\mathbf{F(x)} = \begin{bmatrix} -1 & 0 \\ 1 & -1 - 3x_2^2 \end{bmatrix}$$

Hence

$$\mathbf{\hat{F}(x)} = \mathbf{F'(x)} + \mathbf{F(x)}$$
$$= \begin{bmatrix} -2 & 1 \\ 1 & -2 - 6x_2^2 \end{bmatrix}$$

Since $\mathbf{\hat{F}(x)}$ is negative definite for all $\mathbf{x} \neq \mathbf{0}$, the equilibrium state $\mathbf{x} = \mathbf{0}$ is asymptotically stable. In addition,

$$\mathbf{f'(x)f(x)} = x_1^2 + (x_1 - x_2 - x_2^3)^2 \longrightarrow \infty$$

as $\|\mathbf{x}\| \longrightarrow \infty$. The equilibrium state $\mathbf{x} = \mathbf{0}$ is, therefore, asymptotically stable in the large.

**Variable-gradient method.** The variable-gradient method is based on the fact that if a particular Liapunov function exists which is capable of proving stability for a given system, then a unique gradient of this $V$ function also exists.

Consider the system described by the following equation:

$$\mathbf{\dot{x}} = \mathbf{f(x, }t)$$

We assume that the equilibrium state in question is the origin of the state space. (As stated earlier, we can always transfer any isolated equilibrium state to the origin.) Let us denote a tentative Liapunov function by $V$. In this function, we assume that $V$ is an explicit function of $\mathbf{x}$ but not an explicit function of $t$. Then

$$\dot{V} = \frac{\partial V}{\partial x_1}\dot{x}_1 + \frac{\partial V}{\partial x_2}\dot{x}_2 + \cdots + \frac{\partial V}{\partial x_n}\dot{x}_n$$

which may be written

$$\dot{V} = (\nabla V)'\mathbf{\dot{x}}$$

where $(\nabla V)'$ is the transpose of $\nabla V$. $\nabla V$, the gradient of $V$, is

$$\nabla V = \begin{bmatrix} \dfrac{\partial V}{\partial x_1} \\ \cdot \\ \cdot \\ \cdot \\ \dfrac{\partial V}{\partial x_n} \end{bmatrix} = \begin{bmatrix} \nabla V_1 \\ \cdot \\ \cdot \\ \cdot \\ \nabla V_n \end{bmatrix}$$

$V$ is obtained as a line integral of $\nabla V$ as

$$V = \int_0^x (\nabla V)'\, d\mathbf{x} \qquad (15\text{-}15)$$

The upper limit of integration here does not imply that $V$ is a vector quantity but rather that the integral is a line integral to an arbitrary point $(x_1, x_2, \ldots, x_n)$ in the state space. This integral can be made independent of the path of integration. The simplest path is indicated by the expanded form of Eq. (15-15):

$$V = \int_0^{x_1, (x_2 = x_3 = \cdots = x_n = 0)} \nabla V_1 \, dx_1 + \int_0^{x_2, (x_1 = x_1, x_3 = x_4 = \cdots = x_n = 0)} \nabla V_2 \, dx_2 + \cdots$$

$$+ \int_0^{x_n, (x_1 = x_1, x_2 = x_2, \cdots, x_{n-1} = x_{n-1})} \nabla V_n \, dx_n \tag{15-16}$$

where $\nabla V_i$ is the component of $\nabla V$ in the $x_i$ direction.

Let us define

$$\mathbf{e}_1 = \begin{bmatrix} 1 \\ 0 \\ 0 \\ \cdot \\ \cdot \\ \cdot \\ 0 \end{bmatrix}, \quad \mathbf{e}_2 = \begin{bmatrix} 0 \\ 1 \\ 0 \\ \cdot \\ \cdot \\ \cdot \\ 0 \end{bmatrix}, \quad \cdots, \quad \mathbf{e}_n = \begin{bmatrix} 0 \\ 0 \\ 0 \\ \cdot \\ \cdot \\ \cdot \\ 1 \end{bmatrix}$$

The integral given by Eq. (15-16) states that the path starts from the origin and moves along the vector $\mathbf{e}_1$ to $x_1$. From this point, the path moves in the direction of the vector $\mathbf{e}_2$ to $x_2$. Then, the path moves in the direction of the vector $\mathbf{e}_3$ to $x_3$. In this way the path finally reaches point $(x_1, x_2, \ldots, x_n)$.

For a scalar function $V$ to be obtained uniquely from a line integral of a vector function $\nabla V$, the following matrix $\mathbf{F}$ formed by taking $\partial \nabla V_i / \partial x_j$

$$\mathbf{F} = \begin{bmatrix} \dfrac{\partial \nabla V_1}{\partial x_1} & \dfrac{\partial \nabla V_1}{\partial x_2} & \cdots & \dfrac{\partial \nabla V_1}{\partial x_n} \\[2mm] \dfrac{\partial \nabla V_2}{\partial x_1} & \dfrac{\partial \nabla V_2}{\partial x_2} & \cdots & \dfrac{\partial \nabla V_2}{\partial x_n} \\ \cdot & \cdot & & \cdot \\ \cdot & \cdot & & \cdot \\ \cdot & \cdot & & \cdot \\ \dfrac{\partial \nabla V_n}{\partial x_1} & \dfrac{\partial \nabla V_n}{\partial x_2} & \cdots & \dfrac{\partial \nabla V_n}{\partial x_n} \end{bmatrix}$$

must be symmetric. Thus we require

$$\frac{\partial \nabla V_i}{\partial x_j} = \frac{\partial \nabla V_j}{\partial x_i} \qquad (i, j = 1, 2, \ldots, n)$$

The total number of such equations is $n(n-1)/2$. For example, in the case of $n = 3$, we have three equations

$$\frac{\partial \nabla V_2}{\partial x_1} = \frac{\partial \nabla V_1}{\partial x_2}, \qquad \frac{\partial \nabla V_3}{\partial x_1} = \frac{\partial \nabla V_1}{\partial x_3}, \qquad \frac{\partial \nabla V_3}{\partial x_2} = \frac{\partial \nabla V_2}{\partial x_3}$$

The condition on the matrix $\mathbf{F}$ is thus a generalized curl requirement for the $n$-dimensional case. The problem of determining a function $V$ which satisfies Liapunov's theorem is then transformed into the problem of finding a $\nabla V$ such that the $n$-dimensional curl of $\nabla V$ equals zero. Further, $V$ and $\dot{V}$ determined from

$\nabla V$ must be sufficient to prove stability; that is, they must satisfy Liapunov's theorem. We first set $\nabla V$ equal to an arbitrary column vector

$$\nabla V = \begin{bmatrix} a_{11}x_1 + a_{12}x_2 + \cdots + a_{1n}x_n \\ a_{21}x_1 + a_{22}x_2 + \cdots + a_{2n}x_n \\ \cdot \\ \cdot \\ \cdot \\ a_{n1}x_1 + a_{n2}x_2 + \cdots + a_{nn}x_n \end{bmatrix} \tag{15-17}$$

The $a_{ij}$ are completely undetermined quantities. The $a_{ij}$ may be constant or functions of time $t$ and/or functions of the state variables. It is convenient, however, to choose $a_{nn}$ as a constant or as a function of time $t$. Several of the $a_{ij}$ may be chosen zero, or they are obvious from the constraints on $\dot{V}$ imposed by the investigator, or they may be determined from the curl equations.

If the equilibrium state $\mathbf{x} = 0$ of a nonlinear system is asymptotically stable, then we may be able to obtain a Liapunov function by following this procedure:

1. Assume $\nabla V$ of the form of Eq. (15-17).
2. From $\nabla V$, determine $\dot{V}$.
3. Constrain $\dot{V}$ to be negative definite or at least negative semidefinite.
4. Use the $n(n-1)/2$ curl equations implied by the statement that $\mathbf{F}$ must be symmetric to determine the remaining unknown coefficients in $\nabla V$.
5. Recheck $\dot{V}$, as the addition of terms required as a result of Step 4 may alter $\dot{V}$.
6. Determine $V$ by Eq. (15-16).
7. Examine the region of asymptotic stability.

Note that failure to obtain a suitable Liapunov function by this method does not imply instability of the equilibrium state. (In fact no conclusions about stability can be made.)

---

*Example 15-9.* Let us construct a Liapunov function of the following system:

$$\dot{x}_1 = -x_1 + 2x_1^2 x_2$$
$$\dot{x}_2 = -x_2$$

by use of the variable-gradient method.

Let the gradient of $V$ be

$$\nabla V = \begin{bmatrix} a_{11}x_1 + a_{12}x_2 \\ a_{21}x_1 + 2x_2 \end{bmatrix}$$

The derivative of $V$ is then

$$\begin{aligned} \dot{V} &= \nabla V' \dot{\mathbf{x}} \\ &= (a_{11}x_1 + a_{12}x_2)\dot{x}_1 + (a_{21}x_1 + 2x_2)\dot{x}_2 \\ &= -a_{11}x_1^2 + 2a_{11}x_1^3 x_2 - a_{12}x_1 x_2 + 2a_{12}x_1^2 x_2^2 - a_{21}x_1 x_2 - 2x_2^2 \end{aligned} \tag{15-18}$$

Let us put, as a trial,

$$a_{11} = 1, \qquad a_{12} = a_{21} = 0$$

then Eq. (15-18) becomes

$$\dot{V} = -x_1^2(1 - 2x_1x_2) - 2x_2^2$$

$\dot{V}$ is negative definite if

$$1 - 2x_1x_2 > 0 \tag{15-19}$$

Thus, Eq. (15-19) is a constraint on $x_1$ and $x_2$. The gradient $\nabla V$ is

$$\nabla V = \begin{bmatrix} x_1 \\ 2x_2 \end{bmatrix}$$

Note that

$$\frac{\partial \nabla V_1}{\partial x_2} = \frac{\partial \nabla V_2}{\partial x_1} = 0$$

We see that the curl equation is satisfied. $V$ is

$$V = \int_0^{x_1,\,(x_2=0)} x_1\,dx_1 + \int_0^{x_2,\,(x_1=x_1)} 2x_2\,dx_2$$

$$= \frac{x_1^2}{2} + x_2^2 \tag{15-20}$$

From this Liapunov function, we can say that the system is asymptotically stable in the region

$$1 > 2x_1x_2$$

To show that the Liapunov function given by Eq. (15-20) is not the only one possible, let us choose

$$a_{11} = \frac{2}{(1 - x_1x_2)^2}, \qquad a_{12} = \frac{-x_1^2}{(1 - x_1x_2)^2}, \qquad a_{21} = \frac{x_1^2}{(1 - x_1x_2)^2}$$

Then

$$\dot{V} = -2x_1^2 - 2x_2^2$$

$\dot{V}$ is negative definite in the entire state plane. $\nabla V$ becomes

$$\nabla V = \begin{bmatrix} \dfrac{2x_1}{(1 - x_1x_2)^2} - \dfrac{x_1^2 x_2}{(1 - x_1x_2)^2} \\[2ex] \dfrac{x_1^3}{(1 - x_1x_2)^2} + 2x_2 \end{bmatrix}$$

Since

$$\frac{\partial \nabla V_1}{\partial x_2} = \frac{3x_1^2 - x_1^3 x_2}{(1 - x_1x_2)^3}$$

$$\frac{\partial \nabla V_2}{\partial x_1} = \frac{3x_1^2 - x_1^3 x_2}{(1 - x_1x_2)^3}$$

the curl equation is satisfied. $V$ is

$$V = \int_0^{x_1,\,(x_2=0)} \left[ \frac{2x_1}{(1 - x_1x_2)^2} - \frac{x_1^2 x_2}{(1 - x_1x_2)^2} \right] dx_1 + \int_0^{x_2,\,(x_1=x_1)} \left[ \frac{x_1^3}{(1 - x_1x_2)^2} + 2x_2 \right] dx_2$$

$$= x_2^2 + \frac{x_1^2}{1 - x_1x_2} \tag{15-21}$$

From this Liapunov function, the origin of the system is seen to be asymptotically stable in the region

$$1 > x_1x_2$$

It is clear that the Liapunov function defined by Eq. (15-20) gives a smaller asymptotically stable region than the Liapunov function given by Eq. (15-21). Hence the latter function is a better choice than the former.

## 15-7  CONCLUDING COMMENTS

In concluding this chapter, we shall summarize important points to remember:

1. A Liapunov function has the following properties:
   (a) It is a scalar function.
   (b) It is positive definite, at least in the neighborhood of the origin.
   (c) Its time derivative is nonpositive.
   (d) It is not a unique function for a given system.

2. An advantage of the second method of Liapunov lies in its ability to give stability information in the large for both linear and nonlinear systems.

3. If a Liapunov function can be found in a region $\Omega$ including the origin of the state space, then the stability or asymptotic stability of the origin is verified. This, however, does not necessarily mean that the trajectories starting from a state outside the region $\Omega$ approach infinity since the second method establishes only sufficient conditions for stability. (Remember that failure to find a Liapunov function to show stability, asymptotic stability, or instability of an equilibrium state does not mean anything.) Note also that for nonlinear systems some Liapunov functions are superior to others in that the region of asymptotic stability is larger or $\dot{V}$ is negative definite rather than negative semidefinite, etc.

4. If the origin of the system is stable or asymptotically stable, then Liapunov functions with the required properties always exist, although finding one may be quite difficult. This difficulty is the major drawback of stability analysis based on the second method of Liapunov.

5. In addition to providing a stability criterion, Liapunov functions may be used for the transient-response analysis of simple systems. (See Section 15-5.) Liapunov functions may also be used for solving optimization problems based on quadratic performance indexes. (See Section 16-5.)

EXAMPLE PROBLEMS AND SOLUTIONS

**PROBLEM A-15-1.** Determine whether or not the following quadratic form is negative definite.

$$Q = -x_1^2 - 3x_2^2 - 11x_3^2 + 2x_1x_2 - 4x_2x_3 - 2x_1x_3$$

**Solution.** The given quadratic form $Q$ can be written

$$Q = [x_1 \quad x_2 \quad x_3] \begin{bmatrix} -1 & 1 & -1 \\ 1 & -3 & -2 \\ -1 & -2 & -11 \end{bmatrix} \begin{bmatrix} x_1 \\ x_2 \\ x_3 \end{bmatrix}$$

Applying Sylvester's criterion, we find

$$-1 < 0, \quad \begin{vmatrix} -1 & 1 \\ 1 & -3 \end{vmatrix} > 0, \quad \begin{vmatrix} -1 & 1 & -1 \\ 1 & -3 & -2 \\ -1 & -2 & -11 \end{vmatrix} < 0$$

The quadratic form is negative definite.

**PROBLEM A-15-2.** Consider the system

$$\dot{\mathbf{x}} = \mathbf{A}\mathbf{x}$$

where the eigenvalues of $\mathbf{A}$ are distinct and have negative real parts. Assume that the transformation

$$\mathbf{x} = \mathbf{P}\mathbf{y}$$

transforms the coefficient matrix into diagonal form:

$$\dot{\mathbf{y}} = \mathbf{P}^{-1}\mathbf{A}\mathbf{P}\mathbf{y} = \mathbf{D}\mathbf{y}$$

Obtain a Liapunov function for this system.

**Solution.** Consider the following $V(\mathbf{y})$:

$$V(\mathbf{y}) = \mathbf{y}^*\mathbf{y} = \text{positive definite}$$

The time derivative of $V(\mathbf{y})$ is

$$\begin{aligned} \dot{V}(\mathbf{y}) &= \dot{\mathbf{y}}^*\mathbf{y} + \mathbf{y}^*\dot{\mathbf{y}} \\ &= \mathbf{y}^*\mathbf{D}^*\mathbf{y} + \mathbf{y}^*\mathbf{D}\mathbf{y} \\ &= 2\mathbf{y}^*\mathbf{D}\mathbf{y} \end{aligned}$$

Since the eigenvalues of $\mathbf{D}$ and those of $\mathbf{A}$ are identical and the eigenvalues of $\mathbf{A}$ have negative real parts, $\dot{V}(\mathbf{y})$ is negative definite. Hence the function $V(\mathbf{y})$ considered here is a Liapunov function.

**PROBLEM A-15-3.** Consider the second-order system

$$\dot{\mathbf{x}} = \mathbf{A}\mathbf{x}$$

where

$$\mathbf{x} = \begin{bmatrix} x_1 \\ x_2 \end{bmatrix}, \quad \mathbf{A} = \begin{bmatrix} a_{11} & a_{12} \\ a_{21} & a_{22} \end{bmatrix} \quad (a_{ij} = \text{real})$$

Find the real symmetric matrix $\mathbf{P}$ which satisfies

$$\mathbf{A}'\mathbf{P} + \mathbf{P}\mathbf{A} = -\mathbf{I}$$

Then find the condition that $\mathbf{P}$ is positive definite. (Note that $\mathbf{P}$ being positive definite implies that the origin $\mathbf{x} = \mathbf{0}$ is asymptotically stable in the large.)

**Solution.** The equation

$$\begin{bmatrix} a_{11} & a_{21} \\ a_{12} & a_{22} \end{bmatrix}\begin{bmatrix} p_{11} & p_{12} \\ p_{12} & p_{22} \end{bmatrix} + \begin{bmatrix} p_{11} & p_{12} \\ p_{12} & p_{22} \end{bmatrix}\begin{bmatrix} a_{11} & a_{12} \\ a_{21} & a_{22} \end{bmatrix} = \begin{bmatrix} -1 & 0 \\ 0 & -1 \end{bmatrix}$$

yields the following three simultaneous equations:

$$2(a_{11}p_{11} + a_{21}p_{12}) = -1$$
$$a_{11}p_{12} + a_{21}p_{22} + a_{12}p_{11} + a_{22}p_{12} = 0$$
$$2(a_{12}p_{12} + a_{22}p_{22}) = -1$$

Solving for the $p_{ij}$, we obtain

$$\mathbf{P} = \frac{1}{2(a_{11} + a_{22})|\mathbf{A}|} \begin{bmatrix} -(|\mathbf{A}| + a_{21}^2 + a_{22}^2) & a_{12}a_{22} + a_{21}a_{11} \\ a_{12}a_{22} + a_{21}a_{11} & -(|\mathbf{A}| + a_{11}^2 + a_{12}^2) \end{bmatrix}$$

$\mathbf{P}$ is positive definite if

$$p_{11} = -\frac{|\mathbf{A}| + a_{21}^2 + a_{22}^2}{2(a_{11} + a_{22})|\mathbf{A}|} > 0$$

$$|\mathbf{P}| = \frac{(a_{11} + a_{22})^2 + (a_{12} - a_{21})^2}{4(a_{11} + a_{22})^2 |\mathbf{A}|} > 0$$

from which we obtain

$$|\mathbf{A}| > 0, \qquad a_{11} + a_{22} < 0$$

as the conditions that $\mathbf{P}$ is positive definite.

**PROBLEM A-15-4.** Consider the motion of a space vehicle about the principal axes of inertia. The Euler equations are

$$A\dot{\omega}_x - (B - C)\omega_y\omega_z = T_x$$
$$B\dot{\omega}_y - (C - A)\omega_z\omega_x = T_y$$
$$C\dot{\omega}_z - (A - B)\omega_x\omega_y = T_z$$

where $A$, $B$, and $C$ denote the moments of inertia about the principal axes: $\omega_x$, $\omega_y$, and $\omega_z$ denote the angular velocities about the principal axes; and $T_x$, $T_y$, and $T_z$ are the control torques.

Assume that the space vehicle is tumbling in orbit. It is desired to stop the tumbling by applying control torques which are assumed to be

$$T_x = k_1 A\omega_x$$
$$T_y = k_2 B\omega_y$$
$$T_z = k_3 C\omega_z$$

Determine sufficient conditions for asymptotically stable operation of the system.

**Solution.** Let us choose the state variables as

$$x_1 = \omega_x, \qquad x_2 = \omega_y, \qquad x_3 = \omega_z$$

Then the system equations become

$$\dot{x}_1 - \left(\frac{B}{A} - \frac{C}{A}\right)x_2x_3 = k_1x_1$$

$$\dot{x}_2 - \left(\frac{C}{B} - \frac{A}{B}\right)x_3x_1 = k_2x_2$$

$$\dot{x}_3 - \left(\frac{A}{C} - \frac{B}{C}\right)x_1x_2 = k_3x_3$$

or

$$\begin{bmatrix} \dot{x}_1 \\ \dot{x}_2 \\ \dot{x}_3 \end{bmatrix} = \begin{bmatrix} k_1 & \frac{B}{A}x_3 & -\frac{C}{A}x_2 \\ -\frac{A}{B}x_3 & k_2 & \frac{C}{B}x_1 \\ \frac{A}{C}x_2 & -\frac{B}{C}x_1 & k_3 \end{bmatrix} \begin{bmatrix} x_1 \\ x_2 \\ x_3 \end{bmatrix}$$

The equilibrium state is the origin, or $\mathbf{x} = \mathbf{0}$. If we choose

$$V(\mathbf{x}) = \mathbf{x}'\mathbf{P}\mathbf{x} = \mathbf{x}' \begin{bmatrix} A^2 & 0 & 0 \\ 0 & B^2 & 0 \\ 0 & 0 & C^2 \end{bmatrix} \mathbf{x}$$

$$= A^2 x_1^2 + B^2 x_2^2 + C^2 x_3^2$$

$$= \text{positive definite}$$

then the time derivative of $V(\mathbf{x})$ is

$$\dot{V}(\mathbf{x}) = \dot{\mathbf{x}}'\mathbf{P}\mathbf{x} + \mathbf{x}'\mathbf{P}\dot{\mathbf{x}}$$

$$= \mathbf{x}' \begin{bmatrix} k_1 & -\dfrac{A}{B}x_3 & \dfrac{A}{C}x_2 \\[2mm] \dfrac{B}{A}x_3 & k_2 & -\dfrac{B}{C}x_1 \\[2mm] -\dfrac{C}{A}x_2 & \dfrac{C}{B}x_1 & k_3 \end{bmatrix} \begin{bmatrix} A^2 & 0 & 0 \\ 0 & B^2 & 0 \\ 0 & 0 & C^2 \end{bmatrix} \mathbf{x}$$

$$+ \mathbf{x}' \begin{bmatrix} A^2 & 0 & 0 \\ 0 & B^2 & 0 \\ 0 & 0 & C^2 \end{bmatrix} \begin{bmatrix} k_1 & \dfrac{B}{A}x_3 & -\dfrac{C}{A}x_2 \\[2mm] -\dfrac{A}{B}x_3 & k_2 & \dfrac{C}{B}x_1 \\[2mm] \dfrac{A}{C}x_2 & -\dfrac{B}{C}x_1 & k_3 \end{bmatrix} \mathbf{x}$$

$$= \mathbf{x}' \begin{bmatrix} 2k_1 A^2 & 0 & 0 \\ 0 & 2k_2 B^2 & 0 \\ 0 & 0 & 2k_3 C^2 \end{bmatrix} \mathbf{x} = -\mathbf{x}'\mathbf{Q}\mathbf{x}$$

For asymptotic stability, the sufficient condition is that $\mathbf{Q}$ be positive definite. Hence we require

$$k_1 < 0, \qquad k_2 < 0, \qquad k_3 < 0$$

If the $k_i$ are negative, then noting that $V(\mathbf{x}) \longrightarrow \infty$ as $\|\mathbf{x}\| \longrightarrow \infty$, we see that the equilibrium state is asymptotically stable in the large.

**PROBLEM A-15-5.** For physical systems, complex state variables must occur in conjugate pairs. Consider the system

$$\dot{\mathbf{x}} = \mathbf{f}(\mathbf{x})$$

Suppose that the state variables $x_1$ and $x_2$ are complex-conjugate pairs. Then $\partial f_1/\partial x_1$ is the complex conjugate of $\partial f_2/\partial x_2$, or

$$\frac{\partial f_1}{\partial x_1} = \overline{\frac{\partial f_2}{\partial x_2}}$$

also

$$\frac{\partial f_1}{\partial x_2} = \overline{\frac{\partial f_2}{\partial x_1}}$$

Using this fact, show that

$$[\mathbf{x}^*\mathbf{F}(\mathbf{x})\mathbf{x}]^* = \mathbf{x}^*\mathbf{F}(\mathbf{x})\mathbf{x}$$

Assume that $\mathbf{x}$ is a two-dimensional vector.

**Solution**

$$\mathbf{x}^*\mathbf{F}(\mathbf{x})\mathbf{x} = [\bar{x}_1 \quad \bar{x}_2]\begin{bmatrix} \dfrac{\partial f_1}{\partial x_1} & \dfrac{\partial f_1}{\partial x_2} \\[2mm] \dfrac{\partial f_2}{\partial x_1} & \dfrac{\partial f_2}{\partial x_2} \end{bmatrix}\begin{bmatrix} x_1 \\[2mm] x_2 \end{bmatrix}$$

$$= [\bar{x}_1 \quad \bar{x}_2]\begin{bmatrix} \dfrac{\partial f_1}{\partial x_1}x_1 + \dfrac{\partial f_1}{\partial x_2}x_2 \\[2mm] \dfrac{\partial f_2}{\partial x_1}x_1 + \dfrac{\partial f_2}{\partial x_2}x_2 \end{bmatrix}$$

$$= \frac{\partial f_1}{\partial x_1}\bar{x}_1 x_1 + \frac{\partial f_1}{\partial x_2}\bar{x}_1 x_2 + \frac{\partial f_2}{\partial x_1}\bar{x}_2 x_1 + \frac{\partial f_2}{\partial x_2}\bar{x}_2 x_2 \qquad (15\text{-}22)$$

The conjugate transpose of $\mathbf{x}^*\mathbf{F}(\mathbf{x})\mathbf{x}$ is

$$[\mathbf{x}^*\mathbf{F}(\mathbf{x})\mathbf{x}]^* = \overline{\frac{\partial f_1}{\partial x_1}}x_1\bar{x}_1 + \overline{\frac{\partial f_1}{\partial x_2}}x_1\bar{x}_2 + \overline{\frac{\partial f_2}{\partial x_1}}x_2\bar{x}_1 + \overline{\frac{\partial f_2}{\partial x_2}}x_2\bar{x}_2$$

Noting that

$$x_1 = \bar{x}_2, \qquad x_2 = \bar{x}_1$$

$$\frac{\partial f_1}{\partial x_1} = \overline{\frac{\partial f_2}{\partial x_2}}, \qquad \frac{\partial f_1}{\partial x_2} = \overline{\frac{\partial f_2}{\partial x_1}}$$

we obtain

$$[\mathbf{x}^*\mathbf{F}(\mathbf{x})\mathbf{x}]^* = \frac{\partial f_2}{\partial x_2}\bar{x}_2 x_2 + \frac{\partial f_2}{\partial x_1}\bar{x}_2 x_1 + \frac{\partial f_1}{\partial x_2}\bar{x}_1 x_2 + \frac{\partial f_1}{\partial x_1}\bar{x}_1 x_1 \qquad (15\text{-}23)$$

Comparing Eqs. (15-22) and (15-23), we obtain

$$[\mathbf{x}^*\mathbf{F}(\mathbf{x})\mathbf{x}]^* = \mathbf{x}^*\mathbf{F}(\mathbf{x})\mathbf{x}$$

**PROBLEM A-15-6.** Consider the system

$$\frac{Y(s)}{U(s)} = \frac{4}{(s+2)(s^2+2s+2)}$$

which can be rewritten

$$\frac{Y(s)}{U(s)} = \frac{2}{s+2}\left[\frac{j}{s+1+j} + \frac{-j}{s+1-j}\right]$$

The block-diagram representation of this system is shown in Fig. 15-5. Let us choose the state variables as shown there. Assuming that $u = 0$, show that

$$[\mathbf{x}^*\mathbf{F}(\mathbf{x})\mathbf{x}]^* = \mathbf{x}^*\mathbf{F}(\mathbf{x})\mathbf{x}$$

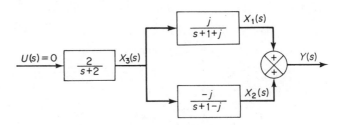

**Fig. 15-5.** Control system.

**Solution.** From the block diagram, we obtain

$$\frac{X_1}{X_3} = \frac{j}{s+1+j}$$

$$\frac{X_2}{X_3} = \frac{-j}{s+1-j}$$

$$\frac{X_3}{U} = \frac{2}{s+2}$$

or

$$\dot{x}_1 = -(1+j)x_1 + jx_3$$
$$\dot{x}_2 = -(1-j)x_2 - jx_3$$
$$\dot{x}_3 = -2x_3 + 2u$$

Since $u$ is assumed to be zero, we obtain

$$\dot{x}_1 = -(1+j)x_1 + jx_3 = f_1(\mathbf{x})$$
$$\dot{x}_2 = -(1-j)x_2 - jx_3 = f_2(\mathbf{x})$$
$$\dot{x}_3 = -2x_3 = f_3(\mathbf{x})$$

The Jacobian matrix for this system is

$$\mathbf{F}(\mathbf{x}) = \begin{bmatrix} -(1+j) & 0 & j \\ 0 & -(1-j) & -j \\ 0 & 0 & -2 \end{bmatrix}$$

Note that

$$\mathbf{x} = \begin{bmatrix} x_1 \\ x_2 \\ x_3 \end{bmatrix} = \begin{bmatrix} \bar{x}_2 \\ \bar{x}_1 \\ x_3 \end{bmatrix}$$

($x_3$ is a real state variable or $\bar{x}_3 = x_3$). Now compute $\mathbf{x}^*\mathbf{F}(\mathbf{x})\mathbf{x}$.

$$\mathbf{x}^*\mathbf{F}(\mathbf{x})\mathbf{x} = [\bar{x}_1 \quad \bar{x}_2 \quad x_3] \begin{bmatrix} -(1+j) & 0 & j \\ 0 & -(1-j) & -j \\ 0 & 0 & -2 \end{bmatrix} \begin{bmatrix} x_1 \\ x_2 \\ x_3 \end{bmatrix}$$

$$= [\bar{x}_1 \quad \bar{x}_2 \quad x_3] \begin{bmatrix} -(1+j)x_1 + jx_3 \\ -(1-j)x_2 - jx_3 \\ -2x_3 \end{bmatrix}$$

$$= -(1+j)\bar{x}_1 x_1 + j\bar{x}_1 x_3 - (1-j)\bar{x}_2 x_2 - j\bar{x}_2 x_3 - 2x_3^2 \qquad (15\text{-}24)$$

Now take the complex conjugate of $\mathbf{x}^*\mathbf{F}(\mathbf{x})\mathbf{x}$. From Eq. (15-24),

$$[\mathbf{x}^*\mathbf{F}(\mathbf{x})\mathbf{x}]^* = -(1-j)x_1\bar{x}_1 - jx_1 x_3 - (1+j)x_2\bar{x}_2 + jx_2 x_3 - 2x_3^2$$

Noting that $x_1 = \bar{x}_2$, $x_2 = \bar{x}_1$, we obtain

$$[\mathbf{x}^*\mathbf{F}(\mathbf{x})\mathbf{x}]^* = -(1-j)\bar{x}_2 x_2 - j\bar{x}_2 x_3 - (1+j)\bar{x}_1 x_1 + j\bar{x}_1 x_3 - 2x_3^2 \qquad (15\text{-}25)$$

Comparing Eqs. (15-24) and (15-25), we see that

$$[\mathbf{x}^*\mathbf{F}(\mathbf{x})\mathbf{x}]^* = \mathbf{x}^*\mathbf{F}(\mathbf{x})\mathbf{x}$$

**PROBLEM A-15-7.** Show that if

$$\hat{\mathbf{F}}(\mathbf{x}) = \mathbf{F}^*(\mathbf{x}) + \mathbf{F}(\mathbf{x})$$

where

$$\mathbf{F}(\mathbf{x}) = \begin{bmatrix} \dfrac{\partial f_1}{\partial x_1} & \dfrac{\partial f_1}{\partial x_2} & \cdots & \dfrac{\partial f_1}{\partial x_n} \\ \cdot & \cdot & & \cdot \\ \cdot & \cdot & & \cdot \\ \cdot & \cdot & & \cdot \\ \dfrac{\partial f_n}{\partial x_1} & \dfrac{\partial f_n}{\partial x_2} & \cdots & \dfrac{\partial f_n}{\partial x_n} \end{bmatrix}$$

is negative definite, then the determinant of $\mathbf{F}(\mathbf{x})$ is nonzero for $\mathbf{x} \neq \mathbf{0}$, or

$$|\mathbf{F}(\mathbf{x})| \neq 0 \qquad \text{for } \mathbf{x} \neq \mathbf{0}$$

**Solution.** By use of the relationship

$$[\mathbf{x}^*\mathbf{F}(\mathbf{x})\mathbf{x}]^* = \mathbf{x}^*\mathbf{F}(\mathbf{x})\mathbf{x}$$

we have

$$\begin{aligned} \mathbf{x}^*\hat{\mathbf{F}}(\mathbf{x})\mathbf{x} &= \mathbf{x}^*[\mathbf{F}^*(\mathbf{x}) + \mathbf{F}(\mathbf{x})]\mathbf{x} \\ &= \mathbf{x}^*\mathbf{F}^*(\mathbf{x})\mathbf{x} + \mathbf{x}^*\mathbf{F}(\mathbf{x})\mathbf{x} \\ &= [\mathbf{x}^*\mathbf{F}(\mathbf{x})\mathbf{x}]^* + \mathbf{x}^*\mathbf{F}(\mathbf{x})\mathbf{x} \\ &= 2\mathbf{x}^*\mathbf{F}(\mathbf{x})\mathbf{x} \end{aligned}$$

Since $\hat{\mathbf{F}}(\mathbf{x})$ is negative definite, $\mathbf{F}(\mathbf{x})$ is also negative definite. Hence

$$|\mathbf{F}(\mathbf{x})| \neq 0 \qquad \text{for } \mathbf{x} \neq \mathbf{0}$$

**PROBLEM A-15-8.** Determine the stability of the origin of the linear time-varying system described by

$$\dot{\mathbf{x}} = \mathbf{A}(t)\mathbf{x}$$

where

$$\mathbf{x} = \begin{bmatrix} x_1 \\ x_2 \end{bmatrix}, \qquad \mathbf{A}(t) = \begin{bmatrix} 0 & 1 \\ -\dfrac{1}{t+1} & -10 \end{bmatrix} \qquad (t \geq 0)$$

Use the variable-gradient method.

**Solution.** Let the gradient of $V$ be defined by

$$\nabla V = \begin{bmatrix} a_{11}x_1 + a_{12}x_2 \\ a_{21}x_1 + a_{22}x_2 \end{bmatrix}$$

Then the derivative of $V$ is

$$\dot{V} = (\nabla V)'\dot{\mathbf{x}}$$

$$= (a_{11}x_1 + a_{12}x_2)x_2 + (a_{21}x_1 + a_{22}x_2)\left(-\frac{1}{t+1}x_1 - 10x_2\right)$$

Let us choose $a_{12} = a_{21} = 0$, then

$$\dot{V} = a_{11}x_1x_2 + a_{22}x_2\left(-\frac{1}{t+1}x_1 - 10x_2\right)$$

By choosing $a_{11} = 1$ and $a_{22} = t + 1$, $\nabla V$ can be written

$$\nabla V = \begin{bmatrix} x_1 \\ (t+1)x_2 \end{bmatrix}$$

Since

$$\int_0^{x_1, (x_2=0)} x_1 \, dx_1 + \int_0^{x_2, (x_1=x_1)} (t+1)x_2 \, dx_2 = \frac{1}{2}[x_1^2 + (t+1)x_2^2]$$

let us choose

$$V = \frac{1}{2}[x_1^2 + (t+1)x_2^2]$$

Then

$$\dot{V} = x_1\dot{x}_1 + \frac{x_2^2}{2} + (t+1)x_2\dot{x}_2$$

$$= -(10t + 9.5)x_2^2$$

$\dot{V}$ is thus negative semidefinite. Noting that $\dot{V}$ is identically equal to zero only at the origin $x_1 = x_2 = 0$, by Theorem 15-2, we conclude that the equilibrium state at the origin $\mathbf{x} = \mathbf{0}$ is asymptotically stable in the large.

## PROBLEMS

**PROBLEM B-15-1.** Determine whether or not the following quadratic form is positive definite.

$$Q = x_1^2 + 4x_2^2 + x_3^2 + 2x_1x_2 - 6x_2x_3 - 2x_1x_3$$

**PROBLEM B-15-2.** Find a Liapunov function for the following system:

$$\begin{bmatrix} \dot{x}_1 \\ \dot{x}_2 \end{bmatrix} = \begin{bmatrix} -1 & 1 \\ 2 & -3 \end{bmatrix} \begin{bmatrix} x_1 \\ x_2 \end{bmatrix}$$

**PROBLEM B-15-3.** Determine the stability of the equilibrium state of the following system:

$$x_1(k+1) = x_1(k) + 3x_2(k)$$
$$x_2(k+1) = -3x_1(k) - 2x_2(k) - 3x_3(k)$$
$$x_3(k+1) = x_1(k)$$

**PROBLEM B-15-4.** Determine the stability of the origin of the following system:

$$\dot{x}_1 = x_2$$
$$\dot{x}_2 = -g(x_1) + x_2$$

where $g(x_1)/x_1 > 0$ for $x_1 \neq 0$.

**PROBLEM B-15-5.** Consider the system described by

$$\dot{x}_1 = -g_1(x_1) + g_2(x_1, x_2)$$
$$\dot{x}_2 = g_3(x_2)$$

where $g_1(0) = g_3(0) = 0$, $g_2(0, x_2) = 0$. Obtain sufficient conditions for the stability of the origin of the system.

**PROBLEM B-15-6.** Determine the stability of the origin of the following system:

$$\dot{x}_1 = x_2$$
$$\dot{x}_2 = -x_1^3 - x_2$$

**PROBLEM B-15-7.** Determine the stability of the origin of the following system:

$$\dot{x}_1 = -x_1 + x_2 + x_1(x_1^2 + x_2^2)$$
$$\dot{x}_2 = -x_1 - x_2 + x_2(x_1^2 + x_2^2)$$

**PROBLEM B-15-8.** Determine the stability of the origin of the following system:

$$\dot{x}_1 = x_1 - x_2 - x_1^3$$
$$\dot{x}_2 = x_1 + x_2 - x_2^3$$

# 16

# OPTIMAL AND ADAPTIVE
# CONTROL SYSTEMS

## 16-1  INTRODUCTION

**Optimal control systems.** Problems of optimal control have received a great deal of attention during the past decade owing to increasing demand for systems of high performance and to the ready availablility of the digital computer.

The concept of control system optimization comprises a selection of a performance index and a design which yields the optimal control system within limits imposed by physical constraints. Such an optimal control system differs from an ideal one in that the former is the best attainable in the presence of physical constraints whereas the latter may well be an unattainable goal.

**Performance indexes.** In solving problems of optimal control systems, we may have the goal of finding a rule for determining the present control decision, subject to certain constraints which will minimize some measure of a deviation from ideal behavior. Such a measure is usually provided by a criterion of optimization, or performance index. The performance index which was defined in Chapter 7 is a function whose value indicates how well the actual performance of the system matches the desired performance. In most practical cases, system behavior is optimized by choosing the control vector in such a way that the performance index is minimized (or maximized).

The performance index is important because it, to a large degree, determines the nature of the resulting optimal control. That is, the resulting control may be linear, nonlinear, stationary or time-varying, depending on the form of the per-

formance index. The control engineer formulates this index based on the requirements of the problem. Thus, he influences the nature of the resulting system.
The requirements of the problem usually include not only performance requirements
but also restrictions on the form of the control to ensure physical realizability.

The optimization process should provide not only control policies, parameter
configurations which are optimal, but also a measure of the degradation in performance by the departure of the performance index function from its minimum
(or maximum) value which results from the use of nonoptimal control policies.

To a considerable degree, use of optimization theory in system design has been
much hampered by the conflict between analytic feasibility and practical utility
in the selection of the performance index. It is desirable that the criteria for optimal
control originate not from a mathematical but from an applicational point of view.
In general, however, the choice of a performance index involves a compromise
between a meaningful evaluation of the system performance and a tractable mathematical problem.

Choosing the most appropriate performance index for a given problem is very
difficult, especially in complex systems. For example, consider the problem of the
maximization of a payload of a space vehicle. The payload may be considered the
difference between the vehicle weight after accomplishing the mission and the resi-
dual components of the vehicle, such as supporting structures, communications,
and power and altitude control equipment. Maximizing the payload, therefore,
will involve an optimization of both the thrust program and mission design for
minimum propellant expenditures, as well as an optimal design of the components
of the vehicle. In space-vehicle applications, other possible performance specifications may be minimum fuel expenditure, minimum target miss, minimum time,
etc. In civilian, as differentiated from military, applications of control, the prime
considerations are usually economic.

**Formulation of optimization problems.** The quantities appearing in control
system optimization problems are state variables, control variables, and system
parameters. Consider, for example, a vehicle which is assumed to be a point mass
traveling through space. The state variables of this system may be the three position
coordinates of the vehicle, the three velocity coordinates, and the instantaneous
mass of the vehicle. These state variables are generated from a set of differential
equations which, in this example, may be simply Newton's equations of motion
and a continuity equation relating the propellant flow to the mass loss rate of the
vehicle. The control variables for this example might be the thrust magnitude of
the vehicle and a set of angles defining the thrust direction. The system parameters
are constants describing certain properties of the problem. Such parameters might
be the exhaust velocity of the propulsion system or a prespecified time-of-thrust
termination. For systems using ionic propulsion, these parameters might be the
values of the exhaust velocity and the size of the power plant carried by the vehicle.

In general, the problem of optimization of control systems may be formulated
if the following information is given:

1. system state equation and output equation
2. control vector

3. constraints of the problem
4. performance index
5. system parameters

An optimal control problem is to determine the optimal control vector $\mathbf{u}(t)$ within the class of allowable control vectors. This vector $\mathbf{u}(t)$ usually depends on

1. initial state or initial output
2. desired state or desired output
3. nature of the constraints
4. nature of the performance index

Except for special cases, the problem may be so complicated for analytic solution that a computational solution must be obtained.

In this chapter, we shall discuss time-optimal control systems and optimal control systems based on quadratic performance indexes.

**Time-optimal control systems.** In Section 16-4, we shall discuss two time-optimal control problems. One is, given the discrete-time system

$$\mathbf{x}((k+1)T) = \mathbf{G}(T)\mathbf{x}(kT) + \mathbf{H}(T)\mathbf{u}(kT)$$

where the norm of $\mathbf{u}(kT)$ is not bounded, find the control vector $\mathbf{u}(kT)$ which will bring any initial state to the origin of the state space in the minimum number of sampling periods.

The other time-optimal control problem is, given the plant described by

$$\overset{(n)}{y} + a_1 \overset{(n-1)}{y} + \cdots + a_{n-1}\dot{y} + a_n y = b_0 \overset{(n)}{u} + b_1 \overset{(n-1)}{u} + \cdots + b_{n-1}\dot{u} + b_n u$$

determine the control function $u$ which will transfer the output $y$ of the system from a given intial condition to zero in minimum time $T$. The control function $u$ must satisfy a magnitude constraint and the output and its derivatives are to remain at zero for time $t \geq T$ if no subsequent disturbances are applied.

Obtaining the explicit solution for the time-optimal control of an $n$th-order plant is extremely difficult. Therefore, we shall discuss only a simple case where the plant is described by

$$\ddot{y} + a_1\dot{y} + a_2 y = bu$$

That is, the transfer function of the plant involves two poles and no zeros.

**Optimal control systems based on quadratic performance indexes.** In many practical control systems, we desire to minimize some function of the error signal. For example, given the system

$$\dot{\mathbf{x}} = \mathbf{A}\mathbf{x} + \mathbf{B}\mathbf{u}$$

we may desire to minimize a generalized error function such as

$$J = \int_0^T [\boldsymbol{\xi}(t) - \mathbf{x}(t)]^*\mathbf{Q}[\boldsymbol{\xi}(t) - \mathbf{x}(t)]\, dt$$

where $\boldsymbol{\xi}(t)$ represents the desired state, $\mathbf{x}(t)$ the actual state, [thus, $\boldsymbol{\xi}(t) - \mathbf{x}(t)$ is the error vector], **Q** a positive-definite (or positive-semidefinite) matrix, and the time interval $0 \leq t \leq T$ is either finite or infinite.

In addition to considering errors as a measure of system performance, however, we usually must pay attention to the energy required for control action. Since $u(t)$ may have the dimension of force or torque, the control energy is proportional to the integral of $[u(t)]^2$. If the errors are minimized regardless of the energy required, then a design may result which calls for overly large values of $u(t)$. This is undesirable since all physical systems are subject to saturation. Large-amplitude control signals are ineffective outside the range determined by saturation. Thus, practical considerations place a constraint on the control vector, for example,

$$\int_0^T \mathbf{u}^*(t)\mathbf{R}\mathbf{u}(t)\, dt = K$$

where **R** is a positive-definite matrix and $K$ is a positive constant. The performance index of a control system over the time interval $0 \leq t \leq T$ may then be written, with the use of a Lagrange multiplier $\lambda$, as follows:

$$J = \int_0^T [\boldsymbol{\xi}(t) - \mathbf{x}(t)]^* \mathbf{Q}[\boldsymbol{\xi}(t) - \mathbf{x}(t)]\, dt + \lambda \int_0^T \mathbf{u}^*(t)\mathbf{R}\mathbf{u}(t)\, dt \quad (0 \leq t \leq T) \quad (16\text{-}1)$$

The Lagrange multipier $\lambda$ is a positive constant indicating the weight of control cost with respect to the minimizing errors.

Finding the optimal control law for the system

$$\dot{\mathbf{x}} = \mathbf{A}\mathbf{x} + \mathbf{B}\mathbf{u}$$

subject to the performance index given by Eq. (16-1), has a practical significance that the resulting system compromises between minimizing the integral error and minimizing the control energy. In Section 16-5, we consider a special case where $T = \infty$, the desired state $\boldsymbol{\xi}$ is the origin, or $\boldsymbol{\xi} = \mathbf{0}$, and the state vector is real. Under these conditions, the quadratic performance index given by Eq. (16-1) can be expressed as

$$J = \int_0^\infty [\mathbf{x}'(t)\mathbf{Q}\mathbf{x}(t) + \mathbf{u}'(t)\mathbf{R}\mathbf{u}(t)]\, dt$$

where we have included $\lambda$ in the positive definite matrix **R**. Note that in this problem $\mathbf{u}(t)$ is unconstrained. We shall see later that such a quadratic performance index leads to convenient mathematical operations in the solution of optimal control problems.

**Comments on optimal control systems.** The system which minimizes (or maximizes) the selected performance index is, by definition, optimal. It is evident that the performance index, in reality, determines the system configuration. It is important to point out that an optimal control system under a given performance index is, in general, not optimal under other performance indexes. In addition, hardware realization of a particular optimal control law may be quite difficult and expensive. Hence it may be pointless to devote too much expense to implementing an optimal controller which is the best in some narrow, individualistic sense. A control system is seldom designed to perform a single task completely specified beforehand. Instead, it is designed to perform a task selected at random from a complete repertory of possible tasks. In practical systems, it then may be more sensible to seek approximate optimal control laws which are not rigidly tied to a single performance index.

Strictly speaking, we should realize that a mathematically obtained optimal control system gives, in most practical situations, the ultimate performance limitation under the given performance index and is more a measuring stick than a practical goal. Therefore, before we decide whether to build the optimal control system or something inferior but simpler, we should carefully evaluate a measure of the degree to which the performance of the complex, optimal control system exceeds that of a simpler, suboptimal one. Unless the optimal control system can be justified, we shall not build extremely complicated, optimal control systems.

Once the ultimate performance limitation is found by use of optimal control theory, we should make efforts to design a simple system that is close to optimal. Keeping this in mind, we build a prototype physical system, test it and modify it until a satisfactory system is obtained which has performance characteristics close to the optimal control system synthesized by use of optimal control theory.

**Questions concerning the existence of the solution to optimal control problems.** It has been stated that the optimal control problem, given any initial state $\mathbf{x}(t_0)$, is to find an allowable control vector $\mathbf{u}(t)$ that transfers the state to the desired region of the state space and for which the performance index is minimized.

It is important to mention that in some cases a particular combination of plant, desired state, performance index, and constraints make optimal control impossible. This is a matter of requiring performance beyond the physical capabilities of the system.

Questions regarding the existence of an optimal control system are important since they serve to inform the designer whether or not a proposed performance index is realistic for a given system and set of constraints. Two of the most important among these questions are those of controllability and observability. In the following, we shall very briefly explain what is meant by controllability and observability.

**Controllability and observability.** A system is said to be controllable at time $t_0$ if it is possible by means of an unconstrained control vector to transfer the system from any initial state $\mathbf{x}(t_0)$ to any other state in a finite interval of time.

A system is said to be observable at time $t_0$ if, with the system in state $\mathbf{x}(t_0)$, it is possible to determine this state from the observation of the output over a finite time interval.

The concepts of controllability and observability were introduced by Kalman. They play an important role in the optimal control of multivariable systems. In fact, the conditions of controllability and observability may govern the existence of a complete solution to the optimal control problem.

**Outline of the chapter.** This chapter introduces the reader to some of the more important optimal and adaptive control systems. Section 16-2 presents the conditions for complete controllability and Section 16-3 gives those for complete observability. Section 16-4 discusses the time-optimal control of discrete-time and continuous-time systems.

Optimal control systems based on quadratic performance indexes are presented in Section 16-5. Here the Liapunov approach is presented. Section 16-6 briefly discusses model reference control systems. The final two sections, Sections 16-7 and 16-8, present introductory discussions of adaptive control systems. Such systems continually measure the system performance and adjust their parameters in order to maintain near-optimal performance.

## 16-2  CONTROLLABILITY

The solution to an optimal control problem may not exist if the system considered is not controllable. Although most physical systems are controllable and observable, corresponding mathematical models may not possess the property of controllability and observability. Then it is necessary to know the conditions under which a system is controllable and observable.

In this section, we shall first define the linear independence of vectors and then derive the conditions for complete state controllability. We shall begin our discussion with discrete-time systems since they are simpler to analyze than continuous-time systems. Then we derive the conditions for complete state controllability for continuous-time systems. Finally, we shall discuss complete output controllability.

**Linear independence of vectors.** The vectors $\mathbf{x}_1$, $\mathbf{x}_2, \ldots, \mathbf{x}_n$ are said to be linearly independent if

$$c_1\mathbf{x}_1 + c_2\mathbf{x}_2 + \cdots + c_n\mathbf{x}_n = \mathbf{0}$$

where $c_1, c_2, \ldots, c_n$ are constants, implies

$$c_1 = c_2 = \cdots = c_n = 0$$

Conversely, the vectors $\mathbf{x}_1, \mathbf{x}_2, \ldots, \mathbf{x}_n$ are said to be linearly dependent if and only if $\mathbf{x}_i$ can be expressed as a linear combination of $\mathbf{x}_j$ ($j = 1, 2, \ldots, n; j \neq i$), or

$$\mathbf{x}_i = \sum_{\substack{j=1 \\ j \neq i}}^{n} c_j\mathbf{x}_j$$

for some set of constants $c_j$. This means that if $\mathbf{x}_i$ can be expressed as a linear combination of the other vectors in the set, it is linearly dependent upon them or it is not an independent member of the set.

---

*Example 16-1.* The vectors

$$\mathbf{x}_1 = \begin{bmatrix} 1 \\ 2 \\ 3 \end{bmatrix}, \quad \mathbf{x}_2 = \begin{bmatrix} 1 \\ 0 \\ 1 \end{bmatrix}, \quad \mathbf{x}_3 = \begin{bmatrix} 2 \\ 2 \\ 4 \end{bmatrix}$$

are linearly dependent since

$$\mathbf{x}_1 + \mathbf{x}_2 - \mathbf{x}_3 = \mathbf{0}$$

The vectors

$$\mathbf{y}_1 = \begin{bmatrix} 1 \\ 2 \\ 3 \end{bmatrix}, \qquad \mathbf{y}_2 = \begin{bmatrix} 1 \\ 0 \\ 1 \end{bmatrix}, \qquad \mathbf{y}_3 = \begin{bmatrix} 2 \\ 2 \\ 2 \end{bmatrix}$$

are linearly independent since

$$c_1\mathbf{y}_1 + c_2\mathbf{y}_2 + c_3\mathbf{y}_3 = \mathbf{0}$$

implies

$$c_1 = c_2 = c_3 = 0$$

Note that if an $n \times n$ matrix is nonsingular (i.e., the matrix is of rank $n$ or the determinant is nonzero), then $n$ column (or row) vectors are linearly independent. If the $n \times n$ matrix is singular (i.e., the rank of the matrix is less than $n$ or the determinant is zero), then $n$ column (or row) vectors are linearly dependent. To demonstrate this, notice that

$$[\mathbf{x}_1 \,|\, \mathbf{x}_2 \,|\, \mathbf{x}_3] = \begin{bmatrix} 1 & 1 & 2 \\ 2 & 0 & 2 \\ 3 & 1 & 4 \end{bmatrix} = \text{singular}$$

$$[\mathbf{y}_1 \,|\, \mathbf{y}_2 \,|\, \mathbf{y}_3] = \begin{bmatrix} 1 & 1 & 2 \\ 2 & 0 & 2 \\ 3 & 1 & 2 \end{bmatrix} = \text{nonsingular}$$

**Complete state controllability of discrete-time systems.** Consider the discrete-time system described by

$$\mathbf{x}((k + 1)T) = \mathbf{G}\mathbf{x}(kT) + \mathbf{H}u(kT) \tag{16-2}$$

where

$\mathbf{x}(kT) = $ state vector ($n$-dimensional vector)

$u(kT) = $ control signal

$\mathbf{G} = n \times n$ nonsingular matrix

$\mathbf{H} = n \times 1$ matrix

$T = $ sampling period

Note that $u(kT)$ is constant for $kT \leq t < (k + 1)T$. Without loss of generality, we can assume that the initial state is arbitrary and that the final state is the origin of the state space.

The discrete-time system given by Eq. (16-2) is state controllable if there exists a piecewise-constant control signal $u(kT)$ defined over a finite sampling interval $0 \leq kT < nT$ such that, starting from any initial state, the state $\mathbf{x}(kT)$ can be made zero for $kT \geq nT$. If every state is controllable, then the system is said to be completely state controllable.

We shall derive the condition for complete state controllability by using the fact that if a system is completely state controllable, then a piecewise-continuous control signal exists which will transfer any initial state to the origin in a finite number of sampling periods.

The solution of Eq. (16-2) is

$$\mathbf{x}(kT) = \mathbf{G}^k\mathbf{x}(0) + \sum_{j=0}^{k-1} \mathbf{G}^{k-j-1}\mathbf{H}u(jT)$$

If the system is controllable, then, starting from an arbitrary $\mathbf{x}(0)$, we can bring the state to the origin or $\mathbf{x}(kT) = \mathbf{0}$ for $k \geq n$ by the application of some $u(0)$, $u(T), \ldots, u((n-1)T)$. Thus,

$$\mathbf{G}^n\mathbf{x}(0) + \sum_{j=0}^{n-1} \mathbf{G}^{n-j-1}\mathbf{H}u(jT) = \mathbf{0}$$

or

$$\mathbf{x}(0) = -\sum_{j=0}^{n-1} \mathbf{G}^{-(j+1)}\mathbf{H}u(jT)$$
$$= -[\mathbf{G}^{-1}\mathbf{H}u(0) + \mathbf{G}^{-2}\mathbf{H}u(T) + \cdots + \mathbf{G}^{-n}\mathbf{H}u((n-1)T)] \quad (16\text{-}3)$$

Since $\mathbf{G}$ is an $n \times n$ nonsingular matrix and $\mathbf{H}$ is an $n \times 1$ matrix, clearly $\mathbf{G}^{-1}\mathbf{H}$, $\mathbf{G}^{-2}\mathbf{H}, \ldots, \mathbf{G}^{-n}\mathbf{H}$ are $n \times 1$ matrices or column vectors. Let us define

$$\mathbf{G}^{-1}\mathbf{H} = \begin{bmatrix} a_{11} \\ a_{21} \\ \cdot \\ \cdot \\ \cdot \\ a_{n1} \end{bmatrix}, \quad \mathbf{G}^{-2}\mathbf{H} = \begin{bmatrix} a_{12} \\ a_{22} \\ \cdot \\ \cdot \\ \cdot \\ a_{n2} \end{bmatrix}, \quad \ldots, \quad \mathbf{G}^{-n}\mathbf{H} = \begin{bmatrix} a_{1n} \\ a_{2n} \\ \cdot \\ \cdot \\ \cdot \\ a_{nn} \end{bmatrix}$$

Then Eq. (16-3) can be rewritten as the following $n$ simultaneous algebraic equations:

$$x_1(0) = -a_{11}u(0) - a_{12}u(T) - \cdots - a_{1n}u((n-1)T)$$
$$x_2(0) = -a_{21}u(0) - a_{22}u(T) - \cdots - a_{2n}u((n-1)T)$$
$$\cdot \quad \cdot \quad \cdot$$
$$x_n(0) = -a_{n1}u(0) - a_{n2}u(T) - \cdots - a_{nn}u((n-1)T)$$

If these equations are to give a solution to any given set of values $x_1(0)$, $x_2(0)$, $\ldots, x_n(0)$, then we require the $n \times n$ matrix $(a_{ij})$ to be nonsingular. (Otherwise the solution may not exist.) The condition that the $n \times n$ matrix $(a_{ij})$ is nonsingular implies that each column of the matrix $(a_{ij})$ is linearly independent, or the matrix $(a_{ij})$ is of rank $n$.

From the preceding analysis, we may state the conditions for complete state controllability of the system of Eq. (16-2) as follows: The system is completely state controllable if and only if the vectors $\mathbf{G}^{-1}\mathbf{H}, \mathbf{G}^{-2}\mathbf{H}, \ldots, \mathbf{G}^{-n}\mathbf{H}$ are linearly independent or the $n \times n$ matrix

$$[\mathbf{G}^{-1}\mathbf{H} \mid \mathbf{G}^{-2}\mathbf{H} \mid \cdots \mid \mathbf{G}^{-n}\mathbf{H}]$$

is of rank $n$. Since $\mathbf{G}$ is nonsingular, this condition can be stated that the rank of the matrix

$$[\mathbf{H} \mid \mathbf{G}^{-1}\mathbf{H} \mid \cdots \mid \mathbf{G}^{-(n-1)}\mathbf{H}]$$

is $n$. This condition can also be stated that the rank of the matrix

$$[\mathbf{H} \mid \mathbf{GH} \mid \cdots \mid \mathbf{G}^{n-1}\mathbf{H}]$$

is $n$.

Note that if the system of Eq. (16-2) is completely state controllable, then we can transfer any initial state to the origin in at most $n$ sampling periods. [Remember that this is true if and only if the magnitude of $u(kT)$ is unbounded. If the magnitude of $u(kT)$ is bounded, it may take more than $n$ sampling periods.]

If the system is described by

$$\mathbf{x}((k+1)T) = \mathbf{Gx}(kT) + \mathbf{Hu}(kT)$$

where $\mathbf{u}(kT)$ is an $r$-dimensional vector, then it can be proved that the condition for complete state controllability is that the following $n \times nr$ matrix

$$[\mathbf{G}^{-1}\mathbf{H} \mid \mathbf{G}^{-2}\mathbf{H} \mid \cdots \mid \mathbf{G}^{-n}\mathbf{H}]$$

is of rank $n$. This condition is the same as the condition that the rank of the matrix

$$[\mathbf{H} \mid \mathbf{G}^{-1}\mathbf{H} \mid \cdots \mid \mathbf{G}^{-(n-1)}\mathbf{H}]$$

is $n$, or the rank of the matrix

$$[\mathbf{H} \mid \mathbf{GH} \mid \cdots \mid \mathbf{G}^{n-1}\mathbf{H}]$$

is $n$.

**Complete state controllability of continuous-time systems.** Consider the continuous-time system

$$\dot{\mathbf{x}} = \mathbf{Ax} + \mathbf{B}u \tag{16-4}$$

where

$\mathbf{x}$ = state vector ($n$-dimensional vector)
$u$ = control signal
$\mathbf{A}$ = $n \times n$ matrix
$\mathbf{B}$ = $n \times 1$ matrix

The system described by Eq. (16-4) is said to be state controllable at $t = t_0$ if it is possible to construct an unconstrained control signal which will transfer an initial state to any final state in a finite time interval $t_0 \leq t \leq t_1$. If every state is controllable, then the system is said to be completely state controllable.

We shall now derive the condition for complete state controllability. Without loss of generality, we can assume that the final state is the origin of the state space and that the initial time is zero, or $t_0 = 0$.

The solution of Eq. (16-4) is

$$\mathbf{x}(t) = e^{\mathbf{A}t}\mathbf{x}(0) + \int_0^t e^{\mathbf{A}(t-\tau)}\mathbf{B}u(\tau)\,d\tau$$

Applying the definition of complete state controllability just given, we have

$$\mathbf{x}(t_1) = \mathbf{0} = e^{\mathbf{A}t_1}\mathbf{x}(0) + \int_0^{t_1} e^{\mathbf{A}(t_1-\tau)}\mathbf{B}u(\tau)\,d\tau$$

or

$$x(0) = -\int_0^{t_1} e^{-\mathbf{A}\tau}\mathbf{B}u(\tau)\,d\tau \qquad (16\text{-}5)$$

Note that $e^{-\mathbf{A}\tau}$ can be written

$$e^{-\mathbf{A}\tau} = \sum_{k=0}^{n-1} \alpha_k(\tau)\mathbf{A}^k \qquad (16\text{-}6)$$

Substituting Eq. (16-6) into Eq. (16-5) gives

$$x(0) = -\sum_{k=0}^{n-1} \mathbf{A}^k\mathbf{B}\int_0^{t_1} \alpha_k(\tau)u(\tau)\,d\tau \qquad (16\text{-}7)$$

Let us put

$$\int_0^{t_1} \alpha_k(\tau)u(\tau)\,d\tau = \beta_k$$

Then Eq. (16-7) becomes

$$x(0) = -\sum_{k=0}^{n-1} \mathbf{A}^k\mathbf{B}\beta_k$$

$$= -[\mathbf{B} \mid \mathbf{AB} \mid \cdots \mid \mathbf{A}^{n-1}\mathbf{B}]\begin{bmatrix} \beta_0 \\ \hline \beta_1 \\ \hline \cdot \\ \cdot \\ \cdot \\ \hline \beta_{n-1} \end{bmatrix} \qquad (16\text{-}8)$$

If the system is completely state controllable, then, given any initial state $x(0)$, Eq. (16-8) must be satisfied. This requires that the rank of the $n \times n$ matrix

$$[\mathbf{B} \mid \mathbf{AB} \mid \cdots \mid \mathbf{A}^{n-1}\mathbf{B}]$$

be $n$.

From this analysis, we can state the condition for complete state controllability as follows: The system given by Eq. (16-4) is completely state controllable if and only if the vectors $\mathbf{B}, \mathbf{AB}, \ldots, \mathbf{A}^{n-1}\mathbf{B}$ are linearly independent, or the $n \times n$ matrix

$$[\mathbf{B} \mid \mathbf{AB} \mid \cdots \mid \mathbf{A}^{n-1}\mathbf{B}]$$

is of rank $n$.

The result just obtained can be extended to the case where the control vector $\mathbf{u}$ is $r$-dimensional. If the system is described by

$$\dot{\mathbf{x}} = \mathbf{Ax} + \mathbf{Bu}$$

where $\mathbf{u}$ is an $r$-dimensional vector, then it can be proved that the condition for complete state controllability is that the following $n \times nr$ matrix

$$[\mathbf{B} \mid \mathbf{AB} \mid \cdots \mid \mathbf{A}^{n-1}\mathbf{B}]$$

is of rank $n$, or contains $n$ linearly independent column vectors.

*Example 16-2.* Consider the system given by

$$\begin{bmatrix} \dot{x}_1 \\ \dot{x}_2 \end{bmatrix} = \begin{bmatrix} 1 & 1 \\ 0 & -1 \end{bmatrix} \begin{bmatrix} x_1 \\ x_2 \end{bmatrix} + \begin{bmatrix} 1 \\ 0 \end{bmatrix} [u]$$

Since

$$[\mathbf{B} \mid \mathbf{AB}] = \begin{bmatrix} 1 & 1 \\ 0 & 0 \end{bmatrix} = \text{singular}$$

the system is not completely state controllable.

*Example 16-3.* Consider the system given by

$$\begin{bmatrix} \dot{x}_1 \\ \dot{x}_2 \end{bmatrix} = \begin{bmatrix} 1 & 1 \\ 2 & -1 \end{bmatrix} \begin{bmatrix} x_1 \\ x_2 \end{bmatrix} + \begin{bmatrix} 0 \\ 1 \end{bmatrix} [u]$$

For this case

$$[\mathbf{B} \mid \mathbf{AB}] = \begin{bmatrix} 0 & 1 \\ 1 & -1 \end{bmatrix} = \text{nonsingular}$$

The system is therefore completely state controllable.

**Alternate form of the condition for complete state controllability.** Consider the system defined by

$$\dot{\mathbf{x}} = \mathbf{Ax} + \mathbf{Bu} \tag{16-9}$$

where

$\mathbf{x}$ = state vector ($n$-dimensional vector)
$\mathbf{u}$ = control vector ($r$-dimensional vector)
$\mathbf{A}$ = $n \times n$ matrix
$\mathbf{B}$ = $n \times r$ matrix

If the eigenvectors of $\mathbf{A}$ are distinct, then it is possible to find a transformation matrix $\mathbf{P}$ such that

$$\mathbf{P}^{-1}\mathbf{AP} = \mathbf{D} = \begin{bmatrix} \lambda_1 & & & 0 \\ & \lambda_2 & & \\ & & \cdot & \\ & & & \cdot \\ 0 & & & \lambda_n \end{bmatrix}$$

Note that if the eigenvalues of $\mathbf{A}$ are distinct, then the eigenvectors of $\mathbf{A}$ are distinct; however, the converse is not true. For example, an $n \times n$ real symmetric matrix having multiple eigenvalues has $n$ distinct eigenvectors. Note also that each column of the $\mathbf{P}$ matrix is an eigenvector of $\mathbf{A}$ associated with $\lambda_i$ ($i = 1, 2, \ldots, n$).

Let us define

$$\mathbf{x} = \mathbf{Pz} \tag{16-10}$$

Substituting Eq. (16-10) into Eq. (16-9), we obtain

$$\dot{z} = \mathbf{P}^{-1}\mathbf{A}\mathbf{P}z + \mathbf{P}^{-1}\mathbf{B}u \qquad (16\text{-}11)$$

By defining

$$\mathbf{P}^{-1}\mathbf{B} = \mathbf{F} = (f_{ij})$$

we can rewrite Eq. (16-11) as

$$\dot{z}_1 = \lambda_1 z_1 + f_{11}u_1 + f_{12}u_2 + \cdots + f_{1r}u_r$$
$$\dot{z}_2 = \lambda_2 z_2 + f_{21}u_1 + f_{22}u_2 + \cdots + f_{2r}u_r$$
$$\cdot \quad \cdot \quad \cdot$$
$$\dot{z}_n = \lambda_n z_n + f_{n1}u_1 + f_{n2}u_2 + \cdots + f_{nr}u_r$$

If the elements of any one row of the $n \times r$ matrix $\mathbf{F}$ are all zero, then the corresponding state variable cannot be controlled by any of the $u_i$. Hence, the condition of complete state controllability is that if the eigenvectors of $\mathbf{A}$ are distinct, then the system is completely state controllable if and only if no row of $\mathbf{P}^{-1}\mathbf{B}$ has all zero elements. It is important to note that to apply this condition for complete state controllability, we must put the matrix $\mathbf{P}^{-1}\mathbf{A}\mathbf{P}$ in Eq. (16-11) in diagonal form.

If the $\mathbf{A}$ matrix in Eq. (16-9) does not possess distinct eigenvectors, then diagonalization is impossible. In such a case, we may transform $\mathbf{A}$ into the Jordan canonical form. If, for example, $\mathbf{A}$ has eigenvalues $\lambda_1, \lambda_1, \lambda_1, \lambda_4, \lambda_4, \lambda_6, \ldots,$ $\lambda_n$ and has $n - 3$ distinct eigenvectors, then the Jordan canonical form of $\mathbf{A}$ is

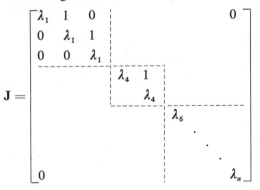

The $3 \times 3$ and $2 \times 2$ submatrices on the main diagonal are called Jordan blocks.

Suppose that we can find a transformation matrix $\mathbf{S}$ such that

$$\mathbf{S}^{-1}\mathbf{A}\mathbf{S} = \mathbf{J}$$

If we define a new state vector $\mathbf{z}$ by

$$\mathbf{x} = \mathbf{S}\mathbf{z} \qquad (16\text{-}12)$$

then substitution of Eq. (16-12) into Eq. (16-9) yields

$$\dot{z} = \mathbf{S}^{-1}\mathbf{A}\mathbf{S}z + \mathbf{S}^{-1}\mathbf{B}u$$
$$= \mathbf{J}z + \mathbf{S}^{-1}\mathbf{B}u \qquad (16\text{-}13)$$

The condition for complete state controllability of the system of Eq. (16-9) may then be stated as follows: The system is completely state controllable if and only if (1) no two Jordan blocks in $\mathbf{J}$ of Eq. (16-13) are associated with the same eigenvalues, (2) the elements of any row of $\mathbf{S}^{-1}\mathbf{B}$ that correspond to the last row of each Jordan block are not all zero, and (3) the elements of each row of $\mathbf{S}^{-1}\mathbf{B}$ that correspond to distinct eigenvalues are not all zero.

---

*Example 16-4.* The following systems are completely state controllable:

$$\begin{bmatrix} \dot{x}_1 \\ \dot{x}_2 \end{bmatrix} = \begin{bmatrix} -1 & 0 \\ 0 & -2 \end{bmatrix} \begin{bmatrix} x_1 \\ x_2 \end{bmatrix} + \begin{bmatrix} 2 \\ 5 \end{bmatrix} [u]$$

$$\begin{bmatrix} \dot{x}_1 \\ \dot{x}_2 \\ \dot{x}_3 \end{bmatrix} = \begin{bmatrix} -1 & 1 & 0 \\ 0 & -1 & 0 \\ 0 & 0 & -2 \end{bmatrix} \begin{bmatrix} x_1 \\ x_2 \\ x_3 \end{bmatrix} + \begin{bmatrix} 0 \\ 4 \\ 3 \end{bmatrix} [u]$$

$$\begin{bmatrix} \dot{x}_1 \\ \dot{x}_2 \\ \dot{x}_3 \\ \dot{x}_4 \\ \dot{x}_5 \end{bmatrix} = \begin{bmatrix} -2 & 1 & 0 & & 0 \\ 0 & -2 & 1 & & \\ 0 & 0 & -2 & & \\ \hline & & & -5 & 1 \\ 0 & & & 0 & -5 \end{bmatrix} \begin{bmatrix} x_1 \\ x_2 \\ x_3 \\ x_4 \\ x_5 \end{bmatrix} + \begin{bmatrix} 0 & 1 \\ 0 & 0 \\ 3 & 0 \\ 0 & 0 \\ 2 & 1 \end{bmatrix} \begin{bmatrix} u_1 \\ u_2 \end{bmatrix}$$

The following systems are not completely state controllable:

$$\begin{bmatrix} \dot{x}_1 \\ \dot{x}_2 \end{bmatrix} = \begin{bmatrix} -1 & 0 \\ 0 & -2 \end{bmatrix} \begin{bmatrix} x_1 \\ x_2 \end{bmatrix} + \begin{bmatrix} 2 \\ 0 \end{bmatrix} [u]$$

$$\begin{bmatrix} \dot{x}_1 \\ \dot{x}_2 \\ \dot{x}_3 \end{bmatrix} = \begin{bmatrix} -1 & 1 & 0 \\ 0 & -1 & 0 \\ 0 & 0 & -2 \end{bmatrix} \begin{bmatrix} x_1 \\ x_2 \\ x_3 \end{bmatrix} + \begin{bmatrix} 4 & 2 \\ 0 & 0 \\ 3 & 0 \end{bmatrix} \begin{bmatrix} u_1 \\ u_2 \end{bmatrix}$$

$$\begin{bmatrix} \dot{x}_1 \\ \dot{x}_2 \\ \dot{x}_3 \\ \dot{x}_4 \\ \dot{x}_5 \end{bmatrix} = \begin{bmatrix} -2 & 1 & 0 & & 0 \\ 0 & -2 & 1 & & \\ 0 & 0 & -2 & & \\ \hline & & & -5 & 1 \\ 0 & & & 0 & -5 \end{bmatrix} \begin{bmatrix} x_1 \\ x_2 \\ x_3 \\ x_4 \\ x_5 \end{bmatrix} + \begin{bmatrix} 4 \\ 2 \\ 1 \\ 3 \\ 0 \end{bmatrix} [u]$$

**Condition for complete state controllability in the $s$ plane.** The condition for complete state controllability can be stated in terms of transfer functions or transfer matrices.

A necessary and sufficient condition for complete state controllability is that no cancellation occurs in the transfer function or transfer matrix. If cancellation occurs, the system cannot be controlled in the direction of the canceled mode.

---

*Example 16-5.* Consider the following transfer function:

$$\frac{X(s)}{U(s)} = \frac{s + 2.5}{(s + 2.5)(s - 1)}$$

Clearly cancellation of the factor $(s + 2.5)$ occurs in the numerator and denominator of this transfer function. (Thus one degree of freedom is lost.) Because of this cancellation, this system is not completely state controllable.

The same conclusion can, of course, be obtained by writing this transfer function in the form of state equations. A state-space representation is

$$\begin{bmatrix} \dot{x}_1 \\ \dot{x}_2 \end{bmatrix} = \begin{bmatrix} 0 & 1 \\ 2.5 & -1.5 \end{bmatrix} \begin{bmatrix} x_1 \\ x_2 \end{bmatrix} + \begin{bmatrix} 1 \\ 1 \end{bmatrix} [u]$$

Since

$$[\mathbf{B} \vdots \mathbf{AB}] = \begin{bmatrix} 1 & 1 \\ 1 & 1 \end{bmatrix}$$

the rank of the matrix $[\mathbf{B} \vdots \mathbf{AB}]$ is one. Therefore we arrive at the same conclusion: The system is not completely state controllable.

**Output controllability.** In the practical design of a control system, we may want to control the output rather than the state of the system. Complete state controllability is neither necessary nor sufficient for controlling the output of the system. For this reason, it is desirable to define separately complete output controllability.

Consider the system described by

$$\dot{\mathbf{x}} = \mathbf{Ax} + \mathbf{Bu} \tag{16-14}$$

$$\mathbf{y} = \mathbf{Cx} + \mathbf{Du} \tag{16-15}$$

where

$\mathbf{x}$ = state vector ($n$-dimensional vector)
$\mathbf{u}$ = control vector ($r$-dimensional vector)
$\mathbf{y}$ = output vector ($m$-dimensional vector)
$\mathbf{A} = n \times n$ matrix
$\mathbf{B} = n \times r$ matrix
$\mathbf{C} = m \times n$ matrix
$\mathbf{D} = m \times r$ matrix

The system described by Eqs. (16-14) and (16-15) is said to be completely output controllable if it is possible to construct an unconstrained control vector $\mathbf{u}(t)$ which will transfer any given initial output $\mathbf{y}(t_0)$ to any final output $\mathbf{y}(t_1)$ in a finite time interval $t_0 \leq t \leq t_1$.

It can be proved that the condition for complete output controllability is as follows: The system described by Eqs. (16-14) and (16-15) is completely output controllable if and only if the $m \times (n + 1)r$ matrix

$$[\mathbf{CB} \vdots \mathbf{CAB} \vdots \mathbf{CA^2B} \vdots \cdots \vdots \mathbf{CA^{n-1}B} \vdots \mathbf{D}]$$

is of rank $m$.

## 16-3 OBSERVABILITY

In this section we discuss the observability of linear systems. Consider the unforced system described by the following equations:

$$\dot{\mathbf{x}} = \mathbf{Ax} \tag{16-16}$$

$$\mathbf{y} = \mathbf{Cx} \tag{16-17}$$

where

$\mathbf{x}$ = state vector ($n$-dimensional vector)

$\mathbf{y}$ = output vector ($m$-dimensional vector)

$\mathbf{A} = n \times n$ matrix

$\mathbf{C} = m \times n$ matrix

The system is said to be completely observable if every initial state $\mathbf{x}(0)$ can be determined from the observation of $\mathbf{y}(t)$ over a finite time interval. The system is, therefore, completely observable if every transition of the state eventually affects every element of the output vector. The concept of observability is useful in solving the problem of reconstructing unmeasurable state variables from measurable ones in the minimum possible length of time. (Note that, in practice, the difficulty encountered with optimal control systems is that some of the state variables are not accessible for direct measurement. Then it is necessary to estimate the unmeasurable state variables in order to construct the optimal control signals.)

The reason that we considered the unforced system is as follows: If the system is described by

$$\dot{\mathbf{x}} = \mathbf{Ax} + \mathbf{Bu}$$
$$\mathbf{y} = \mathbf{Cx}$$

then

$$\mathbf{x}(t) = e^{\mathbf{A}t}\mathbf{x}(0) + \int_0^t e^{\mathbf{A}(t-\tau)}\mathbf{Bu}(\tau)\, d\tau$$

and $\mathbf{y}(t)$ is

$$\mathbf{y}(t) = \mathbf{C}e^{\mathbf{A}t}\mathbf{x}(0) + \mathbf{C}\int_0^t e^{\mathbf{A}(t-\tau)}\mathbf{Bu}(\tau)\, d\tau$$

Since the matrices $\mathbf{A}$, $\mathbf{B}$, and $\mathbf{C}$ are known and $\mathbf{u}(t)$ is also known, the integral term on the right-hand side of this last equation is a known quantity. Therefore, it may be subtracted from the observed value of $\mathbf{y}(t)$. Hence, for investigating a necessary and sufficient condition for complete observability, it suffices to consider the system described by Eqs. (16-16) and (16-17).

**Complete observability of discrete-time systems.** Consider the system described by

$$\mathbf{x}((k+1)T) = \mathbf{Gx}(kT) \tag{16-18}$$
$$\mathbf{y}(kT) = \mathbf{Cx}(kT) \tag{16-19}$$

where

$\mathbf{x}(kT)$ = state vector ($n$-dimensional vector)

$\mathbf{y}(kT)$ = output vector ($m$-dimensional vector)

$\mathbf{G} = n \times n$ matrix

$\mathbf{C} = m \times n$ matrix

$T$ = sampling period

This system is completely observable if, given the output $\mathbf{y}(kT)$ for finite sampling periods, it is possible to determine the initial state vector $\mathbf{x}(0)$.

In the following, we derive the condition for complete observability of the discrete-time system described by Eqs. (16-18) and (16-19). Since the solution $\mathbf{x}(kT)$ is

$$\mathbf{x}(kT) = \mathbf{G}^k\mathbf{x}(0)$$

we obtain

$$\mathbf{y}(kT) = \mathbf{CG}^k\mathbf{x}(0)$$

Complete observability means that, given $\mathbf{y}(0), \mathbf{y}(T), \ldots, \mathbf{y}(NT)$, we can determine $x_1(0), x_2(0), \ldots, x_n(0)$. To determine $n$ unknowns, we need only $n$ values of $\mathbf{y}(kT)$. Hence, $N = n - 1$. For a completely observable system, given

$$\mathbf{y}(0) = \mathbf{Cx}(0)$$
$$\mathbf{y}(T) = \mathbf{CGx}(0)$$
$$\cdot \quad \cdot \quad \cdot$$
$$\mathbf{y}((n - 1)T) = \mathbf{CG}^{n-1}\mathbf{x}(0)$$

we must be able to determine $x_1(0), x_2(0), \ldots, x_n(0)$. Noting that $\mathbf{y}$ is an $m$ vector, the preceding $n$ simultaneous equations yield $nm$ equations, all involving $x_1(0)$, $x_2(0), \ldots, x_n(0)$. To obtain a unique set of solutions $x_1(0), x_2(0), \ldots, x_n(0)$ from these $nm$ equations, we must be able to write exactly $n$ linearly independent equations among them. This requires that the $nm \times n$ matrix

$$\begin{bmatrix} \mathbf{C} \\ \hline \mathbf{CG} \\ \hline \cdot \\ \cdot \\ \cdot \\ \hline \mathbf{CG}^{n-1} \end{bmatrix}$$

be of rank $n$.

Noting that the rank of a matrix and that of the conjugate transpose of the matrix are the same, we can state the condition for complete observability as follows: The system described by Eqs. (16-18) and (16-19) is completely observable if and only if the $n \times nm$ matrix

$$[\mathbf{C}^* \mid \mathbf{G}^*\mathbf{C}^* \mid \cdots \mid (\mathbf{G}^*)^{n-1}\mathbf{C}^*]$$

is of rank $n$, or has $n$ linearly independent column vectors.

**Complete observability of continuous-time systems.** Consider the system described by Eqs. (16-16) and (16-17), rewritten

$$\dot{\mathbf{x}} = \mathbf{Ax}$$
$$\mathbf{y} = \mathbf{Cx}$$

The output vector $\mathbf{y}(t)$ is

$$\mathbf{y}(t) = \mathbf{C}e^{\mathbf{A}t}\mathbf{x}(0)$$

Noting that

$$e^{\mathbf{A}t} = \sum_{k=0}^{n-1} \alpha_k(t)\mathbf{A}^k$$

we obtain

$$\mathbf{y}(t) = \sum_{k=0}^{n-1} \alpha_k(t)\mathbf{C}\mathbf{A}^k\mathbf{x}(0)$$

or

$$\mathbf{y}(t) = \alpha_0(t)\mathbf{C}\mathbf{x}(0) + \alpha_1(t)\mathbf{C}\mathbf{A}\mathbf{x}(0) + \cdots + \alpha_{n-1}(t)\mathbf{C}\mathbf{A}^{n-1}\mathbf{x}(0) \qquad (16\text{-}20)$$

If the system is completely observable, then given the output $\mathbf{y}(t)$ over a time interval $0 \le t \le t_1$, $\mathbf{x}(0)$ is uniquely determined from Eq. (16-20). It can be shown that this requires the rank of the $nm \times n$ matrix

to be $n$.

From this analysis, we can state the condition for complete observability as follows: The system described by Eqs. (16-16) and (16-17) is completely observable if and only if the $n \times nm$ matrix

$$[\mathbf{C}^* \mid \mathbf{A}^*\mathbf{C}^* \mid \cdots \mid (\mathbf{A}^*)^{n-1}\mathbf{C}^*]$$

is of rank $n$, or has $n$ linearly independent column vectors.

---

*Example 16-6.* Consider the system described by

$$\begin{bmatrix} \dot{x}_1 \\ \dot{x}_2 \end{bmatrix} = \begin{bmatrix} 1 & 1 \\ -2 & -1 \end{bmatrix} \begin{bmatrix} x_1 \\ x_2 \end{bmatrix} + \begin{bmatrix} 0 \\ 1 \end{bmatrix} [u]$$

$$y = \begin{bmatrix} 1 & 0 \end{bmatrix} \begin{bmatrix} x_1 \\ x_2 \end{bmatrix}$$

Is this system controllable and observable?

Since the rank of the matrix

$$[\mathbf{B} \mid \mathbf{A}\mathbf{B}] = \begin{bmatrix} 0 & 1 \\ 1 & -1 \end{bmatrix}$$

is two, the system is completely state controllable.

For output controllability, let us find the rank of the matrix $[\mathbf{C}\mathbf{B} \mid \mathbf{C}\mathbf{A}\mathbf{B}]$. Since

$$[\mathbf{C}\mathbf{B} \mid \mathbf{C}\mathbf{A}\mathbf{B}] = \begin{bmatrix} 0 & 1 \end{bmatrix}$$

the rank of this matrix is one. Hence the system is completely output controllable.

To test the observability condition, examine the rank of $[\mathbf{C}' \mid \mathbf{A}'\mathbf{C}']$. Since

$$[\mathbf{C}' \mid \mathbf{A}'\mathbf{C}'] = \begin{bmatrix} 1 & 1 \\ 0 & 1 \end{bmatrix}$$

the rank of $[\mathbf{C}' \mid \mathbf{A}'\mathbf{C}']$ is two. Hence the system is completely observable.

**Conditions for complete observability in the $s$ plane.** The conditions for complete observability can also be stated in terms of transfer functions or transfer matrices. The necessary and sufficient condition for complete observability is that no cancellation occurs in the transfer function or transfer matrix. If cancellation occurs, the canceled mode cannot be observed in the output.

---

*Example 16-7.* Show that the following system is not completely observable.

$$\dot{\mathbf{x}} = \mathbf{A}\mathbf{x} + \mathbf{B}u$$
$$y = \mathbf{C}\mathbf{x}$$

where

$$\mathbf{x} = \begin{bmatrix} x_1 \\ x_2 \\ x_3 \end{bmatrix}, \quad \mathbf{A} = \begin{bmatrix} 0 & 1 & 0 \\ 0 & 0 & 1 \\ -6 & -11 & -6 \end{bmatrix}, \quad \mathbf{B} = \begin{bmatrix} 0 \\ 0 \\ 1 \end{bmatrix}$$
$$\mathbf{C} = [4 \quad 5 \quad 1]$$

Note that the control function $u$ does not affect the complete observability of the system. In order to examine complete observability, we may simply set $u = 0$. For this system, we have

$$[\mathbf{C}' \mid \mathbf{A}'\mathbf{C}' \mid (\mathbf{A}')^2\mathbf{C}'] = \begin{bmatrix} 4 & -6 & 6 \\ 5 & -7 & 5 \\ 1 & -1 & -1 \end{bmatrix}$$

Note that

$$\begin{vmatrix} 4 & -6 & 6 \\ 5 & -7 & 5 \\ 1 & -1 & -1 \end{vmatrix} = 0$$

Hence the rank of the matrix $[\mathbf{C}' \mid \mathbf{A}'\mathbf{C}' \mid (\mathbf{A}')^2\mathbf{C}']$ is less than three. Therefore, the system is not completely observable.

In fact in this system, cancellation occurs in the transfer function of the system. The transfer function between $X_1(s)$ and $U(s)$ is

$$\frac{X_1(s)}{U(s)} = \frac{1}{(s+1)(s+2)(s+3)}$$

and the transfer function between $Y(s)$ and $X_1(s)$ is

$$\frac{Y(s)}{X_1(s)} = (s+1)(s+4)$$

Therefore the transfer function between the output $Y(s)$ and the input $U(s)$ is

$$\frac{Y(s)}{U(s)} = \frac{(s+1)(s+4)}{(s+1)(s+2)(s+3)}$$

Clearly the two factors $(s+1)$ cancel each other. This means that there are nonzero initial states $\mathbf{x}(0)$ which cannot be determined from the measurement of $y(t)$.

**Relationships between controllability, observability, and transfer functions.** The transfer function has no cancellation if and only if the system is completely state controllable and completely observable. This means that the canceled transfer function does not carry along all the information characterizing the dynamic system.

**Alternate form of the condition for complete observability.** Consider the system described by Eqs. (16-16) and (16-17), rewritten

$$\dot{\mathbf{x}} = \mathbf{Ax} \qquad (16\text{-}21)$$

$$\mathbf{y} = \mathbf{Cx} \qquad (16\text{-}22)$$

Suppose that the transformation matrix $\mathbf{P}$ transforms $\mathbf{A}$ into a diagonal matrix, or

$$\mathbf{P}^{-1}\mathbf{AP} = \mathbf{D}$$

where $\mathbf{D}$ is a diagonal matrix. Let us define

$$\mathbf{x} = \mathbf{Pz}$$

Then Eqs. (16-21) and (16-22) can be written

$$\dot{\mathbf{z}} = \mathbf{P}^{-1}\mathbf{APz} = \mathbf{Dz}$$

$$\mathbf{y} = \mathbf{CPz}$$

Hence,

$$\mathbf{y}(t) = \mathbf{CP}e^{\mathbf{D}t}\mathbf{z}(0)$$

or

$$\mathbf{y}(t) = \mathbf{CP}\begin{bmatrix} e^{\lambda_1 t} & & & 0 \\ & e^{\lambda_2 t} & & \\ & & \cdot & \\ & & & \cdot \\ & & & \cdot \\ 0 & & & e^{\lambda_n t} \end{bmatrix} \mathbf{z}(0) = \mathbf{CP}\begin{bmatrix} e^{\lambda_1 t}z_1(0) \\ e^{\lambda_2 t}z_2(0) \\ \cdot \\ \cdot \\ \cdot \\ e^{\lambda_n t}z_n(0) \end{bmatrix}$$

The system is completely observable if none of the columns of the $m \times n$ matrix $\mathbf{CP}$ consists of all zero elements. This is because if the $i$th column of $\mathbf{CP}$ consists of all zero elements, then the state variable $z_i(0)$ will not appear in the output equation and therefore cannot be determined from observation of $\mathbf{y}(t)$. Thus, $\mathbf{x}(0)$, which is related to $\mathbf{z}(0)$ by the nonsingular matrix $\mathbf{P}$, cannot be determined. (Remember that this test applies only if the matrix $\mathbf{P}^{-1}\mathbf{AP}$ is in diagonal form.)

If the matrix $\mathbf{A}$ cannot be transformed into a diagonal matrix, then by use of a suitable transformation matrix $\mathbf{S}$, we can transform $\mathbf{A}$ into the Jordan canonical form, or

$$\mathbf{S}^{-1}\mathbf{AS} = \mathbf{J}$$

where $\mathbf{J}$ is in the Jordan canonical form.

Let us define

$$\mathbf{x} = \mathbf{Sz}$$

Then Eqs. (16-21) and (16-22) can be written

$$\dot{\mathbf{z}} = \mathbf{S}^{-1}\mathbf{ASz} = \mathbf{Jz}$$

$$\mathbf{y} = \mathbf{CSz}$$

Hence

$$y(t) = \mathbf{CS}e^{\mathbf{J}t}\mathbf{z}(0)$$

The system is completely observable if (1) no two Jordan blocks in $\mathbf{J}$ are associated with the same eigenvalues, (2) no columns of $\mathbf{CS}$ that corresspond to the first row of each Jordan block consist of zero elements, and (3) no columns of $\mathbf{CS}$ that correspond to distinct eigenvalues consist of zero elements.

To clarify condition (2) above, in Example 16-8, we have encircled by dotted lines the columns of $\mathbf{CS}$ that correspond to the first row of each Jordan block.

---

*Example 16-8.* The following systems are completely observable.

$$\begin{bmatrix} \dot{x}_1 \\ \dot{x}_2 \end{bmatrix} = \begin{bmatrix} -1 & 0 \\ 0 & -2 \end{bmatrix} \begin{bmatrix} x_1 \\ x_2 \end{bmatrix}, \qquad y = \begin{bmatrix} 1 & 3 \end{bmatrix} \begin{bmatrix} x_1 \\ x_2 \end{bmatrix}$$

$$\begin{bmatrix} \dot{x}_1 \\ \dot{x}_2 \\ \dot{x}_3 \end{bmatrix} = \begin{bmatrix} 2 & 1 & 0 \\ 0 & 2 & 1 \\ 0 & 0 & 2 \end{bmatrix} \begin{bmatrix} x_1 \\ x_2 \\ x_3 \end{bmatrix}, \qquad \begin{bmatrix} y_1 \\ y_2 \end{bmatrix} = \begin{bmatrix} 3 & 0 & 0 \\ 4 & 0 & 0 \end{bmatrix} \begin{bmatrix} x_1 \\ x_2 \\ x_3 \end{bmatrix}$$

$$\begin{bmatrix} \dot{x}_1 \\ \dot{x}_2 \\ \dot{x}_3 \\ \dot{x}_4 \\ \dot{x}_5 \end{bmatrix} = \begin{bmatrix} 2 & 1 & 0 & & 0 \\ 0 & 2 & 1 & & \\ 0 & 0 & 2 & & \\ & & & -3 & 1 \\ 0 & & & 0 & -3 \end{bmatrix} \begin{bmatrix} x_1 \\ x_2 \\ x_3 \\ x_4 \\ x_5 \end{bmatrix}, \qquad \begin{bmatrix} y_1 \\ y_2 \end{bmatrix} = \begin{bmatrix} 1 & 1 & 1 & 0 & 0 \\ 0 & 1 & 1 & 1 & 0 \end{bmatrix} \begin{bmatrix} x_1 \\ x_2 \\ x_3 \\ x_4 \\ x_5 \end{bmatrix}$$

The following systems are not completely observable.

$$\begin{bmatrix} \dot{x}_1 \\ \dot{x}_2 \end{bmatrix} = \begin{bmatrix} -1 & 0 \\ 0 & -2 \end{bmatrix} \begin{bmatrix} x_1 \\ x_2 \end{bmatrix}, \qquad y = \begin{bmatrix} 0 & 1 \end{bmatrix} \begin{bmatrix} x_1 \\ x_2 \end{bmatrix}$$

$$\begin{bmatrix} \dot{x}_1 \\ \dot{x}_2 \\ \dot{x}_3 \end{bmatrix} = \begin{bmatrix} 2 & 1 & 0 \\ 0 & 2 & 1 \\ 0 & 0 & 2 \end{bmatrix} \begin{bmatrix} x_1 \\ x_2 \\ x_3 \end{bmatrix}, \qquad \begin{bmatrix} y_1 \\ y_2 \end{bmatrix} = \begin{bmatrix} 0 & 1 & 3 \\ 0 & 2 & 4 \end{bmatrix} \begin{bmatrix} x_1 \\ x_2 \\ x_3 \end{bmatrix}$$

$$\begin{bmatrix} \dot{x}_1 \\ \dot{x}_2 \\ \dot{x}_3 \\ \dot{x}_4 \\ \dot{x}_5 \end{bmatrix} = \begin{bmatrix} 2 & 1 & 0 & & 0 \\ 0 & 2 & 1 & & \\ 0 & 0 & 2 & & \\ & & & -3 & 1 \\ 0 & & & 0 & -3 \end{bmatrix} \begin{bmatrix} x_1 \\ x_2 \\ x_3 \\ x_4 \\ x_5 \end{bmatrix}, \qquad \begin{bmatrix} y_1 \\ y_2 \end{bmatrix} = \begin{bmatrix} 1 & 1 & 1 & 0 & 0 \\ 0 & 1 & 1 & 0 & 0 \end{bmatrix} \begin{bmatrix} x_1 \\ x_2 \\ x_3 \\ x_4 \\ x_5 \end{bmatrix}$$

**Principle of duality.** We shall now discuss the relation between controllability and observability. We shall introduce the principle of duality, due to Kalman, to clarify apparent analogies between controllability and observability.

Consider the system $S_1$ described by

$$\dot{\mathbf{x}} = \mathbf{A}\mathbf{x} + \mathbf{B}\mathbf{u}$$
$$\mathbf{y} = \mathbf{C}\mathbf{x}$$

where

$\quad$ $\mathbf{x} =$ state vector ($n$-dimensional vector)

$\quad$ $\mathbf{u} =$ control vector ($r$-dimensional vector)

$\quad$ $\mathbf{y} =$ output vector ($m$-dimensional vector)

$\quad$ $\mathbf{A} = n \times n$ matrix

$\quad$ $\mathbf{B} = n \times r$ matrix

$\quad$ $\mathbf{C} = m \times n$ matrix

and the dual system $S_2$ defined by

$$\dot{\mathbf{z}} = \mathbf{A}^*\mathbf{z} + \mathbf{C}^*\mathbf{v}$$
$$\mathbf{n} = \mathbf{B}^*\mathbf{z}$$

where

$\quad$ $\mathbf{z} =$ state vector ($n$-dimensional vector)

$\quad$ $\mathbf{v} =$ control vector ($m$-dimensional vector)

$\quad$ $\mathbf{n} =$ output vector ($r$-dimensional vector)

$\quad$ $\mathbf{A}^* =$ conjugate transpose of $\mathbf{A}$

$\quad$ $\mathbf{B}^* =$ conjugate transpose of $\mathbf{B}$

$\quad$ $\mathbf{C}^* =$ conjugate transpose of $\mathbf{C}$

The principle of duality states that system $S_1$ is completely state controllable (observable) if and only if system $S_2$ is completely observable (state controllable).

To verify this principle, let us write down the necessary and sufficient conditions for complete state controllability and complete observability of systems $S_1$ and $S_2$.

For system $S_1$:

1. A necessary and sufficient condition for complete state controllability is that the rank of the $n \times nr$ matrix

$$[\mathbf{B} \mid \mathbf{AB} \mid \cdots \mid \mathbf{A}^{n-1}\mathbf{B}]$$

is $n$.

2. A necessary and sufficient condition for complete observability is that the rank of the $n \times nm$ matrix

$$[\mathbf{C}^* \mid \mathbf{A}^*\mathbf{C}^* \mid \cdots \mid (\mathbf{A}^*)^{n-1}\mathbf{C}^*]$$

is $n$.

For system $S_2$:

1. A necessary and sufficient condition for complete state controllability is that the rank of the $n \times nm$ matrix

$$[\mathbf{C}^* \mid \mathbf{A}^*\mathbf{C}^* \mid \cdots \mid (\mathbf{A}^*)^{n-1}\mathbf{C}^*]$$

is $n$.

2. A necessary and sufficient condition for complete observability is that the rank of the $n \times nr$ matrix

$$[\mathbf{B} \mid \mathbf{AB} \mid \cdots \mid \mathbf{A}^{n-1}\mathbf{B}]$$

is $n$.

By comparing these conditions, the truth of this principle is apparent. This principle, of course, equally applies to discrete-time systems. By use of this principle, the observability of a given system can be checked by testing the state controllability of its dual.

## 16-4  TIME-OPTIMAL CONTROL SYSTEMS

In this section, we first consider the time-optimal control of discrete-time systems and then consider the time-optimal control of continuous-time systems.

**Time-optimal control of discrete-time systems.** If the $n$th-order system is discrete-time, then an arbitrary initial state can be brought to a prescribed state in at most $n$ sampling periods if the system is completely state controllable and the norm of the control vector is not bounded.

Consider the following discrete-time system:

$$\mathbf{x}((k+1)T) = \mathbf{G}(T)\mathbf{x}(kT) + \mathbf{H}(T)\mathbf{u}(kT), \qquad \|\mathbf{u}(kT)\| \leq \infty$$

where

$\mathbf{x}$ = state vector ($n$-dimensional vector)

$\mathbf{u}$ = control vector ($r$-dimensional vector)

$\mathbf{G}(T) = n \times n$ constant matrix

$\mathbf{H}(T) = n \times r$ constant matrix

(Note that the sampling period $T$ is a constant.) If this system is completely state controllable, then we can transfer an arbitrary initial state $\mathbf{x}(0)$ to the origin in $n$ steps, or

$$\mathbf{x}(nT) = \mathbf{G}^n(T)\mathbf{x}(0) + \sum_{k=0}^{n-1} \mathbf{G}^{n-k-1}(T)\mathbf{H}(T)\mathbf{u}(kT) = \mathbf{0}$$

The required control vector $\mathbf{u}(kT)$ can be determined by solving the following $n$ simultaneous vector-matrix equations:

$$\mathbf{x}(T) = \mathbf{G}(T)\mathbf{x}(0) + \mathbf{H}(T)\mathbf{u}(0)$$
$$\mathbf{x}(2T) = \mathbf{G}(T)\mathbf{x}(T) + \mathbf{H}(T)\mathbf{u}(T)$$

$$\cdot \quad \cdot \quad \cdot$$

$$\mathbf{x}(nT) = \mathbf{G}(T)\mathbf{x}((n-1)T) + \mathbf{H}(T)\mathbf{u}((n-1)T) = \mathbf{0}$$

To be a practical solution, the control vector $\mathbf{u}(kT)$ must be determined as a function of the state vector $\mathbf{x}(kT)$.

Note that since we assumed the norm of $\mathbf{u}(kT)$ to be unbounded, we can make the response time shorter by making the sampling period shorter. Note also that if the control signal is bounded, then the number of sampling periods required to transfer the arbitrary initial state to the origin may increase.

We shall illustrate the determination of the required $\mathbf{u}(kT)$ by use of an example.

*Example 16-9.* Consider the following discrete-time system:

$$\begin{bmatrix} x_1((k+1)T) \\ x_2((k+1)T) \end{bmatrix} = \begin{bmatrix} 1 & 1-e^{-T} \\ 0 & e^{-T} \end{bmatrix} \begin{bmatrix} x_1(kT) \\ x_2(kT) \end{bmatrix} + \begin{bmatrix} e^{-T}+T-1 \\ 1-e^{-T} \end{bmatrix} [u(kT)] \quad (16\text{-}23)$$

Note that this is the discrete version of the continuous-time system described by

$$\begin{bmatrix} \dot{x}_1 \\ \dot{x}_2 \end{bmatrix} = \begin{bmatrix} 0 & 1 \\ 0 & -1 \end{bmatrix} \begin{bmatrix} x_1 \\ x_2 \end{bmatrix} + \begin{bmatrix} 0 \\ 1 \end{bmatrix} [u]$$

Let us rewrite Eq. (16-23) as

$$\mathbf{x}((k+1)T) = \mathbf{G}(T)\mathbf{x}(kT) + \mathbf{H}(T)u(kT)$$

We assume that $u(kT)$ is unbounded, or

$$-\infty \leq u(kT) \leq \infty$$

Clearly, this system is completely state controllable. We shall solve the problem of transferring an arbitrary initial state to the origin by using a piecewise-constant control signal.

Since the given system is of second order, we need at most two sampling periods to transfer any initial state $\mathbf{x}(0)$ to the origin. Noting that

$$\mathbf{x}(T) = \mathbf{G}(T)\mathbf{x}(0) + \mathbf{H}(T)u(0) \quad (16\text{-}24)$$
$$\mathbf{x}(2T) = \mathbf{0} = \mathbf{G}(T)\mathbf{x}(T) + \mathbf{H}(T)u(T)$$
$$= \mathbf{G}^2(T)\mathbf{x}(0) + \mathbf{G}(T)\mathbf{H}(T)u(0) + \mathbf{H}(T)u(T)$$

we obtain

$$\mathbf{x}(0) = -\mathbf{G}^{-1}(T)\mathbf{H}(T)u(0) - \mathbf{G}^{-2}(T)\mathbf{H}(T)u(T) \quad (16\text{-}25)$$

Substituting Eq. (16-25) into Eq. (16-24), we obtain

$$\mathbf{x}(T) = -\mathbf{G}^{-1}(T)\mathbf{H}(T)u(T) \quad (16\text{-}26)$$

Noting that

$$\mathbf{G}^{-1}(T)\mathbf{H}(T) = \begin{bmatrix} 1 & 1-e^{T} \\ 0 & e^{T} \end{bmatrix} \begin{bmatrix} e^{-T}+T-1 \\ 1-e^{-T} \end{bmatrix} = \begin{bmatrix} T-e^{T}+1 \\ e^{T}-1 \end{bmatrix}$$

$$\mathbf{G}^{-2}(T)\mathbf{H}(T) = \begin{bmatrix} 1 & 1-e^{2T} \\ 0 & e^{2T} \end{bmatrix} \begin{bmatrix} e^{-T}+T-1 \\ 1-e^{-T} \end{bmatrix} = \begin{bmatrix} T+e^{T}-e^{2T} \\ e^{2T}-e^{T} \end{bmatrix}$$

we obtain from Eqs. (16-25) and (16-26)

$$\begin{bmatrix} x_1(0) \\ x_2(0) \end{bmatrix} = -\begin{bmatrix} T-e^{T}+1 \\ e^{T}-1 \end{bmatrix} [u(0)] - \begin{bmatrix} T+e^{T}-e^{2T} \\ e^{2T}-e^{T} \end{bmatrix} [u(T)] \quad (16\text{-}27)$$

$$\begin{bmatrix} x_1(T) \\ x_2(T) \end{bmatrix} = -\begin{bmatrix} T-e^{T}+1 \\ e^{T}-1 \end{bmatrix} [u(T)] \quad (16\text{-}28)$$

Combining Eqs. (16-27) and (16-28) gives

$$\begin{bmatrix} x_1(0) & x_1(T) \\ x_2(0) & x_2(T) \end{bmatrix} = \begin{bmatrix} -T+e^{T}-1 & -T-e^{T}+e^{2T} \\ -e^{T}+1 & -e^{2T}+e^{T} \end{bmatrix} \begin{bmatrix} u(0) & u(T) \\ u(T) & 0 \end{bmatrix}$$

Hence

$$\begin{bmatrix} u(0) & u(T) \\ u(T) & 0 \end{bmatrix} = \begin{bmatrix} -T+e^{T}-1 & -T-e^{T}+e^{2T} \\ -e^{T}+1 & -e^{2T}+e^{T} \end{bmatrix}^{-1} \begin{bmatrix} x_1(0) & x_1(T) \\ x_2(0) & x_2(T) \end{bmatrix} \quad (16\text{-}29)$$

Equation (16-29) defines the optimal control sequence $u(0)$ and $u(T)$. Note that $u(T)$ can be expressed in terms of $\mathbf{x}(0)$ only or $\mathbf{x}(T)$ only.

To illustrate the solution $u(kT)$ in terms of $\mathbf{x}(kT)$, let us assume $T$ equals 1 sec. Then

$$
\begin{bmatrix} u(0) & u(1) \\ u(1) & u(0) \end{bmatrix} = \begin{bmatrix} -1.58 & -1.24 \\ 0.58 & 0.24 \end{bmatrix} \begin{bmatrix} x_1(0) & x_1(1) \\ x_2(0) & x_2(1) \end{bmatrix} \tag{16-30}
$$

from which we obtain

$$
u(k) = -1.58x_1(k) - 1.24x_2(k) \qquad (k = 0, 1)
$$

With this control law, any initial state $\mathbf{x}(0)$ can be transferred to the origin in 2 sec. Figure 16-1 is the block diagram of the optimal control system.

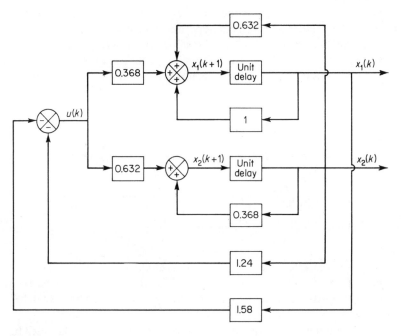

**Fig. 16-1.** Block diagram of the optimal control system considered in Example 16-9.

Let us now find the initial states from which we can transfer the system to the origin in one sampling period. To do so, put $\mathbf{x}(T) = \mathbf{0}$ in Eq. (16-24). Then

$$
\mathbf{x}(T) = \mathbf{G}(T)\mathbf{x}(0) + \mathbf{H}(T)u(0) = \mathbf{0}
$$

from which we obtain

$$
\mathbf{x}(0) = -\mathbf{G}^{-1}(T)\mathbf{H}(T)u(0)
$$

$$
= -\begin{bmatrix} T - e^T + 1 \\ e^T - 1 \end{bmatrix} u(0)
$$

For $T = 1$,

$$
\begin{bmatrix} x_1(0) \\ x_2(0) \end{bmatrix} = \begin{bmatrix} 0.72 \\ -1.72 \end{bmatrix} [u(0)] \tag{16-31}
$$

From Eq. (16-31), we find that if the initial state lies on the line

$$1.72x_1(0) + 0.72x_2(0) = 0$$

then it can be transferred to the origin in one sampling period, or 1 sec. Otherwise we require two sampling periods, or 2 sec, to transfer the initial state to the origin.

Finally, note that if we wish to reduce the response time to at most 0.5 sec, this can be done by choosing $T = 0.25$ sec. However, this will increase the magnitude of the required control signal.

**Time-optimal control of continuous-time systems with bounded control signals.** By plausible arguments and experimentation in the past it was believed that if a control system is being operated under limited power, then the system can be moved from one state to another in the shortest time by at all times properly utilizing all available power. This hypothesis is called the "bang-bang principle." Although many time-optimal control systems with limited control power are of the bang-bang type, it is not true that all such systems are of this type. (For example, if the transfer function of a plant includes zeros, it can be shown that the strict bang-bang solution in which the control signal switches between the constant values $+u_0$ and $-u_0$ is not optimal.)

Consider a plant described by

$$\overset{(n)}{y} + a_1 \overset{(n-1)}{y} + \cdots + a_{n-1}\dot{y} + a_n y = bu \tag{16-32}$$

or

$$\frac{Y(s)}{U(s)} = \frac{b}{s^n + a_1 s^{n-1} + \cdots + a_{n-1}s + a_n}$$

where the magnitude of $u$ is limited, or

$$-u_0 \leq u \leq u_0$$

It can be proved (References B-2 and L-4) that, for the time-optimal control of the plant given by Eq. (16-32), the control law is of the bang-bang type; that is, the control signal takes the maximum value, the sign of which switches at most $n - 1$ times.

In the following analysis, we limit our discussion to plants described by Eq. (16-32). We shall not prove the bang-bang principle here. Instead, we shall use the proved fact that the time-optimal control of systems described by Eq. (16-32) is of the bang-bang type and apply this to the solution of the specific problem where $n = 2$.

For second-order systems, the phase-plane approach is the most convenient in determining the switching curve on which the sign of the control signal must be changed. For higher-order systems, however, whatever approach is used, it is quite difficult to find the switching surfaces in $n$-dimensional state space. In most cases, constructing explicitly the switching surfaces in $n$-dimensional state space is almost impossible.

---

*Example 16-10.* Consider the time-optimal control of the plant described by $1/Js^2$. The optimal controller is of the bang-bang, or maximum effort, type. That is, the controller actuates the motor to supply full positive torque for accelerating the load or full negative torque for decelerating the load, depending on whether the error is positive or negative. Assuming that the system is subjected only to changes in initial conditions, determine the switching curve in the phase plane.

Referring to Fig. 16-2, we see that the equation for the system is

$$J\ddot{c} = u = \pm T \tag{16-33}$$

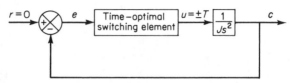

**Fig. 16-2.** Block diagram of a time-optimal control system.

Since the system is subjected only to changes in initial conditions, we set $r = 0$ for $t > 0$. Then we obtain $c = -e$, and Eq. (16-33) becomes

$$\ddot{e} = -\frac{u}{J} = \mp\frac{T}{J}$$

This last equation may be rewritten

$$\dot{e}\frac{d\dot{e}}{de} = \mp\frac{T}{J}$$

from which we obtain

$$\dot{e}^2 = \mp2\frac{T}{J}e + K$$

where $K$ is a constant. This equation describes two families of parabolas in the $e$-$\dot{e}$ plane, depending on whether $u = T$ or $u = -T$.

Once we set $u = T$ or $u = -T$, the particular parabola is determined by the value of $K$, which depends on the initial conditions $e(0)$ and $\dot{e}(0)$. The only parabolas which pass through the origin are parabola $AOA'$ and parabola $BOB'$, shown in Fig. 16-3.

Referring to Fig. 16-3, we see that to reach the origin a representative point in the $e$-$\dot{e}$ plane must follow the trajectory $AO$ or $BO$. This means that, given the initial state represented by point $P$ in Fig. 16-4, we must use $u = T$ until the representative point,

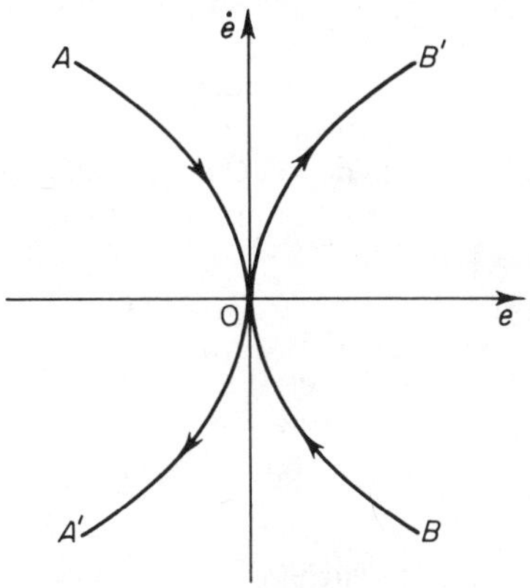

**Fig. 16-3.** Phase-plane diagram showing trajectories passing through the origin.

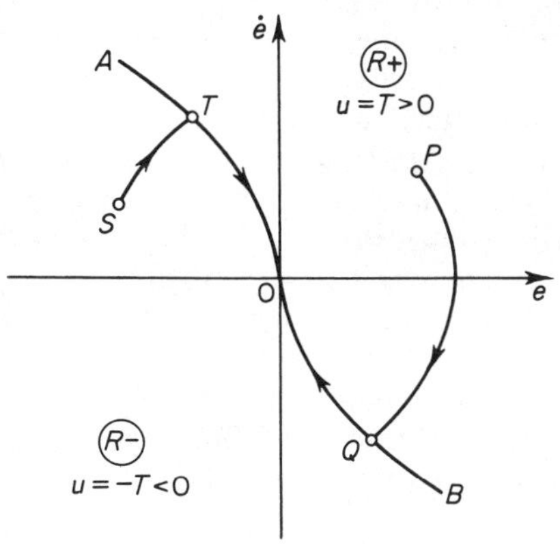

**Fig. 16-4.** Phase-plane diagram showing time-optimal trajectories.

following the parabolic path, reaches point $Q$, where the control signal must be switched to $u = -T$. Then the representative point moves along the path $QO$. This will give the minimum response time, or time-optimal response.

To see this graphically, refer to Fig. 16-5. Let us compare the two paths $PQO$ and $PQRSTUO$. Noting that the time required to travel from point $Q$ to point $O$ is the same as that from point $S$ to point $T$, we see that

$$(t_f - t_i)_{PQO} < (t_f - t_i)_{PQRSTUO}$$

where $t_f - t_i$ is the time required to travel from point $P$ to the origin.

Similarly, comparing the two trajectories $PQO$ and $PRSTO$ in Fig. 16-6, we see that

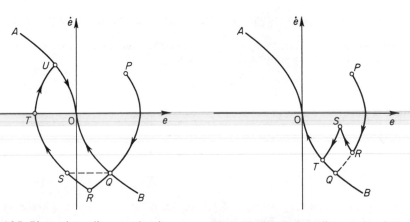

**Fig. 16-5.** Phase-plane diagram showing optimal and nonoptimal trajectories.

**Fig. 16-6.** Phase-plane diagram showing optimal and nonoptimal trajectories.

it takes more time to travel from point $R$ to point $T$ via point $S$ than via point $Q$ because the average velocity is larger for path $RQT$ than for path $RST$. Hence

$$(t_f - t_i)_{PQO} < (t_f - t_i)_{PRSTO}$$

Thus, if switching occurs at points other than on parabola $AOB$, the path requires extra time to reach the origin and cannot be time optimal.

Similarly, if the initial state is point $S$ in Fig. 16-4, then the optimal path is $STO$. Thus, for time-optimal response, a representative point must remain on its initial parabola until it reaches parabola $AO$ or parabola $BO$. The control signal then switches the sign, and the representative point follows the path along parabola $AO$ or parabola $BO$ until the origin is reached.

In summary, the time-optimal control of the plant whose transfer function is given by $1/Js^2$ has been considered. We have found that the switching curve separates the $e$-$\dot{e}$ plane into regions $R_+$ and $R_-$, as shown in Fig. 16-4, where $R_+$ is the set of points to the right and $R_-$ the set of points to the left of the switching curve $AOB$. A formal statement of the control law for this problem may be given as follows: The time-optimal control law as a function of state is given by

$$u(t) = T \qquad \text{for } (x_1, x_2) \text{ in region } R_+ \text{ or on the switching curve } AO$$

$$u(t) = -T \qquad \text{for } (x_1, x_2) \text{ in region } R_- \text{ or on the switching curve } BO$$

It is important to observe that, regardless of the initial state, we require at most one switching of the control signal.

Finally, from a practical point of view, we note the following:

1. Such effects as friction will modify the switching curve, and inaccurate switchings may lead to a path which will not approach the origin of the phase plane. For this reason, a region of linear operation may be provided in the vicinity of the origin. Such a system is often called a dual-mode time-optimal control system.

2. To simplify the problem of mechanizing the optimal controller, we may approximate the parabolic switching curve by a series of straight-line segments. The resulting system then is not time optimal, but it is close to it. (See Problem A-16-5.)

**A few comments on the time-optimal control of higher-order systems with bounded control signals.** Consider the following higher-order system with bounded control signal:

$$\overset{(n)}{y} + a_1 \overset{(n-1)}{y} + \cdots + a_{n-1}\dot{y} + a_n y = bu$$

or in the state-space representation

$$\begin{bmatrix} \dot{x}_1 \\ \dot{x}_2 \\ \cdot \\ \cdot \\ \cdot \\ \dot{x}_n \end{bmatrix} = \begin{bmatrix} 0 & 1 & 0 & \cdots & 0 \\ 0 & 0 & 1 & \cdots & 0 \\ \cdot & \cdot & \cdot & & \cdot \\ \cdot & \cdot & \cdot & & \cdot \\ \cdot & \cdot & \cdot & & \cdot \\ -a_n & -a_{n-1} & -a_{n-2} & \cdots & -a_1 \end{bmatrix} \begin{bmatrix} x_1 \\ x_2 \\ \cdot \\ \cdot \\ \cdot \\ x_n \end{bmatrix} + \begin{bmatrix} 0 \\ 0 \\ \cdot \\ \cdot \\ \cdot \\ b \end{bmatrix}[u]$$

$$y = [1 \quad 0 \quad \cdots \quad 0] \begin{bmatrix} x_1 \\ x_2 \\ \cdot \\ \cdot \\ \cdot \\ x_n \end{bmatrix}$$

The problem of the time-optimal control of this system is to transfer the state from a given point in the state space to the origin (or some other specified point) in the minimum time that is consistent with the given constraints. Since the initial conditions may be arbitrary, any point in the state space can be the initial point. Therefore, it is necessary to determine the control law which minimizes the response time as a function of the state variables. This requires the determination of the loci of points in state space at which the control signal is to be switched to the opposite direction.

The use of a matrix method will yield the solution of the switching problem of such higher-order systems in general terms. It should be mentioned, however, that in practice the real problem is not to find the solution in general matrix form but to find a method for solving several or more simultaneous equations involving transcendental terms within a reasonable amount of time. When working with higher-order systems, we find that the computations involved become quite laborious as the number of variables or the order of the system increases. (The matrix approach does not offer any help in solving this real problem.)

### 16-5 OPTIMAL CONTROL SYSTEMS BASED ON QUADRATIC PERFORMANCE INDEXES

In this section we shall consider the design of optimal control systems based on quadratic performance indexes. The control system which we shall consider may be represented by

$$\dot{\mathbf{x}} = \mathbf{A}\mathbf{x} + \mathbf{B}\mathbf{u}$$

where

$\mathbf{x}$ = state vector ($n$-dimensional real vector)

$\mathbf{u}$ = control vector ($r$-dimensional real vector)

$\mathbf{A} = n \times n$ constant matrix

$\mathbf{B} = n \times r$ constant matrix

In designing control systems, we are often interested in choosing the control vector $\mathbf{u}(t)$ so that a given performance index is minimized. It can be proved that the quadratic performance indexes where the limits of integration are 0 and $\infty$, such as

$$J = \int_0^\infty L(\mathbf{x}, \mathbf{u})dt$$

where $L(\mathbf{x}, \mathbf{u})$ is a quadratic function of $\mathbf{x}$ and $\mathbf{u}$, will yield linear control laws, that is,

$$\mathbf{u}(t) = -\mathbf{K}\mathbf{x}(t)$$

where $\mathbf{K}$ is an $r \times n$ matrix, or

$$\begin{bmatrix} u_1 \\ u_2 \\ \cdot \\ \cdot \\ \cdot \\ u_r \end{bmatrix} = -\begin{bmatrix} k_{11} & k_{12} & \cdots & k_{1n} \\ k_{21} & k_{22} & \cdots & k_{2n} \\ \cdot & \cdot & & \cdot \\ \cdot & \cdot & & \cdot \\ \cdot & \cdot & & \cdot \\ k_{r1} & k_{r2} & \cdots & k_{rn} \end{bmatrix}\begin{bmatrix} x_1 \\ x_2 \\ \cdot \\ \cdot \\ \cdot \\ x_n \end{bmatrix}$$

Therefore, the design of optimal control systems based on such quadratic performance indexes boils down to the determination of the elements of the matrix $\mathbf{K}$.

In the following, we shall first discuss the parameter-optimization problem, i.e., determine the optimal values of system parameters. Specifically, we consider an initially displaced system

$$\dot{\mathbf{x}} = \mathbf{A}\mathbf{x}, \qquad \mathbf{x}(0) = \mathbf{c}$$

where $\mathbf{A}$ has an adjustable parameter or parameters. It is desired to transfer any initial state to the origin while minimizing the quadratic performance index

$$J = \int_0^\infty \mathbf{x}'\mathbf{Q}\mathbf{x}\, dt$$

where $\mathbf{Q}$ is a positive-definite (or positive-semidefinite) real symmetric matrix. The problem thus becomes that of determining the value(s) of the adjustable parameter(s), so as to minimize the performance index.

We shall next consider the optimal control problem based on quadratic performance indexes and determine the control law. That is, we consider the problem of determining the optimal control vector $\mathbf{u}(t)$ for the system described by

$$\dot{\mathbf{x}} = \mathbf{Ax} + \mathbf{Bu}$$

and the performance index given by

$$J = \int_0^\infty (\mathbf{x'Qx} + \mathbf{u'Ru})\, dt$$

where $\mathbf{Q}$ is a positive-definite (or positive-semidefinite) real symmetric matrix, $\mathbf{R}$ is a positive-definite real symmetric matrix, and $\mathbf{u}$ is unconstrained.

There are many different approaches to the solution of these two types of problems. In this section, we present one approach based on the second method of Liapunov. (Recall that in Chapter 7 we discussed a classical approach to the solution of similar optimization problems.)

**Control system optimization via the second method of Liapunov.** Classically, control systems are first designed and then their stability is examined. An approach different from this is where the conditions for stability are formulated first and then the system is designed within these limitations. For a large class of control systems, a direct relationship can be shown between Liapunov functions and the generalized quadratic performance indexes used in the synthesis of optimal control systems. If the second method of Liapunov is utilized to form the basis for the design of an optimal controller, then we are assured that the system will work, i.e., that the system output will be continually driven toward its desired value. Thus the designed system has a configuration with inherent stability characteristics.

**Parameter-optimization problems solved by the second method of Liapunov.** In the following, we shall discuss a direct relationship between Liapunov functions and generalized quadratic performance indexes and solve the parameter-optimization problem using this relationship. Let us consider the system:

$$\dot{\mathbf{x}} = \mathbf{Ax}$$

where all eigenvalues of $\mathbf{A}$ have negative real parts, or the origin $\mathbf{x} = \mathbf{0}$ is asymptotically stable. (We shall call such a matrix $\mathbf{A}$ a *stable* matrix.) It is desired to minimize the quadratic performance index defined by

$$J = \int_0^\infty \mathbf{x'Qx}\, dt$$

where $\mathbf{Q}$ is a positive-definite (or positive-semidefinite) real symmetric matrix.

We shall show that a Liapunov function can effectively be used in the solution of this problem. Let us assume that

$$\mathbf{x'Qx} = -\frac{d}{dt}(\mathbf{x'Px})$$

where $\mathbf{P}$ is a positive-definite real symmetric matrix. Then we obtain

$$\mathbf{x}'\mathbf{Q}\mathbf{x} = -\dot{\mathbf{x}}'\mathbf{P}\mathbf{x} - \mathbf{x}'\mathbf{P}\dot{\mathbf{x}} = -\mathbf{x}'\mathbf{A}'\mathbf{P}\mathbf{x} - \mathbf{x}'\mathbf{P}\mathbf{A}\mathbf{x} = -\mathbf{x}'(\mathbf{A}'\mathbf{P} + \mathbf{P}\mathbf{A})\mathbf{x}$$

By the second method of Liapunov, we know that for a given $\mathbf{Q}$ there exists $\mathbf{P}$, if $\mathbf{A}$ is stable, such that

$$\mathbf{A}'\mathbf{P} + \mathbf{P}\mathbf{A} = -\mathbf{Q} \tag{16-34}$$

Hence we can determine the elements of $\mathbf{P}$ from this equation.

The performance index $J$ can be evaluated as

$$J = \int_0^\infty \mathbf{x}'\mathbf{Q}\mathbf{x}\,dt = -\mathbf{x}'\mathbf{P}\mathbf{x}\Big|_0^\infty = -\mathbf{x}'(\infty)\mathbf{P}\mathbf{x}(\infty) + \mathbf{x}'(0)\mathbf{P}\mathbf{x}(0)$$

Since all eigenvalues of $\mathbf{A}$ have negative real parts, we have $\mathbf{x}(\infty) \longrightarrow \mathbf{0}$. Therefore, we obtain

$$J = \mathbf{x}'(0)\mathbf{P}\mathbf{x}(0)$$

Thus the performance index $J$ can be obtained in terms of the initial condition $\mathbf{x}(0)$ and $\mathbf{P}$, which is related to $\mathbf{A}$ and $\mathbf{Q}$ by Eq. (16-34). If, for example, a system parameter is to be adjusted so as to minimize the performance index $J$, then it can be accomplished by minimizing $\mathbf{x}'(0)\mathbf{P}\mathbf{x}(0)$ with respect to the parameter in question. Since $\mathbf{x}(0)$ is the given initial condition and $\mathbf{Q}$ is also given, $\mathbf{P}$ is a function of the elements of $\mathbf{A}$. Hence this minimization process will result in the optimal value of the adjustable parameter.

It is important to note that the optimal value of this parameter depends, in general, upon the initial condition $\mathbf{x}(0)$. However, if $\mathbf{x}(0)$ involves only one nonzero component; i.e., $x_1(0) \neq 0$, and other initial conditions are zero, then the optimal value of the parameter does not depend on the numerical value of $x_1(0)$. (See the following example.)

---

*Example 16-11.* Consider the system shown in Fig. 16-7. Determine the value of the damping ratio $\zeta > 0$ so that when the system is subjected to a unit-step input $r(t) = 1(t)$, the following performance index is minimized:

$$J = \int_0^\infty \mathbf{x}'(t)\mathbf{Q}\mathbf{x}(t)\,dt$$

where

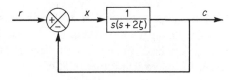

**Fig. 16-7.** Control system.

$$\mathbf{x} = \begin{bmatrix} x_1 \\ x_2 \end{bmatrix} = \begin{bmatrix} x \\ \dot{x} \end{bmatrix}, \qquad \mathbf{Q} = \begin{bmatrix} 1 & 0 \\ 0 & \mu \end{bmatrix} \quad (\mu > 0)$$

The system is assumed to be at rest initially.

From Fig. 16-7, we obtain the following equation for the system:

$$\ddot{c} + 2\zeta\dot{c} + c = r$$

Noting that $x = r - c, r(t) = 1(t)$, and the initial conditions are equal to zero, we have

$$\ddot{x} + 2\zeta\dot{x} + x = 0 \quad (t > 0)$$

The state-space representation of this last equation becomes

$$\begin{bmatrix} \dot{x}_1 \\ \dot{x}_2 \end{bmatrix} = \begin{bmatrix} 0 & 1 \\ -1 & -2\zeta \end{bmatrix}\begin{bmatrix} x_1 \\ x_2 \end{bmatrix}, \qquad \begin{bmatrix} x_1(0) \\ x_2(0) \end{bmatrix} = \begin{bmatrix} 1 \\ 0 \end{bmatrix}$$

or

$$\dot{x} = Ax$$

where

$$A = \begin{bmatrix} 0 & 1 \\ -1 & -2\zeta \end{bmatrix}$$

Since $A$ is a stable matrix, the value of $J$ is given by

$$J = x'(0)Px(0)$$

where $P$ is determined from

$$A'P + PA = -Q$$

or

$$\begin{bmatrix} 0 & -1 \\ 1 & -2\zeta \end{bmatrix}\begin{bmatrix} p_{11} & p_{12} \\ p_{12} & p_{22} \end{bmatrix} + \begin{bmatrix} p_{11} & p_{12} \\ p_{12} & p_{22} \end{bmatrix}\begin{bmatrix} 0 & 1 \\ -1 & -2\zeta \end{bmatrix} = \begin{bmatrix} -1 & 0 \\ 0 & -\mu \end{bmatrix}$$

This equation results in the following three equations:

$$-2p_{12} = -1$$
$$p_{11} - 2\zeta p_{12} - p_{22} = 0$$
$$2p_{12} - 4\zeta p_{22} = -\mu$$

Solving these three equations for the $p_{ij}$, we obtain

$$P = \begin{bmatrix} p_{11} & p_{12} \\ \\ p_{12} & p_{22} \end{bmatrix} = \begin{bmatrix} \zeta + \dfrac{1+\mu}{4\zeta} & \dfrac{1}{2} \\ \dfrac{1}{2} & \dfrac{1+\mu}{4\zeta} \end{bmatrix}$$

Thus the performance index $J$ becomes

$$J = x'(0)Px(0)$$
$$= \left(\zeta + \frac{1+\mu}{4\zeta}\right)x_1^2(0) + x_1(0)x_2(0) + \frac{1+\mu}{4\zeta}x_2^2(0)$$

Substituting the initial conditions $x_1(0) = 1$, $x_2(0) = 0$ into this last equation, we obtain

$$J = \zeta + \frac{1+\mu}{4\zeta}$$

To minimize $J$ with respect to $\zeta$, we set $\partial J/\partial\zeta = 0$, or

$$\frac{\partial J}{\partial\zeta} = 1 - \frac{1+\mu}{4\zeta^2} = 0$$

This yields

$$\zeta = \frac{\sqrt{1+\mu}}{2}$$

Thus the optimal value of $\zeta$ is $\sqrt{1+\mu}/2$. For example, if $\mu = 1$, then the optimal value of $\zeta$ is $\sqrt{2}/2$ or 0.707.

**Optimal control systems based on quadratic performance indexes.** We shall now consider the optimal control problem that, given the system equation

$$\dot{\mathbf{x}} = \mathbf{A}\mathbf{x} + \mathbf{B}\mathbf{u} \tag{16-35}$$

determine the matrix $\mathbf{K}$ of the optimal control vector

$$\mathbf{u}(t) = -\mathbf{K}\mathbf{x}(t) \tag{16-36}$$

so as to minimize the performance index

$$J = \int_0^\infty (\mathbf{x}'\mathbf{Q}\mathbf{x} + \mathbf{u}'\mathbf{R}\mathbf{u})\, dt \tag{16-37}$$

where $\mathbf{Q}$ is a positive-definite (or positive-semidefinite) real symmetric matrix and $\mathbf{R}$ is a positive-definite real symmetric matrix. Note that the second term on the right-hand side of Eq. (16-37) accounts for the expenditure of the energy of the control signals. The matrices $\mathbf{Q}$ and $\mathbf{R}$ determine the relative importance of the error and the expenditure of this energy. In this problem, we assume that the control vector $\mathbf{u}(t)$ is unconstrained.

As mentioned earlier, without proof, the linear control law given by Eq. (16-36) is the optimal control law. Therefore, if the unknown elements of the matrix $\mathbf{K}$ are determined so as to minimize the performance index, then $\mathbf{u}(t) = -\mathbf{K}\mathbf{x}(t)$ is optimal for any initial state $\mathbf{x}(0)$. The block diagram showing the optimal configuration is shown in Fig. 16-8.

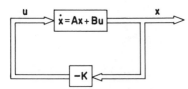

**Fig. 16-8.** Optimal control system.

Now let us solve the optimization problem. Substituting Eq. (16-36) into Eq. (16-35), we obtain

$$\dot{\mathbf{x}} = \mathbf{A}\mathbf{x} - \mathbf{B}\mathbf{K}\mathbf{x} = (\mathbf{A} - \mathbf{B}\mathbf{K})\mathbf{x}$$

In the following derivations, we assume that the matrix $\mathbf{A} - \mathbf{B}\mathbf{K}$ is stable, or that the eigenvalues of $\mathbf{A} - \mathbf{B}\mathbf{K}$ have negative real parts.

Substituting Eq. (16-36) into Eq. (16-37) yields

$$J = \int_0^\infty (\mathbf{x}'\mathbf{Q}\mathbf{x} + \mathbf{x}'\mathbf{K}'\mathbf{R}\mathbf{K}\mathbf{x})\, dt$$

$$= \int_0^\infty \mathbf{x}'(\mathbf{Q} + \mathbf{K}'\mathbf{R}\mathbf{K})\mathbf{x}\, dt$$

Following the discussion given in solving the parameter-optimization problem, we set

$$\mathbf{x}'(\mathbf{Q} + \mathbf{K}'\mathbf{R}\mathbf{K})\mathbf{x} = -\frac{d}{dt}(\mathbf{x}'\mathbf{P}\mathbf{x})$$

Then we obtain

$$\mathbf{x}'(\mathbf{Q} + \mathbf{K}'\mathbf{R}\mathbf{K})\mathbf{x} = -\dot{\mathbf{x}}'\mathbf{P}\mathbf{x} - \mathbf{x}'\mathbf{P}\dot{\mathbf{x}}$$

$$= -\mathbf{x}'[(\mathbf{A} - \mathbf{B}\mathbf{K})'\mathbf{P} + \mathbf{P}(\mathbf{A} - \mathbf{B}\mathbf{K})]\mathbf{x}$$

Comparing both sides of this last equation and noting that this equation must hold true for any $\mathbf{x}$, we require that

$$(\mathbf{A} - \mathbf{B}\mathbf{K})'\mathbf{P} + \mathbf{P}(\mathbf{A} - \mathbf{B}\mathbf{K}) = -(\mathbf{Q} + \mathbf{K}'\mathbf{R}\mathbf{K}) \tag{16-38}$$

By the second method of Liapunov if $\mathbf{A} - \mathbf{BK}$ is a stable matrix, there exists a positive definite matrix $\mathbf{P}$ which satisfies Eq. (16-38). Then, noting that $\mathbf{x}(\infty) = \mathbf{0}$, the performance index can be written

$$J = \mathbf{x}'(0)\mathbf{Px}(0) \qquad (16\text{-}39)$$

The design steps may be stated as follows:

1. Determine the matrix $\mathbf{P}$ which satisfies Eq. (16-38) as a function of $\mathbf{K}$.
2. Substitute the matrix $\mathbf{P}$ into Eq. (16-39). Then the performance index becomes a function of $\mathbf{K}$.
3. Determine the elements of $\mathbf{K}$ so that the performance index $J$ is minimized. The minimization of $J$ with respect to the elements $k_{ij}$ of $\mathbf{K}$ can be accomplished by setting $\partial J/\partial k_{ij}$ equal to zero and solving for the optimal values of $k_{ij}$.

For details of this design approach, see Problem A-16-8. When the number of the elements $k_{ij}$ is not small, this approach is not convenient.

A better approach to the design of the optimal control system is available. Since $\mathbf{R}$ has been assumed a positive definite real symmetric matrix, we can write

$$\mathbf{R} = \mathbf{T'T}$$

where $\mathbf{T}$ is a nonsingular matrix. Then Eq. (16-38) can be written as

$$(\mathbf{A}' - \mathbf{K'B'})\mathbf{P} + \mathbf{P}(\mathbf{A} - \mathbf{BK}) + \mathbf{Q} + \mathbf{K'T'TK} = \mathbf{0}$$

which can be rewritten

$$\mathbf{A'P} + \mathbf{PA} + [\mathbf{TK} - (\mathbf{T'})^{-1}\mathbf{B'P}]'[\mathbf{TK} - (\mathbf{T'})^{-1}\mathbf{B'P}] - \mathbf{PBR}^{-1}\mathbf{B'P} + \mathbf{Q} = \mathbf{0}$$

The minimization of $J$ with respect to $\mathbf{K}$ requires the minimization of

$$\mathbf{x}'[\mathbf{TK} - (\mathbf{T'})^{-1}\mathbf{B'P}]'[\mathbf{TK} - (\mathbf{T'})^{-1}\mathbf{B'P}]\mathbf{x}$$

with respect to $\mathbf{K}$. Since this value is nonnegative, the minimum occurs when it is zero, or when

$$\mathbf{TK} = (\mathbf{T'})^{-1}\mathbf{B'P}$$

Hence

$$\mathbf{K} = \mathbf{T}^{-1}(\mathbf{T'})^{-1}\mathbf{B'P} = \mathbf{R}^{-1}\mathbf{B'P} \qquad (16\text{-}40)$$

Equation (16-40) gives the optimal matrix $\mathbf{K}$. The matrix $\mathbf{P}$ in Eq. (16-40) must satisfy Eq. (16-38) or the following reduced equation:

$$\mathbf{A'P} + \mathbf{PA} - \mathbf{PBR}^{-1}\mathbf{B'P} + \mathbf{Q} = \mathbf{0} \qquad (16\text{-}41)$$

Equation (16-41) is called the reduced-matrix Riccati equation. The design steps may be stated as follows:

1. Solve Eq. (16-41), the reduced-matrix Riccati equation, for the matrix $\mathbf{P}$.
2. Substitute this matrix $\mathbf{P}$ into Eq. (16-40). The resulting matrix $\mathbf{K}$ is the optimal one.

A design example based on this latter approach is given in Example 16-12.

If the matrix $\mathbf{A} - \mathbf{BK}$ is a stable one, the present method always gives the correct result. Kalman has shown that the requirement of $A - BK$ being a stable matrix is equivalent to that of the rank of

$$[S' \mid A'S' \mid \cdots \mid (A')^{n-1}S']$$  (16-42)

being $n$, where the matrix $S$ is defined by

$$S'S = Q$$

This rank condition may conveniently be applied to check if the matrix $A - BK$ is a stable one.

It is important to note, however, that even if $A - BK$ is not a stable matrix, the present method will still give correct results in a special case where the matrix $P$ determined from Eq. (16-41) becomes positive semidefinite; nevertheless it yields $x'(\infty)Px(\infty) = 0$.

Finally, note that if the performance index is given in terms of the output vector rather than the state vector, i.e.,

$$J = \int_0^\infty (y'Qy + u'Ru)\, dt$$

then the index can be modified by using the output equation

$$y = Cx$$

to

$$J = \int_0^\infty (x'C'QCx + u'Ru)\, dt$$

and the same procedure as we discussed here can be applied to obtain the optimal matrix $K$.

---

*Example 16-12.* Consider the system shown in Fig. 16-9. Assuming the control signal to be

$$u(t) = -Kx(t)$$

design the optimal feedback gain matrix $K$ such that the following performance index is minimized:

$$J = \int_0^\infty (x'Qx + u'u)\, dt$$

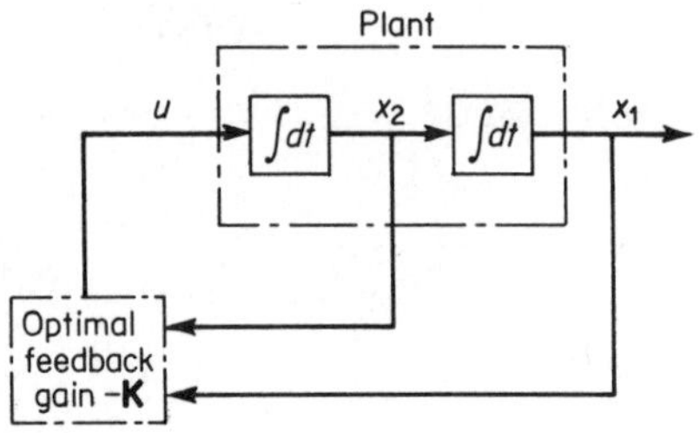

where

**Fig. 16-9.** Control system.

$$Q = \begin{bmatrix} 1 & 0 \\ 0 & \mu \end{bmatrix} \quad (\mu \geq 0)$$

From Fig. 16-9, we see that the equation for the plant is

$$\begin{bmatrix} \dot{x}_1 \\ \dot{x}_2 \end{bmatrix} = \begin{bmatrix} 0 & 1 \\ 0 & 0 \end{bmatrix} \begin{bmatrix} x_1 \\ x_2 \end{bmatrix} + \begin{bmatrix} 0 \\ 1 \end{bmatrix} [u]$$

Noting that

$$Q = \begin{bmatrix} 1 & 0 \\ 0 & \sqrt{\mu} \end{bmatrix} \begin{bmatrix} 1 & 0 \\ 0 & \sqrt{\mu} \end{bmatrix} = S'S$$

we find the rank of the matrix given by (16-42), or

$$[\mathbf{S}' \mid \mathbf{A}'\mathbf{S}'] = \begin{bmatrix} 1 & 0 & 0 & 0 \\ 0 & \sqrt{\mu} & 1 & 0 \end{bmatrix}$$

to be two. Thus, $\mathbf{A} - \mathbf{BK}$ is a stable matrix, and the Liapunov approach presented in this section yields the correct result.

We shall demonstrate the use of the reduced-matrix Riccati equation in the design of the optimal control system. Let us solve Eq. (16-41), rewritten as

$$\mathbf{A}'\mathbf{P} + \mathbf{PA} - \mathbf{PBR}^{-1}\mathbf{B}'\mathbf{P} + \mathbf{Q} = \mathbf{0}$$

or

$$\begin{bmatrix} 0 & 0 \\ 1 & 0 \end{bmatrix}\begin{bmatrix} p_{11} & p_{12} \\ p_{12} & p_{22} \end{bmatrix} + \begin{bmatrix} p_{11} & p_{12} \\ p_{12} & p_{22} \end{bmatrix}\begin{bmatrix} 0 & 1 \\ 0 & 0 \end{bmatrix}$$
$$- \begin{bmatrix} p_{11} & p_{12} \\ p_{12} & p_{22} \end{bmatrix}\begin{bmatrix} 0 \\ 1 \end{bmatrix}[1][0 \quad 1]\begin{bmatrix} p_{11} & p_{12} \\ p_{12} & p_{22} \end{bmatrix} + \begin{bmatrix} 1 & 0 \\ 0 & \mu \end{bmatrix} = \begin{bmatrix} 0 & 0 \\ 0 & 0 \end{bmatrix}$$

This equation can be simplified to

$$\begin{bmatrix} 0 & 0 \\ p_{11} & p_{12} \end{bmatrix} + \begin{bmatrix} 0 & p_{11} \\ 0 & p_{12} \end{bmatrix} - \begin{bmatrix} p_{12}^2 & p_{12}p_{22} \\ p_{12}p_{22} & p_{22}^2 \end{bmatrix} + \begin{bmatrix} 1 & 0 \\ 0 & \mu \end{bmatrix} = \begin{bmatrix} 0 & 0 \\ 0 & 0 \end{bmatrix}$$

from which we obtain the following three equations:

$$1 - p_{12}^2 = 0$$
$$p_{11} - p_{12}p_{22} = 0$$
$$\mu + 2p_{12} - p_{22}^2 = 0$$

Solving these three simultaneous equations for $p_{11}, p_{12}$, and $p_{22}$, requiring $\mathbf{P}$ to be positive definite, we obtain

$$\mathbf{P} = \begin{bmatrix} p_{11} & p_{12} \\ p_{12} & p_{22} \end{bmatrix} = \begin{bmatrix} \sqrt{\mu + 2} & 1 \\ 1 & \sqrt{\mu + 2} \end{bmatrix}$$

The optimal feedback gain matrix $\mathbf{K}$ is then obtained from

$$\mathbf{K} = \mathbf{R}^{-1}\mathbf{B}'\mathbf{P}$$
$$= [1][0 \quad 1]\begin{bmatrix} p_{11} & p_{12} \\ p_{12} & p_{22} \end{bmatrix}$$
$$= [p_{12} \quad p_{22}]$$
$$= [1 \quad \sqrt{\mu + 2}]$$

Thus the optimal control signal is

$$u = -\mathbf{Kx} = -x_1 - \sqrt{\mu + 2}\,x_2 \qquad (16\text{-}43)$$

Note that the control law given by Eq. (16-43) yields an optimal result for any initial state under the given performance index. Figure 16-10 is the block diagram for this system.

### Concluding comments

1. The characteristic of an optimal control law based on a quadratic performance index is that it is a linear function of the state variables, which implies that we need to feed back all state variables. This requires that all such variables be available for feedback. It is, therefore, desirable to represent the

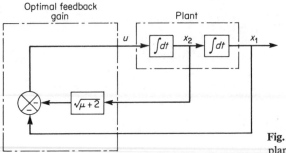

**Fig. 16-10.** Optimal control of the plant shown in Fig. 16-9.

system in terms of measurable state variables. (In complicated systems, it is rather unlikely that we are able to measure all state variables. Then we must estimate the unmeasurable ones and use these estimated values to generate optimal control signals.) It is important to note that we should avoid differentiating a state variable to generate another one. Differentiation of a signal always decreases the signal-to-noise ratio because noise generally fluctuates more rapidly than the command signal. Sometimes the signal-to-noise ratio may be decreased by several times by a single differentiation process. It is generally desirable to avoid differentiation processes; for example, if the acceleration of the output displacement is needed, it is better to measure the acceleration directly by use of an accelerometer rather than differentiating the velocity signal.

2. When the optimal control system is designed in the time domain, it is desirable to investigate the frequency-response characteristics to compensate for noise effects. The system frequency-response characteristics must be such that the system attenuates highly in the frequency range where noise and resonance of components are expected. (To compensate for noise effects we must in some cases either modify the optimal configuration and accept suboptimal performance or modify the performance index.)

3. If the upper limit of integration in the performance index $J$ given by Eq. (16-37) is finite, then it can be shown that the optimal control vector is still a linear function of the state variables and the reference inputs, but with time-varying coefficients. (Therefore, the determination of the optimal control vector involves that of optimal time-varying matrices.)

### 16-6   MODEL-REFERENCE CONTROL SYSTEMS

In Chapter 10, we presented design and compensation techniques of linear time-invariant control systems. Since all physical plants are nonlinear to some degree, the designed system will behave satisfactorily only over a limited range of operation.

If the assumption of linearity in the plant equation is removed, the design techniques of Chapter 10 are not applicable. In such a case the model-reference approach to the system design presented in this section may be useful.

**Model-reference control systems.** One of the useful methods for specifying system performance is by means of a model which will produce the desired output for a given input. The model need not be actual hardware. It can be only a mathematical model simulated on a computer. In a model-reference control system the output of the model and that of the plant are compared and the difference is used to generate the control signals.

Model-reference control has been used to obtain acceptable performance in some very difficult control situations involving nonlinearity and/or time-varying parameters.

**Design of a controller.*** We shall assume that the plant is characterized by the following state equation:

$$\dot{\mathbf{x}} = \mathbf{f}(\mathbf{x}, \mathbf{u}, t) \tag{16-44}$$

where

$\mathbf{x}$ = state vector ($n$-dimensional real vector)

$\mathbf{u}$ = control vector ($r$-dimensional real vector)

$\mathbf{f}$ = vector-valued function

It is desired that the control system follow closely some model system. Our design problem here is to synthesize a controller which always generates a signal which forces the plant state toward the model state. Figure 16-11 is the block diagram showing the system configuration.

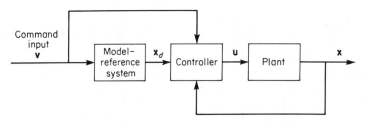

**Fig. 16-11.** Model-reference control system.

We shall assume that the model-reference system is linear and described by

$$\dot{\mathbf{x}}_d = \mathbf{A}\mathbf{x}_d + \mathbf{B}\mathbf{v} \tag{16-45}$$

where

$\mathbf{x}_d$ = state vector of the model ($n$-dimensional real vector)

$\mathbf{v}$ = input vector ($r$-dimensional real vector)

$\mathbf{A}$ = $n \times n$ constant matrix

$\mathbf{B}$ = $n \times r$ constant matrix

We assume that the eigenvalues of $\mathbf{A}$ have negative real parts so that the model-reference system has an asymptotically stable equilibrium state.

*References M-8 and V-1.

Let us define the error vector **e** by

$$\mathbf{e} = \mathbf{x}_d - \mathbf{x} \tag{16-46}$$

In the present problem, we wish to reduce the error vector to zero by a suitable control vector **u**. From Eqs. (16-44), (16-45), and (16-46), we obtain

$$\dot{\mathbf{e}} = \dot{\mathbf{x}}_d - \dot{\mathbf{x}} = \mathbf{A}\mathbf{x}_d + \mathbf{B}\mathbf{v} - \mathbf{f}(\mathbf{x}, \mathbf{u}, t)$$
$$= \mathbf{A}\mathbf{e} + \mathbf{A}\mathbf{x} - \mathbf{f}(\mathbf{x}, \mathbf{u}, t) + \mathbf{B}\mathbf{v} \tag{16-47}$$

Equation (16-47) is a differential equation for the error vector.

We now design a controller such that at steady state $\mathbf{x} = \mathbf{x}_d$ and $\dot{\mathbf{x}} = \dot{\mathbf{x}}_d$, or $\mathbf{e} = \dot{\mathbf{e}} = \mathbf{0}$. Thus the origin $\mathbf{e} = \mathbf{0}$ will be an equilibrium state.

A convenient starting point in the synthesis of the control vector **u** is the construction of a Liapunov function for the system given by Eq. (16-47).

Let us assume that the form of the Liapunov function is

$$V(\mathbf{e}) = \mathbf{e}'\mathbf{P}\mathbf{e}$$

where **P** is a positive-definite real symmetric matrix. Taking the derivative of $V(\mathbf{e})$ with respect to time gives

$$\dot{V}(\mathbf{e}) = \dot{\mathbf{e}}'\mathbf{P}\mathbf{e} + \mathbf{e}'\mathbf{P}\dot{\mathbf{e}}$$
$$= [\mathbf{e}'\mathbf{A}' + \mathbf{x}'\mathbf{A}' - \mathbf{f}'(\mathbf{x}, \mathbf{u}, t) + \mathbf{v}'\mathbf{B}']\mathbf{P}\mathbf{e} + \mathbf{e}'\mathbf{P}[\mathbf{A}\mathbf{e} + \mathbf{A}\mathbf{x} - \mathbf{f}(\mathbf{x}, \mathbf{u}, t) + \mathbf{B}\mathbf{v}]$$
$$= \mathbf{e}'(\mathbf{A}'\mathbf{P} + \mathbf{P}\mathbf{A})\mathbf{e} + 2M \tag{16-48}$$

where

$$M = \mathbf{e}'\mathbf{P}[\mathbf{A}\mathbf{x} - \mathbf{f}(\mathbf{x}, \mathbf{u}, t) + \mathbf{B}\mathbf{v}] = \text{scalar quantity}$$

The assumed $V(\mathbf{e})$ function is a Liapunov function if

1. $\mathbf{A}'\mathbf{P} + \mathbf{P}\mathbf{A} = -\mathbf{Q}$ is a negative-definite matrix.
2. The control vector **u** can be chosen to make the scalar quantity $M$ nonpositive.

Then, noting that $V(\mathbf{e}) \rightarrow \infty$ as $\|\mathbf{e}\| \rightarrow \infty$, we see that the equilibrium state $\mathbf{e} = \mathbf{0}$ is asymptotically stable in the large. Condition 1 can always be met by a proper choice of **P**, since the eigenvalues of **A** are assumed to have negative real parts. The problem here is to choose an appropriate control vector **u** so that $M$ is either zero or negative.

We shall illustrate the application of the present approach to the design of a nonlinear controller by means of an example.

---

*Example 16-13.* Consider a nonlinear time-varying plant described by

$$\begin{bmatrix} \dot{x}_1 \\ \dot{x}_2 \end{bmatrix} = \begin{bmatrix} 0 & 1 \\ -b & -a(t)x_2 \end{bmatrix} \begin{bmatrix} x_1 \\ x_2 \end{bmatrix} + \begin{bmatrix} 0 \\ 1 \end{bmatrix} [u]$$

where $a(t)$ is time-varying and $b$ is a positive constant. Assuming the reference model equation to be

$$\begin{bmatrix} \dot{x}_{d1} \\ \dot{x}_{d2} \end{bmatrix} = \begin{bmatrix} 0 & 1 \\ -\omega_n^2 & -2\zeta\omega_n \end{bmatrix} \begin{bmatrix} x_{d1} \\ x_{d2} \end{bmatrix} + \begin{bmatrix} 0 \\ \omega_n^2 \end{bmatrix} [v] \tag{16-49}$$

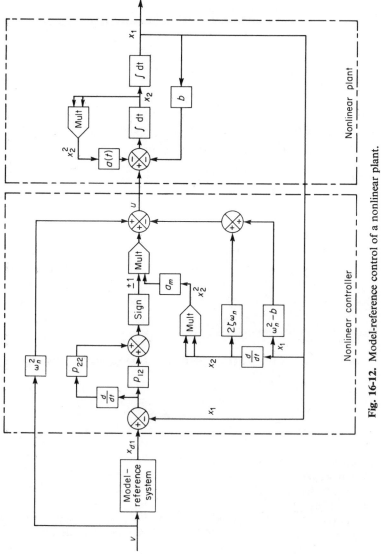

**Fig. 16-12.** Model-reference control of a nonlinear plant.

design a nonlinear controller which will give stable operation of the system.

Define the error vector by

$$\mathbf{e} = \mathbf{x}_d - \mathbf{x}$$

and a Liapunov function by

$$V(\mathbf{e}) = \mathbf{e}'\mathbf{P}\mathbf{e}$$

where $\mathbf{P}$ is a positive-definite real symmetric matrix. Then, referring to Eq. (16-48) we obtain $\dot{V}(\mathbf{e})$ as

$$\dot{V}(\mathbf{e}) = \mathbf{e}'(\mathbf{A}'\mathbf{P} + \mathbf{P}\mathbf{A})\mathbf{e} + 2M$$

where

$$M = \mathbf{e}'\mathbf{P}[\mathbf{A}\mathbf{x} - \mathbf{f}(\mathbf{x}, \mathbf{u}, t) + \mathbf{B}\mathbf{v}]$$

By identifying the matrices $\mathbf{A}$ and $\mathbf{B}$ from Eq. (16-49) and choosing the matrix $\mathbf{Q}$ to be

$$\mathbf{Q} = \begin{bmatrix} q_{11} & 0 \\ 0 & q_{22} \end{bmatrix} = \text{positive definite}$$

we obtain

$$\dot{V}(\mathbf{e}) = -(q_{11}e_1^2 + q_{22}e_2^2) + 2M$$

where

$$M = [e_1 \quad e_2] \begin{bmatrix} p_{11} & p_{12} \\ p_{12} & p_{22} \end{bmatrix} \left\{ \begin{bmatrix} 0 & 1 \\ -\omega_n^2 & -2\zeta\omega_n \end{bmatrix} \begin{bmatrix} x_1 \\ x_2 \end{bmatrix} \right.$$
$$- \begin{bmatrix} 0 & 1 \\ -b & -a(t)x_2 \end{bmatrix} \begin{bmatrix} x_1 \\ x_2 \end{bmatrix} - \begin{bmatrix} 0 \\ u \end{bmatrix} + \left. \begin{bmatrix} 0 \\ \omega_n^2 v \end{bmatrix} \right\}$$
$$= (e_1 p_{12} + e_2 p_{22})[-(\omega_n^2 - b)x_1 - 2\zeta\omega_n x_2 + a(t)x_2^2 + \omega_n^2 v - u]$$

If we choose $u$ so that

$$u = -(\omega_n^2 - b)x_1 - 2\zeta\omega_n x_2 + \omega_n^2 v + a_m x_2^2 \operatorname{sign}(e_1 p_{12} + e_2 p_{22}) \qquad (16\text{-}50)$$

where

$$a_m = \max|a(t)|$$

then

$$M = (e_1 p_{12} + e_2 p_{22})[a(t) - a_m \operatorname{sign}(e_1 p_{12} + e_2 p_{22})]x_2^2 = \text{nonpositive}$$

With the control function $u$ given by Eq. (16-50), the equilibrium state $\mathbf{e} = \mathbf{0}$ is asymptotically stable in the large. Thus Eq. (16-50) defines a nonlinear control law that will yield an asymptotically stable operation. The block diagram for the present control system is shown in Fig. 16-12.

Note that the rate of convergence of the transient response depends on the matrix $\mathbf{P}$, which in turn depends on the matrix $\mathbf{Q}$ chosen at the start of the design.


## 16-7  ADAPTIVE CONTROL SYSTEMS

In recent years, interest in adaptive control systems has increased rapidly along with interest and progress in control topics in general. The term *adaptive system* has a variety of specific meanings, but it usually implies that the system is capable of accommodating unpredictable environmental changes, whether these changes arise within the system or external to it. This concept has a great deal of appeal to the systems designer since a highly adaptive system, besides accommodating environmental changes, would also accommodate moderate engineering design errors or uncertainties and would compensate for the failure of minor system components, thereby increasing system reliability.

We shall first present some basic concepts of adaptive control systems and explain what such systems are. Then we shall discuss the necessary functions a controller must perform in order that it be called adaptive. Finally, we shall introduce some concepts of learning systems.

**Introduction.** In most feedback control systems, small deviations in parameter values from their design values will not cause any problem in the normal operation of the system, provided these parameters are inside the loop. If plant parameters vary widely according to environmental changes, however, then the control system may exhibit satisfactory response for one environmental condition but may fail to provide satisfactory performance under other conditions. In certain cases, large variations of plant parameters may even cause instability.

In the simplest analysis, one may consider different sets of values of the plant parameters. It is then desirable to design a control system which works well for all sets. As soon as this demand is formulated, the strict optimal control problem loses its importance. By asking for good performance over a range, we have to abandon the best performance for one parameter set.

If the plant transfer function can be identified continuously, then we can compensate variations in the transfer function of the plant simply by varying adjustable parameters of the controller and thereby obtain satisfactory system performance continuously under various environmental conditions. Such an adaptive approach is quite useful to cope with a problem where the plant is normally exposed to varying environments so that plant parameters change from time to time. (Since changes are not predictable in most practical cases, a fixed-parameter or preprogrammed time-varying controller cannot provide an answer.)

**Definition of adaptive control systems.** Adaptation is a fundamental characteristic of living organisms since they attempt to maintain physiological equilibrium in the midst of changing environmental conditions. An approach to the design of adaptive systems is then to consider the adaptive aspects of human or animal behavior and to develop systems which behave somewhat analogously.

There are different definitions of adaptive control systems now in use in the literature. The vagueness surrounding most definitions and classifications of adaptive systems is due to the large variety of mechanisms by which adaptation may be achieved and to a failure to differentiate between the external mainfestations of adaptive behavior and the internal mechanisms used to achieve it. Primarily, different definitions arise because of the various classifications and delineations which divide control systems into those which are adaptive and those which are not. (The small degree of adaptivity required by most system specifications could be achieved by the use of familiar feedback techniques with fixed gains, compensators, and, in some cases, nonlinearities.)

We shall find it necessary to define adaptive system characteristics which are fundamentally different from those of conventional feedback systems so that we may restrict our attention to only those unique aspects of adaptive system behavior and design. In this book we define adaptive control systems as follows:

*Definition.* An adaptive control system is one which continuously and automatically measures the dynamic characteristics (such as the transfer function) of the plant, compares them with the desired dynamic characteristics, and uses the difference to vary adjustable system parameters (usually controller characteristics) or to generate an actuating signal so that the optimal performance can be maintained regardless of the environmental changes; alternatively, such a system may continuously measure its own performance according to a given performance index and modify, if necessary, its own parameters so as to maintain optimal performance regardless of the environmental changes.

To be called an adaptive system, self-organizing features must exist. If the adjustment of the system parameters is done only by direct measurement of the environment, the system is not adaptive.

A seemingly adaptive system is exemplified by the aircraft autopilot, which is designed to adjust its loop gains as a function of altitude in order to compensate for corresponding changes in aircraft parameters. The adjustment is based on direct information about the environment (in this case, atmospheric pressure) and not on a self-organizing scheme. These systems do not possess any self-organizing features and therefore are essentially conventional closed-loop systems.

Another example of systems which may seem to be adaptive but really are not exists in the field of model-reference systems. Some of these systems (such as the one considered in Example 16-13) merely use the difference between model response and plant response as an input signal to the plant, as shown in Fig. 16-13 (a). These systems cannot be considered truly adaptive since block-diagram

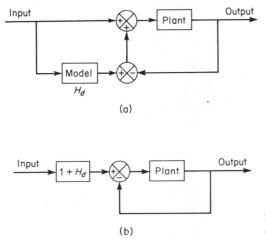

(a)

(b)

Fig. 16-13. (a) Model-reference system; (b) simplified block diagram.

manipulation will, in this case, reduce the configuration to that of Fig. 16-13 (b), which is simply a basic feedback loop with a prefilter. (Note that some authors call this type of control *model adaptation.* The model is either a physical model or a simulated system on the computer. The model has no variable parameters.)

**Performance indexes.** The very basis of adaptive control rests in the premise that there is some condition of operation or performance for the system which is better than any other. Thus it becomes necessary to define what constitutes optimal performance. In adaptive control systems, such performance is defined in terms of the performance index, which we must decide upon after stating our goals. These goals will be as diverse as the systems to which they are applied, but it is usually possible to generalize the object of optimization to that of minimizing the cost of operation, or maximizing the profit.

Some general characteristics which are usually considered desirable are

1. reliability
2. selectivity
3. applicability

Hence, the performance index should be reliable, or it should be a uniform measure of "goodness" for systems of all orders. It should be selective, or it should involve a sharply defined optimum as a function of system parameters. It should not have local optima or saddle points. The performance index should be easily applicable to practical systems and should be readily measurable.

If the performance index assumes a value of zero at the optimal operating condition, rather than a maximum or a minimum, then it can be used as the adaptive-loop error signal and can be used for feedback directly in some systems.

Note that, in general, all the mathematically tractable performance indexes have one serious drawback in common; although they specify the cost of system operation in terms of error and energy, they do not give us information about the transient-response characteristics of the system. Thus a system that is designed to operate optimally from the standpoint of maximum "profit" may have undesirable transient characteristics or may even be unstable. Therefore, to assure satisfactory response characteristics, we may need secondary criteria relating to response characteristics in order to influence the choice of cost weighting elements.

Finally, remember that the performance index used in an adaptive control system defines an optimal performance for that system. This means that the performance index essentially gives the upper limit of the performance of the system. Therefore, the selection of a proper performance index is most important.

**Adaptive controllers.** An adaptive controller may consist of the following three functions:

1. Identification of dynamic characteristics of the plant.
2. Decision making based on the identification of the plant.
3. Modification or actuation based on the decision made.

If the plant is imperfectly known, perhaps because of random time-varying parameters or because of the effects of environmental changes on the plant dynamic characteristics, then the initial identification, decision, and modification procedures will not be sufficient to minimize (or maximize) the performance index. It then becomes necessary to carry out these procedures continuously or at intervals of time, depending upon how fast the plant parameters are changing. This constant "self-redesign" or self-organization of the system to compensate for unpredictable changes in the plant is the aspect of performance which is usually considered in defining an adaptive control system.

A block diagram representation of an adaptive control system is shown in Fig. 16-14. In this system, the plant is identified and the performance index measured

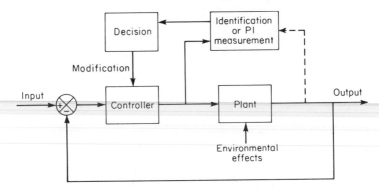

**Fig. 16-14.** Block diagram representation of an adaptive control system.

continuously or periodically. Once these have been accomplished, the performance index is compared with the optimal and a decision is made based on the findings as to how to modify the actuating signal. Since the plant is identified within the system itself, adjustment of the parameters is a closed-loop operation. Note that in such closed-loop adaptation the question of stability may arise.

In the following, we shall explain in some detail the three functions: identification, decision, and modification.

**Identification of the dynamic characteristics of the plant.** The dynamic characteristics of the plant must be measured and identified continuously or at least very frequently. This should be accomplished without affecting the normal operation of the system. To identify the characteristics of a system, we must perform a test and analyze the results. (For a control system, this entails imposing a control signal on the plant and analyzing system response.) Identification may be made from normal operating data of the plant or by use of test signals, such as sinusoidal ones of small amplitude or various stochastic signals of small amplitude. In practice, no direct application of step or impulse inputs can be made. (Except in certain special cases, the plant will be in normal operation during the test so that the test signals imposed should not unduly disturb normal outputs; furthermore, normal inputs and system noise should not disturb and confuse the test.) Normal inputs are ideal as test signals since no difficulties with undesired outputs or confusing inputs will arise. However, identification with normal inputs is only possible when they have adequate signal characteristics (bandwidth, amplitude, etc.) for proper identification.

Stochastic test signals are quite convenient in certain applications. By using crosscorrelation techniques, we may analyze the output as a function of the stochastic input in order to determine the response characteristics. With a stochastic input, the excitation energy can be spread over a band of frequencies making the effect tolerable. Also, since the crosscorrelator can be designed to closely correlate inputs and outputs, the level of the test signal may be kept low.

Identification must not take too long since if it does, further variations of the plant parameter may occur. Identification time should be sufficiently short compared with the rate of environmental changes. With time for identification limited, it is usually impossible to identify the plant completely; the best one can expect is only partial identification.

It is important to note that not all adaptive systems require identification explicitly. Some systems have already been identified to the extent that the measurement of the value of the performance index can indicate what controller parameters must be changed. That is, the system is fairly well known so that a measurement of the performance index completes the identification.

On the other hand, if plant identification is very difficult, we need to measure directly the performance index and build an adaptive controller based on it. If identification is not called for and adaptivity is based on measurements of the performance index alone, the control system is called an optimalizing control system. Since self-organization is achieved with this approach, we shall consider such a system adaptive.

The difficulty of making a realistic identification will depend upon how much information about the plant is required and upon the amount of prior knowledge of the plant. In general, these are also the factors which will determine whether to use an identification approach or a direct search of the controller parameter space as a function of the performance index, which is discussed later under optimalizing control systems.

**Decision making based on the identification of the plant.** Decision here refers to one made on the basis of plant characteristics which have been identified and on the performance index computed.

Once the plant has been identified, it is compared with the optimal characteristics (or optimal performance), and then a decision must be made as to how the adjustable parameters (controller characteristics) should be varied in order to maintain optimal performance. The decision is accomplished by a computer.

**Modification based on the decision made.** Modification refers to the change of control signals according to the results of the identification and decision. In most schemes, the decision and modification are conceptually a single operation with the modification consisting of a means of mechanizing the transformation of a decision output signal into the control signal (the input to the plant).

This control signal, or the input signal to the plant, can be modified in two ways. The first approach is to adjust the controller parameters in order to compensate for changes in the plant dynamics. This is called controller parameter modification. The second approach is to synthesize the optimal control signal, based on the plant transfer function, performance index, and desired transient response. This is called control-signal synthesis.

The choice between controller parameter modification and control-signal synthesis is primarily a hardware decision since the two approaches are conceptually equivalent. Where reliability is very important, as in aerospace applications, the use of parameter change adaptation is often favored over the use of control-signal synthesis. (This is because the system can operate even after the failure of the adaptive loop if the control signal is not entirely dependent upon the adaptive portion of the system.) Improvements in components, however, especially in complex electronic equipments, may swing the preference to signal synthesis.

**Optimalizing control systems.** Optimalizing control systems heavily rely on optimization techniques. In general, optimization consists of searching the space of variable controller parameters as a function of some performance index to determine where the performance index is maximized or minimized. Implicit in the previous statement is the fact that a scalar performance index which is a function of system outputs can be defined so that its extremum represents the best possible system performance. This is generally possible and necessary for any adaptive control system.

The methods of finding the optimal operating point are basically trial-and-error procedures. In the steepest-descent approach, the gradient of the performance index surface is measured by noting the effects of small changes in the variable parameters. (This may be called the derivative sensing method.) The parameter vector is then moved in the direction of maximum slope, either a fixed amount or else an amount determined by the surface gradient. Where the parameters are slowly varying, the gradient may be calculated relatively infrequently. Under conditions of more rapid parameter changes, however, a procedure known as alternate biasing is superior to the derivative-sensing method. With alternate biasing, the system is never operated at the optimal condition but is alternately operated a fixed distance on either side of the calculated optimum, and a new optimum is calculated from the difference in the values of the performance index.

Probably the greatest advantage of the optimalizing control approach is that no restrictions have been placed on the plant. It may be nonlinear, mutliple-input–multiple-output, time-varying, etc. A major difficulty of this optimization approach is that no satisfactory method has been found for discriminating against local extremum. Therefore, this approach is useful for any physical process whose performance surface has a single optimum and whose variations are slow enough for the control system to accommodate them.

**Learning systems.** One significant difference between a skilled human operator and the adaptive controller discussed above is that the human operator recognizes familiar inputs and can use his past learned experiences in order to react in an optimal manner. Adaptive control systems are designed to modify the control signal as the system environment changes so that performance is always optimal.

A system which is capable of recognizing the familiar feature and patterns of a situation and which uses its past learned experiences in behaving in an optimal fashion is called a *learning system.*

A learning system is a higher-level system than an adaptive system. Gibson divides the space of all control systems into four basic hierarchal levels:

1. open loop
2. closed loop
3. adaptive loop
4. learning loop

where each level is responsive to a performance index or control error measured at the next lower level and where levels higher than the fourth will exist for more complex environments.

A learning system responding to a familiar situation will not require identification of the system. The approach to the design of such a system is to "teach" the system the best choice for each situation. Once the system has learned the optimal control law for each possible situation, it may operate near the optimal condition regardless of environmental changes.

A learning system, when subjected to a new situation, learns how to behave by an adaptive approach. If the system experiences the same situation it has learned before, it will recognize it and behave optimally without going through the same adaptive approach. Models of human behavior which are being developed by many researchers will undoubtedly yield useful results for application to learning systems.

**Concluding comments.** Recent developments of high performance aerospace vehicles and high efficiency manufacturing plants impose more and more stringent requirements on their associated control systems. In designing such control systems, we are concerned with the development of systems which will meet the specifications imposed by the users under the anticipated operating conditions. Most control systems which require exacting performance over a wide range of operating conditions will necessarily be adaptive to some extent. When high adaptivity is clearly called for, most present-day requirements will be met by an identification-decision-modification system with either sequential or continuous modification, depending upon the rate of change of the varying parameters.

The most fascinating developments in adaptive control systems lie in the areas of pattern recognition and learning systems. Pattern recognition techniques may someday answer the need for a general identification procedure. When tied in with learning approaches, they may appreciably reduce the "time barrier" which presently plagues many adaptive control systems.

This section has presented only an outline of adaptive control systems. The reader interested in such systems should refer to recent research results available in the literature.

EXAMPLE PROBLEMS AND SOLUTIONS

**PROBLEM A-16-1.** Show that the following system

$$\dot{x} = Ax + Bu$$
$$y = Cx$$

where

$$
\mathbf{x} = \begin{bmatrix} x_1 \\ x_2 \\ x_3 \end{bmatrix}, \qquad
\mathbf{A} = \begin{bmatrix} -1 & -2 & -2 \\ 0 & -1 & 1 \\ 1 & 0 & -1 \end{bmatrix}, \qquad
\mathbf{B} = \begin{bmatrix} 2 \\ 0 \\ 1 \end{bmatrix}
$$

$$
\mathbf{C} = \begin{bmatrix} 1 & 1 & 0 \end{bmatrix}
$$

is completely state controllable and completely observable.

**Solution.** For complete state controllability, the vectors $\mathbf{B}$, $\mathbf{AB}$, and $\mathbf{A}^2\mathbf{B}$ must be linearly independent. For this system, we obtain

$$
\mathbf{B} = \begin{bmatrix} 2 \\ 0 \\ 1 \end{bmatrix}, \qquad
\mathbf{AB} = \begin{bmatrix} -4 \\ 1 \\ 1 \end{bmatrix}, \qquad
\mathbf{A}^2\mathbf{B} = \begin{bmatrix} 0 \\ 0 \\ -5 \end{bmatrix}
$$

Since

$$
\begin{vmatrix} 2 & -4 & 0 \\ 0 & 1 & 0 \\ 1 & 1 & -5 \end{vmatrix} = -10
$$

the rank of the matrix $[\mathbf{B} \mid \mathbf{AB} \mid \mathbf{A}^2\mathbf{B}]$ is three, and the system is completely state controllable.

For complete observability, the vectors $\mathbf{C}'$, $\mathbf{A}'\mathbf{C}'$, and $(\mathbf{A}')^2\mathbf{C}'$ must be linearly independent. For this system,

$$
\mathbf{C}' = \begin{bmatrix} 1 \\ 1 \\ 0 \end{bmatrix}, \qquad
\mathbf{A}'\mathbf{C}' = \begin{bmatrix} -1 \\ -3 \\ -1 \end{bmatrix}, \qquad
(\mathbf{A}')^2\mathbf{C}' = \begin{bmatrix} 0 \\ 5 \\ 0 \end{bmatrix}
$$

Note that

$$
\begin{vmatrix} 1 & -1 & 0 \\ 1 & -3 & 5 \\ 0 & -1 & 0 \end{vmatrix} = 5
$$

Therefore the rank of the matrix $[\mathbf{C}' \mid \mathbf{A}'\mathbf{C}' \mid (\mathbf{A}')^2\mathbf{C}']$ is three, and the system is completely observable.

**PROBLEM A-16-2.** Show that the continuous-time system described by

$$
\begin{bmatrix} \dot{x}_1 \\ \dot{x}_2 \end{bmatrix} = \begin{bmatrix} 0 & 1 \\ -1 & 0 \end{bmatrix} \begin{bmatrix} x_1 \\ x_2 \end{bmatrix} + \begin{bmatrix} 0 \\ 1 \end{bmatrix} [u] \tag{16-51}
$$

$$
y = \begin{bmatrix} 1 & 0 \end{bmatrix} \begin{bmatrix} x_1 \\ x_2 \end{bmatrix} \tag{16-52}
$$

is completely state controllable and completely observable.

The discrete-time system which is equivalent to the continuous-time system of Eqs. (16-51) and (16-52) is given by

$$
\begin{bmatrix} x_1((k+1)T) \\ x_2((k+1)T) \end{bmatrix} = \begin{bmatrix} \cos T & \sin T \\ -\sin T & \cos T \end{bmatrix} \begin{bmatrix} x_1(kT) \\ x_2(kT) \end{bmatrix} + \begin{bmatrix} 1 - \cos T \\ \sin T \end{bmatrix} [u(kT)]
$$

$$
y(kT) = \begin{bmatrix} 1 & 0 \end{bmatrix} \begin{bmatrix} x_1(kT) \\ x_2(kT) \end{bmatrix}
$$

where $T$ is the sampling period. Show that the discrete-time system is completely state controllable and completely observable if and only if $T \neq n\pi$ $(n = 0, 1, 2, \ldots)$.

**Solution.** For the continuous-time system described by Eqs. (16-51) and (16-52),

$$[\mathbf{B} \mid \mathbf{AB}] = \begin{bmatrix} 0 & 1 \\ 1 & 0 \end{bmatrix}$$

The rank of $[\mathbf{B} \mid \mathbf{AB}]$ is two and the system is completely state controllable. Now

$$[\mathbf{C}' \mid \mathbf{A}'\mathbf{C}'] = \begin{bmatrix} 1 & 0 \\ 0 & 1 \end{bmatrix}$$

The rank of $[\mathbf{C}' \mid \mathbf{A}'\mathbf{C}']$ is also two and the system is completely observable.

For the discrete-time system, we have

$$[\mathbf{H} \mid \mathbf{GH}] = \begin{bmatrix} 1 - \cos T & 1 + \cos T - 2\cos^2 T \\ \sin T & -\sin T + 2\cos T \sin T \end{bmatrix}$$

The rank of the matrix $[\mathbf{H} \mid \mathbf{GH}]$ is two if and only if $T \neq n\pi$ $(n = 0, 1, 2, \ldots)$. The rank of

$$[\mathbf{C}' \mid \mathbf{G}'\mathbf{C}'] = \begin{bmatrix} 1 & \cos T \\ 0 & \sin T \end{bmatrix}$$

is two if and only if $T \neq n\pi$ $(n = 0, 1, 2, \ldots)$. Hence the discrete-time system is completely state controllable and completely observable if and only if $T \neq n\pi$ $(n = 0, 1, 2, \ldots)$.

It is important to point out that the sampling process impairs the controllability and observability. (Physically, in the present case the continuous output signal is a sinusoidal signal, but the sampled output signal is zero if $T = n\pi$.) Note that we can always avoid the loss of controllability and observability by choosing the sampling period sufficiently small compared with the smallest time constant of the system.

**PROBLEM A-16-3.** Consider the discrete-time system discussed in Example 16-9. It was stated that any initial state can be brought to the origin in at most two sampling periods if the magnitude of $u(kT)$ is not bounded. If, however, the magnitude of $u(kT)$ is bounded, then some initial states cannot be transferred to the origin in two sampling periods. They may require three, four, or more sampling periods.

Suppose that the magnitude of $u(kT)$ is bounded, or

$$|u(kT)| \leq 1$$

Find the regions of initial states in the $x_1$-$x_2$ plane which can be transferred to the origin in one sampling period and two sampling periods, respectively. Assume that $T = 1$ sec.

**Solution.** If we require $\mathbf{x}(T) = \mathbf{0}$, then from Eq. (16-31), we have

$$\begin{bmatrix} x_1(0) \\ x_2(0) \end{bmatrix} = \begin{bmatrix} 0.72 \\ -1.72 \end{bmatrix}[u(0)]$$

Since $|u(0)| \leq 1$, we obtain

$$|x_1(0)| \leq 0.72, \qquad |x_2(0)| \leq 1.72$$

Hence if the initial state lies on the line segment

$$1.72x_1(0) + 0.72x_2(0) = 0, \qquad -0.72 \leq x_1(0) \leq 0.72$$

it can be brought to the origin in one sampling period, or 1 sec. This line segment is shown in Fig. 16-15.

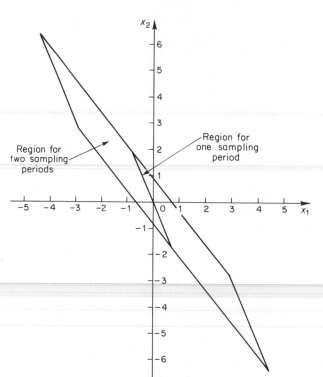

**Fig. 16-15.** Regions from which initial states can be brought to the origin in one or two sampling periods.

If we require $\mathbf{x}(2) = \mathbf{0}$, then from Eq. (16-30) we obtain

$$u(0) = -1.58x_1(0) - 1.24x_2(0)$$
$$u(1) = 0.58x_1(0) + 0.24x_2(0)$$

Since $|u(0)| \leq 1$ and $|u(1)| \leq 1$, we obtain the following four relationships:

$$1.58x_1(0) + 1.24x_2(0) \leq 1 \tag{16-53}$$
$$1.58x_1(0) + 1.24x_2(0) \geq -1 \tag{16-54}$$
$$0.58x_1(0) + 0.24x_2(0) \leq 1 \tag{16-55}$$
$$0.58x_1(0) + 0.24x_2(0) \geq -1 \tag{16-56}$$

The region bounded by Eqs. (16-53) through (16-56) is shown in Fig. 16-15. If the initial state lies within this region, then it can be transferred to the origin in two sampling periods, or 2 sec.

**PROBLEM A-16-4.** Consider the time-optimal control system shown in Fig. 16-2. Referring to Fig. 16-4, obtain the equation for the switching curve $AOB$. If the maximum torque is increased, how is the shape of the switching curve altered?

Assuming that $T/J = 2$ and the initial condition is in the first quadrant in the $e$-$\dot{e}$ plane, plot typical $e$ versus $t$ and $u$ versus $t$ curves.

**Solution.** The equation for the curve $AO$ in Fig. 16-4 is

$$\dot{e}^2 + 2\frac{T}{J}e = 0$$

The equation for the curve $BO$ is

$$\dot{e}^2 - 2\frac{T}{J}e = 0$$

The equation for the switching curve can be written, by combining the preceding two equations, as follows:

$$|\dot{e}|\dot{e} + 2\frac{T}{J}e = 0$$

The change in the shape of the switching curve as the maximum torque is increased may be described as follows: The larger the maximum torque is, the closer the switching curve is to the ordinate in the phase plane. Figure 16-16 shows two switching curves for two values of $T$.

Finally, Fig. 16-17 shows a typical time-optimal trajectory in the $e$-$\dot{e}$ plane and the corresponding $e$ versus $t$ and $m$ versus $t$ curves.

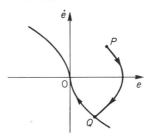

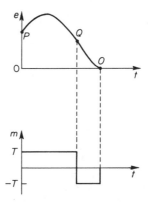

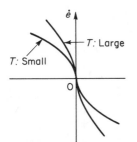

**Fig. 16-16.** Switching curves for two values of maximum torque $T$.

**Fig. 16-17.** Typical time-optimal trajectory in the $e$-$\dot{e}$ plane and the corresponding $e$ versus $t$ and $m$ versus $t$ curves.

**PROBLEM A-16-5.** The switching curve for the time-optimal control system shown in Fig. 16-2 is the parabola $AOB$, as shown in Fig. 16-4. The switching from positive to negative torque can be accomplished by means of a specialized computer. However, a linear approximation of the switching curve gives, in many cases, such a good result that there is not much justification for complicating the controller by introducing computing devices to determine the optimal switching. Thus, in many practical cases, it may be sufficient to have switchings occur along a straight line rather than a parabola.

Consider the system shown in Fig. 16-18. Plot the switching line and a typical trajectory.

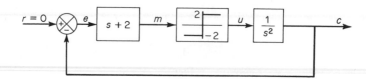

**Fig. 16-18.** Control system.

**Solution.** From the block diagram, the equation for the switching line $AOB$ can be written

$$\dot{e} + 2e = 0$$

The equation describing the plant is

$$\ddot{c} = u = 2 \qquad \text{for } \dot{e} + 2e > 0$$
$$\phantom{\ddot{c} = u} = -2 \qquad \text{for } \dot{e} + 2e < 0$$

Noting that $r = 0$, we have $c = -e$. Hence

$$\ddot{e} = -2 \qquad \text{for } \dot{e} + 2e > 0$$
$$\phantom{\ddot{e}} = 2 \qquad \text{for } \dot{e} + 2e < 0$$

The equation for the trajectories is

$$\dot{e}^2 = \pm 4e + K$$

where $K$ is a constant. This equation describes two families of parabolas in the $e$-$\dot{e}$ plane. Figure 16-19 shows the switching line and a typical trajectory. Note that one might think that the trajectory stops at point $P$. In practice, however, due to the existence of small time lags in switching, deviations occur from the switching line, and the trajectory converges to the origin of the phase plane, as shown in Fig. 16-19. (A system having such a straight switching line is not time optimal. But the response time is reasonably short and adequate for practical purposes.)

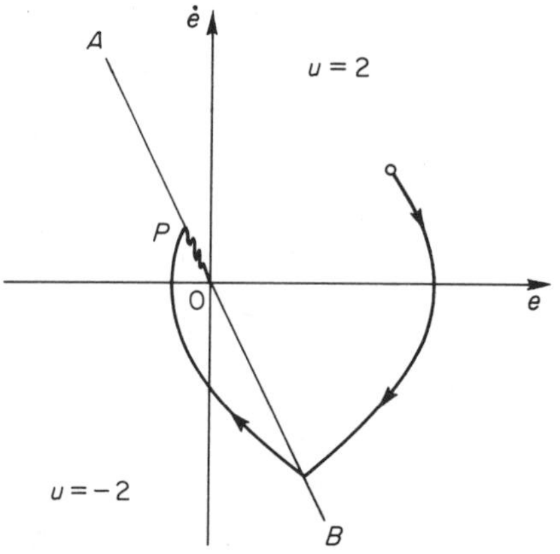

**Fig. 16-19.** Switching line and a typical trajectory.

**PROBLEM A-16-6.** Consider the system

$$\dot{\mathbf{x}} = \mathbf{A}\mathbf{x} + \mathbf{B}\mathbf{u}$$

where $\mathbf{x}$ and $\mathbf{u}$ are real vectors and $\mathbf{A}$ and $\mathbf{B}$ are real matrices. Show that if the control vector $\mathbf{u}$ is given by

$$\mathbf{u} = -\mathbf{B}'\mathbf{P}\mathbf{x} \qquad\qquad (16\text{-}57)$$

where **P** is a positive-definite real symmetric matrix satisfying the condition that

$$\mathbf{A'P} + \mathbf{PA} = -\mathbf{I}$$

then the origin of the system is asymptotically stable in the large.

**Solution.** Let us choose the following $V$ as a possible Liapunov function:

$$V = \mathbf{x'Px}$$

Then

$$\dot{V} = \dot{\mathbf{x}}'\mathbf{Px} + \mathbf{x'P}\dot{\mathbf{x}}$$
$$= \mathbf{x'}(\mathbf{A'P} + \mathbf{PA})\mathbf{x} + \mathbf{u'B'Px} + \mathbf{x'PBu}$$
$$= -\mathbf{x'Ix} + 2\mathbf{x'PBu} \tag{16-58}$$

If Eq. (16-57) is substituted into Eq. (16-58), we see that

$$\dot{V} = -\mathbf{x'Ix} - 2\mathbf{x'PBB'Px}$$

Since **PBB'P** is always positive definite or positive semidefinite, $\dot{V}$ is always negative definite. Since $V$ is positive definite, $\dot{V}$ negative definite, and $V \to \infty$ as $\|\mathbf{x}\| \to \infty$, the origin of the system is asymptotically stable in the large.

**PROBLEM A-16-7.** Consider the control system described by

$$\dot{\mathbf{x}} = \mathbf{Ax} + \mathbf{B}u \tag{16-59}$$

where

$$\mathbf{x} = \begin{bmatrix} x_1 \\ x_2 \end{bmatrix}, \qquad \mathbf{A} = \begin{bmatrix} 0 & 1 \\ 0 & 0 \end{bmatrix}, \qquad \mathbf{B} = \begin{bmatrix} 0 \\ 1 \end{bmatrix}$$

Assuming the linear control law

$$u = -\mathbf{Kx} = -k_1 x_1 - k_2 x_2 \tag{16-60}$$

determine the constants $k_1$ and $k_2$ so that the following performance index is minimized:

$$J = \int_0^\infty \mathbf{x'x}\, dt$$

Consider only the case where the initial condition is

$$\mathbf{x}(0) = \begin{bmatrix} c \\ 0 \end{bmatrix}$$

Choose the undamped natural frequency to be 2 rad/sec.

**Solution.** Substituting Eq. (16-60) into Eq. (16-59), we obtain

$$\dot{\mathbf{x}} = \mathbf{Ax} - \mathbf{BKx}$$

or

$$\begin{bmatrix} \dot{x}_1 \\ \dot{x}_2 \end{bmatrix} = \begin{bmatrix} 0 & 1 \\ 0 & 0 \end{bmatrix}\begin{bmatrix} x_1 \\ x_2 \end{bmatrix} + \begin{bmatrix} 0 \\ 1 \end{bmatrix}[-k_1 x_1 - k_2 x_2]$$
$$= \begin{bmatrix} 0 & 1 \\ -k_1 & -k_2 \end{bmatrix}\begin{bmatrix} x_1 \\ x_2 \end{bmatrix} \tag{16-61}$$

Thus

$$\mathbf{A} - \mathbf{BK} = \begin{bmatrix} 0 & 1 \\ -k_1 & -k_2 \end{bmatrix}$$

Elimination of $x_2$ from Eq. (16-61) yields

$$\ddot{x}_1 + k_2 \dot{x}_1 + k_1 x_1 = 0$$

Since the undamped natural frequency is specified as 2 rad/sec, we obtain

$$k_1 = 4$$

Therefore

$$\mathbf{A} - \mathbf{BK} = \begin{bmatrix} 0 & 1 \\ -4 & -k_2 \end{bmatrix}$$

$\mathbf{A} - \mathbf{BK}$ is a stable matrix if $k_2 > 0$. Our problem now is to determine the value of $k_2$ so that the performance index

$$J = \int_0^\infty \mathbf{x}'\mathbf{x}\, dt = \mathbf{x}'(0)\mathbf{P}\mathbf{x}(0)$$

is minimized, where the matrix $\mathbf{P}$ is determined from

$$(\mathbf{A} - \mathbf{BK})'\mathbf{P} + \mathbf{P}(\mathbf{A} - \mathbf{BK}) = -\mathbf{I}$$

or

$$\begin{bmatrix} 0 & -4 \\ 1 & -k_2 \end{bmatrix} \begin{bmatrix} p_{11} & p_{12} \\ p_{12} & p_{22} \end{bmatrix} + \begin{bmatrix} p_{11} & p_{12} \\ p_{12} & p_{22} \end{bmatrix} \begin{bmatrix} 0 & 1 \\ -4 & -k_2 \end{bmatrix} = \begin{bmatrix} -1 & 0 \\ 0 & -1 \end{bmatrix}$$

Solving for the matrix $\mathbf{P}$, we obtain

$$\mathbf{P} = \begin{bmatrix} p_{11} & p_{12} \\ p_{12} & p_{22} \end{bmatrix} = \begin{bmatrix} \dfrac{5}{2k_2} + \dfrac{k_2}{8} & \dfrac{1}{8} \\ \dfrac{1}{8} & \dfrac{5}{8k_2} \end{bmatrix}$$

The performance index is then

$$J = \mathbf{x}'(0)\mathbf{P}\mathbf{x}(0)$$

$$= \begin{bmatrix} c & 0 \end{bmatrix} \begin{bmatrix} p_{11} & p_{12} \\ p_{12} & p_{22} \end{bmatrix} \begin{bmatrix} c \\ 0 \end{bmatrix} = p_{11}c^2$$

$$= \left(\frac{5}{2k_2} + \frac{k_2}{8}\right)c^2 \tag{16-62}$$

In order to minimize $J$, we differentiate $J$ with respect to $k_2$ and set it equal to zero as follows:

$$\frac{\partial J}{\partial k_2} = \left(\frac{-5}{2k_2^2} + \frac{1}{8}\right)c^2 = 0$$

Hence

$$k_2 = \sqrt{20}$$

With this value of $k_2$ we have $\partial^2 J/\partial k_2^2 > 0$. Thus the minimum value of $J$ is obtained by substituting $k_2 = \sqrt{20}$ into Eq. (16-62), or

$$J_{\min} = \frac{\sqrt{5}}{2}c^2$$

The designed system has the control law

$$u = -4x_1 - \sqrt{20}x_2$$

The designed system is optimal in that it results in a minimum value for the performance index $J$ under the assumed initial condition.

**PROBLEM A-16-8.** Consider the control system shown in Fig. 16-9. The equation for the plant is

$$\dot{\mathbf{x}} = \mathbf{Ax} + \mathbf{B}u \qquad (16\text{-}63)$$

where

$$\mathbf{x} = \begin{bmatrix} x_1 \\ x_2 \end{bmatrix}, \qquad \mathbf{A} = \begin{bmatrix} 0 & 1 \\ 0 & 0 \end{bmatrix}, \qquad \mathbf{B} = \begin{bmatrix} 0 \\ 1 \end{bmatrix}$$

Assuming the linear control law

$$u = -\mathbf{Kx} = -k_1 x_1 - k_2 x_2 \qquad (16\text{-}64)$$

determine the constants $k_1$ and $k_2$ so that the following performance index is minimized:

$$J = \int_0^\infty (\mathbf{x}'\mathbf{x} + u'u)\, dt$$

**Solution.** By substituting Eq. (16-64) into Eq. (16-63), we obtain

$$\dot{\mathbf{x}} = \mathbf{Ax} - \mathbf{BKx} = (\mathbf{A} - \mathbf{BK})\mathbf{x} = \begin{bmatrix} 0 & 1 \\ -k_1 & -k_2 \end{bmatrix} \mathbf{x}$$

If we assume $k_1$ and $k_2$ to be positive constants, then $\mathbf{A} - \mathbf{BK}$ becomes a stable matrix and $\mathbf{x}(\infty) = \mathbf{0}$. Hence the performance index can be written

$$J = \int_0^\infty (\mathbf{x}'\mathbf{x} + \mathbf{x}'\mathbf{K}'\mathbf{Kx})\, dt$$

$$= \int_0^\infty \mathbf{x}'(\mathbf{I} + \mathbf{K}'\mathbf{K})\mathbf{x}\, dt$$

$$= \mathbf{x}'(0)\mathbf{P}\mathbf{x}(0)$$

where $\mathbf{P}$ is determined from

$$(\mathbf{A} - \mathbf{BK})'\mathbf{P} + \mathbf{P}(\mathbf{A} - \mathbf{BK}) = -(\mathbf{I} + \mathbf{K}'\mathbf{K})$$

or

$$\begin{bmatrix} 0 & -k_1 \\ 1 & -k_2 \end{bmatrix} \begin{bmatrix} p_{11} & p_{12} \\ p_{12} & p_{22} \end{bmatrix} + \begin{bmatrix} p_{11} & p_{12} \\ p_{12} & p_{22} \end{bmatrix} \begin{bmatrix} 0 & 1 \\ -k_1 & -k_2 \end{bmatrix} = -\begin{bmatrix} 1 & 0 \\ 0 & 1 \end{bmatrix} - \begin{bmatrix} k_1^2 & k_1 k_2 \\ k_1 k_2 & k_2^2 \end{bmatrix}$$

This matrix equation results in the following three equations in $p_{ij}$:

$$-2k_1 p_{12} = -1 - k_1^2$$
$$p_{11} - k_2 p_{12} - k_1 p_{22} = -k_1 k_2$$
$$2p_{12} - 2k_2 p_{22} = -1 - k_2^2$$

Solving these three equations for the $p_{ij}$, we obtain

$$\mathbf{P} = \begin{bmatrix} p_{11} & p_{12} \\ p_{12} & p_{22} \end{bmatrix} = \begin{bmatrix} \dfrac{1}{2}\left(\dfrac{k_2}{k_1} + \dfrac{k_1}{k_2}\right) + \dfrac{k_1}{2k_2}\left(\dfrac{1}{k_1} + k_1\right) & \dfrac{1}{2}\left(\dfrac{1}{k_1} + k_1\right) \\[2ex] \dfrac{1}{2}\left(\dfrac{1}{k_1} + k_1\right) & \dfrac{1}{2}\left(\dfrac{1}{k_2} + k_2\right) + \dfrac{1}{2k_2}\left(\dfrac{1}{k_1} + k_1\right) \end{bmatrix}$$

Now

$$J = \mathbf{x}'(0)\mathbf{P}\mathbf{x}(0)$$

$$= \left[\frac{1}{2}\left(\frac{k_2}{k_1} + \frac{k_1}{k_2}\right) + \frac{k_1}{2k_2}\left(\frac{1}{k_1} + k_1\right)\right]x_1^2(0) + \left(\frac{1}{k_1} + k_1\right)x_1(0)x_2(0)$$

$$+ \left[\frac{1}{2}\left(\frac{1}{k_2} + k_2\right) + \frac{1}{2k_2}\left(\frac{1}{k_1} + k_1\right)\right]x_2^2(0)$$

To minimize $J$, we set $\partial J/\partial k_1 = 0$ and $\partial J/\partial k_2 = 0$, or

$$\frac{\partial J}{\partial k_1} = \left[\frac{1}{2}\left(\frac{-k_2}{k_1^2} + \frac{1}{k_2}\right) + \frac{k_1}{k_2}\right]x_1^2(0) + \left(\frac{-1}{k_1^2} + 1\right)x_1(0)x_2(0) + \left[\frac{1}{2k_2}\left(\frac{-1}{k_1^2} + 1\right)\right]x_2^2(0) = 0$$

$$\frac{\partial J}{\partial k_2} = \left[\frac{1}{2}\left(\frac{1}{k_1} - \frac{k_1}{k_2^2}\right) + \frac{-k_1}{2k_2^2}\left(\frac{1}{k_1} + k_1\right)\right]x_1^2(0) + \left[\frac{1}{2}\left(\frac{-1}{k_2^2} + 1\right) - \frac{1}{2k_2^2}\left(\frac{1}{k_1} + k_1\right)\right]x_2^2(0) = 0$$

For any given initial conditions $x_1(0)$ and $x_2(0)$, the value of $J$ becomes minimum when

$$k_1 = 1, \qquad k_2 = \sqrt{3}$$

Note that $k_1$ and $k_2$ are positive constants as we assumed in the solution. Hence for the optimal control law

$$\mathbf{K} = [k_1 \quad k_2] = [1 \quad \sqrt{3}]$$

The block diagram of this optimal control system is shown in Fig. 16-10 with $\mu = 1$.

**PROBLEM A-16-9.** Consider the system

$$\dot{\mathbf{x}} = \mathbf{f}(\mathbf{x}, \mathbf{u})$$

which may be linear or nonlinear. It is desired to determine the optimal control law $\mathbf{u} = \mathbf{g}(\mathbf{x})$ such that the following performance index

$$J = \int_0^\infty L(\mathbf{x}, \mathbf{u})\, dt$$

is minimized, where $\mathbf{u}$ is unconstrained.

If the origin of the system described by

$$\dot{\mathbf{x}} = \mathbf{f}(\mathbf{x}, \mathbf{g}(\mathbf{x}))$$

is asymptotically stable, and thus a Liapunov function $V(\mathbf{x})$ exists such that $\dot{V}(\mathbf{x})$ is negative definite, then show that a sufficient condition for a control vector $\mathbf{u}_1$ to be optimal is that $H(\mathbf{x}, \mathbf{u})$, where

$$H(\mathbf{x}, \mathbf{u}) = \frac{dV}{dt} + L(\mathbf{x}, \mathbf{u}) \tag{16-65}$$

is minimum with $\mathbf{u} = \mathbf{u}_1$, or

$$\min_{\mathbf{u}} H(\mathbf{x}, \mathbf{u}) = \min_{\mathbf{u}} \left[\frac{dV}{dt} + L(\mathbf{x}, \mathbf{u})\right]$$

$$= \frac{dV}{dt}\bigg|_{\mathbf{u}=\mathbf{u}_1} + L(\mathbf{x}, \mathbf{u}_1) \tag{16-66}$$

and

$$\frac{dV}{dt}\bigg|_{\mathbf{u}=\mathbf{u}_1} = -L(\mathbf{x}, \mathbf{u}_1) \tag{16-67}$$

**Solution.** Let us integrate both sides of Eq. (16-67). Then

$$V(\mathbf{x}(\infty)) - V(\mathbf{x}(0)) = -\int_0^\infty L(\mathbf{x}(t), \mathbf{u}_1(t))\, dt \tag{16-68}$$

Since the origin of the system is asymptotically stable, $\mathbf{x}(\infty) = \mathbf{0}$ and $V(\mathbf{x}(\infty)) = 0$. Then Eq. (16-68) becomes

$$V(\mathbf{x}(0)) = \int_0^\infty L(\mathbf{x}(t), \mathbf{u}_1(t))\, dt \tag{16-69}$$

To prove that $\mathbf{u}_1(t)$ is optimal, assume that $\mathbf{u}_1(t)$ is not optimal and that the control vector $\mathbf{u}_2(t)$ will yield a smaller value of $J$. Then

$$\int_0^\infty L(\mathbf{x}(t), \mathbf{u}_2(t)) \, dt < \int_0^\infty L(\mathbf{x}(t), \mathbf{u}_1(t)) \, dt$$

Note that from Eq. (16-66), the minimum value of $H(\mathbf{x}, \mathbf{u})$ occurs at $\mathbf{u} = \mathbf{u}_1$. Note also that from Eq. (16-67) this minimum value is equal to zero. Hence

$$H(\mathbf{x}, \mathbf{u}) \geq 0$$

for all $\mathbf{u}$. Therefore

$$H(\mathbf{x}, \mathbf{u}_2) = \frac{dV}{dt}\bigg|_{\mathbf{u}=\mathbf{u}_2} + L(\mathbf{x}, \mathbf{u}_2) \geq 0$$

Integrating both sides of this inequality from 0 to $\infty$, we obtain

$$V(\mathbf{x}(\infty)) - V(\mathbf{x}(0)) \geq -\int_0^\infty L(\mathbf{x}(t), \mathbf{u}_2(t)) \, dt$$

Since $V(\mathbf{x}(\infty)) = 0$, we have

$$V(\mathbf{x}(0)) \leq \int_0^\infty L(\mathbf{x}(t), \mathbf{u}_2(t)) \, dt \tag{16-70}$$

Then from Eqs. (16-69) and (16-70)

$$\int_0^\infty L(\mathbf{x}(t), \mathbf{u}_1(t)) \, dt \leq \int_0^\infty L(\mathbf{x}(t), \mathbf{u}_2(t)) \, dt$$

This is a contradiction. Hence $\mathbf{u}_1(t)$ is the optimal control vector.

**PROBLEM A-16-10.** Consider the system

$$\begin{bmatrix} \dot{x}_1 \\ \dot{x}_2 \end{bmatrix} = \begin{bmatrix} 0 & 1 \\ 0 & 0 \end{bmatrix} \begin{bmatrix} x_1 \\ x_2 \end{bmatrix} + \begin{bmatrix} 0 \\ 1 \end{bmatrix} [u]$$

It is desired to find the optimal control function $u$ such that the following performance index

$$J = \int_0^\infty (\mathbf{x}'\mathbf{Q}\mathbf{x} + u'u) \, dt, \qquad \mathbf{Q} = \begin{bmatrix} 1 & 0 \\ 0 & \mu \end{bmatrix}$$

is minimized.

Referring to the sufficient condition for the optimal control vector presented in Problem A-16-9, determine the optimal control signal $u(t)$.

**Solution.** Let us define

$$V = \mathbf{x}'\mathbf{P}\mathbf{x}$$

where

$$\mathbf{P} = \begin{bmatrix} p_{11} & p_{12} \\ p_{12} & p_{22} \end{bmatrix}$$

Then

$$V = p_{11}x_1^2 + 2p_{12}x_1x_2 + p_{22}x_2^2$$

and

$$\dot{V} = 2p_{11}x_1x_2 + 2p_{12}x_2^2 + (2p_{12}x_1 + 2p_{22}x_2)u$$

Equation (16-65) becomes

$$H(\mathbf{x}, u) = \frac{dV}{dt} + L(\mathbf{x}, u)$$

$$= 2p_{11}x_1x_2 + 2p_{12}x_2^2 + (2p_{12}x_1 + 2p_{22}x_2)u + x_1^2 + \mu x_2^2 + u^2 \quad (16\text{-}71)$$

The optimal control signal $u(t)$ is the one which minimizes $H(\mathbf{x}, u)$, Therefore, we differentiate $H(\mathbf{x}, u)$ with respect to $u$ and set the resulting equation equal to zero, or

$$\frac{\partial H}{\partial u} = 2p_{12}x_1 + 2p_{22}x_2 + 2u = 0$$

from which we obtain

$$u = -p_{12}x_1 - p_{22}x_2 \quad (16\text{-}72)$$

By substituting Eq. (16-72) into Eq. (16-71) and setting the result equal to zero, we obtain

$$x_1^2(1 - p_{12}^2) + x_1x_2(2p_{11} - 2p_{12}p_{22}) + x_2^2(\mu + 2p_{12} - p_{22}^2) = 0$$

This last equation must hold for any $x_1$ and $x_2$. Hence we require

$$1 - p_{12}^2 = 0$$

$$p_{11} - p_{12}p_{22} = 0$$

$$\mu + 2p_{12} - p_{22}^2 = 0$$

(Note that these three equations in the $p_{ij}$ are identical with those obtained in Example 16-12.) Solving these three simultaneous equations for $p_{11}, p_{12}$, and $p_{22}$, requiring that $\mathbf{P}$ be positive definite, we obtain

$$p_{11} = \sqrt{\mu + 2}, \qquad p_{12} = 1, \qquad p_{22} = \sqrt{\mu + 2}$$

The optimal control law is then given by

$$u = -x_1 - \sqrt{\mu + 2}\, x_2$$

With this law, the system state equation becomes

$$\begin{bmatrix} \dot{x}_1 \\ \dot{x}_2 \end{bmatrix} = \begin{bmatrix} 0 & 1 \\ 0 & 0 \end{bmatrix} \begin{bmatrix} x_1 \\ x_2 \end{bmatrix} + \begin{bmatrix} 0 \\ 1 \end{bmatrix} [-x_1 - \sqrt{\mu + 2}\, x_2]$$

or

$$\begin{bmatrix} \dot{x}_1 \\ \dot{x}_2 \end{bmatrix} = \begin{bmatrix} 0 & 1 \\ -1 & -\sqrt{\mu + 2} \end{bmatrix} \begin{bmatrix} x_1 \\ x_2 \end{bmatrix}$$

Note that the coefficient matrix in this last equation is a stable one. Thus the origin of the system is asymptotically stable and the present approach to the solution is valid.

## PROBLEMS

**PROBLEM B-16-1.** Determine the state controllability and observability of the system described by

$$\begin{bmatrix} \dot{x}_1 \\ \dot{x}_2 \\ \dot{x}_3 \end{bmatrix} = \begin{bmatrix} 0 & 1 & 0 \\ 0 & 0 & 1 \\ -6 & -11 & -6 \end{bmatrix} \begin{bmatrix} x_1 \\ x_2 \\ x_3 \end{bmatrix} + \begin{bmatrix} 0 \\ 0 \\ 1 \end{bmatrix} [u]$$

$$y = [4 \quad 5 \quad 1] \begin{bmatrix} x_1 \\ x_2 \\ x_3 \end{bmatrix}$$

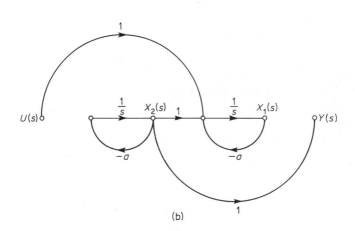

**Fig. 16-20.** Signal flow graphs of systems.

**PROBLEM B-16-2.** Are the systems shown in Figs. 16-20 (a) and (b) controllable and observable?

**PROBLEM B-16-3.** Consider the system described by

$$\begin{bmatrix} \dot{x}_1 \\ \dot{x}_2 \\ \dot{x}_3 \end{bmatrix} = \begin{bmatrix} \lambda_1 & 1 & 0 \\ 0 & \lambda_1 & 1 \\ 0 & 0 & \lambda_1 \end{bmatrix} \begin{bmatrix} x_1 \\ x_2 \\ x_3 \end{bmatrix} + \begin{bmatrix} a \\ b \\ c \end{bmatrix} [u]$$

Determine the conditions on $a$, $b$, and $c$ for complete state controllability.

**PROBLEM B-16-4.** Consider the system described by

$$\begin{bmatrix} \dot{x}_1 \\ \dot{x}_2 \end{bmatrix} = \begin{bmatrix} a & b \\ c & d \end{bmatrix} \begin{bmatrix} x_1 \\ x_2 \end{bmatrix} + \begin{bmatrix} 1 \\ 1 \end{bmatrix} [u]$$

$$y = [1 \quad 0] \begin{bmatrix} x_1 \\ x_2 \end{bmatrix}$$

Determine the conditions on $a$, $b$, $c$, and $d$ for complete state controllability and complete observability.

**PROBLEM B-16-5.** Consider the system

$$\begin{bmatrix} \dot{x}_1 \\ \dot{x}_2 \end{bmatrix} = \begin{bmatrix} -1 & 0 \\ 0 & -1 \end{bmatrix} \begin{bmatrix} x_1 \\ x_2 \end{bmatrix} + \begin{bmatrix} 1 \\ -1 \end{bmatrix} [u]$$

$$y = [1 \quad 1] \begin{bmatrix} x_1 \\ x_2 \end{bmatrix}$$

Show that the output $y$ is independent of the control function $u$.

**PROBLEM B-16-6.** Assume that a mechanical vibratory system is described by

$$\dot{\mathbf{x}} = \mathbf{Ax} + \mathbf{Bu}$$

where

    $\mathbf{x}$ = state vector ($n$-dimensional vector)

    $\mathbf{u}$ = input vector ($r$-dimensional vector)

    $\mathbf{A}$ = $n \times n$ constant matrix

    $\mathbf{B}$ = $n \times r$ constant matrix

Find the condition on $\mathbf{A}$ and $\mathbf{B}$ that all modes of oscillation can be excited.

**PROBLEM B-16-7.** Consider the time-optimal control system shown in Fig. 16-21. Obtain the equation of the switching curve. Plot the switching curve on the $e$-$\dot{e}$ plane and show a few typical time-optimal trajectories.

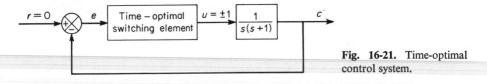

Fig. 16-21. Time-optimal control system.

**PROBLEM B-16-8.** Plot the phase-plane diagram of the system shown in Fig. 16-22.

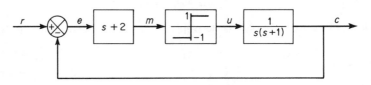

Fig. 16-22. Control system.

**PROBLEM B-16-9.** Consider the system described by

$$\dot{\mathbf{x}} = \mathbf{Ax}, \qquad \mathbf{x}(0) = \mathbf{c}$$

where

$$\mathbf{x} = \begin{bmatrix} x_1 \\ x_2 \\ x_3 \end{bmatrix}, \qquad \mathbf{A} = \begin{bmatrix} 0 & 1 & 0 \\ 0 & 0 & 1 \\ -1 & -2 & -a \end{bmatrix}, \qquad \mathbf{c} = \begin{bmatrix} c_1 \\ 0 \\ 0 \end{bmatrix}$$

$a$ = adjustable parameter $> 0$

Determine the value of the parameter $a$ so as to minimize the following performance index:

$$J = \int_0^\infty \mathbf{x}'\mathbf{x} \, dt$$

**PROBLEM B-16-10.** Consider the system shown in Fig. 16-23. Determine the value of the gain $K$ so that the damping ratio $\zeta$ of the closed-loop system is equal to 0.5. Then compute the undamped natural frequency $\omega_n$ of the closed-loop system. Assuming that $e(0) =$

1 and $\dot{e}(0) = 0$, compute

$$\int_0^\infty e^2(t)\, dt$$

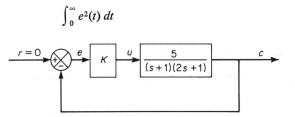

**Fig. 16-23.** Control system.

**PROBLEM B-16-11.** Determine the optimal control function $u$ for the system described by

$$\dot{\mathbf{x}} = \mathbf{Ax} + \mathbf{B}u$$

where

$$\mathbf{x} = \begin{bmatrix} x_1 \\ x_2 \end{bmatrix}, \qquad \mathbf{A} = \begin{bmatrix} 0 & 1 \\ 0 & -1 \end{bmatrix}, \qquad \mathbf{B} = \begin{bmatrix} 0 \\ 1 \end{bmatrix}$$

such that the following performance index is minimized:

$$J = \int_0^\infty (\mathbf{x'x} + u'u)\, dt$$

# REFERENCES

A-1   Andronow, A. A., and C. E. Chaikin, *Theory of Oscillations*, Princeton, N.J.: Princeton University Press, 1949.

A-2   Aseltine, J. A., *Transform Methods in Linear System Analysis*, New York, N.Y.: McGraw-Hill Book Company, Inc., 1958.

A-3   Athanassiades, M., and P. L. Falb, "Time Optimal Control for Plants with Numerator Dynamics," *IRE Trans. Automatic Control* (1962), pp. 46-7.

A-4   Athans, M., and P. L. Falb, *Optimal Control: An Introduction to the Theory and Its Applications*, New York, N.Y.: McGraw-Hill Book Company, Inc., 1965.

B-1   Bayliss, L. E., *Living Control Systems*, London, England: English Universities Press Limited, 1966.

B-2   Bellman, R., I. Glicksberg, and O. Gross, "On the Bang-Bang Control Problem," *Quart. Appl. Math.*, **14** (1956), pp. 11-8.

B-3   Brown, G. S., and D. P. Campbell, *Principles of Servomechanisms*, New York, N.Y.: John Wiley & Sons, Inc., 1955.

B-4   Buland, R. N., "Analysis of Nonlinear Servos by Phase Plane-Delta Method," *J. Franklin Inst.*, 257 (1954), pp. 37-48.

B-5   Bushaw, D. W., "Differential Equations with a Discontinuous Forcing Term," *Stevens Inst. Tech. Experimental Towing Tank Report* **469**, Hoboken, N.J., Jan., 1953.

B-6   Bushaw, D. W., "Optimal Discontinuous Forcing Terms," in S. Lefschetz (ed.), *Contributions to the Theory of Nonlinear Oscillations*, **4**, Princeton, N.J.: Princeton University Press, 1958, pp. 29–52.

B-7   Butman, S., and R. Sivan (Sussman), "On Cancellations, Controllability and Observability," *IEEE Trans. Automatic Control* (1964), pp. 317–8.

C-1   Campbell, D. P., *Process Dynamics*, New York, N.Y.: John Wiley & Sons, Inc., 1958.

C-2   Chandaket, P., C. T. Leondes, and E. C. Deland, "Optimum Non-linear Bang-Bang Control Systems with Complex Roots," *AIEE Trans. Part II*, **80** (1961), pp. 82–102.

C-3   Chang, S. S. L., *Synthesis of Optimum Control Systems*, New York, N.Y.: McGraw-Hill Book Company, Inc., 1961.

C-4   Cheng, D. K., *Analysis of Linear Systems*, Reading, Mass.: Addison-Wesley Publishing Company, Inc., 1959.

C-5   Chestnut, H., and R. W. Mayer, *Servomechanisms and Regulating System Design*, vol. 1, New York, N.Y.: John Wiley & Sons, Inc., 1959.

C-6   Coddington, E. A., and N. Levinson, *Theory of Ordinary Differential Equations*, New York, N.Y.: McGraw-Hill Book Company, Inc., 1955.

C-7   Cosgriff, R. L., *Nonlinear Control Systems*, New York, N.Y.: McGraw-Hill Book Company, Inc., 1958.

C-8   Cunningham, W. J., *Introduction to Nonlinear Analysis*, New York, N.Y.: McGraw-Hill Book Company, Inc., 1958.

D-1   D'Azzo, J. J., and C. H. Houpis, *Feedback Control System Analysis and Synthesis*, 2nd ed., New York, N.Y.: McGraw-Hill Book Company, Inc., 1966.

D-2   Del Toro, V., and S. R. Parker, *Principles of Control System Engineering*, New York, N.Y.: McGraw-Hill Book Company, Inc., 1960.

D-3   DeRusso, P. M., R. J. Roy, and C. M. Close, *State Variables for Engineers*, New Yrok, N.Y.: John Wiley & Sons, Inc., 1965.

D-4   Dorf, R,, *Time Domain Analysis and Design of Control Systems*, Reading, Mass.: Addison-Wesley Publishing Company, Inc., 1965.

E-1   Eckman, D. P., *Automatic Process Control*, New York, N.Y.: John Wiley & Sons, Inc., 1958.

E-2   Enns, M., J. R. Greenwood, III, J. E. Matheson, and F. T. Thompson, "Practical Aspects of State-Space Methods Part I: System Formulation and Reduction," *IEEE Trans. Military Electronics* (1964), pp. 81–93.

E-3   Evans, W. R., "Graphical Analysis of Control Systems," *AIEE Trans. Part II*, **67** (1948), pp. 547–51.

E-4   Evans, W. R., "Control System Synthesis by Root Locus Method," *AIEE Trans. Part II*, **69** (1950), pp. 66–9.

E-5 Evans, W. R., "The Use of Zeros and Poles for Frequency Response or Transient Response," *ASME Trans.*, **76** (1954), pp. 1335–44.

F-1 Flügge-Lotz, I., "Synthesis of Third-order Contactor Control Systems," *Proc. First Intern, Cong. IFAC*, Moscow, 1960; *Automatic and Remote Control*, London, England: Butterworths & Company, Ltd., 1961, pp. 390–7.

F-2 Flügge-Lotz, I., and T. Ishikawa, "Investigation of Third-order Contactor Control Systems with Two Complex Poles without Zeros," *NASA Tech. Note D248*, 1960.

F-3 Freeman, H., *Discrete-Time Systems*, New York, N.Y.: John Wiley & Sons, Inc., 1965.

F-4 Friedland, B., "The Structure of Optimum Control Systems," *ASME J. Basic Engineering*, ser. D, **84** (1962), pp. 1–12.

F-5 Fuller, A. T., "Phase Space in the Theory of Optimum Control," *J. Elec. Control*, ser. 1 **VIII** (1960), pp. 381–400.

G-1 Gardener, M. F., and J. L. Barnes, *Transients in Linear Systems*, vol 1, New York, N.Y.: John Wiley & Sons, Inc., 1942.

G-2 Gibson, J. E., *Nonlinear Automatic Control*, New York, N.Y.: McGraw-Hill Book Company, Inc., 1963.

G-3 Gilbert, E. G., "Controllability and Observability in Multivariable Control Systems," *J. SIAM Control*, ser. A, **1** (1963), pp. 128–51.

G-4 Gille, J. C., M. J. Pelegrin, and P. Decaulne, *Feedback Control Systems*, New York, N.Y.: McGraw-Hill Book Company, 1959.

G-5 Graham, D., and R. C. Lathrop, "The Synthesis of Optimum Response: Criteria and Standard Forms," *AIEE Trans. Part II*, **72** (1953), pp. 273–88.

G-6 Graham, D., and D. McRuer, *Analysis of Nonlinear Control Systems*, New York, N.Y.: John Wiley & Sons, Inc., 1961.

G-7 Gupta, S. C., *Transform and State Variable Methods in Linear Systems*, New York, N.Y.: John Wiley & Sons, Inc., 1966.

H-1 Hahn, W., *Theory and Application of Liapunov's Direct Method*, Englewood Cliffs, N.J.: Prentice-Hall, Inc., 1963.

H-2 Harrison, H. L., and J. G. Bollinger, *Introduction to Automatic Controls*, Scranton, Pa.: International Textbook Company, Inc., 1963.

H-3 Harvey, C. A., "Determining the Switching Criterion for Time Optimal Control," *J. Math. Anal. Appl.*, **5** (1962), pp. 245–57.

H-4 Harvey, C. A., and E. B. Lee, "On the Uniqueness of Time-optimal Control for Linear Processes," *J. Math . Anal. Appl.*, **5** (1962), pp. 258–68.

H-5 Henke, R. W., "Digital Fluidics Works Now," Control Engineering, **14**, no. 1, Jan., 1967, pp. 100–4.

H-6 Higdon, D. T., and R. H. Cannon, Jr., "On the Control of Unstable Multiple-Output Mechanical Systems," *ASME Paper no.* **63-WA-148**, 1963.

J-1 Jury, E. I., *Sampled-Data Control Systems*, New York, N.Y.: John Wiley & Sons, Inc., 1958.

J-2 Jury, E. I., "A Simplified Stability Criterion for Linear Discrete Systems," *Proc. IRE*, **50** (1962), pp. 1493–500.

J-3 Jury, E. I., *Theory and Applications of z-Transform Method*, New York, N.Y.: John Wiley & Sons, Inc., 1964.

K-1 Kalman, R. E., "Analysis and Design Principles of Second and Higher Order Saturating Servomechanisms," *AIEE Trans. Part II*, **74** (1955), pp. 294–310.

K-2 Kalman, R. E., "Contributions to the Theory of Optimal Control," *Bol. Soc. Mat. Mex.*, **5** (1960), pp. 102–19.

K-3 Kalman, R. E., "On the General Theory of Control Systems," *Proc. First Intern, Cong. IFAC*, Moscow, 1960; *Automatic and Remote Control*, London, England: Butterworths & Company, Ltd., 1961, pp. 481–92.

K-4 Kalman, R. E., "Canonical Structure of Linear Dynamical Systems," *Proc. Natl. Acad. Sci., USA* **48** (1962), pp. 596–600.

K-5 Kalman, R. E., "When Is a Linear Control System Optimal?" *ASME J. Basic Engineering*, ser. D, **86** (1964), pp. 51–60.

K-6 Kalman, R. E., and J. E. Bertram, "Control System Analysis and Design Via the Second Method of Lyapunov: I Continuous-Time Systems," *ASME J. Basic Engineering*, ser. D, **82** (1960), pp. 371–93.

K-7 Kalman, R. E., Y. C. Ho, and K. S. Narendra, "Controllability of Linear Dynamic Systems," in *Contributions to Differential Equations*, vol. 1, New York, N.Y.: Interscience Publishers, Inc., 1962.

K-8 Korn, G. A., and T. M. Korn, *Electronic Analog Computers*, New York, N.Y.: McGraw-Hill Book Company, Inc., 1956.

K-9 Kreindler, E., "Contributions to the Theory of Time Optimal Control," *J. Franklin Inst.*, **275** (1963), pp. 314–44.

K-10 Kreindler, E., and P. E. Sarachick, "On the Concepts of Controllability and Observability of Linear Systems," *IEEE Trans. Automatic Control* (1964), pp. 129–36.

K-11 Kuo, B. C., *Automatic Control Systems*, Englewood Cliffs, N.J.: Prentice-Hall, Inc., 1962.

K-12 Kuo, B. C., *Analysis and Synthesis of Sampled-Data Control Systems*, Englewood Cliffs, N.J.: Prentice-Hall, Inc., 1963.

L-1 Lago, G., and L. M. Benningfield, *Control System Theory*, New York, N.Y.: Ronald Press Company, 1962.

L-2   LaSalle, J. P., "Time Optimal Control Systems," *Proc. Natl. Acad. Sci.,* USA, **45** (1959), pp. 573–7.

L-3   LaSalle, J. P., "Time Optimal Control," *Bol. Soc. Mat. Mex.,* **5** (1960), pp. 120–4.

L-4   LaSalle, J. P., "The Time-optimal Control Problem," in *Contributions to Differential Equations,* vol. 5, Princeton, N.J.: Princeton University Press, 1960, pp. 1–24.

L-5   LaSalle, J. P., "The Bang-Bang Principle," *Proc. First Intern. Cong. IFAC,* Moscow, 1960; *Automatic and Remote Control,* London, England: Butterworths & Company, Ltd., 1961, pp. 493–7.

L-6   LaSalle, J. P., and S. Lefschetz, *Stability by Liapunov's Direct Method with Applications,* New York, N.Y.: Academic Press, Inc., 1961.

L-7   Lefschetz, S., *Differential Equations: Geometric Theory,* New York, N.Y.: Interscience Publishers, Inc., 1957.

L-8   Lewis, J. B., "The Use of Nonlinear Feedback to Improve the Transient Response of a Servomechanism," *AIEE Trans. Part II,* **71** (1952), pp. 449–53.

L-9   Lindorff, D. P., *Theory of Sampled Data Control Systems,* New York, N.Y.: John Wiley & Sons, Inc., 1965.

M-1   Mason, S. J., "Feedback Theory: Some Properties of Signal Flow Graphs," *Proc. IRE,* **41** (1953), pp. 1144–56.

M-2   Mason, S. J., "Feedback Theory: Further Properties of Signal Flow Graphs," *Proc. IRE,* **44** (1956), pp. 920–6.

M-3   Mason, S. J., and H. J. Zimmerman, *Electronic Circuits, Signals, and Systems,* New York, N.Y.: John Wiley & Sons, Inc., 1960.

M-4   Melbourne, W. G., "Three Dimensional Optimum Thrust Trajectories for Power-Limited Propulsion Systems," *ARS J.,* **31** (1961), pp. 1723–8.

M-5   Melbourne, W. G., and C. G. Sauer, Jr., "Optimum Interplanetary Rendezvous with Power-Limited Vehicles, *AIAA J.,* **1** (1963), pp. 54–60.

M-6   Minorsky, N., *Nonlinear Oscillations,* Princeton, N.J.: D. Van Nostrand Company, Inc., 1962.

M-7   Mirsky, L., *An Introduction to Linear Algebra,* Oxford, England: Clarendon Press, Ltd., 1955.

M-8   Monopoli, R. V., "Controller Design for Nonlinear and Time-Varying Plants," *NASA CR-152,* Jan., 1965.

N-1   Nyquist, H., "Regeneration Theory," *Bell System Tech. J.,* **11** (1932), pp. 126–47.

O-1   Ogata, K., *State Space Analysis of Control Systems,* Englewood Cliffs, N.J.: Prentice-Hall, Inc., 1967.

P-1    Peschon, J., "Learning in Automatic Systems," *Proc. Allerton Conference on Circuit and System Theory*, Univ. of Illinois, Sept., 1964.

P-2    Pipes, L. A., *Matrix Methods for Engineers*, Englewood Cliffs, N.J.: Prentice-Hall, Inc., 1963.

R-1    Ragazzini, J. R., and G. F. Franklin, *Sampled Data Control Systems*, New York, N.Y.: McGraw-Hill Book Company, Inc., 1958.

R-2    Reid, W. T., "A Matrix Differential Equation of Riccati Type," *Amer. J. Math.*, **68** (1946), pp. 237–46.

R-3    Rekasius, Z. V., "A General Performance Index for Analytical Design of Control Systems," *IRE Trans. Automatic Control* (1961), pp. 217–22.

S-1    Schultz, D. G., and J. E. Gibson, "The Variable Gradient Method for Generating Liapunov Functions;" *AIEE Trans. Part II*, **81** (1962), pp. 203–9.

S-2    Schultz, W. C., and V. C. Rideout, "Control System Performance Measures: Past, Present, and Future," *IRE Trans., Automatic Control* (1961), pp. 22–35.

S-3    Shannon, C. E., "Communication in the Presence of Noise," *Proc. IRE*, **37** (1949), pp. 10–21.

S-4    Smith, O. J. M., *Feedback Control Systems*, New York, N.Y.: McGraw-Hill Book Company, Inc., 1958.

S-5    Staats, P. F., "A Survey of Adaptive Control Topics'" *Plan B paper*, Dept. of Mech. Eng., Univ. of Minn., Mar., 1966.

S-6    Stallard, D. V., "A Series Method of Calculating Control System Transient Responses from the Frequency Response," *AIEE Trans., Part II*, **74** (1955), pp. 61–4.

T-1    Thaler, G. J., and R. G. Brown, *Analysis and Design of Feedback Control Systems*, New York, N.Y.: McGraw-Hill Book Company, Inc., 1960.

T-2    Thaler, G. J., and M. P. Pastel, *Nonlinear Control Systems*, New York, N.Y.: McGraw-Hill Book Company, Inc., 1962.

T-3    Tou, J. T., *Digital and Sampled-data Control Systems*, New York, N.Y.: McGraw-Hill Book Company, Inc., 1959.

T-4    Tou, J. T., *Modern Control Theory*, New York, N.Y.: McGraw-Hill Book Company, Inc., 1964.

T-5    Tsien, H. S., *Engineering Cybernetics*, New York, N.Y.: McGraw-Hill Book Company, Inc., 1958.

T-6    Turnbull, H. W., and A. C. Aitken, *An Introduction to the Theory of Canonical Matrices*, London, England: Blackie and Son, Ltd., 1932.

V-1    Van Landingham, H. F., and W. A. Blackwell, "Controller Design for Nonlinear and Time-Varying Plants," *Educational Monograph*, College of Engineering, Oklahoma State Univ., 1967.

W-1  Wadel, L. B., "Describing Function as Power Series," *IRE Trans. Automatic Control* (1962), p. 50.

W-2  Waltz, M. D., and K. S. Fu, "A Learning Control System," *Proc. Joint Automatic Control Conference*, 1964, pp. 1–5.

W-3  Wilcox, R. B., "Analysis and Synthesis of Dynamic Performance of Industrial Organizations—The Application of Feedback Control Techniques to Organizational Systems," *IRE Trans. Automatic Control* (1962), pp. 55–67.

W-4  Wojcik, C. K., "Analytical Representation of the Root Locus," *ASME J. Basic Engineering*, ser. D, **86** (1964), pp. 37–43.

Z-1  Zadeh, L. A., and C. A. Desoer, *Linear System Theory; The State Space Approach*, New York, N.Y.: McGraw-Hill Book Company, Inc., 1963.

# INDEX

## B

## C